Essential Ion Channel Methods

Reliable Lab Solutions

Essential Ion Channel Methods

Reliable Lab Solutions

Edited by

P. Michael Conn

Director, Office of Research Advocacy (OHSU)
Senior Scientist, Divisions of Reproductive Sciences
and Neuroscience (ONPRC)

Professor, Departments of Pharmacology and Physiology,
Cell and Developmental Biology, and Obstetrics and Gynecology (OHSU)
Beaverton, Oregon

ELSEVIER

AMSTERDAM • BOSTON • HEIDELBERG • LONDON
NEW YORK • OXFORD • PARIS • SAN DIEGO
SAN FRANCISCO • SINGAPORE • SYDNEY • TOKYO
Academic Press is an imprint of Elsevier

Academic Press is an imprint of Elsevier
30 Corporate Drive, Suite 400, Burlington, MA 01803, USA
525 B Street, Suite 1900, San Diego, CA 92101-4495, USA
32 Jamestown Road, London NW1 7BY, UK
Linacre House, Jordan Hill, Oxford OX2 8DP, UK

Material in the work originally appeared in Volumes 293
and 294 of *Methods in Enzymology* (1998, 1999, Elsevier Inc.)

Notice
No responsibility is assumed by the publisher for any injury and/or damage to persons
or property as a matter of products liability, negligence or otherwise, or from any use
or operation of any methods, products, instructions or ideas contained in the material
herein. Because of rapid advances in the medical sciences, in particular, independent
verification of diagnoses and drug dosages should be made

ISBN: 978-0-12-382204-8

For information on all Academic Press publications
visit our website at www.elsevierdirect.com

Transferred to Digital Printing in 2013

CONTENTS

4. Tight-Seal, Whole-Cell Patch Clamping of *C. elegans* Neurons

S. R. Lockery and M. B. Goodman

5. Gating Currents

Francisco Bezanilla and Enrico Stefani

6. Determining Ion Channel Permeation Properties

Ted Begenisich

PART IV   Expression Systems

PART V Model Simulations

PART VI Physical

PART VII   Purification and Reconstitution

PART VIII   Second Messengers and Biochemical Approaches

PART X   Toxins and Other Membrane Active Compounds

CONTRIBUTORS

Numbers in parentheses indicate the pages on which the authors' contributions begin.

Peter Agre (487), Department of Biological Chemistry, John Hopkins University School of Medicine, Baltimore, MD, USA

L. Aguilar-Bryan (449), Department of Cell Biology, Baylor College of Medicine, One Baylor Plaza, Houston, Texas, USA

Michael Akong (211), Cell Biology, SIBIA Neurosciences, Inc., La Jolla, California, USA

O. S. Andersen (315), Department of Physiology and Biophysics, Weill Cornell Medical College, New York, USA

Sarah M. Assmann (411), Department of Biology, Pennsylvania State University, University Park, Pennsylvania

Christine E. Bear (337), Programme in Molecular Structure and Function, Research Institute, Hospital for Sick Children, Toronto, Canada

Ted Begenisich (101), Department of Physiology, University of Rochester Medical Center, Rochester, New York, USA

Heinrich Betz (473), Max-Plank Institute, Hirnforschung, Deutschordenstrasse, Frankfurt, Germany

Francisco Bezanilla (81), Department of Anesthesiology, UCLA, Los Angeles, California, USA

J. Bryan (449), Department of Cell Biology, Baylor College of Medicine, One Baylor Plaza, Houston, Texas, USA

Gary Buell (111), Geneva Biomedical Research Institute, Geneva, Switzerland

F. Chudziak (449), Department of Cell Biology, Baylor College of Medicine, One Baylor Plaza, Houston, Texas, USA

J. P. Clement IV (449), Department of Cell Biology, Baylor College of Medicine, One Baylor Plaza, Houston, Texas, USA

Lourdes J. Cruz (511), Department of Biology, University of Utah, Salt Lake City, Utah, USA

Norman Davidson (135), Division of Biology, California Institute of Technology, Pasadena, California, USA

Carol Deutsch (3), Department of Physiology, University of Pennsylvania, Philadelphia, Pennsylvania, USA

Paul D. W. Eckford (337), Programme in Molecular Structure and Function, Research Institute, Hospital for Sick Children, Toronto, Canada

Alison L. Eertmoed (155), Department of Neurobiology, University of Chicago, Chicago, Illinois, USA

Markus U. Ehrengruber (135), Division of Biology, California Institute of Technology, Pasadena, California, USA

Cecilia Farre (297), Department of Chemistry, Stanford University, Stanford, California, USA

Isabelle Favre (359), Department of Biology, Clarkson University, Potsdam, New York, USA

Harvey A. Fishman (297), Department of Chemistry, Stanford University, Stanford, California, USA

Kevin Galley (337), Programme in Molecular Structure and Function, Research Institute, Hospital for Sick Children, Toronto, Canada

Elizabeth Garami (337), Programme in Molecular Structure and Function, Research Institute, Hospital for Sick Children, Toronto, Canada

Maria L. Garcia (529), Membrane Biochemistry & Biophysics, Merck Research Labs R80N-C31, Rahway, New Jersey, USA

G. Gonzalez (449), Department of Cell Biology, Baylor College of Medicine, One Baylor Plaza, Houston, Texas, USA

M. B. Goodman (65), Department of Molecular and Cellular Physiology, Stanford University, Stanford, California, USA

M. Goulian (315), Center for Studies in Physics and Biology, The Rockefeller University, New York, USA

Michael E. Green (179), Department of Chemistry, City College of the City University of New York, New York, NY 10031

William N. Green (155), Department of Neurobiology, University of Chicago, Chicago, Illinois, USA

Owen P. Hamill (465), University of Texas Medical Branch, 301 University Blvd, Galveston, Texas, USA

Markus Hanner (529), Membrane Biochemistry & Biophysics, Merck Research Labs R80N-C31, Rahway, New Jersey, USA

John W. Hanrahan (41), Department of Physiology, McIntyre Medical Science, McGill University, Montreal, Quebec, Canada

Ian Hart (561), Neurosciences Group, Institute of Molecular Medicine, John Radcliffe Hospital, Oxford, United Kingdom

Stephen D. Hess (211), Cell Biology, SIBIA Neurosciences, Inc., La Jolla, California, USA

Richard L. Huganir (379), HHMI, Department of Neuroscience, John Hopkins University School of Medicine, Baltimore, Maryland, USA

Ingemar Jacobson (297), Department of Chemistry, Stanford University, Stanford, California, USA

Kent Jardemark (297), Department of Chemistry, Stanford University, Stanford, California, USA

Mark C. Jasek (135), Division of Biology, California Institute of Technology, Pasadena, California, USA

Yanlin Jia (41), Department of Physiology, McIntyre Medical Science, McGill University, Montreal, Quebec, Canada

Edwin C. Johnson (211), Cell Biology, SIBIA Neurosciences, Inc., La Jolla, California, USA

Gregory J. Kaczorowski (529), Membrane Biochemistry & Biophysics, Merck Research Labs R80N-C31, Rahway, New Jersey, USA

Bruce L. Kagan (545), Department of Psychiatry, UCLA, Los Angeles, California, USA

Sunjeev Kamboj (379), HHMI, Department of Neuroscience, John Hopkins University School of Medicine, Baltimore, Maryland, USA

David B. Kantor (135), Division of Biology, California Institute of Technology, Pasadena, California, USA

Hans-Günther Knaus (529), Membrane Biochemistry & Biophysics, Merck Research Labs R80N-C31, Rahway, New Jersey, USA

R. E. Koeppe II (315), Department of Chemistry and Biochemistry, University of Arkansas, Fayetteville, Arkansas, USA

Zie Kone (41), Department of Physiology, McIntyre Medical Science, McGill University, Montreal, Quebec, Canada

Markus Lanzrein (135), Division of Biology, California Institute of Technology, Pasadena, California, USA

Bodo Laube (473), Max-Plank Institute, Hirnforschung, Deutschordenstrasse, Frankfurt, Germany

Henry A. Lester (135), Division of Biology, California Institute of Technology, Pasadena, California, USA

Canhui Li (337), Programme in Molecular Structure and Function, Research Institute, Hospital for Sick Children, Toronto, Canada

Sheri J. Lillard (297), Department of Chemistry, Stanford University, Stanford, California, USA

Paul Linsdell (41), Department of Physiology, McIntyre Medical Science, McGill University, Montreal, Quebec, Canada

S. R. Lockery (65), Institute of Neuroscience, University of Oregon, Eugene, Oregon, USA

J. A. Lundbæk (315), Department of Neuropharmacology, Novo-Nordisk A/S, Måløv, Denmark

Jiexin Luo (41), Department of Physiology, McIntyre Medical Science, McGill University, Montreal, Quebec, Canada

Monica M. Lurtz (237), Department of Molecular Physiology and Biophysics, Baylor College of Medicine, Houston, Texas, USA

A. M. Maer (315), Department of Physiology and Biophysics, Weill Cornell Medical College, New York, USA

Andrew L. Mammen (379), HHMI, Department of Neuroscience, John Hopkins University School of Medicine, Baltimore, Maryland, USA

John C. Mathai (487), Department of Biological Chemistry, John Hopkins University School of Medicine, Baltimore, MD, USA

Ceri J. Mathews (41), Department of Physiology, McIntyre Medical Science, McGill University, Montreal, Quebec, Canada

Don W. McBride, Jr. (465), University of Texas Medical Branch, 301 University Blvd, Galveston, Texas, USA

J. Michael McIntosh (511), Department of Biology, University of Utah, Salt Lake City, Utah, USA

Tajib A. Mirzabekov (545), Department of Psychiatry, UCLA, Los Angeles, California, USA

Alok K. Mitra (255), Department of Cell Biology, The Scripps Research Institute, La Jolla, California, USA

Edward Moczydlowski (359), Department of Biology, Clarkson University, Potsdam, New York, USA

Alexander Moscho (297), Department of Chemistry, Stanford University, Stanford, California, USA

C. Nielsen (315), Department of Physiology and Biophysics, Weill Cornell Medical College, New York, USA

Martin Niethammer (21), Howard Hughes Medical Institute, Mass General Hospital, Boston, Massachusetts, USA

Baldomero M. Olivera (511), Department of Biology, University of Utah, Salt Lake City, Utah, USA

Owe Orwar (297), Department of Chemistry, Stanford University, Stanford, California, USA

U. Panten (449), Department of Cell Biology, Baylor College of Medicine, One Baylor Plaza, Houston, Texas, USA

Rao V. L. Papineni (237), Department of Molecular Physiology and Biophysics, Baylor College of Medicine, Houston, Texas, USA

Steen E. Pedersen (237), Department of Molecular Physiology and Biophysics, Baylor College of Medicine, Houston, Texas, USA

Ashwin Pinto (561), Neurosciences Group, Institute of Molecular Medicine, John Radcliffe Hospital, Oxford, United Kingdom

Gregory M. Preston (487), Department of Biological Chemistry, John Hopkins University School of Medicine, Baltimore, MD, USA

Kathryn Radford (111), Geneva Biomedical Research Institute, Geneva, Switzerland

Mohabir Ramjeesingh (337), Programme in Molecular Structure and Function, Research Institute, Hospital for Sick Children, Toronto, Canada

Lisa Romano (411), Department of Biology, Pennsylvania State University, University Park, Pennsylvania

Erin M. Schuman (135), Division of Biology, California Institute of Technology, Pasadena, California, USA

C. Schwanstecher (449), Department of Cell Biology, Baylor College of Medicine, One Baylor Plaza, Houston, Texas, USA

M. Schwanstecher (449), Department of Cell Biology, Baylor College of Medicine, One Baylor Plaza, Houston, Texas, USA

Jason B. Shear (297), Department of Chemistry, Stanford University, Stanford, California, USA

Morgan Sheng (21, 397), Howard Hughes Medical Institute, Mass General Hospital, Boston, Massachusetts, USA

ZuFang Sheng (3), Department of Physiology, University of Pennsylvania, Philadelphia, Pennsylvania, USA

Anatoly Y. Silberstein (545), Department of Psychiatry, UCLA, Los Angeles, California, USA

Robert Slaughter (529), Membrane Biochemistry & Biophysics, Merck Research Labs R80N-C31, Rahway, New Jersey, USA

Barbara L. Smith (487), Department of Biological Chemistry, John Hopkins University School of Medicine, Baltimore, MD, USA

Kenneth A. Stauderman (211), Cell Biology, SIBIA Neurosciences, Inc., La Jolla, California, USA

Enrico Stefani (81), Department of Anesthesiology, UCLA, Los Angeles, California, USA

F. Anne Stephenson (561), Neurosciences Group, Institute of Molecular Medicine, John Radcliffe Hospital, Oxford, United Kingdom

Ye-Ming Sun (359), Department of Biology, Clarkson University, Potsdam, New York, USA

Vinzenz M. Unger (255), Department of Cell Biology, The Scripps Research Institute, La Jolla, California, USA

Yolanda F. Vallejo (155), Department of Neurobiology, University of Chicago, Chicago, Illinois, USA

Mark A. Varney (211), Cell Biology, SIBIA Neurosciences, Inc., La Jolla, California, USA

Gönül Veliçelebi (211), Cell Biology, SIBIA Neurosciences, Inc., La Jolla, California, USA

Angela Vincent (561), Neurosciences Group, Institute of Molecular Medicine, John Radcliffe Hospital, Oxford, United Kingdom

Yanchun Wang (337), Programme in Molecular Structure and Function, Research Institute, Hospital for Sick Children, Toronto, Canada

Michael Wyszynski (397), Howard Hughes Medical Institute, Mass General Hospital, Boston, Massachusetts, USA

Youfeng Xu (135), Division of Biology, California Institute of Technology, Pasadena, California, USA

Mark Yeager (255), Department of Cell Biology, The Scripps Research Institute, La Jolla, California, USA

Richard N. Zare (297), Department of Chemistry, Stanford University, Stanford, California, USA

PREFACE

The rapid growth of interest and research activity in ion channels is indicative of their fundamental importance in the maintenance of the living state. This volume was prepared with a view to providing a sampling of the range of molecular and physical methods that are significant for the study of ion channels.

The authors were selected from the contributors of previous Methods in Enzymology volumes on that topic on the basis of their significant research contributions in the area about which they have written. They have been encouraged to make use of graphics and comparisons with other methods, and to provide tricks and approaches that make it possible to adapt methods to other systems.

The authors were encouraged to present these methods in a fashion that allows their replication by individuals new to the field, yet providing valuable information for seasoned investigators.

I express my appreciation to the contributors for revising their contributions and to the staff of Academic Press for helpful input and maintaining outstanding production standards.

P. Michael Conn

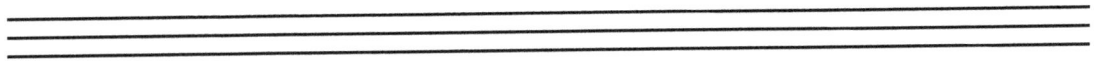

PART I

Assembly

CHAPTER 1

Assembly of Ion Channels

ZuFang Sheng and Carol Deutsch

Department of Physiology
University of Pennsylvania
Philadelphia
Pennsylvania
USA

I. Introduction

Most ion channels are multisubunit conglomerates. Because synthesis and as-sembly of many different types of pore-forming subunits occur in a single cell, how do the right subunits find each other to give the correct stoichiometry and avoid scrambling to channel homogeneity? This problem is even more striking if we consider the vast number of nonchannel transmembrane proteins made simulta-neously in a cell. Assembly is a multistep process that requires specific intersubunit recognition events. Each of these steps may include intermediate folded conforma-tions of subunits and/or intermediate subunit stoichiometries. Such possibilities have not been explored for most types of ion channels, including K^+ channels, nor is it known which regions of the subunits actually interact during each assembly step.

In some cases, the NH_2-terminal domains of ion channels can function as specific recognition motifs between subunits (Babila *et al.*, 1994; Li *et al.*, 1992; Shen *et al.*, 1993; Verrall and Hall, 1992; see also Xu and Li, 1998, this volume), but it is not clear that such elements contribute to stabilization of the mature multi-meric protein or whether additional subunit–subunit interactions between transmembrane segments provide the energy to shift the equilibrium in a lipid bilayer toward multimerization and the final, mature channel that functions in the plasma membrane. Most voltage-gated K^+ channels are homotetrameric membrane proteins, each subunit containing six putative transmembrane segments, S1–S6. It is not clear what holds the tetramer together; intersubunit covalent linkages do not appear to be responsible (Boland *et al.*, 1994). In these channels the cytoplasmic NH_2 terminus contains a recognition domain, T1 ("first tetramerization"), that tetramerizes *in vitro* and confers subfamily specificity (Li *et al.*, 1992; Shen and Pfaffinger, 1995; Shen *et al.*, 1993; Xu *et al.*, 1995). However, in the native channel there are also intramembrane association (IMA) sites in the central core of voltage-gated K^+ channels that provide sufficient recognition and stabilization interactions for channel assembly, and disruption of one or more of these interactions may suppress channel formation (Sheng *et al.*, 1997; Tu *et al.*, 1996). The relative contributions of different domain interactions (e.g., T1 and IMA) may vary from channel isoform to isoform. What are these T1 and IMA domains in the native full-length K^+ channel, and what are their relative contributions to channel formation?

Identification of the recognition and stabilization motifs in the primary sequence of channel proteins is a good beginning to understanding channel assembly; however, it still leaves many questions unanswered. How specific are these intersubunit interactions? How strong are they? At which stage in assembly are subunits integrated into the membrane? What are the spatial and temporal events involved in channel assembly? What is the subunit stoichiometry of the channel? What is the history of the subunits during assembly? Is recruitment of subunits a random event? What is the nature of the subunit pool? Where is it located? When are subunits recruited into multimeric channels, and where? We can address these issues both biochemically and biophysically, as described in the next section, using a variety of *in vitro* translation systems and *in vivo* expression systems.

The *in vitro* translation systems include rabbit reticulocyte lysate (RRL) and wheat germ agglutinin (WGA) systems, which contain cellular components necessary for protein synthesis (tRNA, ribosomes, amino acids, and initiation, elongation, and termination factors) and are capable of a variety of posttranslational processing activities (acetylation, isoprenylation, proteolysis, and some phosphorylation activity). Signal peptide cleavage and core glycosylation can be reconstituted and studied by adding canine pancreatic microsomal membranes to the translation reaction. These systems permit studies, for example, of transcriptional and translational control, association of proteins, and their membrane integration. However, the translation efficiency of high molecular weight proteins ($>100,000$) is relatively poor, and it is not clear that all aspects of *in vivo* processing have been reconstituted. Thus, caution must be used in extrapolating findings with the *in vitro* system to *in vivo* events.

The *in vivo* expression system most used for study of channel function and assembly has been *Xenopus* oocytes (Rudy and Iverson, 1992). Mammalian cells are also used frequently and involve DNA transfection techniques (Rudy and

Iverson, 1992). Oocytes typically require injection of channel mRNA (typically 50 nl/oocyte; <0.1–100 ng mRNA/oocyte). This system is an intact cell system that expresses at high levels for both electrophysiological and biochemical measurements, which can be done simultaneously in parallel samples. Both the oocyte and a mammalian T-cell expression system are described later, as well as the methods used to study channel protein synthesis, integration into membranes, and oligomerization.

Broadly defined, assembly also involves trafficking, posttranslational modification, and localization of channel proteins in specific subcellular compartments, as well as the aforementioned processes of recognition and association (oligomerization). This chapter, however, focuses only on strategies and methods that can be used (1) to identify regions of a protein that are potentially involved in intersubunit interactions during assembly of the pore-forming unit of ion channels, (2) to determine the strength, kinetics, spatial, and temporal characteristics of the intersubunit interactions, and (3) to determine the subunit stoichiometry and history of subunits during assembly. For some cases we illustrate the approaches by describing experiments in our laboratory involving a voltage-gated K^+ channel, Kvl.3. However, these strategies and methods can be, and have been, used for other multimeric channels.

II. Strategies and Methods

The strategies used to address the issues just stated entail either direct or indirect determinations of various aspects of subunit association. The former category includes primarily biochemical approaches; the latter makes use of functional readouts. These strategies are protein based, yet each can have additional strategies at the DNA level. For example, strategies that entail constructing genes that link multiple channel domains in tandem, swapping channel domains to create chimeras, and/or deleting or mutating domains can be combined with the protein assays to elucidate mechanisms of channel assembly.

A. Identification of Putative Regions Involved in Intersubunit Interactions

Intersubunit association can be assessed by direct and indirect methods as described in the following subsections. To discover which regions of the channel interact across subunit boundaries, physical association between channel subunits or between peptide fragments of a channel and the full-length channel protein must be demonstrated. This can be done directly by (1) immunoprecipitation of one member of a complex by antibody against the other member, (2) cross-linking interacting proteins using bifunctional reagents, or (3) binding assays of interacting peptides. Such binding assays have been employed to show that K^+ channel subunits, or parts of these subunits, multimerize both *in vitro* and *in vivo* (Babila *et al.*, 1994; Li *et al.*, 1992; Shen and Pfaffinger, 1995; Shen *et al.*, 1993). But these studies have been concerned primarily with cytoplasmic NH_2-terminal interactions. We describe one of these methods used in our laboratory, namely, immunoprecipitation. One important caveat concerning the association of peptide fragments of a channel with the channel protein is that it is not clear that such association faithfully reflects native associations between full-length subunits *in situ*. For instance, constraints imposed on a segment of the channel in the context of the full-length folded protein may lead to different interactions with another subunit compared with the isolated truncated channel peptide fragment. Therefore, for a transmembrane segment, it is ultimately important to determine not only whether these interactions occur in the native protein but also the topology and orientation of the peptide fragment.

1. Immunoprecipitation

This method requires the use of antibodies (antisera) to a protein or a peptide construct. If the antibodies to native epitopes are not sufficiently good, an epitope tag may be used; c-*myc* (MEQKLI-SEEDL) (Evans *et al.*, 1985) is excellent for this purpose. Such nonnative epitopes, however, should be inserted into a primary sequence at a nonperturbing distance (>15 amino acids) from putative topogenic determinants. The first step in this approach involves making the appropriate plasmid DNA either for use in transfections for subsequent *in vivo* expression, or for *in vitro* transcription to produce mRNA for subsequent use in either *in vivo* or *in vitro* experiments. Standard methods of restriction enzyme analysis, agarose gel electrophoresis, and bacterial transformation are used for these studies. Plasmid DNA are purified using Qiagen columns (Valencia, CA), and capped mRNA is synthesized *in vitro* from linearized templates using Sp6 or T7 RNA polymerase (Promega, Madison, WI).

For *in vitro* immunoprecipitation experiments, proteins are translated *in vitro* with [^{35}S]methionine (2 μl/25 μl translation mixture; ~10 μCi/μl Dupont/NEN Research Products, Boston, MA) in RRL (commercial preparations are available from Promega, and from MBI Fermentas, Amherst, NY; laboratory preparations can be made according to Jackson and Hunt, 1983; Walter and Blobel, 1983) in the presence (1.8 μl membrane suspension/25 μl translation mixture) or absence of canine pancreatic microsomal membranes (Promega or MBI Fermentas), according to the Promega *Protocol and Application Guide*. Two proteins that are proposed to interact are then cotranslated. Relative mRNA concentrations should be determined from the efficiencies of each construct to yield protein ratios that are desired. To maximize coimmunoprecipitation, microsomal membranes should be used in limiting concentration compared with the total mRNA concentration. The translation reaction can be visualized and quantitated using SDS–PAGE and phosphor imaging.

To perform immunoprecipitation from an *in vitro* translation system (RRL, microsomal membranes), 1–5 μl of cell-free translation products is mixed in 400 μl of buffer A [0.1 M NaCl, 0.1 M Tris (pH 8.0), 10 mM EDTA, and 1% (v/v)

Triton X-100] containing 0.1 mM phenylmethylsulfonyl fluoride (PMSF). Ascites fluid (9E10 to *myc* epitope; 1 μl) or channel antisera (4 μl) are added and samples incubated at 4 °C for 30 min. Protein A Affi-Gel beads (10–20 μl; Bio-Rad, Richmond, CA) are added and the suspension mixed continuously at 4 °C for 6–15 h with constant mixing. The beads are centrifuged and washed three times with buffer A and two times with 0.1 M NaCl, 0.1 M Tris (pH 8.0) prior to SDS–PAGE and fluorography. Where relevant, counts per minute in immunoprecipitated proteins should be corrected by the efficiency of recovering precipitated protein from the translation mixture. For example, for a *myc*-labeled peptide the correction factor can be calculated as the ratio of *myc*-peptide in the translation mixture to *myc*-peptide in the immunoprecipitate, as measured using the anti-*myc* antibody. The efficiency of immunoprecipitation will probably range from 10% to 25%. Each batch of membranes must be titrated for each protein or peptide construct to determine the maximal coimmunoprecipitation conditions, that is, the proper mRNA-to-membranes ratio. Also, a study of the time course of addition of membranes indicates that maximal coimmunoprecipitation occurs when membranes are added 5–10 min after translation has begun (Andrews, 1996; Sheng and Deutsch, 1997).

To perform immunoprecipitation experiments from oocytes, the *in vitro* transcribed mRNA is mixed with [^{35}S]methionine/cysteine [10× concentration; Tran ^{35}S-label (ICN, Irvine, CA) 20 μCi/μl; 90% methionine, 10% cysteine] and injected directly into oocytes (50 nl/oocyte) as described later. Coinjections of mRNA encoding channel proteins that are proposed to interact are followed by incubation of the oocytes in 1.5-ml Eppendorf tubes at 18 °C for 4 h in methionine/cysteine-free medium. The medium is then changed to a 1 mM methionine, 1 mM cysteine medium. For each time point, five to seven oocytes are frozen (−80 °C). Later, all samples are thawed and homogenized in 5× volumes (35 μl) of homogenization buffer (0.25 M sucrose, 50 mM Tris, pH 7.5, 50 mM potassium acetate, 5 mM $MgCl_2$, 1 mM dithiothreitol) while kept on ice. To the oocyte homogenate, 1.2 ml of buffer A (0.1 M NaCl, 0.1 M Tris (pH 8.0), 10 mM EDTA, and 1% Triton X-100) containing 0.1 mM PMSF is added. The solution is mixed continuously at 4 °C for 6 h and centrifuged at 14,000 × g at 4 °C for 15 min and the supernatant removed. Antibody (the amount will be determined by the titer) is added to the supernatant and incubated for 30 min at 4 °C. Protein A Affi-Gel beads (15 μl) are added, and the suspension is mixed continuously at 4 °C overnight. The beads are centrifuged and washed three times with buffer A and twice with 0.1 M NaCl. 0.1 M Tris (pH 8.0) prior to SDS–PAGE and fluorography.

2. Yeast Two-Hybrid System

We can use another approach, the yeast two-hybrid method (Fields and Song, 1989), which relies on a functional readout, to learn whether two proteins interact *in vivo*. This method is a genetic assay based on the fact that eukaryotic transcriptional activators are bifunctional, containing discrete functional domains. One domain binds to DNA while the other activates transcription. Two fusion proteins are generated *in vivo* (in yeast), one from a plasmid containing the DNA-binding domain and protein X, and the other from a plasmid containing the activation domain and protein Y. The two-hybrid genes are expressed in a yeast host strain containing latent reporter genes, typically *lacZ* or *His3*. Expression of the reporter gene, therefore, indicates possible interaction between proteins X and Y. This technique is restricted to detecting interactions between cytoplasmic regions of channel proteins and cannot be used to detect interactions between transmembrane segments of proteins. It has been used to identify and define the role of T1 recognition domains in voltage-gated K^+ channels (Xu *et al.*, 1995). A more detailed discussion of this technique is given by Xu and Li (1998, this volume).

3. Dominant Negative Suppression

To learn which regions of channel subunits may associate, including transmembrane segments, a dominant negative suppression strategy can be used. Evidence for protein–protein association is obtained as follows. A full-length channel is coexpressed with a fragment of the channel that putatively interacts across subunit boundaries. If the fragment associates with the channel subunit, then it will result in scavenging of available monomers, competitively inhibiting association of a full-length subunit in the multimer, and/or associating with a multimer. In any of these cases, the result may be suppression of a fully functional channel. Thus suppression, measured by a variety of readouts, may be interpreted as evidence of protein–protein interaction, with the additional caveat that it must then be shown that such interactions also occur between the fragment region and channel protein in the full-length channel tetramer *in situ*. Some of these readouts are functional tests (e.g., current measurements), and others are biochemical (e.g., ligand binding, immunoprecipitation assays of channel formation; see earlier discussion). Ideally, several criteria should be experimentally demonstrated in order to use this strategy to infer putative sites. First, it should be shown that protein is being made and is stable *in vivo*. Second, if suppression occurs, it must be shown not to be a consequence of inhibition of transcription or translation, but rather due to physical interaction of the peptide fragment with channel protein. Third, it must be shown that if these peptide fragments suppress current, they are specific. Fourth, the orientation and topology of channel subunits and peptide fragments should be known. This strategy has been used in oocytes and mammalian cells to identify multimerization domains in hydrophilic NH_2-terminal segments (Babila *et al.*, 1994; Li *et al.*, 1992; Tu *et al.*, 1996; Verrall and Hall, 1992; Xu and Li, 1998, this volume) and to probe for IMA sites within the hydrophobic core containing transmembrane segments (Sheng *et al.*, 1997; Tu *et al.*, 1996).

To determine experimentally which regions of channel subunits may be interacting, oocyte expression and electrophysiology are convenient and relatively simple tools for this purpose. While standard methods are used, some attention must be paid to conditions for suppression experiments. Specifically, the level of current expressed and the time after injection of the oocytes should be chosen to optimize detection of suppression (see later discussion). Oocytes can be isolated from *Xenopus laevis* females (Xenopus I, Ann Arbor, MI) as described previously (Goldin, 1992). Stage V–VI oocytes are selected and microinjected with mRNA (usually 0.1–10 ng) encoding for the channel of choice. The amount of mRNA should be adjusted to produce an appropriate current amplitude (see later discussion). Where applicable coinjections can be made in specified mole ratios, for example, for suppression experiments, we use 1:2 mole ratios of channel mRNA to truncated K^- channel mRNA or transmembrane control mRNA, respectively. K^+ currents from mRNA-injected oocytes are measured with a two-microelectrode voltage clamp after 24–48 h, at which time currents should be 2–10 μA. This level of expressed current is optimal for observing suppression because it avoids voltage-clamp artifacts that would mask true maximum current levels for control currents and therefore underestimate the extent of suppression. Electrodes (<1 MΩ) typically contain 3 M KCl, while the bath Ringer's solution contains (in mM): 116 NaCl, 2 KCl, 1.8 $CaCl_2$, 2 $MgCl_2$, 5 HEPES (pH 7.6). The appropriate holding potential for Kv1.3 is -100 mV. Data should be presented as box plots, which represent the central tendency of the measured current. This is a better means of displaying such data than bar graphs of the mean value because suppression data are quite variable and often non-Gaussian, and this method of analysis permits the entire range of data to be presented. The box and the bars indicate the 25–75 and 10–90 percentiles of the data, respectively. The horizontal line inside the box represents the median of the data. It may be necessary to carry out a statistical analysis of these results to determine whether experimental and control values are different. A nonparametric test such as a Mann–Whitney rank sum test is sufficient.

B. Characterization of Intersubunit Interactions

1. Avidity

In those cases in which the avidity (strength of association) between interacting channel proteins or peptides is to be determined by titrating the association with increasing amounts of detergent, sodium dodecyl sulfate (0–1.0% SDS in 0.1 M Tris, pH 7.8), or sodium *N*-dodecanoylsarcosinate (0–0.1% sarkosyl in 50 mM NaCl, 0.1 M Tris, pH 7.5) should be added (100 μl) to the translation products (3.5 μl) and incubated for 30 min at 4 °C before diluting 10× with buffer A (+PMSF) and continuing according to the procedures described earlier for immunoprecipitation.

2. Membrane Versus Nonmembrane Compartment

It is important to determine whether association requires membranes and, if so, whether proteins must first be integrated into the membranes for association to occur, and moreover, whether the proteins must be integrated into the same membrane. Finally, what is the temporal relationship between synthesis of associating proteins, membrane integration, and complex formation?

To demonstrate whether membranes are required for association of subunits and/or channel peptides, proteins are cotranslated in the presence and absence of microsomal membranes and then immunoprecipitated. The extent of membrane integration of each protein can be measured by extracting translation products made in the presence of microsomal membranes with either Tris buffer or carbonate buffer and comparing the pellet and supernatant content of ^{35}S-labeled protein. *In vitro* translation products (1–5 μl) are diluted in 750 μl of either sodium carbonate (0.1 M Na_2CO_3, pH 11.5) or Tris (0.25 M sucrose, 0.1 M Tris, pH 7.5) solution. Samples are incubated on ice for 30 min prior to centrifugation at 70,000 rpm (208 kG) TLA 100.3 rotor for 30 min. Supernatants are removed and 10% (v/v) trichloroacetic acid (TCA) added to precipitate protein, which is resuspended in SDS–PAGE sample buffer. Membrane pellets are then dissolved directly in SDS–PAGE sample buffer. The pellet contains the membrane fraction of protein. As a measure of the physical and functional integrity of the microsomal membranes, control mRNA should be simultaneously translated in the assay, including one that encodes a known transmembrane protein and one that encodes a secreted protein. The former should remain in the pellet fraction at pH 7.5 and 11.5, whereas the latter will be in the pellet fraction (microsomal lumen) at pH 7.5 and in the supernatant at pH 11.5. The ratio of pellet to pellet plus supernatant for extractions done at pH 7.5 gives the fraction of pellet associated with microsomes, and for extractions done at pH 11.5, this ratio is the fraction of protein integrated into the microsomal membrane.

To determine whether two putatively interacting proteins can associate if they are integrated into different membranes, the two proteins are translated separately in the presence or absence of membranes, then mixed together, and immunoprecipitated. In this case, puromycin (1 mM, 15 min) is added to terminate translation prior to combining the two reaction mixtures. A control should be included to show that the channel proteins are actually being translated on the membranes. The ideal control is one in which one of the monitored proteins is itself glycosylated. Thus, synthesis and translocation (glycosylation) of this protein will verify whether the membranes are functioning properly.

To determine whether synthesis, membrane integration, and association occur sequentially or concurrently, and which of these steps is rate limiting, a study of the relative time course of the synthesis, membrane integration, and association of the interacting proteins can be done by measuring ^{35}S incorporation, carbonate extraction, and coimmunoprecipitation, respectively. In the case of Kvl.3, such a comparison has shown that synthesis and integration of channel protein are rapid and that the association step itself is the rate-determining, membrane-delimited step in complex formation (Sheng *et al.*, 1997).

3. Complex Size

To determine the size of the associated channel complex, sucrose gradient experiments can be used. This method is based on the hydrodynamic properties of proteins and protein complexes. The fractional migration of a protein through a gradient is determined by the sedimentation characteristics of the protein or complex. Limitations of this method include experimental tailing of bands due to self-aggregation of proteins, and resolution that is often >30 kDa. However, the choice of detergent and salt concentration can have a significant impact on aggregation and resolution. *In vitro* translated protein (50 μl) is loaded on a 100-μl sucrose cushion (0.5 M sucrose, 100 mM KCl, 50 mM HEPES (pH 7.5), 5 mM $MgCl_2$, 1 mM DTT) and spun at 117 kG using a TLA rotor for 5 min to obtain pellets that contain the membrane fraction only. These pellets are solubilized for 30 min on ice in 100 μl of 0.05% dodecylmaltoside ($C_{12}M$), 50 mM NaCl, 50 mM Tris, 1 mM EDTA, pH 7.5, or in 1.5% CHAPS (Sigma, St. Louis, MO) buffer solution, 150–200 mM NaCl, 50 mM Tris–HCl (pH 8.0), 1 mM EDTA. Sometimes 0.015% phosphatidyl choline can be added to prevent aggregation. The solubilized proteins are then centrifuged for 1 h at 4 °C, 139,000 $\times$ g (60,000 rpm) in a Beckman (Columbia, MD) TL-100 centrifuge to remove insoluble material. Twelve milliliters of 5–20% linear sucrose gradient (in 0.05% $C_{12}M$ or in 1% CHAPS, buffer as above) is poured into Nalgene (Fisher Scientific, Philadelphia, PA) ultratubes, using an HBI gradient maker. The solubilized protein is loaded on top of the gradient and sedimented at 164,000 $\times$ g (36,000 rpm) in a SW-40 Ti rotor on a Beckman L8-70M ultracentrifuge for 20 h at 4 °C. Fractions (either 0.25 or 1 ml) are collected and either used directly in immunoprecipitation assays or precipitated with TCA (10%) for 1 h at 0 °C, and spun in an Eppendorf centrifuge (14,000 rpm) at 4 °C for 30 min to pellet the protein. The supernatant is removed and the pellet dried. SDS sample buffer is added to the pellet for separation on SDS–PAGE gels.

C. Determination of Subunit Stoichiometry and History During Assembly

1. Mass Tagging

To determine subunit stoichiometry, molecular weight markers can be engineered into subunits (Heginbotham *et al.*, 1997; Sakaguchi *et al.*, 1997). This is known as *mass tagging* (Heginbotham *et al.*, 1997). Identical subunits can be labeled with additional peptide chains that will shift the molecular weights of monomer and multimer detected on SDS–PAGE. If assembly of subunits is random and stable, then the number of distinctly different protein bands observed on SDS–PAGE will indicate the subunit stoichiometry of the assembled channel. This approach has been used to biochemically demonstrate the tetrameric stoichiometry of the influenza virus M_2 channel (Sakaguchi *et al.*, 1997) and of a prokaryotic K^+ channel (Heginbotham *et al.*, 1997). In each case, two plasmids, each containing different engineered lengths of the C terminus of the channel were expressed, isolated, and run on SDS–PAGE. If two different species of monomers associate randomly to form a tetramer, then the relative proportion of the five resultant channel types will be described by a binomial distribution (see later section). Five distinct bands, representing each of the five possible channels subunit compositions, appeared on SDS–PAGE (Heginbotham *et al.*, 1997; Sakaguchi *et al.*, 1997).

This method can be adapted for use in low-yield expression systems by labeling the translated subunits with [^{35}S]methionine. However, this method does require that multimeric species be stable under conditions of SDS–PAGE, which is rarely the case. Moreover, quantitative analysis using binomial distributions may not be practical because the stability of each mass-tagged multimeric complex may not be identical, hence skewing the distribution for noncombinatoric reasons (Heginbotham *et al.*, 1997). An alternative means to mass tagging a polypeptide is to engineer a glycosylation site into the subunit. In this case the cell itself labels the subunit *in vivo*. (Such strategies are used routinely to determine topologies of polytopic transmembrane proteins; Galen and Oslwald, 1995; Holman *et al.*, 1994.) However, the creation or deletion of consensus sites must be experimentally confirmed because consensus sites may not always be glycosylated. Sometimes specific lengths and amino acid sequences in the flanking regions are required for glycosylation (Galen and Ostwald, 1995).

2. Functional Tagging

In addition to the direct methods already described, subunit stoichiometry can also be determined using a functional readout. A mutant is made that has altered conductance, pharmacology, and/or kinetics of gating. We refer to this approach as *functional tagging*. It has been used to determine not only subunit stoichiometry (MacKinnon, 1991) but also the contribution of individual subunits to specific K$^+$ channel functions (Kavanaugh *et al.*, 1992; MacKinnon *et al.*, 1993; Ogielska *et al.*, 1995; Panyi *et al.*, 1995). Moreover, one can engineer a particular function of a channel to reveal its prior history, namely, to learn something about the assembly of individual subunits. In this regard the strategy and methods described later can be used to answer the following questions: Does synthesis and assembly of different channel subunits occur in the same shared compartment? Are subunits recruited randomly or preferentially? Does multimer formation occur in the plasma membrane? Are channel monomers and multimers in equilibrium in the plasma membrane? Is channel diversity temporally or spatially regulated? Any functional property of the channel can be used for this purpose. The major criterion that must be met, however, is that an order of magnitude difference in the chosen functional parameters for the wild-type (WT) and mutant (MUT) subunits must exist, regardless of which parameter is being studied, whether it be time constants of gating, binding constants of some ligand, or single-channel conductances.

The functional tagging experiments are performed either by heterologously coexpressing a WT subunit with a MUT subunit or by heterologously expressing a MUT subunit in a cell expressing endogenous WT channels. The appropriately modified function is measured and an analysis, as described later, is performed.

$$\boxed{\text{Gating kinetics:} \qquad I(t) = \sum_{m \, 0}^{N} B(N, p, m) \, Y_m(t)}$$

$$\boxed{\text{Affinity of channel blocker:} \qquad \frac{I([bk])}{I(0)} = \sum_{m \, 0}^{N} B(N, p, m) F_{unbk, m}([bk])}$$

$$\boxed{\text{Single channel conductance:} \qquad \frac{\textit{Number of openings to level } i_m}{\textit{Total number of openings}} = B(N, p, m)}$$

$$B(N, p, m) = \frac{N!}{m!(N-m)!} \, p^m (1-p)^{N\,m}$$

Fig. 1 Readouts of functionally tagged subunits. Equations that describe gating kinetics, open-channel block, and single-channel conductance. In each case, B (N, p, m) represents the binomial distribution, as described in the text, along with the functions and symbols used in these equations.

The method of analysis in functional tagging studies assumes a binomial distribution for the random formation of heteromultimeric channels. For a multimer of N subunits, the fraction of channels with exactly m MUT subunits will be

$$B(N, p, m) = \frac{N!}{m!(N-m)!} p^m (1-p)^{N-m}$$

where p is the fraction of MUT subunits in the membrane. The WT homomultimer is represented by $m = 0$, whereas $m = N$ represents the homomultimeric MUT channel. If the biophysical properties of each member of this population are known; it is possible to estimate both p and the validity of the underlying assumption, namely, that WT and MUT subunits assemble randomly. Biophysical properties that can be quantified in this way include the kinetics of inactivation, the affinity of an open-channel blocker, and the single-channel conductance.

Three similar equations (Fig. 1) can be used to fit the data obtained from a cell expressing both endogenous and heterologous subunits, depending on the biophysical parameter to be measured. In the case of gating kinetics, $I(t)$ is the current at time t and $Y_m(t)$ is a function describing the gating kinetics for a channel with m MUT subunits (see later example). In the case of blocker affinity, $I([bk])$ is the current in the presence of blocking agent, $I(0)$ is the current in the absence of blocking agent, bk is the blocker molecule, $[bk]$ is the blocker concentration, and $F_{unbk,m}([bk])$ is the fraction of unblocked current for a channel with m MUT subunits. In the case of single-channel conductance, i_m is the single-channel current for a channel with m MUT subunits.

We introduce an approach that provides a general two-step test for possible sources of nonrandomness in channel assembly that can give us insights into the prior history of channel subunits. As outlined in Fig. 2, the first test determines the relative affinities of mixed subunits. WT and MUT subunits are heterologously coexpressed in a cell devoid of the channel in question, and the resulting channel

Test 1: Mix and heterologously express WT and MUT

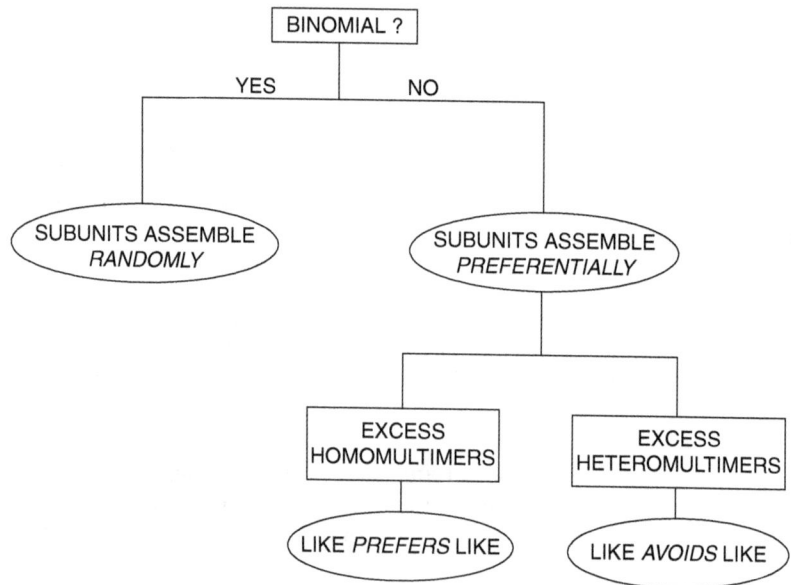

BINOMIAL ?

YES   NO

SUBUNITS ASSEMBLE *RANDOMLY*

SUBUNITS ASSEMBLE *PREFERENTIALLY*

EXCESS HOMOMULTIMERS

EXCESS HETEROMULTIMERS

LIKE *PREFERS* LIKE

LIKE *AVOIDS* LIKE

Test 2: Express MUT heterologously in cell with endogenous WT

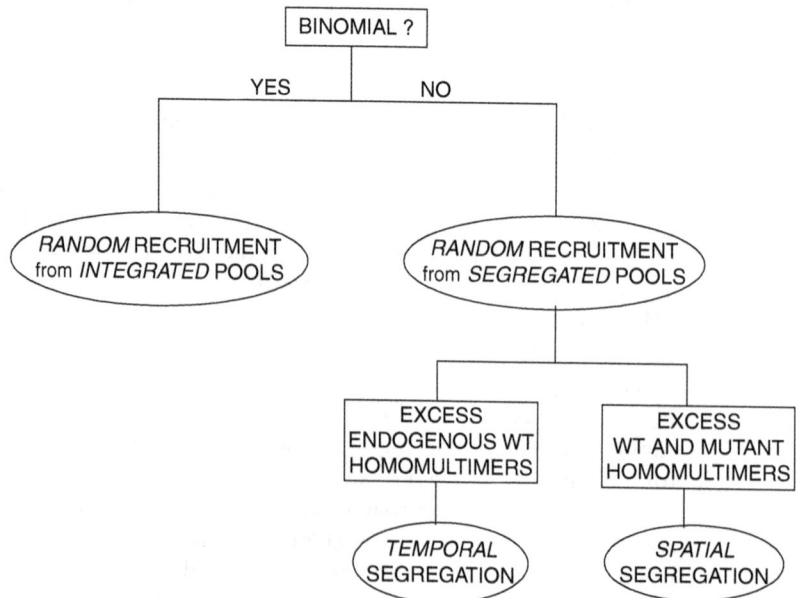

BINOMIAL ?

YES   NO

RANDOM RECRUITMENT from *INTEGRATED* POOLS

RANDOM RECRUITMENT from *SEGREGATED* POOLS

EXCESS ENDOGENOUS WT HOMOMULTIMERS

EXCESS WT AND MUTANT HOMOMULTIMERS

TEMPORAL SEGREGATION

SPATIAL SEGREGATION

population either conforms to or fails to conform to a binomial distribution. If it conforms, this means that WT and MUT subunits are recruited randomly and independently with the same probability. Failure means that subunits are not selected randomly, but since they are present at the same time in the same heterologous cell compartment, failure means there is some cooperativity, positive ("like prefers like") or negative ("like avoids like"). This will be manifest as an excess of homomultimers or an excess of heteromultimers, respectively, compared to a binomial distribution.

The second test determines whether subunits are recruited from integrated or segregated monomer pools and what must be the nature of the segregation. It should be used after verifying that the first test shows no cooperativity. MUT subunits are heterologously expressed in a cell that already has endogenous WT subunits. A binomial distribution can be interpreted as evidence that random recruitment of subunits occurred from an integrated pool of subunits. This test can only fail the binomial distribution if there is some segregation, assuming that these subunits did not fail the first test (i.e., did not show cooperativity). Segregation will be manifest either as an excess of endogenous WT channels or as an excess of both WT and MUT channels compared to the binomial distribution. The first case indicates temporal segregation of subunits (i.e., multimers were formed irreversibly at different times), while the second indicates spatial segregation (i.e., WT and MUT homomultimers were formed in spatially separate compartments).

To illustrate a specific application of this approach, we describe our studies of Kv1.3. The WT homotetramer Kv1.3 inactivates with a time constant of 200 ms, and a point mutation in the S6 segment of this channel produces a MUT homotetramer that inactivates with a time constant of 4 ms, $50\times$ faster than the WT channel (Panyi et al., 1995). When we expressed this functionally tagged subunit along with the WT subunit in a cytotoxic T cell (CTLL), which is devoid of endogenous channels, we found that the rate of inactivation depended exponentially on the number of MUT subunits (m). We could therefore account for the inactivation kinetics of a population of heterotetramers by a binomially weighted sum of inactivation rates when WT and MUT subunits were coexpressed in a cell line devoid of endogenous voltage-gated K^+ channels (Panyi et al., 1995), indicating a random association of MUT and WT subunits with no cooperative assembly. The equation for gating kinetics (Fig. 1) was used for this analysis. Here $Y_m(t)$ is a first-order process and equals $[k_{i,m}/(k_{i,m} + k_{r,m})e^{(k_{i,m}+k_{r,m})t} + k_{r,m}/(k_{i,m} + k_{r,m})]I_{peak}$, where I_{peak} is the peak current. The inactivation and recovery rate constants for a channel with m MUT subunits are $k_{i,m}$ and $k_{r,m}$, respectively. The values for the

Fig. 2 Prior history of channel subunits. Flow diagram illustrating tests for subunit preferences in association (test 1) and for segregation of subunits (test 2). The results (rectangles) and interpretations (ovals) of heterologous coexpression of wild-type (WT) and mutant (MUT) subunits in a cell are shown in test 1. The results and interpretations of heterologous expression of MUT in a cell already expressing endogenous WT are shown in test 2, where it is known that WT and MUT subunits show no preferential association.

rate constants were determined from the particular model invoked for the inactivation of Kv1.3 and from the measured parameters for the WT and MUT homomultimers. When we applied this method of analysis to the resulting current obtained from MUT-transfected Jurkat T cells, which also express endogenous WT channels, we found that the inactivation kinetics did not conform to a simple binomial distribution of channel types, but rather to a binomial distribution plus an additional term for preformed endogenous WT channels in the membrane, that is, excess WT channels compared to a binomial distribution of channel types, consistent with temporal, but not spatial segregation of the WT and MUT subunits (Panyi and Deutsch, 1996) (Fig. 2).

In addition to the strategies and methods described, a variety of techniques have been used to define protein topology, that is, the location of a protein or region of protein with respect to a membrane-bound compartment. These methods include protease digestion, cell surface labeling with epitope-specific antibodies, cross-linking with sulfhydryl specific reagents, or cysteine accessibility. We refer the reader to a recent review by Skach (in press) that describes assays of protein topology and the insights they provide regarding protein folding and general mechanisms of polytopic protein biogenesis. These approaches are designed to investigate the mechanisms of protein folding and assembly in the endoplasmic reticulum membrane, and are complementary to the approaches described herein.

References

Andrews, D. (1996). Unpublished data.

Babila, T., Moscucci, A., Wang, H., Weaver, F. E., and Koren, G. (1994). *Neuron* **12,** 615.

Boland, L. M., Jurman, M. E., and Yellen, G. (1994). *Biophys. J.* **66,** 694.

Evans, G., Lewis, G., Ramsay, G., and Bishop, M. J. (1985). *Mol. Cell Biol.* **5,** 3610.

Fields, S., and Song, O. (1989). *Nature* **340,** 245.

Galen, Z. G., and Ostwald, R. E. (1995). *J. Biol. Chem.* **270,** 2000.

Goldin, A. (1992). *Methods Enzymol.* **207,** 266.

Heginbotham, L., Odessey, E., and Miller, C. (1997). *Biochemistry* **36,** 10355–110342.

Holman, M., Maron, C., and Heinemann, S. (1994). *Neuron* **13,** 1331.

Jackson, R. J., and Hunt, T. (1983). *Methods Enzymol.* **96,** 50.

Kavanaugh, M. P., Hurst, R. S., Yakel, J., Adelman, J. P., and North, R. A. (1992). *Neuron* **8,** 493.

Li, M., Jan, Y. N., and Jan, L. Y. (1992). *Science* **257,** 1225.

MacKinnon, R. (1991). *Nature* **500,** 232.

MacKinnon, R., Aldrich, R. W., and Lee, A. W. (1993). *Science* **262,** 757.

Ogielska, E. M., Zagotta, W. N., Hoshi, T., Heinemann, S. H., Haab, J., and Aldrich, R. W. (1995). *Biophys. J.* **69,** 2449.

Panyi, G., and Deutsch, C. (1996). *J. Gen. Physiol.* **107,** 409.

Panyi, G., Sheng, Z., Tu, L., and Deutsch, C. (1995). *Biophys. J.* **69,** 896.

Rudy, B., and Iverson, L. E. (eds.), (1992). *Methods Enzymol.* **207,** 225.

Sakaguchi, T., Tu, Q., Pinto, L. H., and Lamb, R. A. (1997). *Proc. Natl. Acad. Sci. USA* **94,** 5000.

Sheng, Z., and Deutsch, C. (1997). Unpublished data.

Sheng, Z., Skach, W., Santarelli, V., and Deutsch, C. (1997). *Biochemistry* **36,** 15501–15513.

Shen, N. V., Chen, X., Boyer, M. M., and Pfaffinger, P. (1993). *Neuron* **11,** 67.

Shen, N. V., and Pfaffinger, P. J. (1995). *Neuron* **14,** 625.

Skach, W. (in press). *Methods Enzymol.* **292,** 265–278.

Tu, L., Santarelli, V., Sheng, Z., Skach, W., Pain, D., and Deutsch, C. (1996). *J. Biol. Chem.* **271,** 18904.

Verrall, S., and Hall, Z. W. (1992). *Cell* **68,** 23.

Walter, P., and Blobel, G. (1983). *Methods Enzymol.* **96,** 84.

Xu, J., and Li, M. (1998, this volume). *Methods Enzymol.* **293,** Chapter 1.

Xu, J., Yu, W., Jan, J. N., Jan, L., and Li, M. (1995). *J. Biol. Chem.* **270,** 24761.

PART II

Genetics

CHAPTER 2

Identification of Ion Channel-Associated Proteins Using the Yeast Two-Hybrid System

Martin Niethammer and Morgan Sheng

Howard Hughes Medical Institute
Mass General Hospital
Boston, Massachusetts
USA

I. Introduction

A wide variety of ion channels are involved in generating and regulating the electrical behavior of a neuronal cell. Typically, these ion channels are not randomly distributed on the cell surface but rather localized at specific subcellular sites. Good examples include ligand-gated ion channels (ionotropic neurotransmitter receptors) at postsynaptic sites and voltage-gated sodium channels at nodes of Ranvier. Presumably, the targeted localization of these ion channels depends on protein–protein interactions between ion channel subunits and the cytoskeleton. In addition, it is likely that ion channels interact with intracellular proteins that are involved in their modulation (e.g., protein kinases such as the tyrosine kinase Src (Holmes *et al.*, 1996; Yu *et al.*, 1997) or protein kinase C (Tsunoda *et al.*, 1997) or that are involved in downstream signaling mechanisms activated by the ion channel (Brenman *et al.*, 1996; Tsunoda *et al.*, 1997). Thus, knowledge about intracellular proteins associated with ion channels is critical to our understanding of signal transduction by ion channels and of mechanisms of ion channels clustering at specific subcellular sites. Currently, few such proteins have been identified.

At the neuromuscular junction, 43K/rapsyn is involved in ACh receptor clustering and postsynaptic specialization (Froehner, 1993; Gautam *et al.*, 1995). In the spinal chord, glycine receptors are clustered due to association with gephyrin, which may link receptor subunits to tubulin (Kirsch *et al.*, 1993). Both rapsyn and gephyrin were initially identified biochemically as proteins that copurified with their respective receptors.

Biochemical copurification of associated proteins may not be feasible for many ion channels, especially those of low abundance and low solubility (such as NMDA receptors). An alternative approach is the yeast two-hybrid system. The two-hybrid system is a widely used yeast-based genetic approach for detecting protein–protein interactions and provides a powerful way of isolating ion channel-associated proteins in an *in vivo* setting. It can be used both to assay known interactor proteins and for screening of cDNA libraries for novel proteins, giving immediate access to the primary DNA sequence of the interacting protein, as well as allowing for relatively easy determination of the binding domains involved. Since ion channels are transmembrane proteins, using them in a two-hybrid screen is not trivial, even though it has been done recently with some success.

In this chapter, we describe the yeast two-hybrid assay we have used to identify proteins binding to Shaker-type channels and NMDA receptor subunits. We also discuss some general and specific considerations about using ion channel proteins in the two-hybrid system.

II. Principles of Two-Hybrid System

The yeast two-hybrid system (Chien *et al.*, 1991; Fields and Song, 1989) makes use of the modular nature of eukaryotic transcription factors. Transcription activators typically consist of a DNA binding domain, which recognizes specific sequence motifs in the regulatory region of genes, and a distinct activator domain (Keegan *et al.*, 1986). Typically, the DNA binding domain and the transcription activating domain function independently; that is, hybrids between different DNA binding domains and activation domains form fully functional transcription factors (Brent and Ptashne, 1985). Furthermore, the two domains need not be

Table I
Ion-Channel Baits Used in Two-Hybrid Screens

Ion channel	Residues[a]	Vector	Results	Reference
Kv1.4	568–655[a]	pBHA	PSD-95	Kim et al. (1995)
			SAP97	
			Chapsyn-110	
NMDA-R1	834–938[a]	pBHA	α-Actinin-2	Wyszynski et al. (1997)
NMDA-R2A	834–1464[a]	pGBT9	PSD-95	Kornau et al. (1995)
	1349–1464[a]	pBHA	PSD-95	Niethammcr et al. (1996)
			SAP97	
NMDA-R2B	1361–1482[a]	pBHA	PSD-95	Niethammcr et al. (1996)
			SAP97	
GluR2	834–883[a]	pPC97	GRIP	Dong et al. (1997)

[a]C-terminal residue of the ion channel.

covalently linked (Ma and Ptashne, 1988; McKnight et al., 1987). This latter property is exploited in the yeast two-hybrid system.

A plasmid construct encoding a fusion protein of a known protein (the "bait") is made with a DNA binding domain, and cotransfected into reporter yeast strains with a plasmid encoding a second fusion protein, consisting of a known protein or a random cDNA fragment derived from a library, fused to a transcriptional activation domain. Thus, by using an entire library of cDNA fragments for the second construct, a screen for unknown interactors can be performed. Neither of the two fusions alone can activate the reporter genes; only when the expressed fusion proteins bind each other within the yeast nucleus is the transcription activator reconstituted. This is detected by activation of one or more reporter genes whose upstream regulatory region carries the appropriate DNA sequence elements for binding of the bait-DNA binding domain.

Since its inception, the two-hybrid system has been used to identify numerous protein–protein interactions in a variety of systems. More recently, several two-hybrid screens have successfully identified proteins that interact with ion channels (Table I) (Dong et al., 1997; Kim et al., 1995; Kornau et al., 1995; Niethammer et al., 1996; Wyszynski et al., 1997). It is noteworthy that all of these proteins, for reasons discussed later, interact with the intracellular carboxyterminal region of the channel proteins. A number of other ion channels have also been shown to interact with known proteins in the two-hybrid system (such as Kv1.1–1.3) (Kim et al., 1995); these channel parts have not, however, been tested as bait in library screens.

III. Materials and Methods

A variety of vectors and reporter strains have been designed for use in the two-hybrid system. The most commonly used DNA binding domains are derived from the *Saccharomyces cerevisiae* transcription factor Gal4p and the *Escherichia coli* repressor *lex A*, while the most commonly used activation domains stem from Gal4p and the herpes simplex virus protein VP16 (Bartel and Fields, 1995). Reporter genes usually encode for yeast nutritional marker genes or the *E. coli* gene β-galactosidase. These vectors have been described in detail elsewhere (Bartel and Fields, 1995), and we thus restrict this chapter to the system used in our laboratory, which utilizes the *lexA* DNA binding domain and the Gal4p activation domain.

Formulations for solutions and media needed for two-hybrid screens are given in the Appendix.

A. Vectors and Reporter Yeast Strains

Most two-hybrid vectors share some common features such as bacterial *ori* (for prokaryotic replication) and *bla* (for ampicillin resistance) sequences, as well as yeast origin sequences such as 2μ or ARS (for replication in yeast). Expression of the selectable nutritional markers (e.g., *TRP1* or *LEU2*) from these vectors is usually controlled by an ADH1-derived yeast promoter. We use two vectors in our system. The bait protein is inserted into vector pBHA [a derivative of pBTM116, with a hemagglutinin (HA) tag inserted between the LexA sequence and the multiple cloning site]; and the library is fused into vector pGAD10. The pGAD10 vector is part of the commercially available matchmaker two-hybrid system (Clontech Laboratories, Palo Alto, CA). We use the yeast host strain L40 (genotype *MATa trpl-901 leu2-3,112 his3Δ200 ade2 lys2-801am LYS2::(lexAop)₄-HIS3 URA3::(lexAop)₈-lacZ*), which is mutant for the chromosomal copies of *LEU2*, *TRP1*, and *HIS3*. Two independent reporter genes *HIS3* and *lacZ*, with upstream *lexA* binding sites, are stably integrated into the genome (Hollenberg *et al.*, 1995). Activation of both of these reporter genes confers histidine-independent growth and β-galactosidase activity, thereby allowing nutritional selection and blue-white X-Gal screening for yeast harboring interacting bait and activation domain fusions. pBHA contains the *TRP1* gene, and pGAD10 contains the *LEU2* gene as selectable markers. We can thus select for cotransformed yeast by growing the yeast on leucine/tryptophan-deficient media. Interaction of the two fusion proteins can be selected for by additionally making the plates histidine deficient. Some leaky His3p expression usually occurs even in the absence of interaction of the fusion proteins. This can be overcome by adding the competitive inhibitor 3-amino-1,2,4-triazole (3-AT) to the selection plates (we usually use a concentration of 2.5–5 mM). 3-AT can also help to suppress any intrinsic low-level transcriptional activating activity that a bait may exhibit.

B. Making Bait Constructs

Generally, we find that the following types of baits tend to be more successful in the two-hybrid system: (1) defined modular domains of proteins, (2) C-terminal ends, and (3) whole proteins. Presumably, modular domains and C termini will more likely fold in a native confirmation as fusion proteins and thus make the screen more "realistic." Using domains requires prior knowledge about the bait protein, such as the presence of recognized domains that can fold in varying contexts, such as SH3 or PDZ domains. Using the whole protein is often impractical for a number of reasons—the proteins may be large, and thus hard to subclone and transform, and, for reasons that are not understood, large baits often display transcriptional activation by themselves. However, if these problems do not occur, whole proteins are advantageous because they make no presumption about domain structure and may be more likely to fold correctly. Even more worrisome for screening with ion channels or other transmembrane proteins is that these proteins will not fold correctly when not membrane inserted and will likely never even reach the yeast nucleus due to hydrophobic transmembrane regions. The whole-protein approach thus usually is only feasible for small, intracellular proteins and not for transmembrane proteins.

Generally, an intracellular C terminus seems to be the best suited part of ion channels for two-hybrid screens. It is probably no coincidence that all successful two-hybrid screens with ion channels have used the intracellular C-terminal region as a bait (Table I), and to date we know of no instance of successful yeast two-hybrid screens with N termini or extra- or intracellular loops of ion channels. Extracellular regions are often glycosylated *in vivo*. If this glycosylation is involved in the binding of other proteins or the tertiary structure of the channel, the interaction is not likely to show up in the two-hybrid system. While intracellular N termini should behave similarly to C termini in the two-hybrid system (with respect to folding), they at present have an additional problem. All vectors commercially available to this point fuse the bait to the C-terminal end of *lexA* or Gal4 DNA binding proteins, thus likely masking interactions that might take place with a free N terminus. An N-terminal fusion vector has been reported (Béranger *et al.*, 1997); however, to our knowledge it has not yet been used in any ion channel screens.

Construction of the bait is done by standard molecular techniques (Sambrook *et al.*, 1989). The DNA encoding the channel part of interest is subcloned into the multicloning site of pBHA (Fig. 1). pBHA has stop codons in all three frames downstream of the multicloning site. Nevertheless, when making a C-terminal bait it is important to include the stop codon at the C terminus. The easiest way to generate the insert (especially of shorter fragments) is by PCR (polymerase chain reaction) using specific primers, adding the appropriate restriction sites at the ends. Typically, we insert the bait in frame beginning at the *Eco*RI site. Because frameshifts that result in faulty bait proteins are a major concern, it is essential to sequence the final bait plasmid before using it in the two-hybrid system. For sequencing of the *lexA*-bait fusion junction we use the primer 5′-CTT CGT CAG CAG AGC TTC ACC ATT G-3′ which corresponds to amino acids 181–188 of the lexA protein ($\sim$100 nucleotides upstream of the *Eco*RI site). Once the sequence is confirmed, it is of equal importance to test any bait for self-activation, by cotransforming the bait plasmid with an insert-free pGAD10 and testing reporter gene activation.

HA tag

```
                 Tyr Pro Tyr Asp Val Pro Asp Tyr Ala
     GAA TTG TAC CCA TAC GAC GTC CCA GAC TAC GCT

     GAA TTC CCG GGG ATC CGT CGA CCT GCA GCC AAG

     EcoRI   Smal    BamHI   SalI    PstI

     CTA ATT CCG GGC GAA TTT CTT ATG ATT TAT GAT
       *                               *         *

     TTT TAT TAT TAA ATA AGT TAT AAA AAT AAG TGT
                   *       *           *       *

     ATA CAA ATT TTA AAG TGA CTC
                         *
```

Fig. 1 Multicloning site of pBHA. The hemagglutinin tag (HA) and unique restriction sites are indicated. Downstream stop codons in all three frames are marked by asterisks (*). The LexA sequence ends directly 5′ of the sequence shown here.

C. What to Do When Bait is Self-Activating

Self-activation is a poorly understood process. Many baits, probably for different reasons, will activate the reporter genes to variable degrees even when cotransfected with an empty pGAD10 vector. For instance, this may be due to the bait behaving like a true transcription activator, binding to a transcription factor, or being nonspecifically able to recruit transcriptional machinery. Self-activation is a common problem with two-hybrid baits, and will make a successful screen next to impossible. Since it is impossible to predict if a bait is activating, all bait constructs must be tested by cotransfection with empty library vector. Basically, two steps can be taken if a bait does activate transcription. If the self-activation is weak (i.e., β-Gal activity and growth on His⁻ plates are weak), it may still be possible to detect strong interaction above that background (especially if His⁻ growth is suppressed with 3-AT). The other option is to alter the bait. Often subtracting or adding a few amino acids on either side of the bait can abolish self-activation. This may be because these modifications allow for proper folding of the bait, thus avoiding nonspecific binding of an incorrectly folded bait to proteins of the transcriptional machinery, for instance. In other cases, larger bits may have to be added or deleted from the bait. Unfortunately, this is not a predictable process. We routinely make a series of alternative overlapping baits to minimize this problem. Generally, we find that longer baits are more likely to be self-activating. For instance, we originally attempted to use the C-terminal 350 amino acids of NR2A (residues 1115–1464) for a two-hybrid screen. That construct proved to be self-activating. (Note that self-activation is system dependent, and a bait may not be self-activating in the context of one vector, but strongly in the context of another.) By contrast, a shorter construct restricted to the C-terminal 116 amino acids of NR2A (Niethammer et al., 1996) displays virtually no self-activation of reporter genes.

If the two-hybrid system is simply used to test for interaction between specific known proteins (rather than for a library screen), a third option should be considered. Simply reversing the two proteins between DNA binding and activation domain vectors can, and most of the time will, abolish the problem. That is to say, it is unlikely that both interacting proteins will self-activate reporter gene transcription.

Another approach has been reported by Cormack and Somssich (1997), adding a specific repressor sequence to the bait construct. With this method they were able to repress self-activation that even high levels of 3-AT could not suppress. Although this approach remains to be tested in a screen situation, it may provide a last ditch solution when all else fails.

D. Sequence of Events for Library Screen

All the protocols needed for a library screen are given below. The following summary of the events will help to plan the experiments. It takes about 2–3 weeks to complete an entire screen. First, yeast are cotransformed with a bait plasmid and cDNA library and plated on trp⁻leu⁻his⁻ plates. Three days later, the yeast colonies that grow independently of histidine are further assayed for expression of the second reporter gene (β-galactosidase activity). Positive colonies need to be restreaked again on trp⁻leu⁻his⁻ plates. We streak the colonies with a sterile loop onto sectors of 150 mm plates. This helps with the isolation of pure clones and also confirms that colonies picked were truly showing histidine-independent growth and β-galactosidase activity. This step is analogous to purification of a positive plaque from a conventional hybridization screen. After 2 days, these restreaked colonies are again assayed for β-galactosidase activity. If it is still not possible to isolate single colonies, another restreaking is necessary. Plasmid DNA is isolated from the yeast and transformed into bacteria by electroporation. The isolated library plasmids are then transformed one more time into yeast together with the original bait plasmid in order to check if this reconfers the interaction phenotype. This step is needed to confirm isolation of truly interacting plasmids from the library.

E. Yeast Transformation

To transform yeast, we use a method described by Bartel and Fields (1995). This method is a modified version of the lithium acetate method developed by Ito *et al.* (1983) and improved by Schiestl and Gietz (1989) and Hill *et al.* (1986). Dimethyl sulfoxide (DMSO) is added to improve transformation efficiency (Hill *et al.*, 1991). All the volumes and amounts given here are for testing interactions between two known proteins. For library screens, these must be adjusted as described at the end of the protocol.

1. Grow overnight culture of yeast in 20–40 ml YPD (yeast extract, peptone, dextrose) medium (inoculated from a recently streaked YPD plate grown for 2–3 days at 30 °C) in a sterile 250-ml flask, in shaking water bath at 30 °C.

2. Dilute overnight culture into fresh YPD (~50- to 100-fold) in a sterile flask (the YPD volume should never exceed more than 20% of total flask volume; ideally it should be less than 10%). The total final volume should be about 10 ml culture per intended transformation. Grow for another 3–4 h with shaking at 30 °C (the final OD_{600} should be about 0.20–0.25). Some protocols such as that described by Bartel and Fields (1995) grow the yeast until an OD_{600} of 0.5 is reached. Generally, we find reduced efficiency with ODs under 0.2; for most transformations any OD between 0.2 and 0.8 seems to work well.

Time saver: The procedure outlined in steps 1 and 2 will give optimal transformation efficiency. This is really only needed for library screens. For standard two-hybrid assays testing interaction between two known proteins, transformation efficiency is less important. In this case the yeast can be diluted the night before and grown overnight to an OD of 0.2–0.8 (assume a 2-h doubling time). We usually try to err on the side of overgrowth, since OD values higher than 0.25 are easily tolerated, whereas lower OD values will diminish the transformation efficiency. For this procedure it is useful to grow the 20 ml culture an extra day to saturation, before the final dilution.

3. Harvest the cells by centrifuging 5–10 min at 1500–2000 × *g* at 4 °C, in 50-ml sterile conical tubes. Decant supernatant and wash the cells by resuspending the pellet in 10 ml sterile water per 50-ml tube and pelleting again at 1500–2000 × *g* for 5–10 min. At this stage, if several 50-ml tubes have been used, pool the cells after resuspending each pellet in water.

4. Resuspend the washed yeast cell pellet in freshly made 1× LiAc/1× TE, pH 7.5, buffer at a density of 100 μl per transformation (i.e., per 10 ml of the original culture).

5. Preload transforming DNAs into 1.5-ml microfuge tubes. This can be done toward the end of the yeast growth phase, and the DNA kept on ice. Use 200 ng to 1 μg of each plasmid for the cotransformation, with ~100 μg of carrier single-stranded salmon sperm DNA (10 mg/ml stock). Whenever feasible, use master mixes to minimize tube-to-tube variability. If the total volume of DNA is greater than 10 μl, it is worthwhile to add the appropriate amounts of 10× LiAc and 10× TE to bring contents to 1× buffer concentrations. However, for most applications, it is sufficient to mix 2 μl of each plasmid (assuming typical minipreparation concentrations of 100–300 ng/μl) and 6 μl (60 μg) of salmon sperm DNA. Add 100 μl of yeast cells (resuspended in step 4) to each tube of DNA.

6. Add 0.6 ml of freshly made 40% polyethylene glycol (PEG) solution in 1× LiAc/TE (i.e., 0.8 volume of 50% PEG, 0.1 volume 10× LiAc, 0.1 volume 10× TE) to each transformation. Invert the tube several times and vortex *briefly* to mix contents. Because 40% PEG is quite viscous, take care to mix well but without excess agitation of cells.

7. Incubate 30 min in 30 °C water bath, shaking moderately. Mix occasionally.

8. Add 70 μl of DMSO. Vortex briefly or invert several times.

9. Heat shock for 15 min in a 42 °C water bath. No mixing is required during this stage.

10. Spin 5–10 s in microfuge (high speed) and carefully aspirate supernatant.

11. Resuspend pellets well in 1 ml YPD medium (the pellet will be rather clumpy because of the PEG), and incubate 2–3 h in 30 °C shaking water bath, to allow for *HIS3* expression. This rather long outgrowth is due to the slow doubling time of yeast, as well as the fact that for successful *HIS3* expression both the bait and the library plasmid need to be expressed before activation of the *HIS3* gene can occur.

12. Spin cells 30 s in microfuge. With a quick shake of the wrist, decant most of the YPD from the microtube. Resuspend the pellet in the remaining YPD (about 50 μl). Plate half of the resuspension on each of the selection plates (trp⁻leu⁻his⁻ and trp⁻leu⁻). If only plating on trp leu plates, no outgrowth is needed. Just directly resuspend the heat shocked cells in 50–100 μl YPD and plate. The yeast can either be plated on 60-mm plates or on sectors of 150-mm plates. Grow 2–3 days at 30 °C and perform X-Gal assay. Library screens are always grown 3 days.

For the screening of a cDNA library, the following adjustments need to be made: if a library is screened, the preceding protocol needs to be scaled up. A single transformation (to be plated on a 150-mm plate) is done with 5–10 μg of each plasmid (bait and library), 200 μg of carrier DNA, and 200 μl of yeast (still diluted at 100 μl per 10 ml of original culture). Double the volumes of PEG and DMSO as well. We normally perform a library screen with the equivalent of 5–10 such 200-μl transformations. In this case, the transformation can be pooled in 50-ml conical tubes or Falcon 2059 tubes rather than doing separate microtubes. In the final step, the outgrowth is then resuspended into 400 μl YPD per 150-mm petri dish (trp⁻leu⁻his⁻ selection, with 2.5–5 mM 3-aminotriazole). In addition, do not forget to make a 1:200–1:400 dilution (i.e., 2–4 μl of the resuspension into 100 μl YPD) and plate these on 100-ml leu⁻trp⁻ plates (containing His!) to test for transformation efficiency. This step is critical to evaluate the efficiency of the screen. We get routinely between 50,000 and 150,000 transformants per 150-mm plate using the amount of yeast and DNA described earlier (i.e., for a 10-plate screen 500,000–1,500,000 colonies total).

F. Reporter Gene Activation Assays

Usually, only a small fraction of transformed yeast grows on his⁻ plates. Once the yeast colonies grow up (usually after 2 or 3 days), they can be tested for β-galactosidase activity. The His selection even in the presence of 3-AT is not always complete (and sometimes spontaneous mutations can make the yeast His⁺ again), so β-galactosidase activation serves as an independent secondary screen to ensure that colonies growing on the trp⁻leu⁻his⁻ plates contain interacting fusion proteins. When testing for interaction of already isolated proteins, we usually plate the yeast on trp⁻leu⁻his⁻ and trp⁻leu⁻ plates. The number of colonies growing without histidine compared to the growth with histidine correlates well with β-galactosidase activity, and can serve as an additional measure of strength of interaction.

There are a variety of ways to assay for β-galactosidase activity. We use a filter assay, both for isolation of colonies in a screen and to test interactions. The relative times it takes for colonies to turn blue in our experience correlates well with the more quantitative colorimetric assays (Kim *et al.*, 1995) and has the advantage of convenience (protocols of quantitative liquid colorimetric assays are described elsewhere; Bartel and Fields, 1995; Bartel *et al.*, 1993).

1. X-Gal is kept as a stock solution of 20 mg/ml in *N,N*-dimethylformamide at −20 °C. When performing the assay, add 1.67 ml X-Gal stock per 100 ml Z buffer (see Appendix). Some protocols will also add 0.27 ml 2-mercaptoethanol to this, but this is usually unnecessary. Put a filter paper of appropriate size (e.g., Whatman 1, Clifton, NJ) into a petri dish (1 per yeast plate), and add X-Gal solution. Be careful not to add too much (about 3.5–4 ml for a 150-mm plate); there should be no excess pooling of solution.

2. For library screens we use nitrocellulose membranes (e.g., Protran, Schleicher & Schuell, Keene, NH, pore size 0.45 μm). Colonies stick better to nitrocellulose and stay more discretely separate from each other. Since nitrocellulose is more expensive, filter paper (e.g., Whatman 1) can be substituted when testing yeast two-hybrid interactions between known proteins.

3. Place the filter onto the agar plate (make sure to avoid air bubbles) until completely moist. Mark the filter in some way that will allow for orientation of filter to agar. Grab with forceps, and pull off in a smooth, rapid motion. Be careful not to drag the filter/nitrocellulose over the plate, as colonies may get smeared. Submerge in liquid nitrogen for about 10–20 s (be careful when using nitrocellulose as it gets very brittle during freezing; supported nitrocellulose circles are more sturdy, e.g., Optitran, Schleicher & Schuell). This freeze–thaw step permeabilizes the yeast.

4. Place the filter onto the X-Gal soaked filter, *yeast colony side up*. Again make sure that no air gets trapped underneath. It is best to hold the filter with forceps and lower it gently from one side as it thaws.

5. Start a timer and wait for the yeast colonies to turn blue. Some protocols place the yeast at 30 °C during this time. We keep it at room temperature, which seems to work well in our experience. Since X-Gal time can be somewhat variable depending on room or bench temperature and colony density, it is important to have a positive control included in every assay (including library screens). Weak interactors can take quite a while; it is thus advisable to add a negative control as well because many weakly self-activating baits will cause the yeast to turn blue after many hours (we routinely watch for at least 4 h for a library screen).

6. Whether performing a library screen or restreaking colonies, one can just pick the positive colonies with a sterile toothpick directly from the filters (there is sufficient viable yeast left after freezing for subsequent growth).

The filter X-Gal assay is very reliable. It is, in addition, semiquantitative, in that it assays the relative time it takes the different colonies to turn blue. Unlike colorimetric assays, the threshold of "blueness" is set by the observer—this may cause some variability between different people. Usually, the differences seem to be not large, and the results are consistent and reproducible, especially when compared with a positive control as internal reference. It is also not linear, in that it sets a lower limit for the speed at which the assay can work (usually around 12–15 min at room temperature). On the other hand, it does have the advantage of great sensitivity.

G. Recovery of Plasmids from Yeast

There are several protocols for the isolation of plasmid DNA from yeast. Generally, it is not a trivial procedure, because of the tough yeast cell wall. Also, DNA yields are low, and the DNA is contaminated with genomic DNA and cellular proteins. We use glass beads to disrupt the cell walls mechanically. A variation of this method described in the Clontech yeast protocol handbook (PT3024-1, Clontech Laboratories) additionally uses an enzyme preparation (lyticase) to weaken the cell walls. We do not find this to be necessary, but it may increase the yield for some low copy number plasmids. The resulting DNA is very impure (it is not useful for sequencing or restriction digests), but sufficient for electroporation into bacteria. A cleaner, higher yield DNA preparation is then obtained from bacteria by standard plasmid preparation methods.

1. Inoculate single yeast colonies in 2 ml of the appropriate selection medium (leu⁻) and incubate at 30 °C overnight.

2. Transfer cultures to microtubes and pellet in microfuge (5 s). Decant the supernatant and resuspend the pellet in the residual liquid by vortexing.

3. Add 0.2 ml of yeast lysis buffer, 0.2 ml of neutralized phenol/chloroform/isoamyl alcohol (25:24:1) and about 0.3 g (it is not critical to be very exact here) acid-washed glass beads (Celis, 1994) (Sigma, St. Louis, MO).

4. Vortex for 2 min at room temperature. This step is critical to disrupting the yeast cell walls.

5. Spin 5 min at room temperature in microfuge. Transfer the supernatant to a clean tube and ethanol precipitate the DNA. Wash with 70% (v/v) ethanol and dry under vacuum.

6. Resuspend the pellet in a small volume of TE (25 μl).

7. Transform *E. coli* with electroporation.

Bartel and Fields (1995) describe essentially the same method, except they directly lyse a single yeast colony without overnight growth by adding 50 μl of lysis solution and phenol/chloroform, and 0.1 g of glass beads.

The final DNA is rather crude, but works well for electroporation. We usually get 20–200 colonies of *E. coli* with 1.5 μl of the yeast plasmid prep. Of course, this preparation will result in isolation of both the bait and the library plasmid. In a library screen it is essential to recover the library plasmid vector (pGAD10) rather than the known bait construct. Fortunately, the *LEU2* marker carried by pGAD10 complements the bacterial *leuB* marker. This can be exploited by using a bacterial strain with a *leuB* mutation, such as HB101 (Bolivar and Backman, 1979; Boyer and Roulland-Dussoix, 1969) or MH4 (Hall *et al.*, 1984). After electroporation, the cells are plated on M9 minimal plates (with 50 μg/ml ampicillin) containing all amino acids except leucine. As a backup, we usually also plate some of the cells on LB (50 μg/ml Amp) plates. This will not select for pGAD10 library plasmids but the cells usually grow better, which sometimes helps when transformation efficiencies are very low. The disadvantage is that they will then need to be rescreened on M9 plates to distinguish between library plasmids and bait plasmid.

Library plasmid DNA is then prepared from the transformed bacteria by standard methods. It is advisable at this stage to perform some crude restriction mapping, especially if a large number of clones have been isolated. This may give hints about multiple hits, an important factor in judging the veracity of the results (see below).

Since it is possible for multiple copies of pGAD10 to get transformed into the same yeast cells, clones may be isolated during this procedure that were not responsible for the original reporter gene activation. We thus usually pick at least two colonies from each bacterial transformation. Also, it is absolutely essential to retransform the isolated library plasmid into yeast together with the original baits. These retransformants are then assayed for histidine-independent growth and β-galactosidase activity. Failure to reconfer such activity eliminates clones that falsely showed up as positive in the original screen, a common occurrence. It is not unusual for half or more of the clones to drop out at this point. At this stage, it is also sensible to retransform the isolated clones with an entirely unrelated bait, as well as empty bait vector, which can eliminate pGAD10 constructs that can activate transcription by themselves or nonspecifically interact with different baits.

H. Judging Quality of Library Screen

The number of transformants (as calculated from leu⁻trp⁻ plating efficiency) is a measure of completeness of the library screen. Ideally, the total number of screened colonies should exceed the number of independent clones of the library. Note that in a random fragment library, only one-sixth of clones will generate true in-frame fusions with the GAL4 activation domain.

If positive clones are isolated, the main concerns are to eliminate false positives, that is, clones not binding to the bait, or clones where the binding has no *in vivo* biological relevance. The first type of false positives, clones not binding (or nonspecifically binding) to the bait but spuriously isolated during the screen, is eliminated by retransformation with the original bait vector. This class of false positives will not induce *HIS3* or β-galactosidase activity when coexpressed with the original bait in fresh yeast, or they will do so by interacting indiscriminately with various unrelated baits.

The other type of false positives is much more difficult to exclude—clones that indeed bind specifically to the bait in yeast, but whose interaction is of no biological significance *in vivo*. For example, the isolated "interacting" protein might never be in the same cellular or subcellular localization as the protein used for the bait.

So how is a screen assessed for the potential validity of the isolated clones? Generally, outcomes of screens can be grouped into three categories: (1) no positive clones are isolated, that is, even if there were blue colonies in the original screen, the recovered library plasmids do not reconfer *HIS3* and β-galactosidase activity with the original bait; (2) several identical or overlapping fragments of the same gene, or fragments of related genes are isolated, with very little else; or (3) many different, unrelated fragments are isolated.

In the case of category 1, if a large enough number of colonies was screened, the bait may either not "work" in the two-hybrid system or there may be no interacting clones of high enough affinity represented in the library. Either way, not much time is wasted. Category 2 represents the ideal scenario—not only were the fragments isolated several times (which at least shows a robust interaction), but it may even be possible to map the region of interaction based on common regions shared by overlapping cDNA fragments of the same gene. This scenario generally predicts that the isolated clones will be interesting to study and are true interactors *in vivo*. Situation 3, on the other hand, can be difficult. Unless there is a compelling reason to follow up some of the clones (such as previously suspected interactors), it becomes a game of chance. It may be advisable to modify the bait and perform another screen. We often perform more than one screen in parallel with slightly differing baits or different libraries, thus possibly isolating the same protein from more than screen. Using more than one library also increases the complexity of the screen.

Some clones such as heat shock proteins or ribosomal proteins seem to pop up frequently in two-hybrid screens (Hengen, 1997). The laboratory of Erica Golemis at the Fox Chase Cancer Center (Philadelphia, PA) maintains a web page that lists some of the most common false-positive clones isolated in two-hybrid screens; this may be useful in excluding some of the screen results (http://www.fccc.edu/research/labs/golemis/InteractionTrapInWork.html).

If any of the clones are judged to be useful, that is, the interaction in the two-hybrid system is specific, an independent assay should be performed to confirm the interaction in an independent manner, such as an *in vitro* filter overlay assay using fusion proteins (Li *et al.*, 1992). It may also be possible to confirm the interaction in recently developed mammalian and bacterial two-hybrid systems (Dove *et al.*, 1997; Luo *et al.*, 1997).

The problem of proving the validity of the interaction *in vivo* is the major challenge, and is discussed in Wyszynski and Sheng (1998).

IV. Limitations, Perspectives, and Outlooks

When doing a two-hybrid screen, it is important to keep in mind that many protein–protein interactions, even if they do occur *in vivo*, are simply undetectable by this method. Examples are baits or interactions that are toxic when expressed in yeast, or proteins that do not express, do not fold or modify properly, are degraded, or never enter the nucleus. Expression can be verified by methods described elsewhere (Langlands and Prochownik, 1997; Printen and Sprague, 1994), but some of the other problems may be unsolvable, even by altering the bait. For reasons outlined earlier, ion channels are likely to be susceptible to the folding problem, especially with loops flanked by transmembrane domains. It may thus be fundamentally impossible to detect some type of interactions between ion channels and other proteins using the two-hybrid system. Nevertheless, we feel that the yeast two-hybrid system is a powerful approach for identification of novel interactions. Particularly, the speed of the screen makes it a very viable option because it allows for the testing of many different parts of ion channels. Our understanding of ion channel-associated protein complexes has rapidly expanded in recent years, thanks to the two-hybrid system, and this will be a major factor in dissecting functional aspects of synapses in the future.

Appendix: Solutions

For convenience, we provide recipes for most of the solutions and media needed for the two-hybrid screen.

YPD Medium

Per liter, to 950 ml of deionized water, add:

Bacto-peptone	20 g
Bacto-yeast extract	10 g
Glucose	20 g

Shake until dissolved, adjust to 1 l, autoclave for 20 min at 15 lb/in^2 on liquid cycle. (*Note:* Some of the glucose will caramelize during autoclaving, giving the medium a tea-colored appearance. This is not a problem—just do not leave the medium in the autoclave for a long time after the cycle is finished.) For YPD plates add 15 g/l Bacto-agar before autoclaving.

Minimal Plates and Media

Yeast minimal plates. Per liter, to 800 ml of deionized water, add:

Yeast nitrogen base without amino acids	6.7 g
Bacto-agar	15 g

Adjust volume to 860 ml, autoclave for 20 min at 15 lb/in^2 on liquid cycle. Let cool to about 45 °C (water bath). Add 40 ml 50% glucose (filter sterilized) and 100 ml 10× dropout (HLT$^-$, see below). For LT$^-$ plates, 10 ml 100× His (see below). (*Note:* It is advantageous to add a stir-bar into the flask before autoclaving. It will allow you to mix in the dropout and glucose without creating bubbles, which are a nuisance during plate pouring.)

If background depression on HLT$^-$ plates is desired, add an appropriate amount of 1 M 3-aminotriazole (usually to a final concentration of 2.5 or 5 mM).

Selection medium is made the same way, except without Bacto-agar.

10× dropout HLT$^-$. This solution is deficient for His/Leu/Trp. Per liter, to 950 ml of deionized water, add:

Adenine	200 mg
L-Arginine hydrochloride	200 mg
L-Aspartic acid	1000 mg
L-Glutamic acid	1000 mg
L-Isoleucine	300 mg
L-Lysine hydrochloride	300 mg
L-Methionine	200 mg
L-Phenylalanine	500 mg
L-Serine	1350 mg
L-Threonine	2000 mg
L-Tyrosine	300 mg
Uracil	200 mg
L-Valine	1500 mg

Adjust to 1 l, filter sterilize (0.2 μm), store at 4 °C.

100× histidine/100× tryptophan. Per 100 ml, dissolve 200 mg L-histidine (monohydrate) or 200 mg L-tryptophan, filter sterilize (0.2 μm), store at 4 °C.

1 M 3-aminotriazole. Dissolve 3-amino-1,2,4-triazole in deionized water (8.4 g/ 100 ml). 3-AT takes a long time to dissolve and may in fact not go completely into solution. Filter sterilize the final solution (0.2 μm); this will also remove any remaining particles. Store at 4 °C. (*Note:* At temperatures even slightly below 4 °C, 3-AT may fall out of solution. If that occurs put it at room temperature until it is redissolved. Protect from light.)

M9 minimal medium. Per liter, to 650 ml of autoclaved deionized water (less than 50 °C), add:

5× M9 salts	200 ml
1 M MgSO$_4$ (autoclaved or filter sterilized)	2 ml
50% glucose (filter sterilized)	8 ml
1 M CaCl$_2$ (filter sterilized)	0.1 ml
10 mg/ml proline (filter sterilized)	4 ml
1 M thiamine hydrochloride (filter sterilized)	1 ml
100 mg/ml ampicillin	0.5 ml

If necessary, supplement with appropriate dropouts, for example, for L$^-$ medium:

10× HLT$^-$ dropout	100 ml
100× histidine	10 ml
100× tryptophan	10 ml

Adjust to 1 l with sterile deionized water. (*Note:* Make sure to add the MgSO$_4$ and CaCl$_2$ after dilution of the M9 salts. Ideally, add CaCl$_2$ at the very last, i.e., after dropouts have been added as well.) M9 is not a very rich medium, so the final concentration of ampicillin is only 50 μg/ml. For M9 plates, add 15 g Bacto-agar to the 650 ml water before autoclaving.

The 5× M9 salts is made by dissolving the following salts in deionized water to a final volume of 1 l:

Na$_2$HPO$_4$ • 7H$_2$O	64 g (or 33.9 g of anhydrous Na$_2$HPO$_4$)
KH$_2$PO$_4$	15 g
NaCl	2.5 g
NH$_4$Cl	5 g

Autoclave the salt solution or filter sterilize.

50% glucose. Dissolve glucose to 50% (w/v). (*Note:* Add glucose slowly to water, and not the other way around, otherwise the glucose will solidify. Slight heating may be necessary to get the glucose into solution.) Filter sterilize (0.2 μm). Store at 4 °C.

Yeast Transformation Solutions

The $10\times$ LiAc is 1 M lithium acetate dihydrate (102 g/l). Adjust pH with acetic acid to 7.5. The $10\times$ TE is 100 mM Tris, 10 mM EDTA. Adjust pH to 7.5. The 50% PEG is 50 g/100 ml polyethylene glycol 3350 in deionized water. Stir overnight to dissolve. Filter sterilize (0.2 μm) all three solutions, store at room temperature.

Yeast Lysis Buffer

Triton X-100	2% (v/v)
SDS	1% (w/v)
NaCl	100 mM
Tris, pH 8.0	10 mM
EDTA, pH 8.0	1 mM

After preparing the solution, filter sterilize (0.2 μm) and store at room temperature.

X-Gal Assay

Z-buffer for X-Gal staining. Per liter, add:

$Na_2HPO_4 \cdot 7H_2O$	16.1 g (8.53 g anhydrous Na_2HPO_4)
$NaH_2PO_4 \cdot H_2O$	5.5 g
KCl	0.75 g
$MgSO_4 \cdot 7H_2O$	0.25 g

Adjust to pH 7.0 with NaOH. Autoclave and store at room temperature.

X-Gal stock solution. X-Gal (5-bromo-4-chloro-3-indolyl-β-D-galactopyranoside) is kept as a stock solution of 20 mg/ml in *N,N*-dimethylformamide at $-20\,°$C in the dark.

Acknowledgments

We thank Elaine Aidonidis for assistance with the manuscript, Roland Roberts for helpful comments, and Cheng-Ting Chien for help in teaching us the yeast two-hybrid system. M. S. is an Assistant Investigator of the Howard Hughes Medical Institute.

References

Bartel, P. L., and Fields, S. (1995). *Methods Enzymol.* **254**, 241.
Bartel, P. L., Chien, C. T., Sternglanz, R., and Fields, S. (1993). *In* "Cellular Interactions in Development: A Practical Approach," (D. A. Hartley, ed.), p. 153. Oxford University Press, Oxford.

Béranger, F., Aresta, S., de Gunzburg, J., and Camonis, J. (1997). *Nucleic Acids Res.* **25,** 2035.

Bolivar, F., and Backman, K. (1979). *Methods Enzymol.* **68,** 245.

Boyer, H. W., and Roulland-Dussoix, D. (1969). *J. Mol. Biol.* **41,** 459.

Brenman, J. E., Chao, D. S., Gee, S. H., McGee, A. W., Craven, S. E., Santillano, D. R., Wu, Z., Huang, F., Xia, H., Peters, M. F., Froehner, S. C., and Bredt, D. S. (1996). *Cell* **84,** 757.

Brent, R., and Ptashne, M. (1985). *Cell* **43,** 729.

Celis, J. E. (ed.) (1994). *In* "Cell Biology." Academic Press, San Diego, CA.

Chien, C. T., Bartel, P. L., Sternglanz, R., and Fields, S. (1991). *Proc. Natl. Acad. Sci. USA* **88,** 9578.

Cormack, R. S., and Somssich, I. E. (1997). *Anal. Biochem.* **248,** 184.

Dong, H., O'Brien, R. J., Fung, E. T., Lanahan, A. A., Worley, P. F., and Huganir, R. L. (1997). *Nature* (*London*) **386,** 279.

Dove, S. L., Joung, J. K., and Hochschild, A. (1997). *Nature* (*London*) **386,** 627.

Fields, S., and Song, O. K. (1989). *Nature* (*London*) **340,** 245.

Froehner, S. C. (1993). *Annu. Rev. Neurosci.* **16,** 347.

Gautam, M., Noakes, P. G., Mudd, J., Nichol, M., Chu, G. C., Sanes, J. R., and Merlie, J. P. (1995). *Nature* (*London*) **377,** 232.

Hall, M. N., Hereford, L., and Herskowitz, I. (1984). *Cell* **4,** 1057.

Hengen, P. N. (1997). *Trends Biochem. Sci.* **22,** 33.

Hill, J. E., Myers, A. M., Koerner, T. J., and Tzagoloff, A. (1986). *Yeast* **2,** 163.

Hill, J., Donald, K. A., and Griffiths, D. E. (1991). *Nucleic Acids Res.* **19,** 5791.

Hollenberg, S. M., Sternglanz, R., Cheng, P. F., and Weintraub, H. (1995). *Mol. Cell Biol.* **15,** 3813.

Holmes, T. C., Fadool, D. A., Ren, R., and Levitan, I. B. (1996). *Science* **274,** 2089.

Ito, H., Fukuda, Y., Murata, K., and Kimura, A. (1983). *J. Bacteriol.* **153,** 163.

Keegan, L., Gill, G., and Ptashne, M. (1986). *Science* **14,** 699.

Kim, E., Niethammer, M., Rothschild, A., Jan, Y. N., and Sheng, M. (1995). *Nature* (*London*) **378,** 85.

Kirsch, J., Wolters, I., Triller, A., and Betz, H. (1993). *Nature* (*London*) **366,** 745.

Kornau, H. C., Schenker, L. T., Kennedy, M. B., and Seeburg, P. H. (1995). *Science* **269,** 1737.

Langlands, K., and Prochownik, E. V. (1997). *Anal. Biochem.* **249,** 250.

Li, M., Jan, Y. N., and Jan, L. Y. (1992). *Science* **257,** 1225.

Luo, Y., Batalao, A., Zhou, H., and Zhu, L. (1997). *BioTechniques* **22,** 350.

Ma, J., and Ptashne, M. (1988). *Cell* **55,** 443.

McKnight, J. L., Kristie, T. M., and Roizman, B. (1987). *Proc. Natl. Acad. Sci. USA* **84,** 7061.

Niethammer, M., Kim, E., and Sheng, M. (1996). *J. Neurosci.* **16,** 2157.

Printen, J. A., and Sprague, G. F., Jr. (1994). *Genetics* **138,** 609.

Sambrook, J., Fritsch, E. F., and Maniatis, T. (eds.) (1989). *In* "Molecular Cloning: A Laboratory Manual." Cold Spring Harbor Laboratory Press, New York.

Schiestl, R. H., and Gietz, R. D. (1989). *Curr. Genet.* **16,** 339.

Tsunoda, S., Sierralta, J., Sun, Y., Bodner, R., Suzuki, E., Becker, A., Socolich, M., and Zuker, C. S. (1997). *Nature* (*London*) **388,** 243.

Wyszynski, M., and Sheng, M. (1998, this volume). *Methods Enzymol.* **293,** Chapter 20.

Wyszynski, M., Lin, J., Rao, A., Nigh, E., Beggs, A. H., Craig, A. M., and Sheng, M. (1997). *Nature* (*London*) **385,** 439.

Yu, X. M., Askalan, R., Keil, G. J., and Salter, M. W. (1997). *Science* **275,** 674.

PART III

Electrophysiology

CHAPTER 3

Patch–Clamp Studies of Cystic Fibrosis Transmembrane Conductance Regulator Chloride Channel

John W. Hanrahan, Zie Kone, Ceri J. Mathews, Jiexin Luo, Yanlin Jia, and Paul Linsdell

Department of Physiology
McIntyre Medical Science
McGill University
Montreal, Quebec, Canada

DOI: 10.1016/B978-0-12-382204-8.00003-5

I. Introduction

The cystic fibrosis transmembrane conductance regulator (CFTR) has two sets of six membrane-spanning regions (TM1–TM12), two nucleotide-binding folds (NBF1, NBF2), and a regulatory domain containing numerous potential sites for phosphorylation by protein kinases (Riordan *et al.*, 1989). It belongs to an important superfamily of ATP-binding cassette (ABC) transport proteins that has members in bacteria, yeast, and higher eukaryotes. Although it may have multiple functions, it is generally accepted that CFTR functions as an ATP-dependent, phosphorylation-activated Cl^- channel (Anderson *et al.*, 1991a,b; Bear *et al.*, 1992; Gadsby *et al.*, 1995; Kartner *et al.*, 1991). In this chapter we describe methods that we have found useful for studying the channel activity of CFTR. The reader is referred to Vol. 207 of this series for more general information on patch-clamp techniques (Rudy and Iverson, 1992).

II. Choice of Expression Systems

Functional studies of endogenous CFTR have been carried out using epithelial (Champigny *et al.*, 1990; Gray *et al.*, 1988, 1990; Haws *et al.*, 1994; Tabcharani *et al.*, 1990) and cardiac cells (Baukrowitz *et al.*, 1994; Hart *et al.*, 1996; Nagel *et al.*, 1992). However, because it is often difficult to obtain gigaohm seals on native epithelial cells, heterologous expression systems are generally preferred for biophysical studies. Many cell lines have been used successfully for transient and stable transfection, including Chinese hamster ovary (CHO) cells, fibroblasts, baby hamster kidney cells (BHK), and human enbryonic kidney (HEK 293) cells. The choice of preparations, of course, depends on the purpose of the experiment. We find transient expression of CFTR in *Spodoptera frugiperda* fall armyworm ovary (Sf9) (Kartner *et al.*, 1991) and *Xenopus* oocytes (Bear *et al.*, 1991) to be inconvenient for detailed patch-clamp studies mainly due to the difficulty of controlling the level of expression. In our experience, the additional work involved in the preparation of stable lines is worthwhile (see Chang *et al.*, 1993). Hi-5 insect cells are reportedly (Yang *et al.*, 1996) superior to Sf9 cells because they do not have other channel types that could potentially be confused with CFTR (Gabriel *et al.*, 1992). Single CFTR channels can be recorded on *Xenopus* oocytes, but this requires careful titration of the amount of RNA injected into the oocytes (or DNA for nuclear injections) to achieve low expression.

The ideal expression system for investigation of the unitary properties of CFTR would have a low but consistent level of CFTR expression so that, on average, one channel is observed per patch. This dream is seldom realized but patches containing single channels are sometimes obtained on CHO cells that have been stably transfected with pNUT-CFTR. We find 3–10 channels to be more typical

(Tabcharani *et al.*, 1991). The main disadvantage of CHO cells is the presence of endogenous channels, which have higher conductance and become active once membrane patches are excised. This problem is not serious when only short recordings are needed, but it becomes a severe limitation when long (10- to 20-min) recordings from excised patches are desired, such as when studying modulation (Jia *et al.*, 1997a). CFTR channel activity runs down rapidly when membrane patches are excised from CHO or epithelial cells into bath solution lacking protein kinase A (PKA; <100 s at 20 °C, and <15 s at 37 °C). This rundown can be advantageous when studying the pharmacology of the membrane-associated phosphatase, which presumably dephosphorylates and inactivates CFTR after patch excision (Becq *et al.*, 1994). We have found that the membrane-associated phosphatase activity of CHO cells can be used to remove constitutive phosphorylation on CFTR simply by holding excised, inside-out patches for at least 10 min before testing exogenous kinases (see below). However, because stable activity is not achieved, rundown prevents studies of CFTR regulation by exogenous phosphatases. Examining the direct effects of low ATP concentrations on channel gating is also problematic, because MgATP must be present as a substrate for PKA in order to maintain channel phosphorylation.

Fortunately, CFTR channels do not usually run down in patches excised from BHK cells, perhaps because CFTR expression levels are extremely high compared to those of endogenous phosphatases. As discussed later, BHK cells express up to 7000 channels per patch when transfectants are selected with 500 μM methotrexate (Seibert *et al.*, 1995). This high density may enable most channels to escape dephosphorylation by endogenous phosphatases. These patches have very large currents and little rundown, therefore they are useful for macroscopic selectivity measurements (Linsdell and Hanrahan, 1996) and for examining regulation by exogenous phosphatases (Luo *et al.*, 1998).

III. General Patch-Clamp Methods

Pipettes are prepared from borosilicate glass using a two-stage vertical puller (PP-83, Narishige Instrument Laboratory, Tokyo). For single-channel recording we use pipettes that have resistances of 4–6 MΩ when filled with 150 mM NaCl solution. For macroscopic current recording, a lower heat setting is used during the second pull so that pipettes have 1- to 2-MΩ resistance. We do not routinely coat pipettes (e.g., with Sylgard; Dow Corning, Midland, MI) because our experiments on the voltage-independent CFTR channel do not require rapid changes in membrane potential. Furthermore, because our CFTR-expressing cell lines have very good sealing properties, we do not find it necessary to firepolish the pipettes. The bath solution is always grounded through an agar bridge having the same ionic composition as the pipette solution. Single-channel currents are amplified (Axopatch 1B-C or 200 A-B, Axon Instruments, Inc., Foster City, CA), recorded on video cassette tape by pulse-coded modulation-type recording adapters (e.g., DR384, Neurodata Instrument Co., New York) and low-pass filtered during playback using eight-pole Bessel-type filters (900 LPF, Frequency Devices, Haverhill, MA). Alternatively, data are filtered and then sampled by a standard Digidata 2000 interface at frequencies that are two- to fivefold higher than the filter setting.

IV. CFTR Channel Permeation

A. Selectivity

CFTR is highly selective for anions over monovalent cations and has anion permeability ratios consistent with a lyotropic sequence (Linsdell *et al.*, 1997a,b; Tabcharani *et al.*, 1997). Low iodide permeability is commonly used as a diagnostic property of CFTR, but this criterion must be used with care since I^- permeability depends on the experimental protocol used. Single-channel studies indicate that P_I/P_{Cl} is initially high when first exposed to iodide ($P_I/P_{Cl} \sim 1.9$) but falls within tens of seconds to a very low value (Fig. 1). This switch leads to a hysteresis in the current–voltage relationship and is accelerated by holding the membrane patch at potentials that would drive Cl^- into the pore. The precise mechanisms of this switch are not presently known (Tabcharani *et al.*, 1997). The switch to low iodide permeability does not occur under biionic conditions if the ionic strength is elevated to 400 mM (Tabcharani *et al.*, 1992). Also, when only a fraction of the Cl^- is replaced by iodide, the reversal potential shifts in the direction expected when $P_I > P_{Cl}$. However, in most preparations, exposure to iodide eventually leads to a reduction in macroscopic conductance and $P_I < P_{Cl}$.

Because I^- exposure alters the permeability of the channel to I^- itself, the steady-state value of P_I/P_{Cl} after the switch should not be compared with the permeability ratios obtained for other ions in the absence of iodide. CFTR shows a lyotropic or weak field strength selectivity sequence (Linsdell *et al.*, 1997b; Tabcharani *et al.*, 1997), meaning that ions that are relatively easily dehydrated tend to be more permeant than those that retain their waters of hydration more strongly. P_I/P_{Cl} estimated before the switch is exactly the value expected based on the ratios for other halides (Tabcharani *et al.*, 1997).

Selectivity is usually determined by exposing channels to different ions and measuring the resulting change in the zero current (reversal) potential. The simplest situation is to have equal concentrations of different ions present on either side. Under these biionic conditions, with Cl^- in the extracellular solution and a test anion X^- in the intracellular solution, the permeability of X^- relative to that of Cl^- (assuming zero cation permeability) is given by the Goldman–Hodgkin–Katz voltage equation:

$$\frac{P_X}{P_{Cl}} = \exp\left(\frac{E_{rev}F}{RT}\right) \tag{1}$$

A

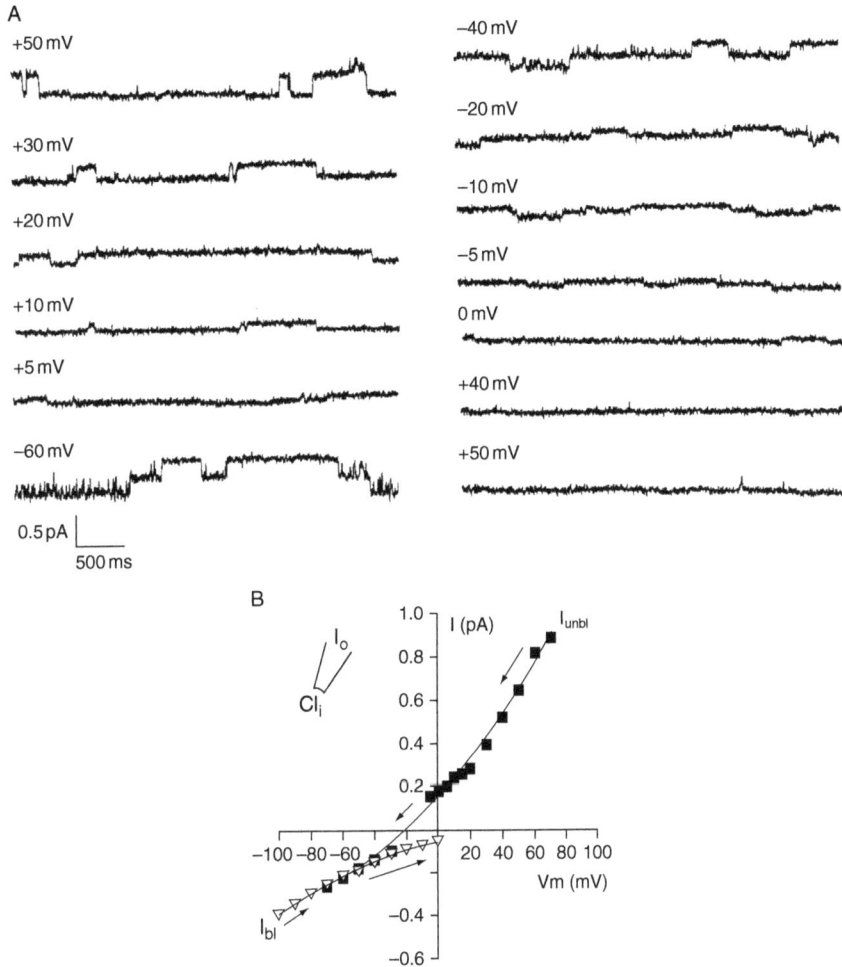

Fig. 1 Single CFTR channel currents recorded under biionic conditions showing both high (solid squares) and low (inverted triangles) permeability to iodide. The criterion of low iodide permeability needs to be used with caution when identifying CFTR's contribution to macroscopic Cl⁻ conductance. With permission from Tabcharani *et al.* (1997).

where E_{rev} is the reversal potential, F is Faraday's constant, R is the gas constant, and T is the temperature in Kelvins. Because recordings made under biionic conditions involve different pipette and bath solutions, a liquid junction potential will occur between these two dissimilar solutions that must be corrected. Liquid junction potentials are determined for different anions relative to Cl⁻ as described previously (Neher, 1992) and are given in Table I.

Table I

Calculated Minimum Dimensions and Measured Liquid Junction Potentials for Different Anions[a]

Anion	Ion dimensions (Å)	Liquid junction potential (mV)
Chloride	$3.62 \times 3.62 \times 3.62$	0
Fluoride	$2.72 \times 2.72 \times 2.72$	-3
Bromide	$3.90 \times 3.90 \times 3.90$	0
Iodide	$4.32 \times 4.32 \times 4.32$	0
Cyanate (OCN⁻)	$3.56 \times 3.56 \times 5.68$	ND[b]
Au(CN)$_4^-$	$3.56 \times 3.56 \times 8.49$	ND
Thiocyanate (SCN⁻)	$3.60 \times 3.60 \times 3.60$	-1
SeCN	$3.80 \times 3.80 \times 6.61$	ND
Nitrate (NO$_3^-$)	$3.10 \times 4.76 \times 5.17$	-1
Formate	$3.40 \times 4.62 \times 4.82$	-3
N(CN)$_2^-$	$3.65 \times 4.47 \times 7.71$	ND
Bicarbonate	$3.40 \times 4.95 \times 5.96$	-7
Acetate	$3.99 \times 5.18 \times 5.47$	-5
Perchlorate (ClO$_4^-$)	$4.63 \times 4.64 \times 4.91$	-3
Propanoate	$4.12 \times 5.23 \times 7.05$	-6
Benzoate	$3.55 \times 6.30 \times 8.57$	-7
Pyruvate	$4.09 \times 5.73 \times 6.82$	-6
Hexafluorophosphate (PF$_6^-$)	$4.92 \times 4.98 \times 5.16$	-2
Ni(CN)$_4$	$3.56 \times 7.39 \times 7.39$	ND
Ethane sulfonate	$4.94 \times 5.37 \times 7.44$	-6
Methane sulfonate	$5.08 \times 5.43 \times 5.54$	-5
C(CN)$_3$	$3.85 \times 7.22 \times 7.85$	ND
Glutamate	$4.67 \times 6.52 \times 10.78$	-8
Isethionate	$5.35 \times 5.79 \times 7.60$	-8
Gluconate	$4.91 \times 6.86 \times 12.09$	-10
Glucoheptonate	$5.50 \times 6.51 \times 13.54$	-11
Glucuronate	$5.23 \times 7.73 \times 9.43$	-12
MES	$6.00 \times 6.99 \times 9.65$	-9
Galacturonate	$6.51 \times 6.66 \times 8.10$	-10
HEPES	$6.64 \times 6.64 \times 12.86$	-11
TES	$6.63 \times 6.75 \times 11.35$	-10
Lactobionate	$7.57 \times 9.32 \times 13.11$	-12

[a]Nonhalide anions are given in order of increasing size (mean diameter, calculated as described in the text). Liquid junction potentials were measured as described (Neher, 1992) with 154 mM of the test ion in the intracellular (bath) solution and 154 mM Cl⁻ in the extracellular (pipette) solution.

[b]ND, not determined.

B. Pore Size

CFTR also discriminates among anions on the basis of their size. The dimensions of the largest permeant anions presumably give some indication of the dimensions of the most constricted part of the pore. To estimate the pore diameter of CFTR we determine the unhydrated diameter of permeant ions, which can be related to permeability ratios according to an excluded volume effect

(Dwyer *et al.*, 1980). Slight permeability to the external organic anions formate, bicarbonate, and acetate, and the apparent impermeability of the larger external anions propanoate, pyruvate, methane sulfonate, ethane sulfonate, and gluconate, suggested that the narrowest region of the CFTR pore has a functional diameter of ~5.3 Å (Linsdell *et al.*, 1997). In that study the functional diameter of an unhydrated ion was taken to be the geometric mean of its two smallest dimensions, based on the hypothesis that the ability of elongated, cylindrical shaped ions to pass through the channel would be relatively insensitive to their length (Linsdell *et al.*, 1997). Table I lists the minimum unhydrated dimensions of a large number of different anions, which we have estimated using Molecular Modeling Pro computer software (WindowChem Software Inc., Fairfield, CA).

C. Barrier Models

Permeating ions are thought to interact briefly with sites in the open pore. According to rate theory models, the free energies and positions within the transmembrane electric field of binding sites and energy barriers control ion permeability and conductance. We have previously generated such models for Cl^- and gluconate permeation in CFTR using the AJUSTE computer program developed by Dr. Osvaldo Alvarez and colleagues (Alvarez *et al.*, 1992). The design and use of the program have been described in detail (Alvarez *et al.*, 1992; Linsdell *et al.*, 1997a). Models for CFTR permeation obtained using this program when the channel was assumed to have either two or three ion-binding sites are shown in Fig. 2. Both models can reproduce the permeation properties of CFTR under a range of ionic conditions (Linsdell *et al.*, 1997b). The three-site model also explains the anomalous conductance of CFTR in Cl^-/SCN^- mixtures, and the loss of this behavior on mutation of a positively charged, pore lining amino acid. These models require many assumptions that are probably invalid. For example, anion-binding sites are assumed to be at fixed positions regardless of the presence of other anions in the pore. Nevertheless, such models provide a useful framework for thinking about anion permeation and the effects of mutations.

D. Macroscopic Current Recording from Excised Patches

When single-channel currents are too small to resolve (e.g., when studying relatively impermeant ions, fast channel blockers, or CFTR mutants having low conductance), we exploit the high expression of CFTR in BHK cells by recording macroscopic currents as described earlier (Fig. 3). This approach has been used to study CFTR channel block (Linsdell and Hanrahan, 1996; Linsdell *et al.*, 1997a) and selectivity (Linsdell and Hanrahan, 1997; Linsdell *et al.*, 1997c). The macroscopic selectivity sequence for small anions is similar to that measured for single channels (Linsdell *et al.*, 1997b; Tabcharani *et al.*, 1997). Reversal potentials are also obtained for large intracellular organic anions such as propanoate, which are not measurable as unitary currents (Fig. 4; see also Linsdell *et al.*, 1997b). Large

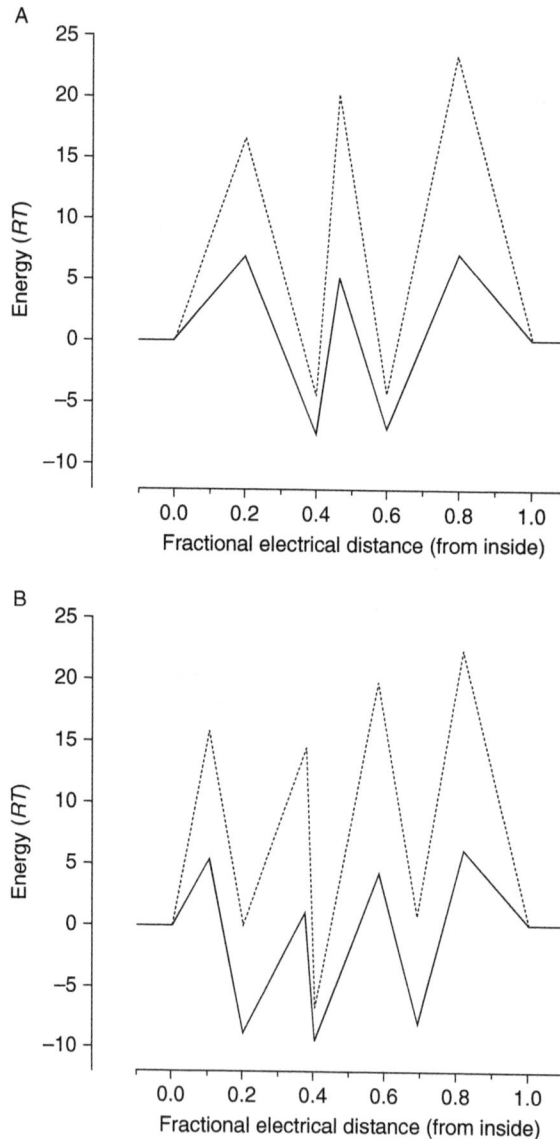

Fig. 2 Best-fit free-energy profiles for Cl^- (solid line) and gluconate (dashed line) movement in the CFTR Cl^- channel pore, for both (A) the three-barrier, two-site model and (B) the four-barrier, three-site model. See Linsdell *et al.* (1997a) for parameters.

anions carry small unitary currents and also act as open channel blockers of Cl^- permeation (Linsdell and Hanrahan, 1996).

After obtaining a gigohm seal on the BHK cell, the patch is normally excised into bath solution containing MgATP but lacking PKA. CFTR channels are

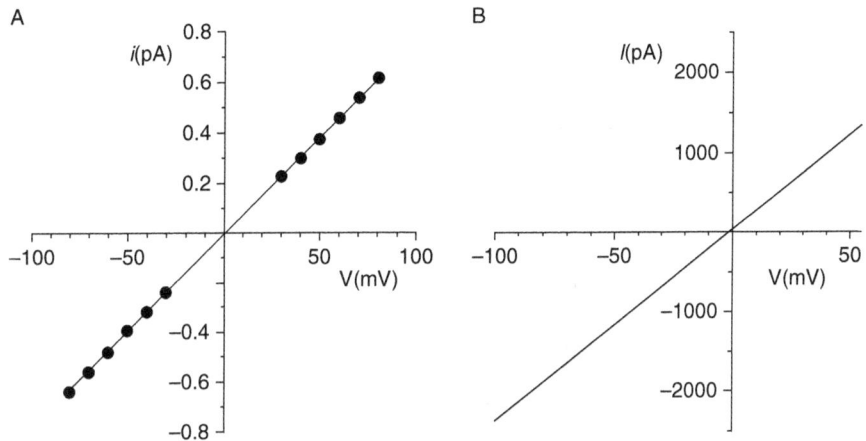

Fig. 3 Comparison of current–voltage relationship for (A) single-channel CFTR channel and (B) for the leak-subtracted macroscopic conductance of a BHK membrane patch. The slopes indicate conductances of 7.8 pS and 24.1 nS, respectively. Assuming similar P_0 values (0.4), these data indicate that the macroscopic current is carried by ~7700 channels.

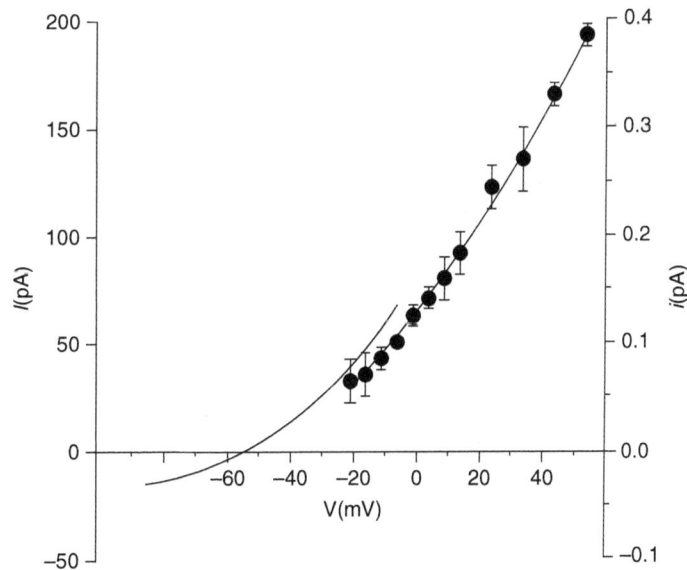

Fig. 4 Illustration of the usefulness of macroscopic selectivity measurements. The solid points indicate the current–voltage relationship obtained with external Cl⁻ and cytoplasmic propanoate. The single-channel currents were too small to allow calculation of a reversal potential (note scale on right). The solid curve without symbols shows a macroscopic I–V relationship (scale on left) that demonstrates finite propanoate permeability.

inactive under these conditions, which allows the background or leak current to be measured. Current–voltage (*I–V*) relationships are then obtained by applying a slow (37.5–100 mV/s) depolarizing voltage ramp to the patch (Linsdell and Hanrahan, 1996). Adding PKA to the bath rapidly activates a large current, which is mediated exclusively by CFTR (Linsdell and Hanrahan, 1996; Linsdell and Hanrahan, unpublished observations, 1998). The leak current can be subtracted digitally to give a macroscopic *I–V* relationship for CFTR as in Figs. 3 and 4. This *I–V* curve usually has the same shape as the single-channel *I–V* relationship; however, this needs to be confirmed for mutants because there are instances in which mutations in the membrane-spanning segments can induce voltage-dependent gating.

E. Whole–Cell Currents

For whole-cell studies, pipette capacitance is canceled using the internal circuitry of the patch clamp while still in the cell-attached configuration. Excess suction is applied to break the patch, whole-cell capacitance is nulled, and currents recorded as described earlier. Current is measured while holding V_m at 0 mV, and during 1-s alternating pulses to ±60 mV (1-s duration). Under these ionic conditions, an increase in the outward current at 0 mV indicates elevation of Cl^- conductance. *I–V* relationships are generated by voltage steps from −100 to +100 mV in 10-mV increments (500-ms duration). Currents are plotted using Clampfit and the average current values between 400 and 500 ms are plotted against membrane potential. Currents are normalized to cell capacitance (pA/pF) to allow comparison of data from different cells.

Standard whole-cell patch-clamp methods are used when recording CFTR currents in transfected BHK cells. The standard bath (extracellular) solution contains (in mM): 150 NaCl, 10 *N*-tris[hydroxymethyl]methyl-2-aminoethanesulfonic acid, 2 $MgCl_2$ (pH 7.40), and is supplemented with 50 mM sucrose to prevent development of swelling-activated Cl current. The pipette (intracellular) solution contains (in mM): 110 *N*-methyl-D-glucamine (NMDG)-aspartate, 30 NMDG-Cl, 1 $MgCl_2$, 10 TES, and 0.1 BAPTA [1,2-bis-(*o*-aminophenoxy)ethane-*N*,*N*,*N'*,*N'*-tetraacetic acid; tetrasodium salt] (pH 7.20) supplemented with 1 mM MgATP (from a 100 mM stock in buffer). Symmetrical 150 mM Cl^- solutions can also be used so that whole-cell *I–V* relationships can be compared with those for single channels; however, cell viability is compromised when intracellular chloride concentration is elevated. NMDG solutions have a short shelf-life and should be refrigerated, filtered (0.22 μm) frequently, and replaced every few days. Whole-cell CFTR currents are activated by adding cpt-cAMP or 8-bromo-cAMP (100 μM final concentration) to the bath solution from 50 mM stock solutions in water. The stocks are stable for several months when stored in small aliquots at −70 °C.

V. Regulation of CFTR Chloride Channel

A. Studying PKA Activation and Nucleotide Dependence of CFTR

PKA catalyzes phosphorylation of CFTR at multiple sites (Cheng *et al.*, 1991; Cohn *et al.*, 1992; Picciotto *et al.*, 1992) and activates the channel (Bear *et al.*, 1991; Berger *et al.*, 1991; Kartner *et al.*, 1991; Rommens *et al.*, 1991; Tabcharani *et al.*, 1991). The major PKA sites phosphorylated *in vivo* are serines at positions 660, 700, 737, 795, and 813; however, studies of full-length CFTR and R domain peptides indicate that additional sites (namely, serine-422, -712, -753, and -768) can also be phosphorylated *in vitro*. Minor discrepancies between the sites phosphorylated in various studies probably reflect the different techniques and reagents used; for example, serine at position 422 was phosphorylated in an NBF1-R domain peptide40 but not in full-length CFTR (Cheng *et al.*, 1991); whereas serine-753 was phosphorylated in full-length CFTR (Seibert *et al.*, 1995) but not in the peptide (Townsend *et al.*, 1996). PKA phosphorylation sites on CFTR have distinct functions. Sites with distinct stimulatory and inhibitory effects have been identified (Wilkinson *et al.*, 1997). We have recently suggested that one site, or set of sites, regulates burst duration (PKAb) while another controls interburst duration (PKAi) (Luo *et al.*, 1997). A distinction has also been made between modulatory and activating PKA sites (Gadsby and Nairn, 1994). Some PKA sites capable of activating CFTR remain to be identified (the "cryptic" sites) because 10–15% responsiveness persists in a mutant (15SA) lacking serine-422 and all mono- and dibasic PKA sites on the R domain (F. Seibert, unpublished observation, 1997). In defense of mutagenesis, this loss of activation by PKA is probably not due to a general disruption caused by mutating the serines, because tyrosine phosphorylation causes very robust activation of the mutant (Jia *et al.*, 1997b).

The phosphorylation state of CFTR must be controlled when studying ATP dependence of channel gating (Li *et al.*, 1996; Mathews *et al.*, 1998). Phosphorylation of the dibasic PKA sites on CFTR reduces the EC_{50} for ATP-dependent gating and increases the maximum P_0 that can be achieved at high ATP concentrations (Mathews *et al.*, 1998). Variations in the level of phosphorylation may help explain the wide range of EC_{50} values reported for ATP dependence in the literature. Also, demonstrating the strict dependence of CFTR gating on ATP may require exposure of patches to flowing ATP-free solution. Channel activity can persist for many minutes when patches are excised into stagnant bath solution that is nominally ATP free, even if the chamber is rinsed intermittently.

We use the catalytic subunit of type II bovine cardiac PKA ($\sim$0.3 mg/ml) to activate CFTR channels (Tabcharani *et al.*, 1991). PKA is stable for more than 2 years at -70 °C when stored in buffer containing (in mM): 150 KCl (pH 7.0), 30 KH_2PO_4, 1 dithiothreitol (DTT), and 1 ethylenediaminetetraacetic acid (EDTA; disodium salt); however, it should not be thawed and refrozen. Its potency declines after approximately 1 h at room temperature ($\sim$23 °C); therefore, small aliquots are thawed and kept at 4 °C until used, typically within 2–3 weeks. In North America,

Promega (Madison, WI) or Upstate Biotechnology Inc. (UBI, Lake Placid, NY) are reliable, though expensive, sources. Lyophilized PKA is avoided because of its low activity. MgATP and other nucleotides are stored at $-70\ °C$ as 100 mM stock solutions in buffer (pH 7.4), and fresh aliquots are thawed each day. Note that all commercially available nucleotide preparations contain contaminants (nucleoside mono- and diphosphates).

B. Burst Analysis in Patches with Multiple CFTR Channels

The open probability of CFTR depends primarily on alterations in its burst kinetics. Open burst duration may reflect the NBF–NBF interactions that control ATP turnover at NBF1 (Baukrowitz *et al.*, 1994; Gunderson and Kopito, 1994; Hwang *et al.*, 1994; Li *et al.*, 1996; discussed in Mathews *et al.*, 1998). Unfortunately, most patches contain multiple channels, which precludes use of the threshold crossing method for estimating open and closed times. To evaluate kinetics in multichannel patches, the mean number of channels open is determined by measuring the fraction of time spent at each multiple of the unitary current. The single-channel open probability (P_0) is calculated from:

$$P_0 = \sum_{i=1}^{N} \frac{t_i}{TN} \tag{2}$$

where t_i is the time spent above a threshold i set at 0.5, 1.5, 2.5, …, times the single-channel amplitude, N is the number of channels locked open at the end of the experiment using AMP–PNP (see below), and T is the duration of the segment (typically >120 s). The number of opening transitions during each segment is counted (see below) and used to estimate the mean burst (τ_{open}) and interburst (τ_{closed}) durations from the "cycle time" according to:

$$\tau_{open} = \frac{(NP_0)T}{n} \tag{3}$$

$$\tau_{closed} = \frac{(N - NP_0)T}{n - 1} \tag{4}$$

where N is the number of channels locked open by AMP–PNP (see below), T is the duration of the segment, and n is the number of identifiable bursts.

This method assumes there is one open and one closed state, although it is clear by inspection of recordings that CFTR has at least two closed states (interburst closures and flickery closures within bursts; see, e.g., Fischer and Machen, 1994; Haws *et al.*, 1992). Fortunately, brief closures within bursts have little impact on P_0 and do not depend on nucleotides. When they are excluded from the analysis, the kinetics of CFTR are well described by long openings (actually bursts) and closings (interbursts). To estimate the number of functional channels in patches (N), we add the nonhydrolyzable nucleotide adenylyl imidodiphosphate (AMP–PNP;

tetralithium salt) to the bath from a 100 mM stock solution at the end of each experiment (final concentration of 1 mM). When P_0 is low (e.g., at low [ATP] or when channels are partially phosphorylated), the value of N obtained by locking channels with AMP–PNP exceeds that estimated from the maximum number that open simultaneously during long recordings. Although AMP–PNP does not cause locking at physiologic temperatures, it still increases P_0 and therefore improves the estimate of N.

C. Studying Regulation by PKC

In addition to the well-known activation of CFTR by PKA, it has recently become clear that PKC phosphorylation also has profound effects on CFTR and is required for activation by PKA (Jia *et al.*, 1997a). Patch-clamp studies indicate that CFTR is constitutively phosphorylated by PKC (and PKA) in several mammalian cell types. This dependence must therefore be considered when designing experiments to study activation by other protein kinases. CFTR channel activity was weakly stimulated (10%) when freshly excised, multichannel patches were exposed to PKC (Tabcharani *et al.*, 1991). However, more recent work has revealed that PKC phosphorylation alone has no effect after prolonged rundown (Jia *et al.*, 1997a), when membrane-associated phosphatases have removed constitutive phosphorylation (Fig. 5). We now interpret the stimulation by PKC observed previously as resulting from phosphorylation of permissive PKC sites, and reactivation of a few channels that still had residual PKA phosphorylation. Since the level of constitutive phosphorylation by PKC undoubtedly varies among cell lines and under different experimental conditions, PKC regulation must be considered when assessing the effects of other kinases or when comparing the responsiveness of CFTR in different preparations, particularly if negative results are obtained.

To study the role of PKC, the inhibitors chelerythrine chloride and Gö6976 are prepared in dimethyl sulfoxide (DMSO; 10 and 5 mM stocks, respectively) and stored in tightly sealed containers under nitrogen gas at $-20\,°C$. Fresh aliquots are thawed each day and diluted immediately before use to yield final concentrations of 1 μM for chelerythrine, and 500 nM for Gö6976. Preincubating cells with chelerythrine in culture medium in a 5% CO_2 incubator at 37 °C for 30–120 min prior to patch-clamp recording abrogates PKA stimulation of CFTR channels in 80% of membrane patches. Gö6976 is routinely added to both the bath and pipette solutions, and several minutes are allowed to elapse after entering the whole-cell configuration to ensure adequate diffusion of Gö6976 into the cell before currents are recorded (see Marty and Neher, 1995). For studies of exogenous PKC we use calcium and phospholipid-dependent PKC II from rat brain (Tabcharani *et al.*, 1991), although other isoforms give similar results (Berger *et al.*, 1993). After thawing an aliquot, the PKC is kept at 4 °C and can be used over a period of 1–2 weeks. We also add 1,2-dioctanoylglycerol (8:0) DiC_8 to the bath solution. The DiC_8 is dissolved in chloroform at 10 mM, stored under nitrogen gas in a

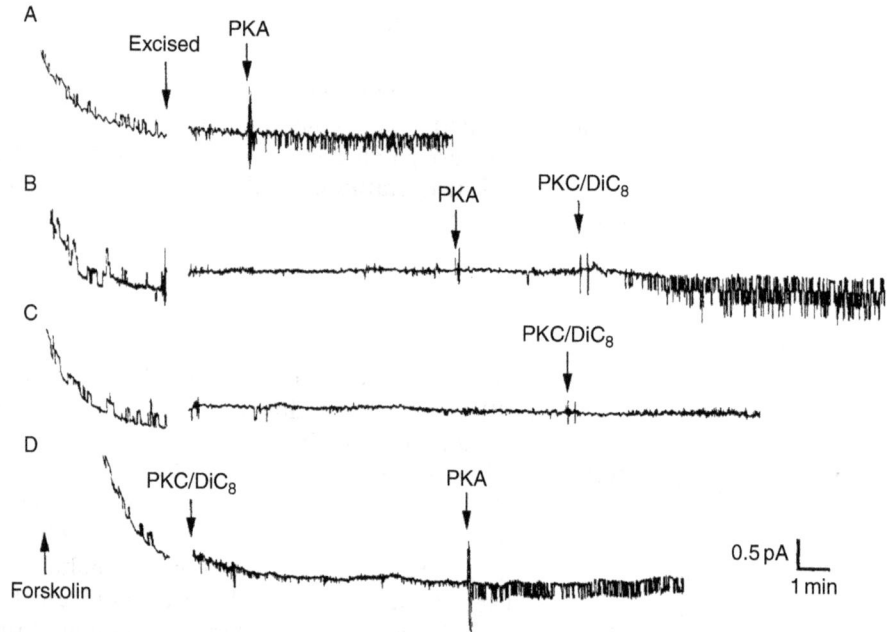

Fig. 5 The responsiveness of CFTR channels to PKA catalytic subunit under various conditions reveals the importance of constitutive phosphorylation. PKA activates CFTR channels when added soon after excision (A), but not after 10 min, although subsequent addition of PKC and the lipid activator DiC_8 does allow restimulation (B). PKC does not stimulate CFTR channels in the absence of PKA phosphorylation (C), but when present preserves responsiveness to PKA (D). With permission from Jia *et al.* (1997a).

tightly sealed container at $-20\,°C$, and diluted in bath solution immediately before being added at $5\,\mu M$ final concentration.

D. Studying Regulation by Src

Tyrosine phosphorylation is not detectable on CFTR under control conditions; however, the tyrosine kinase p60c-Src is a potent channel activator when added to excised patches and phosphorylates immunoprecipitated CFTR protein *in vitro* (Jia *et al.*, 1998). Moreover, tyrosines on CFTR become phosphorylated *in vivo* when v-Src is coexpressed with CFTR. These results have implications for both the specificity and mechanism of channel regulation by phosphorylation. Although the physiologic relevance of tyrosine phosphorylation remains to be determined, at the very least, the results indicate that many kinases other than PKA and PKC may participate in regulating CFTR. To study CFTR regulation in excised patches, we add 30 U/ml of p60c-Src from UBI to the cytoplasmic side. Control experiments are carried out using p60c-Src, which has been heated to 100 °C for 10 min to destroy its enzymatic activity. When cells are transiently transfected with an expression plasmid containing v-Src, CFTR channels can be activated in excised patches simply by exposing the patches to 1 mM MgATP. This activity is greatly enhanced by the protein tyrosine phosphatase (PTP) inhibitor dephostatin; therefore, there is an endogenous, membrane-associated PTP, which can dephosphorylate tyrosine residues on CFTR.

E. Studying Regulation by Protein Phosphatases

CFTR channel activity declines rapidly when patches are excised from CHO or human airway cells. This rundown is mediated by a membrane-associated phosphatase activity that is not sensitive to okadaic acid or calyculin A nor is it dependent on calcium and calmodulin (Becq *et al.*, 1994; Tabcharani *et al.*, 1991), but it does require magnesium consistent with PP2C (Luo *et al.*, 1998). As mentioned earlier, this rundown does not usually occur in patches from BHK cells expressing very high levels of CFTR. Adding PP2A, PP2C, or alkaline phosphatase individually to BHK patches with stable CFTR activity reduces P_0 by more than 90% but does not abolish it. PP1 and PP2B have no effect on channel activity (Berger *et al.*, 1993; Luo *et al.*, 1998). Deactivation by PP2C is more rapid than by PP2A or alkaline phosphatase and has a time course resembling the spontaneous rundown, which is sometimes observed even in membrane patches excised from BHK cells. Deactivation by exogenous PP2A is associated with dramatic shortening of the mean burst duration. However, burst duration does not change after addition of PP2C or during spontaneous rundown (when it occurs). Thus, functionally distinct PKA sites may differ in their sensitivities to protein phosphatases; sites controlling burst duration (PKAb) are susceptible to PP2A but not PP2C, whereas those regulating interburst duration (PKAi) can be dephosphorylated by either phosphatase.

PP1, PP2A, and PP2B are available from Promega and UBI. UBI also sells antibodies to PP2A, recombinant PP2Cα protein, and a polyclonal antibody to PP2Cα, although they are expensive. We have used PP2Cα, which is isolated from chicken gizzard (Pato and Adelstein, 1983a,b). Commercially available PP2A is avoided because of its low activity. Regardless of the source, the phosphatase activity of any enzyme preparation should be assayed prior to use, preferably under conditions that approximate those present during patch-clamp experiments. Full-length phospho-CFTR would be the most appropriate substrate for such studies but is difficult to purify in sufficient quantities for measuring release of radiolabeled phosphate. R domain peptide can be expressed in bacteria as a glutathione *S*-transferase fusion protein or with a histidine tag, purified, and then phosphorylated *in vitro*. Alternatively, phosphorylated myosin light chains or other phosphatase substrates can be used, although the relative efficiency of dephosphorylation by the phosphatases will depend on substrate used; for example, PP2C dephosphorylates myosin light chains much more effectively than phosphocasein.

The protein phosphatases are stable for at least 6 months when stored as small aliquots at −20 °C; however, PP1 activity declines two- to threefold once thawed and stored for 2–3 h on ice. Protein phosphatases can be diluted in 50% glycerol and stored without freezing at $-20\,°C$, although the glycerol may cause problems later when added during patch-clamp experiments. Patch-clamp solutions usually contain 150 mM NaCl; however, we found that this salt partially inhibits protein phosphatases relative to the same buffer lacking NaCl; that is, PP1 activity is reduced by 72%, PP2A by 66%, PP2B by 38%, and PP2C by 20% (Luo *et al.*, 1997). We elevate the Mg^{2+} concentration of the bath to 12 mM when studying regulation by PP2C, which requires ~2 mM Mg^{2+} for half maximal activity. Phosphatase inhibitors such as vanadate and fluoride must be used with care because they may act directly on CFTR. The PP1 and PP2A inhibitor calyculin A is more effective than okadaic acid when added to intact cells, although both have IC_{50} values in the 1- to 10-nM range *in vitro*. We typically use 10–100 nM of calyculin A or 1–10 μM of okadaic acid. Calyculin A concentrations higher than 100 nM are toxic to the human colonic epithelial cell line T_{84}. Neither inhibitor affects PP2C (at least up to 1 mM okadaic acid). Microcystin permeates poorly through cell membranes but is useful for whole-cell and inside-out patches (Hwang *et al.*, 1993).

VI. DRSCAN: A pCLAMP-Compatible Program for Analyzing Long Records

The pCLAMP suite of programs (Axon Instruments Inc.) copes adequately with well-behaved data. However, in our studies of CFTR (<10 pS) we found that setting the optimum threshold was sometimes difficult because the cursor position could not be set for individual traces. Moreover, experiments often last more than 20 min and involve several interventions; which require repeated switching to the General Parameters menu to enter start and end times of segments for each analysis. More serious problems arise when patches contain more than one channel, which is usually the case. To overcome these limitations while retaining the many useful features of pClamp, we have developed a companion program with a user-friendly graphical interface. The structure of this program, which we call DRSCAN, is shown in Fig. 6. It reads pClamp Axon Binary Format (.ABF) data files that are acquired using FETCHEX in gap-free mode. The gain and duration of the displayed data are easily altered, and threshold level(s) can be set for individual screens using cursors. To aid threshold positioning, an all-points histogram is automatically drawn adjacent to the current trace (Fig. 6). DRSCAN performs threshold-crossing analysis for up to 99 cursors and calculates a running value for NP0, mean burst duration, and mean interburst duration as described earlier. The results of DRSCAN analyses are output in Axon.EVL format and also in text format, which can be imported into graphing and statistical programs such as Origin 4.10 (Microcal, Northampton, MA). The program also calculates current amplitudes for selected segments of data using Gaussian fits to the all-points histogram, allowing the generation of $I–V$ relationships (Fig. 7). A brief overview

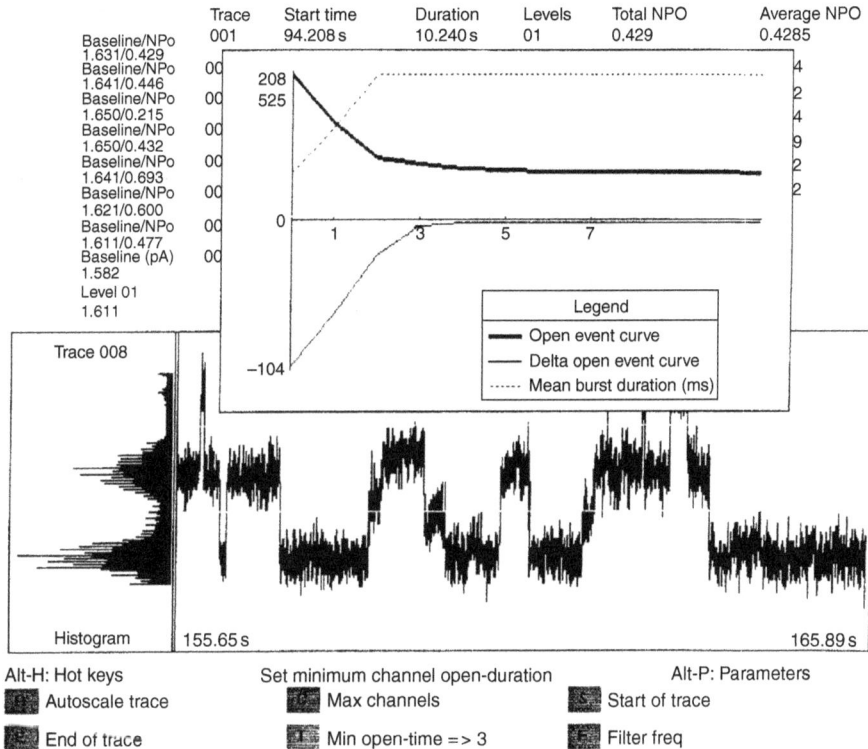

Fig. 6 The "event analysis" display provided by DRSCAN. The thin white lines correspond to event levels set by the user. Curves in the pop-up window guide the selection of durations for accepting or rejecting events.

of DRSCAN is given in the Appendix. Similar features could be incorporated into other programs intended for analyzing long continuous recordings of CFTR.

A. Output of Single-Channel Current Records for Production of Figures

Whether DRSCAN or pCLAMP is used for data analysis, routine plotting of current records is easily achieved using the Export Plot commands of the FETCHAN program. Although the output is not easily edited or of high quality, we find that Origin 4.10 (e.g., 16-bit version with the optional pCLAMP plug-in; Microcal) is a convenient and versatile way to plot and edit Axon binary data files, and yields publication-quality figures with multiple panels. A digital (Gaussian) filter can be applied to the data in FETCHAN by specifying a low-pass cutoff frequency from the General Parameters menu. The built-in fast Fourier transform (FFT) filter of Origin 4.1 also works well and improves the clarity of recordings.

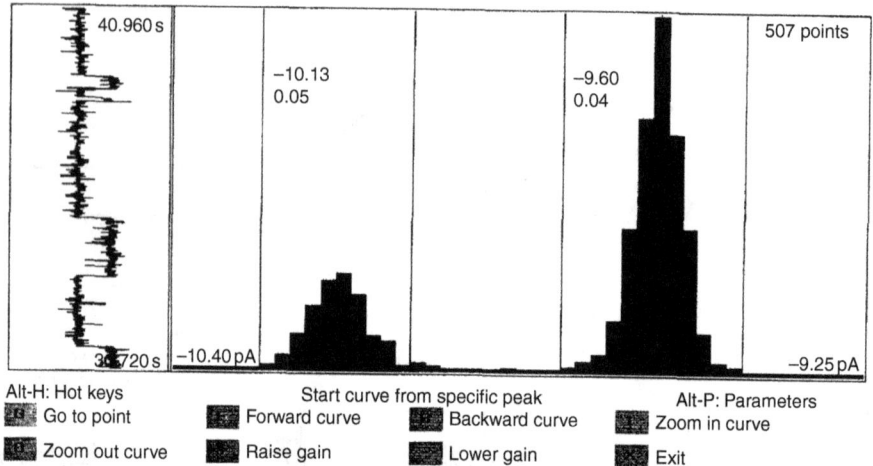

Fig. 7 Current–voltage analysis using Gaussian fits of all-points histograms. One-sided fits to the peaks or manual cursor positioning can be used when the histograms are skewed due to filtering.

Appendix

Hardware and Software Requirements

DRSCAN runs on stand-alone PCs with minimal hardware requirements. It requires a 486 or higher processor with at least 8 Mb of RAM and 1 Mb of free hard disk space. It was written using the Borland TURBO C++ for DOS compiler, and uses the TURBO C++ graphics library; therefore, it also requires at least an enhanced graphics adapter. An Axon Instruments compatible hardware interface is required for data acquisition. To prevent aliasing, data are digitized at a rate which is at least twice the analog filter cutoff frequency (−3 dB; eight-pole Bessel filter). DRSCAN requires MS-DOS 3.31 or higher, although it can also be run as a DOS command from within any version of Microsoft Windows. It is a companion to the pCLAMP suite of programs, which is also required for data acquisition.

Program Description

DRSCAN is compiled as a project (TURBO C++.PRJ) file consisting of both source code modules and object modules. The object modules are precompiled TURBO C++ graphics files (.OBJ files) linked with a set of specialized modular source code files (Fig. 8). These handle application file I/O routines (FIOROUT), DOS file system interface (DROUTINE), graphical user interface (GINTFACE), Axon binary file interface (ABFROUT), histogram generation algorithms and plotting routines (HISTROUT), and event list analysis algorithms (ANALYSIS, EVNTANAL). Finally, there are two display formatting modules, one for the text mode interface (TINTFACE) and another for the graphical mode interface (GRAFROUT).

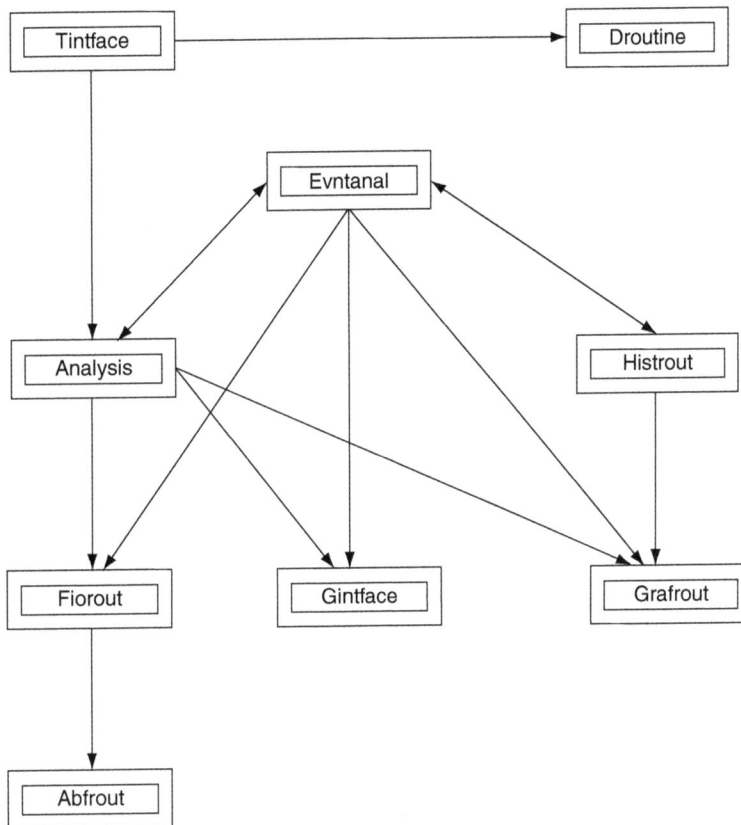

Fig. 8 Scheme showing the interrelationships between DRSCAN modules. An arrow from module M_1 to M_2 means that module M_2's computations are triggered by a service request issued in module M_1.

Text Mode Interface (TINTFACE) Module

TINTFACE is a navigation system between submenus. It takes user requests in the form of menu item selections and communicates with the DROUTINE module, which specializes in reading and writing to and from the DOS file system. TINT-FACE allows users to view the application output files in text format. On receiving a user request for an I–V analysis or an event list analysis, TINTFACE builds the necessary user application files and passes control to the ANALYSIS module. On running DRSCAN, the program parameters are validated by this module. The minimal parameter for starting DRSCAN are the program name and a user name, as in "DRSCAN User_Name." The following files are automatically built:

1. A \DRSCAN\Data\User_Name-directory is created (if none exists). All user files are kept here, separate from the application files located in the \DRSCAN root directory.

2. A user profile parameter file is created if none exists, to store the user's last data and analysis files source directories, display preferences, etc.

3. A Trace_Log file is created, which records chronological actions, as the user analyzes records in the open data file.

4. A circular event list file is created consisting of the FETCHEX data file name with extension ".EV?" where "?" can be either "L" or a number between 1 and 9. This allows users to have up to 10 different analyses on the same data file with results stored in a meaningful, easy-to-retrieve file-naming convention. The same holds for the *I–V* analysis, although with different extension conventions. Event list files are Axon ".EVL" file format compatible and therefore can be read by other programs such as Axon's READEVNT or PSTAT.

DOS File System Interface (DROUTINE) Module

This module specializes in handling the DOS file system. It maintains and displays the application directory entries, services formatting requests issued by TINTFACE, and organizes users' working spaces by keeping them apart from one another and separate from the application-wide system files. It adds flexibility to file naming and buffers DOS I/O calls to improve system performance.

ANALYSIS Module

This module is the driver for both the *I–V* and event list analyses. It is also the repository for application utilities (such as sorting and help routines). It is activated by TINTFACE, and it determines whether the user has requested *I–V* or event list analysis. It then performs a system initialization and loads the right table (help, menu, hot keys, short-cut keys, etc). It orders the correct screen resolution for both the histogram and trace curves. Requests are made to EVNTANAL for loading corresponding algorithms, to GINTFACE for setting the user graphical display resolution, to HISTROUT for processing and displaying the trace histogram, to GRAFROUT for displaying the trace, and to FIOROUT for reading Axon FETCHEX data files.

EVNTANAL Module

This module handles the algorithmic implementation of the DRSCAN application. It allows the user to select records in the data file to be processed, to set baselines, to correct baseline drift, and to set event level cursors at the desired place on the trace window, EVNTANAL also computes the NP_0 for the segment of data selected. This quantity is weighted by the duration of the given interval. The mean NP_0, total running NP_0, and "NP_0 so far" values are also computed (Fig. 6). Event files created by EVNTANAL reflect the true behavior of the data, as recorded by

FETCHEX. This means that a typical event file created by EVNTANAL may contain events of zero length when there is a transition across two or more cursor levels.

The weighted NP_0 computation is based on Eq. (2), except that T is the duration of the trace in the window. At the end of an analysis, EVNTANAL can build a "reprocessed" event list file, which excludes events of duration less than some user specified parameter. The default is two data points. To help the user choose an optimal minimum duration to be excluded, a "*Delta Open Event Curve*" curve is also generated showing the number of open events computed during the segment of data as a function of event durations excluded (Fig. 6). The inflection point of this curve is taken as the optimal "minimum event length" that eliminates the most false events without excluding true ones. Analytically, this point corresponds to the null value of the first derivative of the function f, such that $f(x) = y_i - y_{i-1}$ ($i \geq 2$) and y_i is the number of events whose duration expands over i data points. In addition to ignoring spurious brief events, the event list reprocessing routine can calculate mean burst (τ_{open}) and interburst durations (τ_{closed}) according to Eqs. (3) and (4).

EVNTANAL also outputs the following files:

1. *An event list file, whose format matches the Axon ".EVL" binary file.* Event list files created by EVNTANAL record all events as they appear in the experiment file. All events are detected and written to the file whenever the current level crosses the threshold set by the user at the beginning of each trace analysis.

2. *One or more reprocessed event list files.* Reprocessed event list files rebuild the event list file created above by eliminating false events due to noise in the experiment file. This noise-elimination process is carried out empirically using the first derivative of the "*Delta Open Event*" curve as described earlier.

3. *An analysis log file.* The analysis log file saves on disk the contents of the user graphical display screen. In the case of an event analysis this would contain information concerning the portion of the experiment records being processed. The analysis log file in this case would specify a trace starting time and a duration, the number of event levels detected in that window, and the NP_0 values for the trace, that is, "Trace NP_0," "Total Analysis NP_0 so far," and "Mean NP_0 so far." In the case of an I–V analysis, the log file would give the number of current intervals for each trace (also called "Peaks"), the range of current values in each peak, and the mean current value and its standard deviation.

HISTROUT Module

This module constructs the all-points amplitude histogram for the displayed current trace. The user adjusts the length and start time of the current trace to be analyzed according to the number of transitions and severity of baseline drift. HISTROUT allows the split screen to be toggled between "active" mode, which allows interactive cursor setting, and "passive" mode, in which the all-points amplitude histogram is simply displayed to the left of the trace (Fig. 6). Histogram bin width is calculated automatically based on the horizontal resolution of the screen and the current amplitude range of the selected data trace. The mean and variance of each peak is obtained by fitting data between two cursors with a Gaussian curve and these are displayed on the histogram. For example, Fig. 7 shows two peaks and user-defined vertical cursors. The mean currents and variances are -10.13 and -9.60 pA, and 0.05 and 0.04 pA, respectively.

HISTROUT is very precise in computing these parameters because it processes the file records point by point without unnecessary estimation. As a result most DRSCAN algorithms run in quadratic complexity time $O(n^2)$, with the exception of the event level detection, which runs in cubic complexity time $O(n^3)$.

File I/O (FIOROUT) Module

This module serves as an interface to the Axon binary file system, the ABFR-OUT. FIOROUT receives calls from EVNTANAL and ANALYSIS for reading Axon FETCHEX acquisition data files. It routes the same call with the appropriate Axon ".ABF" parameters. Selected channel reading and channel multiplexing reading routines in the Axon ABFFILES.C program are used to receive and service function calls from FIOROUT. ABFFILES.C is a public domain "ABF" I/O utility program that comes with the distribution diskettes of pClamp6. We have modified some functions in that program to handle the reading and formatting of acquisition file records, which are then returned to the ANALYSIS or EVNTANAL modules.

ABFROUT Module

This is the collection of Axon ABFFILES.C I/O routines that deal with reading data records from data files.

Graphical Interface (GINTFACE) Module

This module handles the graphical user interface, receiving user keystrokes and translating them into application procedure calls. GINTFACE contains display resolution computation. It loads font files and interacts with the screen display aspect by detecting the hardware in use and managing the display area of the screen. All graphical menu item selections are serviced by this module. As such, it interacts with ANALYSIS (for displaying application global help topics and hot keys) and manages navigation among the various modules that generate user interface processes. GINTFACE also interacts with EVNTANAL for displaying the cursor levels set by the user (in Fig. 6, one such level cursor is shown as a thin clear line in the middle of the trace). It also displays the trace records read from the acquisition file and processed by GRAFROUT, and the histogram generated by HISTROUT and passed to EVNTANAL.

GRAFROUT Module

This module prepares the trace to be displayed by selecting meaningful data points that can fit in the display area. Data selection is based on a heuristic that filters the outliers in a cloud of points. The cloud of points has a size proportional to the (number of data points in the trace)/(horizontal display resolution). When the user sets a cursor level, this module changes pixels in the range of the cursor and performs textual display of the amplitudes and times specified by cursors.

References

Alvarez, O., Villarroel, A., and Eisenman, G. (1992). *Methods Enzymol.* **207**, 816.

Anderson, M. P., Gregory, R. J., Thompson, S., Souza, D. W., Paul, S., Mulligan, R. C., Smith, A. E., and Welsh, M. J. (1991a). *Science* **253**, 202.

Anderson, M. P., Rich, D. P., Gregory, R. J., Smith, A. E., and Welsh, M. J. (1991b). *Science* **251**, 679.

Baukrowitz, T., Hwang, T. C., Nairn, A. C., and Gadsby, D. C. (1994). *Neuron* **12**, 473.

Bear, C. E., Duguay, F., Naismith, A. L., Kartner, N., Hanrahan, J. W., and Riordan, J. R. (1991). *J. Biol. Chem.* **266**, 19142.

Bear, C. E., Li, C., Kartner, N., Bridges, R. J., Jensen, T. J., Ramjeesingh, M., and Riordan, J. R. (1992). *Cell* **68**, 809.

Becq, F., Jensen, T. J., Chang, X. B., Savoia, A., Rommens, J. M., Tsui, L. C., Buchwald, M., Riordan, J. R., and Hanrahan, J. W. (1994). *Proc. Natl. Acad. Sci. USA* **91**, 9160.

Berger, H. A., Anderson, M. P., Gregory, R. J., Thomson, S., Howard, P. W., Maurer, R. A., Mulligan, R., Smith, A. E., and Welsh, M. J. (1991). *J. Clin. Invest.* **88**, 1422.

Berger, H. A., Travis, S. M., and Welsh, M. J. (1993). *J. Biol. Chem.* **268**, 2037.

Champigny, G., Verrier, B., Gérard, C., Mauchamp, J., and Lazdunski, M. (1990). *FEBS Lett.* **259**, 263.

Chang, X. B., Tabcharani, J. A., Hou, Y. X., Jensen, T. J., Kartner, N., Alon, N., Hanrahan, J. W., and Riordan, J. R. (1993). *J. Biol. Chem.* **268**, 11304.

Cheng, S. H., Rich, D. P., Marshall, J., Gregory, R. J., Welsh, M. J., and Smith, A. E. (1991). *Cell* **66**, 1027.

Cohn, J. A., Nairn, A. C., Marino, C. R., Melhus, O., and Kole, J. (1992). *Proc. Natl. Acad. Sci. USA* **89**, 2340.

Dwyer, T. M., Adams, D. J., and Hille, B. (1980). *J. Gen. Physiol.* **75**, 469.

Fischer, H., and Machen, T. E. (1994). *J. Gen. Physiol.* **104**, 541.

Gabriel, S. E., Price, E. M., Boucher, R. C., and Stutts, M. J. (1992). *Am. J. Physiol. (Cell Physiol.)* **263**, C708.

Gadsby, D. C., and Nairn, A. C. (1994). *TIBS* **19**, 513.

Gadsby, D. C., Nagel, G., and Hwang, T. C. (1995). *Annu. Rev. Physiol.* **57**, 387.

Gray, M. A., Greenwell, J. R., and Argent, B. E. (1988). *J. Membr. Biol.* **105**, 131.

Gray, M. A., Pollard, C. E., Harris, A., Coleman, L., Greenwell, J. R., and Argent, B. E. (1990). *Am. J. Physiol. (Cell Physiol.)* **259**, C752.

Gunderson, K. L., and Kopito, R. R. (1994). *J. Biol. Chem.* **269**, 19349.

Hart, P., Warth, J. D., Levesque, P. C., Collier, M. L., Geary, Y., Horowitz, B., and Hume, J. R. (1996). *Proc. Natl. Acad. Sci. USA* **93**, 6343.

Haws, C., Krouse, M. E., Xia, Y., Gruenert, D. C., and Wine, J. J. (1992). *Am. J. Physiol. (Lung Cell. Mol. Physiol.)* **263**, L692.

Haws, C., Finkbeiner, W. E., Widdicombe, J. H., and Wine, J. J. (1994). *Am. J. Physiol. (Lung Cell. Mol. Physiol.)* **266**, L502.

Hwang, T. C., Horie, M., and Gadsby, D. C. (1993). *J. Gen. Physiol.* **101**, 629.

Hwang, T. C., Nagel, G., Nairn, A. C., and Gadsby, D. C. (1994). *Proc. Natl. Acad. Sci. USA* **91**, 4698.

Jia, Y., Mathews, C. J., and Hanrahan, J. W. (1997a). *J. Biol. Chem.* **272**, 4978.

Jia, Y., Seibert, F., Chang, X. B., Riordan, J. R., and Hanrahan, J. W. (1997b). *Pediatr. Pulmonol.* (Suppl. 14), 214, (Abstract).

Jia, Y., Loo, M. A., Seibert, F., Jensen, T. J., Hou, Y. X., Cui, L., Chang, X. B., Clarke, D. M., Riordan, J. R., and Hanrahan, J. W. (1998). *J. Biol. Chem.*

Kartner, N., Hanrahan, J. W., Jensen, T. J., Naismith, A. L., Sun, S., Ackerley, C. A., Reyes, E. F., Tsui, L. C., Rommens, J. M., Bear, C. E., and Riordan, J. R. (1991). *Cell* **64,** 681.

Linsdell, P., and Hanrahan, J. W. (1996). *J. Physiol. (Lond.)* **496,** 687.

Linsdell, P., and Hanrahan, J. W. (1997). *Pediatr. Pulmonol.* (Suppl. 14), 215, (Abstract).

Linsdell, P., Tabcharani, J. A., and Hanrahan, J. W. (1997a). *J. Gen. Physiol.* **110,** 365.

Linsdell, P., Tabcharani, J. A., Rommens, J. M., Hou, X. Y., Chang, X. B., Tsui, L. C., Riordan, J. R., and Hanrahan, J. W. (1997b). *J. Gen. Physiol.* **110,** 355.

Linsdell, P., Zheng, S. X., and Hanrahan, J. W. (1997c). *Pediatr. Pulmonol.* (Suppl. 14), 214, (Abstract).

Luo, J., Pato, M. D., Seibert, F. S., Chang, X. B., Riordan, J. R., and Hanrahan, J. W. (1997). *Pediatr. Pulmonol.* (Suppl. 14), 217, (Abstract).

Marty, A., and Neher, E. (1995). *In* "Single Channel Recording," (B. Sakmann, and E. Neher, eds.), **2,** pp. 31–52. Plenum Press, New York.

Mathews, C. J., Tabcharani, J. A., Chang, X. B., Riordan, J. R., and Hanrahan, J. W. (1998). *J. Physiol. (Cambr.)* **508,** 365.

Nagel, G., Hwang, T. C., Nastiuk, K. L., Nairn, A. C., and Gadsby, D. C. (1992). *Nature* **360,** 81.

Neher, E. (1992). *Methods Enzymol.* **27,** 123.

Pato, M. D., and Adelstein, R. S. (1983a). *J. Biol. Chem.* **258,** 7047.

Pato, M. D., and Adelstein, R. S. (1983b). *J. Biol. Chem.* **258,** 7055.

Picciotto, M. R., Cohn, J. A., Bertuzzi, G., Greengard, P., and Nairn, A. C. (1992). *J. Biol. Chem.* **267,** 12742.

Riordan, J. R., Rommens, J. M., Kerem, B. S., Alon, N., Rozmahel, R., Grzelczak, Z., Zielenski, J., Lock, S., Plavsic, N., Chou, J. L., Drumm, M. L., Iannuzzi, M. C., *et al.* (1989). *Science* **245,** 1066.

Rommens, J. M., Dho, S., Bear, C. E., Kartner, N., Kennedy, D., Riordan, J. R., Tsui, L. C., and Foskett, J. K. (1991). *Proc. Natl. Acad. Sci. USA* **88,** 7500.

Rudy, B., and Iverson, L. E. (eds.), (1992). *Methods Enzymol.* **207.**

Seibert, F. S., Tabcharani, J. A., Chang, X. B., Dulhanty, A. M., Mathews, C. J., Hanrahan, J. W., and Riordan, J. R. (1995). *J. Biol. Chem.* **270,** 2158.

Tabcharani, J. A., Low, W., Elie, D., and Hanrahan, J. W. (1990). *FEBS Lett.* **270,** 157.

Tabcharani, J. A., Chang, X. B., Riordan, J. R., and Hanrahan, J. W. (1991). *Nature* **352,** 628.

Tabcharani, J. A., Chang, X. B., Riordan, J. R., and Hanrahan, J. W. (1992). *Biophys. J.* **62,** 1.

Tabcharani, J. A., Linsdell, P., and Hanrahan, J. W. (1997). *J. Gen. Physiol.* **110,** 341.

Townsend, R. R., Lipniunas, P. H., Tulk, B. M., and Verkman, A. S. (1996). *Protein Sci.* **5,** 1865.

Wilkinson, D. J., Strong, T. V., Mansoura, M. E., Wood, D. L., Smith, S. S., Collins, F. S., and Dawson, D. C. (1997). *Am. J. Physiol. (Lung Cell. Mol. Physiol.)* **273,** L127.

Yang, I. C. H., Cheng, T. H., Wang, F., Price, E. M., and Hwang, T. C. (1996). *Am. J. Physiol.* **272,** C142.

CHAPTER 4

Tight-Seal, Whole-Cell Patch Clamping of *C. elegans* Neurons

S. R. Lockery[*] and M. B. Goodman[†]

[*]Institute of Neuroscience
University of Oregon
Eugene, Oregon, USA

[†]Department of Molecular and Cellular Physiology
Stanford University
Stanford, California, USA

DOI: 10.1016/B978-0-12-382204-8.00004-7

I. Introduction

The nematode *Caenorhabditis elegans* is widely used to study the relationship between genes, neurons, and behavior. The adult hermaphrodite has a compact nervous system of only 302 neurons and the synaptic connections between these cells have been described completely (White *et al.*, 1986). The neural circuits for many of its behaviors have been delineated (Bargmann, 1993). In addition, hundreds of genes affecting behavior have been identified (Hodgkin *et al.*, 1995) and a variety of promoters active in small subsets of neurons are available for labeling cells of interest in transgenic animals. In the 10 years that have passed since the first *in vivo* whole-cell patch-clamp recordings were reported (Goodman *et al.*, 1998), the number of identified neurons that have been analyzed using this approach has grown from 1 to 11 (Faumont *et al.*, 2006; Mellem *et al.*, 2002, 2008; Nickell *et al.*, 2002; O'Hagan *et al.*, 2005; Pierce-Shimomura *et al.*, 2001; Ramot *et al.*, 2008a; Ward *et al.*, 2008). These studies have analyzed sensory transduction pathways, glutamate receptors, ion channels, and other neuronal phenomena in their native cellular contexts. Other studies have investigated *C. elegans* neurophysiology by recording from dissociated, cultured neurons (Bianchi and Driscoll, 2006).

C. elegans presents a formidable challenge for electrophysiology, however. The animals are only 0.25–1.2 mm long and the cell bodies of *C. elegans* neurons are typically 2 μm in diameter (White *et al.*, 1986). In addition, the body is protected by a tough, pressurized cuticle that explodes when dissected (Tattar *et al.*, 1977).

Here we present an updated, reliable method for making tight-seal, whole-cell patch-clamp recordings from intact neurons in *C. elegans* at all developmental stages. By combining this technique with cell-specific expression of green fluorescent protein (GFP) (Chalfie *et al.*, 1994), whole-cell recordings can be made from identified neurons. Thus, it is now possible to describe the electrical properties of particular neurons and to find out how these properties are altered by mutations affecting neuronal development and behavior. Both the technique and the equipment available for patch-clamp electrophysiology have changed in the past 10 years; key updates are indicated with bullet points following our original text.

II. Overview

Recordings were made using simple modifications of existing techniques and equipment. Standard methods to immobilize worms during DNA injection (Fire, 1986), which dehydrate the animal, were modified for electrophysiology.

This involved immobilizing the animal with a waterproof glue (cyanoacrylate) under moist conditions. Because the worms are quite small, dissections were done with fine glass dissecting needles, based on the procedure for dissecting chromosomes (Brown and Carey, 1994). Rupturing the membrane patch to obtain the whole-cell configuration (breaking in) in *C. elegans* neurons appeared to be more difficult than in many other cells. To solve this problem we developed a way to make extremely blunt pipettes with submicron openings. In addition, we reduced movements produced by suction pulses, thermal drift, and mechanical drift to a fraction of a micron by stabilizing the pipette mechanically. A commercially available patch-clamp amplifier was modified to accommodate the small capacitance and high input resistance of *C. elegans* neurons. With these modifications, the tight-seal, whole-cell patch-clamp recording configuration can be achieved with a success rate of 56%.

III. Preparing *C. elegans* for *In Situ* Electrophysiology

A. Animals

We used approximately synchronous cultures of larval or adult worms. These were obtained by isolating eggs from gravid adults and allowing them to hatch on either sterile agar plates (NGM, Stiernagle, 2006) (which produces animals that arrest at the L1 stage) or plates seeded with bacteria (*Escherichia coli* OP50). Animals were collected from growth plates by washing them off with M9 or distilled water (~1.0–1.5 ml), and pelleting them in a mini centrifuge.

B. Gluing

An agarose pad (25 μm thick) was formed by pressing 10 μl of molten agarose (1.5%, medium EEO, Fisher Scientific, BP161) between two No. 0 coverslips (24 × 60 mm) and removing the top coverslip. The pH of the agarose from which the pad was made was adjusted to 9.0 (30 mM Na TAPS) to accelerate polymerization of the glue. Worms were transferred immediately to the surface of the pad in a 0.35-μl sample from the pellet. The coverslip was sealed with beeswax over a hole in a glass plate to form a recording chamber.

The fluid in which worms were transferred must either absorb into the agarose or evaporate, since excess fluid interferes with the glue. Absorption and evaporation were accelerated by spreading the fluid with a stream of air from a mouth tube. This also helped disperse worms across the agarose pad. When ready for gluing, worms were anesthetized by placing the chamber on the surface of a water-filled tissue-culture flask (50 ml, Falcon 3014) stored at 4 °C.

Anesthetized worms were immobilized with cyanoacrylate glue (Nexaband Quick Seal™, Veterinary Products Laboratory, Phoenix, AZ) applied with a small pipette held in a micromanipulator and viewed under a dissecting microscope. Glue pipettes (tip diameter 17–19 μm) were pulled from thin-walled borosilicate capillary tubes (0.94 mm ID, 1.20 mm OD, Corning 8250, Garner Glass, Claremont, CA). The diameter of the glue drop (~50 μm) was set by adjusting the height of the column of glue at the time the pipette was filled (the taller the column, the larger the glue drop). To compensate for variations in worm size and posture, additional adjustments were made by sucking or blowing very gently on a mouth tube connected to the glue pipette. Worms were glued along one side, leaving the other side clear for dissection and recording. About 25 worms could be glued in 5–10 min. Gluing was done as quickly as possible, since the agarose pad tended to dry out, and worms glued to a dry pad became dehydrated and unsuitable for recording.

The chamber was rinsed with distilled water to remove unglued worms, then filled with extracellular saline. Air pockets appeared on the glue after rinsing. Air pockets made the preparation hard to see, so we removed them with a stream of air bubbles from a mouth tube with a 0.4-mm opening. The recording chamber was then transferred to the stage of a fixed-stage microscope (either inverted or upright).

Electrical recordings were made during the first hour after gluing. Pharyngeal pumping and movements of the nose and tail of glued worms could be observed for several hours after gluing, but the vigor of these movements declined steadily. At the same time, glued worms developed a Clr phenotype (Clark et al., 1992) and vacuoles sometimes appeared in the head. Unglued worms in the same external saline were unaffected, suggesting these effects are due to the glue.

1. Updates

- When using an upright microscope, No. 1 coverslips can be used. They are thicker and less fragile than the No. 0 coverslips required for use on inverted microscopes.

- Worms can be transferred "semidry" by pipetting them (4 μl) onto a filter paper disk and inverting the disk onto the surface of the agarose pad. The disks are formed by using a hole punch to cut circles from filter paper (Whatman No. 1).

- We recently identified a new, improved cyanoacrylate glue: WORMGLU™, which is 80% octyl–20% butyl cyanoacrylate ($r = 5.75$ cP) and available from GluStitch (http://www.glustitch.com, Roberts, WA).

- Agar formed from physiological saline (pH 7.4) is sufficiently basic to support polymerization of the glue.

C. Dissection

Dissection proceeded in two steps. First, internal pressure was relieved by puncturing the cuticle at the level of the gonad primordium with a glass dissecting needle (cutter) held in a micromanipulator mounted on the microscope stage. In adults, pressure is best relieved adjacent to the bend in the anterior arm of the gonad, liberating both a loop of intestine and a gonad arm. Cutters were made from glass rods on a microelectrode puller. Second, cell bodies of neurons in the head were exposed using the cutter to open a slit in the cuticle. This was done by orienting the cutter tangent to the cuticle, working the tip just under the cuticle surface, then moving the cutter along the tangent line, like opening an envelope with a letter opener. Neurons emerged through the slit, forming a hemispherical bouquet of 10–20 cell bodies (Fig. 1). This is called the "slit-worm preparation" (Avery *et al.*, 1995). Cell bodies were easier to see and record from in younger animals (L1–L3) on an inverted microscope. If an upright microscope is used cell bodies on the top surface are easily visualized. Inspection of GFP-labeled neurons in the bouquet showed that exposed cell bodies remained attached to their processes. Moreover, measured capacitance in an identified neuron matched capacitance predicted from its surface area in undissected animals (Goodman *et al.*, 1998), suggesting that exposed neurons remained intact.

2. Updates

- Including a low-molecular-weight fluorescent dye in the recording pipette (10 μM sulforhodamine 101) allows the experimenter to verify neuronal identity and to assess the extent of fluid access to the neuron (O'Hagan *et al.*, 2005; Ramot *et al.*, 2008b)

- Recordings can be obtained with either an inverted microscope or a fixed-stage, upright microscope.

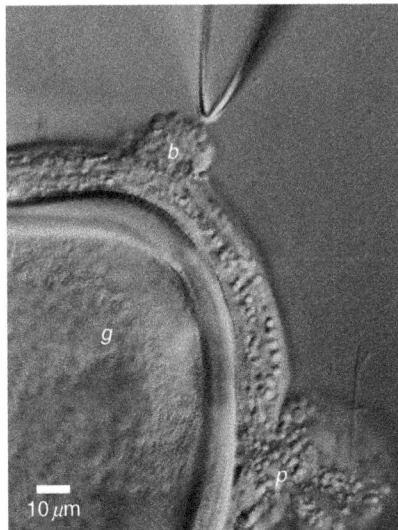

Fig. 1 DIC micrograph of a dissected L1 animal. The pipette is retracted slightly to visualize the membrane connecting the cell and the pipette. The worm's nose is out of view to the left. Symbols: p, point of pressure release; g, glue; b, bouquet of neuronal cell bodies.

IV. Fabricating Patch Pipettes with Submicron Openings

Because *C. elegans* neuronal cell bodies are only 2–5 μm in diameter, recording pipettes must have submicron tip openings. We developed a technique called pressure polishing to fabricate such pipettes having resistances between 5 and 15 MΩ. Consult Goodman and Lockery (2000) for a detailed description of the method; a visualized experiment demonstrating the technique is also available (Johnson *et al.*, 2008).

A. Updates

- Pressure-polishing (Goodman and Lockery, 2000; Johnson *et al.*, 2008) is used to create blunt pipettes from thick-walled, borosilicate glass.
- Equipment to transform a simple inverted microscope into a pressure-polishing microforge is commercially available (CPM-2 with IPH-THP accessory, ALA Scientific Instruments).
- A long-working distance, high-magnification (100×) objective is an essential feature of the polishing apparatus. Such objectives are available from several manufacturers, typically as part of their metallurgical line (e.g., Leica PL Fluotar 100×/0.75 ∞/0).

V. Solutions

The ionic composition of the fluid surrounding *C. elegans* neurons *in vivo* is unknown. Candidate extracellular salines were assayed previously (Avery *et al.*, 1995) by observing pharyngeal pumping and its electropharyngeogram in dissected animals. We used a generic extracellular saline that contained (in mM): 145 NaCl, 5 KCl, 1 CaCl$_2$, 5 MgCl$_2$, 10 HEPES, 20 D-glucose, pH 7.2. The apparent health of exposed neurons depended on the osmolarity of the external saline. By trial and error, we settled on a value of 315–325 mOsm.

The ionic composition of *C. elegans* neuronal cytoplasm is also unknown. We used a generic intracellular saline composed of (in mM): 125 potassium gluconate, 18 KCl, 4 NaCl, 0.6 CaCl$_2$, 1 MgCl$_2$, 10 potassium EGTA, 10 potassium HEPES (pH 7.2), 310–315 mOsm. To make K$^+$-free saline, K$^+$ was replaced with *N*-methyl-D-glucamine.

VI. Patch–Clamp Setup

Because the cell bodies of *C. elegans* neurons are extremely small, recording from them requires higher magnification, greater mechanical stability, and better contrast enhancement in the patch-clamp setup than recording from many other types of neurons. The following steps were taken to improve optics and mechanical stability.

A. Optics

Regardless of whether an upright or an inverted microscope is used, the ability to simultaneously view animals in transmitted light with DIC (differential interference contrast) optics and in fluorescence is essential. Fluorescence imaging is used to identify the GFP-tagged neuron of interest, whereas DIC imaging is used to determine the position of the cell body with sufficient accuracy to obtain a seal. Recordings performed on an inverted microscope (Lockery lab) used a 63×, oil immersion objective with a numerical aperture of 1.4. With this objective, the distance between the top of a No. 0 coverslip and the lens is 160 μm. The diameter of the worm ranges from 15 (L1) to 80 μm (adult). Thus, the agarose pad must be quite thin, between 80 and 145 μm from top to bottom, depending on the size of the animal. We have also used a long working distance, water-immersion objective (Zeiss 40×/0.75, 1.9 mm working distance) (Avery *et al.*, 1995). We prefer the 63×/1.4 objective, however, because of its greater resolving power, which was important for watching the membrane patch while trying break-in. An upright microscope offers the advantage of using optical components optimized for water-immersion. Recordings collected in the Goodman laboratory (O'Hagan *et al.*, 2005; Ramot *et al.*, 2008a) use a water immersion objective on a fixed-stage, upright microscope (Nikon CFI Fluor 60×/1.0, 2.0 mm working distance).

1. Updates

• Recordings can be obtained in larvae and adults. Because of their thickness, adults are best studied using a fixed-stage, upright microscope (e.g., Nikon FN-1).

B. Mechanics

Recording chambers made of acrylic (e.g., Plexiglas), while adequate for recording from large neurons, cannot be used for recording from *C. elegans* neurons. This is because acrylic chambers flex in response to small thermal fluctuations, causing significant movements along the optical axis of the microscope. To avoid this problem, we used a chamber made from a glass plate (76 × 76 × 1 mm), since the coefficient of thermal expansion of glass is much less than acrylic. A hole (18 mm in diameter) was cut in the center of the plate. The coverslip holding the worms was sealed to the underside of the plate with beeswax with the worms centered in the hole.

To minimize drift of the recording pipette, we mounted the amplifier headstage on a motorized manipulator. The manipulator was mounted to either the microscope stage or to a stable, fixed platform that also held the specimen. This approach minimizes vibration and relative movement between the pipette and the preparation.

To obtain whole-cell recordings, we found it was critical to prevent the pipette tip from moving in response to the suction pulse used to break in. This was done by three modifications of the usual arrangement of the pipette holder, air pressure line, and headstage.

(1) The hole in the end cap of the pipette holder (EH-U1, E.W. Wright, Guilford, CT) was drilled to match the outside diameter of the pipette as closely as possible. The final hole size fitted the pipette snugly but allowed it to slide down the hole without jamming.

(2) Suction was delivered by a stainless steel airline connected to the pipette holder and anchored to the manipulator and the microscope. The steel airline was interrupted with a short section of flexible tubing so the manipulator could move.

(3) The pipette was clamped to a stable support attached to the manipulator (Sachs, 1995).

2. Updates

• Using a pipette holder with a conical rather than cylindrical hole stabilizes the back end of the pipette regardless of small variations in pipette outer diameter. This additional stability helps to eliminate the need for a pipette clamp.

• In our current setups, the pipette holder itself is firmly clamped to the headstage and micromanipulator.

• We also recommend outfitting the pipette holder with a cap-seal that is slightly smaller than the pipette's outer diameter (i.e., 1.2 mm for 1.5 mm OD pipettes).

C. Electronics

Several commercial patch-clamp amplifiers are suitable for recording from *C. elegans* neurons without further modification. These include the Axopatch 200B (MDS; http://www.moleculardevices.com) and EPC-10 (HEKA; http://www.heka.com).

VII. Recording from Neurons Labeled with GFP

C. elegans neurons can be identified by size and position inside the animal but these cues are lost when neurons are exposed for recording. We were able to identify exposed neurons, however, by using strains of worms in which single neurons or small sets of neurons expressed GFP. In these experiments, the microscope was equipped for epifluorescence with a 50-W Hg lamp and standard filter set for visualizing GFP (Chalfie *et al.*, 1994). The pipette was brought up to the labeled cell under simultaneous transmitted (DIC) and epifluorescent illumination. To visualize the pipette and the labeled cell optimally, we adjusted the intensity of the transmitted illumination.

VIII. Tight-Seal, Whole-Cell Recording

A. Sealing and "Breaking In"

To keep the pipette tip clean, we applied positive pressure (0.4 kPa) continuously to the back end of the pipette. Immediately after adjusting the pipette capacitance compensation and testing for stability in current clamp, the pipette was brought into contact with the cell body of the target neuron. Contact was detected as an increase in pipette resistance of about 10%. At the moment of contact, positive pressure was released. In most cases, a gigaohm seal forms spontaneously. In others gentle suction was used, alone or in combination with a negative voltage command (-60 mV) to the pipette. Typical seal resistances were 4–11 GΩ. As noted previously (Avery *et al.*, 1995), single-channel activity was apparent in some patches.

The whole-cell configuration was achieved by applying stronger suction together with an electrical zap (0.9 ms, 900 mV) to rupture the membrane patch. GFP-labeled neurons were about twice as easy to break into as unlabeled neurons. In the strain we used most extensively [OH3192 *gcy-5(ntIs1)*], 56% ($n = 42$) of the seals we obtained led to successful whole-cell recordings. The success rate for unlabeled cells was 31% ($n = 100$).

The perforated patch technique (Rae *et al.*, 1991) is an obvious alternative to rupturing the patch when trying to record from small cells. Recordings can be made from *C. elegans* neurons using amphotericin or Nystatin (S. Faumont and S. R. Lockery, unpublished). Using this technique, however, it can take tens of minutes to gain access to the cell and access resistances are impractically high.

B. Whole-Cell Capacitance and Access Resistance

Although *C. elegans* neurons have little capacitance, the transition from the on-cell to the whole-cell recording configuration coincided with a detectable change in the capacity transient elicited by a voltage-clamp pulse (Fig. 2A). The residual capacitance of the pipette and electrode holder (1.16 pF $\pm$ 0.25, $n = 10$, mean $\pm$ S.D.) was comparable to the whole-cell capacitance of the L1 larval neurons (0.5–3 pF), however. Thus, whole-cell capacitance (C_{in}) and access resistance (R_a) were estimated from the difference current obtained by subtracting the current elicited by a 20 mV test pulse in the on-cell configuration from the current recorded in the whole-cell configuration (Fig. 4B). C_{in} was calculated by dividing the integral of the difference current by 20 mV. R_a was estimated as τ/C_{in}, where τ is the time constant of the decay of the difference current. R_a was 19 MΩ (S.D. $= 9, n = 41$).

C. Membrane Current and Membrane Potential

An example of a whole-cell recording obtained from a neuron in the head of a wild-type L1 animal is shown in Fig. 2A. A decaying outward current was observed at potentials greater than about 6 mV. Hyperpolarizing steps elicited little change in net membrane current. Figure 2B shows the voltage response in the same

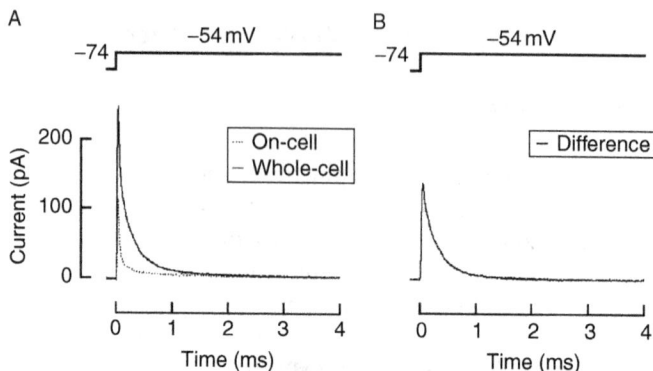

Fig. 2 Voltage-clamp capacity-current transients. (A) Transients recorded in the on-cell (dotted line) and whole-cell (solid line) configurations in response to a 10 ms, 20 mV step from −74 mV. The transient recorded in the on-cell mode represents the pipette capacitance that remained after compensation. The transition from the on-cell to the whole-cell configuration coincided with a detectable change in the capacity transient. For clarity, only the first 4 ms is shown. An equal-amplitude, opposite-polarity transient was apparent on repolarization to −74 mV. (B) The difference current obtained by subtracting the on-cell transient from the whole-cell transient in (A) to eliminate uncompensated pipette capacitance. This current was used to calculate whole-cell capacitance, access resistance, and clamp speed. Reprinted with permission from Goodman *et al.* (1998).

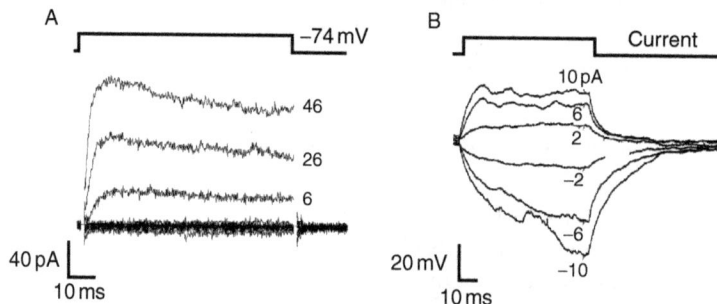

Fig. 3 Whole-cell voltage-clamp and current-clamp recordings from the cell shown in Fig. 2. (A) Membrane current recorded in response to 11 80-ms voltage steps between −154 and +46 mV in 20 mV increments. The holding potential was −74 mV. An outward current was apparent in response voltage steps to more than +6 mV; no change in membrane current was apparent for voltage steps less than +6 mV. (B) Membrane voltage recorded in response to six 75-ms current steps between −10 and +10 pA in 4 pA increments. Membrane voltage is expressed relative to the zero-current potential which was −9 mV. In the trace marked "−2," a voltage artifact produced by the perfusion pump was excised, producing a gap in the trace.

cell to a family of current pulses. As expected from the absence of hyperpolarizing-activated membrane current (Fig. 3A), the amplitude of the response to hyperpolarizing current pulses was proportional to pulse amplitude and its time course was approximately exponential. The response to depolarizing current pulses was also

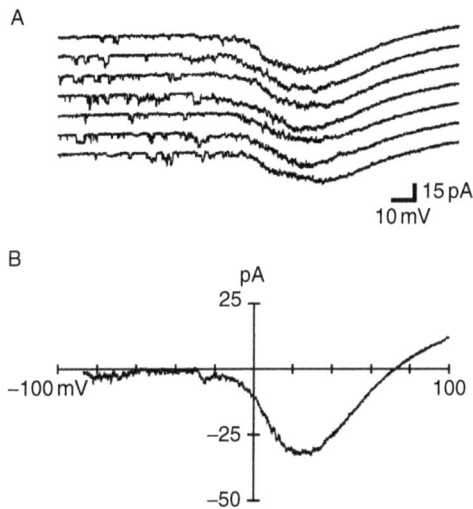

Fig. 4 Whole-cell and putative single-channel currents recorded in the same cell. (A) Current in response to a voltage ramp from −112 to 88 mV (300 ms duration). The holding potential between ramps was −92 mV. The step from the holding potential to −112 mV elicited a capacity transient that was omitted from the traces. The pipette contained K^+-free saline and the cell was superfused by a solution containing (in mM): 110 $BaCl_2$, 10 CsHEPES, pH 7.2. (B) Average of the seven traces in (A). Average current was well fit by the Boltzmann equation, $I = G_{max}(V − E_r)/(1 + exp[(V_0 − V)/V_s])$, with $G_{max} = 0.82$ nS, $E_r = 71$ mV, $V_0 = 11.6$ mV, and $V_s = 8.2$ mV. Reprinted with permission from Goodman *et al.* (1998).

graded with amplitude, but the change in membrane potential was smaller than that evoked by a hyperpolarizing pulse of equal amplitude. This is likely to reflect activation of outward current by depolarization.

Putative single-channel currents were occasionally observed during whole-cell recordings. These events were most apparent under conditions that minimized outward currents. One such experiment is shown in Fig. 4A. The pipette contained K^+-free saline and the cell was superfused with an external solution designed to block K^+ currents and enhance Ca^{2+} currents. At voltages hyperpolarized to −50 mV, single-channel openings could be resolved. At higher voltages the traces became noisy, suggesting that the number of open channels increased. Averaging these traces produced the macroscopic current–voltage curve shown in Fig. 4B.

IX. Interpreting Patch–Clamp Recordings from Small Cells

There are advantages and disadvantages to recording from small cells (Barry and Lynch, 1991). These are discussed briefly below in the context of recording from *C. elegans* neurons.

A. Voltage Clamp

1. Voltage Errors Due to Series Resistance

One advantage of recording from small cells is that the membrane currents are also typically small. This reduces the discrepancy between membrane voltage and command voltage when current flows across R_a. Voltage errors in our recordings were at most 10 mV even without series resistance compensation.

2. Clamp Speed

Without series resistance compensation, the time constant for charging the membrane capacitance in our recordings was 7–118 μs, with a median of 23 μs ($n = 39$). This indicates that in most cases we had rapid control of membrane voltage even in uncompensated recordings.

3. Filtering of Whole-Cell Voltage–Clamp Currents

In voltage clamp, R_a and C_{in} produce a low-pass filter with a cutoff frequency, F_c, of approximately $1/(2\pi R_a C_{in})$. Because C_{in} in C. elegans neurons is small, currents should be filtered less than in larger cells with comparable R_a. In our recordings, F_c for an average C. elegans neuron ($R_a = 20$ MΩ, $C_{in} = 0.8$ pF) was 10 kHz. Higher values of F_c could be achieved with whole-cell capacitance compensation. Thus, it should be possible to resolve both fast macroscopic and single-channel currents even in whole-cell recordings.

B. Current Clamp

In current clamp, R_a, and uncompensated pipette capacitance, C_p, produce a low-pass filter with $F_c \approx 1/(2\pi R_a C_p)$. In large neurons, where $C_{in} \gg C_p$, filtering by the pipette is small compared to filtering by the cell itself. In C. elegans neurons, however, $C_{in} \approx C_p$, so filtering by the pipette could be significant. In our recordings, we estimate F_c was about 8 kHz. F_c could be increased by reducing pipette capacitance. Steps that reduce pipette capacitance include using borosilicate or quartz glass, applying heavier Sylgard coatings, and reducing the depth of the recording bath.

C. Estimating Passive Membrane Properties of Small Neurons

Input resistance, membrane potential, and membrane time constant are difficult to measure in small cells. This is because one cannot make the usual assumptions that input resistance, R_{in}, is much greater than seal resistance, R_s, or that input capacitance, C_{in}, is much greater than uncompensated pipette capacitance, C_p (Barry and Lynch, 1991).

1. Input Resistance

Any estimate of R_{in} in a *C. elegans* neuron is at best a lower bound. This is because R_s is of the same order of magnitude as R_{in}, and R_s and R_{in} act in parallel to determine the response of the cell to a voltage or current pulse. A more accurate estimate may be possible, however, using the time course of single-channel currents in cell-attached patches (Barry and Lynch, 1991).

2. Membrane Potential

A related effect of the similarity between R_{in} and R_s is a reduction in the magnitude of the cell's zero-current (resting) potential, V_m. Because R_s and R_{in} act in parallel, the apparent V_m is the weighted average of the true V_m and the reversal potential associated with R_s, which is approximately 0 mV.

3. Membrane Time Constant

The quantity τ_m is often estimated from the voltage response to a current impulse or step (charging curve). In the case of an isopotential cell recorded with an ideal electrode ($R_{in} \gg R_a$, $C_{in} \gg C_p$), $\tau_m = R_{in}C_{in}$. For *C. elegans* neurons, however, $C_{in} \approx C_p$, so the apparent τ_m is increased to $R_{in}(C_{in} + C_p)$. At the same time, however, the effective value of R_{in} is decreased by R_s, tending to decrease the apparent τ_m. Similar considerations apply to nonisopotential cells (Major *et al.*, 1993). Thus, estimates of τ_m obtained from charging curves in *C. elegans* neurons are unreliable.

X. Spatial Control of Voltage

Recordings from the unbranched, bipolar neuron ASER indicated that this cell could be well-space clamped at all larval stages (Goodman *et al.*, 1998). Most neurons in the head ganglia of *C. elegans* have processes with the same dimensions as those in ASER and are either bipolar or monopolar. Assuming that the electrical properties of ASER are not unique, these two types of neurons should also be well-space clamped. By contrast, the touch receptor neurons, whose neurites extend for one-half the body length (500 μm in an adult), have space constants that are only $\sim$100 μm (O'Hagan *et al.*, 2005), indicating an imperfect ability to space clamp such neurons. Whether a third type of neuron, with a single process running the length of the body, can also be well-space clamped remains to be seen.

XI. Prospects

A. Single Channels in Whole–Cell Recordings

Single-channel currents are rarely resolved in whole-cell recordings. In *C. elegans* neurons, however, we found conditions that revealed single-channel events in whole-cell recordings. This means it should be possible to study the microscopic and macroscopic properties of a current in a single recording. Also, because *C. elegans* neurons are likely to be well-space clamped, it may be possible to record single-channel currents originating at synapses and other sites distant from the cell body.

B. Identified Neurons

A great advantage of invertebrate systems is the existence of identifiable neurons. Identification can be laborious, however, since neurons often must be identified by several criteria at once, including their intrinsic electrical properties, morphology, and connections to other neurons. In GFP-labeled strains of *C. elegans*, neurons can be identified by direct observation. This saves time and makes it possible to manipulate a neuron's membrane currents in ways that obscure the intrinsic electrical properties by which the neuron would otherwise be identified. It is difficult to assess the effects GFP might have on neuronal physiology in *C. elegans* because we did not record from the same neuron with and without GFP expression. In mammalian expression systems, however, the biophysical properties of exogenously expressed ion channels were unaffected by the presence of GFP (Doevendans *et al.*, 1996; Marshall *et al.*, 1995). In *C. elegans*, the amplitude and time course of current in GFP-labeled neurons were not unusual when compared with unlabeled neurons, suggesting that the effects of GFP, if any, are not dramatic.

XII. Conclusion

This chapter describes a method for recording from neurons in *C. elegans* nematodes. By similar means, we have also exposed for recording the cell bodies of neurons in the ventral cord and tail ganglia. It is possible, therefore, to record from almost any neuron in *C. elegans*.

The ability to record form identified neurons in *C. elegans* opens new avenues of research in a system already well known for its complete neuronal wiring diagram (White *et al.*, 1986), developmental cell lineage (Sulston and Horvitz, 1977), and genomic sequence (The *C. elegans* Sequencing Consortium, 1998). For example, it is now possible to combine both classical and molecular genetics with electrical recordings from identified neurons to study the genetic control of neural function and its development. In addition, we can now study how identified neurons function in the neural circuits underlying particular behaviors. It is possible, therefore, to trace the connections between genes and behavior in *C. elegans* more completely than before.

Acknowledgments

We thank D. Raizen, K. Breedlove, and L. Avery for first demonstrating that on-cell patch-clamp recordings can be made from *C. elegans* neurons, J. Thomas for the idea of using a laser beam to break into cells, D. Garbers, S. Yu, and C. Bargmann for the gift of GFP-labeled strains. Portions of the text have been published previously (Goodman *et al.*, 1998) and are reprinted here with permission. This work began in the laboratory of T. Sejnowski. Work in the Lockery and Goodman labs has been supported by NSF, NIMH, NINDS, ONR, The Sloan Foundation, The McKnight Foundation, The Baxter Foundation, The Klingenstein Fund, and The Searle Scholars Program.

References

Avery, L., Raizen, D., and Lockery, S. R. (1995). Electrophysiological methods. *In* "*C. elegans*: Modern Biological Analysis of an Organism," (H. F. Epstein, and D. C. Shakes, eds.). Academic Press, Orlando, FL.

Bargmann, C. I. (1993). Genetic and cellular analysis of behavior in *C. elegans*. *Annu. Rev. Neurosci.* **16**, 47–71.

Barry, P. H., and Lynch, J. W. (1991). Liquid junction potentials and small cell effects in patch-clamp analysis. *J. Membr. Biol.* **121**, 101–117.

Bianchi, L., and Driscoll, M. (2006). Culture of embryonic *C. elegans* cells for electrophysiological and pharmacological analyses. *WormBook* 1–15.

Brown, S. D., and Carey, A. H. (1994). Chromosome dissection and cloning. *Methods Mol. Biol.* **29**, 425–436.

Chalfie, M., Tu, Y., Euskirchen, G., Ward, W. W., and Prasher, D. C. (1994). Green fluorescent protein as a marker for gene expression. *Science* **263**, 802–805.

Clark, S. G., Stern, M. J., and Horvitz, H. R. (1992). *C. elegans* cell-signalling gene sem-5 encodes a protein with SH2 and SH3 domains. *Nature* **356**, 340–344.

Doevendans, P. A., Becker, K. D., An, R. H., and Kass, R. S. (1996). The utility of fluorescent in vivo reporter genes in molecular cardiology. *Biochem. Biophys. Res. Commun.* **222**, 352–358.

Faumont, S., Boulin, T., Hobert, O., and Lockery, S. R. (2006). Developmental regulation of whole cell capacitance and membrane current in identified interneurons in *C. elegans*. *J. Neurophysiol.* **95**, 3665–3673.

Fire, A. (1986). Integrative transformation of *Caenorhabditis elegans*. *EMBO J.* **5**, 2673–2680.

Goodman, M. B., and Lockery, S. R. (2000). Pressure polishing: A method for re-shaping patch pipettes during fire polishing. *J. Neurosci. Methods* **100**, 13–15.

Goodman, M. B., Hall, D. H., Avery, L., and Lockery, S. R. (1998). Active currents regulate sensitivity and dynamic range in *C. elegans* neurons. *Neuron* **20**, 763–772.

Hodgkin, J., Plasterk, R. H., and Waterston, R. H. (1995). The nematode *Caenorhabditis elegans* and its genome. *Science* **270**, 410–414.

Johnson, B. E., Brown, A. L., and Goodman, M. B. (2008). Pressure-polishing pipettes for improved patch-clamp recording. *J. Vis. Exp.*

Major, G., Evans, J. D., and Jack, J. J. (1993). Solutions for transients in arbitrarily branching cables: I. Voltage recording with a somatic shunt. *Biophys. J.* **65**, 423–449.

Marshall, J., Molloy, R., Moss, G. W., Howe, J. R., and Hughes, T. E. (1995). The jellyfish green fluorescent protein: A new tool for studying ion channel expression and function. *Neuron* **14**, 211–215.

Mellem, J. E., Brockie, P. J., Zheng, Y., Madsen, D. M., and Maricq, A. V. (2002). Decoding of polymodal sensory stimuli by postsynaptic glutamate receptors in *C. elegans*. *Neuron* **36**, 933–944.

Mellem, J. E., Brockie, P. J., Madsen, D. M., and Maricq, A. V. (2008). Action potentials contribute to neuronal signaling in *C. elegans*. *Nat. Neurosci.* **11**, 865–867.

Nickell, W. T., Pun, R. Y., Bargmann, C. I., and Kleene, S. J. (2002). Single ionic channels of two *Caenorhabditis elegans* chemosensory neurons in native membrane. *J. Membr. Biol.* **189**, 55–66.

O'Hagan, R., Chalfie, M., and Goodman, M. B. (2005). The MEC-4 DEG/ENaC channel of *Caenorhabditis elegans* touch receptor neurons transduces mechanical signals. *Nat. Neurosci.* **8**, 43–50.

Pierce-Shimomura, J. T., Faumont, S., Gaston, M. R., Pearson, B. J., and Lockery, S. R. (2001). The homeobox gene lim-6 is required for distinct chemosensory representations in *C. elegans*. *Nature* **410**, 694–698.

Rae, J. L., Cooper, K., Gates, P., and Watsky, M. (1991). Low access resistance perforated patch recordings using amphotericin B. *J. Neurosci. Methods* **37**, 15–26.

Ramot, D., MacInnis, B. L., and Goodman, M. B. (2008a). Bidirectional temperature-sensing by a single thermosensory neuron in *C. elegans*. *Nat. Neurosci.* **11**, 908–915.

Ramot, D., MacInnis, B. L., Lee, H. C., and Goodman, M. B. (2008b). Thermotaxis is a robust mechanism for thermoregulation in *Caenorhabditis elegans* nematodes. *J. Neurosci.* **28**, 12546–12557.

Sachs, F. (1995). A low drift micropipette holder. *Pflugers Arch.* **429**, 434–435.

Stiernagle, T. (2006). Maintenance of *C. elegans*. *WormBook* 1–11.

Sulston, J. E., and Horvitz, H. R. (1977). Post-embryonic cell lineages of the nematode, *Caenorhabditis elegans*. *Dev. Biol.* **56**, 110–156.

Tattar, T. A., Stack, J. P., and Zuckerman, B. M. (1977). Apparent nondestructive penetration of *Caenorhabditis elegans* by microelectrodes. *Nematologica* **23**, 267–269.

The C. elegans Sequencing Consortium, B. M. (1998). Genome sequence of the nematode *C. elegans*: A platform for investigating biology. *Science* **282**, 2012.

Ward, A., Liu, J., Feng, Z., and Xu, X. Z. (2008). Light-sensitive neurons and channels mediate phototaxis in *C. elegans*. *Nat. Neurosci.* **11**, 916–2922.

White, J. G., Southgate, E., Thomson, J. N., and Brenner, S. (1986). The structure of the nervous system of the nematode *C. elegans*. *Philos. Trans. R. Soc. B* **314**, 1–340.

CHAPTER 5

Gating Currents

Francisco Bezanilla and Enrico Stefani

Department of Anesthesiology
UCLA, Los Angeles, California, USA

I. Introduction

Several membrane transport mechanisms are voltage dependent. These include voltage-dependent ion channels and some transporters and pumps. Regardless of the structure of the molecule, the membrane voltage exerts its effect by acting on a

DOI: 10.1016/B978-0-12-382204-8.00005-9

"voltage sensor." Hodgkin and Huxley (1952) predicted the existence of a voltage sensor when they stated: "It seems difficult to escape the conclusion that the changes in ionic permeability depend on the movement of some component of the membrane which behaves as though it had a large charge or dipole moment." Thus, the voltage sensor is the machinery that mediates the transduction between the membrane potential and whatever change may occur in the membrane to initiate or terminate ion conduction.

A simple view of the voltage sensor consists of an embedded charge in the channel protein that can move in response to changes in the membrane electric field, and that movement leads to the opening or closing of the conduction pathway. An immediate consequence of the charge movement is that it generates a current and, because this current is connected to the channel gating process, it is called the *gating current*. The most general property of the gating current is that it is transient because the charge that moves is confined in the membrane field, and for long periods will remain in a stable position. In this regard, gating currents share the properties of capacitive currents and may be considered to be displacement currents produced by dielectric relaxation.

At the single-molecule level, gating currents represent the rate of displacement of individual charges or dipole reorientation. In the original Hodgkin and Huxley (1952) formulation, the opening of the potassium channel occurs when four independent gating particles are located simultaneously in the active position. Furthermore, each particle can only be in one of two stable positions, the rate of translocation from the resting to the active position is increased by depolarization, and the reverse rate is increased by hyperpolarization. Figure 1 illustrates a simulation based on these assumptions. Because each charged particle may only dwell in one of two positions, the transition between them is instantaneous, producing an infinitely large current shot (δ function, see Fig. 1, single-channel gating shots). The integral of this current is the total charge moved times the fraction of the field it traverses. In this simulation, the result of a single step of membrane potential started from a very negative value (-120 mV) to a positive value ($+50$ mV) is shown for the single gating shots and the resultant single-channel current. The ensemble current of shots for 1000 pulses is shown as the gating current and the resultant macroscopic current as the macroscopic ionic current. This type of simulation sets the stage for the conditions to record these different types of current from real cells because, depending on the kinetics and the amount of gating charge per channel, the relative magnitudes of the different currents vary widely.

II. Studying Gating Currents

When the voltage-dependent channel opens, ionic current flows can be detected during voltage clamp. If the channel had only two physical states—closed or open—the information obtained from recording the macroscopic ionic current would be enough to infer the voltage dependence of the transition. However,

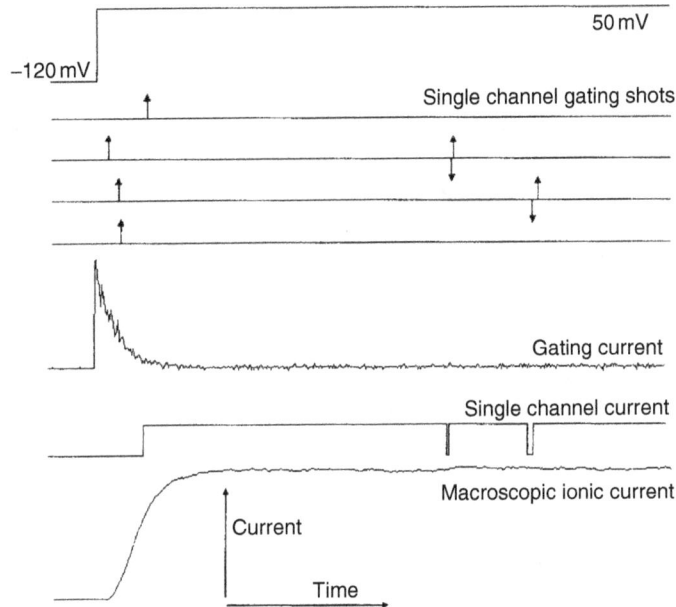

Fig. 1 The predicted ionic and gating events for a channel with four independent gating particles, each obeying first-order kinetics with voltage-dependent rate constants. The pulse goes from −120 to 50 mV, and the single-channel gating shots show the sequence of events for one trial. The resultant single-channel current is shown for the same trial. The average of the gating shots for 1000 trials is shown in the trace labeled "Gating current," while the average of the single-channel openings for the same number of trials is shown in the "Macroscopic ionic current" trace.

channels exhibit many physical states, most of which are several steps away from the open state. In that case, the information obtained from the macroscopic currents or even from single-channel recordings will be incomplete and inferring the properties of transitions between closed states will be more difficult the further that transition occurs from the open state. Gating currents, on the other hand, include the direct contributions of those far-removed charge-carrying transitions and their characterization helps in elucidating the physical states and transitions the channel undergoes from the closed to the open states. In fact, most of the charge-carrying transitions occur before the actual opening of the channel; therefore, the study of gating currents is necessary if we are to obtain a full characterization of the voltage-dependent process.

III. Charge per Channel

The time integral of the gating current measures the product $z \cdot \delta$, where z is the charge displaced and δ is the fraction of the field it traverses. The voltage dependence of a channel will depend on the total number of charges displaced, and this number has been found recently for several types of channels using two different methods. The first method makes the measurement of the total charge (Q_t) by measuring the asymptotic values of the charge versus potential ($Q–V$) curve from the time integral of the gating currents. The charge per channel is computed by dividing Q_t by N, the number of channels present, normally estimated by noise analysis of the ionic currents (Schoppa et al., 1992) or counting the channels with a specific toxin (Aggarwal and MacKinnon, 1996). The second method estimates the charge per channel from the derivative with respect to voltage of the logarithm of the open probability versus voltage relation ($P_0–V$ curve). This estimation is only accurate when the derivative is computed at very low values of P_0, which can be done by single-channel recording (Hirshberg et al., 1996) or by extending the $P_0–V$ curve using the $Q–V$ curve (Seoh et al., 1996; Sigg and Bezanilla, 1997). The values measured for Na^+ and K^+ channels vary between 11 and 13 e_0.

IV. Detection Problem

In the idealized case, the magnitude of the single-channel shot (see simulation of Fig. 1) is predicted to be a delta (δ) function. In practice, the transition will not be instantaneous, but it will appear so to the observer due to bandwidth limitations. In fact, the shots should be recorded as the impulse response of the filter used, and the predicted magnitude is about 1 fA for a 5-kHz bandwidth, which is below the resolution of present technology, if normal low-pass recording techniques are used. The ensemble of many channels will give a signal whose noise can be analyzed to infer the size of the elementary event (see later discussion).

When the recording is done from a membrane that contains a large number of channels, the contribution of the shots of the individual channels sums to a larger signal called the gating current (see Fig. 1), which can be experimentally recorded. The magnitude of the gating current depends on the number of charges moving per channel, the total number of channels, and the kinetics of the charge transition. These parameters have important consequences for the experimental detection of gating currents because present techniques can only resolve signals that are not extremely fast, due to bandwidth limitations, and not too slow because of uncertainties and noise of the baseline. For a given total charge, fast kinetics will produce a larger gating current than a slower channel. It is not surprising than that the Na^+ gating currents were recorded before the K^+ channel gating currents (Armstrong and Bezanilla, 1973) because the K^+ channel density in squid axon is lower, and its kinetics are about 10 times slower than the Na^+ channel. After increasing the temperature by about 15 °C, which speeds up kinetics roughly by a factor of 10, K^+ gating currents were successfully recorded (Bezanilla et al., 1982). The channel density is a very important factor in detecting gating currents with a good signal-to-noise ratio (S/N) because the noise increases with recording area. The best recordings of gating currents have been obtained from preparations with high channel density such as the squid (Armstrong and Bezanilla, 1973), crayfish (Starkus *et al.*, 1981), and Myxicola (Rudy, 1976) giant axons as well as the eel electroplax (Shenkel and Bezanilla, 1991).

With the advent of molecular cloning techniques, it is now possible to express cloned channels in oocytes or mammalian cells at even higher densities, and the S/N of currents recorded from these expression systems are superior to the native cells. In addition, the cloned channel can be modified by replacing, inserting, or deleting residues that are suspected to be important for the function of the channel. These mutated channels can be expressed and gating currents analyzed to infer how a particular residue that was replaced affected the normal operation of the channel. With this procedure, it has been possible to locate the important residues that make up the voltage sensor in *Shaker* K^+ channels (Aggarwal and MacKinnon, 1996; Seoh *et al.*, 1996).

When the expected size of the gating currents is smaller than the noise of the recording, it is always possible to use signal averaging to improve the S/N. The recording of gating currents lends itself to this technique because the events are time locked with the stimulating voltage pulse, allowing the averaging of the time-locked signal in the computer memory.

V. Separation Problem

Once the charge per channel and the density and kinetics are favorable for detecting gating currents, they must be separated from the membrane capacitive current and the ionic currents. Frequently, these currents are much larger than the gating currents, masking them partially or completely. The ratio of the peak gating current to the ionic current is dependent on the kinetics of gating because, as explained earlier, gating currents will be larger for faster channels. Thus, for Na^+ channels the ratio is about 1:50 to 1:100, while for K^+ channels it is smaller than 1:200. The ionic currents can be abolished by replacing permeant ions on both sides of the membrane by species that do not pass through the channel to be measured and other channels that might be present in the membrane. Typical cation replacements are tetramethylammonium ion, cesium ion, or N-methylglucamine (NMG) ions. Some ions must be tested for effects on gating, such as those seen with tetraethylammonium (TEA) (see later discussion). Details of solutions used can be found in the chapter on the cut-open oocyte voltage clamp (Stefani and Bezanilla, 1998). In addition to impermeant ions, it is possible to use blocking agents, but first it should be ascertained that they do not modify gating.

After the ionic currents have been eliminated, the recorded current will include some residual leakage, capacitive currents, and gating currents. The remaining leakage should be as small as possible and should be examined for time-dependent changes. Frequently, time-dependent leaks occur in very leaky membranes and make the separation of the gating current a very difficult or impossible task. The capacitive current is normally separated by using the property that the gating currents, although capacitive themselves, are nonlinearly related to voltage. This is an expected property of buried charges or dipoles that are confined to the membrane field. In contrast, the charging of the membrane capacitance is not expected to show saturation because the charging and discharging are provided by the separation of charge by the virtually infinite reservoir of ions in the solutions surrounding both sides of the membrane. The logic is simple: a positive going pulse and a negative going pulse of the same magnitude produce capacitive currents of the same magnitude and time course but of opposite polarity, regardless of at which base or holding potential (HP) the pulses are given (see Fig. 2A). Therefore, the addition of those currents, synchronized exactly with the beginning of the pulse, will give a null current. However, if a gating charge moves during the positive pulse, but it does not move during the negative pulse (because at that potential HP, the charge movement is saturated), the addition of the currents produced by P and −P pulses will give a net current that will reflect the movement of the gating charge during the P pulse, canceling completely the capacitive currents of the membrane. This pulse protocol (Fig. 2A) has been called the ±P procedure (Armstrong and Bezanilla, 1973). To obtain a reliable measurement of the gating current, it is imperative that there be no gating charge movement in the negative direction from the potential HP. This condition is rarely met because it has been found that there is a detectable amount of gating charge at potentials more negative than the resting potential, the usual value where the HP is set. To correct this problem, one could move HP to a more negative potential, but when giving a large P pulse, the corresponding −P pulse will reach prohibitive values of membrane potentials, with subsequent breakdown of the membrane. The solution is to step to values of membrane potential where there is no gating charge movement (SH) to give the subtraction pulses for the time needed, and give smaller

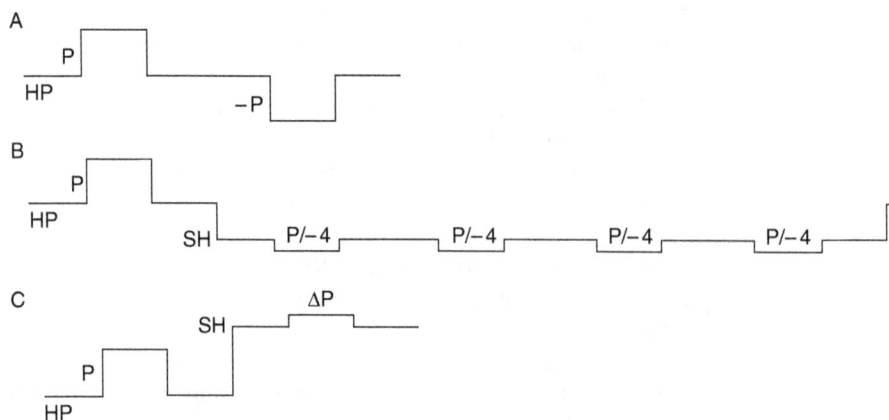

Fig. 2 Subtraction protocols for elimination of the linear capacitive current. HP represents the value of the holding potential: (A) the ±P procedure, (B) the P/−4 procedure with a negative subtracting holding potential (SH), and (C) the variable scaling procedure with a positive SH. For details, see text.

pulses to prevent membrane breakdown. This is the basis of the P/4 procedure (Bezanilla and Armstrong, 1977), which has several variants. Figure 2B shows an example with pulses going in the opposite direction to the test pulse P (the P/−4 procedure). The potential at which subtracting pulses are given can be chosen to be more negative than the HP, or more positive than the HP. One should design the direction of the subtracting pulses so that they occur in the region where no gating charge moves. The reason why the P/4 procedure gives four pulses is because the current produced by a single pulse of one-fourth of the amplitude would have to be multiplied by four before it could be subtracted (or added) to the current produced by the P pulse, increasing the noise of the result. In fact, P/4 does not restore the S/N but it was selected as a good compromise to limit the number of pulses and the time the membrane is taken to the SH potential. The P/4 procedure can be extended to be P/n, and the SH potential may be negative or positive depending on the saturation voltages of the $Q–V$ curve.

Another general subtraction procedure is shown in Fig. 2C, where a pulse of amplitude ΔP is given at an SH (negative or positive, depending on the position of the $Q–V$) and the currents produced by both pulses are stored. Later, the current produced by the ΔP pulse is scaled appropriately to do the subtraction. Frequently, this method is done by recording a single ΔP current and used off-line subtraction of any P-pulse-produced currents. This method has the disadvantage of producing extra noise because of the scaling factor when ΔP is smaller than P. It is recommended that several sizes of ΔP be recorded and signal averaging used to minimize their noise contribution. The recording of ΔP separately from each of the P recordings has the disadvantage that any drift in the time course of the capacitive transient during the experiment will affect the recording of the time course of the gating current. This last problem is not present with the P/n technique because the subtraction is done every time a P pulse is given.

When the gating current is small compared to the capacitive current, the subtraction procedure relies on extracting a small quantity from the subtraction of two large quantities. This could produce serious error because small drifts in the preparation would make the subtraction completely worthless. A typical example where this problem can be pervasive happens when recording in the whole-cell configuration. The time course of the capacity transient depends on the membrane capacity and the series resistance, and the latter depends heavily on the access resistance in the whole-cell configuration. A small change in the access resistance, which frequently occurs in experiments done in that configuration, will make the subtraction invalid.

VI. Voltage Clamp

For the study of gating currents of expressed channels, two major techniques have been used to clamp the membrane: the patch clamp and the cut-open oocyte voltage clamp. The two-microelectrode clamp has also been used, but its poor time resolution makes it less appropriate for studying the kinetics of gating currents.

A. Patch Clamp

In the excised or cell-attached configuration, the patch clamp is an ideal technique because it exhibits low noise and can record currents with high bandwidth. The techniques are described in detail in other chapters in this book (Hilgemann, 1998; Levis, 1998), but for the purposes of studying gating currents there are a few considerations that must be added. First, as the number of channels increases with the area, it is necessary to record from large areas, and the macropatch or the giant patch techniques are the configurations of choice. The coating of the pipettes with Sylgard for macropatches or with the oil–Parafilm mixture for the giant patches is very important to decrease the capacitance and to maintain a stable capacitance during the recording. In addition, the pipettes should be made with short shanks to minimize the access resistance. Best results are obtained with pipettes with resistance lower than 0.6 MΩ. The dynamic range of the input stage of the patch clamp is governed by the feedback resistance of the current-to-voltage (I–V) converter when it is of the resistive type. Capacitive transients, even after compensation, may be much larger than the currents to be recorded. To decrease even further the magnitude of the capacitive transient, the command pulse may be slowed down but if after this is done all of these manipulations still produce saturation, the use of a lower gain for the patch-clamp headstage may be required. This is not recommended because noise will increase, leading to deterioration of the quality of the recordings. A good solution to the dynamic range problem is to use an integrating headstage and to synchronize the resetting of the integrator with a pulse preceding the acquisition period.

The attached patch is generally more stable than the excised patch configuration and, in addition, preserves the time course of the gating current by minimizing excision-induced rundown. The problem of replacing the internal solution with impermeant ions in the attached patch configuration is resolved by immersing the oocyte in the solution with impermeant ions and perforating the surface of the oocyte in several places with a pipette, letting it equilibrate to replace the internal contents.

Although it is quite simple to record gating currents with a bandwidth of 10 kHz, we have recently extended the frequency response of the patch amplifier to about 200 kHz and, using giant patches with access resistance below 0.2 MΩ, we have been able to record gating currents revealing a faster event that precedes the rising phase of the normal gating current. This faster transient has a shallow voltage dependence and carries less than 10% of the total charge (see Stefani and Bezanilla, 1996, 1997).

The patch-clamp method is the only available technique to record gating current fluctuations because the number of channels can be controlled to be within a range that does not exceed the resolution of the elementary event. For the same reason, to count the number of channels to determine charge per channel, the patch clamp is the only technique available.

B. Cut-Open Oocyte Voltage Clamp

This technique is described in detail in another chapter in this volume (Stefani and Bezanilla, 1998). The large recording area of this technique allows the detection of very small gating currents that would escape detection using the patch-clamp technique. This is particularly advantageous when a mutation is introduced in the molecule under study that cuts down the expression of the protein in the membrane. In addition, it has been found that the effect of patch excision, which frequently produces rundown of the currents, is minimized in the cut-open oocyte technique.

VII. Recording Step

Figure 3 shows a generic setup that we will use to analyze the components that can influence the quality of the recorded gating currents.

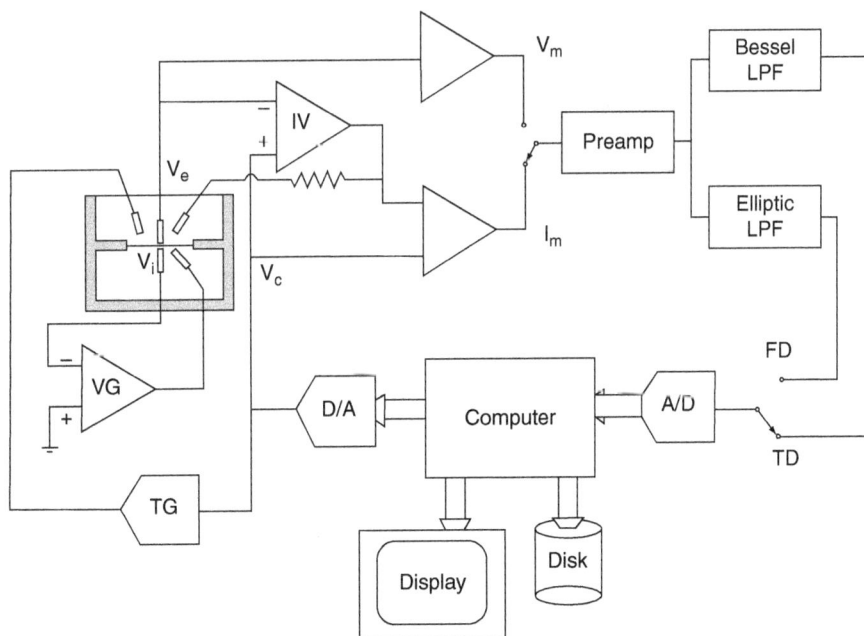

Fig. 3 Schematic diagram of a recording setup to measure gating currents. D/A, digital-to-analog converter; FD, frequency domain mode; *I–V*, current-to-voltage converter; TD, time domain mode; TG, transient generator; V_c, command voltage; V_e, external voltage-sensing electrode; VG, virtual ground control amplifier; V_i, internal voltage-sensing electrode.

A. Membrane and Voltage Clamp

The chamber presented on the left part of Fig. 3 is a general arrangement for a voltage clamp of a piece of membrane. There are many different configurations, depending on the type of cell and voltage clamp used; this particular arrangement is similar to the cut-open oocyte voltage clamp. VG is an amplifier that maintains the inside (bottom chamber) at virtual ground, and I–V is a current-to-voltage converter that imposes the command voltage on the external side of the membrane. Notice that in this configuration, the electrode that feeds back from I–V into the chamber is different from the voltage-sensing electrode V_e, the same is true for the voltage-sensing electrode V_i and the current passing electrode at the output of VG. In a patch configuration, VG is not present and electrode V_e becomes the same as the current passing electrode: this could pose a problem for large currents. In the squid axon voltage clamp, the control is done from the inside, and the outside is maintained at virtual ground by the I–V converter.

B. First Amplifier

It is crucial that the I–V converter amplifier have enough dynamic range so that it will not saturate during the capacity transient. In addition, it should have wide bandwidth and a fast slew rate. Saturation produces different responses for large pulses as compared to small pulses, invalidating the subtraction procedure. Saturation can be prevented by reducing the gain of the amplifier, but it has the disadvantage that the noise increases. The other possibility is to cancel some of the capacitive transient by injecting a similar current of opposite magnitude into the same side of the chamber using a transient generator (TG). Ideally, the TG is capable of generating transients of two or three exponential components that the user can manually change until the capacity transient is abolished. The TG is fed the same pulse that is used as the command to clamp the membrane and, as the clamp is normally slower than the generator itself, it is convenient to add a phase control to match the rising phase of the transient. The output of the TG is fed to the chamber through a different electrode to eliminate interactions with the recording electrode. Again, in a patch-clamp situation, this extra electrode is not implemented. Instead, the TG (built-in in commercial units) will feed the current into the summing junction of the I–V converter.

C. Preamplifiers

The output signal of the I–V converter may be too small to fill the dynamic range of the analog-to-digital (A/D) converter, so an amplifier can be added. This amplifier must have a wide bandwidth, low noise, and excellent linearity characteristics. Saturation should be prevented and carefully checked. For frequency-domain applications it is critical that the frequency response (magnitude and phase) be preserved across the whole range of amplification of the preamplifier.

D. Filters

The filter should be placed after the amplifier to prevent further amplification of the intrinsic noise of the filter. However, in some cases two stages of amplification can be used with the filter between. This arrangement allows preamplification of large transients that still fit within the dynamic range of the filter and use post-amplification after the filter has attenuated the large capacitive transient. In time-domain applications, the filter should be of the linear phase type (Bessel or Gaussian) to prevent ringing of transient responses, but in the frequency domain it is more important to have a sharper cutoff and an elliptic filter can be used. In any case, to prevent aliasing errors, the filter must be set to a 3-dB cutoff frequency no more than one-half the sampling frequency, but because the cutoff is gradual, it is better to use a cutoff at about one-fifth of the sampling frequency. In the frequency domain a value close to one-half of the sampling frequency can be used because the frequency response of elliptic filters is sharper. An exact value of the cutoff to eliminate alisiang can be calculated knowing the frequency response of the filter and matching the frequency at which the attenuation of the signal is just beyond the resolution of the A/D converter.

E. Data Acquisition

Acquisition is carried out with an A/D converter which must be preceded by a sample-and-hold amplifier for typical successive approximation or double ranging converters. This is a critical part of the circuitry and must be selected to be linear, monotonic, and of high resolution. Best results are obtained with 16-bit converters because they allow a larger dynamic range than the typically used 12-bit converters. High resolution is particularly important for subtraction of large transients to extract small gating currents, even after the dynamic range has been enlarged by using a TG, and it is imperative for extracting information from noise analysis of gating currents (see later discussion). The signal should be preamplified to maximize the converter dynamic range.

F. Waveform Generation

The command signal applied to the clamp is generated by a D/A converter. Important characteristics for this converter follow:

1. *Low noise.* Any noise produced by the D/A converter will be differentiated by the membrane capacitance, producing extra noise in the recordings.
2. *Linearity.* Because this is the source for the P and P/*n* pulses, the linearity is critical to achieve good subtraction.
3. *High-frequency response and slew rate.* Again, this is important to match the rising phase of the pulses, regardless of their size, to produce reliable subtraction.

4. *Low glitch power*. Glitches are produced by the asynchronous switching of the bits that form the digital word and they occur with different magnitudes at different values of the input word. Glitches can produce enormous spikes, which may be present in the P pulse and not in the P/n pulse, or *vice versa*, making the subtraction totally unreliable and frequently producing spurious signals that may be confused with gating currents.

G. Computer Control

Typically, the interface between the analog setup and the digital domain is done under program control using a personal computer. The timing of the A/D and D/A converters must be under the control of a master clock that runs independently of the software. This is critical because timing loops in software, especially in multi-tasking operating systems, will make the timing completely unreliable. The clock running the A/D conversion can be different from the clock running the D/A converter, but both must be derived from a common source. It is easy to see that if the pulse generator jitters with respect to the acquisition, even by one clock pulse, it can have disastrous consequences in the subtraction, making it impossible to record gating currents.

H. System Performance

The simplest and most reliable way of testing the setup is to replace the membrane with a model circuit that simulates the membrane resistance, capacitance, series resistance, and electrode resistances. This model circuit is then subjected to the same pulse protocols that are used in the live preparation. Thus, using a P/n procedure it should produce a null trace. Lost sample points at the beginning and end of the pulse should be investigated carefully to trace their origin to saturation or nonlinearities of one or more of the components in the system. By measuring the output of each one of the stages, it is possible to trace where there might be a nonlinearity or saturation. One common pitfall is to look at the output of the filter and decide that is still within its dynamic range, but it is quite common for the input of the filter to be saturated by the incoming signal but look smooth on its output because of its filtering action. Another, more subtle, problem is pickup from the fast rising and fall of the command pulse into one of the inputs of the system. This can happen by capacitive coupling or by imperfect ground return (ground loops) in the system setup. An especially bad situation occurs when the analog return is drained through a ground that drains digital signals because the contribution of the digital noise will not be proportional to the size of the pulse and the P/n procedure will not cancel properly.

VIII. Recording of Gating Currents

In native cells, the recording of gating currents is complicated by the presence of other channels, making it difficult to separate the current of the channel of interest. In this regard, the squid axon is a simple preparation because the two channels present at high density have very different kinetics and temperature changes can be used to separate them (Bezanilla, 1985). In the squid axon the sodium pump is also expressed at very high densities, but its displacement current can be easily dissected from the Na^+ and K^+ gating currents by excluding ATP from the internal medium. The pump transient currents can be studied separately by recording the currents before and after the addition of ouabain, which subtracts the Na^+ and K^+ gating currents (Wagg *et al.*, 1996).

The study of gating currents is easier in expression systems where the channel being studied is made predominant in the membrane. Molecular biology techniques have made possible the ability to put the coding region of channels in expression vectors so that they can be used by cells as DNA or RNA, for synthesis of the protein and insertion into the membrane. By selecting cells that have a very low background of intrinsic channels, it is possible to study the expressed channel in practical isolation, especially when the expression is optimized. Many channels have been expressed in this way, but the gating currents of only a few channels have been studied. Brain sodium channel gating currents expressed in *Xenopus* oocytes were reported by Conti and Stuhmer (1989). *Shaker* K^+ channel gating currents, also expressed in *Xenopus* oocytes, were reported by Bezanilla *et al.* (1991) and calcium channel gating currents were reported by Neely *et al.* (1994).

The *Shaker* K^+ channel expresses at very high densities in *Xenopus* oocytes (Bezanilla *et al.*, 1991) and in HEK cells (Starace *et al.*, 1997). In addition, a mutation in the putative pore region (W434F) renders the channel nonconductive, although the gating currents are still preserved (Perozo *et al.*, 1993). Thus the *Shaker* W434F channel lends itself to detailed studies of the kinetics and steady-state properties of gating charge (Bezanilla *et al.*, 1994; Sigg *et al.*, 1994).

The procedures for recording and analyzing gating currents are illustrated using the *Shaker* H4-IR (inactivation-removed) W434F clone. The expression of this clone is high enough that gating currents can be observed without subtraction. Figure 4 shows a sample experiment using the cut-open oocyte voltage-clamp technique (Stefani and Bezanilla, 1998). The Q–V curve of this clone saturated at 0 mV; therefore, the subtraction can be applied at this potential, and this is the correct potential for compensating the capacitive transient with the analog TG. In Fig. 4A the current recorded for a pulse from 0 to 50 mV is shown, while in Fig. 4C the current shown at higher gain and time resolution has been recorded after adjusting the TG for compensation of most of the capacity current. Figure 4B shows the current recorded for a pulse to −40 mV from an HP of −90 mV. It is clear that the capacitive transient now has a new, slower component in both the ON and OFF positions of the pulse. This new component is the gating current that

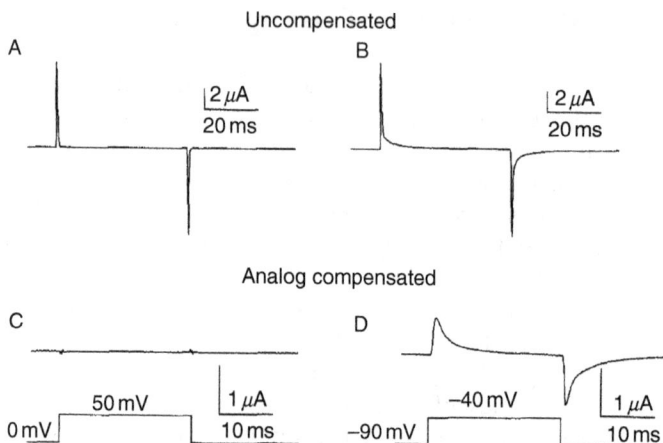

Fig. 4 Gating current recordings with analog compensation of linear components. Traces are records from *Shaker* H4-IR W434F clone. (A) and (B) are membrane currents for pulses from 0 to 50 mV and −90 to −40 mV, respectively. In (C) for the pulse from 0 to 50 mV, the leak and capacity transients were analogically compensated with the capacity and leak compensation control of the cut-open oocyte transient generator. In (D), the same compensation as in part (C), pulse from −90 to −40 mV. Sampling time was 100 μs per point, filter 2 kHz, temperature 22 °C.

contributes as much area to the integral as the capacitive transient alone. By using the same analog settings of the TG shown in Fig. 4C, the current recorded in Fig. 4D shows the gating current with the linear component removed. Thus, in this preparation it is possible to record gating currents without digital subtraction, such as the P/n procedure.

An example of a family of gating current records for the *Shaker* H4-IR W434F mutant recorded at 14 °C is shown in Fig. 5. In this case, the capacity transient was balanced holding the membrane at 0 mV and applying a small 20-mV test pulse. Subsequently, the HP was changed to −90 mV and a series of pulses was applied from a prepulse potential of −130 mV and returning to a postpulse potential of −130 mV. The pulses went from −125 to 30 mV in 5-mV increments. The gating currents during the test pulse (ON gating current) are small for small depolarizations and progressively increase in amplitude for larger depolarizations. Their decay rate is fast at small depolarizations, but at larger depolarizations it slows down and at even larger depolarizations speeds up again. (Notice that the current trace for the 30 mV crosses the current trace for a depolarization to −10 mV.) The ON gating currents exhibit a rising phase, most obvious at large depolarizations, and a decaying phase that has a single exponential component for small and very large depolarizations but has at least two components for intermediate depolarizations (Bezanilla *et al.*, 1994). The components are also obvious in the steady-state properties of the *Q–V* curve. The time integrals of the ON gating current traces shown in Fig. 5 are presented as the filled circles in the lower panel of Fig. 5. A fit to

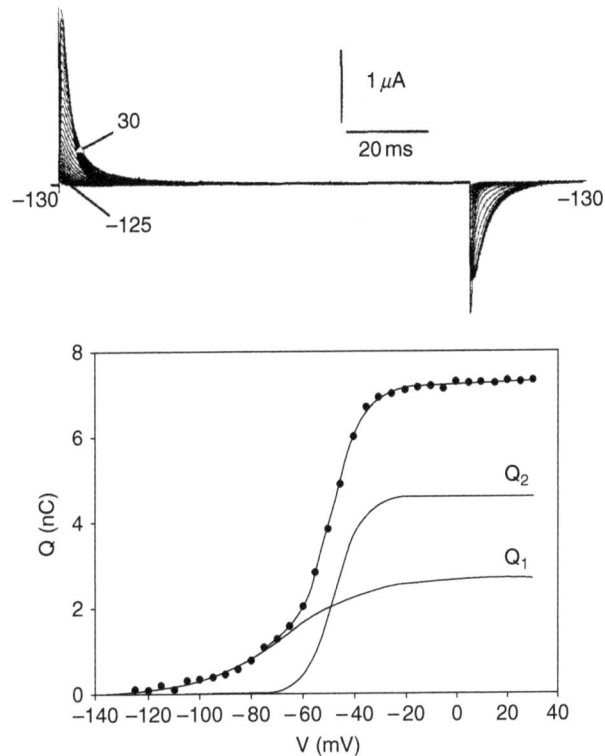

Fig. 5 *Top*: Family of gating current records from *Shaker* H4-IR W434F mutant. Records have been compensated with analog subtraction only (no digital subtraction), which was adjusted at 0 mV. The holding potential was −90 mV and the pulses from −125 to 30 mV were preceded and followed by 50-ms pulses to −130 mV. Sampling time was 50 μs per point and the filter was set to 5 kHz. Temperature was 14 °C. *Bottom*: *Q–V* curve, as measured from the records of the top panel. The continuous line is the fit to the sum of two Boltzmann distributions (Q_1 and Q_2) with the following parameters: $Q_{1max} = 2.69$ nC, $Q_{2max} = 4.59$ nC, $z_1 = 1.61$ e_0, $z_2 = 4.65$ e_0, half points of the distributions: $V_1 = −66.48$ mV and $V_2 = −47.68$ mV.

the sum of two two-state Boltzmann distributions is shown as the continuous curve through the points, and the two components are labeled as Q_1 and Q_2. These two steady-state components follow the same voltage dependence of the kinetic components fitted from gating current traces. The gating currents after the test pulses (OFF gating current) have very different kinetics depending on the depolarization during the preceding pulse. For small depolarizations, the charge returns very quickly, as a single exponential function. As the depolarization during the pulse becomes larger, but smaller than −40 mV, the single exponential decay is maintained, and the peak current increases (largest transient off current shown in Fig. 5). However, for larger depolarizations, OFF gating currents become slower, and at even larger depolarizations (beyond −20 mV) all superimpose (see Fig. 5).

The appearance of the slow component, which eliminates the fast component, has been traced to the opening of the channel, and it can be explained by a stabilization of the open state that slows the return of the charge (Bezanilla *et al.*, 1994; Zagotta *et al.*, 1994).

The return of the charge is even slower in the case of the inactivating (wild-type) *Shaker* channel (Bezanilla *et al.*, 1991) as expected from the extra stabilization induced by the inactivating ball (Hoshi *et al.*, 1990), while it is docked in the channel. In the sodium channel, inactivation also produces charge immobilization (Armstrong and Bezanilla, 1977; Conti and Stuhmer, 1989), which produces an apparent decrease of the OFF charge in relation to the ON charge. In fact, the charge has not disappeared, but it returns much more slowly and becomes undetectable due to baseline noise. By returning to more negative potentials, it is possible to observe the slow mobilization of the missing charge with a time course that resembles the time course of recovery from inactivation (Armstrong and Bezanilla, 1977; Bezanilla *et al.*, 1991). Note that charge immobilization can also occur as a consequence of a blocking effect of ions on the channel. For example, TEA ions in the internal solution can induce complete immobilization of the charge in the inactivation-removed *Shaker* gating current (Bezanilla *et al.*, 1991), but NMG ions do not induce any apparent immobilization. This result points to a careful consideration of the ion replacement to be used in recording gating currents. Thus, charge immobilization should always be looked on first as a possible ion effect that has to be discarded before deducing that there is an intrinsic immobilization of the charge by the channel molecule.

IX. Elementary Gating Event

As discussed earlier, the elementary event carries a small amount of charge that produces a current too small to detect with conventional techniques. However, the elementary event produced fluctuations in the recorded gating currents. Although these fluctuations are small, the low noise achieved with the patch-clamp technique makes it possible to detect it when recording more than about 10,000 channels in the patch. Conti and Stuhmer (1989) succeeded in recording the fluctuations of brain Na^+ gating currents and computed an elementary charge of 2.3 e_0. Crouzy and Sigworth (1993) have published a detailed theory of gating current fluctuations where they point out the effects of limited bandwidth and the interpretation of the calculated elementary charge in terms of a general kinetic model.

The conditions for recording fluctuations of gating currents are a low-noise recording setup and a number of channels that is not too large compared to the expected size of the fluctuations, so that the fluctuations can be resolved by the digitization of the A/D converter. The number of channels recorded with the cut-open oocyte technique is of the order of 10^9, too large in comparison to the expected fluctuations. If the expression is made less efficient, the number of channels may be decreased, but the limiting factor will be the background noise. This implies that the only technique available for this determination is the patch-clamp technique in the cell-attached or excised configuration. For the *Shaker* potassium channel, Sigg *et al.* (1994) found a value of 2.2 e_0 for the elementary charge. The experiments were done with macropatches using the nonconducting W434F version of the *Shaker* H4-IR clone. Figure 6 shows an experiment done with an attached giant patch with an oocyte that was perforated to equilibrate with a solution of isotonic NMG while the patch pipette also contained NMG. The top trace shows the mean, and the second trace shows the variance recorded with a bandwidth of 8 kHz. To eliminate drift effects, the variance was computed by pairs of records. Notice that the variance during the pulse is lower than during the negative potential before the pulse, an expected feature of the shallow Q–V relation at negative potentials of the *Shaker* K^+ channel (see Fig. 5). From the mean–variance plots (Fig. 6, bottom), the estimated elementary charges for the ON and OFF responses were 2.12 e_0, similar to the values reported before. In this case the

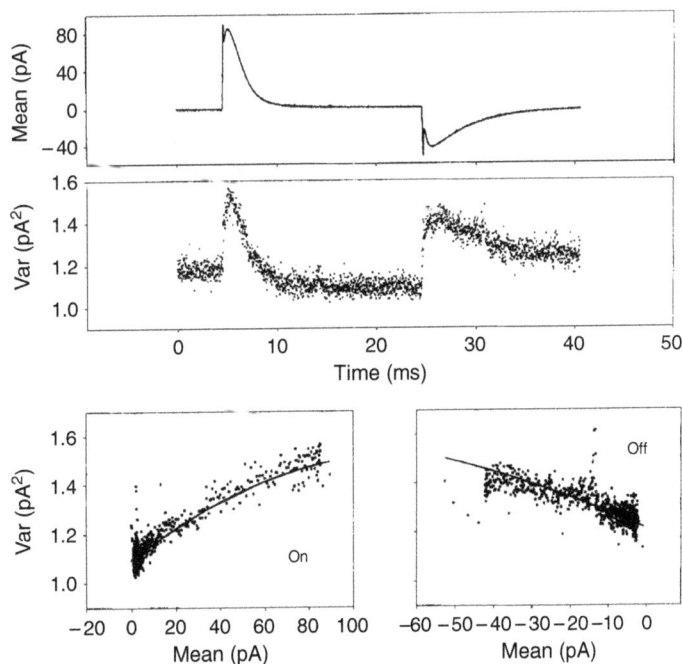

Fig. 6 *Top*: Mean and variance recorded with an attached giant patch from an oocyte expressing *Shaker* H4-IR W434F. Pulse from −120 to 0 mV. The oocyte was perforated in several places and equilibrated with a solution that contained. NMG as the only cation. The pipette contained NMG and calcium. A total of 8000 trials were recorded. Bandwidth was 8 kHz. *Bottom*: Mean–variance plots for the ON and OFF positions of the pulse. The computed elementary charge was the same, 2.12 e_0, for both ON and OFF.

possible contribution of fast-inactivating openings of the W434F mutant, as reported by Yang *et al.* (1997) is not present because there are no permeant ions on either side of the membrane.

X. Recording of Gating Currents in Frequency Domain

As stated earlier, the nonlinear charge movement of the voltage sensor behaves like a voltage-dependent capacitance, which can be studied in finer detail using frequency-domain techniques. The approach followed by Fernandez *et al.* (1982) was to superimpose a small amplitude pseudorandom voltage noise on the command voltage, and after recording the membrane current and the imposed voltage, the admittance was computed. The real part of the capacitance was calculated from the imaginary part of the admittance while the imaginary part of the capacitance was obtained from the real part of the admittance after the membrane leak was subtracted. To apply this technique properly, it is important to subtract the series resistance. The estimation of the series resistance may be obtained from the membrane impedance, which is computed as the reciprocal of the admittance. Details of the theory and methods are in Fernandez *et al.* (1982).

We have applied this technique to the cut-open oocyte voltage clamp in *Xenopus* oocytes expressing the W434F mutant of the *Shaker* H4-IR channel (Bezanilla and Stefani, 1993). In our case, the noise signal was synthesized as the sum of sine waves of voltages of different frequencies and random phase. The admittance was calculated by taking the Fourier transform of the voltage and the current and dividing the transformed current by the transformed voltage. Care should be exercised in using exactly the same pathway for current and voltage amplification, which can be achieved by two identical amplifiers and filters or by switching these devices as indicated in Fig. 3. The Fourier transform is computed on line and the real and imaginary part of the admittance are displayed for immediate evaluation. During the experiment, frequent checks of the properties of the preparation are made by switching to the time domain and recording regular gating currents. An example of the voltage dependence of the capacitance is shown in Fig. 7 for an uninjected oocyte (top) and an oocyte expressing H4-IR W434F. As the noise signal is added to a steady potential, it is found that the maximum capacitance is recorded at around −50 mV, which is an intermediate value of the midpoint of the Q–V curve for an HP of −90 and 0 mV. The shift of the midpoint of the Q–V with depolarized potential is correlated to slow inactivation (Olcese *et al.*, 1997). It is interesting to notice that the real and imaginary parts of the capacitance do not follow a single Debye relaxation. The fitting of the data requires at least two relaxations, as expected from a process that has at least two distinguishable time constants in the time domain. The full characterization of the frequency dependence of the capacitance at different voltages gives a more detailed description of the voltage dependence of the voltage sensor.

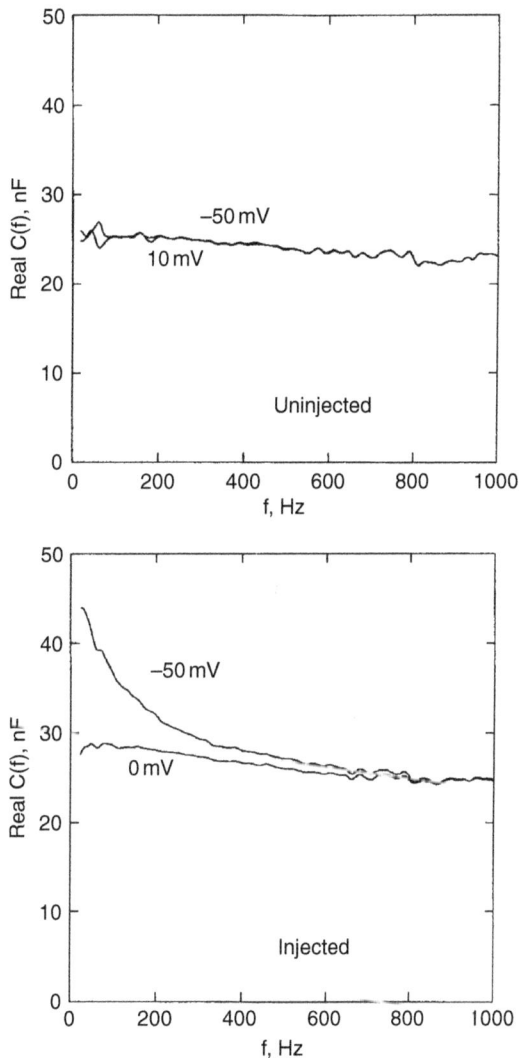

Fig. 7 Frequency dependence of the real part of the capacitance of an uninjected oocyte (top) and injected with H4-IR W434F mutant cRNA (bottom). Voltage dependence of the frequency-dependent capacitance appears as a consequence of channel expression and correlates with the total area of the Q–V curve recorded from the same oocyte.

The study of the charge movement in voltage-dependent channels is normally done by eliminating all of the linear components of the capacitance. The presence of proteins in the membrane should add quasivoltage-independent but frequency-dependent capacitance due to the lossy dielectric properties of buried charges and dipoles. Frequency-domain techniques, which have far better resolution than time-domain techniques, may be used to study this voltage-independent but frequency-dependent capacitance added by the expression of the channels to the oocyte membrane.

Acknowledgments

We thank Abert Cha for reading and commenting on the manuscript. This work is supported by NIH grants GM30376 to F. B. and GM52203 and AR38970 to E. S.

References

Aggarwal, S. K., and MacKinnon, R. (1996). *Neuron* **16**, 1169.
Armstrong, C. M., and Bezanilla, F. (1973). *Nature* **242**, 459.
Armstrong, C. M., and Bezanilla, F. (1977). *J. Gen. Physiol.* **70**, 567.
Bezanilla, F. (1985). *J. Membr. Biol.* **88**, 97.
Bezanilla, F., and Armstrong, C. M. (1977). *J. Gen. Physiol.* **70**, 549.
Bezanilla, F., and Stefani, E. (1993). *Biophys. J.* **64**, A114.
Bezanilla, F., White, M. M., and Taylor, R. E. (1982). *Nature* **296**, 657.
Bezanilla, F., Perozo, E., Papazian, D. M., and Stefani, E. (1991). *Science* **254**, 679.
Bezanilla, F., Perozo, E., and Stefani, E. (1994). *Biophys. J.* **66**, 1011.
Conti, F., and Stuhmer, W. (1989). *Eur. Biophys. J.* **17**, 53.
Crouzy, S. C., and Sigworth, F. J. (1993). *Biophys. J.* **64**, 68.
Fernandez, J. M., Bezanilla, F., and Taylor, R. E. (1982). *J. Gen. Physiol.* **79**, 21.
Hilgemann, D. (1998). *Methods Enzymol.* **293**, 267 (this volume).
Hirshberg, B., Rovner, A., Lieberman, M., and Patlak, J. (1996). *J. Gen. Physiol.* **106**, 1053.
Hodgkin, A. L., and Huxley, A. F. (1952). *J. Physiol. (Lond.)* **117**, 500.
Hoshi, T., Zagotta, W. N., and Aldrich, R. W. (1990). *Science* **250**, 533.
Levis, R. (1998). *Methods Enzymol.* **293**, 218 (this volume).
Neely, A., Wei, X., Olcese, R., Birnbaumer, L., and Stefani, E. (1994). *Science* **262**, 575.
Olcese, R., Latorre, R., Toro, L., Bezanilla, F., and Stefani, E. (1997). *J. Gen. Physiol.* **110**, 579.
Perozo, E., MacKinnon, R., Bezanilla, F., and Stefani, E. (1993). *Neuron* **11**, 353.
Rudy, B. (1976). *Proc. R. Soc. Lond. B.* **193**, 469.
Schoppa, N. E., McCormack, K., Tanouye, M. A., and Sigworth, F. J. (1992). *Science* **255**, 1712.
Seoh, S. A., Sigg, D., Papazian, D. M., and Bezanilla, F. (1996). *Neuron* **16**, 1159.
Shenkel, S., and Bezanilla, F. (1991). *J. Gen. Physiol.* **98**, 465.
Sigg, D., and Bezanilla, F. (1997). *J. Gen. Physiol.* **109**, 27.
Sigg, D., Stefani, E., and Bezanilla, F. (1994). *Science* **264**, 578.
Starkus, J. G., Fellmeth, B. D., and Rayner, M. D. (1981). *Biophys. J.* **35**, 521.
Stefani, E., and Bezanilla, F. (1996). *Biophys. J.* **70**, A143.
Stefani, E., and Bezanilla, F. (1997). *Biophys. J.* **72**, A131.
Stefani, E., and Bezanilla, F. (1998). *Methods Enzymol.* 293, Chapter 17 (this volume).
Wagg, J., Holmgren, M., Gadsby, D. C., Bezanilla, F., Rakowski, R. F., and De Weer, P. (1996). *Biophys. J.* **70**, A19.
Yang, Y., Yan, Y., and Sigworth, F. J. (1997). *J. Gen. Physiol.* **109**, 779.
Zagotta, W. N., Hoshi, T., and Aldrich, R. W. (1994). *J. Gen. Physiol.* **103**, 321.

CHAPTER 6

Determining Ion Channel Permeation Properties

Ted Begenisich

Department of Physiology
University of Rochester Medical Center
Rochester, New York, USA

I. Introduction

Considerable effort is being put into the determination of the mechanism of ion channel permeation. This effort includes many components, but a first step must be to determine which ions can permeate the pore. Additional information can be obtained by determining how many ions may simultaneously occupy the pore and by investigating the ability of small ions to block the pore. These issues are not necessarily independent of one another and, as a result, the proper interpretation of experimental results needs to consider all as part of a whole not as isolated, unrelated pieces. This chapter describes some of the methods used to measure ion channel selectivity, permeation, and blocking and some of the complexities of interpreting the resulting data.

DOI: 10.1016/B978-0-12-382204-8.00006-0

II. Single-Ion Nernst Potential

If a pore is permeable only to a single ion (X) of valence z, the net current through the pore is zero at a potential (V_X) given by the Nernst (or equilibrium) potential for that ion [Eq. (1)]:

$$V_X = \frac{RT}{zF} \ln\left(\frac{a_o}{a_i}\right) \tag{1}$$

where a_o and a_i represent the external and internal activities of ion X, respectively and R, T, and F have their usual thermodynamic meanings; RT/F at 20 °C is approximately 25 mV. This voltage is an equilibrium property and does not depend on the theoretical framework from which it is derived. It is a function only of the ion concentrations and is independent of any impermeant ions and independent of the presence of any fixed charges on the membrane.

The activity terms are related to the ion concentrations by external and internal activity coefficients, γ_o and γ_i. Activity coefficients are used as a simple way to account for the (usually) small interactions among ions in aqueous solutions. Thus, in terms of concentrations, Eq. (1) becomes:

$$V_X = \frac{RT}{zF} \ln\left(\frac{\gamma_o[X]_o}{\gamma_i[X]_i}\right) \tag{2}$$

The activity coefficients depend on the ionic strength of the solution. In normal physiologic solutions, a monovalent ion has an activity coefficient of approximately 0.85, so the ion activity is not much different from the concentration. At higher salt concentrations, the coefficients can become rather small and divalent ions, owing to their higher valence, have rather smaller coefficients than monovalent ions. However, if the internal and external solutions are of similar ionic strength, the activity coefficients in Eq. (2) could be omitted:

$$V_X = \frac{RT}{zF} \ln\left(\frac{[X]_o}{[X]_i}\right) \tag{3}$$

not because they are near unity but rather because they are approximately equal. If the solutions are not sufficiently similar, then these coefficients should not be ignored. With this caution, the remaining discussion will use concentrations rather than activities.

Equation (3) suggests a test that a pore is permeable to a specific ion: measure the channel zero-current (or "reversal") potential with known internal and external concentrations of the test cation and see if this potential is equal to the Nernst potential for the ion. An accurate measurement of the reversal potential requires proper consideration of liquid junction potentials (Neher, 1992).

The use of Eq. (3) as a test for ion selectivity implies knowledge of the intracellular concentration of the ion of interest. Unfortunately, in many experimental situations the internal concentration is not accurately known. This is sometimes true even for the usual whole-cell variant of the patch-clamp technique since it relies on diffusion of the pipette contents into the cell interior (Tang *et al.*, 1992). Thus, rather than compare the expected Nernst potential to a single measurement, the reversal potential can be measured at several external concentrations of the test ion and (often) plotted as a function of the logarithm of the external ion concentration. Equation (3) predicts a linear relationship in such a plot with a $58/z$ mV/decade slope (at 20 °C), and such a finding demonstrates that the pore is highly selective for the tested ion compared to other ions present.

Occasionally, a perfectly linear relationship is found when using physiologic solutions (e.g., Lucero and Pappone, 1989). More often the results are in agreement with Eq. (3) only at high concentrations of the test ion, and at low concentrations the slope is often much less than predicted (e.g., Sah *et al.*, 1988). Such a result suggests that the pore is not perfectly selective for the tested ion.

III. Ion Channel Selectivity

Few, if any, channels are permeant to only a single type of ion, and the expected dependency of pore zero-current (or reversal) potential (V_{rev}) on ion concentration is, in general, both complex and model dependent. However, if two ions (X and Y) of the same valence (z) are considered, a relatively general and simple equation can be derived (see Hille (1992) for some of the details of these computations):

$$V_{rev} = \frac{RT}{zF} \ln \frac{P_X[X]_o + P_Y[Y]_o}{P_X[X]_i + P_Y[Y]_i} \qquad (4)$$

where P_X and P_Y are the permeabilities of ions X and Y, respectively.

The reversal potential of Eq. (4) actually depends only on the relative permeabilities of the two ions:

$$V_{rev} = \frac{RT}{zF} \ln \frac{[X]_o + (P_X/P_Y)[Y]_o}{[X]_i + (P_X/P_Y)[Y]_i} \qquad (5)$$

At high external concentrations of ion X, Eq. (5) predicts a linear relationship between the measured reversal potential and the logarithm of $[X]_o$. Deviations from linearity are expected at lower concentrations owing to the second term in the numerator of Eq. (5). The relative X to Y permeability can be obtained by measuring the channel reversal potential at various ion concentrations and fitting Eq. (5) to the data.

A. Biionic Conditions

While Eq. (5) suggests experiments that can determine the relative ion permeability of a pore, a simpler method can be used if the solution on the cytoplasmic face of the membrane can be accurately controlled. This experimental condition allows the use of a simplified form of Eq. (4) if ions of type X are the only permeant ones in the external solution and Y ions are the only permeant ions in the solution on the cytoplasmic side of the pore:

$$V_{\text{rev}} = \frac{RT}{zF} \ln \left(\frac{P_X [X]_o}{P_Y [Y]_i} \right) \tag{6}$$

Equation (6) can be rewritten in a form that allows computation of the relative ion permeabilities:

$$\frac{P_X}{P_Y} = \left(\frac{[Y]_i}{[X]_o} \right) \exp \left(\frac{z V_{\text{rev}} F}{RT} \right) \tag{7}$$

Thus, a measurement of the current reversal potential with knowledge of the ion concentrations in these biionic conditions allows the determination of the relative ion permeabilities. Chandler and Meves (1965) used this technique to show that the squid axon Na^+ channel is permeable to several other cations include Li^+, K^+, Rb^+, and Cs^+.

B. Internal Ions Not Known But Constant

Hille (1971) showed how relative ion permeabilities can be determined even if the internal medium cannot be controlled but is constant. In this technique, the channel reversal potential is measured twice: once with only ion X in the external solution ($V_{\text{rev,X}}$) and again with only ion Y in the external solution ($V_{\text{rev,Y}}$). Under these conditions, the relative permeability can be computed [using a form of Eq. (4)] from the difference between these two measurements:

$$\frac{P_X}{P_Y} = \frac{[X]_o}{[Y]_o} \exp \left[\frac{zF(V_{\text{rev,X}} - V_{\text{rev,Y}})}{RT} \right] \tag{8}$$

Using this technique, Hille (1972) showed that the selectivity of the Na^+ channel in myelinated nerve is similar to that of the squid axon Na^+ channel and extended the list of permeant ions to include several organic cations. Tsushima et al. (1997) have used this same technique to investigate the role of certain amino acids in determining the selectivity of the rat skeletal muscle isoform of the voltage-gated Na^+ channel.

It is important to note that Eqs. (4)–(8) were derived for ions of the same valence and for a pore that obeys independence (Hille, 1975). Independence means that the movement of one ion is independent of the presence of any others. Ion permeation through most channels is rather complex and seems to have properties consistent with the interaction of the ions with each other while they are in the pore. Consequently, interpretations of data through the use of these equations need to done carefully although if the recording conditions are specified, the computed permeability ratios can be useful empirical parameters.

IV. Classification of Ion Pores

A. One-Ion Pores

In one-ion pores, the presence of a single ion in the permeation pathway prevents entry (perhaps through electrostatic repulsion) of any other ion. Hille (1975) has shown that Eq. (7) applies to one-ion pores even if independence is not obeyed. The permeabilities in Eq. (7) may not be constant for such a pore but, rather, may be functions of membrane potential but not functions of ion concentration. However, in many situations the voltage dependence may be small or nonexistent. Consequently, Eq. (7) is very useful for determining relative selectivities even for one-ion pores.

The current through one-ion pores can be rather sensitive to the type and concentrations of the ions present. Consequently, ion selectivity is best judged not by the relative current carried by different types of ions but rather from measurements of the reversal potential. Ion currents are also affected by many pore-blocking ions (see later discussion) but, providing they are impermeant, they will not contribute to the measured reversal potential, and Eq. (7) can still be used.

B. Multi-Ion Pores

Pores can also be simultaneously occupied by more than a single ion. Not surprisingly, the permeation properties of such multi-ion pores are more complex than those of one-ion pores. Nevertheless, Hille and Schwarz (1978) have shown that Eq. (7) may apply even to these more complicated situations but that the permeability ratio may be a function of ion concentration. Consequently, a finding of concentration-dependent permeabilities in Eq. (7) indicates that the pore under study has multi-ion characteristics. It may be difficult to separate the effects of voltage and concentration because changing ion concentrations will necessarily shift the reversal potential to a new value. Sometimes, however, it is possible to separate these two effects. For example, squid axon Na^+ channel selectivities are concentration but not (significantly) voltage dependent (Begenisich and Cahalan, 1980; Cahalan and Begenisich, 1976).

V. Pore-Blocking Studies

Some aspects of the structure of ion channel pores can be obtained from studies of small ions that block the movement of permeant ions. As described by Woodhull (1974), the dissociation constant for a blocking ion, of valence z, that binds to a site within the membrane electric field is given as

$$K_d(V_m) = K_d(0) \exp\left(\frac{-\delta_z V_m F}{RT}\right) \tag{9}$$

where $K_d(0)$ is the zero-voltage dissociation constant and δ is the location of the binding site as a fraction (measured from the inside in this example) of the membrane voltage. Then, the fraction of current (channels) blocked, $f(V_m)$, will not only be a function of the concentration of blocking ions, [B], but also a function of voltage:

$$f(V_m) = \frac{[B]}{[B] + K_d(V_m)} \tag{10}$$

A determination of the fraction of current blocked as a function of membrane potential will, through the application of Eqs. (9) and (10), yield a value for the electrical distance to the blocking site. In one of the best studies of this type, Miller (1982) used monovalent and bisquaternary ammonium compounds to determine the spatial dependence of the electric potential in part of the pore of a K^+ channel from sarcoplasmic reticulum.

Occupancy of a pore by multiple ions significantly complicates the use of Eq. (10) for interpreting the results of blocking studies. A good example of the difficulty of obtaining the fractional field location of a blocking site can be found in the work of Adelman and French (1978). A literal application of Eq. (10) to Cs^+ block of K^+ channels reveals that the blocking site reached by external Cs^+ is more than 100% across the membrane. This nonsensical result can be a consequence of a multi-ion pore (Hille and Schwarz, 1979) and, indeed, the data of Adelman and French can be reproduced by a specific multi-ion pore model (Begenisich and Smith, 1984). For block to occur in a multi-ion pore, the permeant ion must vacate the blocking ion binding site. Consequently, the voltage dependence of block will reflect not only the voltage-dependent binding of the blocking ion to this site but also the voltage-dependent evacuation of the pore by permeant ions (Hille and Schwarz, 1978).

VI. Determining Pore Occupancy

Mathematical descriptions of selectivity and block of multi-ion pores require the specification of the number and location within the membrane electrical field of both permeant and blocking ions. Doing this accurately for any ion channel certainly approaches the limits of current methodological capabilities. However, certain types of unidirectional flux measurements can provide lower estimates for the maximum number of ions that may simultaneously occupy the pore. Hodgkin and Keynes (1955) measured the unidirectional influx and efflux of K^+ through the voltage-gated, delayed rectifier K^+ channels in cuttlefish axons and compared their results with the prediction made by Ussing (1949) for independent movement of a single type of ion:

$$\frac{m_e}{m_i} = \frac{[K]_i}{[K]_o} \exp\left(\frac{V_m F}{RT}\right) \qquad (11)$$

where m_i and m_e are the unidirectional K^+ influx and efflux, respectively; or, in terms of the K^+ equilibrium potential, V_K:

$$\frac{m_e}{m_i} = \exp\left[\frac{(V_m - V_K)F}{RT}\right] \qquad (12)$$

Hodgkin and Keynes (1955) found that this equation provided a poor description of their data but a good fit was obtained if the right-hand side of Eq. (11) was raised to a power, n':

$$\frac{m_e}{m_i} = \exp\left[\frac{n'(V_m - V_K)F}{RT}\right] \qquad (13)$$

which, to best describe their data, required a value near 2.5. To understand the meaning of the n' parameter, Hodgkin and Keynes (1955) investigated several theoretical descriptions of ion permeation and showed that an expression like Eq. (13) is expected if K^+ channels simultaneously contain several ions. Subsequently, a variety of theoretical approaches (Bek and Jakobsson, 1994; Hille and Schwarz, 1978; Kohler and Heckmann, 1979; Levitt, 1986; Schumaker and MacKinnon, 1990) have all led to the same conclusion: the value of n' in Eq. (13) can approach, but not exceed, the number of ions simultaneously occupying the pore. The 2.5 value of Hodgkin and Keynes (1955) means there can be at least three K^+ ions in the cuttlefish channel.

Measurements of unidirectional ion fluxes (Stampe and Begenisich, 1998) have been done with cloned K^+ channels and the results suggest that *Shaker* K^+ channels, one member of the voltage-gated family, can be simultaneously occupied by at least four K^+ ions. Stampe and Begenisich (1996) and Neyton and Miller (1988) used quite different techniques including an investigation of the K^+ sensitivity of Ba^{2+} block and demonstrated the existence of four ion binding sites in the pore of Ca^{2+}-activated K^+ channels.

VII. Conclusion

Many ion channels have complex permeation properties best understood in terms of simultaneous pore occupancy by several ions. The interaction of ions within the channel pore complicates the interpretation of certain types of experiments and it is useful to consider a variety of approaches with special attention to the multi-ion nature of channel pores. The pore in different types of K^+ channels appears to allow occupancy by at least four ions and it remains a challenge to find where, within the pore, the ions are located and to determine how so much electrostatic charge can be confined to such a relatively small space.

References

Adelman, W. J., Jr., and French, R. J. (1978). *J. Physiol. (Lond.)* **276**, 13.

Begenisich, T., and Cahalan, M. D. (1980). *J. Physiol. (Lond.)* **307**, 217.

Begenisich, T., and Smith, C. (1984). "Current Topics in Membranes and Transport," Vol. **22**, Academic Press, New York.

Bek, S., and Jakobsson, E. (1994). *Biophys. J.* **66**, 1028.

Cahalan, M. D., and Begenisich, T. (1976). *J. Gen. Physiol.* **68**, 111.

Chandler, W. K., and Meves, H. (1965). *J. Physiol. (Lond.)* **180**, 788.

Hille, B. (1971). *J. Gen. Physiol.* **58**, 599.

Hille, B. (1972). *J. Gen. Physiol.* **59**, 647.

Hille, B. (1975). *In* "Membranes—A Series of Advances. Lipid Bilayers and Biological Membranes: Dynamic Properties," (G. Eisenman, ed.), Vol. **4**, p. 255. Dekker, New York.

Hille, B. (1992). "Ionic Channels of Excitable Membranes," 2nd edn., p. 345. Sinauer Associates, Sunderland, MA.

Hille, B., and Schwarz, W. (1978). *J. Gen. Physiol.* **72**, 409.

Hille, B., and Schwarz, W. (1979). *Brain Res. Bull.* **4**, 159.

Hodgkin, A. L., and Keynes, R. D. (1955). *J. Physiol. (Lond.)* **128**, 61.

Kohler, H.-H., and Heckmann, K. (1979). *J. Theor. Biol.* **79**, 381.

Levitt, D. G. (1986). *Ann. Rev. Biophys. Biophys. Chem.* **15**, 29.

Lucero, M. T., and Pappone, P. A. (1989). *J. Gen. Physiol.* **94**, 451.

Miller, C. (1982). *J. Gen. Physiol.* **79**, 869.

Neher, E. (1992). *Methods Enzymol.* **207**, 123.

Neyton, J., and Miller, C. (1988). *J. Gen. Physiol.* **92**, 569.

Sah, P., Gibb, A. J., and Gage, P. W. (1988). *J. Gen. Physiol.* **92**, 264.

Schumaker, M. F., and MacKinnon, R. (1990). *Biophys. J.* **58**, 975.

Stampe, P., and Begenisich, T. (1996). *J. Gen. Physiol.* **107**, 449.

Stampe, P., and Begenisich, T. (1998). *Methods Enzymol.* **293**(30).

Tang, J. M., Wang, J., and Eisenberg, R. S. (1992). *Methods Enzymol.* **207**, 176.

Tsushima, R. G., Li, R. A., and Backx, P. H. (1997). *J. Gen. Physiol.* **109**, 463.

Ussing, H. H. (1949). *Acta Physiol. Scand.* **19**, 43.

Woodhull, A. (1974). *J. Gen. Physiol.* **61**, 687.

PART IV

Expression Systems

CHAPTER 7

Expression of Ligand-Gated Ion Channels Using Semliki Forest Virus and Baculovirus

Kathryn Radford and Gary Buell

Geneva Biomedical Research Institute
Geneva, Switzerland

I. Introduction

Recombinant protein synthesis from baculovirus (BV) and Semliki Forest virus (SFV) vectors has been exploited to study the biochemistry, pharmacology, and structure of ligand-gated ion channels. These viral systems allow high infection efficiencies and result in expression levels that exceed those normally obtained by DNA transfection. Here, we briefly review the current status of these systems for both large- and small-scale expression of ligand-gated ion channels and provide relevant protocols. We have used these techniques to study adenosine triphosphate (ATP)-gated ion channels (P2X receptors), but there are many reports of similar methods with other receptors and channels. Table I lists some of the applications of BV or SFV expression that have been published for ligand-gated ion channels.

A. Baculovirus and Semliki Forest Virus as Expression Vectors

The success of baculovirus as an expression vector is based on the replacement of nonessential viral genes that are controlled by strong promoters, by exogenous genes (Smith *et al.*, 1983). Recombinant virus is used to infect insect cells, the natural host, and the foreign gene products are expressed at times and in amounts that depend on the promoter and infection parameters. Ligand-gated ion channels, expressed with this system, have shown ligand-binding activities ranging from 18 to 71 pmol/mg (Lundstrom *et al.*, 1997; Morr *et al.*, 1995) and have been purified to 0.3–3 μg/mg of total cellular protein (Cascio *et al.*, 1993; Green *et al.*, 1995; Sydow *et al.*, 1996).

SFV offers an alternative system for the high-level expression of ligand-gated ion channels in a broader range of cells that includes mammalian (Liljeström and Garoff, 1991). Although fewer examples of ligand-gated ion channels have been expressed with SFV, similar protein levels have been observed (Lundstrom *et al.*, 1997; Michel *et al.*, 1996a,b). The SFV genome is a single-stranded RNA molecule that functions directly as an mRNA. SFV expression vectors consist of modified SFV cDNAs which lack regions that encode viral structural proteins and they serve as templates for the *in vitro* synthesis of recombinant RNA. Unlike BV vectors, the missing viral genes are essential for assembly, and helper RNA, which encodes the structural proteins, must be cotransfected with the recombinant SFV vector RNA. Vector RNA alone is contained in infectious virions because helper RNA lacks the sequence needed for packaging into nucleocapsids. Thus, in contrast with BV, recombinant SFV cannot replicate during infection.

Due to the broad host range of SFV, an additional mutation has been introduced so that only conditionally infectious virus particles are produced. This mutation in the helper RNA prevents natural proteolytic processing of the viral spike protein (Lobigs and Garoff, 1990). The resulting noninfectious virus stocks must be activated by cleavage with chymotrypsin for host cell infection. Following infection with activated recombinant virus, the amounts of ligand-gated ion channels are comparable to those obtained with BV vectors but differ in their expression

Table I
Ligand-Gated Ion Channels Expressed by Semliki Forest Virus and Baculovirus Systems[a]

Channel family	Receptor subtypes expressed	Virus	Application	References
Nicotinic	Rat nAChrα-4, β-2	BV	Binding studies	Wang and Abood (1996)
	Bovine nAChrα	BV	Purification	Iizuka and Fukuda (1993)
	Chick nAChrα	BV	Fluorescence microscopy, binding studies	Atkinson et al. (1996)
5-HT₃	Mouse 5-HT₃	BV	Large-scale expression, purification, electron microscopy	Green et al. (1995)
	Mouse 5-HT₃	SFV	Large-scale expression	Lundstrom et al. (1997)
	Mouse 5-HT₃	SFV	Electrophysiology	Werner et al. (1994)
Glycine	Glycine α-1	BV	Purification	Morr et al. (1995)
	Glycine α-1	BV	Purification	Cascio et al. (1993)
GABA	GABA(A)α-1, β-1	BV	Mutagenisis, electrophysiology	Binir et al. (1997)
	GABA(A)	BV	Subunit composition, infection parameters	Hartnett et al. (1996)
	GABA(A)α-1,2,3,5; β-1,2,3; γ-2	BV	Coexpression, binding studies	Witt et al. (1996)
	GABA(A)α-1, β-1	BV	Infection parameters, electrophysiology	Binir et al. (1995)
	GABA(A)β-1	BV	Stable cell line	Joyce et al. (1993)
	GABA(A)α-1, β-1	BV	Fluorescence studies, electron microscopy	Binir et al. (1992)
AMPA	AMPAα-1, α-2	BV	Expression	Kawamoto et al. (1990)
	AMPAσ-2	BV	Glycosylation, immunofluorescence, binding	Kawamoto et al. (1995a)
	AMPA α-1	BV	Glycosylation, binding	Kawamoto et al. (1994)
	GluRD	BV	Purification	Kuusinen et al. (1995)
	GluRB, GluRD	BV	Binding, electrophysiology	Keinen et al. (1994)
Kainate	GluR6	BV	Palmitoylation, electrophysiology	Pickering et al. (1995)
	GluR6	BV	Cell line comparison	Ross et al. (1994)
	GluR6	BV	Glycosylation, transmembrane topology	Taverna et al. (1994)
	GluR6	BV	Binding, glycosylation	Taverna and Hampson (1994)
NMDA	NMDAR1	BV	Glycosylation, assembly, purification, binding site	Sydow et al. (1996)
	NMDA zeta 1	BV	Binding site	Uchino et al. (1997)
	NMDA zeta 1	BV	Glycosylation, assembly, binding immunofluorescence	Kawamoto et al. (1995b)
	NR1, NR2A, NR2C	BV	Glycosylation	Kopke et al. (1993)
P2X	P2X₂, P2X₃	BV	Binding, electrophysiology, assembly	Radford et al. (1997)
	P2X₁	SFV	Binding	Michel et al. (1996a)
	P2X₂, P2X₁	SFV	Binding	Michel et al. (1996b)
	P2X₂	SFV	Binding, electrophysiology	Lundstrom et al. (1997)

[a]Members of all ligand-gated ion channel families are included. The references were chosen as examples of different applications and are not exhaustive, in particular for GABA receptors that have been extensively studied with the baculovirus system.

kinetics. A comparison of the steps involved in recombinant protein expression by BV and SFV systems is outlined in Fig. 1.

B. Choice of Promoters

Currently, one promoter is used with the SFV expression system. Foreign DNA is inserted under the control of the viral promoter for the 26S subgenomic mRNA that in wild-type virus drives transcription of structural proteins. In contrast, BV vectors that contain a variety of promoters exist and the choice of the promoter depends on the objective. Ion channels have been mostly expressed with vectors that exploit the very late polyhedrin promoter in order to obtain maximal yields for binding or structural studies. Also active during the very late phase of viral infection, the p10 protein promoter is not competitive with the polyhedrin promoter. They can be used together to construct multiple promoter vectors, capable of simultaneously expressing two proteins (Weyer and Possee, 1991).

As opposed to the very late polyhedrin promoter, use of a late promoter such as the basic protein promoter may be of interest for ion channel characterization (Paine et al., 1996). The basic promoter allows maximum protein expression prior to the onset of the lytic period (30 h postinfection) while the host cell membrane integrity is maintained. This facilitates electrophysiologic recordings due to the absence of the large holding currents that are observed after 50 h postinfection when very late promoters generate maximum expression. The amount of foreign protein, made with the late promoter, is less than with very late promoters but is still higher than the quantities from stable cell lines. Generally, earlier promoters of BV may allow better assembly, localization, and function of proteins since the host cell processing pathways remain uncompromised. Both late and very late promoter vectors are commercially available (Pharmingen, San Diego, CA).

Stable insect cell lines also exploit weaker early promoters in the absence of viral replication and efficient membrane targeting of ligand-gated ion channel receptors has been reported (Atkinson et al., 1996; Joyce et al., 1993; Shotkoski et al., 1996). The chick nicotinic acetylcholine receptor cDNA was integrated randomly into an insect cell genome under the control of the immediate early gene promoter IE1, and efficient membrane targeting was demonstrated by fluorescence microscopy and ligand binding (Atkinson et al., 1996). Stable cell lines remain an attractive alternative if maximal expression is not essential.

C. Incorporation of Protein Tags

Characterization and purification of proteins including ion channels is facilitated by the use of epitope tags, fused to the recombinant polypeptide. Commercially available BV vectors encoding fusion tags such as glutathione transferase (GST) or multiple histidine residues (His_6) generate proteins that can be subsequently purified on glutathione or Ni-NTA agarose beads, respectively (Chen et al., 1993; Davies and Jones, 1991). Other epitope tag sequences may be

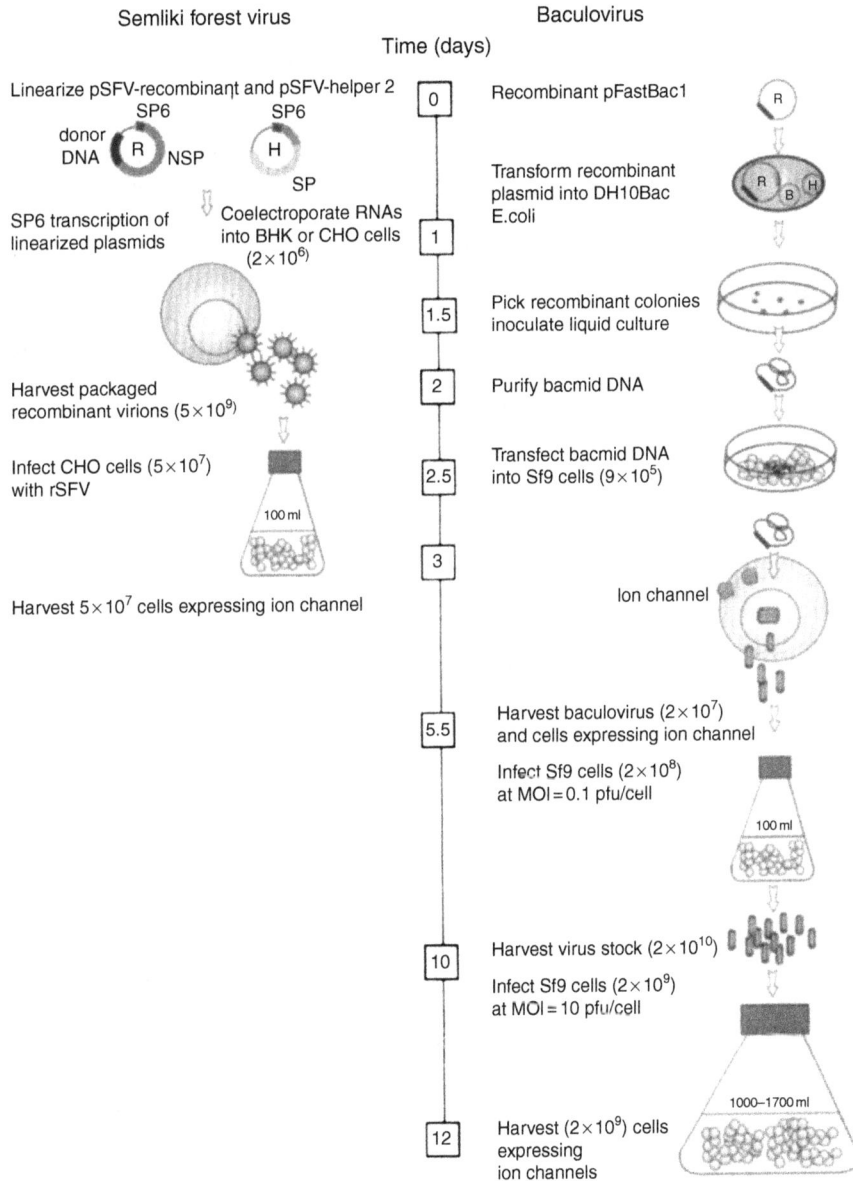

Fig. 1 Schematic representation and timescale of virus construction and expression of ion channels by Semliki Forest virus and baculovirus systems. (Left) Recombinant SFV plasmid (R) containing foreign DNA and helper SFV plasmid (H) are linearized with *Spe*I. RNAs transcribed by SP6 RNA polymerase are cotransfected into BHK or CHO cells and packaged virions are used to infect CHO cells in suspension to produce recombinant ion channels. (Right) Recombinant BV plasmid (R) is transformed into DH10Bac cells that contain a helper plasmid (H) and bacmid (B). Transposition functions provided by the helper plasmid allow transfer between the recombinant BV plasmid and the target bacmid, which is capable of replicating in *Escherichia coli* and transfecting insect cells to produce infectious baculovirus. Baculovirus stock is amplified by one to two successive infections of insect cell cultures at low multiplicity. For ion channel expression, insect cells are infected with baculovirus at 10 pfu/cell and cells are harvested after 50 h.

incorporated directly into the cDNA. Human m_2 muscarinic acetylcholine receptors, expressed in insect cells and tagged at the carboxy terminus with histidine residues have been purified with Ni^{2+} or Co^{2-} immobilized gel (Hayashi and Haga, 1996). Their function was shown to be unaffected by the histidine tag. 5-Hydroxy-tryptamine (5-HT$_3$) and P2X ion channel receptors have been successfully expressed in both SFV and BV with His$_6$ or epitopes specific for monoclonal antibodies (Green et al., 1995; Lundstrom et al., 1997; Radford et al., 1997c). Whole-cell patch-clamp recordings for epitope-tagged P2X receptors were indistinguishable from those for the wild-type receptors. However, His$_6$-tagged 5-HT$_3$ receptors desensitized 50 times more slowly than the corresponding wild-type receptor (Lundstrom et al., 1997). Alternative tag positions may have to be tested to maintain the native channel properties. Epitope tags also serve as indicators of full-length protein expression and stability. Proteolytic cleavage of the C-terminal His$_6$-tagged cholecystokinin receptor expressed in BV-infected insect cells has allowed determination of the optimal time of harvest of the full-length receptor (Gimpl et al., 1996).

D. Optimization of Expression

Despite careful consideration of the vector, the efficiency of heterologous gene expression is largely dictated by the intrinsic nature of the encoded protein. Greater than 1000-fold differences in yields have been reported for BV expressed proteins but there is little evidence that the expression level for a given gene can be influenced more than fivefold by optimizing its sequence (O'Reilly et al., 1994). In our laboratory comparison of P2X receptor, subtypes expressed by SFV or BV have largely shown the same relative ligand binding, regardless of the expression system employed (Fig. 2). Optimization of cell culture and infection parameters can, however, improve the yields of ligand-gated ion channels. The key variables are the multiplicity of infection (MOI) and the cell density at the time of infection (TOI); they control the balance of nutrient utilization between cell growth and product expression and can be manipulated to ensure maximum yields and reproducible measurements (Radford et al., 1997a,b). The MOI is particularly important for the optimal expression of ion channels by SFV (Blasey et al., 1997). By increasing the MOI, the number of RNA molecules per cell that actively transcribes the recombinant protein is increased, resulting in more channels. In contrast, BV is capable of replicating within the host cell and its initial MOI has little effect on the final product yield (Radford et al., 1997a).

E. Glycosylation

The expected putative sites of N-linked glycosylation in P2X ion channels have been observed for BV expressed protein (K. Radford, unpublished results, 1996), but the extent of glycosylation varied. The expression of multiple glycosylation variants introduces sample heterogeneity that can, for example, prevent the accurate determination of the molecular weight of the fully assembled ion channel.

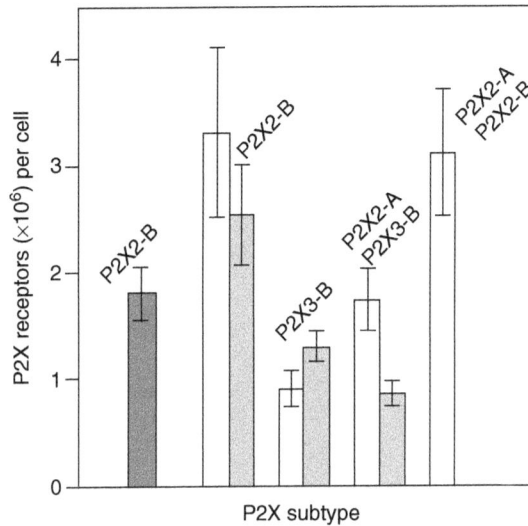

Fig. 2 Expression of P2X$_2$ and P2X$_3$ receptors with SFV or BV systems; comparison of simple and coinfection of cells and culture volume. Each bar shows the average of three determinations of the specific binding of [^{35}S]ATPγS to P2X receptors per cell. Open bars represent baculovirus expression at 150 ml and black bars at 24 l. Hatched bars show expression with SFV in 13-ml cultures. Sf9 cell cultures were grown and infected with baculovirus in suspension cultures and CHO cell cultures were grown and infected with SFV in 75-cm^3 tissue culture flasks. Letters A and B refer to independent epitopes that were used to tag the receptors (Radford *et al.*, 1997c).

Glycosylation also affects the localization of ion channels but may not be essential for their function once they are fully assembled. Inhibition of glycosylation has been shown to effect the surface expression and binding of glutamate receptors (Kawamoto *et al.*, 1994, 1995b; Sydow *et al.*, 1996) expressed in BV-infected insect cells. This loss of function could be due to the inability of the nonglycosylated receptors to bind transiently to molecular chaperones. The simultaneous expression of folding pathway chaperones has previously improved expression levels of secreted recombinant proteins (Hsu *et al.*, 1994), but similar experiments have yet to be performed with ion channels.

Posttranslational modifications have been studied less extensively with SFV; however, it is expected that the broader range of host cells will allow modifications identical to those of native receptors. Short-term expression of proteins in cultured neurons has been achieved with SFV in which the expressed proteins showed appropriate intracellular transport (Olkkonen *et al.*, 1993).

F. Receptor Assembly

Baculovirus has been used to express both heteromeric and homomeric protein complexes (Fig. 3). Successful generation of active heteromeric channels was obtained by the coinfection of insect cells with multiple BVs that encoded different

Infected with	P2X$_2$-A		P2X$_2$-A and P2X$_3$-B		P2X3-B		P2X$_2$-A and P2X$_3$-B	
Immunoprecipitated with	anti-A	anti-B	anti-B	anti-A	anti-B	anti-A	anti-B	anti-A
Detected with	anti-A	anti-A	anti-A	anti-A	anti-B	anti-B	anti-B	anti-B

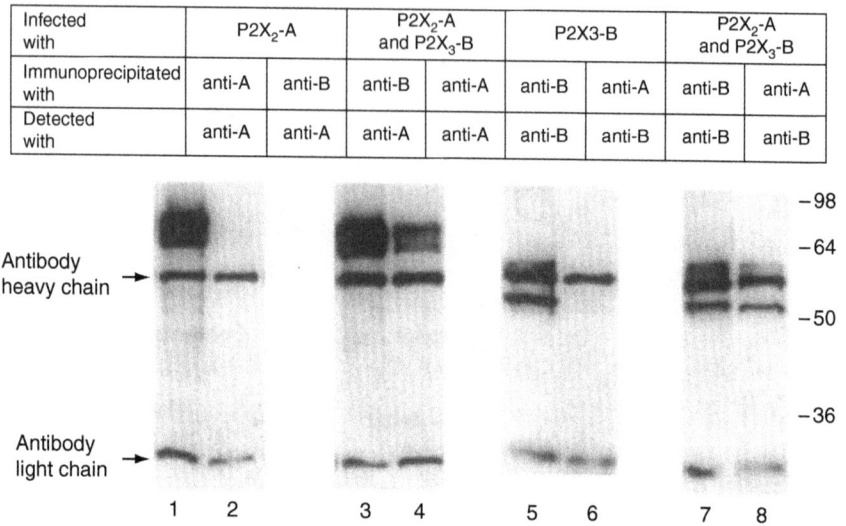

Fig. 3 Western blot analysis of P2X$_2$ and P2X$_3$ coimmunoprecipitation. Cells were infected with baculoviruses encoding a P2X$_2$ receptor tagged with an A epitope, a P2X$_3$ receptor tagged with a B epitope, or both. Cells were harvested at 52 h, solubilized with 1% Triton X-100, and the supernatants immunoprecipitated with antiepitope peptide monoclonal antibodies as indicated. These antibodies were also used to detect the immunoprecipitated channel subunits by Western blotting. P2X$_3$ comigrated with antibody heavy chain bands. Adapted from Radford and Buell (1997).

subunit components (Hartnett *et al.*, 1996; Radford *et al.*, 1997c). Evidence of interactions between other proteins, coexpressed by BVs, has been established using immunoprecipitation (Barr *et al.*, 1997). Although heteromeric ion channel expression has not been reported using SFV, this system has certain advantages. First, multiple viruses encoding subtype variants can be generated rapidly using the SFV system. Second, the nonreplicative nature of the recombinant virus, leading to the direct relationship between MOI and expression yield, could facilitate the study of heteromeric channel stoichiometry.

The BV and SFV systems, therefore, can produce large amounts of ligand-gated ion channels for purification and in addition be used to study the assembly, interaction, and function of the component subunits.

II. Generation of Viral DNA or RNA

A. Recombinant Semliki Forest Virus Production

Two plasmids, the recombinant vector with inserted sequence and pSFV-Helper2 plasmid, are linearized with *Spe*I. Recombinant RNAs are generated from these templates, using SP6 RNA polymerase, and the two *in vitro* transcripts are coelectroporated into Baby Hamster kidney (BHK) or Chinese Hamster ovary (CHO) cells. SFV structural proteins, encoded by the helper RNA, form virions that package only the recombinant vector RNA and this virus can infect other cells without subsequent replication. pSFV-Helper2 should be used to prevent the low-frequency generation of replicon-competent SFV.

1. Protocol 1.1: *In Vitro* Transcription of Viral RNAs

1. 2.5 μg of pSFV1 DNA with foreign insert and pSFV-Helper2 DNA (Life Technologies, Inc., Rockville, MD) are separately digested with *Spe*I. The insert must not contain any *Spe*I restriction sites because this enzyme is used to linearize the recombinant vector plasmid. To ensure that the template is free of ribonucleases after digestion, it is extracted with phenol and precipitated with ethanol.

2. The linearized vector and helper DNAs are used as templates for *in vitro* transcription by incubating with the following components at 37 °C for 1 h in 50 μl [40 mM HEPES/KOH, pH 7.4, 6 mM magnesium acetate, 20 mM spermidine hydrochloride, 1 mM m7G(5')ppp(5')G, 5 mM dithiothreitol (DTT)], rNTP mix (1 mM rATP, 1 mM rCTP, 1 mM rUTP, and 0.5 mM rGTP), 50 U RNasin (Promega, Madison, WI), and 30 U SP6 RNA polymerase.

3. 1 μl of each transcript is analyzed by agarose (0.6%) gel electrophoresis to assess the quality and amount of RNA. The transcripts normally migrate as a single band, more slowly than the linearized plasmid.

2. Protocol 1.2: *In Vivo* Packaging of Recombinant SFV Virions

This protocol produces sufficient virus to infect 2×10^8 cells at an MOI of 100 virus/cell.

1. BHK (ATCCCCL-10, Rockville, MD) or CHO (ATCC CCL-61) cells grown in Ham's F-12 medium: Iscove's (1:1) medium (Seromed, Basle, Switzerland), 20 ml at 5×10^5 cell/ml, are centrifuged at $1000 \times g$ for 5 min, washed, and resuspended in 1 ml phosphate-buffered saline (PBS).

2. Cells (0.8 ml) and 20 μl recombinant vector and helper RNA transcripts are mixed rapidly and electroporated with two pulses at 25 μF, 850 V, and maximum resistance (Bio-Rad, Richmond, CA; Gene Pulser).

3. Electroporated cells are immediately diluted 20-fold in prewarmed, complete growth medium and incubated for 36 h at 37 °C in 5% (v/v) CO_2.

4. Recombinant SFV is harvested by centrifugation at $1000 \times g$ for 5 min and can be stored at 4 °C for several weeks. For long-term storage the virus can be frozen in liquid nitrogen and stored at −80 °C; however, on thawing infective virus titers will be lower.

Compared to BV, the generation of SFV stock is more rapid, but is also limited by the quantity of virus that is formed per electroporation. Repetitive generation of small-scale SPV stocks can introduce variability and necessitates the characterization of each new preparation. For these reasons we have preferred SFV for the small-scale expression of receptor subtypes and mutational analysis while BV has been more useful for large-scale expression.

B. Recombinant Baculovirus Production

Classical construction of recombinant BVs involves the cotransfection of insect cells with baculoviral DNA and a recombinant transfer plasmid that contains the foreign gene. Insertion of the gene into the BV genome occurs via homologous recombination, and the resulting mixture of progeny viruses (recombinant or parent) is separated by several cycles of plaque purification. Linearized BV DNA (BaculoGold DNA, Pharmingen) contains a lethal deletion in the *polh* gene, which can be complemented by recombination with the appropriate transfer plasmid so that only recombinant viruses are able to form plaques. Typically, a single plaque purification is required to ensure the absence of the parental BV.

More recently, recombinant BVs have been made by site-specific transposition of foreign genes into a bacmid that is propagated in *Escherichia coli* (FastBac, Life Technologies). Using this technique, recombinant viral DNA, isolated from independent colonies, is free from parental nonrecombinant virus and plaque purification is not necessary. However, regardless of the system that is chosen, clonal isolation of virus by plaque purification helps to ensure homogeneous expression.

1. Protocol 1.3: Recombinant Virus Construction Using Linearized Baculovirus DNA

1. Seed 2 ml of Sf9 (*Spodoptera frugiperda* fall army worm ovary) cells (Invitrogen, San Diego, CA; 1×10^6 cell/ml) grown in Sf900II (Life Technologies) into 60-mm tissue culture plates and allow to attach for 1 h at 27 °C.

2. Mix 0.5 μg of viral DNA (BaculoGold DNA, Pharmingen) with 5 μg of recombinant baculovirus transfer vector. Use highly purified DNA that is less than 1 month old. After a 5-min incubation at room temperature, add 1 ml of transfection buffer (25 mM HEPES, pH 7.1, 125 mM $CaCl_2$, 140 mM NaCl).

3. Vacuum aspirate medium from cell monolayer and add DNA/transfection buffer mixture drop by drop while rocking the plate. A fine calcium phosphate/DNA precipitate should form. Incubate plates at 27 °C for 5 h.

4. Remove cotransfection mixture by vacuum aspiration. Wash monolayer gently with 3 ml of SF900II. Vacuum aspirate wash medium and add 3 ml of fresh SF900II. Incubate at 27 °C for 5 days.

5. Examine cells at 5 days postinfection. Infected cells are larger with a granular nucleus, 80–90% of the entire cell size. Harvest 3 ml of cotransfection supernatant and store at 4 °C.

6. Plaque purification is performed as for the plaque assay described later with slight modifications. Dilutions of virus from 10^{-2} to 10^{-4} are sufficient and neutral red is not required to stain plaques. Instead, 200 μg/ml X-Gal (5-bromo-4-chloro-

3-indolyl-β-D-galactopyranoside) is added to the agarose/medium overlay immediately prior to use. After 5 days of incubation, white recombinant viral plaques are distinguished from the small percentage ($<$1%) of the blue ($lacZ^+$) nonrecombinant viral plaques.

7. Pick a recombinant plaque with a sterile Pasteur pipette and transfer the entire agarose plug to 1 ml of SF900II medium. Virus is eluted from the agarose by agitation overnight at 4 °C. At this stage the virus can be used for large-scale amplification of virus stock.

2. Protocol 1.4: Recombinant Virus Construction Using pFastBAC

1. Subclone the gene of interest into the pFastBAC1 donor vector (Life Technologies). Digest vector and foreign DNA with the selected endonucleases, dephosphorylate the vector, and purify the vector and foreign insert DNA. Ligate the prepared vector and insert fragment to produce the pFastBAC-recombinant.

2. Incubate 0.1 ml of DH10B competent cells (Life Technologies) with 1 ng of the pFastBAC-recombinant on ice for 30 min. Transform by heatshock for 45 s at 42 °C. Cool 2 min on ice, add 0.9 ml rich medium, and shake at 37 °C for 4 h.

3. Plate 0.1 ml of transformation mix at various dilutions ($10^0, 10^{-1}, 10^{-2}$) onto LB plates containing 50 μg/ml kanamycin sulfate, 10 μg/ml tetracycline, 7 μg/ml gentamicin, 100 μg/ml X-Gal, and 40 μg/ml isopropylthio-β-D-galactoside and incubate 24 h at 37 °C.

4. Select 10 white colonies and streak a fraction of each colony onto fresh plates to verify the recombinant phenotype. Inoculate the same colonies in 2 ml LB with the antibiotics used above. Incubate at 37 °C with agitation for 16 h.

5. Extract and purify DNA from 1 ml of a liquid culture, confirmed to be recombinant. A DNA extraction kit (Qiagen, Hilden, Germany) can be used and provides DNA of adequate purity for transfection. The DNA is dissolved in 40 μl of TE buffer and can be stored at -20 °C.

6. Sf9 cells (9×10^5) are inoculated into 2 ml of Sf900II medium containing antibiotics (50 U/ml penicillin, 50 μg/ml streptomycin), plated onto sterile 35-mm culture dishes, and allowed to attach for 1 h at 27 °C. The cell condition at the time of transfection is critical and cells grown to greater than 2×10^6 cell/ml should not be used.

7. Minipreparation bacmid DNA (5 μl) in 0.1 ml of SF900II is mixed with 6 μl of CellFECTIN reagent (Life Technologies) in 0.1 ml Sf900II and incubated for 45 min at room temperature.

8. Sf9 cell monolayer is rinsed with 2 ml of Sf900II and vacuum aspirated. Sf900II medium (0.8 ml) is added to the transfection mixture and 1 ml of the diluted mixture is overlaid onto the cells. The plates are incubated for 5 h at 27 °C after which the residual transfection mixture is replaced with 2 ml of Sf900II medium containing antibiotics. The transfected cells are incubated for 48 h at 27 °C.

9. Supernatant is harvested for virus titration and subsequent amplification to 100 ml. At this stage it may be possible to detect expression of the ion channel proteins via immunoblot or activity assay of the residual cells.

III. Choice and Cultivation of Host Cells

The host insect cells most commonly used with BV vectors are Sf9, Sf21, and Tn5-High Five cells (Invitrogen). These cell lines grow well in suspension in serum-free medium (SF900II for *S. frugiperda*: Sf9; Sf21 and Excell-401 for *Trichoplusia ni*: Tn5). Although neither cell line is superior for the expression of all ion channels, we have found better yields of functional P2X receptors using Sf9 than with Tn5 cells. Differences between cell lines have also been reported for other receptor families. The human muscarinic m_2 receptors have been expressed using BV in several insect cell lines: Sf9, Sf21, and Tn5 (Heitz *et al.*, 1997). The expression level was slightly increased in Sf21 cells versus Sf9 cells, and Sf9 and Tn5 cells showed significant proteolysis compared to Sf21 cells. The Sf21 and Sf9 cells are smaller and have a higher surface area-to-volume ratio which could improve the expression of membrane proteins. Increased synthesis of toxic metabolites (ammonia and lactate) in Tn5 cell cultures has been reported and may decrease productivity (Yang *et al.*, 1996). Regardless of the cell line employed, the nutrient composition of the medium potentially imposes a limitation. Most examples of ion channels, expressed in insect cells, have used serum containing media. However, the glucose and amino acid content of the original basal medium component of serum containing formulations such as Graces medium, TNMH, or IPL41 media is very low, compared to current serum-free formulations and do not support maximal insect cell growth or optimal protein expression (Radford *et al.*, 1997b).

A. Insect Cell Culture

Cultivation of large volumes of insect cells as monolayers is impractical due to the high surface area-to-volume ratio that is required for adequate oxygenation and to the disruption of the monolayer for monitoring channel expression. Since the application of pluronic polyol F-68 (Life Technologies), a shear protection agent (Murhammer and Goochee, 1990) insect cell cultures of 30–1700 ml are most efficiently cultivated in suspension. Culture volumes should not exceed 50% of the maximum capacity of Erlenmeyer flasks. For larger cultures additional oxygen or air must be supplied so that dissolved oxygen tension is maintained at 50% of that in air. In stirred tank bioreactors (2–10 l), this is achieved by a combination of sparging and stirring. Fixed-angle rather than flat-blade impellers are employed to increase the mixing without excessive shear stress. Large-scale airlift reactors (20–100 l) can oxygenate and mix by sparging alone and the dimensions of these reactors (high height-to-surface area ratio) can minimize cell damage due to bubble bursting at the medium surface.

Due to the diversity of culture vessels and bioreactors, the following protocols describe expression only in readily available flasks. Cultures of 30–1700 ml provide sufficient material for most biochemical, pharmacologic, and preliminary structural studies.

1. Protocol 2.1: Routine Subculture of Insect Cells

1. Inoculate Erlenmeyer shaker flasks (250 cm^3, Schott) with 125 ml of Sf9 cells at 2×10^5 cell/ml in SF900II medium supplemented with 0.1% Pluronic F-68 and rotate at 170 rpm at 27 °C with lids unscrewed.

2. Subculture at 2×10^5 cell/ml into prewarmed medium (27 °C) every 3 or 4 days while cells are in early exponential growth. Sf9 cells should have an average doubling time of 21 h and should be replaced if a doubling time greater than 24 h is observed routinely. Do not use cells for subculture that have exceeded 5×10^6 cell/ml without prior medium exchange. Discard cultures exceeding a density of 7×10^6 cell/ml.

3. Cell density and viability are quantified using a hemocytometer and trypan blue (0.2%) exclusion. Count samples in triplicate, where a single count includes both sides of the hemocytometer and sample is diluted so that approximately 200 cells are counted per side. The mean of triplicate estimations $\pm95\%$ confidence interval can be calculated from the variance estimated by the method of Nielsen *et al.* (1991).

4. Cultures can be scaled up in volume at time of subculture to 1700 ml and are typically infected for expression at a density of 1×10^6 to 5×10^6 cell/ml. Two weeks of reproducible growth kinetics should be observed before infection, especially after thawing cells from liquid nitrogen. New cell cultures from frozen parent stocks should be thawed after Sf9 cells have exceeded 50 passages.

B. SFV Host Cell Line

The preferred hosts for SFV are mammalian cell lines, although the virus is capable of infecting a wide variety of primary cells and most animal cell lines. BHK and CHO cells are most commonly used but CHO cells have been suggested to be less efficient for SFV expression (Blasey *et al.*, 1997). BHK cells in suspension tend to aggregate and are difficult to maintain without specialized bioreactors or nonadherent clones. We, therefore, use CHO cells for medium-scale expression (50–1700 ml) but prefer BHK for small-scale (maximum 50 ml) monolayer cultures. These mammalian cells typically reach maximum cell densities of $1–2 \times 10^6$ cell/ml. This is 5- to 10-fold less than for Sf9 cultures, but BHK and CHO cells grow faster, doubling every 15–18 h.

CHO cell manipulations are performed in Ham's F-12 and Iscove's medium (1:1) supplemented with 4 mM glutamine and 10% fetal calf serum (FCS).

Mammalian cell lines have lower oxygen requirements than insect cells; however, the oxygen demand increases dramatically after infection in both systems and CHO suspension cultures should be rotated at 120 rpm. Unlike insect cells, pH control is critical for optimal growth and ion channel expression by CHO or BHK cells and all incubations should be performed in the presence of 5% (v/v) CO_2 as sodium bicarbonate is used as a buffering component in the medium.

1. Protocol 2.2: Routine Subculture of CHO Cell

1. Inoculate Erlenmeyer shaker flasks (250 cm^3) (Schott) with 125 ml of CHO cells at 4×10^4 cell/ml in Ham's F-12; Iscove's (1:1) medium supplemented with 4 mM glutamine, 10% FCS, and 0.2% pluronic F-68 and rotate at 120 rpm at 37 °C in 5% CO_2 with lids unscrewed.
2. Quantify cell density and viability as explained for insect cells in Protocol 2.1.
3. Allow cells to grow and subculture at 4×10^4 cell/ml into prewarmed medium (37 °C) every 2 or 3 days while cells are in early exponential growth. They are ready for infection with SFV at 5×10^5 cells/ml. Methods for optimization of expression parameters are given in Protocol 4.2.

CHO cells should have an average doubling time of 16 h and should be replaced if a doubling greater than 20 h is observed routinely or they have exceeded 2×10^6 cell/ml during subculturing.

IV. Amplification and Titration of Virus Stocks

The quantity of BV, generated by transfection or plaque purification, is adequate for small-scale receptor expression. Virus can be amplified by repeated infections of insect cell cultures at increasing scale, but the number of serial infections should be kept to a minimum (2–3) and a low MOI employed [0.1–0.2 plaque-forming units (pfu)/cell]. Extensive serial passage of virus or high MOI will produce virus stock, dominated by defective interfering particles, which have extensive mutations and are helper virus dependent (Kool *et al.*, 1990). These particles interfere with replication of the normal virus and substantially diminish the titer of viral stocks.

1. Protocol 3.1: Amplification of Baculovirus Stocks

1. Prepare a 20-ml suspension culture of Sf9 cells in SF900II at 1×10^6 cell/ml in a 100-cm^3 flask. Cells may be centrifuged at $1000 \times g$ for 5 min and resuspended in fresh medium to increase the potential virus titer.
2. Inoculate the culture with supernatant from bacmid transfection to give a final MOI of 0.1 pfu/cell or add 1 ml of the eluted virus prepared by plaque purification.

3. Rotate at 170 rpm at 27 °C for 4–5 days or until viability approaches zero as determined via trypan blue exclusion.

4. Centrifuge at $2000 \times g$ for 30 min and store supernatant at 4 °C in the dark for up to 6 months. For long-term storage, virus may be frozen at −80 °C but is unstable at −20 °C. Repeated freeze–thawing will result in a large reduction of virus titer.

5. Determine the virus titer ($\sim 1 \times 10^7$ pfu/ml) by one of the methods suggested later, and repeat the amplification procedure at 1700 ml scale, using an MOI of 0.1 pfu/ml. The 1700 ml stock should contain 2×10^8 pfu/ml of recombinant virus, which is sufficient for expression of the ion channel at the 10-l scale.

A. Baculovirus Titration

Two forms of BV are always produced that have different natural roles in insects. These are budded, infectious virus and multinucleocapsid, noninfectious virus. The multinucleocapsid form, which is normally surrounded by the protective polyhedrin coat, ensures propagation within insect populations and requires ingestion to infect, whereas the infective form permits efficient infection of adjacent cells within an insect and also insect cell cultures. Plaque and end-point dilution methods for determining BV concentrations measure only infective virus and not the total concentration of virus particles, which may be 100-fold higher.

An important parameter influencing either of these assays is the condition and density of the insect cells. Uninfected controls should be included to monitor the condition of the cell monolayers. Virus is visualized in the plaque assay when a single infectious unit replicates and progeny virus infects surrounding cells that lyse and produce clear zones. The end-point dilution method is often performed when the recombinant protein is expressed with a detectable marker, such as β-galactosidase. However, coexpression of nonrelated marker proteins potentially reduces the maximum yield of the ion channel and should be avoided during its large-scale expression. For proteins without markers an alternative end-point assay is described that reliably detects the titer of virus by the degree of postinfection cell growth. End-point dilution methods are simpler to perform than plaque assays and permit large numbers of sample repeats in multiwell screening plates.

1. Protocol 3.2: Plaque Assay

1. Inoculate 35-mm multiwell tissue culture plates with 1.5 ml of Sf21 cell suspensions grown in TNM-FH insect medium (Pharmingen) containing 5% FCS at 8×10^5 cell/ml. Plaque definition is improved if Sf21 cells are used instead of Sf9 cells. Allow cells to attach for at least 1 h at 27 °C.

2. Prepare serial dilutions of test virus (10^{-3} to 10^{-8}) in TNM-FH without serum.

3. Aspirate cell monolayers, rinse with 2 ml TNM-FH without serum, and inoculate with 0.5 ml of diluted virus. At least one well is left uninfected as a cell monolayer control. Plates are tilted immediately to ensure even distribution of virus over the monolayer. Allow virus to absorb by incubation at 27 °C for 2–3 h in a humid environment. Tilt plates at least twice during the incubation period to prevent cells from drying.

4. Prepare overlay medium (2 ml/well) immediately before use by mixing sterile 3% low gelling temperature agarose (cooled to 37 °C), with an equal volume of prewarmed (27 °C) TNM-FH medium containing 10% FCS.

5. After virus adsorption vacuum aspirate monolayers and overlay 2 ml of the agarose mixture without bubbles to each well. Agarose is allowed to set for 3–4 h at 27 °C.

6. Overlay agarose with 1 ml of TNM-FH medium containing 5% FCS and incubate in a humidified environment at 27 °C for 5–6 days.

7. Plates are stained with 1 ml neutral red (0.3% solution diluted 1:20 in PBS) and incubated for 2 h at 27 °C. Excess stain is poured off and the plate is then blotted by inversion onto paper towels for 5 h or overnight at room temperature.

2. Protocol 3.3: End-Point Dilution Virus Assay

This assay requires that the ion channel construct also encodes a marker protein such as β-galactosidase that can be used to visualize the viral dilution end point. The end point is the dilution of virus that would infect 50% of the cultures ($TCID_{50}$) and can be converted according to Poisson distribution to virus concentration as pfu/ml.

1. Serial dilutions of test virus (10^{-1} to 10^{-10}) are made directly into Sf9 cell suspensions (1×10^6 cell/ml) in Sf900II in a final volume of 1 ml. An uninfected control sample is prepared by mixing 0.1 ml of PBS with 0.9 ml of cells.

2. Multiple repeat aliquots (0.1 ml) of virus/cell mixtures are pipetted into 96-well plates and incubated at 27 °C in a humid environment. One column is loaded with the uninfected control culture.

3. 10 μl of 20 mg/ml X-Gal is added to each well at day 5 or 6 and plates are incubated 4 h at room temperature. Blue wells are scored as positive for virus and the dilution that gives 50% infection response is calculated by linear interpolation.

3. Protocol 3.4: MTT End-Point Dilution Virus Assay

1. Serial dilutions of test virus are made into Sf9 cell suspensions as in Protocol 3.3, step 1, but with slightly fewer cells (8×10^5 cell/ml).

2. Aliquots of virus/cell mixtures are loaded into 96-well plates as in step 2 of Protocol 3.3.

3. After 72 h of incubation 20 μl of MTT [3-(4,5-dimethylthiazol-2-yl)2,5-diphenyltetrazolium bromide] (7.5 mg/ml) is added to all wells and incubated for at least 6 h.

4. Plates are read spectrophotometrically at 570 nm and viral end point is determined by scoring the number of virus positive wells (those indicating a lower absorbance than the noninfected control) at each dilution. Error estimates in the virus titer can be calculated by the method of Nielsen *et al.* (1992).

B. SFV Titration

The methods just outlined for BV titration rely on the single virus end point, which can be determined following virus replication. In contrast, SFV does not produce plaques and the absolute amount of infectious virus cannot be determined. The following technique is based on counting the number of infected cells and is similar to the end-point dilution method. Although this method can only estimate the titer, it can be used to compare the relative concentrations of SFV stocks and to improve reproducibility between experiments.

1. Protocol 3.5

1. Test virus is activated by the addition of α-chymotrypsin (Sigma, St. Louis, MO) 200 μg/ml for 15 min at room temperature. This cleaves the viral p62 precursor into E2 and E3 membrane proteins.

2. After 15 min the protease activity is arrested by the addition of aprotinin (Sigma) to a final concentration of 400 μg/ml.

3. Serial dilutions of activated test virus (10^{-1} to 10^{-10}) in culture medium (Ham's F-12 and Iscove's 1:1, 4 mM glutamine, 10% FCS) are used to infect CHO or BHK monolayers (1×10^4 cells/well) in 96-well plates. Multiple repeats are required for each dilution.

4. Infected plates are incubated at 37 °C in a humid 5% CO_2 environment for 20 h after which β-galactosidase expressing cells are stained blue by the addition of X-Gal to 40 μg/ml. Microscopic examination of the cells in each well is used to score the percentage of cells expressing β-galactosidase at each viral dilution.

V. Optimization of Process Parameters

Nutrient exhaustion limits the potential yield of ion channels expressed by BV and correct application of MOI and TOI is required for reproducible expression (Radford *et al.*, 1997b). Substrate limitation arises when cultures, grown to high density, are infected at low MOI; the uninfected cells continue to divide, leaving few nutrients available for product expression. The simplest strategy is to use an MOI (10 pfu/cell) that allows synchronous infection of the population at increasing cell densities. This is also affected by virus titer since the addition of large volumes of virus inoculum, consisting of spent medium, is inhibitory. It is recommended to stay well below the maximal critical cell density of infection and at the highest MOI possible. Once the optimal cell density at TOI has been established, the optimal time of harvest is chosen as a balance between maximum expression and minimal degradation of the receptor.

1. Protocol 4.1: Optimization of the Times of Infection and Harvest for Ion Channels Expressed by Baculovirus

1. Inoculate a 1700-ml shaker flask with 2×10^5 Sf9 cells/ml and allow cultures to grow under conditions described earlier.

2. At various cell densities (10^6 to 6×10^6 cell/ml) remove 100-ml aliquots from the parent culture under sterile conditions and transfer to 250-cm^3 capacity Erlenmeyer flasks.

3. Infect immediately with recombinant BV at a multiplicity of 10 pfu/cell or up to 5% of the culture volume. Incubate shaker flasks as above and sample at 30, 40, 50, 60, 70 h postinfection (5×1 ml) to determine the cell density and activity.

4. Results of the activity assay are expressed on a per-cell basis. Note that total cell density will decrease late in the infection period due to virus-induced cell lysis and estimates at these times cannot be used to calculate the specific activities.

5. The optimal time of harvest for the culture is chosen as the time at which the highest total yield of active receptor is obtained. If activity assays are not available, immunoblot analysis can be used to approximate the optimal cell density at the TOI and harvest, but active and inactive protein is not distinguished. For P2X receptors the optimal time of harvest is between 50 and 65 hpi for cultures infected at high MOI.

Similar methods can be used to optimize ion channel expression by SFV, but these conditions will vary for different cell lines and media. The MOI cannot be optimized in monolayer cultures and the results extrapolated to suspension cultures. Viruses generally adsorb more slowly in suspension than monolayer cultures. Due to the nonreplicative nature of SFV, small changes in culture infection conditions, such as MOI, may result in significant differences in total yield and expression kinetics. For this reason it is necessary to monitor the time course of protein production to determine the optimal time of harvest. The following protocol is used to determine the optimal MOI and associated optimal time of harvest for ligand-gated ion channels expressed with SFV-infected CHO cells.

2. Protocol 4.2: Optimization of Ion Channel Expression Parameters for SFV

1. Recombinant virus (70 ml at 10^9 virus/ml) is activated by the addition of α-chymotrypsin as in Protocol 3.5.

2. CHO cells grown to a density of 5×10^5 cell/ml as 1-l suspension cultures are centrifuged and resuspended at 5% of the original volume in fresh medium (pH adjusted to 6.9 with acetic acid). Volumes of 10 ml are aliquoted into five 250-ml shaker flasks ready for immediate infection.

3. Cultures are infected at various MOIs (1, 10, 50, 100, 500 virus/cell) with activated virus and incubated at 37 °C (120 rpm) in 5% (v/v) CO_2 for 6 h to ensure adequate adsorption of virus. Each culture is sampled at the TOI. Centrifuge 0.5-ml aliquots at $1000 \times g$ for 5 min and store the pellet at -80 °C as a control and the supernatant at 4 °C to assess virus adsorption.

4. At 6 h postinfection the cultures are centrifuged at $1000 \times g$ for 5 min and resuspended in 100 ml fresh medium. A sample (0.5 ml) of the supernatant is stored at 4 °C to confirm complete virus adsorption.

5. Cultures are sampled (5×1 ml) at 6, 12, 24, 48, and 60 h postinfection and the cell density and ion channel expression levels are determined.

We routinely use SDS–PAGE and Western blot analysis to monitor P2X expression in these cases, but electrophysiologic activity can also be used to determine the optimal MOI and its associated optimal time of harvest. Depending on the quality of the virus stock, maximum expression can occur as early as 20–30 h postinfection with greater than 10^6 receptors/cell.

SFV and BV production are thus optimized by determining the TOI, optimal time of harvest, and, since SFV is nonreplicative, its optimal MOI. Once estimates of these culture and infection parameters have been established, ion channels can be reproducibly expressed by these systems for many function and structure applications.

VI. Applications of Ion Channels Expressed by Baculovirus and SFV

A. Electrophysiologic Recording

The lytic nature of the BV expression system precludes the use of whole-cell recording techniques for determining optimal expression parameters. The infected host cell membrane is compromised late during infection, due to budding of progeny virions and cell lysis. Host cell membrane integrity is usually sufficient for electrophysiologic recording up to 50 h postinfection, after which the holding currents become large and unstable. The use of alternative BV gene promoters for expression of ligand-gated ion channels at earlier times than the very late polyhedrin promoter may offer an advantage for electrophysiologic recording. In general, the expression levels of proteins controlled by earlier promoters such as the basic promoter are lower than those controlled by the polyhedrin promoter but the slightly earlier expression kinetics provide a better separation of the expression and lytic phases of infection.

Mammalian cells infected with SFV are more amenable to electrophysiologic applications due to the earlier onset of expression. Maximum expression of ion channels can be achieved within the first 20 h postinfection, well before cell lysis begins. This is particularly true if a high MOI is employed (MOI > 30). The following protocol for electrophysiologic recording from insect cells infected with BV encoding the P2X$_2$ receptor can also be adapted for BHK cells infected with recombinant SFV by altering the culture conditions, MOI, and time of recording.

1. Protocol 5.1: Preparation of Insect Cells for Whole-Cell Patch Clamp

1. Prepare a 30-ml suspension culture of Sf9 cells and culture under normal conditions at a density of 10^6 cell/ml.
2. Inoculate the culture with virus at MOI of 10 pfu/cell. Synchronous expression is desirable for electrophysiologic recording so that chosen cells are representative of the entire cell population.
3. Rotate infected cultures at 170 rpm at 27 °C for 30 h. Aliquot approximately 2000 cells to polylysine-treated coverslips under sterile conditions and allow to attach at 27 °C for 2 h before electrophysiologic recording.

B. Heteromeric Receptor Expression and Detection

Baculovirus supports the assembly of both homomeric and heteromeric ion channel complexes. Pentameric 5-HT$_3$ receptors, expressed with BV, have been observed by electron microscopy (Green et al., 1995). Coexpression of two subunits, resulting in coassembly and formation active heteromeric channels, has been reported for GABA$_A$ and P2X channels (Hartnett et al., 1996; Pregenzer et al., 1993; Primus et al., 1996; Radford et al., 1997c; Witt et al., 1996). The level of expression for the GABA$_A$ receptor was found to depend on the ratio of infectious virus particles for each subunit in multivirus infections. The stoichiometry of the channel, however, was not affected and only a single type of receptor was produced (Hartnett et al., 1996).

Methods for coexpression of different subunits are essentially identical to those described earlier for a single subtype except that two or more viruses are used to infect a culture simultaneously. The following protocol was used to determine whether epitope-tagged P2X$_2$ and P2X$_3$ subunits assemble to form heteromeric channels (Fig. 3).

2. Protocol 5.2: Coinfection and Coimmunoprecipitation of P2X Receptor Subunits

1. Insect cells at 1.5×10^6 cell/ml grown as 100-ml volumes in 250-cm^3 capacity Erlenmeyer flasks are pooled, centrifuged at $1000 \times g$ for 5 min, and resuspended in fresh medium.

2. Resuspended cultures are inoculated at an MOI of 5 pfu/cell with BV encoding P2X$_2$ (tag A), P2X$_3$ (tag B), or both receptors.

3. Cultures are shaken at 200 rpm at 27 °C for 50 h and aliquots removed at approximately 0, 24, 30, and 50 h and stored as pellets at -80 °C.

4. Cells (2×10^7) harvested at 50 h are lysed on ice in 2 ml of 20 mM Tris, 150 mM NaCl, 1 mM CaCl$_2$, and 1 mM MgCl$_2$ with 1% Triton X-100, and homogenized twice.

5. Lysates are centrifuged at $12,000 \times g$ for 10 min at 4 °C and the supernatant transferred to 200 μl of a 1:1 mixture of washed protein A agarose/protein G agarose (Pharmacia, Piscataway, NJ) for preadsorption of background proteins. Samples are rotated 1 h at 4 °C and beads centrifuged at $12,000 \times g$ for 3 min.

6. Supernatant (0.8 ml) is incubated for 1 h at 4 °C with mouse monoclonal antibodies to either tag (25 μg/ml final concentration), and overnight after the addition of 0.1 ml of washed protein A/protein G agarose.

7. Beads are washed three times in lysis buffer by repeated centrifugation as in step 5. Pellets are resuspended in 50 μl of sample buffer, boiled for 5 min, and 10 μl subjected to SDS–PAGE.

8. Samples immunoprecipitated with tag A antibody are detected by Western blot/ECL (enhanced chemiluminescence) format (Amersham, UK) using the tag B antibody and antimouse IgG as the secondary antibody.

C. Glycosylation

Homogeneous expression of single-receptor subtypes is preferred for applications such as mass spectrometry or crystallography. Expression of ligand-gated ion channels by the BV and SFV systems, however, typically results in the production of multiple molecular weight bands, as seen by SDS–PAGE. This heterogeneity is often the result of varying degrees of glycosylation. Inhibition of glycosylation with tunicamycin or posttranslational removal of carbohydrate moieties with glycosidase will reduce the bands to single species (Kopke *et al.*, 1993).

1. Protocol 5.3: Deglycosylation of Ligand-Gated Ion Channels

1. Disrupt 2×10^6 virus-infected cells (SFV/BHK or baculovirus/Sf9) expressing the channel in 0.5% SDS, 1% 2-mercaptoethanol, and 50 mM sodium citrate, pH 5.5, at 95 °C for 10 min, then chill on ice.

2. High mannose residues are removed with endoglycosidase H (New England Biolabs, Inc., Beverly, MA), added at 10 U/μg of total protein lysate followed by incubation at 4 °C for 4 h.

3. Deglycosylated samples are diluted 1:1 in sample buffer and subjected to SDS–PAGE analysis.

Alternatively, *N*-glycosidase F (PNGase F, New England Biolabs, Inc.) can be used at 40 U/μg protein. This enzyme, which cleaves between the innermost *N*-acetylglucosamine and asparagine residues, is used in 50 mM sodium phosphate, pH 7.5, and the inhibitory activity of SDS in the lysate is neutralized with 1% Nonidet P-40 (NP-40) prior to addition of the enzyme.

2. Protocol 5.4: *In Vitro* Inhibition of Glycosylation

1. Add 30 μl of tunicamycin [stock is filter sterilized at 10 mg/ml in dimethyl sulfoxide (DMSO)] to a 30-ml culture of Sf9 cells at 1×10^6 cell/ml.

2. Incubate the culture for 1 h at 27 °C before infecting with virus at an MOI of 10 pfu/cell.

3. Centrifuge samples (2 ml) of the culture at 0, 24, 36, 48, and 60 h postinfection, and prepare for SDS–PAGE analysis as described in Protocol 5.2. Treated and control cells can also be removed at 24–36 h postinfection for electrophysiologic analysis as in Protocol 5.1.

Tunicamycin may be removed at any time to permit glycosylation of new channels by centrifugation of the culture at $1000 \times g$ and resuspension in fresh medium. The ligand binding and channel activity of control and treated cultures should be examined before large-scale nonglycosylated cultures are produced for structural studies.

VII. Conclusion

This chapter has compared two relatively simple approaches for preparing useful quantities of ligand-gated ion channels. Table II summarizes the important differences between these viral expression systems. Both systems require careful control of the variables that have been considered to ensure reproducible and maximal expression. Attention to these variables will also improve electrophysiologic characterization by allowing comparison between representative sample populations.

Table II
Comparison of Features for Semliki Forest Virus and Baculovirus Expression Systems

Characteristic	SFV	BV
Choice of promoter	One (SP6)	Many
Helper virus	Dependent	Independent
Virus preparation	2 days	10 days
Virus titration	Qualitative	Quantitative
Virus amplification	No	Yes
Host range	Broad	Restricted
Expression time	20 h	2 days
Scale-up	Feasible	Simple

References

Atkinson, A. E., Henderson, J., Hawes, C. R., and King, L. A. (1996). *Cytotechnology* **19**, 37.

Barr, A. J., Brass, L. F., and Manning, D. R. (1997). *J. Biol. Chem.* **272**, 2223.

Binir, B., Tierney, M. L., Howitt, S. M., Cox, G. B., and Gage, P. W. (1992). *Proc. R. Soc. Lond. B: Biol. Sci.* **250**, 307.

Binir, B., Tierney, M. L., Pillai, N. P., Cox, G. B., and Gage, P. W. (1995). *J. Membr. Biol.* **148**, 193.

Binir, B., Tierney, M. L., Lim, M., Cox, G. B., and Gage, P. W. (1997). *Synapse* **26**, 324.

Blasey, H. D., Lundstrom, K., Tate, S., and Bernard, A. R. (1997). *Cytotechnology* **24**, 65.

Cascio, M., Schoppa, N. E., Grodzicki, R. L., Sigworth, F. J., and Fox, R. O. (1993). *J. Biol. Chem.* **268**, 22135.

Chen, X. S., Brash, R., and Funk, C. D. (1993). *Eur. J. Biochem.* **214**, 845.

Davies, A. H., and Jones, L. M. (1991). *Bio/Technology* **11**, 933.

Gimpl, G., Anders, J., Thiele, C., and Fahrenholz, F. (1996). *Eur. J. Biochem.* **237**, 768.

Green, T., Stauffer, K. A., and Lummis, S. C. R. (1995). *J. Biol. Chem.* **270**, 6056.

Hartnett, C., Brown, M. S., Yu, J., Primus, R. J., Meyyappan, M., White, G., Sterling, V. B., Tallman, J. F., Rambhdran, T. V., and Gallager, D. W. (1996). *Receptor Channel* **4**, 179.

Hayashi, M. K., and Haga, T. (1996). *J. Biochem.* **120**, 1232.

Heitz, F., Nay, C., and Guenet, C. (1997). *J. Recept. Signal. Transduct. Res.* **17**, 305.

Hsu, T., Eiden, J. J., Bourgarel, P., Meo, T., and Betenbaugh, M. (1994). *Protein Expr. Purif.* **5**, 595.

Iizuka, M., and Fukuda, K. (1993). *J. Biochem.* **114**, 140.

Joyce, K. A., Atkinson, A. E., Bermudez, I., Beadle, D. J., and King, L. A. (1993). *FEBS Lett.* **335**, 61.

Kawamoto, S., Hattori, S., Oiji, I., Ueda, A., Fukushim, J., Sakimura, K., Mishina, M., and Okuda, K. (1990). *Ann. N. Y. Acad. Sci.* **707**, 460.

Kawamoto, S., Hattori, S., Oiji, I., Hamajima, K., Mishina, M., and Okuda, K. (1994). *Eur. J. Biochem.* **233**, 665.

Kawamoto, S., Hattori, S., Sakimura, K., Mishina, M., and Okuda, K. (1995a). *J. Neurochem.* **64**, 1258.

Kawamoto, S., Uchino, S., Hattori, S., Hamajima, K., Mishina, M., Nakajima-Iijima, S., and Okuda, K. (1995b). *Mol. Brain Res.* **30**, 137.

Keinen, K., Kohr, G., Seeburg, P. H., Laukkanen, M. L., and Oker-Blom, C. (1994). *Bio/Technology* **12**, 802.

Kool, M., Voncken, J. W., van Lier, F. L. J., Tramper, J., and Vlak, J. M. (1990). *Virology* **183**, 739.

Kopke, A. K., Bonk, I., Sydow, S., Menke, H., and Spiess, J. (1993). *Protein Sci.* **2**, 2066.

Kuusinen, A., Arvola, M., Oker-Bloom, C., and Keinanen, K. (1995). *Eur. J. Biochem.* **233**, 720.

Liljeström, P., and Garoff, H. (1991). *Bio/Technology* **9**, 1356.

Lobigs, M., and Garoff, H. (1990). *J. Virol.* **64**, 1233.

Lundstrom, K., Michel, A., Blasey, H., Bernard, A. R., Hovius, R., Vogel, H., and Surprenant, A. (1997). *J. Recept. Signal. Transduct. Res.* **17**, 115.

Michel, A. D., Lundstrom, K., Buell, G. N., Surprenant, A., Valera, S., and Humphrey, P. P. (1996a). *Br. J. Pharmacol.* **117,** 1254.

Michel, A. D., Lundstrom, K., Buell, G. N., Surprenant, A., Valera, S., and Humphrey, P. P. (1996b). *Br. J. Pharmacol.* **118,** 1806.

Morr, J., Rundstrom, N., Betz, H., Langosch, D., and Schmitt, B. (1995). *FEBS Lett.* **368,** 495.

Murhammer, D. W., and Goochee, C. F. (1990). *Biotechnol. Progr.* **6,** 391.

Nielsen, L. K., Smyth, G. K., and Greenfield, P. F. (1991). *Biotechnol. Progr.* **7,** 560.

Nielsen, L. K., Smyth, G. K., and Greenfield, P. F. (1992). *Cytotechnology* **8,** 231.

Olkkonen, V. M., Lilijeström, P., Garoff, H., Simons, K., and Dotti, C. G. (1993). *J. Neurosci. Res.* **35,** 445.

O'Reilly, D. R., Miller, L. K., and Luckow, V. A. (1994). Baculovirus Expression Vectors: A Laboratory Manual Oxford University Press, New York.

Paine, M. J., Gilham, D., Roberts, G. C., and Wolf, C. R. (1996). *Arch. Biochem. Biophys.* **328,** 143.

Pickering, D. S., Taverna, F. A., Salta, M. W., and Hampson, D. R. (1995). *Proc. Natl. Acad. Sci. USA* **92,** 12090.

Pregenzer, J. F., Im, W. B., Carter, D. B., and Thomsen, D. R. (1993). *Mol. Pharmacol.* **43,** 801.

Primus, R. J., Yu, J., Xu, J., Hartnett, C., Meyyappan, M., Kostas, C., Rambhdran, T. V., and Gallager, D. W. (1996). *J. Pharmacol. Exp. Ther.* **276,** 882.

Radford, K. M., D. W., and Buell, G., D. W. (1997). Expression of ligand-gated ion channels using Semliki Forest virus and baculovirus. *Methods Enzymol.*

Radford, K. M., Cavegn, C., Bertrand, M., Bernard, A. R., Reid, S., and Greenfield, P. F. (1997a). *Cytotechnology* **24,** 73.

Radford, K. M., Reid, S., and Greenfield, P. F. (1997b). *Biotechnol. Bioeng.* **56,** 32.

Radford, K. M., Virginio, C., Surprenant, A., North, R. A., and Kawashima, E. H. (1997c). *J. Neurosci.* **17,** 6529.

Ross, S. M., Taverna, F. A., Pickering, D. S., Wang, L. Y., MacDonald, J. F., Pennefather, P. S., and Hampson, D. R. (1994). *Neurosci. Lett.* **173,** 139.

Shotkoski, F., Zhang, H. G., Jackson, M. B., and French-Constant, R. H. (1996). *FEBS Lett.* **380,** 257.

Smith, G. E., Summers, M. D., and Fraser, M. J. (1983). *Mol. Cell. Biol.* **3,** 2156.

Sydow, S., Kopke, A. I. E., Blank, T., and Spiess, J. (1996). *Mol. Brain Res.* **41,** 228.

Taverna, F. A., and Hampson, D. R. (1994). *Eur. J. Pharmacol.* **266,** 181.

Taverna, F. A., Wang, L. Y., MacDonald, J. F., and Hampson, D. R. (1994). *J. Biol. Chem.* **269,** 14159.

Uchino, S., Nakajima-Iijima, S., Okuda, K., Mishina, M., and Kawamoto, S. (1997). *Neuroreport* **8,** 445.

Wang, D. X., and Abood, L. G. (1996). *J. Neurosci. Res.* **44,** 350.

Werner, P., Kawashima, E., Hussy, N., Lundstrom, K., Buell, G., Humbert, Y., and Jones, K. A. (1994). *Mol. Brain Res.* **26,** 233.

Weyer, U., and Possee, R. D. (1991). *J. Gen. Virol.* **72,** 2967.

Witt, M. R., Westh-Hansen, S. E., Rasmussen, P. B., Hastrup, S., and Nielsen, M. (1996). *J. Neurochem.* **67,** 2141.

Yang, J. D., Gecik, P., Collins, A., Czarnecki, S., Hsu, H. H., Lasdun, A., Sundaram, R., Muthukumar, G., and Silberklang, M. (1996). *Biotechnol. Bioeng.* **52,** 696.

CHAPTER 8

Recombinant Adenovirus–Mediated Expression in the Nervous System of Genes Coding for Ion Channels and Other Molecules Involved in Synaptic Function

Markus U. Ehrengruber, Markus Lanzrein, Youfeng Xu, Mark C. Jasek, David B. Kantor, Erin M. Schuman, Henry A. Lester, and Norman Davidson

Division of Biology, California Institute of Technology
Pasadena, California, USA

I. Introduction

Recombinant (and thus nonreplicating) adenovirus is at present the gene transfer vector of choice for introducing genes of interest into postmitotic neurons *in vitro*, in slice cultures, and *in vivo*. The per-cell efficiency can be close to 100%. By controlling the multiplicity of infection and choosing a suitable promoter, the level of expression can be controlled. Expression can persist for weeks under favorable conditions. Cytotoxic effects are sometimes a problem, but generally not severe.

Adenovirions enter the cells by a complex pathway, usually by initial attachment to an integrin (Nemerow *et al.*, 1994), followed by endocytosis into endosomes. The viral DNA is transported to the nucleus where it exists as a linear molecule of approximately 36–40 kb, with a covalently attached protein at each end (Graham, 1984). This protein presumably contributes to nuclease resistance of the DNA and thus to the persistence of expression. In proliferating cells, the DNA is diluted during cell division.

Transgenic mice, either deleting (i.e., "knockouts") or overexpressing a gene of interest, are an attractive tool for studying the effects of genes in neurobiology. In the many cases where dominant-negative mutants exist, overexpression of this mutant is a useful alternative to gene knockouts as a method to study the effect of reduced functional activity of an endogenous gene. The recombinant adenovirus system can be established in a somewhat shorter period of time than can a transgenic mouse. If an introduced gene has a lethal embryonic phenotype, the adenovirus system is much more convenient for studying the adult phenotype. On the other hand, the volume of tissue *in vivo* that can be infected with adenovirus is limited by the number of injections and volume per injection that an animal will tolerate, because the virus does not spread appreciably through the extracellular space. Therefore, for a nonsecreted molecule such as a receptor or an ion channel, the broad region of expression achieved with transgenic mice cannot be duplicated by adenovirus.

In this chapter, we focus mainly on the methods and genes that have been used in our laboratory. Our studies have been carried out in hippocampal cells in culture and in acute slices. Till now, we have studied several potassium channels and several nitric oxide synthase constructions.

II. Preparation of Recombinant Adenovirus

The goal is to prepare, by homologous recombination, a virus in which the essential *Ela* and *Elb* gene products have been inactivated by removing the sequence from 455 to 3333 base pairs (bp) and replaced by an expression cassette for the cDNA of interest (van Ormondt *et al.*, 1980). The *Ela* and *Elb* genes are essential for viral replication and for all late gene expression. Human embryonic kidney 293 (HEK293) cells were originally transformed by a fragment of adenovirus including the E1 genes (Graham *et al.*, 1977); these cells can thus supply the gene product in *trans* for recombinant virus replication. However, the E1 deleted viruses are nonreplicating in postmitotic cells such as neurons, although they may slowly replicate in some transformed cell lines such as HeLa cells.

The general strategy used by us and others is depicted in Fig. 1. After exhaustive overdigestion of the wild-type virus with enzymes that cleave 1–1.3 kb from the essential left end, the longer right arm (plus any undigested full-length virus) is separated from the shorter left end by either sucrose gradient centrifugation or gel electrophoresis. The transfer plasmid, either linearized or still circular, contains an

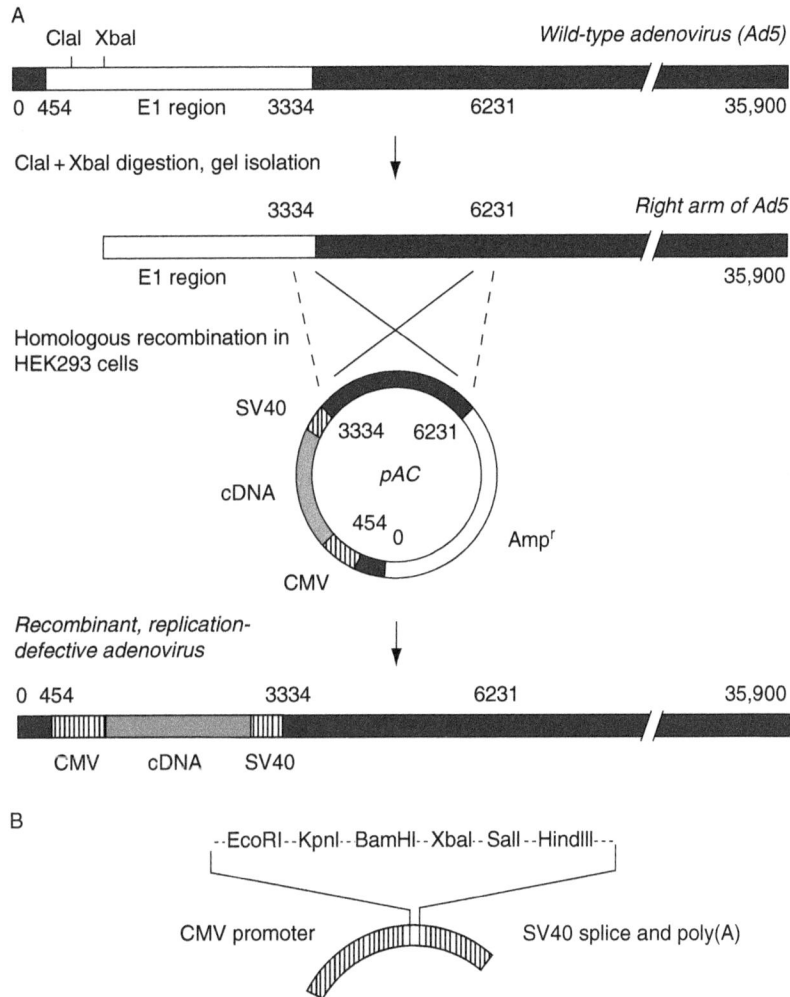

Fig. 1 Schematic representation of the construction of recombinant adenovirus. (A) The adenovirus Ad5 *dl*309 is cut with *Cla*I and *Xba*I; the resulting nonviable right arm is transfected into HEK293 cells together with the pAC adenoviral transfer vector (pACCMVpLpA) (from Gomez-Foix *et al.*, 1992). The plasmid contains adenoviral sequences (nucleotides 0–454 and 3334–6231), and the cDNA of interest under the control of a CMV promoter (760 bp) and upstream of an SV40 splice and poly-adenylation signal (470 bp). On homologous recombination between adenoviral nucleotides 3334–6231 of Ad5 and pAC, recombinant adenovirus is formed; it contains the necessary left-end packaging signal, but is replication-defective as El region genes (within nucleotides 455–3333) have been deleted in the recombination process. The recombinant virus can propagate only in HEK293 cells, which complement the factors encoded by the adenoviral El region (from Graham *et al.*, 1977). (B) Unique restriction sites in the polylinker of pAC.

expression cassette consisting of the cytomegalovirus (CMV) immediate early promoter, the inserted cDNA, and an SV40 splice and polyadenylation sequence. This cassette is flanked by 454 nucleotides of the far left end of the adenovirus and a segment spanning nucleotides 3334–6231 of the virus. The plasmids we used are depicted in Fig. 1 and are denoted pAC plasmids (Gomez-Foix *et al.*, 1992). When cells are cotransfected with pAC plasmid and the isolated long right arm of the virus, sufficient recombination takes place so that some recombinant virus is formed.

Bett *et al.* (1993) report that vectors with inserts resulting in viral DNA close to or less than a net genome size of 105% of that of the wild type, which is 35.9 kb, grow and are relatively stable.

Maps of the viruses we have used for most of our work are shown in Fig. 2. Ad5 *dl*309 has been extensively used by workers in the field (Jones and Shenk, 1979).

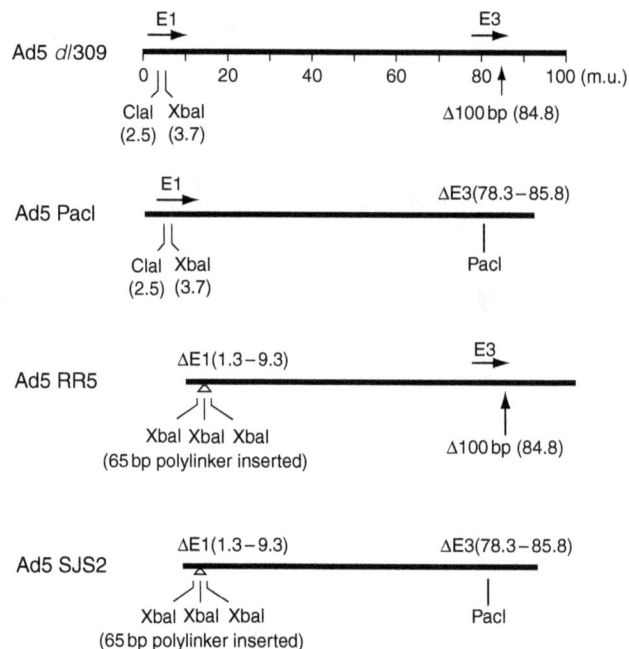

Fig. 2 Structures of adenoviruses used for recombinants. Ad5 *dl*309 is useful because it contains a unique *Xba*I site at 3.7 map units (m.u.) (100 map units = 35.9 kb). The mutation at site 84.8 is due to a spontaneous deletion from 83 to 85 m.u. and partial substitution with a fragment of foreign DNA. Sequencing shows that the net effect is a loss of 104 bp (from Bett *et al.*, 1995). Ad5 *Pac*I (Bett *et al.*, 1994), like Ad5 *dl*309, contains the wild-type E1 region, with an E3 deletion of 2.7 kb replaced by a unique *Pac*I site. In Ad5 RR5, the E1 region from 1.3 to 9.3 m.u. has been replaced by a polylinker, which includes three *Xba*I sites. The left arm of the virus was made by R. D. Gerard by recombination with the plasmid pACESHR. This plasmid contains the same adenovirus sequences as the pACCMV plasmids, with the CMV promoter and the SV40 splice and polyadenylation sequences replaced by the pUC19 polylinker. Ad5 SJS2 was made in our laboratory by S. J. Stary by recombination of the plasmid pACESHR with Ad5 *Pac*I. The right arm of the sequence is from Ad *dl*309.

When used in conjunction with transfer vectors that result in the substitution of the region from 455 to 3333 (Fig. 1) with the expression cassette, the cloning capacity is 4.7 kb. The virus Ad5 *Pac*I (Bett *et al.*, 1994) has an E3 deletion of 2.7 kb, so the cloning capacity is increased to 7.4 kb. Ad5 RR5 and Ad5 SJS2 have the same cloning capacities as Ad5 *dl*309 and Ad5 *Pac*I, respectively. The viruses Ad5 *dl*309 and Ad5 *Pac*I contain the E1 genes; they are therefore replication competent and cytotoxic in most mammalian cells. Ad5 RR5 and Ad5 SJS2 are deleted in the E1 region. They are nonreplicating in many cells; in some transformed cells, such as HeLa, that can supply a substitute for the E1 gene, there is slow replication. Ad5 *Pac*I and Ad5 SJS2 have a unique *Pac*I restriction site in place of the E3 region. As described later, an expression cassette may be introduced at this site by ligation. For example, we have ligated a green fluorescent protein expression cassette of length 0.9 kb into the *Pac*I site of Ad*Pac*I and of Ad5 SJS2. These viruses now have a cloning capacity of 6.5 kb in the E1 region. With such recombinants, one can determine which cells in a culture have been infected by observing the GFP fluorescence.

III. Techniques

A. Safety Considerations

Requirements of the National Institutes of Health (NIH) and Center for Disease Control (CDC) on handling adenovirus and adenoviral recombinants can be found in the current NIH *Guidelines for Recombinant DNA*. The current guidelines can be accessed on the Internet at http://www.ehs.psu.cdu/nih95-1.htm. Adenoviruses are classified in risk group 2 (RG2). This includes agents that are associated with human disease that is rarely serious and for which preventive or therapeutic interventions are often available. In practical terms, adenovirus is regarded as a moderate-risk pathogen.

Recommended precautions for RG2 agents are standard microbiological practices, laboratory coats, decontamination of all infectious wastes, limited access to working areas, protective gloves, posted biohazard signs, and class I or II biological safety cabinets used for mechanical and manipulative procedures that have high aerosol potential.

Of course, any adenovirus recombinant expressing a highly toxic protein should be treated as a special risk.

B. Preparation of Recombinant Adenovirus

1. Preparation of Right Arm of Viral DNA

The separation procedure removes a portion of the 5' end of the viral DNA extending beyond nucleotide 500. The inverted terminal repeat of length 102 nucleotides at each end, and the packaging signal, nucleotides 94–358, are essential for assembling a functional virus.

The viruses Ad5 *dl*309 and Ad5 *Pac*I both have a *Cla*I and an *Xba*I site. Ad5 RR5 and Ad5 SJS2 have three adjacent *Xba*I sites around position 450. Accordingly, these enzymes are used to release the left end.

Twenty micrograms of viral DNA (prepared as described later) is incubated with 100–200 Units of restriction enzyme for 2–3 h. The enzyme is added in two batches, 50–100 Units in the first hour of incubation and the remaining later. This decreases the effects due to thermal degradation of the enzyme. Note that it is very important to digest exhaustively, and that minor nuclease action at the cohesive sites is not deleterious. In our laboratory we prefer to separate the long right arm from the left end by 1% agarose gel electrophoresis. We believe this procedure to be more effective as well as time saving, compared to sucrose gradient centrifugation. For the subsequent gel isolation, we use GeneClean II (Bio101, La Jolla, CA), a silica matrix-based procedure. Vortexing and vigorous pipetting should be avoided to prevent shearing of the long DNA.

2. Cotransfection into HEK293 Cells

HEK293 cells are maintained and propagated under standard conditions [5% humidified CO_2/95% air (v/v) in Dulbecco's modified Eagle's (DME) high-glucose medium (Irvine Scientific, Santa Ana, CA) supplemented with 10% fetal calf serum (FCS), glutamine, and antibiotics (DME-complete)]. For cotransfection, HEK293 cells are grown in 60-mm plastic dishes to 50–70% confluency. Transfection is carried out using Lipofectamine (GIBCO-BRL, Gaithersburg, MD) according to the manufacturer's instructions. Briefly, 1.5 μg plasmid DNA and 1.5 μg right-arm viral DNA (molar ratio of 3–4 to 1) are assembled in 300 μl DME, free of serum and antibiotics (DME). Fifteen microliters of Lipofectamine (30 μg) reagent is diluted in 300 μl DME. The Lipofectamine solution is then added dropwise to the DNA solution.

This mixture is incubated at room temperature for 20 min, diluted to a total volume of 1.5 ml by adding 900 μl DME, and added to the HEK293 cells, previously rinsed once in DME. (This wash step must be done with maximal care because the HEK293 cells do not adhere tightly to the plastic.) After 3–5 h, 3.5 ml DME containing 7% FCS but no antibiotics is added. The cells are incubated overnight and the medium is then changed to DME-complete.

About 7–14 days after transfection, a few round and phase bright cells begin to appear, and will eventually die. This cytopathic effect (CPE) is a consequence of virus production; virions spread to neighboring cells, until all cells eventually become rounded. At this stage of full CPE, the cells and the medium are harvested and freeze–thawed three times to break the cells up and release virus particles. This primary lysate is extracted with an equal volume of chloroform. The chloroform extraction dissociates virion aggregates and removes any possible microbial contamination from the lysate.

3. Plaque Purification

The primary viral lysate from the preceding step will usually still contain wild-type virus. Therefore, it is advisable to purify recombinant virus by isolating single plaques that presumably originate from a single infectious virus particle. In our laboratory, we usually perform two rounds of plaque purification and screen the plaques after each round by PCR (polymerase chain reaction). However, we recommend that, when possible, the primary lysate be tested for functional expression prior to starting the laborious purification procedure. Usually, the primary lysate has a virus titer that is sufficiently high to achieve expression of the transgene in cells other than HEK293 (e.g., CHO) cells. If expression is weak, the proportion of wild-type virus might be high.

For plaque purification, HEK293 cells are seeded into 60-mm dishes (four dishes per sample under test) and grown to 90% confluency. Add 200 μl from a dilution series (four dilutions, 10^4- to 10^7-fold) of the primary lysate in DME-complete to the dishes and incubate for 1 h. The dishes need to be rocked gently every 10 min. Then add 2–3 ml DME-complete and incubate the cells for another 3–5 h before pouring the agarose overlay. Prepare the agarose overlay by melting 1.5% (w/v) low melting point agarose (GIBCO-BRL) in water and mix it with an equal volume of 2× DME-complete (2× DME, 2× glutamine, 2× antibiotics, 10% FCS). Let the mixture cool to 37 °C, remove the medium from the cells, and add 5 ml of the DME–agarose mix to each 60 mm dish. Incubate the dishes at 4 °C for 5–10 min to let the agarose solidify before placing the dishes back into the 37 °C incubator.

Plaques will appear after 4–7 days as small holes (~100-μm diameter) surrounded by round cells. The plaques are marked and harvested with a sterile cotton-plugged Pasteur pipette. This material is suspended in 1 ml of DME-complete, and subjected to three freeze–thaw cycles. Each plaque is amplified in a 60-mm dish containing a confluent monolayer of HEK293 cells by adding 0.5 ml of the plaque suspension. After 3–5 days, when all cells exhibit CPE, cells and medium are harvested, freeze–thawed three times, and chloroform extracted.

4. Screening

Lysates derived from plaques can be screened in various ways. Expression can be tested by infecting cells, provided that an assay is available (e.g., immunostaining or functional tests such as electrophysiologic recording, enzyme assays). Alternatively, virus DNA is isolated from lysates and blotted onto nylon membranes (Southern blots), which are hybridized with a probe specific for the foreign cDNA insert or wild-type virus (e.g., the E1 region). Isolated viral DNA can also be screened by PCR with primers specific for the foreign cDNA insert or wild-type virus. In our laboratory, we routinely screen by PCR analysis.

Viral DNA is isolated by digestion of the lysate with proteinase K. Assemble 0.5 ml of lysate with 5 μl 100× TE (1 M Tris–HCl, 0.1 M EDTA, pH 7.8), 12.5 μl of 20% sodiumdodecyl sulfate (SDS) and 2.5 μl of proteinase K (20 mg/ml,

Boehringer Mannheim, Germany). Incubate at 55 °C for 45 min. Then add 0.5 ml of phenol/chloroform/isoamyl alcohol (25:24:1), vortex thoroughly, and centrifuge in a microfuge at 14,000 rpm for 5 min. Carefully remove the upper phase and transfer it to a fresh tube. Add 0.5 ml of chloroform/isoamyl alcohol (24:1), vortex, and centrifuge as earlier for 5 min. Collect the upper phase and add 50 μl of 3 M sodium acetate and 1.2 ml of 95% (v/v) ethanol. Vortex and centrifuge in a microfuge at 14,000 rpm for 20 min. Wash the pellet once with ∼0.5 ml 70% ethanol and let the pellet dry on the bench for 10 min. Dissolve the pellet in 40 μl of sterile water and use 1 μl for PCR analysis.

For PCR analysis, we use primers specific for the inserted DNA and/or the CMV promoter and the SV40 splice sequence. To check for the presence of wild-type virus (for the cases of Ad5 *dl*309 or Ad5 *Pac*I), primers specific for the E1 region are used. In this case, a faint band can usually be observed even if the recombinant virus is pure. This band presumably originates from trace amounts of DNA from HEK293 cells that contain the E1 genes. If the recombinant virus is pure, the band is about 50–100 times less intense than when wild-type viral DNA is used as template. (Different primers are used to detect Ad5 RR5 and Ad5 SJS2.)

CMV forward and SV40 splice reverse primers used for the insert are 5′-GTG′G-GA′GGT′CTA′TAT′AGC′AG-3′ and 5′-ATC′TCT′GTA′ GGT′AGT′TTG′TCC-3′, respectively. The primers for the E1 region are as follows: forward primer, 5′-GTG′AGT′TCC′TCA′AGA′GGC′C-3′; reverse primer, 5′-ACC′CTC′TTC′A TC′CTC′GTC′G-3′ corresponding to nucleotides 480–498 and 976–958 of the adenovirus genome (GenBank accession no. X02996).

Plateau saturation effects of PCR can be avoided by diluting DNA samples before PCR and/or by reducing the number of cycles of amplification so that ethidium bromide-stained gel bands of intermediate intensity are produced. In this way, a semiquantitative estimate of the amount of recombinant viral DNA and ratio of recombinant to wild-type virus can be achieved.

5. Large-Scale Amplification

Lysate from recombinant viral lysate or wild-type virus is used to infect HEK293 cells grown in large culture flasks (e.g., T-150 cm^2 flasks; Corning, Cambridge, MA). About 300 μl of viral lysate is sufficient to infect a 150-cm^2 flask with a confluent monolayer of cells. After 2–3 days, all the cells exhibit CPE and should then be harvested. A large fraction of the infectious particles is associated with the cells. This provides a convenient way to concentrate the virus: the cells are removed from the bottom of the flask and the cell suspension is centrifuged. The cell pellet is then resuspended in a small volume of DME-complete. One 150-cm^2 flask will produce ∼3 × 10^9 plaque-forming units (pfu) of virus. The cell suspension is freeze–thawed three times and optionally extracted with an equal volume of chloroform. Aliquots should be stored at −80 °C. The presence of cell debris enhances stability for long-term storage. If the virus is chloroform extracted, bovine serum albumin (BSA; to 0.1%) and/or glycerol (to 10%) may be added to increase the stability. We prefer glycerol alone.

6. Virus Purification with a CsCl Isopyknic Gradient

CsCl purification is recommended if highly pure virus is needed (e.g., for *in vivo* injections). In our experience, CsCl-purified virus has fewer toxic effects also on cultured cells and is thus suited for infection of fragile cells such as primary neuronal cultures. The virus lysate used for purification should be as concentrated as possible. Place 3 ml of 3.2 M CsCl ($\rho = 1.4 \text{ g/cm}^3$) into an ultracentrifuge tube (e.g., Beckman polyallomer, 14×89 mm, Beckman Instruments, Fullerton, CA) and overlay with 3 ml of 1.6 M CsCl ($p = 1.2 \text{ g/cm}^3$). Carefully layer 5 ml of chloroform-extracted viral lysate on top, and centrifuge in a swinging-bucket rotor (e.g., Beckman SW41) at 37,000 rpm ($\sim$245,000 $\times$ g) for 16 h at 4 °C.

The band containing the virions appears white opalescent and is located $\sim$1.1 cm from the bottom of the tube. A faint band can be observed $\sim$0.3 cm above the virions; it contains empty particles. Discard the top of the gradient (including the empty particle band) and then remove the band containing the virions. Dilute with an equal volume of Ad solution (137 mM NaCl, 10 mM HEPES, 5 mM KCl, 1 mM $MgCl_2$, pH 7.6). The virus is stable in CsCl for a few months at 4 °C.

CsCl, which is toxic to cells, can be removed by gel filtration using a 9-ml Sephadex G-25 column (Pharmacia PD-10, Piscataway, NJ). Equilibrate the column with Ad solution, apply the virus and elute with Ad solution, collecting 1.5-ml fractions. The fractions can be checked for virus content by measuring absorption at 260 nm. Alternatively, each fraction can be titered as described later. In our experience, fractions 2 and 3, or 3 and 4 contain the bulk of the virus. The virus can be stored in the eluate when glycerol (to 10%) is added. The 1 mM Mg^{2+} in all solutions is important for stabilizing virions.

7. Isolation of Viral DNA

CsCl-purified virus is used to prepare DNA for transfection. The virions are digested with proteinase K, which removes the capsids. Dilute 100$\times$ TE (1 M Tris–HCl, 0.1 M EDTA, pH 7.8) to 1$\times$ in virus solution and add 20% SDS to a final concentration of 0.5% and proteinase K to a final concentration of 100 μg/ml. Incubate at 55 °C for 45 min. Then add an equal volume of phenol/chloroform/isoamyl alcohol (25:24:1), vortex thoroughly, and centrifuge in a microfuge at 14,000 rpm for 5 min. Carefully remove the upper phase and transfer it to a fresh tube. Add an equal volume of chloroform/isoamyl alcohol (24:1), vortex, and centrifuge as above for 5 min. Collect the upper phase and add 0.1 volume of 3 M sodium acetate and 2.2 volumes of 95% ethanol. Vortex and centrifuge in a microfuge at 14,000 rpm for 20 min. Wash the pellet once with 70% ethanol and let the pellet dry on the bench for 10 min. Dissolve the pellet in an appropriate volume of sterile water to a final concentration of $\sim$1 mg/ml.

8. Determination of Viral Titer

a. Plaque Assay

The procedure described earlier for plaque purification can be used to determine the titer of a virus. Several virus dilutions (e.g., 10^7- to 10^9-fold) should be used to infect HEK293 cells in 60-mm dishes, which are then overlaid with agarose. The growing plaques are visible and can easily be counted 5 days after infection.

b. Cell Lysis Assay

This assay may be preferable because it is less laborious and cheaper than the plaque assay. Up to three viruses can be titered simultaneously in triplicate series in a single multiwell plate. In our experience, this assay gives very reproducible results.

Pipette 50 μl of DME-complete into each well of a 96-well plate. Dilute the virus 10^5-fold, add 25 μl to the first well, and mix by pipetting. Take 25 μl from the first well, add it to the second well, and mix. Transfer 25 μl from the second to the third well, etc. Do not add diluted viral solution to the last well; remove 25 μl from the second to last. This dilution series can conveniently be done in triplicate or even more parallel series. Add 50 μl of HEK293 cell suspension ($\sim 10^6$ cells/ml) to each well. The plate is then incubated at 37 °C and cells are fed with 50 μl/well DME-complete at days 3, 6, and 9. After approximately 10 days, the titer can be determined by checking for CPE. It is assumed that the well with the highest dilution of virus that still exhibits CPE, initially contained one infectious particle. The titer is then calculated as $3^n \times 40 \times 10^5$ pfu/ml, where n is the number of wells showing CPE.

9. Cloning into *Pac*I Site of Ad5 *Pac*I and Ad5 SJS2 by Ligation

In these viruses (Fig. 2), a 2.7-kb deletion in E3 has been replaced by a unique *Pac*I site. Thus, in addition to a gene introduced into the E1 region by recombination, a gene can be introduced by ligation into the *Pac*I site.

To prepare the viral vector, 2 μg of Ad5 *Pac*I or SJS2 DNA is digested with 20 Units *Pac*I (New England Biolabs) for 2 h. The enzyme is added in two batches, 10 Units for the first hour of digestion and then the rest. On the second addition of *Pac*I, 5 Units of shrimp alkaline phosphatase (United States Biochemical, Cleveland, OH) is added to the digestion mix to dephosphorylate the ends. The two DNA fragments in the digestion mix are purified by using GeneClean II (Bio101) and eluted in doubly distilled H_2O. Vortexing and vigorous pipetting should be avoided to prevent shearing of the long viral DNA.

The expression cassette consisting of the gene insert with a promoter and poly (A) site needs to be flanked by two *Pac*I sites. This can be achieved by PCR or by linker addition. We prefer the former strategy, which is quick and simple. Two oligonucleotides containing the *Pac*I site plus the 5′ and 3′ sequence of the expression cassette are used as primers. There should be a 5′ extension of five or more nucleotides to ensure subsequent digestion at the *Pac*I sites. PCR is carried out with a high-fidelity PCR kit. We use the Expand High Fidelity PCR System (Boehringer Mannheim). The correct PCR product is purified by the Qiagen PCR purification kit. Two micrograms of the PCR product is digested with 20 Units *Pac*I, then gel purified and eluted in doubly distilled H_2O.

The overnight ligation is carried out by mixing 2 μg DNA insert and 2 μg total of the two dephosphorylated viral vector fragments under ligation conditions at 16 °C. When the high concentration T4 DNA ligase (5 U/μl, Boehringer Mannheim) is used and the ligation volume is small (<20 μl), the ligation mix can be used directly to transfect HEK293 cells without precipitation and desalting.

The transfection and further plaque purification screening procedures are basically the same as previously described for homologous recombination. Individual plaques are screened by PCR and/or *Pac*I digestion. To check for multiple insertions in the *Pac*I site, a long PCR reaction using primers adjacent to the E3 region is used to measure the length of the insert. The primers for the E3 region are as follows: forward primer, 5'-CCT'GAT'TCG'GGA'AGT'TTA'CCC'-3'; reverse primer, 5'-GTG'CTG'CTG'AAT'AAA'CTG'GAC'-3' corresponding to nucleotides 28,030–28,050 and 30,911–30,891 of the wild-type adenovirus genome (GenBank accession no. M73260).

The ligation method can also be applied for inserting genes into the *Xba*I sites of Ad5 RR5 and SJS2. For this purpose, a plasmid such as pZS2 (J. Sheng, personal communication, 1996) which contains the short right arm of adenovirus and an expression cassette with the cDNA of interest terminated by an *Xba*I site is ligated to *Xba*I-digested, dephosphorylated Ad5 RR5 or SJS2.

10. Toxicity Problems

We, and others (Durham *et al.*, 1996; Jordan *et al.*, 1995), have observed the death of neurons and other cells *in vitro*, infected with recombinant adenoviruses. The magnitude of the effect depends on the particular cells and culture conditions, the purity of the virus, the particular gene product being expressed, and its level of expression. Expression level depends on the promoter system being used and on the multiplicity of infection. It is possible but not certain that small amounts of expression of other viral gene products in E1- and E1 plus E3-deleted recombinants contribute to toxicity. If this is the case, the new "gutless" vectors (Parks *et al.*, 1996) will be an improvement. In some cases, addition to the culture medium of a specific inhibitor of the gene product improves viability while permitting accumulation of the gene product. The inhibitor is withdrawn at an appropriate time for observation. Half-lives of infected cells are reported to be reduced so as to range from several days to weeks, depending on all the factors mentioned.

For each particular system, by adjusting the level of expression, one can usually discover a window of time during which the physiologic effects of interest can be observed before cytotoxicity becomes a factor.

IV. Expression of Potassium Channels

To our knowledge, the only ion channel genes that have been studied by electrophysiology after adenovirus-mediated gene transfer are voltage- and G-protein-gated potassium (K^+) channels. The inactivation-removed *Drosophila Shaker B* channel has been expressed *in vitro* and *in vivo* by this method (Johns *et al.*, 1995). We have similarly expressed the *Drosophila Shaker H4* cDNA and several of the heteromeric G-protein-gated inward rectifier K^+ channel (GIRK) cDNAs (Ehrengruber *et al.*, 1997).

A. Preliminary Transfection Tests

Because the preparation of adenovirus is a lengthy procedure, it is useful to test the ion channel gene after cloning into the pAC transfer plasmid before proceeding to make the virus. This is done using standard cotransfection procedures. For example, we use Lipofectamine transfection reagent (GIBCO-BRL) according to the directions of the manufacturer. For a 35-mm Petri dish of semiconfluent CHO cells, we use only 0.25 µg of the GIBCO-BRL (Green Lantern) GFP or 0.1 µg of the Clontech (Palo Alto, CA) pEGFP-1 as a cotransfected reporter gene, and up to 1.5 µg of the necessary plasmids. For example, for a mixture of GIRK1, GIRK2, and muscarinic type 2 acetylcholine receptor (m2AChR) we use 0.5 µg of each. One to 2 days after transfection, 25–50% of the cells are strongly fluorescent; when these fluorescent cells are studied by whole-cell clamping, essentially 100% of them express the expected ion channel type, provided the cDNAs have not undergone mutations during preparation that make them nonfunctional. Recall that GIRKs function best as heteromers; both with plasmids and virus we see good expression with GIRK1 plus GIRK2 or GIRK1 plus GIRK4, but not with GIRK1 alone. Data illustrating typical results are given in Table I.

Because the efficiency of transient transfection into CHO cells is between 20% and 50%, the catalytic activity of nitric oxide synthase constructs and many other enzymes for which there are good assays can be tested in mass cultures. In this case the reporter gene is unnecessary, but useful to indicate that the transfection has been reasonably successful.

Generally speaking, we find it very useful to test all pAC plasmids in this manner before making recombinant virus.

B. Expression of K^+ Channels by Recombinant Adenovirus

Shaker H4 and the cDNAs for GIRKs 1, 2, and 4 were cloned into the pAC vector by standard molecular biology and recombined with Ad5 *dl*309 (*Shaker*; AdH4) or Ad5 *Pac*I (AdGIRK 1, 2, and 4). The serotonin type 1A receptor (5HT$_{1A}$R) was cloned by ligation into Ad5 RR5 (Ad5H$_1$AR). In most cases, the

Table I
Lipofectamine–Mediated Expression of Potassium Channels in CHO Cells[a]

Channel type	Peak current (nA)	Current density (pA/pF, $\mu A/cm^2$)	Channel density per μm^2
Shaker H4[b]	2.9 ± 0.2 (20)	218 ± 42 (20)	2.4
GIRK1 + 2[c]	-0.5 ± 0.1 (10)	-41 ± 10 (10)	3.1
GIRK1 + 4[c]	-1.1 ± 0.1 (39)	-126 ± 13 (39)	9.6

[a]Means $\pm$ S.E.M. (*n*) assayed 1–2 days after transfection. Cells in a 35-mm dish (10–20% confluent) were transfected with 0.13–0.5 μg DNA per pAC plasmid carrying a K^+ channel cDNA. For GIRKs, an equal amount of m2AChR in pMT2 (Kaufman *et al.*, 1989) was cotransfected. In addition, all cells were cotransfected with 0.25 μg Green Lantern (GIBCO-BRL) to allow for selection of successfully transfected, fluorescent cells, which were used for electrophysiologic recording.

[b]*Shaker* peak currents measured at +30 mV as described in Fig. 3 (legend): E_K was −82.9 mV. The channel density was estimated based on a specific membrane capacity of 1 $\mu F/cm^2$ (Hille, 1992), a single-channel conductance of 16 pS, and a channel open probability of 0.5 (from Karschin *et al.*, 1992).

[c]GIRK currents induced by agonists as described in Fig. 4 (legend) at −120 mV and 25 mM $[K^+]_o$ ($E_K \sim$ −44 mV). In the absence of agonist, some CHO cells showed relatively high basal GIRK currents of up to ∼3 nA. The channel density was estimated based on a specific membrane capacity of 1 $\mu F/cm^2$ and values determined for GIRK1 + 5 channels, that is, a single-channel conductance of 40 pS and a channel open probability of 0.043 (from Schreibmayer *et al.*, 1996).

functionality of the construct was confirmed in the plasmids by cotransfection as described earlier.

Figure 3 gives examples of *Shaker* currents induced by infection with AdH4. The inset demonstrates that the level of infection (in atrial and ventricular myocytes in this instance) increases with time postinfection up to 4 days and then levels off.

Several examples of the expression of GIRK currents by mixed cotransfection plus adenoviral infection of CHO cells and by infection of embryonic day 18 (E18) hippocampal neurons are shown in Fig. 4. Figure 4A shows that comparable agonist-activated GIRK currents can be induced for adenovirus-infected CHO cells either by cotransfection with m2AChR (top left) or by infection with Ad5HT$_{1A}$R. Table II summarizes much of these data.

With cultured hippocampal neurons, Lipofectamine-mediated transfection efficiencies are low and thus difficult to use; DOTAP-mediated transfection (Boehringer Mannheim) has about 10% efficiency; however, the viral infection procedure gives close to 100% per cell infection frequency. G-protein stimulation of the GIRKs was via baclofen acting on endogenous GABA$_B$ receptors or by 5-HT stimulation of the 5HT$_{1A}$R introduced by adenovirus. Thus, these experiments show that at least three different transgenes can be effectively introduced into a single cell by adenoviral infection.

As the cloning capacity of Ad *Pac*I is sufficiently large, we have cloned individual expression cassettes for GIRK1 into the E1 region and GIRK4 into the E3 region (by ligation) of Ad *Pac*I and observed a comparable level of expression with a single virus.

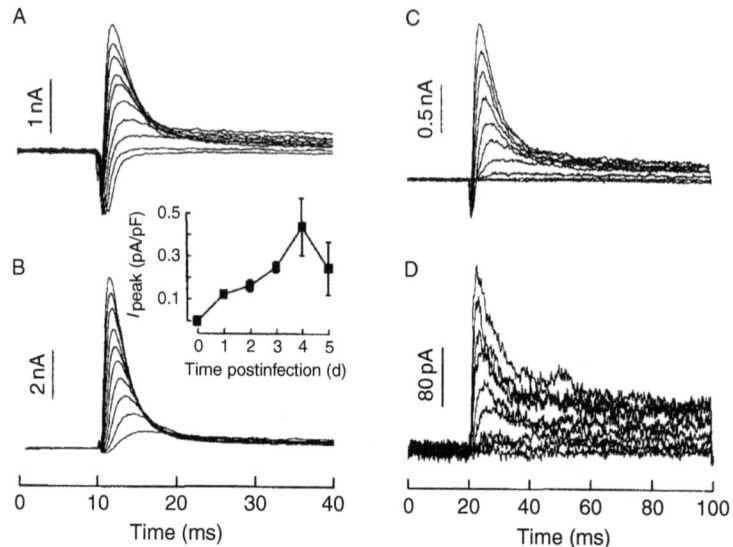

Fig. 3 Adenoviral expression of *Shaker* H4 currents in mammalian cells in primary cultures (A, B) or in mammalian cell lines (C, D). Atrial (A) and ventricular (B) myocytes from 2- to 6-day postnatal rats; CHO cells (C) and pancreatic βTC3 cells (D). Voltage-clamp experiments using a holding potential of -70 mV and leak subtraction, except for ventricular cells (-40 mV, no leak subtraction); after a 10- to 20-m s prepulse to -100 mV, the membrane potential was depolarized in 10-mV steps starting at -50 mV (-40 mV for ventricular myocytes). Currents in βTC3 cells were assayed in 10 mM TEA to suppress delayed outward rectifier K^+ currents. *Shaker* currents in the βTC3 cells are comparatively small because 10 mM TEA, which was added to block other K^+ channels, partially blocks *Shaker* channels. (*Inset*) Time-dependent expression of peak currents at $+30$ mV; means $\pm$ S.E.M. from 3 to 6 myocytes are shown for each postinfection day.

Note that the principal scientific point of our study of GIRK-infected neurons was to analyze quantitatively the mechanisms by which GIRK activation reduces neuronal excitability (Ehrengruber *et al.*, 1997). The large level of expression achieved by adenovirus-mediated overexpression made this possible.

V. Use of Adenovirus in Acute Hippocampal Slice Physiology

The hippocampal slice has been the system of choice in the study of long-term potentiation (LTP) as a model for synaptic modification during learning and memory (Bliss and Collingridge, 1993). Slices are normally used for physiology within a few hours of preparation; however, in appropriately supplemented medium they appear healthy by visual examination and retain their electrophysiologic properties for at least 30 h.

Gene transfer methods for expression of a new gene and for enhancement or inhibition of endogenous genes provide an additional tool in the study of synaptic

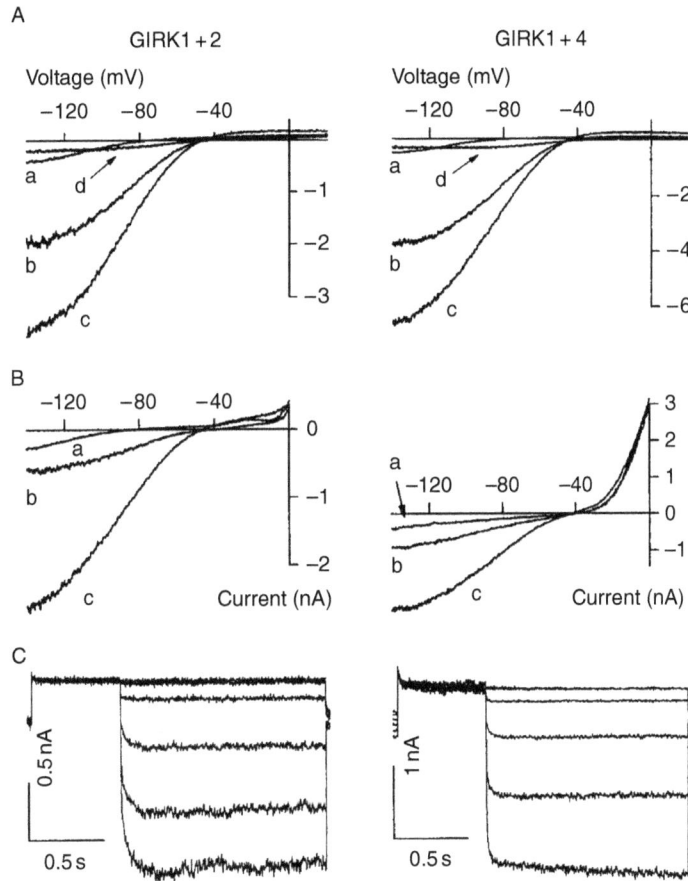

Fig. 4 Adenoviral expression of GIRK channels. GIRK currents in infected cultures of CHO cells (A) and hippocampal neurons from E18 rats (B). Cells were coinfected with AdGIRK1 + 2 (left: for the CHO cell. GIRK1 was cotransfected using Lipofectamine) and AdGIRK1 + 4 (right); CHO cells were additionally transfected with seven transmembrane helix receptor (left, m2AChR; right, coinfection with Ad5HT$_{1A}$R). Currents were assessed, using 2-s voltage ramp protocols, 1–2 days postinfection in 5.4 mM $[K^+]_o$ (a), and 25 mM $[K^+]_o$ both in the presence (c) and absence (b) of agonist (5 μM ACh and 50 μM serotonin for CHO cells, 100 μM baclofen for neurons); 0.5 mM Ba^{2+} blocks the basal and agonist-activated GIRK currents (in 25 mM $[K^+]_o$ (d)). (C) Time-course of agonist-activated GIRK1 + 2 currents in CHO cells (left) and GIRK1 + 4 currents in hippocampal neurons (right). After a 0.6-s prepulse to 0 mV, the membrane potential was stepped to test potentials in 30-mV increments between −120 mV (bottom trace) and 0 mV (top trace). Agonist-activated currents have been isolated by subtracting traces in the absence of agonist.

properties of slices. The technical questions that arise in using recombinant adenovirus for this purpose include the following: (1) Will there be sufficient gene expression from injected virus within the 24- to 30-h lifetime of the slice? (2) Since virus does not spread far through the extracellular space of brain tissue,

Table II

Adenovirus-Mediated Expression of Potassium Channels in Primary and Secondary Cell Cultures[a]

Channel type and host cell	Peak current (nA)	Current density (pA/pF, $\mu A/cm^2$)	Channel density per μm^2
Shaker H4			
Atrial myocytes (rat)	5.2 ± 0.8 (14)	216 ± 37 (14)	2.4
Ventricular myocytes (rat)	4.2 ± 0.8 (9)	354 ± 96 (9)	3.9
βTC3 cells (mouse)[b]	0.9 ± 0.2 (7)	199 ± 61 (7)	2.2
CHO cells (hamster)	0.9 ± 0.3 (6)	87 ± 33 (6)	1.0
GIRK1 + 2			
Hippocampal neurons (rat)[c]	-1.7 ± 0.2 (6)	-76 ± 9 (6)	55
GIRK1 + 4			
Hippocampal neurons (rat)[c]	-1.6 ± 0.5 (4)	-55 ± 17 (4)	4.0
CHO cells (hamster)	-0.6 ± 0.2 (5)	-47 ± 20 (5)	3.5
Endogenous GIRK1 + 4[d]			
Atrial myocytes (rat)	-0.7 ± 0.1 (24)	-15 ± 2 (23)	1.1

[a]Means $\pm$ S.E.M. (*n*) of cells assayed 1–7 days postinfection (10^8–10^9 pfu/ml per virus). Atrial and ventricular myocytes were prepared from 2- to 6-day postnatal rats, and hippocampal neurons from E18 rats. Currents and channel densities were determined as described in Table I.

[b]*Shaker* H4 currents were isolated by a subtraction protocol, that is, inactivation of *Shaker* H4 currents using a holding potential of -30 mV, followed by the test pulses (from Leonard *et al.*, 1989).

[c]$E_K = -40.7$ mV.

[d]For comparison, GIRK1 + 4 currents activated by 5–10 μM ACh at -120 mV and 25 mM $[K^-]_o$ are given ($E_K = -43.5$ mV).

how can one achieve infection and gene expression over a volume sufficiently large for the problem being studied? Our experience so far (Kantor *et al.*, 1996) in these respects is described next.

Slices were infected by microinjection of viral stock (usually about 10^9 pfu/ml) into the extracellular space of the CA1 pyramidal cell layer with glass micropipettes with tips broken off (resistance <1 MΩ). For each of 10–20 injection positions distributed over CA1, multiple pressure injections (5–7 injections per site for 10–20 sites per slice for 100 ms at 5–10 psi) were made by advancing the electrode through the depth of the slice. The approximate volume of solution injected per site is about 0.05 μl, so that ca. 1 μl is needed per slice. For injection of an *Escherichia coli* LacZ reporter virus (AdLacZ), expression was detected within 8 h after injection and was strong by 18–24 h. As shown in Fig. 5A, fairly uniform LacZ expression was observed over a 2 × 1-mm region, which included most of the pyramidal cell layer of CA1; as shown in Fig. 5B, *lacZ* gene product was observed in the dendritic region as well as the pyramidal region. Figures 5C and D show that AdLacZ-infected slices show normal synaptic responses and LTP.

In our work (Kantor *et al.*, 1996), the scientific problem that was addressed was the role of endothelial nitric oxide synthase (eNOS) in generating NO, which functions as a retrograde messenger in some forms of LTP. Endothelial NOS is

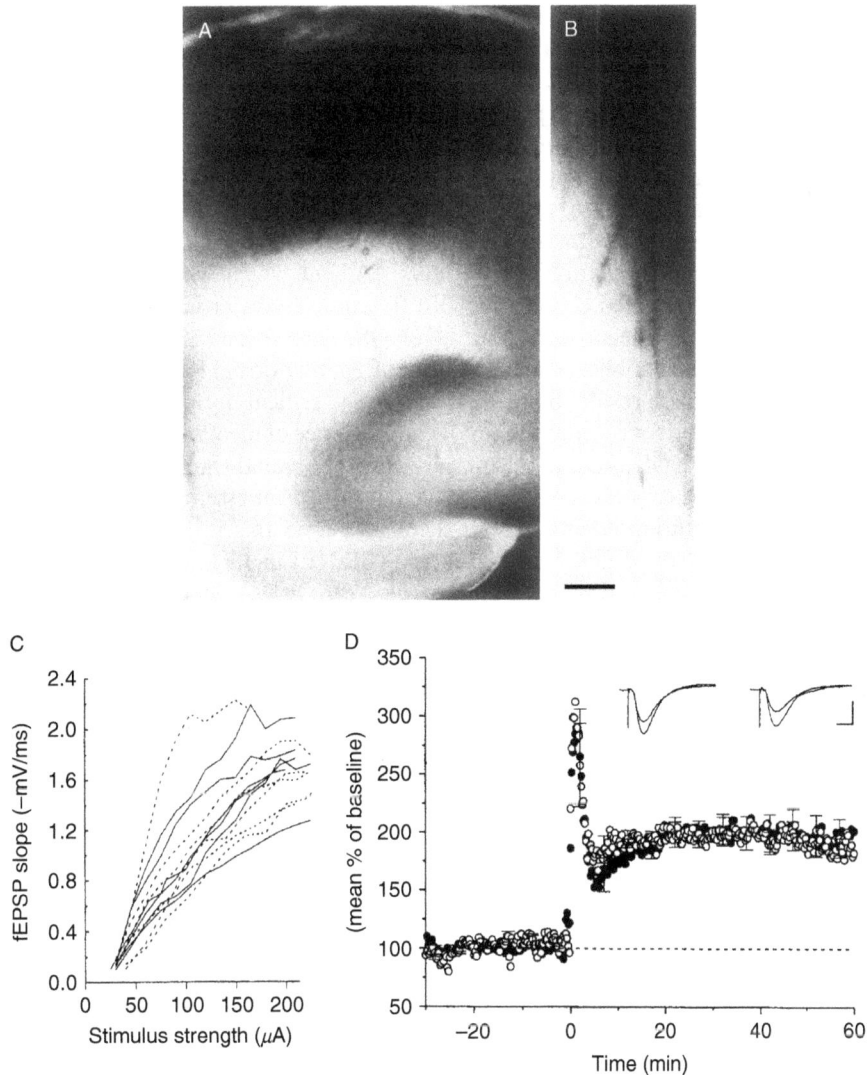

Fig. 5 Hippocampal slices infected with an AdLacZ reporter virus exhibit normal synaptic transmission and plasticity. (A) A slice infected in area CA1 by AdLacZ and stained for X-Gal 24 h after injection. (B) X-Gal labeling of individual pyramidal neurons and their associated dendrites; scale bar: 15 μm. (C) I/O relation for saline-injected (control, solid lines) and AdLacZ-infected (dashed lines) slices. The slope of each line was calculated, and a between-group comparison indicated that they are not significantly different from one another (mean slope $\pm$ S.E.M.: AdLacZ, 0.011 $\pm$ 0.002; saline, 0.01 $\pm$ 0.001). (D) Ensemble average of LTP experiments for both control and AdLacZ-infected slices. In control slices (filled circles), the mean field EPSP slope was -0.14 ± 0.01 mV/ms before LTP and -0.27 ± 0.05 mV/ms after LTP ($P < 0.01$). In AdLacZ-infected slices (open circles), the mean field EPSP slope was -0.16 ± 0.01 mV/ms before LTP and -0.28 ± 0.03 mV/ms after LTP ($P < 0.01$). (*Inset*) Two representative field EPSPs from a control slice (left) and an AdLadZ-infected slice (right), taken 10 min before and 50–60 min after the LTP induction protocol; calibration bar; 0.5 mV, 10 ms. [Reprinted with permission from Kantor *et al.* (1996). Copyright © 1997 American Association for the Advancement of Science.]

myristoylated and hence membrane localized. HMA, an inhibitor of myristoylation, blocks LTP. One question then is whether this is due to specific blocking of the
membrane attachment of eNOS, or blocking of membrane localization of other
myristoylated proteins. A fusion protein consisting of the extracellular and transmembrane domain of the T-cell type I membrane protein CD8 fused to a nonmyr-
istoylated mutant of eNOS was constructed and expressed in CA1, thus conferring
an alternate, myristoylation-independent means of membrane association. Slices
injected with this construct exhibited robust LTP, which was not abolished by the
myristoylation inhibitor, HMA, but which was abolished by NOS inhibitors. A
truncated eNOS (TeNOS) functions by heterodimerization with eNOS as a dominant-negative mutant of eNOS. Adenovirus-mediated expression of this mutant
also abolished LTP without affecting normal basal synaptic transmission. Overall,
these results point to eNOS as the critical myristoylated protein required for LTP
induction, and suggest the importance of membrane- rather than cytosolic-localized
eNOS as a critical component of the signal transduction events underlying LTP.

Korte *et al.* (1996) have developed an essentially identical method of adminis-
trating adenovirus to slices and used this approach to study the role of brain-
derived neurotrophic factor (BDNF) in LTP. A similar procedure has also been
described with vaccinia virus (Pettit *et al.*, 1994).

VI. Future Directions

One may anticipate many ways in which future developments will extend the
usefulness of recombinant adenovirus vectors for the study of neurobiological
phenomena. "Gutless" vectors with an enlarged cloning capacity and reduced
immunological reaction *in vivo* and possibly reduced toxicity *in vitro* are being
developed, principally for applications in gene therapy (Parks *et al.*, 1996). However, they will also provide enlarged cloning capacity for research purposes and
may prove to be even less cytotoxic for cells in culture as well as *in vivo*.

Acute slice protocols available now are not practical for the study of the
presynaptic effects of genes expressed in CA3 for synaptic communication between
CA3 and CA1. This is because the projections via the Schaffer collaterals from a
set of CA3 cell bodies to a set of CA1 dendrites are generally not contained in a
single slice; therefore, administration of adenovirus to the CA3 cell bodies of a slice
will not lead to expression in the axons synapsing on CA1 dendrites of the same
slice. One approach to overcome this difficulty may be the use of organotypic
cultures. It is reported that in a single slice of such a culture, 76% of the CA3
pyramidal neurons are synaptically connected to CA1 neurons in the same slice
(Debanne *et al.*, 1995).

Another promising direction is the development of systems providing inducible
gene expression. One such system based on the tetracycline control system
described by Gossen and Bujard (1992) and requiring two adenoviruses has been
described by Yoshida and Hamada (1997).

Acknowledgments

We thank Sheri L. McKinney for the primary cell cultures. Dr. Paulo Kofuji for subcloning of *Shaker* H4 into pAC, Catherine Lin for construction of AdH4 and Ad5HT$_{1A}$R, and S. Jennifer Stary for advice on the adenovirus technique. We are grateful to Drs. Arnold J. Berk, Frank L. Graham, Robert D. Gerard, and Jackie Sheng for much advice. We thank Dr. Berk for Ad5 *dl*309 and AdLacZ, Dr. Graham for Ad5 *Pac*I, Dr. Gerard for pAC, Dr. David E. Clapham for GIRK2 cDNA, Dr. John P. Adelman for GIRK4 cDNA, Dr. Brian Seed for CD8 cDNA, Dr. J. Sheng for AdRR5 and pZS2, and Dr. Simon Efrat for βTC3 cells. This work was supported by the National Institute of Mental Health, National Institute of General Medical Sciences, Human Frontier Science Program, the Swiss National Science Foundation (fellowships 81BE-40054 and 823A-042966 to M. U. E.), and the European Molecular Biology Organization (fellowship ALTF 168-1996 to M. L.).

References

Bett, A. J., Prevec, L., and Graham, F. L. (1993). *J. Virol.* **67**, 5911.

Bett, A. J., Haddara, W., Prevec, L., and Graham, F. L. (1994). *Proc. Natl. Acad. Sci. USA* **91**, 8802.

Bett, A. J., Krougliak, V., and Graham, F. L. (1995). *Virus Res.* **39**, 75.

Bliss, T. V. P., and Collingridge, G. L. (1993). *Nature* **361**, 31.

Debanne, D., Guérineau, N. C., Gähwiler, B. H., and Thompson, S. M. (1995). *J. Neurophysiol.* **73**, 1282.

Durham, H. D., Lochmuller, H., Jani, A., Acsadi, G., Massie, B., and Karpati, G. (1996). *Exp. Neurol.* **140**, 14.

Ehrengruber, M. U., Doupnik, C. A., Xu, Y., Garvey, J., Jasek, M. C., Lester, H. A., and Davidson, N. (1997). *Proc. Natl. Acad. Sci. USA* **94**, 7070.

Gomez-Foix, A. M., Coats, W. S., Baque, S., Alam, T., Gerard, R. D., and Newgard, C. B. (1992). *J. Biol. Chem.* **267**, 25129.

Gossen, M., and Bujard, H. (1992). *Proc. Natl. Acad. Sci. USA* **89**, 5547.

Graham, F. L. (1984). *EMBO J.* **3**, 2917.

Graham, F. L., Smiley, J., Russell, W. C., and Nairn, R. (1977). *J. Gen. Virol.* **36**, 59.

Hille, B. (1992). Ionic Channels of Excitable Membranes. Sinauer Associates, Sunderland, MA.

Johns, D. C., Nuss, H. B., Chiamvimonvat, N., Ramza, B. M., Marban, E., and Lawrence, J. H. (1995). *J. Clin. Invest.* **96**, 1152.

Jones, N., and Shenk, T. (1979). *Cell* **17**, 683.

Jordan, J., Ghadge, G. D., Prehn, J. H. M., Toth, P. T., Roos, R. P., and Miller, R. J. (1995). *Mol. Pharmacol.* **47**, 1095.

Kantor, D. B., Lanzrein, M., Stary, S. J., Sandoval, G. M., Smith, W. B., Sullivan, B. M., Davidson, N., and Schuman, E. M. (1996). *Science* **274**, 1744.

Karschin, A., Thorne, B. A., Thomas, G., and Lester, H. A. (1992). *Methods Enzymol.* **207**, 408.

Kaufman, R. J., Davies, M. V., Pathak, V. K., and Hershey, J. W. B. (1989). *Mol. Cell. Biol.* **9**, 946.

Korte, M., Griesbeck, O., Gravel, C., Carroll, P., Staiger, V., Thoenen, H., and Bonhoeffer, T. (1996). *Proc. Natl. Acad. Sci. USA* **93**, 12547.

Leonard, R. J., Karschin, A., Jayashree-Aiyar, S., Davidson, N., Tanouye, M. A., Thomas, L., Thomas, G., and Lester, H. A. (1989). *Proc. Natl. Acad. Sci. USA* **86**, 7629.

Nemerow, G. R., Cheresh, D. A., and Wickham, T. J. (1994). *Trends Cell Biol* **4**, 52.

Parks, R. J., Chen, L., Anton, M., Sankar, U., Rudnicki, M. A., and Graham, F. L. (1996). *Proc. Natl. Acad. Sci. USA* **93**, 13565.

Pettit, D. L., Perlman, S., and Malinow, R. (1994). *Science* **266**, 1881.

Schreibmayer, W., Dessauer, C. W., Vorobiov, D., Gilman, A. G., Lester, H. A., Davidson, N., and Dascal, N. (1996). *Nature (Lond.)* **380**, 624.

van Ormondt, H., Maat, J., and van Beveren, C. P. (1980). *Gene* **11**, 299.

Yoshida, Y., and Hamada, H. (1997). *Biochem. Biophys. Res. Commun.* **230**, 426.

CHAPTER 9

Transient Expression of Heteromeric Ion Channels

Alison L. Eertmoed, Yolanda F. Vallejo, and William N. Green

Department of Neurobiology
University of Chicago, Chicago
Illinois, USA

I. Update

On the whole, the methods and issues discussed in the original review are current today. Over time, the methods used to transfer ion channel subunit cDNAs into cells have become more refined and standardized. As discussed in the previous introduction, the methods used are usually dictated by the questions to be addressed. At one extreme is the use of gene transfer methods to study ion channel function and structure. Typically, the objective of these types of experiments is to insert the highest possible levels of the ion channel of interest into the plasma membrane. Examples of expression systems developed for this purpose are the baculovirus expression vectors for use in Sf9 insect cells (e.g., Sobolevsky *et al.*, 2009) and mammalian expression vectors for use in human embryonic kidney 293 (HEK 293) cells (e.g., Clayton *et al.*, 2009). Special variants of the HEK 293 cells have been developed that are glycosylation-deficient, which aids in protein crystallization (Reeves *et al.*, 2002).

At the other possible extreme the gene transfer methods are used to express genetically modified ion channels in an environment that best approximates that of the native channel complex. Ideally, the objective of these types of experiments is to swap the endogenous ion channel subunit of interest with an altered form (i.e., mutated or tagged version) and to avoid either over- or underexpression of the altered subunits. Perhaps the most efficient of these delivery systems are lentiviral vectors developed for protein and RNAi expression in mammalian cells (Lois *et al.*, 2002). Lentiviral vectors have been developed that allow RNAi knockdown of a native protein together with expression another ectopic protein (e.g., Leal-Ortiz *et al.*, 2008). To replace the endogenous proteins with the altered protein, a small hairpin RNA (shRNA) specific for the protein of interest is inserted behind the RNA H1 promoter. Also in the lentivirus vector is a mammalian expression promoter that drives the expression of the altered protein of interest that has been mutated so that the shRNA no longer binds to its mRNA (e.g., Schluter *et al.*, 2006). This strategy should be easily applied to the expression of ion channel subunits in a variety of cell types such as neurons and muscle fibers.

II. Introduction

A major advance in the study of ion channels has been the ability to express channels in foreign host cells after isolation of the channel genes. Initially, the system of choice for rapid expression of ion channel proteins was the *Xenopus* oocyte. Most of the electrophysiologic characterization of newly isolated ion channel subunit cDNAs has been performed using the *Xenopus* oocyte expression system. Unfortunately, the amount of ion channel protein produced using this system is relatively small. For biochemical and cell biological analysis, ion channels are generally expressed in cultured cell lines where production can be generated in enormous numbers of cells. The transfer of ion channel genes to cultured cells allows the synthesis of large amounts of these proteins, which previously could only be purified in minute quantities after much effort. Moreover, cultured cells have the additional advantage over *Xenopus* oocytes in providing an environment that is usually closer to that of the expressed ion channel's native environment.

Our objective in writing this chapter is *not* to provide a step-by-step description of the methods for ion channel gene transfer into cultured cell lines. Many excellent and up-to-date texts describe these methods in detail (e.g., Ashley, 1996; Rudy and Iverson, 1992). Instead, our intent is to describe problems and complexities we have encountered using these methods and to discuss changes we have made in the methods in order to counter problems and to achieve higher levels of expression. Because almost all ion channels are heteromeric proteins (see Green and Millar, 1995 for a review), we have devoted most of the results and following discussion to the problems encountered when attempting to express heteromeric ion channels.

Although cultured cell lines are widely used for the expression of ion channel proteins, the methods used are far from standard. Thus, the first problem one is faced with is sorting among the daunting number of options when considering how to express the ion channel of interest. Generally, the options are narrowed by the questions to be addressed. For example, if large quantities of the ion channel are needed for biochemical or structural analysis, a good choice may be to express the channel in Sf9 (*Spodoptera frugiperda* fall armyworm ovary) cells using a baculovirus expression vector. Or, if it is critical that the ion channel be expressed in a specific host cell type, such as a neuron, then one might choose to infect primary cultures of neurons using an adenovirus expression vector.

Usually the next problem encountered is how to introduce your ion channel subunit cDNAs effectively into the chosen cells. In our laboratory, we have been using a variety of mammalian cell lines to transiently express different nicotinic acetylcholine receptors (AChRs). We have tried several different methods to introduce the subunit cDNAs including electroporation (Potter, 1995), retroviral infection (Cepko, 1995), and transfection, both calcium phosphate and lipid mediated (Claudio, 1992). Each method has its advantages and disadvantages. Unfortunately, with each of these methods the percentage of the cells expressing AChRs is well below 100%. At first glance, the inefficiency of transient expression might not appear to pose a significant problem except to lower the levels of expression. In fact, this seems to be the case for ion channels that are homomeric (see later discussion). However, for heteromeric ion channels such as muscle-type AChRs, which are composed of α, β, γ (or ε), and δ subunits, inefficiency of expression has additional complications. These complications appear to arise from an uneven distribution of the subunits among different populations of the expressing cells. At one extreme, one or more subunits are underexpressed or not expressed at all in some cells, which will result in the expression of different combinations of the subunits. As discussed later, the levels of AChR surface expression and efficiency of assembly are both strongly affected by the combination of subunits expressed. At the other extreme, one or more subunits are overexpressed. Overexpression of subunits can be just as deleterious as the underexpression of subunits (see below).

An alternative to transient expression that avoids problems of expression inefficiency is stable expression. A cell line stably expressing an ion channel is established by integrating the correct combination of subunit genes into the genome of a cell. This procedure involves (1) introduction into the cells of a selectable marker along with the subunit genes, (2) isolation of single colonies containing the selectable marker, (3) growth of enough cells for analysis, and (4) further screening of each single colony isolate for the correct combination of subunits. Although stable expression has a number of advantages over transient expression, it is nonetheless a lengthy process, ranging from 2 to 5 months to complete.

Because the method of stable expression involves such a considerable investment of time, we have continued to use transient expression methods in our laboratory. Transient expression is used primarily for the initial characterization of new AChR subunit constructs before the time-consuming procedure of isolating and characterizing stably transfected cell lines. Considerable effort has been expended during the last several years in our laboratory to test the reliability of these methods and to optimize AChR expression using them. In the course of optimizing AChR transient expression, we have made changes and adjustments to the methods, many of which are described later. To further characterize the transient transfection of a heteromeric AChR, we describe how the expression of a heteromeric AChR compares to that of a homomeric AChR, and how heteromeric AChR transient expression compares with stable expression. For most of these experiments, AChR expression was monitored by measuring the total amount of cell-surface expression on a dish of transfected cells using ^{125}I-labeled α-bungarotoxin (Bgt) binding. We chose this assay because it is relatively easy to perform, but also because it is the feature of AChR expression that we most wanted to optimize for many of our experiments. However, as described later, the measurement of AChR surface levels fails to detect some of the more subtle differences between transient and stable expression of the AChR heteromer. To detect these more subtle differences, we have also assayed for intracellular and surface AChR expression using fluorescence microscopy and measured the efficiency of subunit assembly by metabolically labeling the AChR subunits.

III. Methods

A. AChR Constructs and Cell Lines

Three different AChRs are used in this study. Two of the AChRs are muscle-type AChRs from different species: (1) the *Torpedo* AChR, which is composed of α, β, γ, and δ subunits, and (2) the "adult-form" of mouse muscle-type AChR, which is composed of α, β, ε, and δ subunits. The other "AChR" is a homomeric receptor containing $\alpha_7/5HT_3$ chimeric subunits. These subunits consist of the N-terminal half of the chicken α_7 subunit and the C-terminal half of the mouse $5HT_3$ receptor subunit (Eisele *et al.*, 1993). Previous studies of the $\alpha_7/5HT_3$ chimera demonstrate that it has all the pharmacologic properties of an AChR (Eisele *et al.*, 1993). For transient expression, all of the *Torpedo* and mouse subunit cDNAs are subcloned into the pRBG4 expression vector (Lee *et al.*, 1991), and the $\alpha_7/5HT_3$ subunit cDNA is subcloned into the pMT3 vector (Swick *et al.*, 1992). Both expression vectors contain the same cytomegalovirus (CMV) promoter and the simian virus 40 (SV40) origin and polyadenylation sequence.

The cell lines that stably express the *Torpedo* AChR subunits (Claudio *et al.*, 1987), the mouse AChR subunits (Green and Claudio, 1993), and the $\alpha_7/5HT_3$ chimeric subunits (Rangwala *et al.*, 1997) have been previously described. The human embryonic kidney line, tsA 201 (Margolskee *et al.*, 1993), is used for transient transfections. All cell lines are grown at 37 °C, 5% (v/v) CO_2 in Dulbecco's modified Eagle's medium (DMEM; GIBCO, Grand Island, NY) plus 10% calf serum (Hyclone, Logan, UT). The *Torpedo* and mouse subunit cDNAs in the stably expressing cell lines are under the control of SV40 promoters. To enhance expression of the subunits, the medium is supplemented with 20 mM sodium butyrate (Baker, Phillipsburg, NJ) 36–48 h prior to the experiment. The assembly of the *Torpedo* subunits is temperature dependent (Claudio *et al.*, 1987). To allow assembly to occur, the temperature at which the cells are grown is dropped from 37 to 20 °C for the indicated times.

B. Transfection

For the experiments described next, cells are transfected using the calcium phosphate method. This method was chosen because of its relative ease and its low cost.

Stable transfection. Our stable transfection protocol has been described previously in detail in another volume in this series (Claudio, 1992).

Transient transfection. Transient transfection methods are also used to express the three different AChRs. The following protocol presents the details of the transient transfection methods used in our experiments.

1. Protocol 1: Calcium Phosphate Transfection Procedure

All work involving the preparation of the cell cultures and the steps in the protocol should be performed in a sterile, laminar flow hood. *See below for advice on the choice of host cell.* Care should be taken to maintain the most sterile environment possible for the preparation of the DNA and solutions used.

1. The day before transfection, plate cells in 6-cm culture dishes at a density so that cells will be 50–70% confluent the next day.

2. In a sterile tube (polypropylene), add 250 μl of the 2× HEPES solution and the appropriate amount of your DNA. *Note:* As discussed later, the level of expression is highly dependent on the total amount of transfected DNA and, if you are transfecting more than one subunit, on the DNA ratio for the different subunits. Add sterile, deionized, distilled water to the tube to a final volume of 469.5 μl. *Note:* These are the amounts for the transfection of a single 6-cm plate and should be scaled appropriately if more plates are to be transfected.

3. Slowly add 31.5 μl of 2 M $CaCl_2$ *dropwise* to the solution.

4. Wait 30 min for the precipitate to form. The resulting solution should become slightly cloudy during this period.

5. Distribute the solution dropwise over the plated cells.

6. Incubate 5–6 h at 37 °C, 5% CO_2, then change the medium to remove the precipitate.

Maximal expression of the mouse and $\alpha_7/5HT_3$ AChRs occurs 3 days following transfection. For expression of *Torpedo* AChRs, cells are shifted from 37 to 20 °C 24 h after the transfection. Maximal expression occurs 4 days following the temperature shift.

- *Solutions*

  2× HEPES: 8.0 g of NaCl, 0.198 g of $Na_2HPO_4.7H_2O$, and 6.5 g of HEPES. Add sterile, distilled, deionized water to a volume just under 500 ml. Adjust the pH to 7.0 and readjust the final volume to 500 ml. Aliquot and store at −20 °C. Aliquot currently in use is kept at 4 °C.

  2 M CaCl₂: Mix in distilled, deionized water and sterile filter. Can be kept at room temperature.

C. Assays

1. α-Bungarotoxin Binding

The primary assay used for characterizing the expressed AChRs was binding of Bgt to the AChR. α-Bungarotoxin binds to muscle-type and $\alpha_7/5HT_3$ AChRs with high affinity and it has a very long resident time. Because of this slow off rate, Bgt binding to the AChR basically serves as a method to label the receptors. Two different Bgts were used in the experiments: Bgt conjugated with ^{125}I ($[^{125}I]$Bgt) and Bgt conjugated with tetramethylrhodamine (TMR–Bgt). Protocol 2 presents the details of the methods used for $[^{125}I]$Bgt binding to the transfected cells in our experiments, and Protocol 3 presents the details of the TMR–Bgt staining.

2. Protocol 2: Cell-Surface ^{125}I-Bgt Binding Procedure

This protocol is designed for poorly adherent cells, such as HEK 293 and tsA 201 cells, plated on 6-cm culture plates and grown to confluency. Care must be taken since this protocol involves handling radioisotopes. All work involving $[^{125}I]$ Bgt should be monitored with a gamma probe and, as much as possible, should be performed in a fume hood. Gloves and a lab coat should be worn. $[^{125}I]$Bgt-containing solutions should be aspirated into a designated vacuum flask and, along with $[^{125}I]$Bgt-containing dry waste, safely disposed of.

1. Aspirate media from the cell culture plates to be assayed.

2. 700 μl of the 1× phosphate-buffered saline plus EDTA solution (PBS/EDTA) is added to the cell culture plates. The cells, which are poorly adherent, are removed from the plates simply by applying a gentle fluid stream from a pipette. Cells are transferred to a 1.5-ml microfuge tube. Any remaining cells are removed from the plates by the addition of 600 μl PBS/EDTA and added to the microfuge tube.

3. Cells are pelleted in a centrifuge at 1000–$2000 \times g$ for 3–5 min. This centrifuge spin must be gentle so that the cells remain intact. The supernatant above the pellet in the tube is aspirated using a syringe needle (21-gauge 1/2).

4. The cells are washed by addition of 1 ml of PBS/EDTA gently added back to the tube. The cell pellet is gently dispersed. The cells are again pelleted, and the supernatant is aspirated.

5. A PBS/EDTA solution containing 0.1% bovine serum albumin (BSA; w/v) and 4 nM [^{125}I]Bgt (4 μl/ml of the [^{125}I]Bgt stock) is prepared.

6. 750 μl of this PBS/EDTA solution is added to each microfuge tube and the cell pellet is gently dispersed.

7. The microfuge tubes are incubated for 2 h at room temperature on a rotator to keep the cells dispersed.

8. Following the incubation, the cells are pelleted, the supernatant aspirated, and the cells are washed three to four times in the 1× PBS/EDTA solution.

9. After the last wash, the supernatant is aspirated and the [^{125}I]Bgt bound to the cells is counted in a gamma counter.

- *Solutions*

  5× PBS/EDTA: 40.0 g of NaCl, 1 g of KCl, 7.2 g of Na_2HPO_4, 7.2 g of KH_2PO_4, and 1.2 g of EDTA. Add sterile, distilled, deionized water to a volume of just under 1 l. Adjust to pH 7.4 and readjust the final volume to 1 l. Store at room temperature.

  [^{125}I]Bgt stock solution: Add sterile, distilled, deionized water to a 250-mCi lyophilized aliquot of [^{125}I]Bgt (NEN, Boston, MA; 140–170 cpm/fmol) to the appropriate volume in order to make a 1-μM Bgt solution (usually 2–3 ml of water). Store at 4 °C.

3. Fluorescence Microscopy

The efficiency of expression, that is, the percentage of the cells expressing AChR subunits, and the distribution of the subunits among the cells were determined using fluorescence microscopy techniques. To visualize the Bgt sites expressed on the cell surface or intracellularly we used TMR–Bgt. Subunit-specific antibodies were used to visualize the cell-to-cell distribution of the α and δ subunits.

4. Protocol 3: Staining Procedure with Immunofluorescent TMR–Bgt

1. To coat glass slides with alcian blue (Eastman Kodak, Rochester, NY), which is used to immobilize cells for staining, three to four drops of a 1% alcian blue solution (w/v) are placed on the slides. The slides are allowed to set at room temperature for 15 min, rinsed one time with distilled, deionized water, and patted dry with a Kimwipe.

2. The cells to be stained are placed on the slides in medium after removal from the culture dish (see step 2 in Protocol 2). Cells are allowed to set at room temperature for 10 min.

3. Excess medium is aspirated from the slides and three to four drops of a 2% (w/v) formaldehyde solution are placed on the slides, which are allowed to set at room temperature for 10 min.

4. Excess formaldehyde solution is aspirated and three to four drops of Tris-buffered saline (TBS) are placed on the slides and set for 10 min to quench reactive aldehydes.

5. Excess TBS is aspirated. At this point *to permeabilize cells*, TBS plus 0.1% Triton X-100 is added to the slides for 10 min and then aspirated.

6. To block nonspecific binding sites. TBS plus 2 mg/ml BSA is added to the slides for 10 min and then aspirated.

7. 300 nM TMR–Bgt (Molecular Probes, Eugene, OR) in TBS plus 2 mg/ml BSA is added to the slides and the slides are incubated *in the dark* for 1 h at room temperature.

8. The slides are washed four times for 5–10 min per wash with TBS plus 2 mg/ml BSA. For additional staining with 4′,6-diamidino-2-phenylindole dihydrochloride (DAPI: Molecular Probes), DAPI is added to the TBS of the second wash at a 1:1000 dilution, and cells are incubated for 15 min. *Note*: Cells do not have to be permeabilized for the DAPI staining.

9. One drop of mounting medium (Vectashield from Vector Labs, Burlingame, CA) is added to each slide. A coverslip is applied over the cells, sealed with clear nail polish, and allowed to dry.

10. Slides are viewed immediately or stored at −20 °C.

A rabbit polyclonal anti-α-subunit antibody (Ab) and a mouse monoclonal Ab (MAb) anti-δ-subunit, MAb 88b, were used to stain cells with subunit-specific antibodies. In the protocol for the Ab staining the following two steps are substituted for step 7:

7a. To permeabilized cells, add the Abs diluted 1:100 in TBS plus 2 mg/ml BSA for 1 h and wash two to three times for 5–10 min per wash with TBS plus 2 mg/ml BSA.

7b. The secondary antibodies are added *in the dark* for 1 h: fluorescein goat antimouse IgG(H + L) conjugate at 1:500 dilution in TBS/BSA and Lissamine rhodamine (LRSC)-conjugated AffiniPure FAb goat antirabbit IgG(H + L) at 1:200 dilution in TBS/BSA.

- *Solutions*

2% paraformaldehyde/PBS: 1.24 g of paraformaldehyde (Sigma, St. Louis, MO) is added to 4.7 ml of a 0.5 M Na_2HPO_4 solution and 55.8 ml of distilled, deionized water. The solution is stirred and heated at 60 °C until the paraformaldehyde has dissolved. After, 0.36 g of NaCl is added to the solution, followed by 1.5 ml of a 0.5 M NaH_2PO_4 solution. If necessary, adjust the pH to 7.1 and use fresh.

TBS: 1.0 ml of a 1.0 M Tris solution, pH 7.5, is added to 3.0 ml of a 5.0 M NaCl solution and 96 ml of distilled, deionized water. If necessary, adjust the pH to 7.5 and store at 4 °C.

5. Metabolic Labeling and Immunoprecipitation

We have found that metabolic labeling of AChR subunits *in vivo* is the most sensitive assay for analyzing the subunit properties using sodium dodecyl sulfate–polyacrylamide gel electrophoresis (SDS–PAGE). Metabolic labeling has the added advantage of allowing you to pulse label and then follow or "chase" the labeled proteins through their steps of biogenesis. The pulse-chase protocol used in the results as well as the protocol for cell lysis and immunoprecipitation have been described in detail previously (Millar *et al.*, 1996).

IV. Results

A. Dependence on Cell Line

Not surprisingly, we have found that the level of AChR expression is highly dependent on the choice of host cell. For many cell lines, expression levels are poor because of a low efficiency at taking up the DNA. This is the case for both the rat pheochromocytoma cell line, PC12, and the human neuroblastoma cell line, SH-SY5Y, where 1% or less of the cells express AChRs (S. Rakhilin and W. Green, unpublished results, 1997). Other mammalian cell lines are much more efficient at taking up AChR genes. Figure 1 shows the levels of *Torpedo* AChR cell-surface expression obtained in four different cell lines: mouse fibroblast lines, (1) NIH 3T3 and (2) L tk cells, and human embryonic kidney lines, (3) HEK293 and (4) tsA201 cells. For each of these cell lines, we find that the percentage of cells that take up the DNA ranges from 10% to 40% (data not shown, but see Fig. 7 for the tsA201 cells).

The five- to sixfold higher levels of expression observed for the tsA201 cells are not due to a higher expression efficiency. The tsA201 cells were established by stably transfecting HEK293 cells with the SV40 large tumor (T) antigen (Margolskee *et al.*, 1993). The large T antigen initiates replication of expression vectors containing the SV40 origin, resulting in a high copy number for these expression vectors. The higher level of expression of the tsA201 cells is, thus, the result of an amplification of the expression vector once it has entered the cell. In short, the high levels of expression we obtain in tsA201 cells are a function of both the cells' relatively high transfection efficiency and the ability of the cells to amplify vector copy number after its entry into the cells.

B. Expression of Homomeric and Heteromeric AChRs

Different experiments were performed to compare the transient expression of heteromeric mouse $\alpha\beta\varepsilon\delta$ AChRs with that of the homomeric $\alpha_7/5HT_3$ AChRs. One big difference between heteromeric and homomeric AChR expression was that the

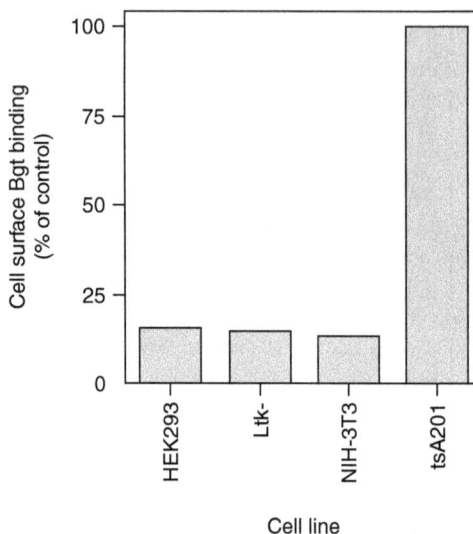

Fig. 1 Cell-surface expression of transiently transfected AChRs is dependent on the host cell line. *Torpedo* AChR subunit cDNAs were transiently transfected into either HEK293, L tk, NIH 3T3, or tsA201 cells. The amount of subunit cDNA transfected was 1 μg α, 0.5 μg β, 2.5 μg γ, 7.5 μg δ per 6-cm plate, and the level of AChR surface expression was determined by [^{125}I]Bgt binding to intact cells. The values represent the mean of [^{125}I]Bgt binding to two 6-cm plates. Background, determined by [^{125}I]Bgt binding to intact cells that were sham transfected with no DNA, was subtracted to obtain the data. The values are plotted as the percentage of the tsA201 value, which was 50 fmol.

level of homomeric AChR expression was considerably higher than that of the heteromeric AChR. As shown in Fig. 2, cell-surface [^{125}I]Bgt binding to the α_7/5HT$_3$ AChRs was approximately fivefold higher than that to the mouse $\alpha\beta\varepsilon\delta$ AChRs, which was about fivefold higher than [^{125}I]Bgt binding to the *Torpedo* AChRs. While the transient expression of the α_7/5HT$_3$ and mouse $\alpha\beta\varepsilon\delta$ AChRs was performed under identical conditions, the conditions were different for *Torpedo* AChR expression. As described in Section III, cells transiently transfected with the *Torpedo* subunit cDNAs were maintained at 20 °C as opposed to 37 °C for the α_7/5HT$_3$ and mouse $\alpha\beta\varepsilon\delta$ AChRs because of the temperature dependence of *Torpedo* AChR assembly. It is likely that the lower temperature is the reason for the lower level of *Torpedo* AChR expression since at 20 °C most of the assembled *Torpedo* AChRs are retained in the endoplasmic reticulum (Ross *et al.*, 1991).

Additional differences between the transient expression of heteromeric and homomeric AChRs were apparent when we measured how varying the amount of DNA affected the level of AChR expression. Figure 3 displays AChR expression, monitored by surface [^{125}I]Bgt binding, as a function of micrograms of DNA added during transfection. The results in Fig. 3A for the homomeric α_7/5HT$_3$ AChR should be compared to those in Fig. 3B for the heteromeric mouse and *Torpedo* AChRs. First, for the homomeric AChR, expression increased linearly

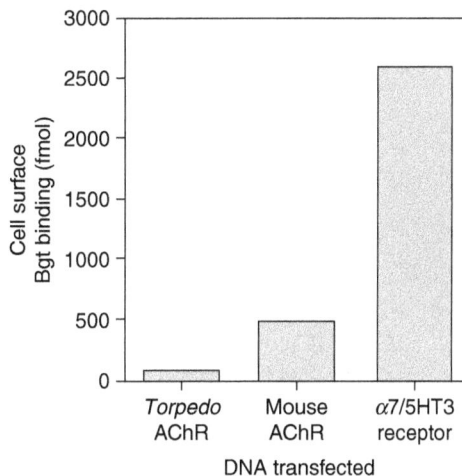

Fig. 2 Cell-surface expression of different AChRs. *Torpedo* AChR α-, β-, γ-, and δ-subunit cDNAs, mouse AChR α-, β-, ε-, and δ-subunit cDNAs, or the chimeric $\alpha_7/5HT_3$ subunit cDNA were transiently transfected into tsA201 cells. The amount of subunit cDNA transfected was 1 μg α, 0.5 μg β, 2.5 μg γ, 7.5 μg δ per 6-cm plate for *Torpedo* AChRs; 5 μg α, 2.5 μg β, 2.5 μg ε, 2.5 μg δ per 6-cm plate for mouse AChRs, and 2.5 μg for the $\alpha_7/5HT_3$ receptor. The level of AChR surface expression, determined as in Fig. 1, was 67 fmol for *Torpedo* AChRs, 500 fmol for mouse AChRs, and 2610 fmol for $\alpha_7/5HT_3$ receptors.

with increasing amounts of transfected DNA, whereas for the heteromeric AChRs expression rose sigmoidally. Second, it only took 2.0 μg of the homomeric AChR DNA for peak expression (Fig. 3A). In contrast, 12.5 μg of total subunit DNA or fivefold higher levels of DNA were needed for peak expression of both heteromeric AChRs (Fig. 3B) even though expression levels for the homomeric AChR were approximately fivefold higher than for the heteromeric AChR (Fig. 1). Finally, homomeric AChR expression plateaued with increasing amounts of transfected DNA (Fig. 3A), whereas expression of the heteromeric AChRs sharply decreased (Fig. 3B). The decrease in expression observed with large amounts of heteromeric AChR DNA is not a nonspecific effect of the increased amount of DNA. If increasing amounts of the expression vector lacking the subunit cDNA were added in the range of 10–30 μg per 6-cm plate, no decrease in AChR expression is observed (data not shown).

C. Complications of Heteromeric AChR Expression

In the previous section, we described how transient expression of heteromeric AChRs is different in a number of ways from that of the homomeric AChRs. To summarize, (1) heteromeric AChR expression is several fold less than that for the homomeric AChR; (2) it requires severalfold more DNA; (3) it displays a sigmoidal rise with increasing DNA concentration as opposed to the linear rise

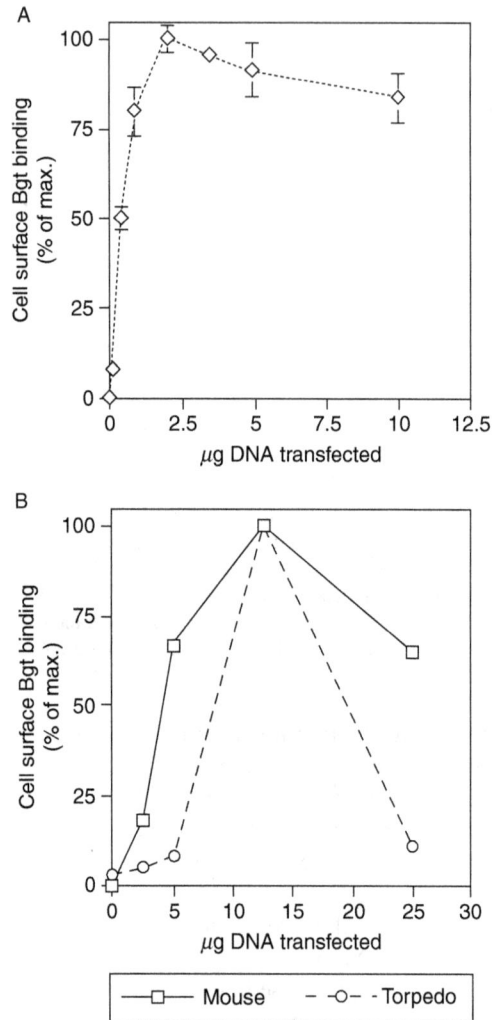

Fig. 3 Dependence of homomeric and heteromeric AChR expression on the amount of transfected DNA. (A) Cell-surface expression of homomeric α_7/5HT$_3$ receptors as a function of the amount of transfected DNA. Maximum levels of expression were obtained by transfecting 2.0 μg of DNA per 6-cm plate. The level of AChR surface expression was determined as in Fig. 1. The values represent the mean ± standard deviation for two separate experiments consisting of two to four 6-cm plates and are plotted as the percentage of maximum binding, which was 2570 fmol. (B) Cell-surface expression of heteromeric *Torpedo* and mouse AChRs as a function of the amount of transfected DNA. Maximum levels of expression for both AChRs were obtained by transfecting 12.5 μg of DNA per 6-cm plate. The level of AChR surface expression was determined as in Fig. 1. The values represent the mean for two 6-cm plates and are plotted as the percentage of maximum binding, which was 3.41 fmol for the α_7/5HT$_3$ AChR, 34 fmol for *Torpedo*, and 357 fmol for the mouse AChR.

observed for homomers; and (4) after reaching peak value, it does not plateau, but declines sharply with increasing amounts of transfected DNA. Obviously, the more complicated process of assembling an AChR consisting of four different subunits is at least part of the explanation for these differences between the expression of AChR heteromers and homomers. However, the differences may also be caused in part by the inefficiency of transient expression. As discussed in Section II, the inefficiency of transient expression would be expected to create more problems for a heteromer than for a homomer. Problems specific to heteromeric expression are likely to be caused by the distribution of the different subunit cDNAs among the cells. An uneven distribution of the different cDNAs could cause either over- or underexpression of different subunits in different cells. In this section, we have tested how varying the amount of different subunit cDNAs affects heteromeric AChR expression.

The uneven distribution of different subunit cDNAs among a large population of cells can have several consequences that affect heteromeric AChR expression. One possibility is that one or more different subunit cDNAs fail to enter a sizable number of the cells and less than the full complement of the subunits is expressed in these cells. In Fig. 4, we have tested how different combinations of the *Torpedo* AChR α, β, γ, and δ subunits affect AChR cell-surface expression. Transient transfection of any combination of subunit cDNAs containing less than the full complement of four subunits resulted in AChR expression levels at or just above background as assayed by cell-surface [^{125}I]Bgt binding. In other experiments,

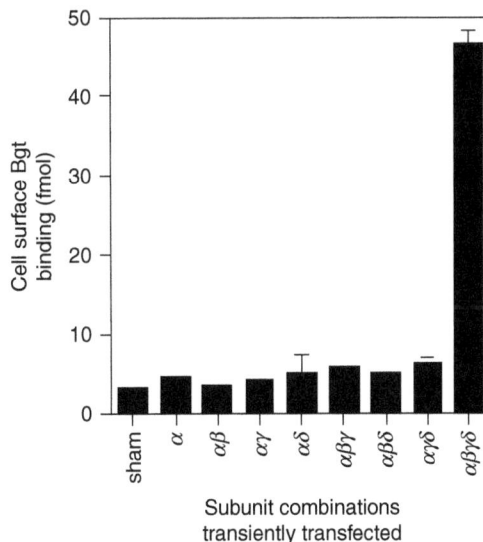

Fig. 4 Cell-surface expression of different combinations of AChR subunits. The indicated combinations of *Torpedo* AChR subunit cDNAs were transiently transfected into tsA201 cells. The amount of subunit cDNA transfected was 1 μg α, 0.5 μg β, 2.5 μg γ, 7.5 μg δ per 6-cm plate. The level of AChR surface expression was determined as in Fig. 1. The values represent the mean $\pm$ standard deviation for three or four 6-cm plates.

we found that only the surface expression levels of the $\alpha\beta\gamma$ and $\alpha\beta\delta$ combinations were significantly above background but extremely small compared to the surface expression levels of all four subunits (data not shown). Similar results were also obtained previously for transient expression of the mouse AChR α, β, γ, and δ subunits in COS cells (Gu *et al.*, 1991). These findings demonstrate that if only one of the four subunit cDNAs fails to enter a cell or if relatively small amounts of one subunit enter, AChR expression on the surface of that cell is greatly diminished or eliminated. Therefore, virtually the only AChR subunit complexes that arrive on the cell surface are AChRs containing all four subunits. Note that the expression of intracellular complexes still occurs in the absence of one or more of the AChR subunits although the number and efficiency of assembly of these complexes appears to be much lower than for all four subunits. Expression of most subunit combinations containing the α subunit (i.e., $\alpha\gamma$, $\alpha\delta$, $\alpha\beta\gamma$, $\alpha\gamma\delta$, $\alpha\beta\gamma$, and $\alpha\beta\delta$) form large numbers of Bgt-binding sites even though there are few or no surface Bgt-binding sites (Blount and Merlie, 1989; Green and Claudio, 1993; Gu *et al.*, 1991; Saedi *et al.*, 1991). Clearly, "quality control" mechanisms exist in the cells that prevent the surface expression of AChRs lacking any of the four subunits.

Another outcome of an uneven distribution of the different subunit DNAs among transfected cells is that one or more subunits are overexpressed relative to the other subunits. The consequence of varying the ratio of one subunit relative to the other subunits was tested as shown in Fig. 5. In this experiment, the amount of

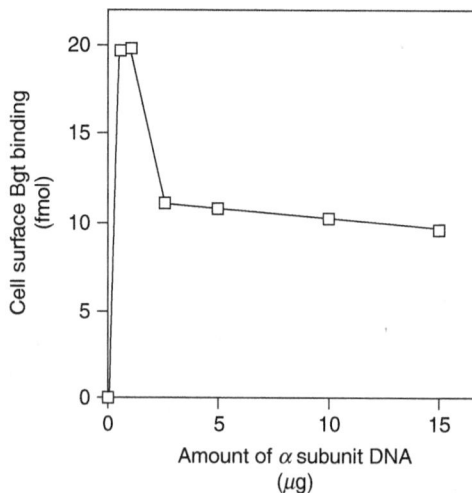

Fig. 5 Dependence of heteromeric AChR expression on varying the amount of transfected DNA for a single subunit. The indicated amount of *Torpedo* AChR α-subunit DNA was transiently transfected into tsA201 cells along with a set amount for the other three subunits (0.1 μg β, 2.5 μg γ, 7.5 μg δ per 6-cm plate). Maximum levels of expression were obtained by transfecting 1.0 μg of α-subunit DNA per 6-cm plate. The level of AChR surface expression was determined as in Fig. 1. The values were obtained from a single 6-cm plate.

transfected α-subunit cDNA was varied while the amount of the other three AChR subunit cDNAs was held constant. Surface expression sharply changed from zero to a peak value at 1 μg of the AChR α-subunit cDNA and then diminished to a plateaued value of half of the peak value at higher DNA concentrations. The dependence of the surface expression on the amount of α-subunit cDNA is different from its dependence on the total amount of subunit cDNA (Fig. 3B) where expression peaked at 12.5 μg and was decreased by 90% at higher DNA concentrations. The much larger decrease in surface expression observed at the higher amounts of total subunit cDNA appears to be caused by increasing the DNA concentration of subunits other than the α subunit. In previous studies, it was shown that AChR surface expression is largely blocked by increasing amounts of mouse δ-subunit cDNA relative to the other three mouse subunits (Gu *et al.*, 1991). We have obtained similar results for both the *Torpedo* γ- and δ-subunit cDNAs (A.L. Eertmoed and W.N. Green, unpublished results, 1997). The relatively small amount of α-subunit cDNA needed to maximize expression reflects the need for much larger amount of other subunits. The dependence on each subunit cDNA, when varied relative to the other three subunits as in Fig. 5, peaked at different values. No expression above background was obtained at a subunit cDNA ratio of 2:1:1:1 (α:β:γ:δ), which is the subunit stoichiometry of the surface AChR and the subunit cDNA ratio we first tried. *Torpedo* AChR expression rose significantly above background after increasing the amount of δ-subunit cDNA relative to the other subunits, and maximum levels of *Torpedo* AChR expression were only obtained when we maintained a subunit cDNA ratio of 1:0.5:2.5:7.5 (α:β:γ:δ).

We have not determined why deviations from a subunit cDNA ratio of 1:0.5:2.5:7.5 (α:β:γ:δ) reduces AChR surface expression and, presumably, the assembly efficiency of fully assembled AChRs. The most likely explanation is suggested by the data of Fig. 4. Over- or underexpression of subunits would be expected to result in the assembly of subunit complexes containing less than the full complement of subunits. As shown in Fig. 4, the assembly of such subunit complexes would decrease both AChR surface expression and the assembly efficiency of fully assembled AChRs.

D. Assembly Efficiency

In the previous section, we described how transient expression of heteromeric AChRs depended on how the different subunit cDNAs are distributed among the transfected cells. High levels of heteromeric AChR transient expression were only achieved when both the full complement and the correct ratio of subunit cDNAs were transfected. Under these conditions, AChR transient expression equals and can even exceed the levels obtained with stably transfected cell lines. With comparable levels of surface expression obtained by stable and transient expression, can we assume that these two methods of expression are similar? Or could other features of AChR biogenesis differ with these two methods? To further compare stable and transient expression methods, we attempted to measure the efficiency of subunit assembly. For these experiments, the *Torpedo* AChR subunits were used because of the temperature dependence of the assembly of this AChR, which slows the rate of assembly by more than an order of magnitude (Green and Claudio, 1993). To assay subunit assembly, the subunits were metabolically labeled with a 1-h pulse of a mixture of $[^{35}S]$methionine and $[^{35}S]$cysteine, and then immunoprecipitated with the indicated antibodies. For the transiently expressed cells, each of the four subunits was successfully labeled and immunoprecipitated as shown in Fig. 6A. As assayed by the amount of $[^{35}S]$methionine and $[^{35}S]$cysteine signal, the levels of subunit synthesis obtained for the transiently transfected 6-cm cultures in Fig. 6A were considerably higher than the levels obtained for 10-cm cultures of the stably transfected cells (data not shown).

To follow the steps in assembly and measure assembly efficiency, the cells were "chased" after the metabolic labeling, and the labeled subunits were immunoprecipitated with MAb 14. Monoclonal Ab 14 was used because it recognizes only subunits that have been assembled during the latter stages of the assembly process

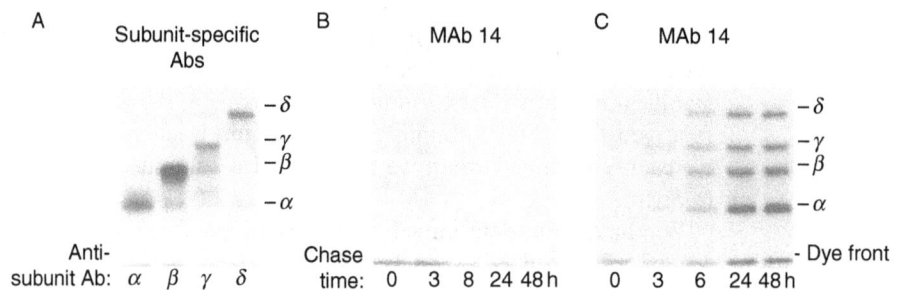

Fig. 6 AChR assembly efficiency for cells either transiently or stably expressing AChR subunits. (A) Metabolic labeling of all four AChR subunits in transiently transfected cells. The four *Torpedo* AChR subunits were transiently expressed in tsA201 cells on 10-cm plates (2 μg α, 1 μg β, 5 μg γ, 15 μg δ). One day after transfection, the cells were pulse labeled for 1 h at 20 °C in methionine-cysteine-free medium supplemented with 333 μCi of an $[^{35}S]$methionine-$[^{35}S]$cysteine mixture (NEN EXPE^{35}S^{35}S). Cells were then lysed and treated with 1% SDS followed by 4% Triton X-100 to eliminate most intersubunit interactions, and the lysate from half a 10-cm plate was immunoprecipitated with the subunit-specific antibodies: polyclonal anti-α Ab, MAb 148 (anti-β), MAb 168 (anti-γ) and polyclonal anti-δ Ab. Labeled subunits were electrophoresed on 7.5% SDS–polyacrylamide gels, fixed, enhanced for 30 min and dried on a gel dryer. Shown is an autoradiogram of the gel. (B) The failure to detect metabolically labeled mature AChR complexes in transiently transfected cells. The four *Torpedo* AChR subunits were transiently expressed in tsA201 cells and metabolically labeled as in part (A) except that following the 1-h pulse the labeled subunits were "chased" (see methods) for the indicated times and immunoprecipitated with MAb 14, which recognizes only subunit complexes that form late in the assembly process. No MAb 14-precipitable complexes were detectable, and we conclude that only a very small fraction of the total number of subunits synthesized assemble into AChRs and are transported to the cell surface. (C) Metabolically labeled mature AChR complexes in stably transfected cells. The four *Torpedo* AChR subunits were stably expressed in mouse L cells on 10-cm plates. Cells were pulse labeled for 30 min at 20 °C and treated as in part (B). In contrast to the transiently transfected cells in part (B), we observed metabolically labeled mature AChR complexes recognized by MAb 14 in cells stably expressing AChRs.

(Green and Claudio, 1993). The MAb 14 precipitation of labeled subunits from stably expressing cells is shown in Fig. 6C. Approximately 30% of the subunits originally labeled during the pulse are precipitated in $\alpha_2\beta\gamma\delta$ complexes 48 h after the pulse. In contrast, we were unable to observe any labeled subunits precipitated with MAb 14 from the transiently transfected cells (Fig. 6B) even though most of the $[^{125}I]$Bgt-bound surface AChRs could be precipitated by this antibody (data not shown). These results demonstrate that the efficiency of AChR assembly in the transiently transfected cells is within the error of our measurements, that is, $\sim 1\%$ or lower. Heteromeric AChR transient expression thus differs from stable transfection in that the efficiency of assembly is at least 30-fold lower. Despite this large difference in assembly efficiency, the levels of AChR surface expression for stably and transiently transfected cells are similar. Transiently transfected cells must synthesize AChR subunits at much higher levels, and the lower assembly efficiency results in similar levels of surface AChRs.

Because transient transfection of the AChR subunits results in such a low efficiency of AChR assembly, we have been unable to carry out detailed studies of AChR assembly using transient expression methods. AChR subunits must be stably expressed in a cell line in order to follow assembly using metabolic labeling of the subunits. Our data suggest that low efficiency of assembly may occur for other heteromeric ion channels, and care should be taken when using transient expression methods to study the properties of ion channels not found on the cell surface.

E. Expression Efficiency and Distribution of Subunits

Why is the efficiency of heteromeric AChR assembly so low when the subunits are transiently transfected compared to when they are stably transfected? To address this question, cells transiently or stably expressing the mouse α, β, ε, and δ subunits were assayed using fluorescence microscopy techniques. The mouse subunits were chosen for study because they expressed higher levels of surface AChRs than the *Torpedo* subunits (see Fig. 2). The efficiency of expression was first tested by staining surface AChRs with TMR–Bgt. Examples of the stained cells are shown in Fig. 7, and the quantification of the results of the staining are presented in Table I. The total number of cells on the slides was visualized by DAPI staining, which stains the nuclei of the cells. As expected, almost all (95%) of the cells stably expressing AChRs were stained on the surface by TMR–Bgt. In contrast, 33% of the cells transiently expressing AChRs were stained. Based on the results of Fig. 4, where only cells transiently transfected with all four subunits expressed surface AChRs at significant levels, it appears that 33% of the transiently transfected cells expressed all four subunits.

Additional assays were performed to test whether AChR subunits are expressed in the 67% of the cells that did not show surface TMR–Bgt staining. The cells were next stained with TMR–Bgt after being permeabilized. This procedure allowed the staining of intracellular as well as the surface Bgt-binding sites, and the results are shown in Fig. 8 and quantified in Table I. If α subunits in combination with only

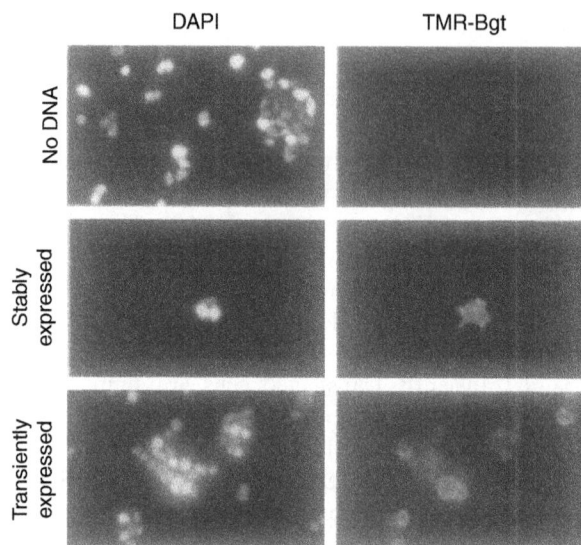

Fig. 7 Detection of cell-surface AChRs using fluorescence microscopy. Intact cells were stained with TMR–Bgt to detect cell-surface AChRs and DAPI to stain the nuclei. Displayed are cells (1) sham transfected with no DNA (top), (2) stably expressing the mouse AChR α, β, ε-, and δ subunits (middle), or (3) transiently expressing the mouse AChR α, β, ε, and δ subunits (bottom). Cells were visualized and photographed, in Germany, with a Carl Zeiss Axioskop using a 40× objective and a 10× eyepiece.

Table I
TMR–Bgt Staining of Cells Transiently and Stably Expressing $\alpha_2\beta\varepsilon\delta$ AChRs

Condition	Fixation	Total number of cells (DAPI stained)	Number stained	Stained (%)
Sham transfected	Intact	354	0	0
(no DNA)	Permeabilized	277	0	0
Stably transfected	Intact	150	142	95
	Permeabilized	154	150	97
Transiently	Intact	651	213	33
transfected	Permeabilized	541	191	35

one or two other AChR subunits are expressed transiently, we would expect to see the formation of intracellular Bgt-binding AChR complexes, but no surface complexes. When the permeabilized cells were stained, the percentage of cells stained increased only to 35% compared to the 33% for the intact cells. These results indicate that AChR subunits were only expressed in the cells that express Bgt sites.

To test further whether the cells that fail to express Bgt sites were expressing AChR subunits, the permeabilized cells were stained with α- and δ-specific antibodies. These results are shown in Fig. 9 and quantified in Table II. Basically, only cells with Bgt sites appeared to be stained by either the α- or δ-specific antibodies

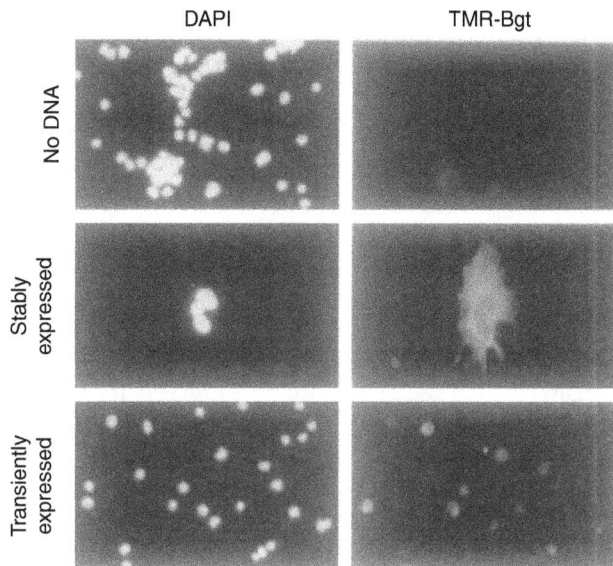

Fig. 8 Detection of cell-surface and intracellular AChRs using fluorescence microscopy. Permeabilized cells were stained and displayed as in Fig. 7.

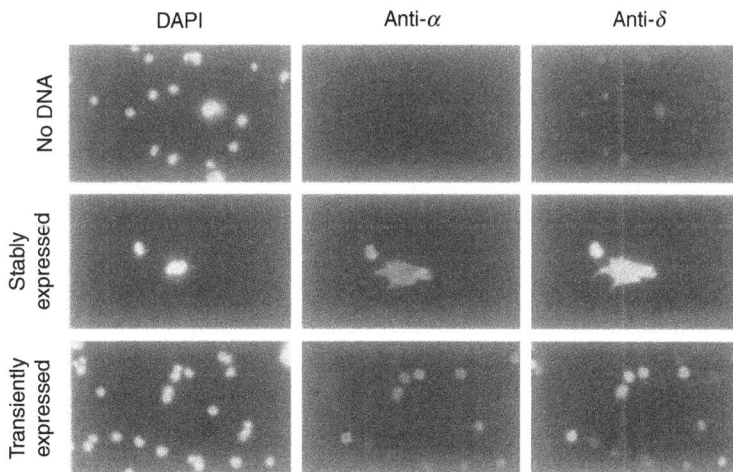

Fig. 9 Detection of AChR α and δ subunits using fluorescence microscopy. Permeabilized cells were stained with polyclonal anti-α Abs and anti-δ MAb 88b (American Type Culture Collection, Rockville, MD) followed by Texas Red and fluorescein isothiocyanate (FITC)-labeled antibodies, respectively. The cells were also stained with DAPI to stain the nuclei and displayed as in Fig. 7.

Table II
Subunit–Specific Staining of Cells Transiently and Stably Expressing $\alpha_2\beta\varepsilon\delta$ AChRs

Condition	Total number of cells (DAPI stained)	Number stained with α-specific Ab	Stained with α-specific Ab (%)	Number stained with α-specific Ab	Stained with α-specific Ab (%)
Stably transfected	203	194	96	202	99.5
Transiently transfected	651	151	36	158	38

since the percentage of permeabilized cells stained with the antibodies was only 36% for the α-specific antibody and 38% for the δ-specific antibody. Based on the results using fluorescence microscopy, we conclude that the transient expression of AChR subunits is inefficient, in the range of 30–40%, but virtually all of the cells that are transfected receive all four subunit cDNAs and, therefore, express all four subunits.

If almost all of the transiently transfected cells express all four AChR subunits, the question remains: Why is the efficiency of heteromeric AChR assembly so low when the subunits are transiently transfected? There is another feature of the staining shown in Figs. 7–9 that distinguishes transiently expressing from stably expressing cells. Although we have made no attempt to quantitate the level of staining in any of the experiments, it is clear that the staining of the transiently expressing cells is very heterogeneous while the staining of the stably expressing cells is homogeneous. For the TMR–Bgt staining of the intact and permeabilized cells, it is the cell-to-cell intensity of the signal that is heterogeneous. For the α- and δ-specific antibody staining of the permeabilized cells, the intensity of both signals also shows cell-to-cell heterogeneity. In addition, although virtually every cell that stains for the α-specific antibody also stains for the δ-specific antibody, the intensity of α-specific antibody signal often does not correlate with the intensity of the δ-specific antibody signal. These results indicate that the ratio of the four subunits varies among the transiently transfected cells.

As already discussed above, it is critical that the four subunit cDNAs are transfected in a set ratio. Deviations from this ratio cause decreases in the level of surface expression. Therefore, altogether the results suggest that cell-to-cell variations in ratio of the four subunits is the cause of low assembly efficiency in the transiently transfected cells. If variations in the subunit ratio are the cause of the low assembly efficiency, then there should be little difference in the assembly efficiency of a homomeric ion channel when either transiently or stably expressed. Although we have not measured the assembly efficiency of the homomeric AChR, a much higher efficiency would explain why the homomeric AChR expression levels are fivefold higher than for the heteromeric AChR (Fig. 2). As we discussed previously, variations in the subunit ratio appear to result in the assembly of subunit complexes containing less than the full complement of subunits. Thus, the low efficiency of heteromeric AChR assembly obtained for transient transfection is likely to be caused by the assembly of large numbers of subunit complexes lacking one or more subunits.

V. Summary

Transient transfection is an excellent method for the expression and study of cell surface, heteromeric ion channels. The cell type, the total amount of DNA, the combination of subunits and the ratio of subunit DNA are all important parameters to consider when attempting to optimize expression. A serious drawback of this method is that the efficiency of subunit assembly is very low in comparison to the efficiency of assembly for stably expressed heteromeric ion channels. The low efficiency of assembly prevents use of transient expression methods for detailed studies of heteromeric AChR assembly, and caution should be taken in the use of these methods for the study of intracellular heteromeric ion channel subunits. After the transient expression of heteromeric AChR subunits, virtually all of the expressing cells contained all four AChR subunits. However, the subunits were heterogeneously distributed among the cells, and the low efficiency of AChR assembly appears to be due to cell-to-cell variations in the ratio of the four subunits.

Acknowledgments

The authors are most grateful to Dr. T. Claudio for the cell lines stably expressing the *Torpedo* and mouse AChRs and the polyclonal anti-α and δ Abs, Dr. S. Sine for the pRBG constructs containing the *Torpedo* and mouse AChR subunit cDNAs, Dr. J.-L. Eisele for the pMT3 construct containing the α_7/5HT$_3$ cDNA, Dr. J. Lindstrom for MAbs 14, 148, and 168, and Dr. J. Kyle for the tsA201 cell line. The authors would also like to thank Christian Wanamaker for help with some of the experiments. This work was supported by grants from the National Institutes of Health and the Brain Research Foundation.

References

Ashley, R. (1996). "Ion Channels: A Practical Approach." IRL Press, London.

Blount, P., and Merlie, J. P. (1989). *Neuron* **3**, 349.

Cepko, C. (1995). *In* "Current Protocols in Molecular Biology," (F. M. Ausubel, R. Brent, R. E. Kingston, D. D. Moore, J. G. Seidman, J. A. Smith, and K. Struhl, eds.), Vol. 1, pp. 9.10.1–9.14.3. Wiley, New York, NY.

Claudio, T. (1992). *Methods Enzymol.* **207**, 391.

Claudio, T., Green, W. N., Hartman, D. S., Hayden, D., Paulson, H. L., Sigworth, F. J., Sine, S. M., and Swedlund, A. (1987). *Science* **238**, 1688.

Clayton, A., *et al.* (2009). Crystal structure of the GluR2 amino-terminal domain provides insights into the architecture and assembly of ionotropic glutamate receptors. *J Mol Biol.* **392**(5), 1125–1132.

Eisele, J. L., Bertrand, S., Galzi, J. L., Devillers, T. A., Changeux, J. P., and Bertrand, D. (1993). *Nature* **366**, 479.

Green, W. N., and Claudio, T. (1993). *Cell* **74**, 57.

Green, W. N., and Millar, N. S. (1995). *Trends Neurosci.* **18**, 280.

Gu, Y., Forsayeth, J. R., Verrall, S., Yu, X. M., and Hall, Z. W. (1991). *J. Cell Biol.* **114,** 799.

Leal-Ortiz, S., *et al.* (2008). Piccolo modulation of Synapsin1a dynamics regulates synaptic vesicle exocytosis. *J. Cell Biol.* **181**(5), 831–846.

Lee, B. S., Gunn, R. B., and Kopito, R. R. (1991). *J. Biol. Chem.* **266,** 11448.

Lois, C., *et al.* (2002). Germline transmission and tissue-specific expression of transgenes delivered by lentiviral vectors. *Science* **295**(5556), 868–872.

Margolskee, R. F., McHendry-Rinde, B., and Horn, R. (1993). *BioTechniques* **15,** 906.

Millar, N. S., Moss, S. J., and Green, W. N. (1996). *In* "Ion Channels: A Practical Approach," (R. Ashley, ed.). IRL Press, London.

Potter, H. (1995). *In* "Current Protocols in Molecular Biology," (F. M. Ausubel, R. Brent, R. E. Kingston, D. D. Moore, J. G. Scidman, J. A. Smith, and K. Struhl, eds.), Vol. 1, pp. 9.3.1–9.3.6. Wiley, New York, NY.

Rangwala, F., Drisdel, R. C., Rakhilin, S., Ko, E., Atluri, P., Harkins, A. B., Fox, A. P., Salman, S. B., and Green, W. N. (1997). *J. Neurosci.* **17,** 8201.

Reeves, P. J., *et al.* (2002). Structure and function in rhodopsin: High-level expression of rhodopsin with restricted and homogeneous N-glycosylation by a tetracycline-inducible N-acetylglucosaminyltransferase I-negative HEK293S stable mammalian cell line. *Proc. Natl. Acad. Sci. USA* **99**(21), 13419–13424.

Ross, A. F., Green, W. N., Hartman, D. S., and Claudio, T. (1991). *J. Cell Biol.* **113,** 623.

Rudy, B., and Iverson, L. E. (1992). *Methods Enzymol.* **207.**

Saedi, M. S., Conroy, W. G., and Lindstrom, J. (1991). *J. Cell Biol.* **112,** 1007.

Schluter, O. M., Xu, W., and Malenka, R. C. (2006). Alternative N-terminal domains of PSD-95 and SAP97 govern activity-dependent regulation of synaptic AMPA receptor function. *Neuron* **51**(1), 99–111.

Sobolevsky, A. I., Rosconi, M. P., and Gouaux, E. (2009). X-ray structure, symmetry and mechanism of an AMPA-subtype glutamate receptor. *Nature* **462**(7274), 745–756.

Swick, A. G., Janicot, M., Cheneval, K. T., McLenithan, J. C., and Lane, M. D. (1992). *Proc. Natl. Acad. Sci. USA* **89,** 1812.

PART V

Model Simulations

CHAPTER 10

Computer Simulations and Modeling of Ion Channels

Michael E. Green

Department of Chemistry
City College of the City University of New York
New York, NY 10031

I. Introduction

Ion channels have been studied by a large variety of experimental techniques during the past half century. The structure and functions of channels are becoming better understood; however, molecular level understanding of the steps in selectivity and gating, as well as relations between the water, the ions, and the protein, are still in need of further study. Because it is difficult to determine experimentally the behavior of individual protein molecules, in spite of the ability to study the physiologic properties of single channels, computer modeling by simulation of all or part of the system has become an increasingly valuable tool in understanding the channels.

In a simulation, one attempts to start with the properties of the molecules that compose the system, and, by following the effects of the molecules on each other as they move, understand the thermodynamics, often the dynamics, and other properties of the system. Other properties may include dielectric constant, and structural parameters such as orientation of the molecules, or their arrangement with respect to boundaries. In addition to intermolecular forces, boundary conditions and external fields may be important. In recent years it has become possible to model the behavior of water and ions in pores defined by either rigid walls, walls to which water has been attached, or even models of proteins. The 1997 state of the art did not quite allow the complete modeling of the protein with the water, a protein with more than 200 amino acids, and more than 50 water molecules in its pore, plus an ion or two, was beyond the reach of the largest computers available in 1997. Computers available in late 2009 are able to handle such simulations, with over 10^5 atoms, thus including essentially all of the protein, plus surrounding lipids and caps of water at either end. However, the accuracy of the simulations may still require further attention. That said, the ability to simulate larger systems does carry with it the potential for much greater understanding of proteins. Considerable progress has begun in that direction with related systems, especially photosynthetic reaction center: that protein has water containing clefts, although it lacks a true channel (Lancaster et al., 1996). In addition, gramicidin, a simple channel composed of just 30 amino acids and with a pore that can hold just seven or eight water molecules, can be modeled in increasing detail (Roux, 1996; Sagnella et al., 1996); in 2009 such a system seems small, and simulations of gramicidin are nearly routine.

Woolf and Roux have carried out an all-atom simulation of gramicidin, with a section of lipid membrane and water caps (Woolf and Roux, 1997). Channels have large electric fields generated by charged amino acids, approaching 10^9 Vm^{-1} and voltage-gated channels generally have moderately large fields, often two orders of magnitude smaller, across their pore, when the membrane is polarized.

Channels have been modeled in the literature in several ways, neglecting different parts of the channel. More recently, complete channels have been modeled. In some work, the water is treated as a dielectric continuum, assigned a dielectric constant of 78 or 80, as appropriate near 25 °C, thereby avoiding the problem of treating the individual water molecules explicitly. We discuss this below, although it is much less common now. In other cases, the protein is replaced by a hydrophobic wall, and the water inside the wall is treated explicitly; this has not been done for some time, as it is no longer necessary, given the possibility now of doing all atom simulations. Various intermediate levels of complexity had been attempted already in 1997. An all-atom simulation of a single transmembrane (TM) segment has been carried out by Shen et al. (1997) on a polyalanine helix in a lipid membrane, surrounded by water; although one TM segment is not a channel, it pointed the way to more complete simulations of channels, which are now (2009) possible. Some principles remain the same for any type of simulation, including the necessity for obtaining intermolecular potentials for all molecules included in the simulation. The general procedures also remain the same. Water, being small, allows a more complete description of its potential than does protein. In part also, the choice of the part of the system to simulate is determined by the present state of knowledge. However, a number of simulations of water in pores have been done. Since 1997, models of channels have been superseded by X-ray structures, as referred to in the introductory paragraph. The first published X-ray structure dates to 1998 (Doyle *et al.*, 1998); it gave the structure of the KcsA channel, a bacterial channel that gated with protons, not voltage, and lacked a voltage sensing domain (VSD).

Since then structures as large as a full voltage-gated channel (with VSD) in the open state has been reported (Long *et al.*, 2005). With the aid of these structures, modeling the channel is essentially unnecessary. Instead, simulations start from the X-ray structures.

Why then would simulations be needed? Obviously not for structure, but there is a great deal that is not contained in the structure, including dynamics, motion of the ion through the channel, hydration, and thermodynamics. The X-ray structures are taken at temperatures well below 220 K, where water undergoes a glass transition. The room temperature behavior is therefore not entirely clear from simply looking at the structure. For this, calculations are needed. Increasing computer power has made possible simulations of the entire channel, with lipid and water, as we shall discuss later. Therefore, although we understand simulations of water in a channel model that may be somewhat too hydrophobic to be rather limited, they are still very informative, easier to follow, and simulations of larger systems will follow the same procedures when increasing computer power and improved protein models become available (as they are beginning over the past decade). Although this is only part of what needs to be simulated, much of this article is devoted to this aspect of the problem.

Now principally of historical interest, modeling of the protein itself was undertaken by several groups, for example, Durell and Guy (1996) and Guy and Durell (1995). Other references are given by Kerr and Sansom (1997) who modeled only the pore: they have done a number of simulations on other models (see later section).

The KcsA channel has since been redone several times (e.g., Domene and Sansom (2003) did an extended simulation of an improved X-ray structure of KcsA (improved structures were published for several years after the first structure was published in 1998)); however, this channel lacks a VSD, gating primarily with a drop in pH, although some voltage dependence apparently remains. At least one complete channel, with VSD, has been reported, in the open state (Long *et al.*, 2005), as have stretch sensitive channels. A complete discussion of the structures would carry us too far a field. These structures now make possible starting the simulations from a reliable set of coordinates. It should be kept in mind, however, that the X-ray structures are determined at temperatures well below 220 K, at which water undergoes a structural transition. Therefore, the room temperature structures may vary somewhat from those determined by X-ray crystallography; the differences are not so large as to cause a major problem.

A. Modeling Water in Pores

To describe the water, it is necessary to use a statistical description. The water has many possible orientations, and these cannot all be separately computed. If an attempt were made to analyze the hydrogen bonds of each of 30 or so water molecules, the number of combinations would be too large for any currently available computer. An attempt to minimize energy would not succeed because the channel is not at zero temperature, and the essential features of the pore water depend on thermal fluctuations. There are also multiple local minima, so that a global minimum would not be found, and a simulation would be needed to weight the various minima correctly. In the discussion of potentials, we consider the effects of neighboring water molecules on hydrogen bond energy.

A completely static picture would also not allow ions to move through the pore. Therefore, one must sample many configurations, all compatible with thermal equilibrium, and average suitably. The average orientation of the water molecules, the variation in density with position, rates of diffusion (to a limited extent, because time-dependent phenomena present additional difficulties), and other properties can be obtained from simulations. It is also possible to put ions in the model, and generate the appropriate electric fields to investigate possible binding sites and forces on the ion, thus obtaining the relevant properties.

Ion channels are unique in that they contain a small pore, with a dielectric boundary, which contains charges itself and in which ions are present; it appears from quantum calculations that in a potassium channel pore there may be only 8–12 water molecules, at most 8 relevant to solvating the ion. The fields in both the pore and the VSD are central to understanding the channels, because they may exceed $3 \times 10^8 \, \mathrm{V \, m^{-1}}$ especially in the VSD. The membrane polarization that keeps voltage-gated channels closed is equivalent to an average field of the order of only $10^7 \, \mathrm{V \, m^{-1}}$ across the entire membrane, or perhaps somewhat higher locally, as the drop across the membrane need not be linear. Therefore, the calculation of the field is a critical element in the determination of the properties of the channel as a whole.

Another major difference with bulk simulations is the presence of a pore wall, which requires that the boundary conditions of the simulation be set for a finite system. In contrast, in bulk simulations, it is typical to use what are called *periodic boundary conditions*, in which the simulation volume is repeated at each boundary of the simulation cell. We describe these in the Section III.C.3.

Intermolecular potentials are needed to describe the interactions of the molecules with each other, with ions, and with the walls. For water, a number of such potentials are available, with parameters derived in part from *ab initio* calculations, but principally characterized by adjustments to obtain agreement with thermodynamic properties of bulk water. Because no potentials are yet (even in 2009) available that have been parameterized specifically for water in confined spaces, there is no choice but to use the standard potentials: these are parameterized for uniform surroundings. Water in pores might still not behave in a manner substantially different from bulk water, in terms of the interactions of the molecules with other molecules, if the molecules do not "feel" the environment to be very different. Models are available that take into account the polarizability of the water; this has the advantage of being somewhat more accurate in high electric fields, but the disadvantage of requiring substantially more computer time. Ions, for which the largest interaction term is their charge, also have established van der Waals parameters, and their polarizabilities may be included. If explicit protein is to be included, then potentials are needed for the atoms of the protein too. Potentials are available for atom groups that can be used for this purpose. Hydrogen bond potentials have been shown to depend on their neighbors, so that neighbor tracking software would be extremely useful. Potentials vary by over $4k_BT$ (k_B is the Boltzmann constant; T, temperature in Kelvins) enough to make a large difference in the behavior of the water and its surroundings.

II. Fundamental Statistical Ideas

A. Definition of Ergodicity

To get thermodynamic averages for a system, it is necessary to average over a large sample of the configurations the system could assume at equilibrium. There is usually little difference in energy between one configuration and another. For example, if a water molecule rotates, the difference in energy may well be less than k_BT, making both configurations about equally likely; however, much larger changes are likely, as just noted for hydrogen bonds. One way to get an adequate sample is by going through many such configurations, sampling, one hopes, enough to provide a good estimate of the thermodynamic average. There must be a method of weighting the sampling according to thermodynamic equilibrium probabilities in order to get an accurate average, or generating them with Boltzmann probability (see Section III.C.1). Alternatively, one can take a single long trajectory as a function of time through the various configurations assumed by a system, and hope that one has sampled the possible configurations adequately. A theorem exists which states that these two possibilities are equivalent, the *ergodic theorem*. Although still not entirely proven in all instances, and of uncertain applicability in a finite system/finite time, we will assume that it holds well enough to allow us to choose either method for sampling a system.

B. Definition of Ensembles

Consider a harmonic oscillator, with energy $E = \frac{1}{2}kx^2 + p^2/2m$, where the first term is the potential energy, the second the kinetic energy: k, spring constant; x, displacement from equilibrium position; p, momentum; m, mass. We can consider a large number of such systems, with energy between E and $E + dE$ (see Fig. 1). Figure 1 shows the region in which all possible configurations of the one-dimensional harmonic oscillator with energy between E and $E + dE$ can be found. A simulation seeking to define the properties of the one-dimensional oscillator would have to sample a reasonable proportion of the shaded area. If, for example, the average position were sought, position left and right of the zero would have to be sampled with about equal frequency, at roughly equal distances from the origin. In this sense, ensemble averages are averages over the possible arrangements of a system in position and momentum. The ensemble itself is, however, a conceptual entity, composed of an indefinitely large number of copies of the system, all defined by the same properties. Three particular types of ensemble are most generally used. The microcanonical ensemble is defined by the number of particles N, the volume V, and the energy E. Figure 1 is an example of a microcanonical ensemble (in other words the energy is considered constant if all systems composing the ensemble have energy in a narrow range dE, not that all systems have identical energy). The canonical ensemble is defined by N, V, and the temperature T. This is more convenient than the microcanonical ensemble if one is interested in thermal equilibrium; temperature is maintained by exchange of heat with a thermal reservoir. The grand canonical ensemble is appropriate for an open system, in which particles as well as heat may be exchanged with V, and T. Then chemical potential is the third quantity that is held constant. It is also possible to do N, p, T ($p =$ pressure) simulations. We will discuss primarily N, V, T simulations.

C. Types of Simulation

Any useful simulation will be of a many-particle system. There are two main types of simulations, Monte Carlo (MC) and molecular dynamics (MD). One must define the ensemble in which one is working in any simulation. However, the two

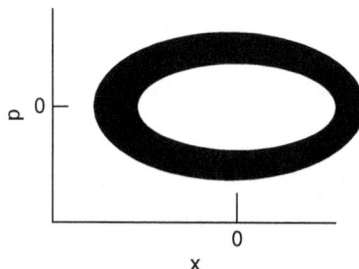

Fig. 1 Ensemble of one-dimensional harmonic oscillators with energy between E and $E + dE$. The inner ellipse is at energy E, the outer ellipse at $E + dE$.

types of simulations differ in how they use the ensemble, and they require ergodicity to be equivalent. MC simulations attempt to sample a large number of configurations, moving randomly from one to another nearby configuration, and repeating until enough of the entire coordinate space has been sampled; one hopes that the sample is adequate to give a good average (but checking is necessary). MD simulations track a system, following Newton's second law, $F = ma$, as the equation of motion, albeit in a much more elaborate fashion. Here, the hope is that the system passes through a large enough sample of configurations that the averages are valid. The equations cannot actually be solved, but particles can be moved according to the force, hence the acceleration: by keeping track of the trajectory and the velocity, one step at a time, the coordinate space is sampled. We consider the consequences of inadequate sampling later. Each step must be so short that there have been no significant motions of the other particles; otherwise, the force would have changed during the step, and the motion calculated for the particle being considered would be incorrect. For the force to be constant, the time step must be so short that even the highest frequency motions of the particles will have hardly changed position. It is therefore necessary to pick a time of approximately 1 fs (10^{-15} s) per step. This is a very short time (e.g., light moves 0.3 μm in 1 fs). Therefore, 10^6 steps produce a track 1 ns long, which is about the upper limit for practical simulations for most moderate size systems in 1997: presumably more powerful computers will be developed in coming years that will make it possible for the technique to be extended to biologically interesting times; in 2009, tracks are approaching 1 μs. It is not quite possible to produce trajectories for systems as large as an ion channel, which continue for times as long as gating, generally in excess of 10 μs.

However, there is a type of technique that randomizes boundary conditions or else forces, stochastic dynamics (SD), which is capable of going to longer times. It is somewhat difficult to avoid errors in complex systems: however, SD has been used successfully, and one such case is examined later. We only mention some other techniques for going to longer times. General references for dynamics simulations as applied to biological macromolecules include McCammon and Harvey (1987) and Brooks et al. (1988).

III. Potential Energy

A. Summary of Forces

A complete potential energy description of the system includes intramolecular and intermolecular energy terms. The former include vibrations, with a term for each normal mode, bending about bond angles and torsion about dihedral angles. All are high-frequency motions, especially the normal mode vibrations. It is to accommodate these motions that MD simulations must take such short time steps. For each quadratic term in the energy, a constant, analogous to a spring constant, is needed (for torsions, usually a trigonometric, not quadratic, dependence appears, e.g., $\cos\theta/a$, where θ is the torsion angle and a is a constant related to the symmetry of the displacement). We will spend little time on the intramolecular terms, which are similar to those in bulk simulations. Water molecules are frequently treated as rigid bodies in bulk simulations, and C–H groups, such as methyl, may be combined into "extended atoms." Among commercial programs in 2009, GROMOS, for example, does this. The groups comprising the protein in ion channels can be parameterized by comparison with other known substances; the properties of the protein are not well known independently.

B. Intermolecular Forces

1. Water

For either MC or MD simulations, it is critical that intermolecular potentials that represent the system as accurately as possible be used; one expects that the degree of realism will increase as experience accumulates and computers improve, understanding that "realism" is a relative term. Several types of forces exist, which are principally either electrostatic or van der Waals forces in origin. There are distributions of charge in chemical bonds, which lead to local dipoles, or charges on atoms, which in turn interact with each other. The van der Waals forces are also well known; for simulations, it is normal to include these forces explicitly in the potentials. Water is best to begin with, as the molecule which illustrates the main features. We do not cover all possible proposed potentials, but limit ourselves to a few commonly used ones. We need to know how the potentials are evaluated. Two types of potentials are available: those which are polarizable and those in which the electrostatic part is represented strictly by point charges. One property that is rarely included, but is almost certainly more important in ion channels than has been allowed for so far is the exchange of groups for ionizable species, especially protons in the channel, these could be represented by allowing dissociative steps, but this is rarely done.

a. Point Charge Models

Several point charge models of water have been proposed. We consider only two of them: the simple point charge (SPC) model (Berendsen *et al.*, 1981), and one of the TIP models (TIP4P) (Jorgensen *et al.*, 1983). These are discussed in Levesque and Weis (1992). Each is characterized by a set of charges on the hydrogen and oxygen atoms (SPC), or near them (TIP4P), plus Lennard–Jones parameters. Given the polarity of the dipoles in the water molecule, the H atoms are, of course, positive, the oxygen negative. In the SPC model, the charges are distributed as shown in Fig. 2A; in TIP4P, as in Fig. 2B. The latter has four points: three for charge, while the oxygen atom is used for the center of van der Waals forces (i.e., Lennard–Jones terms). (We note that TIP3P, a three-point model, is still often used in MD simulations, where the point with no mass found in TIP4P causes a problem, but we use TIP4P, used in MC, in this discussion.) Note that the model includes not only partial charges, but bond lengths and angles as well.

A

$q = -0.82$

$r_{OH} = 1.0\,\text{A}$ $109.47°$

H   H

$q = +0.41$ $q = +0.41$

SPC model

B

O $r_{OM} = 0.15\,\text{A}$

$r_{OH} = 0.9572\,\text{A}$ M $q_M = -1.04$

$104.52°$

H   H

$q = +0.52$ $q = +0.52$

TIP4P model

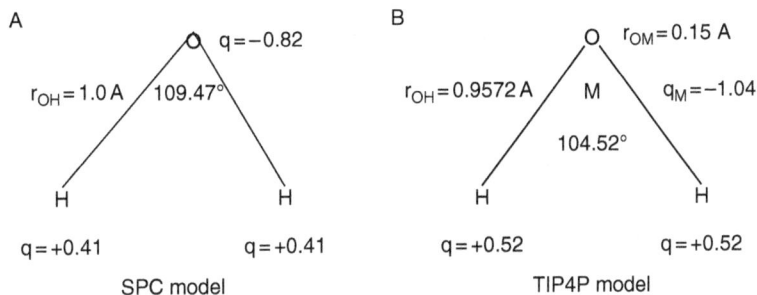

Fig. 2 Parameters of two point charge models of water. The values of q are the electronic charges on the points shown. (B) The negative charge is at point M, not on the O atom. M is 0.15 Å toward the H atoms on the angle bisector. In addition to the charges, there are van der Waals terms centered on the O atom in both models. The charges are in units of the charge on a proton.

Electrostatic interaction energy among molecules can be found from $\Sigma q_i q_j / r_{oo}$, where the sum is over all pairs of interactions between charge sites i,j on different molecules; r_{oo} is the site–site distance. The Lennard–Jones term is $A/r_{oo}^{12} - C/r_{oo}^6$, where for either model oxygen–oxygen distances are used. The values A and C are similar for the two models. Again, it is necessary to sum over each pair of molecules. Therefore, the total energy of interaction can be written as

$$E_{mn} = \sum_i \sum_{j>i} q_i q_j / r_{oo} + A/r_{oo}^{12} - C/r_{oo}^6, \tag{1}$$

In Eq. (1), the energy is for the interaction of molecules m and n; the second sum has molecule indices $j > i$, where i is the first sum index, to avoid double counting. The O–O distances are understood to have the same molecular indices, and there are 9 terms (3×3) for the charges on each molecule–molecule interaction.

The calculation time is dependent approximately on the square of the number of interacting molecules, and thus there is a substantial advantage to cutting off the sum at a finite distance; the computation time then increases only linearly with the number of molecules. Fortunately, this is less of a problem in ion channel simulations, where the channel boundary is less than a reasonable cutoff distance away in at least two dimensions. The cutoff for the Lennard–Jones terms might be at 8 Å; $8^6 = 2.6 \times 10^5$, and the constant, C, is about $4 \times 10^{-18}\,\text{J}\,\text{Å}^6$, in both models, so that the value of the attractive term would be of the order of 10^{-23} J, less than 1% of $k_B T$; the repulsive term is much smaller at large distances (a scaling function, providing a gradual cutoff, is sometimes better). The real problem is with the Coulomb terms; these fall only as $1/r$, and are therefore long range. A variety of schemes exist for obtaining the sum at large distances, and simple cutoffs are not satisfactory. A smoothed cutoff helps; terms beyond may also be included via a reaction field, or sum term (Ewald sum). These too have disadvantages, but it would take too long to discuss them.

Table I
Calculated and Experimental Properties for Liquid Water[a]

Property	SPC	TIP4P	Experimental
d (g cm^{-3})	0.971	0.999	0.997
ΔH_{vap} (kJ mol^{-1})	45.06	44.60	43.97
C_p (J mol^{-1} K^{-1})	97.9	80.7	75.27

[a]Data from Jorgensen *et al.* (1983).

For an ion channel, it is probably simplest to keep the complete set of interactions, as the number of molecules is not great, and the risks of using a cutoff in a system with complicated dielectric boundaries too large. This is especially necessary because of an additional complication: the interaction between charge centers is not simply dependent on distance, but depends also on the relation of position to dielectric boundary. In all atom simulations the dielectric constant can be taken as unity, with the polarization of the atoms included explicitly.

The quality of simulations in bulk water, using these potentials, is good, as shown in Table I. The heat capacity, C_p, is the derivative of the enthalpy, and the error with derivatives is inherently greater as they change sharply with smaller changes in the quantity of which they are the derivative. These data are from MD simulations done at 25 °C and 1 atm. Since this work, from the original TIP4P paper, was published, more extensive work with the same potentials has been carried out; the advent of large computers has meant more extended calculations on larger numbers of molecules. However, the essential points remain unchanged for these potentials. They are good for the bulk liquid, and allow a reasonably good description of its properties, including its radial distribution function (the probability of finding a molecule at a distance r from a central molecule). It is also possible to get a number of nonthermodynamic properties for the water, including properties of the hydrogen bonds. Different sets of parameters produce hydrogen bonds of slightly different length, a matter that also affects the density. As noted above, hydrogen bonds are also dependent on neighboring groups, a point omitted in any of the standard potentials, even in 2009. The importance of this has been noted by Znamenskiy and Green (2007) who used DFT calculations to show that neighbors can raise or lower the energy of a hydrogen bond by as much as $\pm 3k_B T$ (in most cases, $\pm 2k_B T$) with respect to TIP3P energy; as anything greater than $k_B T$ is significant, this implies that the potentials in an ion channel, for instance, must be adjusted for the effects of neighbors. So far this has not been done. Based on unpublished calculations, it appears that salt bridges produce similar dependence on bonding to neighbors.

The average molecular dipole is also important. In the gas phase the dipole moment for the water molecule is 1.85 D (D = Debye, where 1 D = 3.336 × 10^{-30} C m). However, the liquid behaves as if the dipole were about 30% larger. The TIP4P dipole is approximately 2.4 D and the SPC dipole 2.3 D, about correct for the liquid, for which they were parameterized. These models are limited in several ways. For one thing, they assume rigid molecules, thereby losing three vibrational degrees of freedom. Fortunately, the vibrations occur at relatively high energy compared with $k_B T$, so that the contributions of the vibrations are limited (although omitting the vibrations would be a limitation if high accuracy were needed).

b. Polarizable Models

The local charges in a protein produce a fairly large electric field, and this field can in turn distort the electron distribution in the molecule. The problem is more acute in a channel, as the fields are extremely high. For this reason, a model of water in which the water is polarizable would be expected to give improved results. In a polarizable potential, there is an induced addition to the dipole moment of $\mu_{ind} = \alpha \mathbf{F}$, where α is the *polarizability* of the molecule and $\mathbf{F}$ the local electric field. Then there is an additional term in the energy of the form

$$E = -\mu_{ind}\mathbf{F} \tag{2}$$

In Eq. (2), consider α to be a scalar (in a solid it could be a tensor); then one takes the scalar product of the two vectors on the right-hand side. This additional energy can be appreciable; it clearly goes as $\mathbf{F}^2$, as α is a constant. The polarizability is of the order of the size of the molecule, 10^{-24} Å^3 or 10^{-40} C V^{-1} m^2. If $\mathbf{F}^2$ is on the order of 10^{19} V^2 m^{-2}, then the induced energy is of the order of 10^{-21} J for the molecule, or close to $k_B T$; a more precise calculation gives values close to $k_B T$ for several molecules in the channel in each round of simulation. Since, in an MC calculation, the decision on accepting or rejecting a move (see later discussion) depends on quantities of this order, the polarizability does matter. For this reason we will consider one example of a polarizable potential. Several have been proposed, going back at least to the potential of Stillinger and David (1969).

The polarizable SPC (PSPC) potential was proposed by Ahlstrom *et al.* as an extension of the SPC potential. It adds polarizability on the O atom of water, with $\alpha = 0.55 \times 10^{-40}$ C V^{-1} m^2. Calculating the local field is not as simple, because it includes both the external field and a dipole tensor term. The dipole tensor depends on the values of the intermolecular vectors that are being calculated. Therefore, the calculation must be made self-consistent, a substantial increase in computational effort. There are also dipole–charge interactions, and an energy term for creating the dipoles. An additional complication results from the tendency of the polarizability to draw the molecules together, leading to still higher absolute values of the energy, and forcing the molecules to come even closer together; it is necessary to put in some sort of limit on this process to avoid a polarization catastrophe. Ahlstrom *et al.* (1989) added increased repulsion at short distances. Other possibilities include a hard core at an O–O distance of 2.5 Å (Lu and Green, 1997) or a damped field at close approach. The reader is referred to Ahlstrom *et al.* for details. It is in any case possible to prevent the polarization catastrophe. The essential point of including the polarizability is the increased accuracy in the calculation, especially in high field.

In addition to the potentials listed here, a number of other potentials have been proposed (e.g., Chialvo, 1996; Svishchev *et al.*, 1996). Finding an accurate representation of water is an ongoing project, especially when conditions far from 25 °C and 1 atm pressure are to be represented. The most significant extreme condition encountered in ion channels is the high electric field. In addition to polarization, the high field leads to an increase in pressure to some extent due to a phenomenon known as *electrostriction*, but the consequent increase in density is probably not enough to change the required intermolecular potentials. However, it is possible that eventually a more appropriate model of water will be used for higher accuracy simulations of water in channels. In addition the comments on the effects of neighbors on hydrogen bonds should be added to these considerations.

2. Ions

Intermolecular potentials for ions are obviously just as important as those for water, but fortunately much simpler. We are primarily concerned with two ions, K^+ and Na^+, and sometimes Ca^{2+} or Cl^-. A simple potential can be constructed by using the charge on the ion (there is no ambiguity in position for a charge when only one atom constitutes the entire ion), van der Waals parameters, and a polarizability that may be keyed to the water model in use for the simulation. The charge is necessarily one full charge (but a word of caution: some of the charge may be transferred to solvating molecules, a point that can be picked up in high level *ab initio* calculations). In addition, one may observe that for consistency with the surrounding water it would be necessary to include polarizability if a polarizable model of water were used. Since the polarizability is proportional to the molar refraction, the value may be estimated by comparison to the molar refraction data. However, better results can be obtained via *ab initio* calculations. The question of ion water interactions has been considered in the most detail by Aqvist (1990). He gives the following ion–water interaction potential:

$$E_{l-w} = B \sum_j (q_i q_j / r_{ij}) + \sum_j (A_I A_i / r_{ij}^{12} - C_I C_j / r_{ij}^6) \qquad (3)$$

where B is the constant to adjust units and the A_i, C_i, parameters are the SPC values defined earlier, for the van der Waals terms. Then the values of A_I, C_I are the van der Waals parameters associated with the ion. The charges in the first term are the same as in the SPC model. Numerical values can be obtained from Aqvist's paper. The values are obtained by empirical adjustment of the thermodynamic values. *Ab initio* values have improved greatly over the past decade, and now appear to be close to being good enough to rely on. One more generation of supercomputer should suffice to make them highly reliable.

3. Protein

a. Rigid Wall Representation

Channel walls are composed of protein, and must be in some way included in the calculation. The simplest solution is to evade the problem, by substituting a hard wall for the protein; the next more sophisticated solution is to substitute a dielectric medium for the protein. At least two such simulations, using tapered pores to imitate the general structure of a Na^+ or K^+ channel, including both acetylcholine receptor channel and voltage-gated channels (at least most of which are tapered, i.e., hourglass shaped), have been carried out. This approach has been taken by Green and Lu (1995) and Lu and Green (1997). In that model, the hydrophobicity of hard walls was modified by adding fixed water molecules on the channel side of the wall; these molecules, not allowed to translate, replaced a part of the hydrogen bonding ability of the protein that would have made up the channel wall. Since then the same method has been used by Henderson et al. The calculation method is discussed in Section IV.A. In a similar simulation, but without the fixed water molecules, Sansom et al. (1996) used a hydro-phobic wall softened by an S-shaped potential curve at the boundary, with the wall 31 kJ mol^{-1} (5.2×10^{-20} J molecule^{-1} or about $13\,k_BT$) high, and rising to 90% of full height over 1.15 Å. In the latter simulation, no particular dielectric properties were assigned to the wall, and only the hard wall potential was used. An MD simulation was done, while in the former simulation, an MC simulation was used. In the former, the wall was assigned a dielectric constant, and an electric field was calculated. Both concluded that with the taper, the wall aligned molecules significantly; in the Green and Lu simulation, this was true principally in the tapered region, and depended in part on the electric field. In the Sansom et al. simulation the outer layer of the water was oriented throughout the taper. There is little doubt that water is oriented by the geometry of the walls, and that therefore this would affect the dielectric constant in the pore. Furthermore, the orientation would affect the electric field directly, by providing a dipole moment equal to that of a number of aligned molecules. These effects are not surprising (simulations in cylindrical pores have shown some alignment in the water closest to the walls) but are important; it remains to be demonstrated experimentally that they are found in actual channels. However, this type of calcula-tion has been largely superseded. The orientation of water, as shown by quantum calculations (generally DFT type), shows the water solvating the ion when the ion is at the midpoint of the cavity of the channel, and forming a network of hydrogen bonds when the ion moves to the bottom of the selectivity filter. The moral of the story seems to be that clever modeling is no match for a complex natural system.

b. Explicit Representation

It is obviously better to use a complete representation of the protein, and a number of examples had been published (as of 1997). Some are in small channels (alamethicin, gramicidin), formed from peptides of limited size, and not close in structure to the gated channels in which we are interested. One attempt has been made (Singh et al., 1996) to simulate a sodium channel, using a model of the channel proposed by Guy and Durell (1995). As with the models discussed above, the model here has been completely superseded. However, it is also of some interest to look at other proteins, especially if channel-like, such as photosynthetic reaction center. In this section we are concerned with the representation of the protein much more than with the results of the simulation.

The simulation of the pore lining of a model of sodium channel by Singh *et al.* (1996) considered 130 amino acid residues, which line the pore in the model. Just as with water, it was necessary to have intermolecular potentials for the atoms that compose the protein. Since these were represented explicitly, there must be a distribution of charge to go with the atoms. One compromise was made in the choice of potentials; methyl, methylene, and methyne groups were treated as extended atoms, and a single potential assigned to the group, rather than to each C and H atom. The water in the channel was explicitly represented also in this model, with 17- and 20-Å caps at the narrow intracellular end and the wide extracellular end, respectively. In total, 933 water molecules were included, using the SPC model for water. The protein potentials were taken from a commercial package, GROMOS, developed at the University of Groningen (van Gunsteren *et al.*, 1983). It is one of several such, listed in Section V (GROMOS has changed to GROMACS).

While it would take too long to go through the means of parameterizing a protein, it is in principle similar to what is done to parameterize water. In the Singh *et al.* study, in addition to the protein and water, Na^+, K^+, and Cl^- ions were present, and the Aqvist potentials mentioned earlier were used. With all the potentials in hand, an MD study was carried out. Because the potentials are our immediate concern, we postpone consideration of the remainder of the study, which is one of the more interesting in the literature at the present time.

A description of the properties of a set of potentials in the form required for a simulation of a protein is given in McCammon and Harvey. These include the following:

1. *An atom data file.* The properties of each type of atom must be described. This includes partial charges, van der Waals radius, and Lennard–Jones parameters. If an extended atom approach is to be used (H atoms attached to C not explicitly included), the types of atom must include such "extended atoms" as methyl groups, because these will have different parameters than the original atoms. Also, not all atoms of the same element are of the same type. For example, sp^2 and sp^3 C atoms differ in their properties, including in particular bond angles, and must be distinguished.

2. *Coordinate file.* The Cartesian coordinates of the atoms to be used explicitly in the simulation must be included in the data. If an X-ray structure is known, the coordinates may come from the current protein database (in 2009, the URL is www.rcsb.org), for example. If one is starting with a model, as in the Singh *et al.* (1996) calculation, the file must contain the location of the atoms according to the model.

3. *Internal coordinate (molecular topology) file.* From this file, the Cartesian coordinates must be translated into internal coordinates, such that bond lengths and angles are correct. This refers to the list of atoms bonded to each other, the potential

types to use with each, the partial charges, torsion terms, and often the information to build the starting molecule (bond lengths, angles, etc.). If the molecules are to be rigid, these internal coordinates will not change during the simulation. Also we are not generally concerned with cases in which covalent bonds break (but see Sagnella *et al.* (1996) and recall the comments above concerning proton exchange).

Given three translational coordinates for a molecule and three rotational angles, all Cartesian coordinates for atoms in the molecule would be determined if the molecule were rigid. Assuming vibrations are allowed, however, there must be displacement of the atoms from the position they would have if they were all at exactly their equilibrium position.

4. *Energy parameter file.* The Lennard–Jones energy parameters, the charges, and all other inter- and intramolecular interaction energy parameters are given here for each type of atom. If vibrations are allowed, harmonic oscillator terms for internal motions would be included. Three body interactions are not included explicitly in normal simulations, and would go with polarizability calculations, which would require self-consistency, a more elaborate computation.

In addition to the data files listed by McCammon and Harvey (1987), it may be necessary to calculate an electric field for simulations in an ion channel: this may be derived in part from the partial charges in the energy parameter file, if an all-atom simulation is being carried out. However, if dielectric boundaries are involved, the field produced at one atom by another is not simply a function of the distance between the atoms, but depends explicitly on their location with respect to the boundary. We consider this matter separately later.

Once one has the data files, it is possible, in principle, to recalculate the energy in each pass through the simulation, atom pair by atom pair, as a function of distance, with the formulas built into subroutines in the program, or to have large lookup tables as a function of distance. The exact structure of a program, with some lookup tables and some distance dependent computation, has to be decided based on the nature of the computation. We will see a case later in which the electric potential and field computation depended on a preliminary computation of the field, followed by use of the field as a lookup table with interpolation for values at locations between those in the table. This approach is more feasible with ion channels than with general simulations because the volume to be covered is generally more limited. Otherwise lookup tables can quickly exceed the available memory of the largest computers (as of 1997). The use of virtual memory, with continual reference to disk, is slower, but may be unavoidable. However, the memory of computers is no longer so limited (as of 2009), so this is not any longer a major consideration.

C. Simulations

We have already noted that two main types of simulations are of interest to us: MC and MD simulations. We first describe the essential points of MC simulations, then MD. Finally, we add a few comments about SD.

1. Monte Carlo Simulations

In MC simulations, molecules are moved in turn. The change in energy is calculated, following which a decision is made as to whether to accept the change in configuration, or consider the old configuration to also be the new configuration. The point is to discover equilibrium properties. There is no explicit form of time dependence, and it has rarely been attempted to assign time equivalents to the length of an MC simulation. It is not clear that any legitimate way to do so exists. An MC move, as the name implies, is random.

a. Moving the Molecules

To move a water molecule, it is necessary to use a random number generator to get a random change in each coordinate, producing a new location for the molecule. The move may be in more than one coordinate simultaneously. If one is using a rigid water model, like SPC or TIP4P, its position is described by six coordinates: three for the position of some location in the molecule, say, the O atom, and three for rotation. It is possible to move all six simultaneously, but then the move for each should be small enough that the chance of having moved to a much higher energy position (say, within the van der Waals radius of another molecule, or into a channel wall) is small. Translations should be kept to fractions of an angstrom, rotations to a few degrees, up to tens of degrees. It is critical that neither be biased, or the molecules may end up on one side of the volume.

The random number generator may be a matter of some concern, since some random number generators have subtle bias. However, the chance that this will be commensurate with the moves of the molecules, thus biasing the simulation, does not seem very large. As far as the author knows, the question has never been systematically investigated. Generally, one starts with a different random number seed for each simulation, and the sequence of numbers produced by a pseudorandom number generator is very long.

Therefore, different instances of otherwise identical simulations will sample different portions of phase space, and can be considered independent.

b. Acceptance of Moves

A criterion proposed by Metropolis *et al.* (1953) is used to decide whether to accept a move. If the move is to lower energy, it is accepted. If the move is to higher energy, it is tested as follows: for an increase of energy by ΔE, the move is accepted with probability $\exp(-\Delta E/k_B T)$. (Deciding whether to accept the move requires another random number.) If the move is accepted, it becomes the new configuration of that molecule. If the move is rejected, the old configuration becomes the new configuration, and is averaged it again. It is not legitimate to average only accepted moves, because this has the effect of forcing a move on each attempt. In a normal simulation, about 20–50% of the moves should be accepted. If the acceptance rate is too small, it is probable that the steps are too large.

c. Sampling

MC simulations are less than perfectly efficient in sampling the phase space. If too small a portion of coordinate space is sampled, the probability of getting a good estimate of the thermodynamic equilibrium state of the system is small. This can happen if the system is trapped in a local minimum. For example, referring to Fig. 1, suppose the oscillator were always sampled in the upper half of the shaded area, where $p > 0$; some barrier prevented transition to $p < 0$ states. Then the oscillator would appear to be moving continuously to the right, an obvious error. Unfortunately, errors are not always so obvious, and incorrect sampling may lead to conclusions that are wrong but are not discarded. In systems of realistic size, there may be a huge number of minima, and some method of finding a reasonable set is required.

A conceivable alternative is to repeat the simulation several times, and average the results rather than simply extend an individual simulation, an obviously time-consuming procedure. One way around this is to use *simulated annealing*. In this procedure, the system starts at a high temperature, say, 600 K, so that the sampling can cover much more of phase space: the effective depth of any well is inversely proportional to the temperature of the system, from the point of view of a Boltzmann distribution. After the system has passed through many wells, it may be cooled in stages, in the hope that it will settle in the correct minimum, or very nearly so, at the intended simulation temperature of, say, 300 K.

If one has additional information about the system there are several tricks available to introduce this information into the simulation. One method, which has been used for biomolecules, is to introduce a coordinate representing the position of some conformation or other factor that changes the potential, such as the extent of binding a ligand. The simulation is carried out at several values of this coordinate, and the free energy examined as a function of the coordinate. Other thermodynamic properties can be obtained as well. With the coordinate variable, the potential that drives the parameter to the desired range is called the *umbrella potential* and this procedure is called *umbrella sampling*. The normal probability of the system being in some of the positions of coordinate space may be small: by adding the umbrella potential the system may be forced to spend some time in the moderately less probable regions, so that their contribution can be estimated. Then, since the umbrella potential is known, the correct contributions of the various regions of configuration space can be determined with better accuracy than would have been possible in a finite time with normal sampling. Because this is a free-energy difference technique, the umbrella potential in effect being added to the free energy, it is also related to the *potential of mean force*, defined as the derivative with respect to coordinates of the Helmholtz free energy. A method for choosing the potential has been suggested by Mezei (1987). Several other techniques for finding the free energy are also relevant (see the Section IV.C).

More generally, any additional knowledge of the system, say, of the regions of coordinate space which are likely to be more strongly represented among possible configurations, can be incorporated into the simulation to ensure a better sampling of the important regions of coordinate space, by moving the system to entirely different regions of the space to start sampling from there. One example of such knowledge might be large-scale conformational changes, when it is known that more than one set of dihedral angles should be possible for a peptide, and each set must be sampled. Without the large-scale shift, the simulation might never move out of whichever local minimum it finds itself in. New methods of simulation over wide regions in phase space, or at least coguration space, are continually being proposed.

2. Molecular Dynamics Simulations

The basic method by which MD simulations sample phase space is rather different than the method in MC simulations, and is often more efficient in sampling the space. MD actually tracks a system as it moves. Therefore, velocity, acceleration, and time dependence become relevant. The equation of motion of the system is set up to follow Newton's second law, $F = ma$ (F, force; m, mass; a, acceleration). Begin by finding the force on a particle. After this the equations of motion can be integrated. Some of the same considerations apply to MD and to MC. For one thing, the same intermolecular interactions, and energy calculations, apply. Therefore, we need not repeat that part of the discussion.

a. Force Calculation

The energy is determined in the same manner as in the MC calculation. Then the force on the particle can be found from the gradient of the energy:

$$F_i = -\nabla E_i \tag{4}$$

where the gradient is taken with respect to particle i coordinates to obtain the force on that particle. To get useful information from the potential, it needs to be in differentiable form.

b. Integrating the Equations of Motion

To use the force in $F = ma$, observe that $a = d^2x/dt^2$, change in velocity $= \Delta(dx/dt) = (d^2x/dt^2)\Delta t$ and change in position is $\Delta(x) = (dx/dt)\Delta t$. However, the time step is finite, and simply projecting the system forward using the position, momentum, and acceleration particle by particle leads to serious inaccuracies. Another way of looking at this is to expand the position in a Taylor series (we will write it in one dimension for simplicity):

$$x(t + \Delta t) = x(t) + (dx/dt)(\Delta t) + \tfrac{1}{2}(d^2x/dt^2)(\Delta t)^2 + \tfrac{1}{6}(d^3x/dt^3)(\Delta t)^3 + \cdots \tag{5}$$

However, this is an infinite series, and must be truncated, which is done after the square term, the term obtained from the acceleration. A method of avoiding the error introduced by the truncation is needed. The magnitude of the error can be estimated from the fact that energy and momentum must be conserved (all the forces are internal to the system; therefore there is nothing that can change the total energy or momentum). If there are systematic errors, the total energy and momentum of the system will drift from their starting values.

There are in practice a variety of means of correcting the integration of the equation of motion. The most popular are due to Verlet (1967), Gear (1971), and Beeman (1976). A variant called SHAKE (Ryckaert *et al.*, 1977; van Gunsteren and Berendsen, 1977) has also proven very popular. Because these methods are described in the McCammon and Harvey book and its references, we will only mention that these integration algorithms essentially entail more sophisticated methods of compensating for the errors introduced by simply truncating the series in Eq. (5). Observe that the values of x, dx/dt, etc., are the values at the beginning of the time step. It would be better if the average value of these quantities during the step were used. The Verlet method, as an example, looks at the value at the midpoints of the time interval (i.e., $\Delta t/2$ before and after the Δt time step, and then "leapfrogs" the velocity from the value at $t - \Delta t/2$ to the value at $t + \Delta t/2$). The effect is to make the overall time step Δt, but take the average velocity at the midpoint of the interval rather than at the beginning. The other algorithms employ higher order corrections. SHAKE is an improvement that can be applied to an integration algorithm in that it reduces the computation time by freezing certain degrees of freedom, such as C–H bond distances. While the length of the time step is normally determined by the highest frequency motions of the molecule, generally the molecular vibrations, SHAKE sets constraints, typically fixed bond lengths, and takes a longer time step (therefore allowing the computation to proceed through more of space in the same computational time). At each step, the SHAKE algorithm resets the atom positions to fit the constraints within some specified relative margin (say, 10^{-4}). It is done iteratively, treating the positions sequentially, rather than exactly, to avoid solving a nonlinear matrix equation.

c. Sampling

While the sampling efficiency of the MD method is typically greater than that of the MC method, at least where vibrations are included, this does not eliminate the problem of finding the system in a local minimum. It is very nearly impossible to find a global energy minimum with hundreds of degrees of freedom, or more, as in a channel. Sampling a reasonable fraction is extremely important. However, the tricks that could be applied to an MC simulation are essentially the same as the tricks that could be applied to an MD simulation: details differ. Again, it helps tremendously to have some additional information concerning the possible conformations of a system.

3. Boundary Conditions

Very similar questions arise with MC and MD, so there is no need to discuss them separately. The problem arises because a few hundred molecules must stand in for essentially an infinite number, or at any rate a mole, in bulk simulations.

Therefore, a "box" (whatever its actual geometric shape, normally determined by the symmetry of the system) equivalent to about 8 or 10 molecules on a side, has to produce the same thermodynamic quantities as the macroscopic system. There are a couple of ways around this that appear to be successful. However, the most successful does not apply to ion channels. In fact, this is one way in which ion channels differ most seriously from bulk simulations, in that an "infinite" number of molecules need not be represented. Ion channels are finite in size, and must allow water molecules and ions to come into contact with the protein boundary, in whatever way the protein is represented (hard wall, softer wall, or explicit atoms).

The most common method of representing an infinite system with a small number of molecules is to use *periodic boundary conditions* (Binder and Stauffer, 1984; McCammon and Harvey, 1987). The simulation volume ("box") is reproduced on each face by a box containing an identical set of molecules, so that the molecules near one edge see not a hard wall, but another set of molecules, which are in fact identical to those on the opposite side of the simulation box. Only one interaction is allowed with another molecule, that which is at the shortest distance, whether that distance is to the molecule in the original box or to its image in a neighboring box. Thus each molecule, whether in the center or near the edge of the simulation volume, sees an equivalent environment. Any molecule at the left boundary, say, position $(-x, y, z)$, is made a neighbor of molecules on the right boundary, near (x, y, z). This has the effect of keeping the density near any molecule, including those on the boundaries of the simulation volume, essentially the same as the density in the center. In bulk simulations, this is an adequate description, in which the hundreds of molecules in the simulation amount to only a tiny fraction of the solution being simulated. The system, infinite for practical purposes from the viewpoint of any small subvolume, needs to appear infinite in the simulation also. In effect, the boundary disappears, and the system can be translated along any axis. This is not true in a channel, unless the entire protein, including some of the boundary lipid, is included. The electric fields as well as the protein walls are not symmetric. If the model is complete enough, however, and includes enough of the boundary lipid, then the same considerations apply as to any other infinite system.

The physical environment in a channel is also not isotropic. Here, as we discussed in Section III.B, the wall must be represented in some explicit fashion. For example, the boundary conditions appropriate for the channel simulation of water are the protein for at least most of the wall. However, the water at the intracellular and extracellular ends of the channel is in contact with bulk water and must be represented by either a cap of additional water molecules large enough that any edge effects would have relaxed by the edge of the functional part of the simulation volume, or else allow some sort of periodicity in one dimension, along the axis of the channel. The size of a cap would have to be at least on the order of the minimum cutoff distance of intermolecular potentials, say, 9 Å. The techniques vary, but must take into account the specific properties of the model being used.

a. Methods for Longer Time Simulations: Stochastic Dynamics and Others

Stochastic dynamics attempts to speed the simulation approximately 1000-fold by averaging the high-frequency forces. Alternatively, one can allow the boundary conditions to fluctuate, but we do not consider such a case here. To simulate times longer than approximately a nanosecond (in 1997; in 2009, at least 100-fold longer simulations are possible without stretching, and sometimes 1000-fold), it is necessary to have a means of averaging these fast forces. Then large time steps are possible, and times up to hundreds of microseconds can be simulated. The remaining forces are essentially frictional, and a random force must be added. A microsecond is long enough for an ion to pass through a channel. However, SD, in which averaging is applied by way of variable boundaries, is beginning to be applied to biomolecules. Wang *et al.* (1997) have noted that Davis and McCammon (1990) have shown that there must exist, at least in polar liquids, a solvent boundary force, in the form of a pressure on the surface atoms of the solute molecule. Wang *et al.* have extended this work to cyclosporin, a cyclic undecapeptide. A number of workers have used the *boundary element method*, which includes an explicit boundary surface to include surface pressure forces. Wang *et al.* may have carried out the first simulation to include a peptide in solution. Another approach has been suggested by Laakkonen *et al.* (1996) based on a method developed by Guarnieri (1995). This mixes MC and SD; the latter requires the use of the equation of motion, with a random force added:

$$m \, dv/dt = f[x(t)] + R(t) + m\gamma v \qquad (6)$$

where m is the mass of the particle, dv/dt is the acceleration, and forces are $f[x(t)]$ the deterministic force, $m\gamma v$ the frictional force (γ is a friction coefficient), and $R(t)$ a random force with the property that its correlation function is

$$R(t)R(t') = 2m\gamma k_B T \delta(t - t') \qquad (7)$$

In other words, the average value of the product $R(t)R(t')$ is 0 if the times are different; there is no correlation between different times. This is characteristic of a random force that relaxes rapidly compared with the other characteristic times of the system, and is appropriate for simulations that are to be done for a time scale long compared to the time scale of atomic motions. Parameter γ lumps fast degrees of freedom. What is unique in the method used by Laakonen *et al.* (1996) is the combination of SD simulation in the Cartesian space with MC calculations of the torsions in angular space. An MC step, with Metropolis algorithm, intervenes between dynamics steps based on a solution of Eq. (6). The velocities for the succeeding dynamics step (including the random force and frictional force) come from the preceding dynamics step, using the Verlet algorithm discussed earlier, in a form that does not require information at time $t - \Delta t/2$ (the "velocity Verlet algorithm"). The earlier (i.e., at $t - \Delta t/2$) information is unavailable here due to the MC step that intervenes between SD steps: the coordinates depend on the MC step. If the step is accepted, new coordinates are used to compute the position-dependent forces. The method is stated to converge two to three orders of magnitude faster than if "other simulations" are used. Laakonen *et al.* applied it to the TRH receptor-binding pocket; the TRH receptor is a G-protein-coupled receptor (GPCR). Although GPCRs are not the primary subject of this article, they are membrane-spanning proteins and methods applicable to them can in all likelihood be applied to channels.

This is only one example of the attempt to simulate longer time intervals. A variety of techniques involving constraints, multiple time steps, integration of the equation over longer time steps with partial integrals at shorter intervals, and other methods have been tried. None is in common use yet, although the importance of the subject causes continuing effort. A review by Schlick *et al.* (1997) covers the field, with particular reference to details of some methods not covered here.

4. Electric Field Calculations

A critical part of the potential in a channel simulation is the electric field from the charges on the protein, It is also not clear that all amino acids that would be ionized at physiologic pH will be ionized in the channel; a 60-mV potential shifts pK_a by one unit. Potentials of hundreds of millivolts, locally are possible. We will consider two methods of calculating the field. One assumes the water is a continuum, to which a dielectric constant can be assigned, and a concentration of ions. This is entirely appropriate for bulk solution, and almost certainly for the larger volumes at the ends of the channels. It has also been used for the fields in photosynthetic reaction center, in which local potentials as large as 1 V were found. We look at that case first, because it is by far the most popular method, and uses a protein for which a complete structure is known. Second, we consider a method used by Lu and Green for model simulations of a channel, in which the protein was replaced by a dielectric continuum, and a boundary element method was used to find the potential and field. In that set of simulations, the water internal to the channel was represented explicitly.

IV. Nonlinear Poisson–Boltzmann Equation

The Poisson equation gives the potential for a given distribution of charges. Debye and Huckel used it to find the potential around an ion in solution from the distribution of other ions, but only if the solution was so dilute that the interaction energy was less than $k_B T$. The ionic distribution in turn is given to start by the Boltzmann distribution, since the energy of interaction of the ions with the central ion is $q_i \phi / k_B T$, where q_i is the charge on the ion and ϕ the potential at the central ion. Since this is a form of mean field theory, with the average concentration of ions the question of interest, the concentration around the central ion is

$$c = c_0 \exp(-q_i \phi / k_B T) \tag{8}$$

The Poisson equation becomes, with this Boltzmann term (SI units, one radial dimension),

$$d^2\phi/dr^2 = (-1/\varepsilon) \sum_i c_i = (-1/\varepsilon) \sum_i c_0 \exp(-q_i\phi/k_BT) \qquad (9)$$

where $\mathbf{r}$ is the radial distance from the central ion and ε is the dielectric coefficient. This is an obviously nonlinear equation, and is an instance of the nonlinear Poisson–Boltzmann (PB) equation, which Debye and Huckel linearized (Alberty and Silbey, 1997) and then solved, by expanding the exponent to first order. This expansion is correct if other ions are far from the central ion, so that the potential is low enough for $q\phi \ll k_BT$. Therefore, their treatment is valid as an infinite dilution limit, rather than a general result. In the nonlinear PB equation case, the expansion is replaced with a numerical solution. However, Eq. (9) is not complete, because it assumes constant ε. When there is both protein and aqueous solution, the dielectric coefficient varies from place to place in the system. If we also allow charge density to vary, one needs Eq. (9a), the most general form of the equation:

$$\nabla \cdot [\varepsilon(r)\nabla\phi(\mathbf{r})] + \varepsilon(\mathbf{r})\kappa^2 \sinh[\phi(r)] + \rho(\mathbf{r})/\varepsilon_0 = 0 \qquad (9a)$$

where $\varepsilon(\mathbf{r})$ is the dimensionless dielectric coefficient, $\rho(\mathbf{r})$ is total charge density, $\mathbf{r}$ is now the position vector in three dimensions, and κ is the Debye–Huckel parameter, an inverse length: it depends inversely on the square root of ionic strength, the dielectric coefficient, and the temperature. With a 150 mM aqueous solution of 1:1 electrolyte, 25 °C, $\kappa \approx 8$ A^{-1}. A number of possible methods of solving this have been proposed, of which one commonly used example is the DelPhi (Gilson et al., 1987; Honig and Nicholls, 1995; Sharp and Honig, 1990) program. This solves the PB equation by dividing space into small enough segments to allow a local solution. DelPhi was used by Lancaster et al., for example, to solve for the potential and fields in a photosynthetic reaction center molecule with a known configuration. The isolated water molecules fixed in the protein were also represented explicitly, but in one energy-minimized configuration. In more recent work, these water molecules are treated by MC simulation to get the potential more accurately. Lancaster et al. (1996) found potentials up to 1 V: more important, they found potential shifts of hundreds of millivolts near ionizable side chains, which meant that there were pK, shifts as large as 10 units. For a more recent, and more accurate, version (good to about 1 pK unit for a residue in a protein), see Gunner et al. (2006). This has a major effect on how we consider the ionization state of a protein: it is clearly too simple to assume ionization of any basic or acidic residue that would be ionized in water at physiologic pH. There is actually some experimental evidence that bears on the question of high fields. Lockhart and Kim (1992) found, using Stark effect measurements, that the field at the end of a peptide due to its helical dipole was 0.43×10^9 V m^{-1}, quite a substantial value considering that it is measured at a covalently attached residue which is not enclosed in a channel or cleft, but is surrounded by water. This supports the accuracy of the calculation of fields much larger than membrane potentials in the channel.

A. Boundary Charge Method

It is also possible to find the field by determining the effect of a charge at any point in the volume on the potential and field at any other point. If there is a dielectric boundary, a charge anywhere in the volume, whether in the protein or the water, will induce a charge on the boundary. If all the boundary charges can be found, then the dielectric can be removed in favor of the charges induced on the boundary; the potential can then be found by Coulomb's law, summing over the real and the induced charges. To do this, it is necessary to divide the boundary into a set of elements, within which the induced charge is assumed to be constant. The size of the elements may be 1 $\mathring{A}^2$, or less. One gets one linear equation for each element, leading to an $N \times N$ determinant. If this can be solved, all the induced charges are known, and the problem is, for all practical purposes, solved. The problem may be solved, with the source charges on a lattice point, and the resulting potential and field recorded for all other lattice points. For locations between lattice points, interpolation can be used to get the potential and field. This method has been used for simulations of a model channel with explicit water, and protein as dielectric. The method has also been used by Boda *et al.* (2006) (who discovered it independently).

Although the technique is very different from the nonlinear PB equation, potentials and fields of the same order of magnitude are found. These results turn out to be reasonable based on a rough estimate of tile fields to be expected at the distances from charges found in channels. They could be avoided only by having dipoles canceling charges in a concerted fashion, something which could happen only locally. The orientation of the dipoles, of which water molecules are among the most important, can be found by MC or MD simulation, with the former probably being easier.

B. Other Continuum Models

Partenskii *et al.* (1994) attempted to estimate the effect of the carbonyls of a gramicidin-like (in approximate geometry) model in lowering the energy barrier presented by the pore. They attempted to allow for the effect of charges in the pore by introducing a third dielectric constant for the polar part of the channel wall, and solving a nonlinear PB equation. Their work, and references therein, illustrate the difficulty of using continuum models: no really satisfactory choice of dielectric constants in a pore has yet been found for a channel model in which neither the protein nor the water is modeled explicitly.

Eventually, one would want to have a complete model of the channel protein, with all atomic positions defined. In fact, the X-ray crystal structures now (2009) make this possible. Therefore, no dielectric problem arises, but the complete set of charges and dipoles can be used explicitly to obtain the field. However, even this calculation has to be made self-consistent, because the charges at one location alter the pK values elsewhere, and the state of ionization at that location in turn affects dipoles and pK values at the first location. So far, no calculation of this type with pK, shifts has been attempted for a channel.

C. Free-Energy and Pressure Calculations

It is straightforward to get the energy, the density, and other intensive variables by direct averaging. However, free energy and other extensive variables are not as direct, nor is there direct information on the partition function. To obtain the free energy, several special techniques have been devised.

1. Techniques of Free-Energy Calculation

1. *Particle insertion (Widom, 1963 as discussed in Levesque et al., 1984)*. If a test particle is introduced to the system, and the system has an N particle configurational energy of U_N, define $\Delta U_N = U_N - U_{N-1}$; then the chemical potential μ is given by

$$\mu - \mu_0 = k_B T \ln \langle \exp(-U_N/k_B T) \rangle_{N,V,T} \tag{10}$$

where μ_0 is chemical potential of an ideal gas at the same N, V, T. The angle brackets give the canonical average. The added particle tests the energy of its surroundings, hence the value of the free energy. There are substantial sampling difficulties with this procedure, but it has been used successfully in several instances.

2. If the free energy is known in a reference state, *thermodynamic integration* (Ryckaert *et al.*, 1977) can be used. Conceptually, it can be thought of as follows: Because the path by which the system moves from one state to another does not matter for differences in state variables, introduce a variable, λ, such that $0 < \lambda < 1$, where λ is the degree of advancement from the initial state, in which the free energy is known (say, at $T = 0$), to the state at which it is wanted. Then the system can be integrated along the path by increasing λ from 0 to 1, repeating the simulation with enough values of the parameter to provide an accurate integration.

There is an alternate formalism (Partenskii *et al.*, 1994) giving the Helmholtz free energy, A:

$$A_1 - A_0 = -k_B T \ln \langle \exp[-(\psi_1 - \psi_0)/k_B T] \rangle_0 \tag{11}$$

where A_1, A_0 are the Helmholtz free energy for values of the potential before and after perturbation (the NpT ensemble would give the Gibbs free energy). If the perturbation is too large, it can proceed through several simulations with successive values of a parameter λ such that

$$\psi_\lambda = \lambda \psi_1 - (\lambda - 1)\psi_0 \tag{12}$$

where Eq. (12) helps maintain the stability of the motion. The overall free energy is then available from the integral of $d\Delta A(\lambda)/d\lambda$ over λ. Another alternative, when the path would pass through energy differences large compared to $k_B T$, which would prevent adequate sampling, is to go through a *thermodynamic cycle*. This means that the beginning and end points of the change are connected through different, possibly hypothetical species, which can nevertheless be computed. Free energy is path independent, so the difference between initial and final state is the same. The alternate cycle may allow for the cancellation of terms difficult to compute, such as nonideality of the solutions. The method is described in McCammon and Harvey (1987).

Mezei and Beveridge (1986) have given a very thorough, and very reasonable, review of the possibilities of various methods of obtaining the free energy, especially the free-energy perturbation (FEP) method, and several other methods of obtaining the free-energy difference.

2. Pressure

In a liquid characterized by additive pair potentials $\phi(r)$ (r = intermolecular distance), the pressure is normally obtained from the following formula (Binder and Stauffer, 1984):

$$pV/Nk_BT = 1 - (N/6k_BT) \int_0^1 g(r)\mathrm{d}\phi(r)/4\pi^2 \, \mathrm{d}r \qquad (13)$$

where $g(r)$ is the radial distribution function (the probability of finding another molecule at a distance r from the central molecule, another quantity that is typically reported as a result of the simulation). However, a channel, with its major contribution from the wall and the field, would have additional terms, and apparently no attempt to compute directly the pressure in a simulation has been attempted for a channel so far.

D. Examples of Ion Channel Simulations

We conclude with a few examples of ion channel simulations, illustrating the types of information that can be obtained. We begin with channels composed of antibiotics made by bacteria (gramicidin, alamethicin), and then mention a couple of more interesting voltage-gated and acetylcholine receptor examples. The gramicidin simulations are typically all-atom simulations, because the gramicidin molecule is small enough that this is feasible. Roux and Karplus (1991) simulated water with Na^+ and K^+ in an analog of a gramicidin channel, using MD, with CHARMM potentials; the model was constructed to be a periodic helix. To understand the motion of the ion, the potential of mean force was calculated. The energy barrier to ion motion, and the particular water configurations that appeared in the simulation to be responsible, were among the results. Sagnella *et al.* (1996) have examined proton transfer, using Car-Parinello MD, a technique that allows for breaking of covalent bonds, to facilitate consideration of proton motion. Their results indicated that carbonyl solvation was critical for proton transfer, and that water did not move with the proton, which was transferred down the "proton wire." This is one of the few examples of a simulation in which bond breaking and formation is allowed. Breed *et al.* (1996) carried out MD simulations of columns of water in several types of channels, including two types of polyalanine, alamethicin, nicotinic acetylcholine receptor M2 helix, and δ-toxin. In the narrowest channels, the self-diffusion coefficient of water was reduced by an order of magnitude compared to bulk water. Dipole–dipole interactions were strong. Sankararamakrishnan *et al.* (1996) did MD simulations of a model of the pore domain of the nicotinic acetylcholine receptor, relating the behavior of water in the channel to pore diameter and electrostatics. In addition, they carried out nonlinear PB calculations to estimate the electrostatics of the motion of a mono-valent ion down the pore. From these results, they were able to make reasonable suggestions as to the configurations of the open and closed states, and the inter-actions of specific groups with water and with an ion in the pore.

As of 2009, some simulations of channels appear to give very successful results, comparing well with experiments, or carried out in conjunction with experiments (Krepkiy *et al.*, 2009), while others show that there are still substantial difficulties. In some cases differences in choice of potentials leads to very significant changes in the outcome, as in the very thorough simulation of KcsA by Fowler *et al.* (2008). This summary of uses of simulation obviously barely touches the literature on the subject, but does begin to suggest some of the types of information that can be obtained by simulation of a pore, and its water and ions.

V. Summary of Simulation Procedures

We have seen that a simulation requires that the model include the following:

- *Intermolecular potentials.* In a channel, these must include potentials for water, protein, and ions. It may be possible in some simulations to substitute a dielectric medium for either the water or the protein; however, a fully satisfactory model would have to include all three explicitly. Some intramolecular potentials are needed as well.

- *Suitable boundary conditions.* These are likely to be different for a channel, in that periodic boundary conditions are not possible in two dimensions, if the protein model is less than complete. There must be some accommodation of the water as well as lipid to the protein, and the results will depend on the approxima-tion made.

- *External potentials, as electric fields.* These do not normally appear in bulk simulations, but are unavoidable in channels. If omitted, the model is unrealistic for a channel. The fields are large enough that intermolecular potentials which include polarization are likely to be necessary for simulations of channels.

Once one has a model, one can choose a mode of simulation. MC simulations are often simpler than MD, but MD has other advantages. Both provide thermo-dynamic averages and molecular distribution functions as their most important output. MD should produce time dependence, but for times too short generally to be of biological interest. Several techniques have been proposed for simulations that extend to longer times. SD does this, for example, by averaging over the high-frequency motions of the molecules, replacing them with a lumped frictional coefficient: the contribution of these modes is then less well represented for other purposes. For channels, the inherently small sample size implies that errors will be somewhat larger than usual in bulk simulations; repeating simulations, or taking longer runs, may be a way of dealing with this problem.

General references for detailed descriptions of the techniques are particularly useful: for MD, especially for biological macromolecules, McCammon and Harvey (1987) and Brooks *et al.* (1988) for MC, the works edited by Binder and Stauffer (1984), Levesque and Weis (1992), and Widom (1963). Harvey (1989) has reviewed electrostatic calculations for macromolecules, especially the nonlinear PB equation.

There are a number of commercial programs for MD simulations. Among the most commonly used are CHARMm (sic), Amber, GROMACS, and NAMD. This is not a comprehensive list. Each of these programs began in a particular group, but this article cannot contain a historical discussion. There is one MC program, MMC, that is particularly worthy of note, and it is available without charge. Each of these can be found at its own website: The URLs (all preceded by http://) are as follows: CHARMm, www.charmm.org; GROMACS, www.gromacs.org; Amber, ambermd.org (no www); NAMD, www.ks.uiuc.edu/Research/namd; MMC, atlas.physbio.mssm.edu/~mezei/mmc (no www).

Acknowledgment

I am grateful to Dr. Mihaly Mezei for reading the manuscript and providing many helpful comments.

References

Ahlstrom, P., Wallqvist, A., Engstrom, S., and Jonsson, B. (1989). *Mol. Phys.* **68**, 563.

Alberty, R. A., and Silbey, R. J. (1997). *In* "Physical Chemistry," 2nd edn., pp. 230–232. John Wiley, New York. Based on: Debye, P. J. W., and Huckel, E. (1923). *Phys. Z.* **24**, 185, 3(15).

Aqvist, J. (1990). *J. Phys. Chem.* **94**, 8021.

Beeman, D. (1976). *J. Comput. Phys.* **20**, 130.

Berendsen, H. J. C., Postma, J. P. M., van Gunsteren, W. F., and Hermans, J. (1981). *In* "Intermolecular Forces," (B. Pullman, ed.), pp. 331–342. Reidel, Dordrecht, The Netherlands.

Binder, K., and Stauffer, D. (1984). *In* "Applications of the Monte Carlo Method in Statistical Physics," (K. Binder, ed.), pp. 1–36. Springer, Berlin, Germany.

Boda, D., Valisko, M., Eisenberg, B., Nonner, W., Henderson, D., and Gillispie, D. (2006). *J. Chem. Phys.* **125**, 034901/1–11.

Breed, J., Sankararamakrishnan, R., Kerr, I. D., and Sansom, M. S. P. (1996). *Biophys. J.* **70**, 1643.

Brooks, C. L., Karplus, M., and Pettit, B. M. (1988). "Proteins: A Theoretical Perspective of Dynamics, Structure, and Thermodynamics." John Wiley and Sons, New York, NY.

Chialvo, A. A. (1996). *J. Chem. Phys.* **104**, 5240.

Davis, M. E., and McCammon, J. A. (1990). *J. Comput. Chem.* **11**, 401.

Domene, C., and Sansom, M. S. P. (2003). *Biophys. J.* **85**, 2787.

Doyle, D. A., Cabral, J. M., Pfuetzner, R., Kuo, A., Gulbis, J. M., Cohen, S. L., Chait, B. T., and MacKinnon, R. (1998). *Science* **280**, 69.

Durell, S. R., and Guy, H. R. (1996). *Neuropharmacology* **35**, 761.

Fowler, P. W., Tai, K., and Sansom, M. S. P. (2008). *Biophys. J.* **95**, 5062.

Gear, C. W. (1971). "Numerical Initial Value Problems in Ordinary Differential Equations," Prentice-Hall, New York, NY.

Gilson, M. K., Sharp, K. A., and Honig, B. (1987). *J. Comput. Chem.* **9**, 327.

Green, M. E., and Lu, J. (1995). *J. Colloid Interface Sci.* **171**, 117.

Guarnieri, F. (1995). *J. Math. Chem.* **18**, 25.

Gunner, M., Mao, J., Song, Y., and Kim, J. (2006). *Biochem. Biophys. Acta (Bioenergetics)* **1757**, 942.

Guy, H. R., and Durell, S. R. (1995). *In* "Ion Channels and Genetic Diseases," (D. Dawson, ed.), pp. 1–6. Rockefeller University Press, New York, NY.

Harvey, S. C. (1989). *Proteins Struct. Func. Genet.* **5**, 78.

Honig, B., and Nicholls, A. (1995). *Science* **268**, 1144.

Jorgensen, W. L., Chadrasekhar, J., Madura, J. D., Impey, R. W., and Klein, M. L. (1983). *J. Chem. Phys.* **79**, 926.

Kerr, I. D., and Sansom, M. S. P. (1997). *Biophys. J.* **73**, 581.

Krepkiy, D., Mihailescu, M., Alfredo Freites, J., Schow, E. V., Worcester, D. L., Gawrisch, K., Tobias, D. J., White, S. H., and Swartz, K. J. (2009). *Nature* **462**, 473.

Laakkonen, L. J., Guranieri, F., Perlman, J. H., Gershengorn, M. C., and Osman, R. (1996). *Biochemistry* **35**, 7651.

Lancaster, C. R. D., Michel, H., Honig, B., and Gunner, M. R. (1996). *Biophys. J.* **70**, 2469.

Levesque, D., and Weis, J. J. (1992). *In* "The Monte Carlo Method in Condensed Matter Physics," 2nd corrected edn. (K. Binder, ed.), pp. 172–177. Springer, Berlin, Germany.

Levesque, D., Weis, J. J., and Hansen, J. P. (1984). *In* "Applications of Monte Carlo Methods in Statistical Physics," (K. Binder, ed.), p. 37. Springer, Berlin.

Lockhart, D. J., and Kim, P. S. (1992). *Science* 947.

Long, S. B., Campbell, E. B., and MacKinnon, R. (2005). *Science* **309**, 903.

Lu, J., and Green, M. E. (1997). *Prog. Colloid Polym. Sci.* **103**, 121.

McCammon, J. A., and Harvey, S. (1987). "Dynamics of Proteins and Nucleic Acids." Cambridge University Press, Cambridge.

Metropolis, N., Rosenbluth, A. W., Rosenbluth, M. N., Teller, A. H., and Teller, E. (1953). *J. Chem. Phys.* **21**, 1087.

Mezei, M. (1987). *J. Comput. Phys.* **68**, 237.

Mezei, M., and Beveridge, D. (1986). *Ann. N. Y. Acad. Sci.* **482**, 1.

Partenskii, M. B., Dorman, V., and Jordan, P. C. (1994). *Biophys. J.* **67**, 1429.

Roux, B. (1996). *Biophys. J.* **71**, 3177.

Roux, B., and Karplus, M. (1991). *Biophys. J.* **59**, 961.

Ryckaert, J. P., Cicotti, G., and Berendsen, H. J. C. (1977). *J. Comput. Phys.* **23**, 327.

Sagnella, D. E., Laasonen, K., and Klein, M. L. (1996). *Biophys. J.* **71**, 1172.

Sankararamakrishnan, R., Adcock, C., and Sansom, M. S. P. (1996). *Biophys. J.* **71**, 1659.

Sansom, M. S. P., Kerr, I. D., Breed, J., and Sankararamakrishnan, R. (1996). *Biophys. J.* **70**, 693.

Schlick, T., Barth, E., and Mandziuk, M. (1997). *Annu. Rev. Biophys. Biomol. Struct.* **26**, 181.

Sharp, K. A., and Honig, B. (1990). *Ann. Rev. Biophys. Biophys. Chem.* **19**, 301.

Shen, L., Bassolino, D., and Stouch, T. (1997). *Biophys. J.* **73**, 3.

Singh, C., Sankararamakrishnan, R., Subramanian, S., and Jakobsson, E. (1996). *Biophys. J.* **71**, 2276.

Stillinger, F. H., and David, C. W. (1969). *J. Chem. Phys.* **69**, 1473.

Svishchev, I. M., Kusalik, P. G., Wang, J., and Boyd, R. J. (1996). *J. Chem. Phys.* **105**, 4742.

van Gunsteren, W. F., and Berendsen, H. J. C. (1977). *Mol. Phys.* **34**, 1311.

van Gunsteren, W. F., Berendsen, H. J. C., Hermans, J., Hol, W. G. B., and Postma, J. P. M. (1983). *Proc. Natl. Acad. Sci. USA* **80**, 4315.

Verlet, L. (1967). *Phys. Rev.* **159**, 98.

Wang, C. X., Wan, S. Z., Xiang, Z. X., and Shi, Y. Y. (1997). *J. Phys. Chem. B* **101**, 230.

Widom, B. (1963). *J. Chem. Phys.* **39**, 2808.

Woolf, T. B., and Roux, B. (1997). *Biophys. J.* **72**, 1930.

Znamenskiy, V. S., and Green, M. E. (2007). *J. Chem. Theory. Comput.* **3**, 103.

PART VI

Physical

CHAPTER 11

Fluorescence Techniques for Measuring Ion Channel Activity

Gönül Veliçelebi, Kenneth A. Stauderman, Mark A. Varney, Michael Akong, Stephen D. Hess, and Edwin C. Johnson

Cell Biology
SIBIA Neurosciences, Inc.,
La Jolla, California, USA

I. Introduction

Changes in intracellular free calcium concentration ($[Ca^{2+}]_i$) play a crucial role in cellular physiology. A number of cell surface receptors and channels are known to regulate $[Ca^{2+}]_i$ through different molecular mechanisms. Therefore, the functional and pharmacologic properties of many of these cell surface receptors and ion channels can be studied effectively by measuring changes in $[Ca^{2+}]_i$ in intact cells. For our drug discovery efforts, we have targeted several ion channel and receptor systems that play different roles in neuronal physiology and pathophysiology. These molecular targets include voltage- and ligand-gated ion channels: the human neuronal voltage-gated calcium channels (VGCCs) (Brust *et al.*, 1993; Williams *et al.*, 1992a,b, 1994), ligand-gated nicotinic acetylcholine receptor (NAChR) channels (Chavez-Noriega *et al.*, 1997; Elliott *et al.*, 1996), ionotropic *N*-methyl-D-aspartic acid (NMDA) (Hess *et al.*, 1996; Varney *et al.*, 1996), α-amino-3-hydroxy-5-methyl-4-isoxazolepropionic acid (AMPA), and kainate-type excitatory amino acid (EAA) receptor channels. All of these channels mediate elevation of $[Ca^{2+}]_i$ via Ca^{2+} influx from the extracellular medium upon depolarization or activation by agonist.

Our strategy for drug discovery starts with the cloning of cDNAs encoding the specific subtypes of the targeted human ion channels and proceeds to stable expression of functional channels in mammalian host cells, thereby generating subtype-specific cellular targets for drug screening. To facilitate the identification of novel subtype-selective ligands, we sought to develop functional cell-based assays that can detect both competitive agonists and antagonists as well as allosteric activators and inhibitors. First, it was necessary to ascertain that the activity of the targeted receptors and channels in the functional assay displayed the expected pharmacologic profile observed in native systems. Therefore, the assay had to be sensitive and rapid enough to detect the activation or inhibition of ligand- and voltage-gated channels as well as G-protein-coupled receptors. Since the experimental protocol was to be performed with intact cells, it was also desirable that the assay be compatible with testing cells plated on 96-well microtiter dishes. Finally, in order to be a valuable screening tool, the functional assay had to be adaptable to automation so that the assay could be carried out with a higher throughput.

One of the functional assays we have established that meets these criteria measures changes in $[Ca^{2+}]_i$. Calcium-sensing fluorescent dyes offer the most facile means of monitoring changes in $[Ca^{2+}]_i$ both in cell populations and in individual cells. The ability to measure $[Ca^{2+}]_i$ in real time without disrupting cells represents a significant technical advance in cell biology pioneered and led by Tsien (1981). The design, synthesis, and characterization of these fluorescent reagents have been described and reviewed extensively (Tsien, 1988, 1989). The most commonly used calcium-sensitive dyes, Fluo-3, Fura-2, and Indo-1, are structural analogs of the highly selective calcium chelators ethylene glycol-bis(β-aminoethyl ether) *N,N,N',N'*-tetraacetic acid (EGTA) and 1,2-bis(2-aminophenoxy) ethane-*N,N,N'*,

N'-tetraacetic acid (BAPTA). Typically, the acetoxymethyl ester form of the dye readily permeates into the cell, wherein the ester group is hydrolyzed by intracellular esterases, generating the calcium-sensitive form of the free dye that is trapped inside the cell. Fura-2 and Fluo-3 have different affinities for calcium as well as different absorption and fluorescence properties. Fluo-3 is a single excitation wavelength dye, and on binding Ca^{2+} ($K_d = 390$ nM at pH 7.2, 22 °C) (Haugland, 1996) undergoes more than a 100-fold increase in fluorescence without a change in excitation/emission spectra (Haugland, 1996; Kao et al., 1989). The unbound dye is almost nonfluorescent, whereas Ca^{2+}-bound Fluo-3 absorbs and fluoresces in the visible range ($\lambda_{Ex} = 506$ nm and $\lambda_{Em} = 526$ nm) (Haugland, 1996; Minta et al., 1989). Fura-2 is a ratiometric dye that undergoes a shift in its excitation spectrum on binding Ca^{2+} ($K_d = 145$ nM at pH 7.2, 22 °C) (Haugland, 1996). As determined in cuvette-based measurements, the excitation maximum (λ_{Ex}) shifts from 362 nm for the unbound dye to 335 nm for the calcium–dye complex with minimal change in its emission maximum (λ_{Em}) at 510 nm (Grynkiewicz et al., 1985; Haugland, 1996). Because the excitation peak at a saturating concentration of Ca^{2+} exhibits an apparent red shift in microscope-based measurements in Fura-2-loaded cells (Negulescu and Machen, 1990), the shift in excitation maximum is typically monitored by excitation at 350 and 385 nm, and the ratio of fluorescence resulting from these two excitations is used to correct for variations in dye concentration, cell thickness, and cell number.

We have used both Fluo-3 and Fura-2 to develop sensitive and rapid assays to measure changes in $[Ca^{2+}]_i$ in cells expressing recombinant ion channels and receptors. We describe here the experimental methods with particular emphasis on the validation of the assay for human VGCCs, NAChRs, and NMDA receptors.

II. Experimental Procedures

A. Reagents and Instrumentation

Fluo-3-acetoxymethyl ester (Fluo-3 AM) and Fura-2-acetoxymethyl ester (Fura-2 AM) are purchased from Molecular Probes, Inc. (Eugene, OR). The stock solutions of Fluo-3 AM and Fura-2 AM are first prepared in dimethyl sulfoxide (DMSO) and Pluronic F127, then diluted in HEPES-buffered saline (HBS; 125 mM NaCl, 5 mM KCl, 0.62 mM $MgSO_4$, 1.8 mM $CaCl_2$, 20 mM HEPES, 6 mM glucose, pH 7.4) to give final concentrations of 1.0% DMSO and 0.07% Pluronic F127.

The Fluo-3 fluorescence measurements are performed at 0.33-s intervals using a fluorimeter capable of reading a 96-well microtiter plate one well at a time (Cambridge Technology, Inc., Watertown, MA). The fluorimeter is equipped with an excitation band-pass filter of 485 ± 20 nm, and an emission band-pass filter at 530 ± 20 nm. Solutions are added to each well either manually with a pipette or automatically with a Digiflex (Titertek, Huntsville, AL) dispenser. For manual additions, the lid of the fluorimeter is briefly opened to allow access to the microtiter plate; hence, approximately a 5-s gap is introduced into the corresponding fluorescence recording after agonist addition (e.g., Fig. 8, see later section). Automatic additions are performed by a computer-controlled Digiflex apparatus equipped with a dispensing line of polyethylene tubing that is positioned directly above the microtiter well. Additions by the Digiflex facilitate automation of the assay and eliminate the gaps in fluorescence recordings, which can be important for extremely rapid responses.

The Fura-2 fluorescence measurements are performed at 0.25-s intervals per dual excitation (ratio pair) using a customized plate-imaging fluorimeter capable of recording the fluorescence output from all 96 wells simultaneously (SIBIA Neurosciences/Science Applications International Corporation (SIBIA/SAIC), San Diego, CA). The fluorimeter is equipped with two excitation band-pass filters of 350 ± 15 nm and 385 ± 15 nm and an emission filter of 535 ± 50 nm. The solutions are added simultaneously to all wells of the 96-well microplate by an integrated 96-channel pipettor (Carl Creative Systems, Inc., Harbor City, CA). All washing and aspiration procedures are carried out using a 96-well automatic plate washer (Bio-Tek Instruments, Inc., Winooski, VT). Liquid and plate handling are performed in automated fashion using a Zymark robot (Model ZB0021, Zymark Corporation, Hopkinton, MA).

B. Measurement of $[Ca^{2+}]_i$ by Plate-Based Fluo-3/Fura-2 Assay

The following assay protocol was optimized for use with the transfected human embryonic kidney (HEK293 cells) described here. Adaptation to other cell lines may require optimization of several obvious experimental parameters, such as cell number, dye concentration, dye loading time and temperature, cell washing conditions, and instrument settings such as gain and integration time.

1. The cells are plated on poly(D-lysine)-coated 96-well microtiter plates at a density of $1–2 \times 10^5$ cells/well. Twenty-four hours after plating, the culture medium is aspirated and the cells washed twice with 250 μl HBS. Subsequently, 200 μl HBS is added to each well and background fluorescence (F_{bkg}) recorded.

2. The buffer is removed and the cells are incubated with 30–100 μl of 20 μM Fluo-3 AM for 1–2 h at 20 °C, or 1 μM Fura-2 AM for 1 h at room temperature. For Fluo-3, we have calculated that these loading conditions typically result in a maximal fluorescence value (F_{max}) of approximately 15,000 arbitrary units compared to a background of 1000 units. It has previously been shown that if cells are loaded with a Ca^{2+}-sensitive dye (e.g., Fura-2 or Fluo-3) to achieve a final intracellular concentration of approximately 100 μM, a good fluorescence signal can be obtained without excessively buffering cytosolic Ca^{2+} (Neher, 1995).

3. Unloaded dye is removed by aspiration and each well is washed with 250 μl HBS. Next, HBS (180 μl) is added to each well and following a 2-min recovery period, the basal fluorescence (F_b) is recorded for 3 s. The recovery period should be less than 10 min to minimize time-dependent extrusion of the dye from the cells. Next, 20 μl test drug solution is added to each well, and the fluorescence (F_{res}) recorded for 40–60 s (e.g., Figs. 1, 5, and 8) at 0.33- or 0.25-s intervals for Fluo-3 or Fura-2, respectively. This test is carried out to determine if the drug has intrinsic agonist properties.

4. Two minutes after the agonist test, F_b is recorded for another 3 or 5 s before adding a reference agonist to each well and recording F_{res} for another 40–60 s. This test is performed to determine whether the test compound has antagonist properties, because an antagonist will reduce the response to the reference agonist in this test.

5. For Fluo-3 measurements, at the end of the agonist and antagonist tests, F_{max} is determined by adding Triton X-100 to each well to a final concentration of 0.1%. The contents of the wells are mixed by drawing the contents of the well up and down in a pipette tip five times. After an incubation of approximately 2–5 min, F_{max} is recorded. To obtain the minimum fluorescence (F_{min}), $MnCl_2$ is added to a final concentration of 10 mM. Again, the contents of the wells are mixed, and the F_{Mn} value measured after 2 min. For Fura-2 measurements, empirically determined calibration values are used (see below).

6. In those instances where activation of a cell line elicits a very large $[Ca^{2+}]_i$ signal, the peak of fluorescence (F_{res}) can be higher than the F_{max} determined after lysis of the cells with Triton X-100. In part, because the fluorimeter detects the fluorescence emission most efficiently from the bottom of the cell plate, the signal from a layer of cells at the bottom of the plate can be larger than that from a column of cell lysate following solubilization with Triton X-100. We have found that a cocktail of ionomycin/carbonyl cyanide p-(trifluoromethoxy)phenyl hydrazone (FCCP)/carbachol can be used with HEK293 cells to obtain a more precise estimate of F_{max} and F_{Mn}. In this protocol, F_{max} was recorded after the cells were incubated for 2–4 min in HBS, pH 9.0, containing 21.8 mM $CaCl_2$, 5 μM ionomycin, 20 μM FCCP, and 100 μM carbachol. The higher pH and $CaCl_2$ are used to optimize the activity of ionomycin, whereas FCCP and carbachol are added to release Ca^{2+} from intracellular stores and to reduce energy-dependent intracellular buffering. The F_{Mn} values are determined using the same $MnCl_2$ quench procedure described in step 5, except that another 2–4-min incubation period is included after addition of the $MnCl_2$.

C. Data Analysis

1. The peak and basal $[Ca^{2+}]_i$ concentrations for the Fluo-3 results were calculated as described by Kao *et al.* (1989). First, F_{bkg} was subtracted from F_{res}, F_{Mn}, and F_{max}. Next, F_{min} was calculated according to Eq. (1):

$$F_{min} = F_{Mn}/8 \tag{1}$$

Finally, the corrected fluorescence values were used in Eq. (2):

$$[Ca^{2+}](\text{in nM}) = K_d[(F_{res} - F_{min})/(F_{max} - F_{res})] \tag{2}$$

where K_d is 390 nM for Fluo-3 (Haugland, 1996).

2. For Fura-2, the peak and basal $[Ca^{2+}]_i$ concentrations were calculated as described by Grynkiewicz *et al.* (1985) using Eq. (3):

$$[Ca^{2+}](\text{in nM}) = K_d[(R_{res} - R_{min})/(R_{max} - R_{res})](S_{385}) \tag{3}$$

where K_d is 145 nM for Fura-2 (Haugland, 1996); R_{min} and R_{max} represent the fluorescence ratios (F_{350}/F_{385}) determined as described above for F_{min} and F_{max}, and S_{385} represents the ratio of F_{385} for unbound dye to F_{385} for Ca^{2+}-saturated dye.

For our instrument, the R_{min} and R_{max} values were empirically determined from calcium standards as 0.11 and 2.90, respectively. More specifically, two plates were prepared containing, in each well, 180 μl of either 10 mM EGTA or 100 μM CaCl$_2$ in 10 mM Na$_2$HPO$_4$, pH 7.5. Fura-2 (free acid) was added to all wells to a final concentration of 0.1 μM, and the F_{350} and F_{385} values were recorded for all 96 wells in both plates. The R_{min} and R_{max} values were determined from the ratio of average F_{350} to average F_{385} for the EGTA and CaCl$_2$ plates, respectively. The S_{385} value was determined from the ratio of average F_{385} on the EGTA plate to average F_{385} on the CaCl$_2$ plate.

3. For agonist-evoked responses, all fluorescence determinations were performed in four replicate wells. The receptor-mediated $[Ca^{2+}]_i$ responses were quantitated by calculating either the ratio of peak $[Ca^{2+}]_i$ after drug addition to the basal $[Ca^{2+}]_i$ prior to drug addition (P/B), or by calculating the difference between peak $[Ca^{2+}]_i$ and basal $[Ca^{2+}]_i$ (P − B).

D. Development of Stable Cell Lines Expressing Recombinant Human Ion Channels

The stable cell lines expressing VGCCs, NAChRs, and NMDA receptors were established using protocols similar to those described for human metabotropic receptors (Daggett *et al.*, 1995; Lin *et al.*, 1997). Briefly, using the calcium phosphate coprecipitation method (Kingston, 1996), host cells, typically HEK293 cells, were transfected with cDNA expression plasmids encoding one or more individual subunits of human VGCCs, nicotinic, NMDA, AMPA, or kainate receptors. A plasmid encoding the neomycin gene (e.g., pSV2*neo*) was also introduced into the cells to serve as a selectable marker. Two days later, the transfected cells were selected in Dulbecco's modified Eagle's medium containing 10% dialyzed fetal bovine serum, 2 mM L-glutamine, 100 U/ml penicillin G, 100 μg/ml streptomycin, and 500 μg/ml G418. All tissue culture reagents were obtained from GIBCO-BRL (Grand Island, NY). The G418-resistant cells were cloned by two rounds of limiting dilution, and the clones were identified using the $[Ca^{2+}]_i$ assay. The clones exhibiting the most robust $[Ca^{2+}]_i$ responses were characterized further in more detailed pharmacologic studies. The stability of the expression of the recombinant receptor or ion channel in the selected clonal cell line was ascertained by monitoring the magnitude and the pharmacology of the $[Ca^{2+}]_i$ response for 30–40 passages in culture. In addition, we have also expressed most of these channels transiently in HEK293 cells and detected robust $[Ca^{2+}]_i$ signals on activation. An example with the NAChRs is shown later in Fig. 12.

III. Pharmacologic Validation of Assay

The fluorescent dye-based $[Ca^{2+}]_i$ assay was validated in each target receptor system before it was utilized in drug screening. To this end, cell lines stably expressing specific subtypes of the different ion channels were characterized pharmacologically using reference compounds in the Fluo-3- or the Fura-2-based $[Ca^{2+}]_i$ assay. The results obtained in each case were compared with those obtained by measuring inward currents using electrophysiologic techniques. We present selected results here that demonstrate several salient features of the assay. For ease of organization, we have grouped the different voltage- and ligand-gated channels according to the fraction of the inward current carried by $[Ca^{2+}]_i$, based on reported results (Burnashev *et al.*, 1995; Rogers and Dani, 1995). However, additional factors such as number of recombinant receptors expressed per cell, kinetics of the $[Ca^{2+}]_i$ response, desensitization properties, and assay conditions may also affect the absolute magnitude of $[Ca^{2+}]_i$ signals in these cell lines.

IV. Ion Channels with High Fractional Ca^{2+} Response

In this section, we included ion channels in which Ca^{2+} influx accounts for greater than 10% of the inward current.

A. Voltage-Gated Calcium Channels

Neuronal VGCCs are multimeric channel complexes composed of an α_1-, an $\alpha_2\delta$-, and a β-type subunit that are activated by depolarization of the membrane. Several isoforms of the α_1 and the β subunit and several splice variants of genes representing all three subunits have been cloned. Specific subtypes of VGCCs are composed of different combinations of the three types of subunits and exhibit unique biophysical and pharmacologic properties (for review, see Catterall, 1995). We have generated several cell lines stably expressing different combinations of the three human VGCC subunits. Depolarization of the VGCC-expressing cell lines by addition of KCl causes influx of extracellular Ca^{2+}, and the ensuing increase in $[Ca^{2+}]_i$ can be readily detected in real time by an increase in Fluo-3 or Fura-2 fluorescence. Representative kinetic traces for the $\alpha_{1A-2}\alpha_{2b}\delta\beta_{4a}$ (P/Q-type VGCC), $\alpha_{1B-1}\alpha_{2b}\delta\beta_{3a}$ (N-type VGCC), and $\alpha_{1E-3}\alpha_{2b}\delta\beta_{1b}$ (possibly R-type VGCC) subtypes activated by KCl are shown in Fig. 1. The $[Ca^{2+}]_i$ response to KCl addition reached a peak within 10 s in each cell line. The magnitude of the $[Ca^{2+}]_i$ response was comparable in all three cell lines (approximately 500 nM above basal levels), but there were apparent differences in the decay kinetics of the $[Ca^{2+}]_i$ response, with the slowest decay observed in the cell line expressing the $\alpha_{1B-1}\alpha_{2b}\delta\beta_{3a}$ subtype.

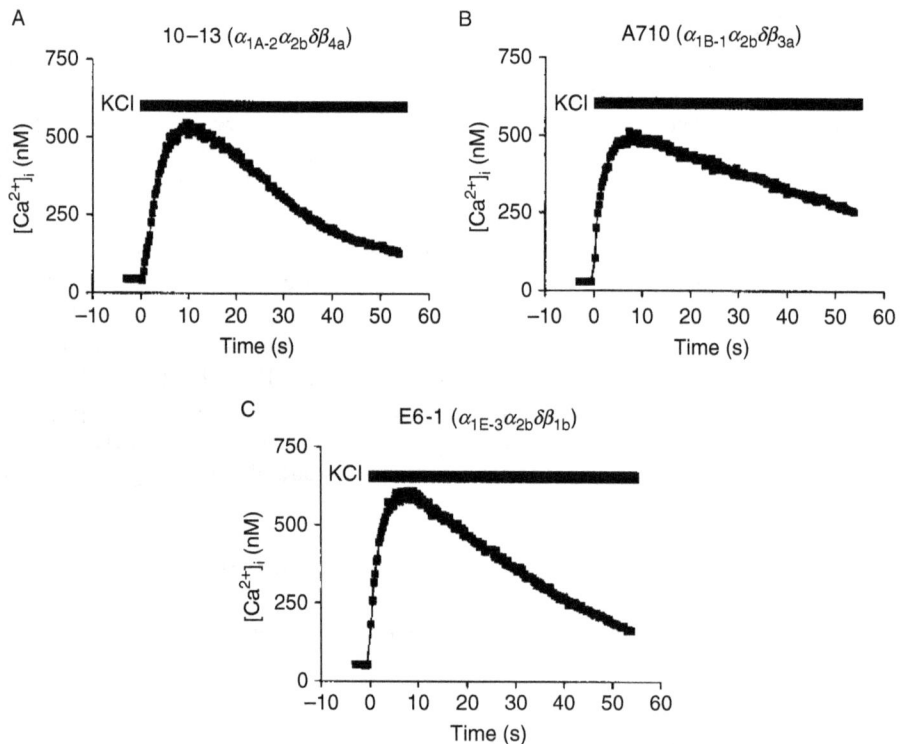

Fig. 1 Time course of the KCl-induced $[Ca^{2+}]_i$ signal in HEK293 cells stably expressing human VGCCs. HEK293 cells were stably transfected with cDNAs encoding the indicated subunits of human VGCCs, and clonal cell lines were established as described in the text. (A) 10–13 cell line expressing $\alpha_{1A-2}\alpha_{2b}\delta\beta_{4a}$. (B) A710 cell line expressing $\alpha_{1B-1}\alpha_{2b}\delta\beta_{3a}$. (C) E6–1 cell line expressing $\alpha_{1E-3}\alpha_{2b}\delta\beta_{1b}$. The cells were loaded with Fluo-3 as described in the text, and the fluorescence was monitored with a plate-reading fluorimeter. Data points were measured at approximately 0.33-s intervals. The first few seconds of data were recorded with cells bathed in HBS. At time zero, depolarization buffer containing 70 mM KCl and 4 mM $CaCl_2$ (final concentrations) was added automatically using the Digiflex dispenser, and the response was monitored for the next 55 s. At the end of the experiment, $[Ca^{2+}]_i$ was calculated from F_{max} and F_{Mn} values determined with the Triton X-100 method. The points represent mean values (error bars omitted for clarity) of quadruplicate determinations from one experiment.

We used the fluorescent dye-based $[Ca^{2+}]_i$ assay to characterize the pharmacologic properties of the three VGCC subtypes. Various reagents and several peptide toxins have been identified that display differential interaction with the α_{1A}, α_{1B}, and α_{1E} subtypes (for review, see Olivera *et al.*, 1994). The results obtained in the fluorescent dye-based $[Ca^{2+}]_i$ assay (Fig. 2) agree with the established pharmacology of the three subtypes. For example, ω-conotoxin GVIA (ω-CgTx-GVIA) specifically inhibited the KCl-induced $[Ca^{2+}]_i$ response in the $\alpha_{1B\text{-}1}\alpha_{2b}\delta\beta_{3a}$-

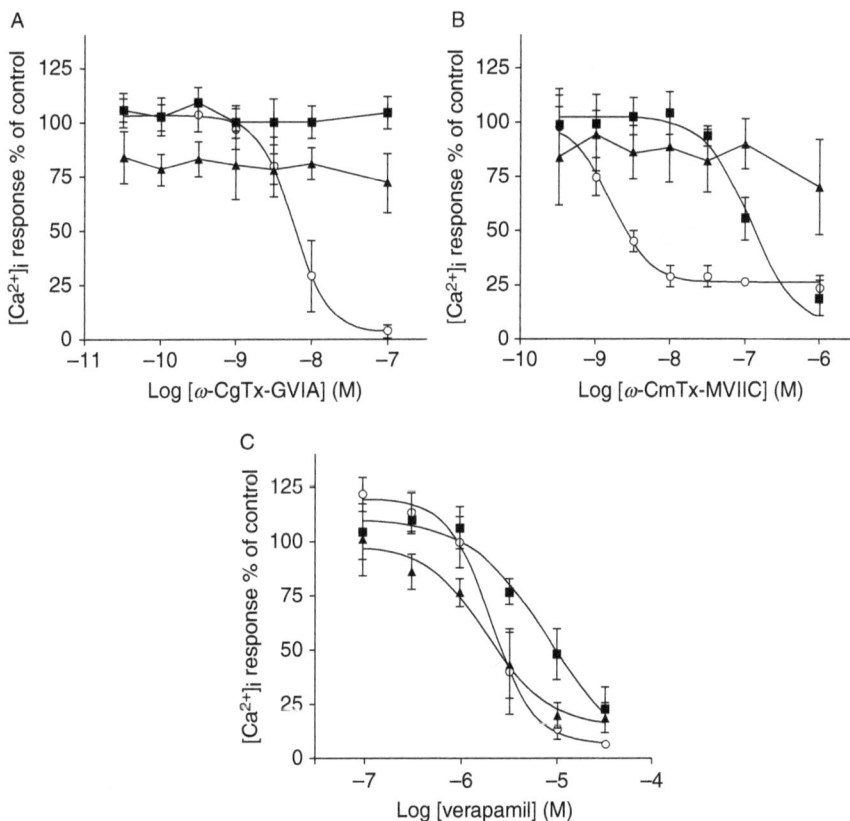

Fig. 2 Pharmacologic characterization of cell lines stably expressing human VGCCs. The cells expressing specific VGCC subtypes were loaded with Fura-2 as described in the text, and the fluorescence was monitored with the SIBIA/SAIC imaging system. The cells were preincubated with ω-CgTx-GVIA (A), ω-CmTx-MVIIC (B), or verapamil (C) for 5–10 min prior to activation with 70 mM KCl. The y-axis shows the response amplitude as a percent of the positive control $[Ca^{2+}]_i$ response to 70 mM KCl with no antagonist added. Points are the mean ± S.D. of quadruplicate determinations from a representative experiment. Curves were fit to points by Prism software (GraphPad, Inc., San Diego, CA). In those instances where curves could not be fit to the points, the points were connected with lines. (■) 10–13 cell line expressing $\alpha_{1A\text{-}2}\alpha_{2b}\delta\beta_{4a}$, (○) A710 cell line expressing $\alpha_{1B\text{-}1}\alpha_{2b}\delta\beta_{3a}$, (▲) E6–1 cell line expressing $\alpha_{1E\text{-}3}\alpha_{2b}\delta\beta_{1b}$.

expressing cell line with an IC_{50} of 8.9 ± 1.4 nM (mean $\pm$ S.E.M.) without any detectable effect on cell lines expressing $\alpha_{1A\text{-}2}\alpha_{2b}\delta\beta_{4a}$ or $\alpha_{1E\text{-}3}\alpha_{2b}\delta\beta_{1b}$ (Fig. 2A). Another peptide toxin, ω-CmTx-MVIIC, blocked the KCl-induced response in cells expressing $\alpha_{1A\text{-}2}\alpha_{2b}\delta\beta_{4a}$ and $\alpha_{1B\text{-}1}\alpha_{2b}\delta\beta_{3a}$ channels with IC_{50} values of 98 ± 12 nM and 5.8 ± 1.7 nM, respectively, without significantly affecting the $[Ca^{2+}]_i$ response in cells expressing $\alpha_{1E\text{-}3}\alpha_{2b}\delta\beta_{1b}$ channels (Fig. 2B). Finally, verapamil, a relatively nonselective blocker, displayed a slight preference for cells expressing $\alpha_{1B\text{-}1}\alpha_{2b}\delta\beta_{3a}$ ($IC_{50} = 3.5 \pm 0.8$ μM) or $\alpha_{1E\text{-}3}\alpha_{2b}\delta\beta_{1b}$ ($IC_{50} = 5.7 \pm 1.4$ μM) channels compared to cells expressing $\alpha_{1A\text{-}2}\alpha_{2b}\delta\beta_{4a}$ channels ($IC_{50} = 32 \pm 4$ μM) (Fig. 2C). These results indicated that the fluorescent dye-based $[Ca^{2+}]_i$ assay was sensitive and rapid enough to detect the changes in $[Ca^{2+}]_i$ resulting from the activation of VGCCs. Furthermore, the $[Ca^{2+}]_i$ response displayed the expected subtype-specific pharmacologic profile observed by whole-cell recording, thus validating the assay as a means to screen unknown compounds for their activity on human VGCC subtypes.

The fluorescent dye-based $[Ca^{2+}]_i$ assay can be more sensitive than electrophysiologic methods for measuring small ionic currents that decay very slowly or incompletely. For example, in HEK293 cells stably expressing $\alpha_{1A\text{-}2}\alpha_{2b}\delta\beta_{3a}$ VGCCs, we were able to detect voltage-gated Ba^{2+} or Ca^{2+} currents in only 5 out of 78 cells, and none of the currents were greater than 120 pA (Fig. 3A). On the other hand, with the fluorescent dye-based $[Ca^{2+}]_i$ assay, we could reliably detect KCl-induced $[Ca^{2+}]_i$ responses of 150–200 nM (Fig. 3B). The higher sensitivity of the fluorescence-based $[Ca^{2+}]_i$ measurements can most easily be explained by the fact that these particular VGCCs inactivate relatively slowly ($\tau_{inact} \approx 690$ ms) and incompletely, with a noninactivating component of approximately 80%. In the fluorescent dye-based $[Ca^{2+}]_i$ assay, the depolarization stimulus, that is, 70 mM KCl, is maintained throughout the measurement, and Ca^{2+} continually enters the cell and accumulates, resulting in a detectable $[Ca^{2+}]_i$ change. In contrast, electrophysiologic measurements monitor only the flux of current across the membrane over a much shorter time scale. Consequently, a very small but incompletely inactivating Ca^{2+} current can be seen as a robust signal with fluorescent dye-based $[Ca^{2+}]_i$ measurements. Neher and Augustine (1992) have described in detail the effects of $[Ca^{2+}]_i$ indicators in prolonging and amplifying the Ca^{2+} current signal by means of Ca^{2+} accumulation and buffering $[Ca^{2+}]_i$ in bovine adrenal chromaffin cells.

One limitation of the fluorescent dye-based $[Ca^{2+}]_i$ assay in the VGCC system may arise from the difficulty in controlling the membrane potential and, therefore, the state of voltage-dependent inactivation of the channels. For example, the $\alpha_{1E\text{-}3}\alpha_{2b}\delta\beta_{1\text{-}3}$-containing neuronal VGCCs undergo steady-state inactivation with membrane depolarization, with approximately one-half of the channels inactivated at -60 mV (Williams *et al.*, 1994). Therefore, if the resting membrane potential of the stable cell line expressing the $\alpha_{1E\text{-}3}\alpha_{2b}\delta\beta_{1\text{-}3}$ channels is more depolarized than -60 mV, the responses to KCl depolarization may be relatively small. Electrophysiologic measurements have revealed that treatment with valinomycin, principally a K^+ ionophore, can hyperpolarize the HEK293 cell membrane by 45–70 mV

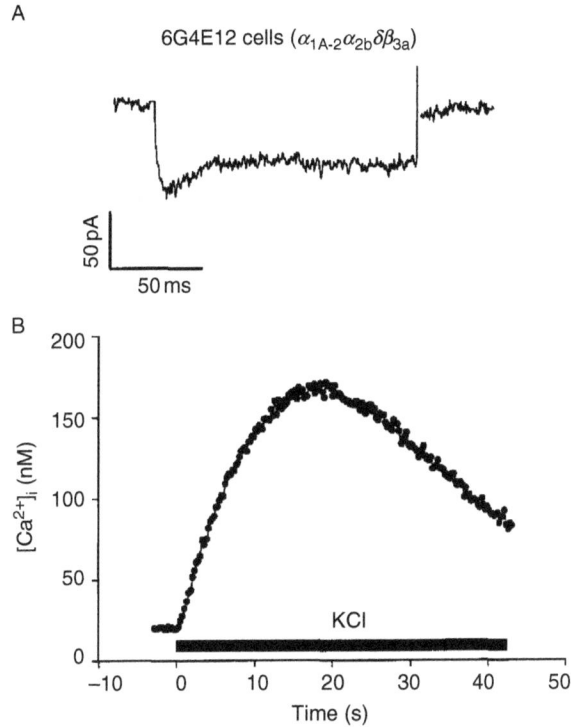

Fig. 3 Comparison of the electrophysiologic current and fluorescent dye-based $[Ca^{2+}]_i$ signal in HEK293 cells stably expressing $\alpha_{1A-2}\alpha_{2b}\delta\beta_{3a}$ VGCCs. In (A), the Ba^{2+} (15 mM) current was recorded from a cell depolarized to +10 mV from a holding potential (V_h) of −90 mV. In (B), cells loaded with Fluo-3 were depolarized with 70 mM KCl.

(data not shown). Therefore, this limitation may be overcome by preincubation of the cells with valinomycin (1–10 μM) before KCl is added. As shown in Fig. 4, in $\alpha_{1E-3}\alpha_{2b}\delta\beta_{1-3}$-expressing HEK293 cells, the $[Ca^{2+}]_i$ response to KCl is higher in valinomycin-treated cells. Because valinomycin acts to hyperpolarize the membrane, it is also possible that some of the increased $[Ca^{2+}]_i$ response is due to the increase in the driving force for Ca^{2+} resulting from the more negative membrane potential.

B. NMDA Receptors

NMDA receptors are ligand-gated cation channels with up to 11% of the inward current carried by Ca^{2+}. These channels are heteromeric complexes composed of at least an R1- and an R2-type subunit and are activated by glutamate and glycine (Burnashev *et al.*, 1995; Hollmann and Heinemann, 1994). NMDA receptors represent a pharmacologically versatile system containing multiple sites of drug

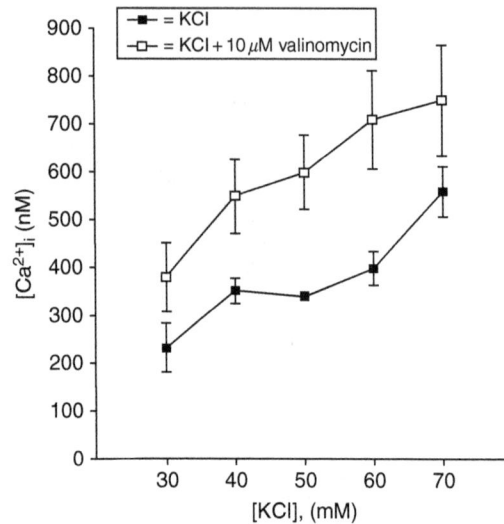

Fig. 4 Effect of valinomycin treatment on the KCl-induced $[Ca^{2+}]_i$ response. The KCl-induced $[Ca^{2+}]_i$ signal was measured in Fluo-3-loaded HEK293 cells expressing $\alpha_{1E-3}\alpha_{2b}\delta\beta_{1b}$-VGCCs. The cells were stimulated with increasing concentrations of KCl either alone or after a 10-min preincubation with 10 μM valinomycin. The points represent mean values ($\pm$S.D.) of quadruplicate determinations from a representative experiment.

interaction (Fig. 5A). HEK293 cells stably expressing the hNMDAR1A/2A and hNMDAR1A/2B subtypes of human NMDA receptors respond to the application of 100 μM glutamate and 30 μM glycine with a rise in $[Ca^{2+}]_i$ of approximately 400 nM that reaches maximal levels 10–20 s after agonist addition and remains elevated for more than 60 s (Fig. 5B).

Using suitable reference compounds that recognize each of these sites, we have demonstrated that the fluorescent dye-based $[Ca^{2+}]_i$ assay can be used to detect effectively the activity of compounds that interact with the receptor through both competitive and noncompetitive mechanisms (Table I). We studied two agonists (glutamate and NMDA) and a competitive antagonist ($\pm$)-3-(2-carboxypiperazin-4-yl)propyl-1-phosphonate (($\pm$)CPP) at the glutamate site (Fig. 6A,–C, Table I). At the glycine site, we were not able to determine the potency of glycine as a coagonist of the NMDA receptor, probably because the contaminating levels of glycine present in the assay are sufficiently high to fully saturate the glycine site (Varney *et al.*, 1996). The cells, rather than the buffer, appear to be the main source of the contaminating glycine. However, the activity of competitive antagonists at the glycine site can be detected. This is demonstrated by the results obtained for 5,7-dichlorokynurenic acid (5,7-DCKA), a competitive glycine-site antagonist, that inhibits the glutamate/glycine-induced $[Ca^{2+}]_i$ signal in hNMDAR1A/2A- and hNMDAR1A/2B-expressing cells with an IC_{50} of approximately 0.2 and

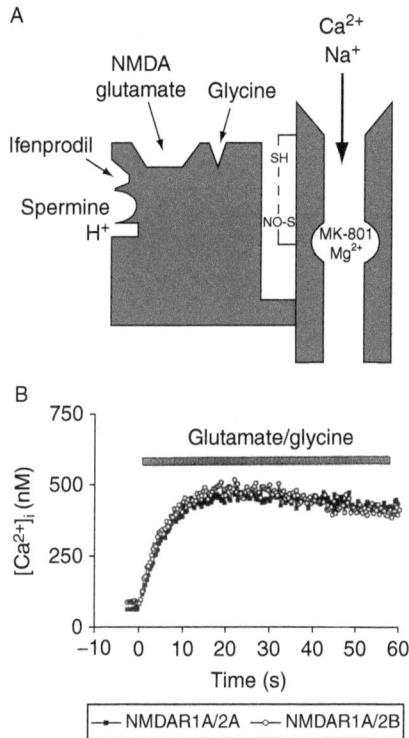

Fig. 5 Agonist-induced $[Ca^{2+}]_i$ responses in HEK293 cells stably expressing human NMDA receptors. (A) Schematic representation of an NMDA receptor. Activation of both the glutamate and glycine site are required for functional activation of the NMDA receptor and subsequent cation entry. In addition, there are several modulatory sites on the receptor. The NMDA receptor channel is inhibited by Mg^{2+} and channel blockers such as MK-801, PCP, memantine, and ketamine. The polyamine site is regulated by spermine, which may overlap with the proton inhibitory site and the ifenprodil binding site (at receptors containing the NMDAR2B subunit). The thiol groups (SH) may react with various forms of nitric oxide to modulate the receptor activity. [This figure is modified from a version published by Lipton and Rosenberg (1994)]. (B) Representative kinetic traces of $[Ca^{2+}]_i$ in response to 100 μM glutamate and 30 μM glycine, measured in Fluo-3-loaded HEK293 cells stably expressing hNMDAR1A/2A or hNMDAR1A/2B. Measurements were performed in nominally Mg^{2+}-free HBS, and data points show mean values from four replicate wells.

0.9 μM, respectively (Fig. 6D, Table I). These values are comparable to IC_{50} values of 0.2 and 0.4 μM determined using two-electrode voltage-clamp recording techniques in *Xenopus* oocytes injected with mRNAs encoding hNMDAR1A/2A or hNMDAR1A/2B, respectively (Hess *et al.*, 1996).

In addition, we examined two noncompetitive antagonists that interact at different sites on the NMDA receptor. Ketamine, a channel blocker, inhibited the glutamate/glycine-induced activation of both subtypes of human NMDA receptors in a concentration-dependent manner, with no apparent selectivity between

Table I

Pharmacology of Recombinant Human NMDA Receptors Stably Expressed in HEK293 Cells[a]

Compound	hNMDAR1A/2A			hNMDAR1A/2B		
	Mean	S.D.	N	Mean	S.D.	N
	EC$_{50}$ or IC$_{50}$ value (μM)					
Agonists						
Glutamate	0.736	0.371	8	0.578	0.222	8
NMDA	8.75	3.78	8	9.43	2.51	8
Antagonists						
5,7-DCKA	0.165	0.06	6	0.943	0.586	8
($\pm$)CPP	1.58	0.94	6	6.43	3.26	8
Ketamine	18.9	15.1	5	10.1	2.96	5
Ifenprodil	>30		8	0.303	0.214	8

[a]Values are expressed as mean $\pm$ S.D. from *N* experiments.

the two subtypes (Fig. 6E, Table I). Another noncompetitive blocker, ifenprodil, displayed marked subtype selectivity, with more than 100-fold greater potency for the NMDAR1A/2B subtype compared to the hNMDAR1A/2A (Fig. 6F, Table I). The ifenprodil results underscore the principal advantage of using a functional assay for screening compounds. In contrast to binding studies that require prior knowledge of a recognition site, the functional assay enables identification of subtype-selective compounds that interact with the receptor at novel sites.

V. Ion Channels with Moderate Functional Ca^{2+} Response

In this section, we included ion channels in which Ca^{2+} influx accounts for 1–10% of the inward current.

A. AMPA/Kainate Receptors

AMPA/kainate receptors, also referred to as non-NMDA ionotropic EAA receptors, are ligand-gated cation channels that respond to glutamate. Both AMPA and kainate receptors are multimeric complexes composed of one or more types of subunits. Burnashev *et al.* (1995) have reported that recombinant homomeric AMPA receptors composed of GluR1, GluR2(Q) or GluR4 or kainate receptors composed of GluR6(V,C,Q) can flux Ca^{2+} with a fractional Ca^{2+} response of 2–4%. Consistent with this, HEK293 cells stably expressing the hGluR3 subtype of human AMPA receptors and the hGluR6(I,Y,Q) subtype of human

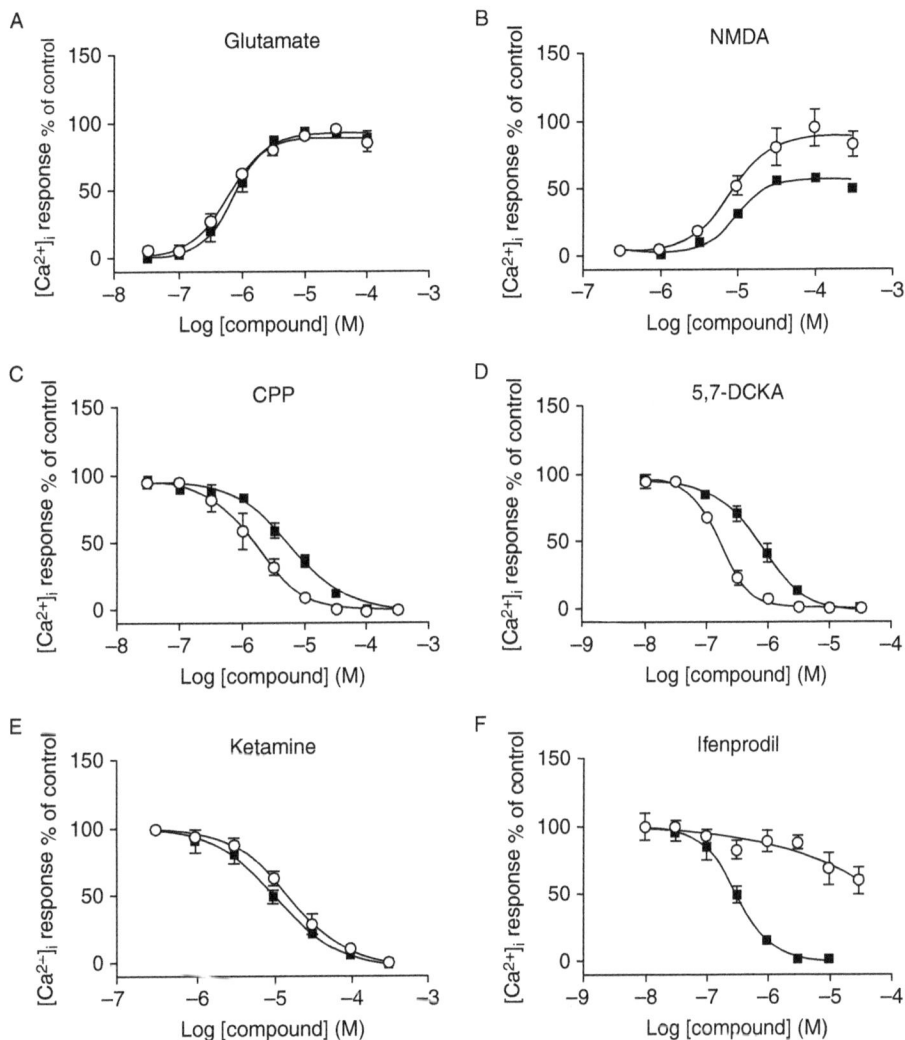

Fig. 6 Pharmacologic profile of HEK293 cells stably expressing hNMDAR1A/2A and hNMDAR1A/2B. Concentration-response curves to (A) glutamate, (B) NMDA, and inhibition curves to (C) ($\pm$)CPP, (D) 5,7-DCKA, (E) ketamine, and (F) ifenprodil were constructed from $[Ca^{2+}]_i$ measurements in Fura-2-loaded HEK293 cells stably expressing hNMDAR1A/2A ($\circ$) or hNMDAR1A/2B ($\blacksquare$). Data represent the mean $\pm$ S.E.M. from five to eight separate experiments, each performed in duplicates. For agonists, the data are normalized to the response elicited by 100 μM glutamate/30 μM glycine. In the antagonist studies, 3 μM glutamate and 3 μM glycine were used.

kainate receptors respond to agonist application with inward currents that desensitize rapidly (Fig. 7A and D). The desensitization can be attenuated by treating the hGluR3- and hGluR6-expressing cells with cyclothiazide (CTZ) or concanavalin A

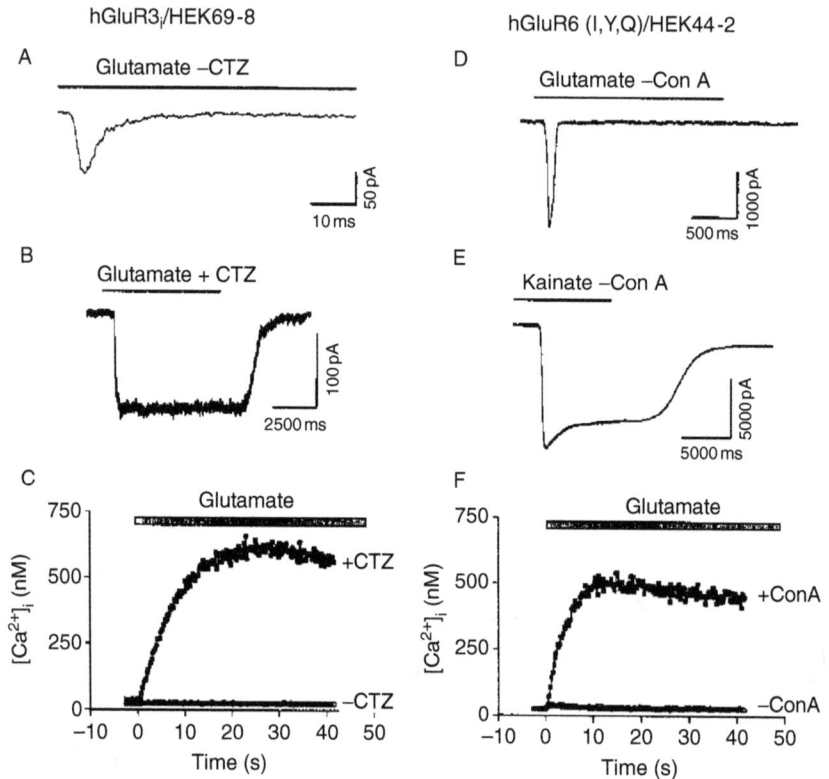

Fig. 7 Comparison of electrophysiologic and fluorescent dye-based $[Ca^{2-}]_i$ measurements of the activation of human AMPA (hGluR3$_i$) or kainate (hGluR6(I,Y,Q)) receptors stably expressed in HEK293 cells. Whole-cell patch-clamp recordings of HEK293 cells stably expressing the human AMPA receptor hGluR3$_i$ (A and B) in the absence (A) and presence (B) of 100 μM CTZ in response to a rapid application of 1 mM glutamate. Recordings from the human kainate receptor hGluR6(I,Y,Q) (D and E) in the absence (D) or following a 5-min pretreatment (E) with 0.3 mg/ml con A in response to a rapid application of 1 mM glutamate (D) or 1 mM kainate (E). Cells were held at a resting membrane potential of −60 mV. Representative kinetic traces of $[Ca^{2+}]_i$ in response to 1 mM glutamate in HEK293 cells stably expressing hGluR3$_i$ (C) and hGluR6(I,Y,Q) (F). Measurements were performed in Fluo-3-loaded cells. AMPA receptor desensitization was inhibited by 100 μM CTZ (C) and kainate receptor desensitization was inhibited by 0.3 mg/ml con A (F) by pretreatment for 5 min with either drug. Data points shown mean values from four replicate wells.

(con A), respectively (Fig. 7B and E). Due to rapid desensitization, glutamate-induced $[Ca^{2+}]_i$ response typically cannot be detected in these cells in the absence of CTZ or con A (Fig. 7C and F). In contrast, in the presence of CTZ or con A, both hGluR3- and hGluR6-expressing cells, respectively, respond to glutamate with a robust increase in $[Ca^{2+}]_i$ of approximately 500 nM above basal levels (Fig. 7C and F).

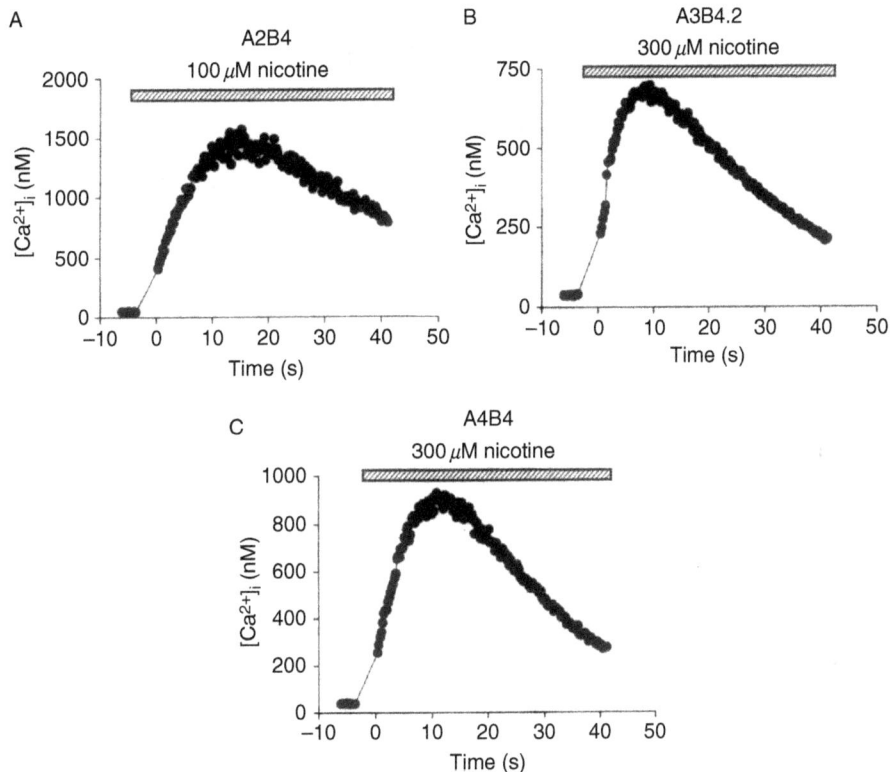

Fig. 8 Nicotine-induced changes of $[Ca^{2+}]_i$ in HEK293 cells stably expressing human NAChR subunits. The $[Ca^{2+}]_i$ measurements were performed in A2B4 (expressing $\alpha_2\beta_4$), A3B4.2 (expressing $\alpha_3\beta_4$), and A4B4 (expressing $\alpha_4\beta_4$) cell lines loaded with Fluo-3. After 10 measurements of basal fluorescence, the lid of the fluorimeter was opened briefly and 40 μl of nicotine was added manually to 160 μl of HBS already in the wells to the indicated final concentration. The 5-s gap at $t = 0$ in each fluorescence record is due to opening and closing of the lid during nicotine addition. Each point represents the mean $[Ca^{2+}]_i$ value (in nM) of four individual wells from a 96-well plate (error bars were omitted for clarity). Representative $[Ca^{2+}]_i$ responses are shown for each cell line.

B. Nicotinic Acetylcholine Receptors

Neuronal NAChRs are also ligand-gated multimeric channel complexes, presumed to be composed of five subunits representing one or more of an α-type and a β-type subunit (McGehee and Role, 1995). Multiple α and β subunits have been cloned and coexpressed stably in different combinations in HEK293 cells (Elliott *et al.*, 1996; Gopalakrishnan *et al.*, 1996; Whiting *et al.*, 1991). Cells stably expressing the binary combinations of $\alpha_2\beta_4$, $\alpha_3\beta_4$, and $\alpha_4\beta_4$ subunits respond to nicotinic agonists with a relatively rapid rise in $[Ca^{2+}]_i$ (Fig. 8). The kinetics and magnitudes of the $[Ca^{2+}]_i$ responses to a maximally effective agonist concentration were compared in the three cell lines. The largest $[Ca^{2+}]_i$ elevation was measured in the $\alpha_2\beta_4$-expressing cell line with a stimulation of more than 1300 nM above basal levels, likely resulting from a higher level of receptor expression in this cell line.

Pharmacologic characterization of the $[Ca^{2+}]_i$ response was performed by determining the potencies of several reference agonists and antagonists in all three cell lines. The concentration–response curves for four nicotinic agonists in the $\alpha_2\beta_4$-expressing cell line are shown in Fig. 9A. In a representative experiment,

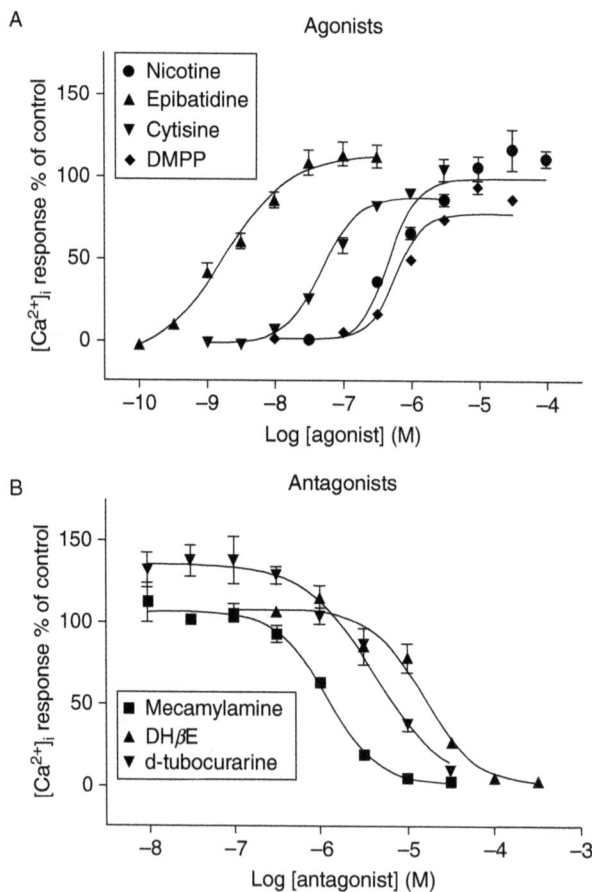

Fig. 9 Pharmacologic profile of A2B4 cells stably expressing human $\alpha_2\beta_4$ neuronal NAChRs. A2B4 cells were loaded with Fura-2 as described in the text, and fluorescence was monitored by the SIBIA/SAIC plate imaging system. The agonist data (A) are shown as a percent of the response to a maximal concentration of nicotine, whereas the antagonist data (B) are shown as a percent of the response to an EC_{80} concentration of nicotine (10 μM). The points are means ($\pm$S.D.) from quadruplicate determinations. The cells were incubated with the antagonists for 5–10 min prior to stimulation with nicotine, and the antagonists did not produce a response by themselves. Note that the compounds were added in a 10× solution containing DMSO, and because the density of this solution is greater than the buffer, the mixture may settle to the bottom of each well, thereby raising the effective concentration of compound. This can alter the apparent potency of the compound by shifting the concentration–response curves to the left.

epibatidine was the most potent agonist on the $\alpha_2\beta_4$ subtype (EC_{50} 1.8 nM), followed by cytisine (EC_{50} 483 nM), nicotine (EC_{50} 50.5 nM), and DMPP (EC_{50} 59.5 nM). In the antagonist studies, we tested mecamylamine, D-tubocurarine, and dihydro-β-erythroidine (DHβE) for their ability to inhibit nicotine-induced $[Ca^{2+}]_i$ responses. Again, in a representative experiment, all three antagonists fully inhibited the response to 10 μM nicotine, with a rank order of mecamylamine (IC_{50} 1.2 μM) > D-tubocurarine (IC_{50} 4.3 μM) > DHβE (IC_{50} 15.5 μM) (Fig. 9B).

The fluorescent dye-based $[Ca^{2+}]_i$ assay can be used to examine the mechanism of action of antagonists using Schild analysis, as shown in Fig. 10 for the three antagonists on the $\alpha_2\beta_4$-expressing cells. Increasing concentrations of mecamyl-amine (Fig. 10A) and D-tubocurarine (Fig. 10B) reduced the efficacy of nicotine without affecting its potency, and thus these two antagonists appeared to act through a noncompetitive mechanism. By contrast, DHβE behaved as a competi-tive antagonist since increasing its concentration reduced the potency of nicotine without significantly reducing its efficacy (Fig. 10C).

C. Cyclic Nucleotide-Gated Channels

Cyclic nucleotide-gated channels (CNGChs) are functionally ligand-gated cat-ion channels that have some structural features of voltage-gated ion channels (Kaupp, 1991). These channels are permeable to both mono- and divalent cations, with 2–8% of the current being carried by Ca^{2+} (Frings et al., 1995). The activation of the CNGChs can be measured using the fluorescent dye-based $[Ca^{2+}]_i$ assay. HEK293 cells stably expressing the rat CNGCh respond to forskolin with a rise in $[Ca^{2+}]_i$ that reaches a peak at approximately 60 s and decays over the next 60 s to an elevated, long-lasting plateau phase (Fig. 11A). The effect of forskolin is concentration dependent, with an EC_{50} of approximately 10 μM (Fig. 11B), similar to the EC_{50} of forskolin to activate adenylyl cyclase.

VI. Ion Channels with Low Fractional Ca^{2+} Response

In this section, we included ion channels in which Ca^{2+} influx accounts for less than 1% of the inward current.

Heteromeric and homomeric AMPA receptors containing the unedited GluR2(R) subunit have relatively low permeability to calcium (Burnashev et al., 1992). In these cases, the activation of the channel can be coupled to a $[Ca^{2+}]_i$ signal through coexpression with a recombinant VGCC subtype. The activation of the ligand-gated channel depolarizes the cell membrane, which in turn activates the VGCC. We tested this approach using one of the channels with moderate relative fractional Ca^{2+} response (the NAChR-$\alpha_3\beta_4$). To this end, we transiently expressed the NAChR-$\alpha_3\beta_4$ in HEK293 cells that stably express the $\alpha_{1B-2}\alpha_{2b}\delta\beta_{1c}$ subtype of human VGCCs (C1–C4 cell line). The nicotine-evoked $[Ca^{2+}]_i$ responses were compared to transient expression of the NAChR-$\alpha_3\beta_4$ in the host HEK293 cell line. The results revealed that

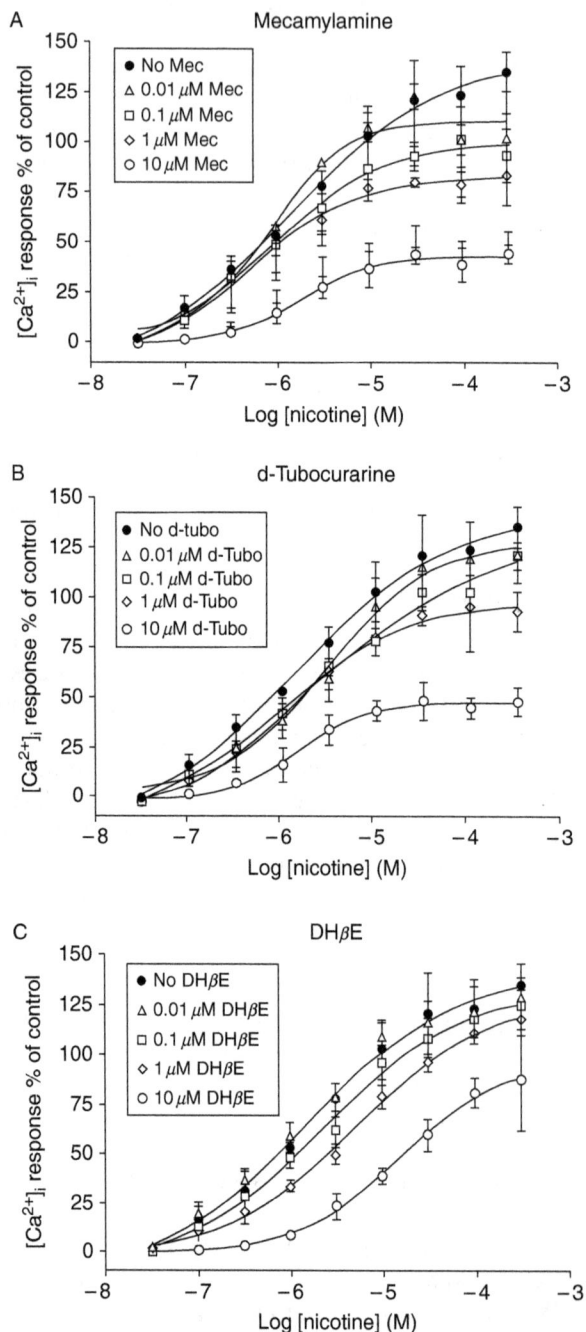

Fig. 10 Schild analysis of the inhibition of $\alpha_2\beta_4$-NAChRs by mecamylamine, D-tubocurarine, or DHβE. A2B4 cells expressing the $\alpha_2\beta_4$-NAChRs were loaded with Fura-2 as described in the text, and fluorescence was monitored by the SIBIA/SAIC plate imaging system. The y-axis represents the response as a percent of the control response to a maximal concentration of nicotine. Nicotine concentration curves were performed in the presence of increasing concentrations of antagonist (5–10 min preincubation) as indicated on the graphs. Points are means ($\pm$S.D.) of quadruplicate determinations.

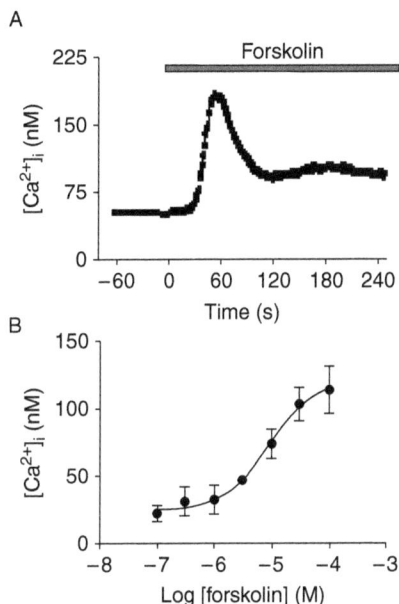

Fig. 11 Time course and pharmacologic characterization of forskolin-induced $[Ca^{2+}]_i$ signal in HEK293 cells stably expressing the rat CNG channel. (A) Representative kinetic trace of $[Ca^{2+}]_i$ in response to 100 μM forskolin in the presence of 1 mM IBMX in Fluo-3-loaded HEK293 cells stably expressing the rat CNG channel. Data points represent mean values from four replicate wells. (B) A concentration–response curve to forskolin in the presence of 1 mM IBMX in Fluo-3-loaded HEK293 cells stably expressing the rat CNG channel. Data points represent the mean $\pm$ S.D. from quadruplicate wells from a single experiment representative of two experiments.

NAChR-$\alpha_3\beta_4$-expressing HEK293 cells responded to the agonist (100 μM DMPP) with approximately a threefold increase in $[Ca^{2+}]_i$ that reached maximal levels after 20 s (Fig. 12A). In comparison, in cells expressing both NAChR-$\alpha_3\beta_4$ and N-type VGCCs, the magnitude of the agonist-induced $[Ca^{2+}]_i$ response was greatly enhanced to approximately 100-fold above basal levels with faster kinetics, reaching maximum levels within 10 s (Fig. 12B). The pharmacology of this response was also compared in both systems. As shown in Fig. 13, the DMPP-induced $[Ca^{2+}]_i$ response in NAChR-$\alpha_3\beta_4$-expressing HEK293 cells was blocked by a nicotinic antagonist (D-tubocurarine) but was not significantly altered by ω-CgTx-GVIA, a specific blocker of the α_{1B}-containing VGCCs. By contrast, in $\alpha_3\beta_4$-expressing C1–C4 cells, the DMPP-induced elevation of $[Ca^{2+}]_i$ was sensitive to both D-tubocurarine and ω-CgTx-GVIA. To compare the pharmacologic properties of the two systems in greater detail, the potencies of nicotine, cytisine, and DMPP were determined using the $[Ca^{2+}]_i$ assay. There were no significant differences in either the potencies of the three reference agonists or their relative efficacies (Table II). Thus, these results indicated that the activity of NAChR channels and other channels with low relative Ca^{2+} permeability can also be quantitated by measurements of the $[Ca^{2+}]_i$ response resulting from the

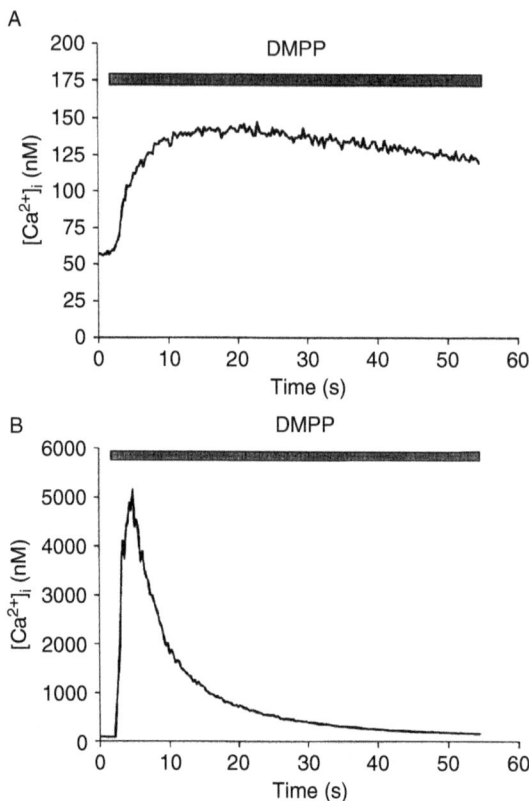

Fig. 12 Coupling the activation of NAChRs to activation of VGCCs by coexpression in HEK293 cells. HEK293 and C1–C4 cells stably expressing the N-type VGCCs ($\alpha_{1B-2}\alpha_{2b}\delta\beta_{1b}$) were transiently transfected with cDNAs encoding the α_3 and β_4 subunits of human NAChRs. Forty-eight hours after the transfection, each group of cells was loaded with Fluo-3, and the fluorescence was monitored by a plate-reading fluorimeter. As indicated on the graphs, the cells were stimulated with the nicotinic agonist DMPP (100 μM). Note the larger amplitude and faster kinetics of the response to DMPP in C1–C4 cells expressing the recombinant VGCCs (B) compared to HEK293 cells (A). The plots show the mean response of quadruplicate wells from a representative experiment.

activation of VGCCs when the two channels are coexpressed in the same host cells. A similar approach has been used to develop a stable cell line coexpressing recombinant rat NAChRs and L-type VGCCs (Stetzer *et al.*, 1996).

Another way to enhance the $[Ca^{2+}]_i$ signal with receptors or ion channels that have low permeability to Ca^{2+} is to increase the extracellular concentration of $CaCl_2$. For example, RD (TE671) cells express human neuromuscular-type NAChRs that have a lower relative permeability to Ca^{2+} than neuronal NAChRs (Bencherif and Lukas, 1991; Vernino *et al.*, 1992). When the RD cells were assayed in normal HBS buffer containing 1.8 mM $CaCl_2$, suberyldicholine stimulated only a small increase in $[Ca^{2+}]_i$

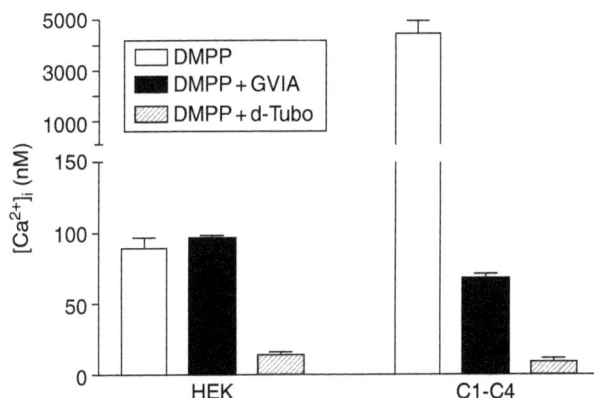

Fig. 13 Comparison of the pharmacology of DMPP-induced $[Ca^{2+}]_i$ signals in HEK293 or in C1–C4 cells transiently expressing $\alpha_3\beta_4$ NAChRs. The data are from the same experiment shown in Fig. 12. The cells were stimulated with 100 μM DMPP alone or after a 10-min preincubation with either 1 μM ω-CgTx-GVIA (GVIA), a selective antagonist of N-type VGCCs, or 10 μM D-tubocurarine (D-Tubo), a NAChR antagonist. Note that ω-CgTx-GVIA had no effect on the DMPP-induced $[Ca^{2+}]_i$ signal in the HEK293 cells and that D-tubocurarine had no effect on the VGCCs expressed in C1–C4 cells (data not shown).

Table II

Comparison of Agonist Activities in Cells Expressing $\alpha_3\beta_4$-NAChR with and without Coexpression of Voltage-Gated Ca^{2+} Channels[a]

Characteristic	Cell line	Nicotine	Cytisine	DMPP
Potency	$\alpha_3\beta_4$	51 ± 5 (3)	46 ± 18 (3)	16 ± 4 (3)
EC$_{50}$ (μM)	VGCC + $\alpha_3\beta_4$	117 ± 74 (3)	38 (2)	30 ± 9 (3)
Efficacy	$\alpha_3\beta_4$	100 (3)	69 ± 5 (3)	58 ± 11 (3)
% of nicotine response	VGCC + $\alpha_3\beta_4$	100 (3)	41 ± 43 (3)	44 ± 39 (3)

[a]Values are expressed as mean $\pm$ S.D. from (N) experiments.

of approximately 25 nM above basal levels compared to 250 nM when assayed in HBS containing 21.8 mM CaCl$_2$ (data not shown). This effect of Ca^{2+} is likely mediated by the increased driving force for Ca^{2+} entry.

VII. Limitations of Fluorescent Dye-Based $[Ca^{2+}]_i$ Assay for Measurement of Ion Channel Activity

The fluorescent dye-based $[Ca^{2+}]_i$ assay is most readily adaptable to those ion channels that flux detectable amounts of Ca^{2+} by the techniques described in this report. However, other fluorescent dyes can potentially be utilized in a similar manner to measure the activity of ion channels that flux other ions. For example, sodium-sensitive dyes, such as SBFI and sodium green (Haugland, 1996) can be used to measure Na^+ flux, and Cl^- flux can be monitored using the fluorescent dye 6-methoxy-N-(3-sulfopropyl)quinolinium (SPQ) (Ehring *et al.*, 1994) while changes in membrane potential can be measured using oxonol dyes (González and Tsien, 1995) and aminonaphthylethenylpyridinium (ANEP) dyes (Loew *et al.*, 1992). As described here for Ca^{2+}-sensing dyes, assays involving other fluorescent dyes would also be optimized and validated to ensure that the particular channel or receptor system of interest displayed the expected pharmacologic characteristics in the assay.

Changes in $[Ca^{2+}]_i$ are not always linear with changes in whole-cell recording currents. Therefore, although the rank order of potency should be the same in both assays, absolute agonist potencies determined from $[Ca^{2+}]_i$ measurements may differ from those determined by electrophysiology. In addition, agonist efficacies may not be the same between the two functional assays, since this will depend on the receptor number and Ca^{2+} permeability. Thus, a compound that is a partial agonist at a receptor when measured by electrophysiology may appear as a full agonist in the Ca^{2+} assay, provided it activates a sufficient number of receptors to elicit a saturating $[Ca^{2+}]_i$ response.

A $[Ca^{2+}]_i$ signal may not be detected if the ion channels desensitize rapidly, that is, AMPA receptors (Fig. 7). However, agents that block or slow receptor desensitization may allow the detection of $[Ca^{2+}]_i$ signals (e.g., CTZ for AMPA receptors). In addition, some agonists elicit different levels of receptor desensitization for the same receptor. For example, glutamate induces a rapid desensitization of AMPA receptors, whereas kainate can evoke nondesensitizing or slowly desensitizing responses (Partin *et al.*, 1994). Consequently, the fluorescent dye-based $[Ca^{2+}]_i$ assay may indicate differences in efficacy between two agonists that may not be detected in whole-cell recordings when desensitization is not blocked.

In $[Ca^{2+}]_i$ assays performed with a population of cells, the resting membrane potential of the cells is not easily controlled, although some manipulation is possible with ionophores such as valinomycin. Therefore, the detection of compounds that interact in a voltage-dependent manner may be compromised (Stocker *et al.*, 1997).

VIII. Advantages of Fluorescent Dye-Based $[Ca^{2+}]_i$ Assays

The studies summarized in this report demonstrate the utility and validity of $[Ca^{2+}]_i$ measurements as a means for evaluating the activity of recombinant ion channels stably expressed in mammalian cells. Although not discussed here, these protocols for $[Ca^{2+}]_i$ measurements are readily adaptable to G-protein-coupled receptors, such as the class I metabotropic glutamate receptors, the activation of which results in increased $[Ca^{2+}]_i$ through the stimulation of the phosphoinositide hydrolysis pathway (Daggett *et al.*, 1995; Lin *et al.*, 1997). Membrane-permeable Ca^{2+}-sensitive fluorescent dyes allow rapid measurement of $[Ca^{2+}]_i$ in whole cells, and detection of the fluorescence output in a plate-reading fluorimeter further facilitates detection of changes in $[Ca^{2+}]_i$ at a resolution of 0.25–0.33 s. The combination of these features with automation has significantly enhanced the capability to screen drugs directly for functional effects on specific receptor subtypes.

The ability to test compounds in a functional assay offers clear advantages over testing compounds in a binding assay. The latter necessitates the availability of a ligand of sufficient potency and selectivity for a particular site on the receptor. In most cases, the binding assay will only detect those compounds that competitively displace the bound reference ligand. By contrast, a functional assay can identify compounds that interact with the receptor either competitively at the same site as the reference agonist or noncompetitively at another site. This, in turn, markedly increases the potential for discovering ligands with novel structures. Additionally, agonists and antagonists can be readily distinguished in a functional assay and their relative efficacies can be determined without the need for secondary assays. Compounds that are detected in a displacement binding assay must subsequently be evaluated in functional assays to discern agonists from antagonists. For the antagonists, $[Ca^{2+}]_i$ measurements can also be used effectively to determine the mechanism of action. Finally, the use of cell lines expressing specific recombinant receptors as the targets for compound screening facilitates the discovery of subtype-selective compounds. In addition to their potential as effective therapeutic agents, subtype-selective compounds also represent valuable pharmacologic tools to study the role of the specific receptor subtypes in normal and pathophysiologic states.

Acknowledgments

We especially thank Dr. Michael Harpold for valuable input and continued support and encouragement throughout this work. In addition, we acknowledge Janis Corey-Naeve, Paul Brust, Alison Gillespie, Fen-Fen Lin, Christine Jachec, Susan Simerson, James Crona, Rhonda Skvoretz, Charlie Deal, Robert Siegel, and Carla Suto for contributions to the establishment and validation of the cell lines; Sandy Madigan for critical review of the manuscript; and Karen Payne for assistance in document preparation.

References

Bencherif, M., and Lukas, R. J. (1991). *Mol. Cell. Neurosci.* **2,** 52.

Brust, P. F., Simerson, S., McCue, A. F., Deal, C. R., Schoonmaker, S., Williams, M. E., Veliçelebi, G., Johnson, E. C., Harpold, M. M., and Ellis, S. B. (1993). *Neuropharmacology* **32,** 1189.

Burnashev, N., Monyer, H., Seeburg, P., and Sakmann, B. (1992). *Neuron* **8,** 189.

Burnashev, N., Zhou, Z., Neher, E., and Sakmann, B. (1995). *J. Physiol.* **485.2,** 403.

Catterall, W. A. (1995). *Annu. Rev. Biochem.* **64,** 493.

Chavez-Noriega, L. E., Crona, J. H., Washburn, M. S., Urrutia, A., Elliott, K. J., and Johnson, E. C. (1997). *J. Pharmacol. Exp. Ther.* **280,** 346.

Daggett, L., Sacaan, A. I., Akong, M., Rao, S., Hess, S. D., Liaw, C., Urruita, A., Jachec, C., Ellis, S. B., Dreessen, J., Knopfel, T., Landwehrmeyer, G. B., *et al.* (1995). *Neuropharmacology* **34,** 871.

Ehring, G. R., Osipchuk, Y. V., and Cahalan, M. D. (1994). *J. Gen. Physiol.* **104,** 1129.

Elliott, K. J., Ellis, S. B., Berckhan, K. J., Urrutia, A., Chavez-Noriega, L. E., Johnson, E. C., Veliçelebi, G., and Harpold, M. M. (1996). *J. Mol. Neurosci.* **7,** 217.

Frings, S., Seifert, R., Godde, M., and Kaupp, U. B. (1995). *Neuron* **15,** 169.

González, J. E., and Tsien, R. Y. (1995). *Biophys. J.* **69,** 1272.

Gopalakrishnan, M., Monteggia, L. M., Anderson, D. J., Molinari, E. J., Piattoni-Kaplan, M., Donnelly-Roberts, D., Arneric, S. P., and Sullivan, J. P. (1996). *J. Pharmacol. Exp. Ther.* **276,** 289.

Grynkiewicz, G., Poenie, M., and Tsien, R. Y. (1985). *J. Biol. Chem.* **260,** 3440.

Haugland, R. P. (1996). *In* "Handbook of Fluorescent Probes and Research Chemicals," (M. T. Z. Spence, ed.). Molecular Probes, Inc., Eugene, OR.

Hess, S. D., Daggett, L. P., Crona, J., Deal, C., Lu, C. C., Urrutia, A., Chavez-Noriega, L., Ellis, S. B., Johnson, E. C., and Veliçelebi, G. (1996). *J. Pharmacol. Exp. Ther.* **278,** 808.

Hollmann, M., and Heinemann, S. F. (1994). *Annu. Rev. Neurosci.* **17,** 31.

Kao, J. P. Y., Harootunian, A. T., and Tsien, R. Y. (1989). *J. Biol. Chem.* **264,** 8179.

Kaupp, U. B. (1991). *Trends Neurosci.* **14,** 150.

Kingston, R. E. (1996). *In* "Current Protocols in Molecular Biology," (F. M. Ausubel, R. Brent, R. E. Kingston, D. D. Moore, J. G. Seidman, J. A. Smith, and K. Struhl, eds.), p. 9.1.4. John Wiley & Sons, New York, NY.

Lin, F. F., Varney, M., Sacaan, A. I., Jachec, C., Daggett, L. P., Rao, S., Whisenant, T., Flor, P., Kuhn, R., Kerner, J. A., Standaert, D., Young, A. B., *et al.* (1997). *Neuropharmacology* **36,** 917.

Lipton, S. A., and Rosenberg, P. A. (1994). *N. Engl. J. Med.* **330,** 613.

Loew, L. M., Cohen, L. B., Dix, J., Fluhler, E. N., Montana, V., Salama, G., and Wu, J. Y. (1992). *J. Membr. Biol.* **130,** 1.

McGehee, D. S., and Role, L. W. (1995). *Annu. Rev. Physiol.* **57,** 521.

Minta, A., Kao, J. P. Y., and Tsien, R. Y. (1989). *J. Biol. Chem.* **264,** 8171.

Negulescu, P. A., and Machen, T. E. (1990). *Methods Enzymol* **192,** 38.

Neher, E. (1995). *Neuropharmacology* **34,** 1423.

Neher, E., and Augustine, G. J. (1992). *J. Physiol.* **450,** 273.

Olivera, B. M., Miljanich, G. P., Ramachandran, J., and Adams, M. E. (1994). *Annu. Rev. Biochem.* **63,** 823.

Partin, K. M., Patneau, D. K., and Mayer, M. L. (1994). *Mol. Pharmacol.* **46,** 129.

Rogers, M., and Dani, J. (1995). *Biophys. J.* **68,** 501.

Stetzer, E., Ebbinghaus, U., Storch, A., Poteur, L., Schrattenholz, A., Kramer, G., Methfessel, C., and Maelicke, A. (1996). *FEBS Lett.* **397,** 39.

Stocker, J. W., Nadasdi, L., Aldrich, R. W., and Tsien, R. W. (1997). *J. Neurosci.* **17,** 3002.

Tsien, R. Y. (1981). *Nature* **290,** 527.

Tsien, R. Y. (1988). *Trends Neurosci.* **11,** 419.

Tsien, R. Y. (1989). *Methods Cell Biol.* **30,** 27.

Varney, M. A., Jachec, C., Deal, C., Hess, S. D., Daggett, L. P., Skvoretz, R., Urcan, M., Morrison, J. H., Moran, T., Johnson, E. C., and Veliçelebi, G. (1996). *J. Pharmacol. Exp. Ther.* **279,** 367.

Vernino, S., Amador, M., Luetje, C. W., Patrick, J., and Dani, J. (1992). *Neuron* **8,** 127.

Whiting, P., Schoepfer, R., Lindstrom, J., and Priestley, T. (1991). *Mol. Pharmacol.* **40,** 463.

Williams, M. E., Feldman, D. H., McCue, A. F., Brenner, R., Veliçelebi, G., Ellis, S. B., and Harpold, M. M. (1992a). *Neuron* **8,** 71.

Williams, M. E., Brust, P. F., Feldman, D. H., Patthi, S., Simerson, S., Maroufi, A., McCue, A. F., Veliçelebi, G., Ellis, S. B., and Harpold, M. M. (1992b). *Science* **257,** 389.

Williams, M. E., Marubio, L. M., Deal, C. R., Hans, M., Brust, P. F., Philipson, L. H., Miller, R. J., Johnson, E. C., Harpold, M. M., and Ellis, S. B. (1994). *J. Biol. Chem.* **269,** 22347.

CHAPTER 12

Ligand Binding Methods for Analysis of Ion Channel Structure and Function

Steen E. Pedersen, Monica M. Lurtz, and Rao V. L. Papineni

Department of Molecular Physiology and Biophysics
Baylor College of Medicine, Houston
Texas, USA

DOI: 10.1016/B978-0-12-382204-8.00012-6

I. Introduction to Revised Article

This chapter focused primarily on binding assays to the muscle type nicotinic acetylcholine from *Torpedo*. Since the original publication, binding assays to nicotinic receptors have increasingly focused on receptors expressed in tissue culture cells, typically HEK293 cells (Wang *et al.*, 2003; Willcockson *et al.*, 2002) or COS cells, and in *Xenopus* oocytes (Xie and Cohen, 2001), to take advantage of analyzing the effects of receptor mutations. The variety of acetylcholine receptors assayed by binding includes not only the muscle type, but also the broad variety of neuronal nicotinic acetylcholine receptors (AChRs). Because of the modest expression efficacy for most nicotinic receptors, binding assays have relied primarily on radioactive ligands: for example, ^{125}I-α-bungarotoxin (α-BgTx) for muscle-type and α7-type neuronal receptors; ^{125}I-epibatidine, [^3H]epibatidine, and [^3H]nicotine for most others. These compounds have also been used extensively to characterize various brain fractions for the presence of nicotinic receptors. The fluorescent assays have remained useful only where sufficiently high concentrations of receptors can be obtained, usually from *Torpedo* electric organ (Andreeva *et al.*, 2006).

A few technical improvements have occurred in binding and its analysis. Assays for mutated expressed receptors could be carried out robotically, permitting relatively rapid analysis of many mutants under a variety of conditions. Double-mutant thermodynamic cycle analysis has found extensive use in analyzing direct ligand–protein interactions in the absence of crystal structures (Willcockson *et al.*, 2002). The extent of interaction between ligand functional groups and receptor side chains was typically expressed in terms of the interaction energy (see below), or its exponential equivalent Ω value: $\Omega = (K_{a1}/K_{a2})/(K_{b1}/K_{b2})$; the labeling is as in Scheme 2.

II. Introduction

The AChR is an ion channel that is opened by the binding of two molecules of acetylcholine on its extracellular surface (Devillers-Thiery *et al.*, 1993). Upon prolonged exposure to acetylcholine, the channel desensitizes and acquires high affinity for acetylcholine. The equilibrium binding to the two sites appears weakly cooperative, but the two sites are distinct, as shown by the binding of various antagonists that preferentially bind one site *versus* the other (Neubig and Cohen, 1979). Two subunits constitute each site: $\alpha\gamma$ and $\alpha\delta$. The heterogeneity of the sites is primarily due to the distinct contributions of the γ- and δ-subunits at each site (Pedersen and Cohen, 1990). The sites are allosterically linked to a binding site located within the channel pore itself, the noncompetitive antagonist site. Binding to this site can alter the conformation in favor of either the resting conformation or the desensitized conformation, depending on the ligand. Desensitization, as induced by noncompetitive antagonists, is also marked by increased affinity for agonist binding to the acetylcholine sites. The ability of many cholinergic ligands to bind all three sites further complicates analyzing the linkage.

An increased awareness of these phenomena has permitted binding to the AChR to be understood in more detail, and binding assays are finding greater use in elucidating the structure of the binding sites. In this chapter we will describe several ligand binding techniques: radioligand binding by centrifugation assay, radio-ligand binding by DE-81 filter binding, and fluorescent ligand binding. In addition, we will discuss how to analyze direct binding measurements, indirect binding by competition, and noncompetitive allosteric effects.

III. Comparison of Methods

A. Radioligand Binding

Measuring ligand binding by radioactively labeled ligands has the advantages of high sensitivity, fewer artifacts than fluorescence, and direct determination of the amount of ligand by scintillation or gamma counting. Counting is substantially more sensitive than fluorescence: fmoles of ligand can be directly detected whereas the limit of detection for fluorescence is near pmoles. There are some drawbacks to using radioactivity: the intrinsic health hazards, detection of beta-emission usually involves handling scintillation cocktails, and all types of radionuclei require significant effort for proper storage and disposal.

B. Fluorescence Binding Measurements

The advantages of fluorescence measurements echo the limitations of using radioactivity. Changes in fluorescence can be followed at any time resolution that still yields a detectable signal, and the data is obtained immediately. There is usually little need for the special handling and disposal of fluorophores. A primary disadvantage of fluorescence is the lower sensitivity, which substantially limits the usable concentration range of fluorescent ligands. A second disadvantage is that the fluorescence signal must be calibrated routinely for quantitative measurements. Fluorescence yield depends on several factors: instrument optics, the fluorophore, the solution, sample geometry, and the detector. Fluorescence measurements also require a sensitive fluorescence spectrophotometer. Some instruments can be obtained for roughly the same price as a scintillation counter, though research grade instruments may be substantially more expensive.

C. Which Ligand Should You Use?

There are a substantial number of radioactive ligands commercially available that bind the acetylcholine binding sites. The most common is ^{125}I-α-BgTx. Tritiated or ^{14}C-labeled ligands available include acetylcholine, epibatidine, and nicotine. For the noncompetitive site, however, there are no radioligands currently available. [^{3}H]Phencyclidine was recently discontinued by Dupont/New England Nuclear (now PerkinElmer). Other ligands that have been used, such as [^{3}H] ethidium or [^{3}H]histrionicotoxin, must be radiolabeled through a radiolabeling service. The fluorescent ligands ethidium, quinacrine, and crystal violet are available through standard sources. Except for derivatives of α-BgTx, fluorescent ligands for the acetylcholine binding sites must be synthesized or obtained from a kind donor.

IV. Methods

A. Preparation of Membranes

The preparation of AChR-rich membranes from *Torpedo* electric organ has been described in this series (Chak and Karlin, 1992). The procedure we follow is essentially that of Sobel *et al.* with some modifications (Pedersen *et al.*, 1986; Sobel *et al.*, 1977). This preparation can be conveniently scaled up to processing 2 kg of electric organ per batch with yields of several hundred mg of membrane protein. The membrane vesicles are near 20% purity in AChR, as assessed by the [^{3}H]acetylcholine binding assay described below (1.5–2 nmol acetylcholine binding sites per mg of protein). For many of the assays, particularly the microcentrifugation assay, it is adequate to use lower specific activity membranes, and often it is desirable to do so. Therefore, we usually save a side fraction from the discontinuous sucrose fractionation which contains membranes with specific activity that varies from 0.1 to 1 nmol acetylcholine binding sites per mg protein.

B. Radioligand Binding Assays

The following procedure describes the microcentrifuge ligand binding assay used for routine binding measurements. It is based on procedures developed and used in Jonathan Cohen's laboratory (Krodel *et al.*, 1979; Neubig and Cohen, 1979; Pedersen, 1995). *Torpedo* AChR-rich membranes are diluted into HTPS (250 mM NaCl, 5 mM KCl, 3 mM CaCl$_2$, 2 mM MgCl$_2$, 0.02% NaN$_3$, 20 mM HEPES, pH 7.0) and centrifuged for 30 min at 15 krpm (~19,000 × g) in a TOMY MTX-150 centrifuge. This initial centrifugation is to remove light membranes that might not sediment during the later centrifugation step. Protein assays show that typically no more than 10–20% of the membranes are lost in this step. The pellet is resuspended by passing the membranes through a 25 gauge syringe needle several times. The membranes are then treated with 1 mM diisopropylfluorophosphonate (DFP) for 1 h at ambient temperature to inactivate acetylcholinesterase. DFP hydrolyzes rapidly in aqueous solution and, therefore, must be diluted from the neat liquid immediately prior to mixing with the membranes. A second

1 h treatment with 0.1 mM DFP is then carried out and the membranes then transferred to ice. Subsequent dilutions and additions are then made in the 0.1 mM DFP solution.

Samples are assembled in eppendorf type microcentrifuge tubes and incubated for 30 min. They are then centrifuged for 30 min at 15 krpm (19,000 × g) in a TOMY microcentrifuge. A sample of the supernatant is retained for counting to determine free ligand and the remainder is removed with a gel-loading pipette tip attached to an aspirator with a trap for collecting the radioactive liquid. Traces of supernatant clinging to the sides of the tubes are adsorbed by a cotton swab. The tubes are then left upside-down on the cotton swab for 15 min to drain any remaining supernatant left on the pellet. The pelleted membranes are dissolved in 100 μl 10% sodium dodecyl sulfate (SDS) by shaking on a vortexer for at least 15 min, and then transferred to a scintillation counting vial; the tube is rinsed with an additional 100 μl 10% SDS and the rinse added to the scintillation vial. The nonspecific binding is determined by parallel samples including excess carbamyl-choline or α-bungarotoxin.

The choice of the ligand's specific radioactivity, its concentration in the binding assays, and the concentration of binding sites must be chosen to optimize the signal with regard to the type of information desired. The concentration of binding sites is limited by the need to form a discrete pellet upon centrifugation. The quality of data deteriorates if less than 50 μg of membranes are used; using 100 μg often improves the data. Thus, the lower limit of binding site concentration is determined by the volume desired and the specific binding activity of the membranes. To minimize the concentration of binding sites, we often use a low specific activity membrane fraction in the assay (0.1–1 nmol [^{3}H]acetylcholine binding sites per mg protein) number. A typical concentration of binding sites is 20–40 nM in final volume of 1 ml.

C. [^{3}H]Acetylcholine Binding

To monitor the conformational changes of the acetylcholine receptors by non-competitive antagonists, a high specific radioactivity [^{3}H]acetylcholine (~30 Ci/mmol; available from American Radiolabeled Chemicals Inc.) can be used at a concentration substantially lower than both the dissociation constant for acetyl-choline and the number of binding sites, as long as only a moderate percentage of the ligand is bound. In this way, the extent of binding is highly sensitive to the conformational equilibrium of the receptor and can be used to determine the conformational preference of the nonradioactive ligand. An example of this is shown in Fig. 1A. The increase in binding is due to the conformational effects of crystal violet that desensitizes the AChR, which is seen as higher affinity binding. At higher concentrations, crystal violet competes directly at the acetylcholine binding sites to reduce binding.

Competitive binding is used to determine the affinity of a ligand that is not radioactive. Nonradioactive ligands are incubated with membranes and an excess

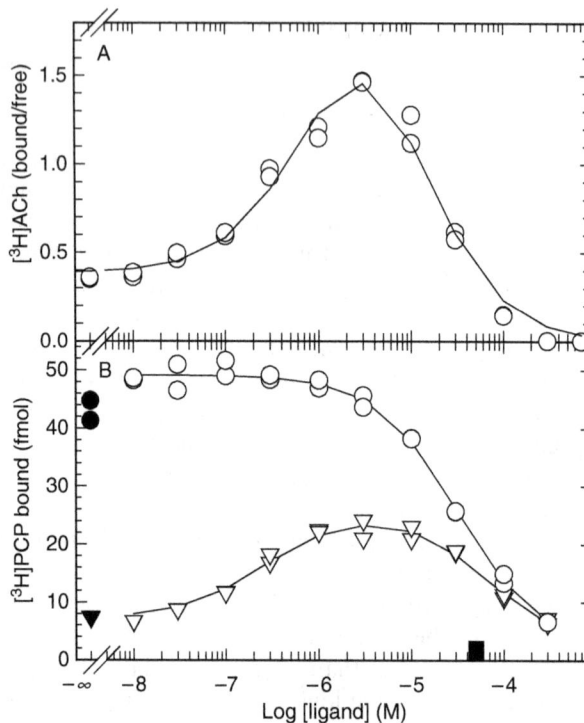

Fig. 1 Allosteric interactions detected by binding assays at low radioligand concentrations. (A) The effect of increasing crystal violet concentrations on [³H]acetylcholine binding was determined using the centrifugation assay described in the text. The assay was carried out in 1 ml volume containing 100 µg AChR-rich membranes (18 nM acetylcholine binding sites) and 1 nM [³H]acetylcholine. The data is plotted as the ratio of the bound ligand to free ligand. Under these conditions, this value is inversely proportional to the K_D. The rising phase shows that crystal violet increases the affinity of the acetylcholine binding sites for acetylcholine. The falling phase at higher concentrations reflects direct competition for binding at the acetylcholine binding sites. (B) The effect of varying concentration of tubocurine, a D-tubocurarine analog, was measured by [³H]phencyclidine binding using the centrifugation assay. AChR-rich membranes (50 µg; 200 µl; 62.5 nM AChR) were incubated with 1 nM [³H]phencyclidine (43 Ci/mmol) and the indicated concentrations of ligand. The rising phase of binding (▽) shows increased affinity for [³H]phencyclidine as a result of desensitization from tubocurine binding to the $\alpha\gamma$ acetylcholine binding site. Tubocurine competitively inhibits the AChR at higher concentrations resulting in the bell-shaped curve. The direct competition by tubocurine at the noncompetitive site is further illustrated by its inhibition in the presence of 100 µM carbamylcholine (○) to block effects at the acetylcholine binding sites. Controls in the absence (▼) and presence (●) of carbamylcholine illustrate the enhanced binding due to the strong desensitization. The binding in the presence of 50 µM proadifen, a noncompetitive antagonist, is also shown (■).

of low specific activity acetylcholine (∼100 mCi/mmol). In this case it is convenient to keep the [³H]acetylcholine concentration higher than both its K_D and the binding site concentration such that the sites are saturated; 100 nM [³H]acetylcholine is typical with 20–40 nM binding sites. The inhibition data will yield a K_{app}, the concentration at which the signal is decreased 50%. This value is easily obtained by fitting the data to an appropriate equation using a nonlinear regression algorithm. It can then be used to calculate the K_D for the inhibitor (see below).

D. Binding to the Noncompetitive Site

For [^{3}H]phencyclidine or [^{3}H]ethidium binding to the noncompetitive antagonist binding site, the assay is carried out as described above except that the DFP incubation steps can be omitted. A low concentration of [^{3}H]phencyclidine is convenient for measuring either competition by other noncompetitive antagonists or measuring conformationally driven effects of agonists or noncompetitive antagonists. Both effects can be seen in Fig. 1B where [^{3}H]phencyclidine binding is altered by the addition of tubocurine: tubocurine binds the acetylcholine binding sites at lower concentrations, resulting in partial conversion to the desensitized conformation of the AChR, which has higher affinity for [^{3}H]phencyclidine as seen by the increase in binding. At higher concentrations, tubocurine competes directly for binding at the noncompetitive antagonist site, producing a loss of [^{3}H]phencyclidine binding. The complete desensitization can be seen by addition of an agonist, carbamylcholine, which increases binding about fivefold over the baseline.

E. Assays Using ^{125}I-α-Bungarotoxin

^{125}I-α-Bungarotoxin is used extensively for measuring binding to the acetylcholine receptor. A variety of assays have been described; the most common one relies on the separation of free from bound ligand by adsorption of the AChR to DE-81 anion exchange filter paper (Schmidt and Raftery, 1973). We describe a variation of this assay to measure binding in physiological buffer. It avoids the use of the detergent Triton X-100, which may contribute to desensitization of the AChR and thereby cause artifacts when assessing the affinity of competing ligands. Because of the intrinsic high affinity of ^{125}I-α-BgTx for the AChR (the K_D is near pM) and its long dissociation time (h), competitive binding assays are not carried out at equilibrium, but rather use an initial rate assay. The assay is carried out for a limited period of time during the linear portion of the association time course. The amount of binding within this time is proportional to the free binding site concentration. Therefore, a competing ligand will reduce the rate of ^{125}I-α-BgTx binding to the extent that it occupies the binding sites. The rate of binding is sensitive to ionic strength (Schmidt and Raftery, 1974), such that an initial rate of binding in low ionic strength buffer can be done in 1–2 min whereas 45 min is necessary for binding in physiological saline.

The initial rate assay relies on maintaining a linear rate of binding of ^{125}I-α-BgTx during the course assay. This requires pseudo-first-order kinetic conditions, such that only a small percentage of the added ^{125}I-α-BgTx is bound and only a minor proportion of the AChR binding sites are bound. A rule of thumb is that the maximum binding should be less than one-half of each site; that is, a quarter of the total binding acetylcholine binding sites.

AChR-rich membranes are suspended to a concentration of 1–2 nM in HTPS supplemented with 0.1% bovine serum albumin and preincubated with the competing ligand for 30 min or longer. ^{125}I-α-BgTx ($\sim$200 Ci/mmol) is then added to a final concentration of 2 nM and the reaction incubated for 45 min. The binding is stopped by dilution with four volumes of 300 nM α-BgTx in 10 mM Tris, 0.1% BSA, 0.1% Triton X-100 (pH 7.4). This serves to isotopically dilute the ^{125}I-α-BgTx and prevent further binding and to lower the ionic strength. The latter is necessary for binding to the DE-81 filters. A 60 μl aliquot of each reaction is spotted on a DE-81 filter pinned to a styrofoam rack. The filters are allowed to sit no more than a few minutes before transfer to a tray with wash buffer: 10 mM Tris, 50 mM NaCl, 0.1% Triton X-100 (pH 7.4). The filters are washed twice, in batches, for 15 min each. The length of washing is not critical. More important is the spotting of the sample onto the filters; the exact time of incubation on the filter before transfer to wash buffer is relatively unimportant as long as the filters remain damp. If the filters begin to dry before washing, the nonspecific binding will increase dramatically.

The background binding in this assay can easily become too high; it appears to be dependent on the batch of DE-81 filters. A simple solution is to increase the amount of binding by using a higher receptor concentration, as long as the conditions for the assay listed above are still met.

F. ^{125}I-α-BgTx Binding to BC$_3$H-1 Cells

BC$_3$H-1 cells express the mouse muscle AChR and provide a convenient system for studying the properties of a mammalian AChR (Sine and Taylor, 1979, 1981). The binding assay can be carried out in 24-well tissue culture plates. Despite the lower receptor number in these cells than are obtained from *Torpedo* membranes, the assay is simpler because the free ^{125}I-α-BgTx can be washed off the cells with quite low background retention.

BC$_3$H-1 cells are maintained in Dulbecco's modified Eagle's medium with 20% fetal bovine serum, 100 units/ml penicillin, and 0.1 mg/ml streptomycin. They are seeded (8000–12,000 cells/well) into 24-well tissue culture plates that have been precoated with gelatin (0.2% porcine gelatin, 500 μl/, for 3 h before plating). The cells are grown until they reach $\sim$70% confluence and then the medium is changed to Dulbecco's modified Eagle's medium with 8% fetal bovine serum, 2% horse serum, 100 units/ml penicillin, and 0.1 mg/ml streptomycin. This initiates differentiation of the cells and expression of the AChR. The cells are ready for assay after 3 days incubation with horse serum.

Like the *Torpedo* membrane assay, this assay relies on the initial rate of binding of ^{125}I-α-BgTx and its slow dissociation rate. The cells are removed from the incubator and equilibrated to room temperature in a hood for 30 min. The remainder of the assay can be carried out on the bench at ambient temperature. Competing ligands are diluted in the media and then applied to the cells (350 μl) after removing the old media. The cells are incubated for 30 min with the competing ligand on a slowly rotating platform. ^{125}I-α-BgTx is then added in a 50 μl aliquot to a final concentration of $\sim$2 nM and allowed to incubate for another 45 min. The solution is then aspirated off the cells and the cells are washed twice with 0.7 ml Dulbecco's modified Eagle's medium. The bound ^{125}I-α-BgTx retained in the well is then transferred to a counting vial after dissolving the cells with 150 μl 1% Triton X-100 for 2 h. The gelatin coating on the wells is necessary for sticking of the cells to the plates during the incubation and washing procedures, but it also slows the dissolution by Triton X-100. It is important to ensure that all the cells are dissolved before transferring the counting vial.

It is important to establish independently that the incubation time with ^{125}I-α-BgTx is in the linear portion of the association rate curve. The rate of binding can be adjusted by varying the ^{125}I-α-BgTx concentration. As mentioned above, a rule of thumb is that no more than 25% of the total binding sites should be bound and only a minor portion of the ^{125}I-α-BgTx should be bound. The total number of receptors can be determined by using longer binding time (3 h) and higher ^{125}I-α-BgTx concentrations. We routinely use ^{125}I-α-BgTx with a specific radioactivity of 200 Ci/mmol, but higher activity ^{125}I-α-BgTx can also be obtained and used in this assay for higher sensitivity.

G. Effects of Noncompetitive Antagonists on Binding to the Agonist Site

Agonists and competitive antagonists, which bind the acetylcholine binding sites, stabilize the desensitized conformation of the AChR to varying degrees. The extent to which this occurs can be measured indirectly by examining the effect of noncompetitive antagonists on the binding of the acetylcholine site ligands. A variety of noncompetitive antagonists stabilize the desensitized conformation. We have used phencyclidine and proadifen. The former is for its high solubility, low partitioning with membranes (Lurtz et al., 1997), and well-characterized binding. The latter for its ability to desensitize to a greater extent (Krodel et al., 1979). Fewer ligands are available that preferentially stabilize the resting conformation of the AChR. The best is tetracaine (Cohen et al., 1986).

Noncompetitive ligands are usually added in competitive radioligand binding assays to a concentration near 30 μM. It is necessary to add sufficient ligand to be well above the noncompetitive ligand's K_D for the resting conformation of the AChR (near 5 μM for proadifen and phencyclidine). However, high concentrations (>100 μM) can sometimes perturb the lipid bilayer and induce nonspecific effects. This can cause problems in obtaining good pellets in the microcentrifuge assay and can cause dissociation of the BC$_3$H-1 cells from the tissue culture dish. High concentrations of some noncompetitive antagonists also inhibit the rate of ^{125}I-α-BgTx binding to Torpedo AChR-rich membranes (Krodel et al., 1979). The exact cause for this effect is not known but may be due to desensitization of the AChR.

V. Fluorescent Ligand Binding Assays

A. Fluorescent Ligands for the Acetylcholine Binding Sites

Only few fluorescent ligands are available that bind the acetylcholine binding sites. A variety of fluorescent derivatives of α-BgTx are commercially available (Haugland, 1996), but are of limited use for binding assays because they do not have substantial changes in fluorescence upon binding. Two agonists, dansyl-C6-choline (Heidmann and Changeux, 1979) and NBD-5-acylcholine (Prinz and Maelicke, 1983), have been characterized extensively. They share the high affinity and rapid binding characteristics of acetylcholine and are excellent real-time monitors of the conformational transitions of the AChR because their quantum yield changes substantially upon binding. The change in signal is enhanced by using energy transfer from tryptophans on the AChR to the bound ligand. Dansyl-choline (Cohen and Changeux, 1973) is an antagonist that binds with somewhat lower affinity but is useful for quantization of receptor specific activity (Neubig and Cohen, 1979; Valenzuela *et al.*, 1992); it is also commercially available (Sigma Chemical Co., St. Louis, MO). These ligands are useful for competitive binding assays for assessing the affinity of unlabeled ligands. Several caveats render competitive inhibition by fluorescence less useful than the radioligand assays. The concentration of receptor needed for an adequate signal is near 100 nM. For competitive inhibition it is desirable to have the ligand in excess, thus requiring several hundred nM concentrations. This will increase the concentration of competing ligand required for complete inhibition. A second caveat is that many competing ligands will absorb at 280–290 nm light, the wavelength required for excitation of these ligands. This adds the complication of having to correct for inner filter effects at higher competitor concentrations.

B. Fluorescent Ligands for the Noncompetitive Site

A variety of fluorophores bind the noncompetitive antagonist site. The best characterized ligand is ethidium bromide (Herz *et al.*, 1987); others include quinacrine (Grünhagen and Changeux, 1976), decidium (Johnson *et al.*, 1987) and crystal violet. The latter compound has a 200-fold increase in fluorescence yield upon binding the ion channel (M. M. Lurtz and S. E. Pedersen, unpublished observations). Ethidium fluorescence enhancement can be used to measure binding to the noncompetitive antagonist site. Titrations with competing ligands can be conveniently performed in a single cuvette to generate inhibition curves.

C. Fluorescent Assay for Noncompetitive Binding

Fluorescent measurements on the AChR require a good quality fluorometer because of the scattering due to membranes. We use research grade instruments (SLM 8000 and ISS PC1); however, less expensive fluorometers are likely to be adequate. For measurements, AChR-rich membranes are sedimented in the ultra-centrifuge and resuspended to 200 nM in acetylcholine binding sites in HTPS (for fluorescence measurements the sodium azide is omitted from the HTPS). Ethidium is added to a final concentration 250 nM and the agonist carbamylcholine is added to 100 μM. Carbamylcholine serves to desensitize the AChR, which improves the affinity for ethidium, and prevents binding of ethidium to the agonist sites. A 2.5–3 ml volume of the suspension is stirred in a 10×10 mm cuvette.

Fluorescence is excited with 340 nm light through a visible-absorbing filter (Oriel 59152) and is measured at 595 nm through a 540 nm cut-on filter (Oriel 59502). The filters improve the signal-to-noise ratio by removing stray light. To measure the affinity of a competing ligand, concentrated solutions can be titrated into the cuvette using a syringe. It is desirable to keep the additions to small volumes to avoid compensating for volume changes and to avoid affecting the equilibrium of the binding of ethidium. It is necessary to wait about 15 min between additions in order to let the slow dissociation of ethidium come to completion. This greatly limits the speed of the assay, but many sets of data can be measured simultaneously in this way.

VI. Analysis of Ligand Binding

The object of this section is to provide some practical guides for the analysis of direct binding data and competitive binding data. The last section will also discuss the use of thermodynamic cycles for interpreting conformational effects and allosteric ligand interactions. The theoretical basis and the derivation for the various equations will not be presented; they can be found elsewhere. The interested reader can find more detail in a number of books and articles (Cantor and Schimmel, 1980; Wyman and Gill, 1990). A good basic primer was recently published by Klotz (1997).

A. Analysis of Binding Isotherms

1. Subtraction of Nonspecific Binding

To analyze binding isotherms, nonspecific binding is subtracted, and the data is then fit by a nonlinear regression algorithm to the equation for single site binding (Eq. (1)). For the centrifugation assays described above, the free ligand concentration is determined directly by counting an aliquot of the supernatant. A common problem in subtracting nonspecific binding in these assays is that the free ligand concentration is not the same in parallel samples for *total binding* and for *nonspecific binding*. In this case, it is incorrect to simply subtract the nonspecific binding from the corresponding samples without inhibitor. If the nonspecific binding is linear with concentration, it can be fit to a line, and the fitted parameters used to calculate the nonspecific for each sample, using the corresponding free

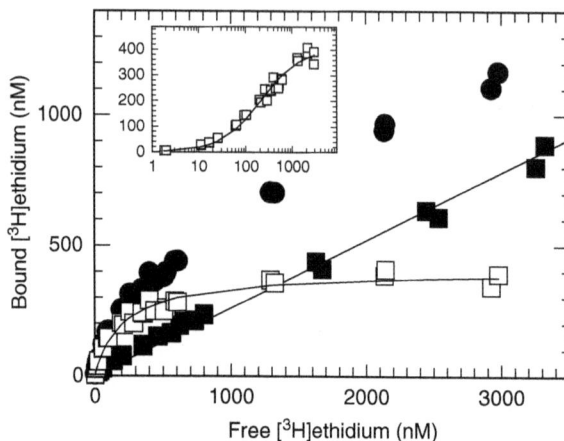

Fig. 2 Subtraction of nonspecific binding by direct determination of free ligand concentrations. AChR-rich membranes (88 μg; 200 μl; 100 nM AChR) were incubated with varying concentrations of [³H]ethidium (0.2 Ci/mmol) in the absence (•) and presence (▪) of 100 μM phencyclidine. After centrifugation of the membranes, the supernatants were counted to determine the free [³H]ethidium concentration. The nonspecific binding (▪), with phencyclidine, was fit to a line. The linear parameters were then used to calculate the nonspecific component of binding using the values of free [³H]ethidium for the samples without phencyclidine. This nonspecific component was then subtracted from the binding value to give the specific binding (□). The specific component of binding was then fit to Eq. (1) (*solid curve*). (Inset) The inset shows the semilog plot of the specific binding. It illustrates the difficulty of adequately demonstrating saturability of binding data. Data that appears to level off in the linear plot does not appear to reach a well-defined plateau value when viewed in a semilog plot. However, obtaining data at higher concentrations is limited by the substantial nonspecific binding.

concentration of ligand. If the *nonspecific binding* is nonlinear, then the values can be fit to other functions; a hyperbolic equation often works well ($B = A \cdot L/(L + K)$). Alternatively, the values can be manually interpolated from the *nonspecific binding* data. This problem is illustrated in Fig. 2.

In cases where the concentrations of receptor are the same or higher than the ligand concentrations, substantial overestimates of K_D values can arise from assuming the free ligand concentration to be equal to the concentration of ligand added. Further error results from directly subtracting the parallel samples that define nonspecific binding. This is a particular problem for fluorescence titration assays where it is inconvenient to directly determine the free ligand for each data point. A correction can improve the estimate: If the amount of receptor is known, then the amount of bound ligand can be estimated and subtracted from the total ligand concentration to provide a corrected free ligand concentration. Replotting the data in terms of this new concentration then gives an improved estimate for the K_D. This ignores any ligand depletion from partitioning of ligand into the membrane, which may also be significant and result in several-fold errors in K_D estimates. When precise determinations of the K_D by fluorescence are desired, the free ligand concentration can be determined independently by removing the bound ligand by centrifugation and measuring the supernatant concentration by fluorescence or by HPLC (Lurtz et al., 1997).

2. Curve Fitting

Prior to the ubiquitous use of computers, binding data was often linearized and plotted in accordance with the Scatchard equation (Scatchard, 1949). This method is still useful for qualitative evaluation of the data, but nonlinear regression of the data is now easy to perform with commonly available programs. We routinely use Sigmaplot (Jandel Scientific). Binding is analyzed using the equation for the binding of a ligand to a single binding site (Eq. (1)), where RL is the measured binding concentration, R_0 is the total binding site concentration, L is the *free* ligand concentration. This equation is simply derived from the definition of the dissociation constant, $K_D = R \cdot L / RL$, and the mass balance equation, $R_0 = R + RL$.

$$RL = \frac{R_0 \cdot L}{L + K_D} \qquad (1)$$

The maximum binding, R_0, and the K_D for binding are extracted from the fit. Most fitting routines will also supply statistical data that indicate the quality of the fit. It is important that the data demonstrate saturability, otherwise both parameters can have substantial error. As discussed by Klotz (1997), the best plot for visualizing whether the binding has reached saturation is a semilog plot of bound ligand *versus* the log of the ligand concentration (see the *inset* in Fig. 2). The quality of the fit can usually be determined by comparing the best fit with the data and looking for systematic deviations. This can also be done by plotting the residuals and looking for a pattern. If the residuals are not scattered evenly about the origin, then the fit is poor. Poor fits may arise from a number of problems. If the data do not saturate, the maximum value will be poorly determined and may deviate significantly from the expected value, resulting in error in the K_D as well. The data may reflect components from multiple sites, cooperativity, or heterogeneous samples that lead to a poor approximation by the single site model.

If the data is presumed to bind more than one site, it must be analyzed by other equations. In the case of cooperativity between sites, the data can often be fit to the Hill equation (Hill, 1910) (Eq. (2)). The value n reflects the degree of apparent cooperativity between two or more sites. In the case of positive cooperativity, the slope of the semilog plot will be steeper than for a single binding site and $n > 1$. For negative cooperativity or for multiple independent sites, $n < 1$. It is difficult to distinguish multiple independent sites from negative cooperativity among the sites by equilibrium binding. If multiple independent sites are suspected, they may also be fit to a sum of terms, each term with the same form as Eq. (1).

$$RL = \frac{R_0}{1 + (K_D/L)^n} \qquad (2)$$

B. Analysis of Competitive Inhibition Data

Competitive inhibition to a single site can generally be fit to Eq. (3), which describes the inhibition of binding of L by increasing concentrations of I. In this equation, RL is the amount bound observed, A is the amplitude of the change in binding, Bcg is the background or nonspecific binding, and K_{app} is the concentration that produces 50% inhibition. To calculate the K_I for the inhibiting ligand (i.e., dissociation constant for the inhibitor I), it is necessary to correct for the concentration of the observed ligand L using Eq. (4), where K_D is the dissociation constant for L. K_D must be determined independently under similar conditions.

$$RL = \frac{A}{1 + I/K_{app}} + Bcg \qquad (3)$$

$$K_I = K_{app} \frac{K_D}{L + K_D} \qquad (4)$$

Competitive inhibition data must have well-defined maximum and minimum values in order to obtain reliable values for K_{app} by fitting to Eq. (3): data should extend 2 orders of magnitude on either side of the K_{app}. For analyzing compounds of unknown affinity, controls should be included using a known ligand at a concentration that gives complete inhibition. Then, if the data do not extend to complete inhibition, the fit can be forced to the Bcg value defined by the control inhibitor.

For fitting data to the acetylcholine binding sites of the AChR, it is often necessary, as well as desirable, to fit the data to inhibition at two distinct sites and obtain a K_{app} for each site. This is accomplished by fitting the data to Eq. (5) which represents inhibition to two independent sites present in equal amounts. At times, it is necessary to fit the data with no prior assumption about the relative amounts of the two sites is assumed; this is described by Eq. (6). This situation occurs, for instance, with inhibition of ^{125}I-α-BgTx binding by 13′-iodo-D-tubocurarine: although the two binding sites are present in equal amounts, the binding of 13′-iodo-D-tubocurarine influences the rate of ^{125}I-α-BgTx to the second site, and it does not appear represented in equal amplitude (Pedersen and Papineni, 1995).

$$RL = A \left[\frac{1}{1 + I/K_{1app}} + \frac{1}{1 + I/K_{2app}} \right] + Bcg \qquad (5)$$

$$RL = \frac{A_1}{1 + I/K_{1app}} + \frac{A_2}{1 + I/K_{2app}} + Bcg \qquad (6)$$

When attempting to distinguish binding to two sites, it is important that K_{1app} and K_{2app} differ sufficiently. The binding to two sites of equal intrinsic affinity can be nearly as well described by binding to two sites that differ fourfold in K_D. Therefore, a separation of K_{1app} and K_{2app} of 10-fold is often required for reliable determination of the individual constants. A good test is to run both single-site and two-site fits to the data. While there will generally be improvement in the residuals because of the increase in the number of parameters, there should be clear evidence from inspection of the graph that the two-site model provides a better fit to the data before it is interpreted as such. As a rule, the best determination of the error is to repeat the experiment.

VII. Thermodynamic Cycle Analysis

Thermodynamic cycles are useful tools for analyzing ligand binding when it is necessary to interpret the effects of ligand structure, receptor mutations, hetero-tropic allosteric effects, or conformational changes. Thermodynamic cycles simply reflect the first law of thermodynamics, the conservation of energy, and the corollary of path independence: because energy is a state function, the difference in energy between two states is independent of the path. Scheme 1 shows a version for analyzing the allosteric effects of first ligand (I) upon a second ligand (L). If the two states are considered to be the free receptor with unbound ligand ($R + I + L$) and the ternary complex (RIL), then it matters not whether L binds first and then I or *vice versa*. The free energy of each step is proportional to the log of the equilibrium constant ($\Delta G = -RT \ln K$). Therefore, taking each path and setting the energies equal and removing the logarithms yields $K_L K_I' = K_I K_L'$. This can be rewritten $K_L / K_L' = K_I / K_I'$, which says that the ratio of the binding constants for L in the absence and presence of I is the same as the ratio of the binding constants for I in the presence and absence of L. For example, if I inhibits the binding of L, then L will inhibit the binding I to the same extent. The equation also shows that if three of the constants are known, then the fourth is also determined.

The scheme does not illustrate the underlying conformational changes, but they are implicit in the changes in the equilibrium constants: each constant reflects the binding to the equilibrium distribution of conformations for the receptor, which may be influenced by the presence of the second ligand. It is clear that long-range interactions between the ligands must take place, often mediated by receptor conformational changes, or else K_L would simply equal K_L'.

This cycle can be used to analyze several scenarios: the effect of varying concentrations of an allosteric ligand I on the binding of the measured ligand L, or the effect of an isotonic concentration of I on the binding affinity of L. The latter case

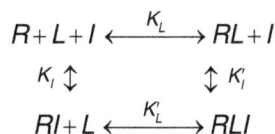

$$R + L + I \xleftrightarrow{\quad K_L \quad} RL + I$$
$$K_I \updownarrow \qquad\qquad \updownarrow K_I$$
$$RI + L \xleftrightarrow{\quad K_L' \quad} RLI$$

Scheme 1 Thermodynamic cycle analysis of allosteric ligand binding.

is described by Eq. (7), where the binding of L is analyzed by Eq. (1) or by Eqs. (3) and (4), depending on whether the data is obtained by direct binding or by competitive inhibition. The K value obtained equals the right-hand term in the denominator of Eq. (7). If a saturating concentration of I is used, the K value reduces to $K_L K_I'/K_I$ which is simply equal to K_L'.

$$RL = R_0 \frac{L}{L + K_L \frac{1+I/K_I}{1+I/K_I'}} \qquad (7)$$

Conversely, if the radioactive (or fluorescent) ligand L is held constant and the effect of varying concentrations of I is measured, then Eq. (7) can be rearranged to give Eq. (8). The data can be fit according to Eq. (3) to yield the maximum, minimum, and K_{app}. Those values can then be interpreted according to Eqs. (9)–(11). In this case, the plateau value at high concentrations of I does not reflect nonspecific binding, but rather reflects the binding of L as affected by I (Eq. (11)). For this reason, it is important to have an independent determination of nonspecific binding using an excess of a ligand known to compete with L. This value should be subtracted before the analysis.

$$\frac{RL}{R_0} = (B_\infty - B_0)\frac{I}{I + K_{app}} + B_0 \qquad (8)$$

$$K_{app} = \frac{K_I K_I'(L + K_L)}{LK_I + K_L K_I'} \qquad (9)$$

$$B_0 = \frac{L}{L + K_L} \qquad (10)$$

$$B_\infty = \frac{L}{L + K_L K_L'/K_I} \qquad (11)$$

A. Double-Mutant Thermodynamic Cycle Analysis

The principle behind the double-mutant thermodynamic cycle analysis has been articulated by Carter *et al.* (1984) and was outlined by Ackers and Smith (1985) for the study of pairwise interactions of residues in proteins. The pairwise analysis overcomes some of the caveats of single mutation analysis on binding energetics and conformational changes. It can be used to determine whether or not changes in homologous ligands are independent and can be used to evaluate interaction among specific loci on ligands and receptors.

To illustrate the method, consider ligands L_1 and a close analog L_2, which differ by a small, discrete structural change, for example, a single methylation in

D-tubocurarine. L_1 and L_2 bind to a receptor site, R_a. A single mutation in R_a will yield the receptor homologue R_b. If the ligand functional group does not interact with the modified receptor amino acid, then the *change* in affinity of the ligand will be *independent* of the receptor modification. If the ligand functional group does interact with the receptor amino acid residue, then the *change* in affinity upon modifying L_1 to L_2 will be *dependent* on whether the receptor residue is also modified.

This explanation can be formalized using the thermodynamic cycle shown in Scheme 2. A binding analysis is performed to measure the equilibrium dissociation constants for both L_1 and L_2 to both R_a and R_b. This will generate four binding constants and the corresponding free energies: ΔG_{a1}, ΔG_{a2}, ΔG_{b1}, and ΔG_{b2}; the subscripts indicate the receptor and the ligand, respectively. Each binding energy is represented at a corner of the cycle. The principle of path independence leads to the same constraints discussed above: all paths from $R_a L_1$ to $R_b L_2$ must have the same energy. Therefore, $\Delta\Delta G_a + \Delta\Delta G_2 = \Delta\Delta G_1 + \Delta\Delta G_b$, where $\Delta\Delta G_x$ represents the *difference* between two binding energies.

If the two ligand and receptor sites are *independent* of each other then the *change* in ligand binding affinities ($\Delta\Delta G_x$) will be the same in each direction; that is, $\Delta\Delta G_a = \Delta\Delta G_b$ and $\Delta\Delta G_1 = \Delta\Delta G_2$. In this case, the change in binding energy caused by modifying the ligand ($\Delta\Delta G_a$ and $\Delta\Delta G_b$) is the same for both receptor types. Conversely, the *change* in binding energy caused by mutating the receptor ($\Delta\Delta G_1$ and $\Delta\Delta G_2$) is the same for either ligand.

If the mutated receptor residue interacts with the altered ligand site, then $\Delta\Delta G_a \neq \Delta\Delta G_b$ and $\Delta\Delta G_1 \neq \Delta\Delta G_2$. However, the cyclic relationship must hold, which can be rewritten as follows: $\Delta\Delta G_a - \Delta\Delta G_b = \Delta\Delta G_1 - \Delta\Delta G_2$. This difference is the interaction energy and reflects the strength of the interaction. The identity of the two differences provides an internal control on the experimental measurements.

This type of analysis has been used to study the role of individual hemoglobin residues in conformational transitions upon ligand binding (Ackers and Smith, 1985). It was also recently applied to study the interaction of scorpion agatoxin 2 with the *Shaker* potassium channel and successfully revealed close electrostatic interactions (Hidalgo and MacKinnon, 1995). This method is likely to find further use in analyzing the structures of binding sites that are inaccessible to direct structural determination.

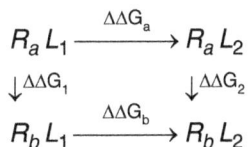

$$R_a L_1 \xrightarrow{\Delta\Delta G_a} R_a L_2$$
$$\downarrow \Delta\Delta G_1 \qquad\qquad\qquad \downarrow \Delta\Delta G_2$$
$$R_b L_1 \xrightarrow{\Delta\Delta G_b} R_b L_2$$

Scheme 2   A double-mutant thermodynamic cycle analysis for ligand-receptor interactions.

References

Ackers, G. K., and Smith, F. R. (1985). *Ann. Rev. Biochem.* **54,** 597–629.

Andreeva, I. E., Nirthana, S., Cohen, J. B., and Pedersen, S. E. (2006). *Biochemistry* **45,** 195–204.

Cantor, C. R., and Schimmel, P. R. (1980). Biophysical Chemistry: Part III, The Behavior of Biological Macromolecules. W. H. Freeman Co., San Francisco, CA.

Carter, P. J., Winter, G., Wilkinson, A. J., and Fersht, A. R. (1984). *Cell* **38,** 835–840.

Chak, A., and Karlin, A. (1992). *Methods Enzymol.* **207,** 546–555.

Cohen, J. B., and Changeux, J. P. (1973). *Biochemistry* **12,** 4855–4864.

Cohen, J. B., Correll, L. A., Dreyer, E. B., Kuisk, I. R., Medynski, D. C., and Strnad, N. P. (1986). *In* "Molecular and Cellular Mechanisms of Anesthetics," (S. H. Roth, and K. W. Miller, eds.), pp. 111–124. Plenum Publishing, New York.

Devillers-Thiery, A., Galzi, J. L., Eisele, J. L., Bertrand, S., and Changeux, J. P. (1993). *J. Membr. Biol.* **136,** 97–112.

Grünhagen, H. H., and Changeux, J. P. (1976). *J. Mol. Biol.* **106,** 517–535.

Haugland, R. P. (1996). Handbook of Fluorescent Probes and Research Chemicals. Molecular Probes, Eugene, OR.

Heidmann, T., and Changeux, J. P. (1979). *Eur. J. Biochem.* **94,** 255–279.

Herz, J. M., Johnson, D. A., and Taylor, P. (1987). *J. Biol. Chem.* **262,** 7238–7247.

Hidalgo, P., and MacKinnon, R. (1995). *Science* **268,** 307–310.

Hill, A. V. (1910). *J. Physiol. (Lond.)* **40**(iv).

Johnson, D. A., Brown, R. D., Herz, J. M., Berman, H. A., Andreasen, G. L., and Taylor, P. (1987). *J. Biol. Chem.* **262,** 14022–14029.

Klotz, I. M. (1997). Ligand-Receptor Energetics: A Guide for the Perplexed. John Wiley & Sons Inc., New York.

Krodel, E. K., Beckman, R. A., and Cohen, J. B. (1979). *Mol. Pharmacol.* **15,** 294–312.

Lurtz, M. M., Hareland, M. L., and Pedersen, S. E. (1997). *Biochemistry* **36,** 2068–2075.

Neubig, R. R., and Cohen, J. B. (1979). *Biochemistry* **18,** 5464–5475.

Pedersen, S. E. (1995). *Mol. Pharmacol.* **47,** 1–9.

Pedersen, S. E., and Cohen, J. B. (1990). *Proc. Natl. Acad. Sci. USA* **87,** 2785–2789.

Pedersen, S. E., and Papineni, R. V. L. (1995). *J. Biol. Chem.* **270,** 31141–31150.

Pedersen, S. E., Dreyer, E. B., and Cohen, J. B. (1986). *J. Biol. Chem.* **261,** 13735–13743.

Prinz, H., and Maelicke, A. (1983). *J. Biol. Chem.* **258,** 10263–10271.

Scatchard, G. (1949). *Ann. N. Y. Acad. Sci.* **51,** 660.

Schmidt, J., and Raftery, M. A. (1973). *Anal. Biochem.* **52,** 349–354.

Schmidt, J., and Raftery, M. A. (1974). *J. Neurochem.* **23,** 617–623.

Sine, S., and Taylor, P. (1979). *J. Biol. Chem.* **254,** 3315–3325.

Sine, S., and Taylor, P. (1981). *J. Biol. Chem.* **256,** 6692–6699.

Sobel, A., Weber, M., and Changeux, J. P. (1977). *Eur. J. Biochem.* **80,** 215–224.

Valenzuela, C. F., Kerr, J. A., and Johnson, D. A. (1992). *J. Biol. Chem.* **267,** 8238–8244.

Wang, H. L., Gao, F., Bren, N., and Sine, S. M. (2003). *J. Biol. Chem.* **278,** 32284–32291.

Willcockson, I. U., Hong, A., Whisenant, R. P., Edwards, J. B., Wang, H., Sarkar, H. K., and Pedersen, S. E. (2002). *J. Biol. Chem.* **277,** 42249–42258.

Wyman, J., and Gill, S. J. (1990). Binding and Linkage: Function Chemistry of Biological Macromolecules. University Science Books, Mill Valley, CA.

Xie, Y., and Cohen, J. B. (2001). *J. Biol. Chem.* **276,** 2417–2426.

CHAPTER 13

Three-Dimensional Structure of Membrane Proteins Determined by Two-Dimensional Crystallization, Electron Cryomicroscopy, and Image Analysis

Mark Yeager, Vinzenz M. Unger, and Alok K. Mitra

Department of Cell Biology
The Scripps Research Institute
La Jolla, California, USA

ESSENTIAL ION CHANNEL METHODS

255

DOI: 10.1016/B978-0-12-382204-8.00013-8

I. Introduction

The Brookhaven Protein Data Bank now has about 5000 atomic structures available for soluble proteins. This compares with about 20 membrane protein structures, many of which are of the same class. Strategies continue to be developed for growing three-dimensional (3D) crystals of membrane proteins (Garavito *et al.*, 1996; Kühlbrandt, 1988; Michel, 1991; Pebay-Peyroula *et al.*, 1997; Reiss-Husson, 1992), and recent progress has been encouraging (Garavito and White, 1997; Ostermeier and Michel, 1997). Nevertheless, the high-resolution structure analysis of membrane proteins is still a formidable task. In addition, no recombinant eucaryotic membrane protein has as yet been amenable to 3D crystallization. Soluble fragments of membrane proteins have been overexpressed, purified, and examined by conventional X-ray crystallography (De Vos *et al.*, 1992; Shapiro *et al.*, 1995; Wang *et al.*, 1990). However, this approach does not allow examination of transmembrane domains, which are involved in signal transduction and transport across membranes. Solid state nuclear magnetic resonance (NMR) spectroscopy has successfully been used to examine the transmembrane domains of membrane proteins (Opella, 1977). However, the protein must be examined in micelles and the molecular weight limit precludes examination of complex polytopic proteins.

An alternative approach is to grow two-dimensional (2D) crystals and use electron cryomicroscopy and image analysis to solve the structure (Amos *et al.*, 1982; Boekema, 1990; Bullough, 1990; Chiu and Schmid, 1993; Chiu *et al.*, 1993; Dorset, 1996; Engel *et al.*, 1992; Jap *et al.*, 1992; Kühlbrandt, 1992; Stewart, 1988). Currently, the structural characterization of the majority of membrane proteins by electron cryocrystallography is indeed a challenging endeavor because the proteins are often present in low abundance in their native membranes, and only a few hundred micrograms of purified recombinant protein may be available. Furthermore, membrane proteins tend to be labile when solubilized in detergent. However, in several cases it has now been possible to grow 2D crystals in lipid bilayer membranes that can be analyzed by electron cryocrystallography. Moreover, the success in obtaining atomic resolution structures for bacteriorhodopsin (Henderson *et al.*, 1990; Kimura *et al.*, 1997) and the light-harvesting complex II (Kühlbrandt *et al.*, 1994) attests that 2D crystallization, electron cryomicroscopy, and image analysis are emerging as powerful approaches for the structural characterization of membrane proteins.

There are several advantages of analyzing 2D crystals by electron cryomicroscopy and image analysis compared with X-ray analysis of 3D crystals. First, a great advantage of this technique is the absence of the phase problem. Because images of the crystals are recorded, phases can be directly calculated by Fourier transformation. Second, the amount of material required for electron crystallography is substantially less than that needed for X-ray crystallography. For instance, our current analysis of gap junction channels is based on membrane specimens that are enriched for the recombinant connexin but still do not show a detectable band on Coomassie-stained sodium dodecyl sulfate (SDS) gels.

This compares with the need for several milligrams of protein required for typical 3D crystallization. Furthermore, purification of such large quantities of protein is quite expensive because of the need for substantial amounts of detergents, protease inhibitors, and tissue culture media. For electron cryomicroscopy of membrane proteins, specimens can be preserved in an unstained, frozen-hydrated state within the lipid bilayer so that the native structure is revealed (Chiu, 1986; Dubochet *et al.*, 1988; Milligan *et al.*, 1984; Taylor and Glaeser, 1976; Unwin, 1986). This compares with the analysis of 3D crystals in the presence of detergents and nonphysiologic precipitants. One would assume that the detergent belt around the hydrophobic perimeter of the membrane protein in the 3D crystal would mimic the lipid bilayer environment, but at least one case has been identified in which a dramatic difference occurred when the protein was examined in bilayers versus 3D crystals. The pore-forming toxin α-hemolysin assembled as a heptamer in 3D crystals (Song *et al.*, 1996) but was hexameric in lipid bilayer membranes (Czajkowsky and Shao, 1997; Olofsson *et al.*, 1988). With the membrane protein embedded in a lipid bilayer, it is also possible to examine functionally important states. For example, Berriman and Unwin (1994) have developed a rapid freezing method that has been successfully used to examine the open conformation of the acetylcholine receptor by exposing the crystals to acetylcholine within milliseconds of plunge-freezing for electron cryomicroscopy (Unwin, 1995).

Similar to X-ray crystallography, the quality of the crystals is a major limitation. To date, most studies using electron cryocrystallography of 2D crystals only allowed analysis at modest $\sim$6-Å resolution. This is sufficient for revealing the packing of transmembrane α-helices within the lipid bilayer, but the connecting loops and β structure are often not delineated at this resolution. To date, only a few proteins have been crystallized in 2D with sufficient order that they diffract to atomic resolution. This problem is in part due to disorder in the crystals. However, besides crystal disorder, instrumental effects are also important. Because the phases are computed directly from images, the microscope must be extremely stable so that movement during imaging is on the order of 1 Å/s or less. Such microscopes are indeed expensive, and setting up a laboratory for electron cryo-crystallography is substantially more expensive than setting up one for X-ray crystallography. The best microscopes available use a field emission gun as an electron source to generate extremely bright and coherent beams, operate at higher voltages (200–400 kV) so that the depth of field is increased to allow examination of thicker specimens, employ stages that are exquisitely stable to vibrations, and operate at liquid nitrogen or even liquid helium temperatures.

In this review we provide a general outline for the steps involved in the structure analysis of membrane proteins by electron cryocrystallography. We present our results on the analysis of 2D crystals of gap junction channels, aquaporin channels, and histidine-tagged Fab molecules, to exemplify respectively the general methods of *in situ* 2D crystallization, *in vitro* 2D crystallization, and lipid monolayer crystallization. Also discussed are the preparation of frozen-hydrated specimens, performance of low-dose transmission electron cryomicroscopy, and image analysis of 2D crystals.

II. Steps in Structural Analysis of Membrane Proteins by Electron Cryocrystallography

Table I outlines the strategy for the structural characterization of membrane proteins by electron cryocrystallography, which involves isolation of membranes containing the desired protein, 2D crystallization, electron cryomicroscopy, and computer image processing (Yeager, 1995).

A. Isolation of Membranes

To our knowledge, our analysis of cardiac gap junction channels is the first example where a polytopic membrane protein has been expressed in a heterologous system and examined by structural methods (Unger *et al.*, 1997c). The overexpression of membrane proteins is not at all routine. The reader is referred to

Table I
General Outline for Structure Analysis of 2D Crystals

Isolation of membranes
   Protein source:
- *High abundance*: obtain tissue
- *Low abundance*: overexpression in insect cells, BHK cells, yeast cells, etc.

   Homogenization and low-speed centrifugations to remove soluble components
   Consider extraction with chaotropic agents (e.g., KI and $Na_2S_2O_3$) to remove extrinsic proteins
   Sucrose gradient ultracentrifugation to enrich the membrane fraction

Two-dimensional crystallization
   In situ method:
- Incubate with low concentrations of nonionic detergents, phospholipases, etc. to remove lipids and concentrate protein in bilayer
- Remove detergent (e.g., by dialysis or Bio-Beads)

   In vitro reconstitution:
- Solubilize protein using high concentrations of nonionic detergents
- Purify protein using size-exclusion, ion-exchange, hydrophobic interaction, or affinity chromatography
- Mix purified protein–detergent complexes with synthetic or native lipids that have been solubilized in detergents
- Remove detergent by dialysis or Bio-Beads

   Lipid monolayer crystallization:
- Isolate and purify water-soluble protein domain:
  - Water-soluble conformation of membrane protein (e.g., annexins)
  - Ectodomain generated by protease cleavage
  - Ectodomain expressed using recombinant DNA methods
- Incubate protein solution in droplet having lipid monolayer surface:
  - Use positively and negatively charged lipids to exploit electrostatic interaction between protein and lipid headgroups
  - Exploit specific affinity between particular headgroup (e.g., nickel lipid) and protein (e.g., hexahistidine tag)

(continues)

Table I (*continued*)

Transmission electron microscopy

    Deposit membranes on grid by centrifugation or drop adherence

    Prepare specimen for cryomicroscopy by plunging the grid into ethane slush or embedding specimen in mordants such as glucose, trehalose, or tannin

    Store grids in liquid nitrogen

    Transfer grid to stage using cryotransfer system

    Perform low-dose cryomicroscopy with specimen at temperature of liquid nitrogen or helium

    Identify large, unbroken membrane sheets, preferably with polygonal edges and with proper ice thickness

    Record image using minimal dose techniques:

- *Search mode*: scan grid in overfocused diffraction mode
- *Focus mode*: at high magnification ($\sim$200,000$\times$) in area just adjacent to region to be photographed located Gaussian focus, select defocus level (e.g., 4000-Å underfocus), correct for astigmatism, and ensure that there is no drift of stage
- *Exposure mode*: record image or diffraction pattern of crystal using dose of $<$10 e$^-$/Å^2

Image appraisal

    Examine EM negatives

    Perform optical diffraction and select images that display sharp, bright reflections to high resolution; optimal defocus; minimal drift

    Digitize crystalline patches identified by optical diffraction

Image processing (see Table II)

    Taper edges of digitized array

    Compute Fourier transform

    Index reflections in computed diffraction pattern

    Refine crystal lattice parameters

    Correct for crystal lattice distortions:

- Mask diffraction pattern and calculate filtered image
- Select reference image
- Calculate cross-correlation map
- Correct for lattice displacements

    Extract corrected amplitudes and phases

    Correct for effects due to CTF and astigmatism

    Evaluate plane group symmetry from images of untilted crystals

    Refine phase origin, tilt axis, and tilt angle

    Merge data from multiple images

    If amplitudes are derived by:

- *Electron diffraction*: subtract background due to inelastic scattering
- *Fourier transformation of images*: apply inverse temperature factor to correct for resolution dependent fall-off of image amplitudes

    Interpolate phase and amplitude data in 3D lattice lines

    Compute 3D map by Fourier inversion with symmetry constraints defined by lattice

    Use graphics software to examine map

Grisshammer and Tate (1995) for a state-of-the-art review. For our analysis of gap junctions, a C-terminal truncation mutant (after lysine 263) of rat heart α_1Cx43 connexin (designated α_1Cx263T) was expressed in a stably transfected baby hamster kidney (BHK) cell line under control of the inducible mouse metallothionein promoter (Kumar *et al.*, 1995). Freeze-fracture, thin-section, and negative stain electron microscopy of the BHK cell membranes demonstrated that the recombinant protein assembles with the characteristic septalaminar morphology of gap junctions. In addition, dye transfer experiments (Unger *et al.*, 1997a) demonstrated that the gap junctions are functional. Of note, however, is that the levels of expression are quite low. Nevertheless, this amount of protein is sufficient to pursue 2D crystallization experiments because the junctions naturally assemble in specialized membrane domains that can be selectively enriched.

B. Procedure for Isolation of Membranes Enriched for Recombinant Gap Junctions

1. Induction of Connexin Expression

   1. Examine the cells by light microscopy and assess the density. When the cells appear almost confluent, they are ready for induction.
   2. Aspirate the medium and replace with fresh medium containing 100 mM zinc acetate.

2. Enrichment of Gap Junctions

   1. Place a 500-ml JA10 centrifuge bottle into an ice bucket and insert a funnel.
   2. Use a cell scraper to dislodge the cells from the bottom of the plates, and decant the media with the suspended cells.
   3. *Spin 1* (2 krpm, 5 min, 4 °C, JA10 rotor, Beckman J-30I centrifuge).
   4. Decant the supernatant.
   5. Carefully pipette 10 ml of ice-cold HEPES buffer (10 mM HEPES, pH 7.5, 0.8% NaCl) into the centrifuge tube so as to not disturb the pellet.
   6. Detach the pellet by swirling the centrifuge bottle.
   7. Pour the contents with the intact pellet into a 50-ml conical plastic tube (Falcon, Corning).
   8. Vortex the Falcon tube to break up the pellet.
   9. Increase the volume to 50 ml with ice-cold HEPES buffer.
   10. *Spin 2* (2 krpm, 5 min, 4 °C, GH-3.8 rotor, Beckman GS-6KR centrifuge).
   11. Decant the supernatant from the Falcon tube, and add 5 ml of ice-cold HEPES buffer containing 140 μg/ml phenylmethylsulfonyl fluoride (PMSF) to the pellet.
   12. Suspend the pellet by vortex mixing, and increase the total volume to 15 ml with HEPES buffer containing 140 μg/ml PMSF.

13. Incubate the Falcon tube for 5 min on ice.

14. Sonicate the same three times for 5 s (Branson sonicator, output setting 6–7, 60% duty cycle) waiting 15 s between each burst. Keep the sonicator tip at the bottom of the tube to minimize foaming. The final suspension should appear milky and homogeneous.

15. Examine a drop of the suspension by light microscopy to ensure that cell lysis is at least 90%. If there are still numerous intact cells, repeat the sonication for 5 s.

16. Increase the volume to 35 ml for ultracentrifugation.

17. *Spin 3* (25 krpm, 45 min, 4 °C, SW28, rotor, Beckman LE-80K ultracentrifuge). Decant the supernatant, and resuspend the pellet in 5 ml of ice-cold HEPES buffer.

18. Add 30 ml of 49% sucrose in HEPES buffer to an SW28 ultracentrifuge tube, and pipette the 5 ml of crude membrane homogenate onto the top of the sucrose cushion.

19. *Spin 4* (25 krpm, 45 min, 4 °C, SW28 rotor, Beckman LE-80K ultracentrifuge).

20. Use a plastic disposable pipette to carefully remove material (e.g., solid PMSF) floating on the top of the tube. Then retrieve the 0/49% interface and transfer to another SW28 tube. The homogenate should appear milky white.

21. Increase the volume to 35 ml with ice-cold sucrose-free HEPES buffer. Use a plastic pipette to mix the solution to dilute the sucrose in the aspirated interface.

22. *Spin 5* (25 krpm, 45 min, 4 °C, SW28 rotor, Beckman LE-80K ultracentrifuge).

23. Decant the supernatant, add 1 ml ice-cold HEPES buffer containing 140 μg/ml PMSF, disrupt the pellet by repeated passage through a plastic Eppendorf pipette and homogenize the sample by sonication for a few seconds.

24. The specimen is stable at 4 °C for ~2 weeks.

C. Two-Dimensional Crystallization

As in any crystallization trial, one must optimize and refine the conditions for crystallization and specimen handling to maximize the degree of order (McPherson, 1982). In general, the variables to be tested to achieve 2D crystallization are similar to those used for 3D crystallization and include evaluation of divalent cations, pH, detergents, lipids, buffers, ionic strength, temperature, precipitants, ligands, and inhibitors. Purity and yield of the protein are maximized, and proteolysis is minimized. We currently use three methods (*in situ* crystallization, *in vitro* reconstitution, and lipid monolayer crystallization) to grow 2D crystals (Fig. 1). Another method is to grow 2D crystals of a detergent-solubilized and purified membrane protein directly on the electron microscope (EM) grid.

In situ crystallization:

In vitro reconstitution:

Lipid monolayer crystallization:

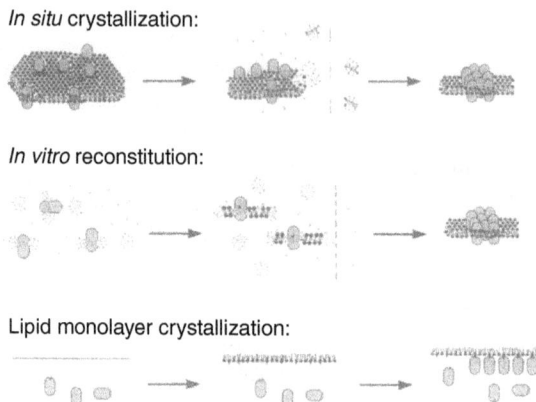

Fig. 1 Schematic diagram for three strategies that have been used to grow 2D crystals. In the method of *in situ crystallization*, the protein is never removed from its native membrane. A requirement for the success of this technique is that the protein must already be fairly concentrated in the membrane. 2D crystallization is induced using gentle conditions that extract membrane lipids without solubilizing the protein. In the method of *in vitro reconstitution*, the protein is solubilized from its native membrane and purified as a complex with detergent micelles. The purified protein–detergent complexes are mixed with lipids that have also been solubilized in detergent. Lipid bilayers are reconstituted by removing the detergent via dialysis. Crystallization is induced by using a lipid:protein ratio that is high enough to promote protein/protein interactions. In the *lipid monolayer crystallization* technique, lipids are spread at an air/water interface so that the aliphatic chains extend into the air, and the lipid headgroups are exposed at the aqueous interface. A soluble protein is present in the aqueous phase. Binding of the protein to the lipid headgroups occurs by electrostatic interactions or by a specific affinity tag. For instance, the lipid monolayer can be doped with a lipid containing a nickel headgroup that will chelate a polyhistidine tag on the protein.

This approach has been successfully used to grow 2D crystals of the H^+-ATPase from *Neurospora crassa* (Auer *et al.*, 1998). Crystallization is induced by using buffer conditions that are similar to those used for 3D crystallization (e.g., high concentrations of PEG or ammonium sulfate). In this way crystallization is induced by interactions of the hydrophilic domains of the protein. Some soluble proteins can also be grown as 2D crystals, such as catalase (Akey *et al.*, 1984; Unwin, 1975), tubulin (Amos and Baker, 1979; Nogales *et al.*, 1997), and VP6, the rotavirus inner capsid protein (Hsu *et al.*, 1997). The progress of 2D crystallization is usually assessed by negative stain electron microscopy.

D. *In Situ* Crystallization

A special advantage of this approach is that the protein is never removed from its native membrane (Engel *et al.*, 1992; Jap *et al.*, 1992; Kühlbrandt, 1992; Yeager, 1994). However, a requirement for the success of this technique is that the protein must already be fairly concentrated in the membrane or have a tendency to self-associate in patches that can be isolated. 2D crystallization is induced using gentle conditions that extract membrane lipids without solubilizing the protein.

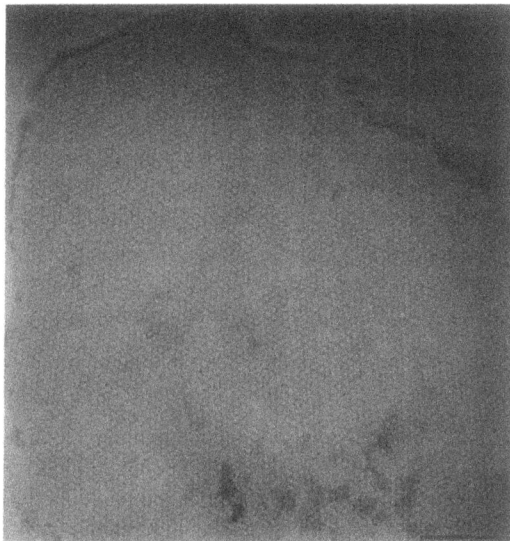

Fig. 2 Electron micrograph of a negatively stained 2D crystal of recombinant gap junction channels formed by a C-terminal truncation mutant of rat heart α_1-connexin (α_1Cx263T). The hexagonal packing of the channels is clearly seen. Such crystals are grown by *in situ* crystallization in which the membranes are incubated in buffer containing Tween 20 and DHPC to extract lipids. The mottled appearance at the bottom of the crystal may represent regions that are partially solubilized by the detergents. Note that the crystal is made of several mosaic domains, and there is a fracture in the crystal that may have occurred during transfer to the supporting carbon substrate. For membrane proteins with hydrophilic ectodomains, we have found that adherence of the specimen to the carbon support is optimal if the grids are rendered hydrophilic by glow discharge (Dubochet *et al.*, 1982). However, strong interactions between the carbon and the crystals may also deform the crystals. We have found that pretreatment of the grid with a detergent such as Tween or octylglucoside can be used to modulate the adherence of the crystals to the carbon. Bar: 1000 Å.

Detergents, enzymes, and chaotropes (Hatefi and Hanstein, 1974) that have been used include deoxycholate (Glaeser *et al.*, 1985; Yeager, 1994; Yeager and Gilula, 1992), dodecylmaltoside (Gogol and Unwin, 1988; Yeager, 1994; Yeager and Gilula, 1992), Lubrol (Zampighi and Unwin, 1979), D_2O (Sikerwar and Unwin, 1988), Tween (Schertler *et al.*, 1993; Unger *et al.*, 1997c), 1,2-diheptanoyl-*sn*-phosphocholine (DHPC) (Unger *et al.*, 1997c), and phospholipases (Mannella, 1984, 1989). For the expressed gap junction channels, the order of naturally occurring 2D arrays was improved by sequential exposure to Tween 20 and DHPC (Fig. 2). In addition, exposure to the detergents enriched the preparation by solubilizing nonjunctional protein.

E. Procedure for *In Situ* 2D Crystallization of Recombinant Gap Junction Channels

1. The protein concentration of the membrane sample enriched for gap junctions is estimated using the DotMetric assay (Geno Technology Inc.).

2. The volume of the enriched gap junction suspension is adjusted so that the protein concentration is 1 mg/ml, after addition of the following agents to the final concentrations that are indicated: 2.8% Tween 20, 200 mM KI, 2 mM sodium thiosulfate, 140 μg/ml PMSF, 50 μg/ml gentamicin in 10 mM HEPES buffer (pH 7.5) containing 0.8% NaCl. Because Tween 20 is contaminated with aldehydes, peroxides, and free acids, it must first be purified by ion-exchange chromatography [resin AG501-X8(D); Bio-Rad, Richmond, CA] immediately before use.

3. Add a magnetic mini-stirring bar to the tube and stir at 27 °C for 12 h.

4. After a 12-h incubation, add solid DHPC (Avanti Polar Lipids, Birmingham, AL) to 13 mg/ml, and stir for an additional 1 h at 27 °C.

5. Add 30 ml 25% sucrose (prepared in 10 mM HEPES, 0.8% NaCl, pH 7.5) to an SW28 centrifuge tube, and carefully add the entire contents on top of the sucrose cushion.

6. *Spin 6* (25 krpm, 1 h, 4 °C, SW28 rotor, Beckman LE-80K ultracentrifuge).

7. Carefully pipette off the sucrose so that the detergent meniscus does not touch the pellet, and wipe the tube clean with a Kimwipe to remove any residual sucrose or detergent.

8. Resuspend the pellet in 600–800 μl of HEPES buffer, and sonicate for a few seconds as above so that the suspension is homogeneous.

9. Dialyze (exclusion limit for the membrane of 100,000) for at least 24 h against 500 ml of the buffer with one change of the buffer (HEPES, plus 5 ml gentamicin/l of buffer) at ~12 h using the Bio-Tech International tubes (Fig. 3).

10. Store the suspension at 4 °C in a 1.5-ml Eppendorf tube.

In the method of *in vitro reconstitution*, the protein is solubilized from its native membrane and purified as a complex with detergent micelles (Engel *et al.*, 1992; Jap *et al.*, 1992; Kühlbrandt, 1992). The purified protein–detergent complexes are mixed with synthetic lipids or extracted native lipids that have also been solubilized in detergent. Lipid bilayers are then reconstituted by removing the detergent by dialysis or by the use of Bio-Beads (Lacapere *et al.*, 1997). Crystallization is induced by using a lipid:protein ratio that is low enough to promote protein/ protein interactions. An important aspect of this approach is that the solubilized protein must be stable for a period that is sufficient to remove detergent by dialysis. A critical variable is the lipid:protein ratio. If the lipid concentration is too high, then the protein will not be sufficiently concentrated in the plane of the lipid bilayer to induce 2D crystallization. Alternatively, if the protein concentration is too high, the protein may aggregate and denature on removal of the detergent. Dedicated glassware should be used for handling all solvents so that they are not contaminated by detergents used for routine glass washing. Lipids to be considered for reconstitution include dimyris-toylphosphatidylcholine (DMPC), dioleylphosphatidylcholine (DOPC), DPOPC, where PC stands for a mixture of phosphatidylcholines from soybean, dilaurylphosphatidylcholine (DLPC), OPPC, POPS,

Fig. 3 Microdialysis chambers available from Bio-Tech International Inc. (Seattle, WA). The bottom of the tubes are fitted with dialysis membranes available in a molecular weight cutoff from 8000 to 500,000 Da. The advantage of this system is that the tube caps can be opened to retrieve small aliquots of the sample during dialysis in order to assess the progression of crystallization. We have cut holes in inert 8.5-cm plastic disks to hold several tubes during dialysis.

DOPE, DPOPE, DOPG, POPG, as well as the lipids from the native membrane. Lipid:cholesterol mixtures are also examined.

For aquaporin I (AQP1, formerly CHIP28, channel-forming integral membrane protein of 28 kDa), the 2D crystals were better ordered if the protein was deglycosylated (Mitra *et al.*, 1994, 1995). For the lipids tested, crystals were obtained with only DMPC and DOPC at 27 °C (Fig. 4). In general, the crystals grown with DOPC were larger and better ordered than for DMPC. In addition, cholesterol had no effect on 2D crystallization. Other groups have had success growing high-resolution 2D crystals of AQP1 using phospholipase treatment (Jap and Li, 1995) as well as reconstitution with *Escherichia coli* lipids (Walz *et al.*, 1994, 1995).

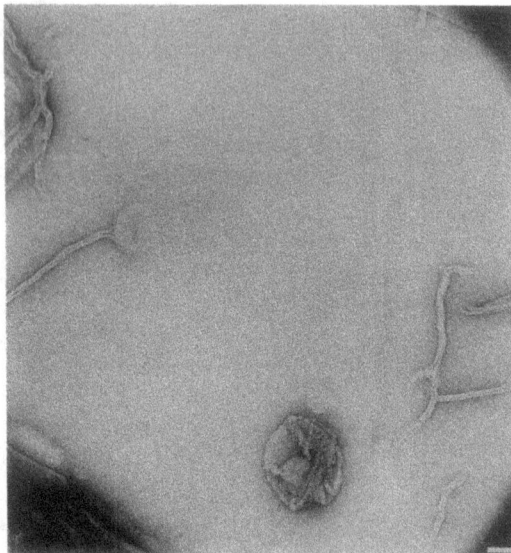

Fig. 4 Electron micrograph of a negatively stained 2D crystal of aquaporin 1 (formerly, CHIP28). Such crystals are grown by *in vitro* reconstitution and are typically 1.5–2.5 μm in diameter. The square lattice is clearly visible, as well as folds in the crystal and a separate vesicle that has collapsed onto the 2D crystal. (Reproduced from Mitra *et al.* (1994) with permission of the American Chemical Society.) Bar: 1000 Å.

F. Procedure for 2D Crystallization of AQP1 by *In Vitro* Reconstitution

1. Lipids are obtained from Avanti Polar Lipids in glass ampules and are prepared as 1% stock solutions in chloroform (Aldrich, Milwaukee, WI, HPLC Grade) and are stored at −30 °C in Reacti-Vials (Pierce, Rockford, IL) that have Teflon caps with a rubber septum that allows removal of the lipid solution without opening the vial. The lipid solution in the vial is flushed with a gentle stream of dry nitrogen gas as the cap is screwed on.

2. To solubilize the lipids in detergent, an aliquot is transferred to a 5-ml glass test tube and dried under a gentle stream of nitrogen. A buffer containing 4% octyl-β-D-glucopyranoside (OG, Anatrace), 20 mM sodium phosphate (pH 7.0), 0.1 mM EDTA, 100 mM NaCl, and 0.025% NaN_3 is added to the dried lipids to give a final detergent concentration of 1% when solubilized.

3. A micromagnetic stirring bar is added, the tube is flushed with nitrogen, and then capped with a cork or sealed with Parafilm. The lipids are solubilized by magnetic stirring overnight at room temperature.

4. Purified human AQP1 is deglycosylated with PNGase F (NEB), and the final purification step involves Q-Sepharose chromatography in a buffer containing 35–80 mM OG, 20–50 mM sodium phosphate (pH 7.2–7.5), 0.1 mM EDTA, and 100 mM NaCl.

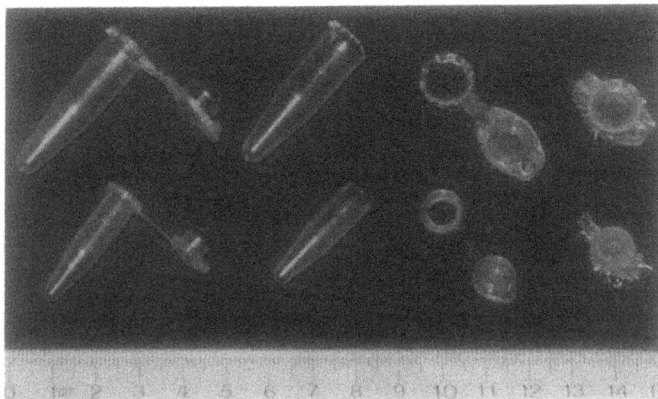

Fig. 5 Microdialysis apparatus constructed from the caps of Eppendorf tubes (Mitra *et al.*, 1994). A razor blade is used to cut the caps off of the tubes. The cap is inverted and the specimen is added to the well ($\sim$10 and $\sim$50 μl for 0.5- and 1.5-ml tube caps, respectively). A piece of dialysis membrane is placed over the well and the ring is snapped closed. The caps will float on the surface of the dialysis buffer. The sample is retrieved by piercing the membrane with a needle and aspirating the sample with a micropipette.

5. The protein concentration is determined by the Lowry method and adjusted to 1–2 mg/ml.

6. For any crystallization experiment, the total sample volume before dialysis is 50 μl, and optimal crystallization occurs at lipid:protein ratios between 1:1 and 1:3 (w/w). Cholesterol at a lipid:cholesterol ratio of 1:0–1:10 has no effect.

7. The solutions are pipetted into microdialysis chambers prepared from the caps of Eppendorf tubes (Fig. 5), and dialysis is performed using SpectraPor membranes with a molecular weight cutoff of 12–14 kDa. The microdialysis caps float on the surface of the buffer.

8. Slow detergent dialysis is performed at 27 °C for 6–8 days against 1-l volumes of buffer, with four to five buffer changes. For any single dialysis experiment, a single buffer and lipid combination can be tested over a range of lipid:protein ratios.

9. Samples are retrieved from the dialysis caps and stored at 4 °C in 1.5-ml Eppendorf tubes.

In the *lipid monolayer crystallization* technique (Brisson *et al.*, 1994; Chiu *et al.*, 1997; Hemming *et al.*, 1995), lipids are spread at an air/water interface so that the aliphatic chains will extend into the air, and the lipid headgroups will be exposed at the aqueous interface. Soluble protein molecules present in the aqueous phase bind to the lipid headgroups via electrostatic interactions or by a specific affinity tag. Negatively charged lipids include stearic and oleic acid (Supelco). Examples of positively charged lipids include octadecylamine (Sigma) and 1,2-diacyl-3-dimethylammonium-propane (Avanti Polar Lipids). Affinity tags can exploit a receptor/ligand or enzyme/substrate interaction (e.g., cholera toxin binding to its receptor (Ribi *et al.*, 1988); avidin binding to a biotinylated lipid (Darst *et al.*, 1991)). A method of

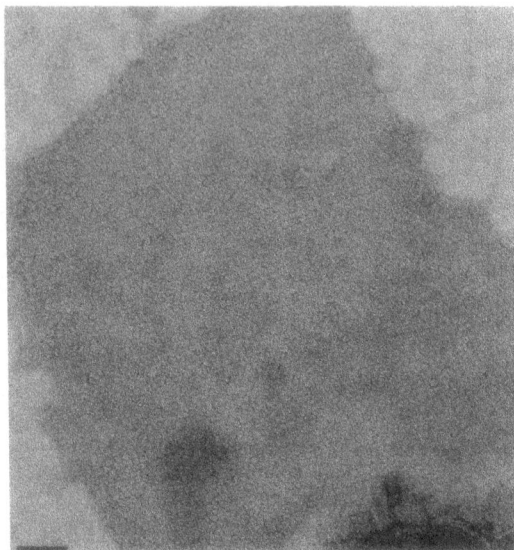

Fig. 6 Electron micrograph of a negatively stained 2D crystal of a histidine-tagged Fab fragment (Adair *et al.*, 2000). Such crystals are grown by lipid monolayer crystallization. The hexagonal lattice is clearly visible, and separate, circular Fab oligomers can be seen at the edges of the crystal. The growth of such crystals exploits the specific attachment of the histidine-tagged Fab molecules to lipids containing a nickel headgroup. Bar: 1000 Å.

potential general utility exploits the strategy of using a histidine-tagged protein (Barklis *et al.*, 1997; Kubalek *et al.*, 1994). The lipid monolayer can be doped with a synthetic lipid containing a chelating nickel headgroup that will bind to a protein with a hexahistidine tag (Fig. 6). This method would certainly be applicable for the 2D crystallization of soluble ectodomains of membrane proteins generated by recombinant DNA technology or by release from the membrane by proteolysis. A recent adaptation of the monolayer crystallization technique has exploited the propensity of some lipids such as galactosylceramide to form unilamellar tubes that provide a substrate for the crystallization of helical arrays of proteins (Ringler *et al.*, 1997; Wilson-Kubalek *et al.*, 1998).

G. Procedure for Lipid Monolayer 2D Crystallization

1. Stock solutions of the special lipid (i.e., charged or liganded) in 10% egg phosphatidylcholine are prepared in chloroform and hexane. For 2D crystallization trials the ratio of the special lipid to the bulk lipid is varied from 0 (as a control) to 1. Egg phosphatidylcholine is typically used because it remains fluid over a wide range of temperatures and compressions.

2. A Teflon block that has wells 3 mm in diameter (the diameter of an EM grid) is thoroughly cleaned by storage in chromic and sulfuric acid (Fisher Cleaning Solution). The block is washed by sonication for 30 min in detergent solution (7X,

Linbro or Versa-Clean, Fisher) and then rinsed for ~30 min in doubly distilled, deionized H_2O.

3. To maintain a hydrated atmosphere, filter paper is placed on the bottom of a 5-cm petri dish and thoroughly wetted with buffer. The Teflon block is then inserted.

4. High concentrations of glycerol or salt in the protein solution may impede crystallization. Hence, the sample may need to be diluted just before addition to the well in the Telfon block.

5. Use an Eppendorf pipette to fill the 0.5-mm-deep well with 10 μl protein solution at a concentration of 50–1000 $\mu g/ml$.

6. Use a glass micropipette to add ~0.5–1 μl of the lipid solution at a concentration of ≤ 0.5 mg/ml.

7. Cover the petri dish and incubate at the desired temperature (e.g., room temperature or 4 $^\circ$C).

8. 2D crystallization on the lipid monolayer can occur quickly (30–60 min) or take days.

9. EM grids with a hydrophobic carbon substrate are used since the attachment occurs via the lipid aliphatic chains.

10. EM forceps are used to carefully place the grid on top of the droplet with the carbon side facing the droplet surface.

11. For some systems, a clue whether crystals have formed is that the grid will not rotate or spin when placed on the droplet.

12. The grid is immediately removed from the droplet, and prepared for microscopy.

13. Alternatively, the surface of the droplet can be picked up with a platinum/paladium wire loop via surface tension (Asturias and Kornberg, 1995). The loop has a diameter slightly larger than the EM grid, so that transfer to the grid is accomplished by carefully passing the grid through the loop. This process presumably reduces stress on the crystal during transfer from the air/water interface, thereby reducing deformation and breakage.

H. Transmission Electron Cryomicroscopy

1. Transfer the 2D crystals to the EM grid:

   a. By centrifugation of an aliquot of the membrane suspension at $1000 \times g$ for 5 min (used for gap junction crystals, Fig. 7).

   b. By pipetting a 5-μl droplet of the sample onto the grid (used for AQP1).

   c. As noted above for lipid monolayer crystallization.

2. Prepare frozen-hydrated specimens. (We use an ethane slush as the cryogen because its freezing rate of about 10^6 deg/s ensures that the buffer will be vitrified (Dubochet et al., 1988).) To prevent evaporation and concentration of salts during

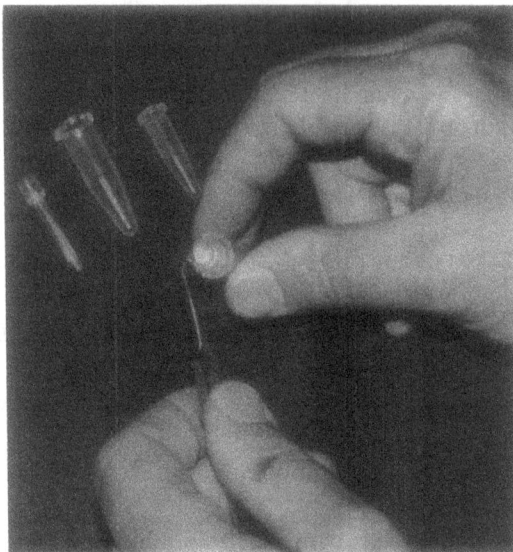

Fig. 7 Plastic buckets for centrifugation of samples onto EM grids. The snap lids of 0.5- and 1.5-ml Eppendorf tubes are removed with a razor blade. The bucket is inserted into the 0.5-ml tube, which is inserted into a 1.5-ml tube, and the system then fits into the rotor of a benchtop microfuge. The buckets accommodate 50–75 μl of sample, and the grid is placed in the bucket as shown.

blotting and plunging, high humidity is maintained by working in a cold room or placing the cryoplunger in a controlled humidity chamber.

a. Attach EM forceps to the drop rod, and adjust the position of the Dewar that holds the ethane cup so that the grid will fall into the center of the cup (Fig. 8). If the cryoplunger is not in an enclosed chamber, the grid should be positioned fairly close to the ethane cup so that evaporation does not occur during free fall of the plunger. Now fill the Dewar with liquid nitrogen.

b. Gently blow a stream of ethane gas via a Pasteur pipette into the cup. The ethane will first liquify and then solidify (Fig. 8B, item 1). (*Warning*: Wear safety glasses and be aware that ethane/air mixtures are explosive! Grid freezing should be conducted away from open flames in a well-ventilated, spark-free environment. Avoid any contact with liquid ethane because it will immediately cause severe burns.)

c. A transfer cup with an attached wire loop (Fig. 8B, item 4) is placed in the first Dewar and filled with liquid nitrogen. This transfer well is used to store the grids temporarily until you are ready to fill the grid boxes.

d. A second Dewar (Fig. 9A, item 5) is filled with liquid nitrogen and is used to hold the 50-ml conical plastic tube in which the grid boxes are stored.

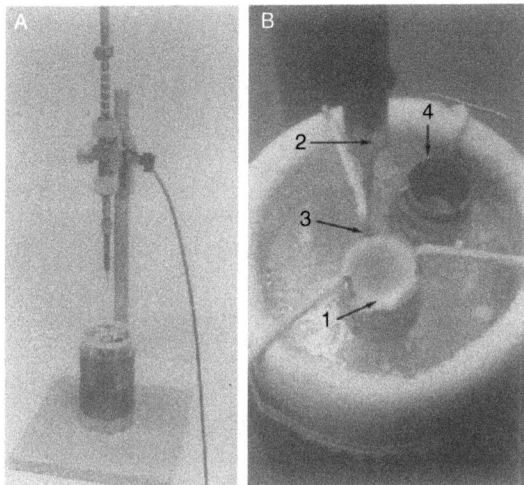

Fig. 8 (A) The original cryoplunger used by Nigel Unwin in his analysis of frozen-hydrated 2D crystals of gap junctions (Unwin and Ennis, 1984) and the acetylcholine receptor (Brisson and Unwin, 1985). (B) The metal cap from a culture tube (1) is inverted and supported by a wire tripod frame in a Dewar filled with liquid nitrogen. Ethane gas is blown into the inverted cap, which liquefies and then solidifies. A metal rod or ethane gas is used to just melt the solidified ethane so that it is slushy. The EM forceps (2) and grid (3) are attached to a weighted rod, which is centered above the ethane cup. The grid is blotted face-on almost to dryness with filter paper. This can be assessed by the capillary spread of the liquid onto the paper. As the wet circle stops growing in diameter, the paper is removed from the grid, and the grid is immediately plunged into the ethane slush via a foot peddle, which releases a lever holding the weighted rod. The frozen grids are placed in a transfer cup (4) before placement in grid boxes for storage. Elaborations of this basic design allow application of ligands or a change in buffer conditions just before the grid enters the ethane slush (Bellare *et al.*, 1988). In this way, dynamic states of macromolecular complexes can be trapped. For instance, this approach has been used to capture the intermediates in the photocycle of bacteriorhodopsin (Subramaniam *et al.*, 1993), the open state of the acetylcholine receptor (Unwin, 1995), conformational changes that occur with pH activation of viral surface fusion glycoproteins (Fuller *et al.*, 1995), and conformational states of myosin (Walker *et al.*, 1995).

    e. Pick up the EM grid that has the sample applied, and attach the forceps to the drop rod.

    f. Melt the solidified ethane by gently inserting a scalpel blade into the ethane cup until the rod or blade touches the bottom of the cup. Alternatively, the ethane can be warmed by blowing a gentle stream of gaseous ethane into the cup using a Pasteur pipette.

    g. If the sample is applied to the EM grid as a droplet, the sample can be pipetted onto the grid with the forceps already attached to the drop rod, and the sample is allowed to settle for ~90 s. For specimens in which the crystals are pelleted onto the grid, add a ~4-μl droplet of buffer to the grid so that it does not dry during positioning of the forceps on the drop rod.

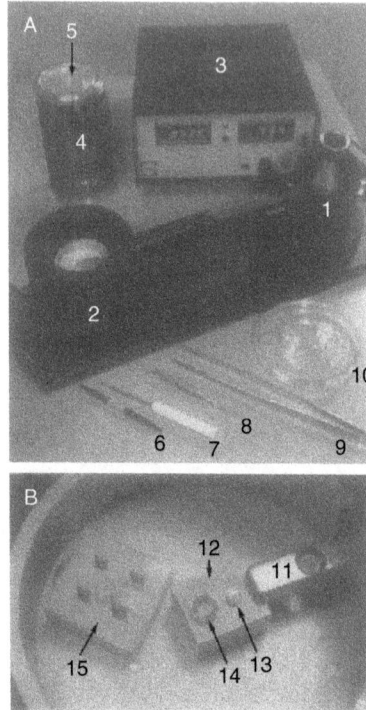

Fig. 9 (A) A Gatan cryostage (1), cryotransfer system (2), and temperature control unit (3). (B) Closeup view of the cryotransfer stage. The numbered items are (4) Dewar filled with liquid nitrogen; (5) 50-ml conical plastic tube for long-term storage of grid boxes; (6) clip ring tool that is used to secure the grid in place at the tip of the cryostage; (7) screwdriver used to loosen and tighten the screw in the lid of the grid box; (8) EM forceps for transferring the grid from the grid box to the cryotransfer system; (9) large forceps for transferring the grid box from the 50-ml conical tube to the cryotransfer system; (10) Plexiglass lid that covers the grid mounting chamber; (11) tip of the cryostage showing the circular well that holds the grid; (12) aluminum block that is positioned under the tip of the cryostage to prevent bubbling of nitrogen into the open well, which makes the grid "dance," thereby impeding rapid deployment of the clip ring; (13) EM grid; (14) clip ring that locks the grid into the well at the tip of the cryostage; (15) boxes that accommodate four grids that are prepared by cutting standard 24-grid boxes from Pellco (Redding, CA). A threaded hole is drilled in the center of the grid box to attach a plastic lid.

h. Blot the grid almost to dryness by pressing a strip of Whatman (Clifton, NJ) #2 filter paper onto the grid. The blotting time will have to be optimized for your specimen in order to have a layer of ice that is thin enough for the electron beam to penetrate but thick enough so that the entire specimen is preserved in vitrified buffer. Note that filter paper may contain trace amounts of chemicals such as divalent cations that may affect the structure of your specimen during blotting. If freezing is done in a cold room, blotting is sometimes optimized by preheating the filter paper using a coffee cup warmer to ensure that the filter paper is dry.

i. Just as you remove the filter paper from the grid, immediately press the foot pedal of the cryoplunger to release the drop rod.

j. Detach the forceps from the rod and quickly insert the grid into the transfer cup filled with liquid nitrogen.

k. After freezing four grids, the transfer cup is emptied into a Styrofoam boat that is filled with liquid nitrogen.

l. The lid of a gridbox (Fig. 9B, item 15) is unscrewed with a screwdriver (Fig. 9A, item 7), and the grid box is placed in the boat.

m. In the Styrofoam boat under liquid nitrogen, EM forceps (Fig. 9A, item 8) are used to transfer the grids into the slots of the grid box.

n. When the four grids have been loaded in the grid box, hold the grid with dissecting forceps (Fig. 9A, item 9), and use a screwdriver to tighten the lid of the grid box lid. Do not overtighten the plastic screw because it is fragile at liquid nitrogen temperatures and will break off.

o. The grid box is then placed in a 50-ml Falcon conical plastic tube (Fig. 9A, item 5) for long-term storage in large ~15-l Dewars. To facilitate retrieval of the tube from the cane of the storage Dewar, two holes are punched in the cap of the tube through which a length of fishing line is tied. Attach a label to the end of the line identifying the specimen.

3. A cryotransfer system is used to transfer the grid to the cold stage. In our laboratory we use a modified Gatan cryotransfer system (Fig. 9).

a. The Dewar of the Gatan cryostage (Fig. 9A, item 1) should be evacuated by pumping at least overnight on a Gatan dry pumping station or equivalent turbomolecular pump.

b. The electron microscope should be aligned and ready for use with the anticontaminator precooled with liquid nitrogen.

c. Use the attached fishing line to transfer the 50-ml conical storage tube with the grid boxes (Fig. 9A, item 5) from the 15-l Dewar flask to a Dewar (Fig. 9A, item 4) filled with liquid nitrogen.

d. Fill a Styrofoam boat with liquid nitrogen.

e. Hold the storage tube with your gloved hand, unscrew the cap, and pour the liquid nitrogen and the grid boxes into the Styofoam boat.

f. Insert the cryoholder into the cryotransfer system (Fig. 9A, item 2), and carefully position a 1-cm aluminum block under the tip of the holder (Fig. 9B, item 12) to support the tip when the grid is secured in position (Fig. 9B, item 11).

g. Fill the well in the cryotransfer system and the Dewar on the cryoholder with liquid nitrogen. Cover the well with the Lucite disk (Fig. 9A, item 10) while the holder cools down.

h. Attach the heater (Fig. 9A, item 3) to the Dewar on the specimen holder and turn on the main switch to monitor the temperature. Be sure to keep the heater switch off.

i. When the temperature has reached −190 °C, unscrew the lid of the grid box and then very quickly transfer the box from the Styrofoam boat to the well in the cryotransfer unit. This transfer should be fast to minimize condensation, which may contaminate the grids.

j. The grid tends to "dance" less in the well at the tip of the cryostage if the liquid nitrogen level is kept just slightly above the tip of the specimen holder. Be careful not to knock the tip of the holder since the metal is soft and the tip is quite brittle when cooled down.

k. Screw the aluminum clip ring (Fig. 9B, item 14) that holds the grid in place onto the threaded end of the clip ring tool (Fig. 9A, item 6), make sure the aluminum cube (Fig. 9B, item 12) is under the tip for support, and press fit the ring over the grid. The tip of the handle of the attachment tool is cooled in liquid nitrogen and is then used to press the ring into place to ensure a snug fit against the grid. (If not snug, the grid may drift during examination.)

l. While waiting for the temperature of the cold stage to stabilize, the tip of the cryoholder is covered with a sliding brass flange to prevent contamination.

m. Cycle the rotary pump on the electron microscope so that it does not turn on during insertion of the cryostage.

4. Insert the cold stage into the microscope (Fig. 10).

a. This step is critical because the cryostage is a very expensive, delicate piece of equipment, and the cryotransfer needs to be done quickly to minimize condensation on the tip of the specimen holder during insertion into the microscope, which will contaminate the grid with cubic and hexagonal ice.

b. For cryostages that are inserted through a side port of the microscope column (e.g., Gatan and Oxford Instruments cryostages), the stage has to be rotated during insertion. During this process liquid nitrogen can spill out of the Dewar of the cold stage. The window on the viewing chamber and the surface of the microscope should therefore be covered with a towel to prevent liquid nitrogen from coming into direct contact with surfaces on the microscope.

c. During insertion of a stage into the microscope, the airlock is evacuated for a defined period of time. We tend to use as short a time as possible to prevent condensation of water which will contaminate the grid. The pumping time must still be sufficient so that the vacuum in the column of the microscope does not deteriorate when the stage is inserted into the column.

d. Liquid nitrogen in the Dewar of the cold stage often boils. To eliminate nitrogen bubbling, a hollow tube equipped with a stopper can be inserted into the nitrogen for a few seconds to bleed off the nitrogen gas.

Fig. 10 Philips CM200FEG transmission electron microscope equipped with a Gatan cold stage (1) for electron cryomicroscopy at about $-180\,^{\circ}$C. The field emission gun provides an extremely bright and coherent source for high-resolution imaging, nominally at 1.2-Å resolution. The control screen (2), the keys, the push buttons and the knobs on the front panel of the microscope give access to the different functions of a computer system which keeps track of all operator commands and directs them to the appropriate device associated with the selected mode. The microcontroller screen (2) serves to provide information about the status of the microscope and also accepts operator commands. XY translation and Z height are controlled with a joystick. The image of the specimen is viewed via binoculars (3) on a fluorescent viewing screen. The settings for beam intensity, magnification, astigmatism, lens currents, etc., are under computer control and can be downloaded from a laptop computer (4) and provide ease of use for low-dose microscopy. The Gatan multiscan 1024×1024 charge-coupled device (CCD) camera (5) is controlled online via computer (6) and is particularly useful for recording electron diffraction patterns (Brink and Chiu, 1994; Sherman *et al.*, 1996). With the current technology the optimal resolution for recording of CCD images from beam sensitive biologic specimens is about 10 Å. Therefore, high-resolution images are still recorded on photographic film. Because the microscope is under computer control, automated data collection is feasible via a remote microscopy server (7).

5. The microscope is set for low-dose cryomicroscopy in which the field being photographed only receives a small dose of electrons (5–10 e$^-$/Å^2) during the photographic exposure. This is accomplished by performing microscopy under three conditions (Williams and Fisher, 1970) (Fig. 11):

    a. *SEARCH mode*: In search mode the grid is scanned with a beam about one-quarter the size of one grid square to locate regions to photograph. Overfocused diffraction mode is preferred over simply viewing the grid at low magnification with a dim beam because the acceptable brightness is much higher without increasing the electron dose. Hence, it is easier to identify the specimen and evaluate the ice thickness and regions contaminated with

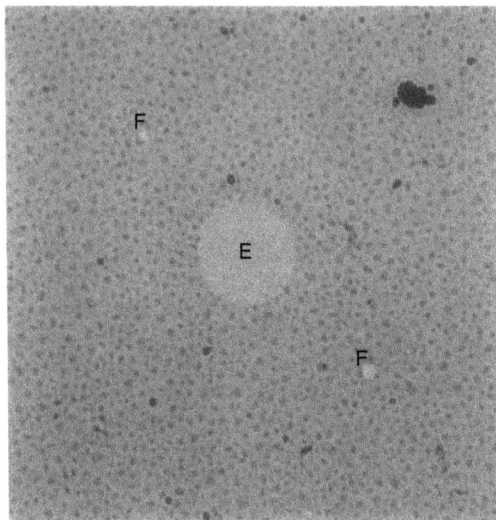

Fig. 11 A low-magnification image of a frozen-hydrated grid showing the technique for low-dose electron microscopy. In focus mode (F) the beam is deflected to a region of carbon on either side of the area to be photographed so that the carbon grain can be examined with a bright beam at high magnification (220,000×) in order to adjust the level of defocus, correct for astigmatism, and check for specimen drift. The translations in focus mode are parallel with the tilt axis of the grid. The exposure (E) is taken at a magnification of 30,000–60,000× so that the electron dose is 5–10 e$^-$/Å^2. The cobblestone appearance in the background is due to boiling of the ice during this high-dose exposure. The black deposits in the background are contaminating ice crystals.

hexagonal or cubic ice. Hexagonal ice can be identified by its characteristic hexagonal crystal habit. In diffraction mode, hexagonal ice will generate sharp diffraction spots. Cubic ice is mosaic in appearance and generates a powder pattern when examined in diffraction mode.

b. *FOCUS mode*: In focus mode, a region of the carbon substrate is examined at high magnification (~220,000×) in order to find Gaussian focus, correct for astigmatism, and to check that there is no specimen drift. In focus mode, the beam is deflected about one beam diameter (~3 μm for an exposure magnification of ~45,000×) from the area to be photographed. For examining tilted crystals, it is critical that the beam translation be set parallel with the tilt axis of the grid. On Philips electron microscopes the focus mode allows translation (designated S1 and S2) to either side of the area to be photographed so that either translation can be selected if you happen to translate the beam over a grid bar. In addition, assessing the level of focus on either side of the area to be photographed allows more precise determination of true focus.

c. *EXPOSURE mode*: Compared to alloys examined in materials science, biologic specimens are exquisitely sensitive to electron radiation damage. This limits the maximum magnification that can be used to ~60,000×, and images are typically recorded at 30,000–60,000×. During low-dose microscopy the specimen is only illuminated during the ~1-s exposure so that the dose is ~5–10 e$^-$/Å^2. For these conditions the optical density on Kodak film SO163 will be about 1 when the film is developed in full-strength D19 developer for about 10 min. Electron diffraction patterns are typically recorded at a camera length of ~1.7 m using ~20 s exposures, with a much lower cumulative dose of ~0.4 e$^-$/Å^2.

6. Scan the grid and record images. The Philips electron microscopes have three touch keys that allow shifting from search mode to focus mode to exposure mode.

a. Press the SEARCH key and scan the grid by viewing the image on the fluorescent screen with binoculars or via an attached television camera.

b. Carefully center an area to be photographed. The pointer in the microscope can be inserted for precise centering.

c. Press the FOCUS key to examine the carbon grain in both S1 and S2 positions (Fig. 11). To locate Gaussian focus, view the grain both under- and overfocused with finer and finer increments of focus. Because contrast is minimized at true focus, the texture of the carbon grain will appear smoothest at true focus (Fig. 12). True focus is located in both the S1 and S2 positions, which may differ slightly. The best estimate for true focus is therefore set at half of the difference between the values for true focus in S1 and S2. Adjustments are also made to correct for astigmatism (Figs. 12 and 13). Having found true focus, the beam is underfocused a desired amount (e.g., −5000 Å using 200-kV electrons). If the microscope is not aligned to be parfocal, an additional defocus difference will occur when shifting from 220,000× in focus mode to the exposure magnification. This shift can sometimes be several thousand ångstoms and must be taken into account when setting the defocus value for the exposure.

d. While still in focus mode, check that the specimen is not drifting (Fig. 13), and only then should the photograph be taken. Because the microscope is sensitive to vibration, keep your hands off the microscope during the exposure, do not talk, and sit still.

e. Illumination of the specimen during the exposure may induce movement, which degrades the resolution of the image. This may be overcome by recording the exposure using a focused beam of ~1000 Å (compared with a flood beam of 2–3 µm), which is scanned over the specimen to record the image (Downing, 1991). Such "spot-scan imaging" can be combined with "dynamic focusing" in which the level of defocus is adjusted during the exposure to correct automatically for the changes in defocus that occur when recording images from tilted specimens (Downing, 1992).

f. The area that you just photographed can now be viewed by pressing the EXPOSURE key.

g. Press the SEARCH key and repeat steps b–e for the next area to be photographed.

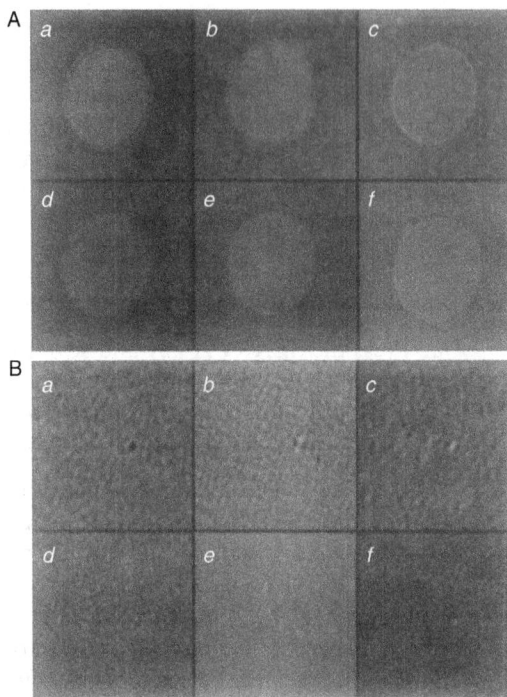

Fig. 12 True focus and astigmatism are determined by evaluation of the carbon grain. Astigmatism is not corrected in *a*, *b*, and *c*, and is corrected in *d*, *e*, and *f*. The images in *a* and *d*, *b* and *e*, and *c* and *f* were recorded at underfocus, true focus, and overfocus, respectively. (A) Astigmatism is easily seen by the asymmetric Fresnel fringes at the edges of a hole in the carbon. Note that the asymmetric arcs are in orthogonal directions when the image is underfocused versus overfocused (compare *a* and *c*). (B) Astigmatism causes the carbon grain to be stretched into "line foci" that are in orthogonal directions when the image is underfocused versus overfocused (compare *a* and *c*). Reproduced with permission of Philips Electron Instruments.

I. Image Appraisal

1. Examine EM negatives on a lightbox to assess whether the ice is too thick, whether there is contamination on the grid, or whether the crystal is broken.

2. Evaluate the negatives using an optical diffractometer (Fig. 14) to check whether the crystal was drifting during the exposure. Select the crystals that display sharp, bright reflections to the highest resolution.

3. Digitize crystalline patches identified by optical diffraction using a step size that is appropriate for the resolution expected in the image (Fig. 15). According to principles of signal processing, the Nyquist limit states that if a continuous signal is to be reproduced with fidelity, then the signal has to be sampled at twice the highest frequency in the signal (Lynn and Fuerst, 1994; Watt and Watt, 1992). This concept has also been referred to as the Whittaker–Shannon sampling theorem

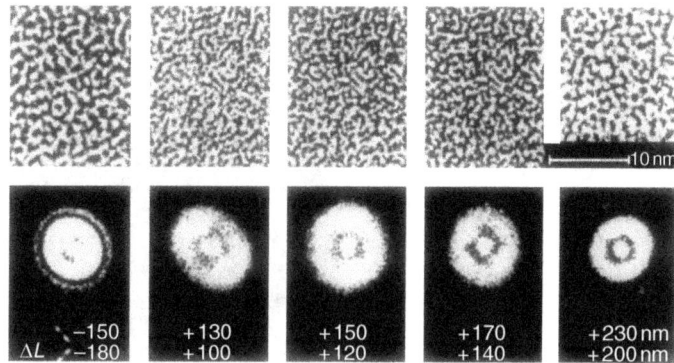

Fig. 13 Electron micrographs and optical transforms of an untilted carbon support film, showing various image defects: (A) specimen drift, (B) misaligned objective aperture generating astigmatism, (C, D) charging of the objective aperture. (Reproduced from Misell (1978) with permission of Springer-Verlag.) Note the corresponding distortions in the Thon rings of the optical diffraction patterns that can be used as a diagnostic aid to assess astigmatism and drift. The Thon rings are a consequence of the contrast transfer function (CTF) of the electron microscope, which results from objective lense aberrations such as astigmatism and spherical aberrations, as well as defocus (Thon, 1966). To achieve contrast in unstained, frozen-hydrated images, micrographs are recorded under conditions where the electron beam is underfocused. The location of the nodes in the Thon rings can be used to determine the level of defocus. Precise determination of the level of defocus is important because phases of the reflections alternate by 180 °C between successive peaks in the CTF. The CTF is a periodic sin function, and the defocus value corresponding to the first node in the CTF will be given by $\Delta F = [C_s v^4 \lambda^3 + 2]/2v^2\lambda$, where C_s is the spherical aberration of the objective lens, λ is the electron wavelength, and v is the spatial frequency. Because the values of C_s and λ are very small (usually 2 mm and 0.037 Å for 100-kV electrons), a simple estimation for the defocus level of a micrograph is given by $1/v^2\lambda$, where the location of the first CTF node (v) in the diffraction pattern can be calculated from the display of the Fourier transform.

(Mellema, 1980; Sayre, 1952). If we anticipate a resolution in the image of ~5 Å, then the image should be sampled with a step size of at least 2.5 Å. For example, micrographs recorded at 45,000× magnification are digitized using a step size of 10 μm, corresponding to 2.17 Å on the specimen [(10 μm/46,000) × (10^4 Å/μm)].

J. Image Processing

Image processing is an approach to derive an enhanced 2D or 3D reconstruction of a structure based on the analysis of images recorded from 2D projections (Amos *et al.*, 1982; Beeston *et al.*, 1972; Misell, 1978; Moody, 1990; Russ, 1992; Stewart, 1988). For the analysis of 2D crystals recorded by electron microscopy, there are four aspects to the enhancement: (1) the images are recorded from 2D crystals that represent several thousand copies of the molecular structure; (2) the spots in the diffraction pattern represent the signal, whereas the noise in between the spots can be removed by filtering; (3) the resolution of the data can be extended by methods to straighten or "unbend" the crystal lattice; and (4) the packing of the

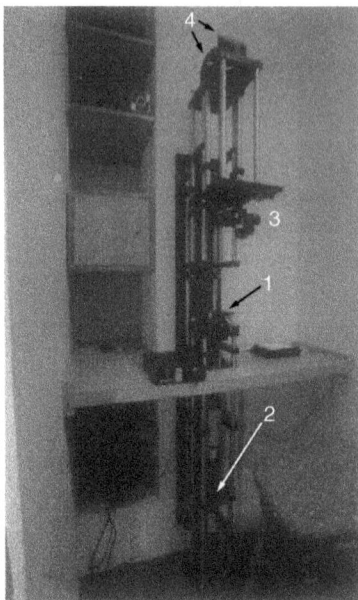

Fig. 14 An optical diffractometer is used to evaluate the quality of the 2D crystals. The photographic negative is placed on the stage (1), which is illuminated with coherent laser light (2). A 35-mm camera (3) can be used to record images or to view the negative via an eyepiece. By removing the objective lens, the diffraction pattern from the negative can be visualized. The mirrors (4) allow a folded design for upright placement in a compact space (Salmon and DeRosier, 1981).

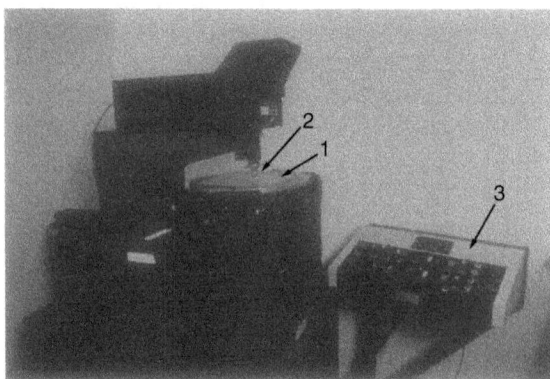

Fig. 15 Images from the best crystals are digitized using a Perkin-Elmer microdensitometer, which generates an XY array of optical densities. The negative is held in place on the glass plate (1) using a suction device. Lenses (2) are located above and below the negative. The instrument has the capability for scanning in 2-μm increments and has a very high dynamic range. The control panel (3) is interfaced to a computer for automated data collection.

molecules in the crystal is based on a certain symmetry, which can be enforced in the density map.

We use the MRC suite of programs for image processing of 2D crystals (Agard, 1983; Baldwin and Henderson, 1984; Crowther and Sleytr, 1977; Crowther *et al.*, 1996; Havelka *et al.*, 1993, 1995; Henderson *et al.*, 1986; Valpuesta *et al.*, 1994) which can be operated on DEC Alpha, SGI, or other workstations. A discussion of each program is beyond the scope of this review, but they are described in general in Table II, and their implementation will be described in detail (Unger *et al.*, in preparation). A general flowchart of the steps with the corresponding programs is shown in Fig. 16, and a summary of the steps is as follows.

Table II
Programs Used for Processing Images of 2D Crystals

Image/map display and general processing	
XIMDISP	General program to display images, Fourier transforms, cross-correlation maps, etc.
HEADER	Prints out information in header record
LABEL	Performs various image manipulations such as pixel averaging to reduce image size
HISTO	Generates histogram of densities in image
FFTRANS	Computes fast Fourier transform
BOXIMAGE	Masks selected area within a digitized image
TAPEREDGE	Tapers density at edge of image to remove central spikes in transform arising from image boundary
TWOFILE	Performs linear combination, or multiplies/divides data in two files
Processing of images of two-dimensional crystals	
EMTILT	Calculates tilt angles from lattice parameters of 2D crystal
MASKTRANA	Masks the Fourier transform in preparation for filtering
AUTOCORRL	Performs an autocorrelation calculation and expansion
QUADSERCHB	Searches cross-correlation map for the position and height of cross-correlation peaks by profile fitting
CCUNBENDD	Corrects image for distortions in the 2D crystal lattice
MMBOX	Provides amplitudes and phases from Fourier transform
TTBOX	Corrects for tilt transfer function and writes out amplitudes and phases
TTMASK	Combines MASKTRAN and TTBOX for processing images of highly tilted crystals
CTFAPPLY	Applies contrast transfer function to phase data from MMBOX
ORIGTILTD	Determines phase origins, refines tilt geometries, and merges data
LATLINED	Performs a least-squares fit to the amplitude and phase data from tilted crystals to derive lattice line
LATLINPRESCALE	Corrects image amplitudes for contrast transfer function
PREPMKMTZ	Removes unreliable lattice line points from the 3D dataset and readjusts figures-of-merit as desired
ALLSPACE	Determines the plane group, origin and beam tilt on a single image of an untilted crystal
AVRGAMPHS	Averages projection amplitudes and phases from multiple images
CALIMAMP3D	Calculates inverse B factors for individual images and rescales image-derived amplitudes

(continues)

Table II (*continued*)

General utilities programs

OMSTATS — Adjusts the figure-of-merit values from AVRGAMPHS for derivations from the theoretical phase (projection data only), calculates statistics in resolution bins and overall phase error, adjusts the phase for twofold constraints, and rejects forbidden reflections

LOTALL — Generates a plot of the phase error for individual Fourier terms of projection data

CHINDEXING — Reindexes structure factor lists obtained by MMBOX (useful for images of more highly tilted images where indexing of the original transform can be ambiguous)

IDEAL — Generates ideal p1 reference dataset for calculating the point spread function

NTSPREAD — Reads the 3D data file and resets the amplitudes to unity and the phases to 0 in order to estimate the resolution

Programs to calculate and contour maps

2MTZ — Converts input data of diverse formats to standard CCP4 *mtz* format

FFT — Crystallographic fast Fourier transform program

EXTEND — Extends maps/images to multiples of the unit cell

-PLUTO — General contouring and atomic model plotting program

O — Software package designed for visualization of density maps and docking atomic models

AVS — Software package from Advanced Visualization Systems for graphics rendering of maps: isosurfaces, transparencies, texture mapping, etc.

1. Use HISTO to generate a histogram of the optical densities in the image.

2. TAPEREDGE is used to wrap around the optical densities at the edges of the image to prevent central spikes in the Fourier transform that arise from the discontinuity in density at the opposite edges of the image.

3. For large images LABEL is used to average adjacent pixels to speed up processing in the refinement of the crystal lattice.

4. FFTRANS is now used to compute the Fourier transform, which can be displayed using XIMDISP (Fig. 17).

5. XIMDISP can also be used to index the reflections in the computed diffraction pattern and iteratively refine the crystal lattice parameters.

6. Correct for crystal lattice distortions (Fig. 18).

   a. Select high signal-to-noise reflections that will be used to generate a filtered image.

   b. MASKTRANA masks the selected diffraction spots from the background, and a filtered image is generated by inverse Fourier transformation.

   c. The image is first filtered tightly using a small box of pixels that includes each spot. The filtered image is then visualized using XIMDISP to select a region that is visually judged to be the most crystalline. This region is later boxed as the reference area for unbending the lattice.

   d. AUTOCORRL is then used to generate an autocorrelation map from a part of the tightly filtered image that is centered around the center of the reference.

Digitize image

|

Calculate fourier transform and determine lattice parameter

Tightly mask transform, calculate filtered image   Mask transform loosely for
                                                           cross-correlation

Calculate autocorrelation profile from filtered
image

Box reference area from filtered image and
calculate reference transform

Calculate fourier transform of the cross-correlation
map and convert into real-space map

Use autocorrelation profile to search real space
cross-correlation map for position and
height of cross-correlation peaks

Use information about peak position to
reinterpolate optical densities of the original
image (="unbend" lattice distortions)

Use information about height of cross-correlation
peaks to box best area of corrected image and generate
transform

Extract amplitudes and phases from transform. If
applicable, repeat whole process using transform of
corrected and boxed image for reference area. Cross-
correlate with loosely masked transform of corrected
but unboxed image and unbend the already corrected
image for a second time.

Correct phases for the effect of the
contract transfer function (CTF)

Determine plane group; merge data from several
images; refine phase origins, tilt geometries and CTF;
average data or fit lattice lines; calculate and display
projection or 3D density maps

Relevant program(s)

FFTRANS, XIMDISP

MASKTRANA
(TTMASK)

LABEL, AUTOCORRL

BOXIMAGE, FFTRANS

TWOFILE, FFTRANS

QUADSERCHB

CCUNBENDD

BOXIMAGE, FFTRANS

MMBOX
(TTBOX)

CTFAPPLY

ALLSPACE, ORIGTILTD,
AVRGAMPHS, LATILINPRESCALE
LATLINED, CCP4, PLUTO, O, AVS

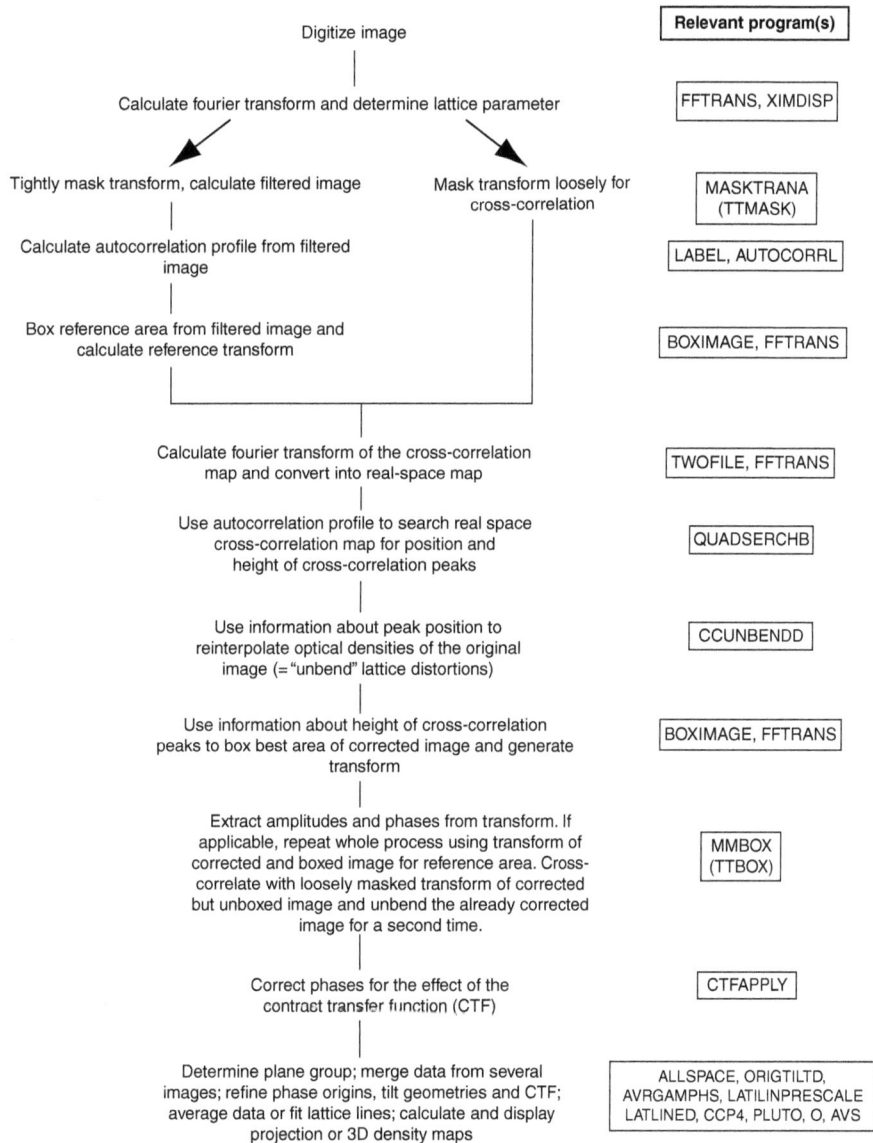

Fig. 16 An outline of the steps used for image processing of 2D crystals. The program titles in the MCR image processing suite are indicated.

e. XIMDISP is used to view the autocorrelation map in order to determine the shape and extent of the profile that will be used for searching the cross-correlation map.

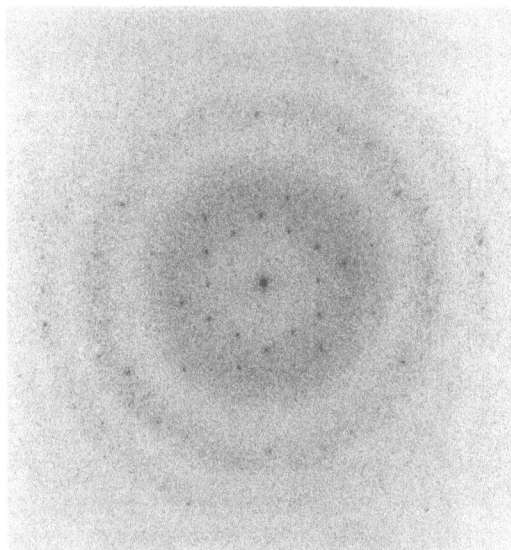

Fig. 17 Computed diffraction pattern from a frozen-hydrated 2D crystal of a recombinant, truncated form of rat heart α_1-connexin (α_1Cx263T). Spots at the outer edge of the pattern correspond to $\sim$9.5-Å resolution. Note the diffuse rings of density in the background, referred to as Thon rings, that are due to the CTF.

f. MASKTRANA is then used a second time to loosely filter the original image, which will contain the information for correcting imperfections in the lattice.

g. The cross-correlation transform is then generated by TWOFILE, which multiplies the loosely masked image Fourier transform by the complex conjugate of the reference Fourier transform. The inverse Fourier transform generates the cross-correlation map.

h. QUADSERCHB is then used to search the cross-correlation map for the location and height of peaks.

i. CCUNBENDD uses the position of the cross-correlation peaks to generate a map of vectors that provides the direction and magnitude for shifting the image density into best agreement with the density in the reference image. A corrected or "unbent" image is generated by applying these translations to the original image.

j. XIMDISP is now used to display the cross-correlation map to identify the region in best agreement with the reference image. The original image is then reboxed by including only those regions that have a certain correlation coefficient (as a general guideline, e.g., $\leq$75% for thick specimens or $\sim$30% for thin specimens) with the reference image.

7. The process from steps 3 to 6 can be repeated in several variations to correct even finer lattice deviations in the crystal.

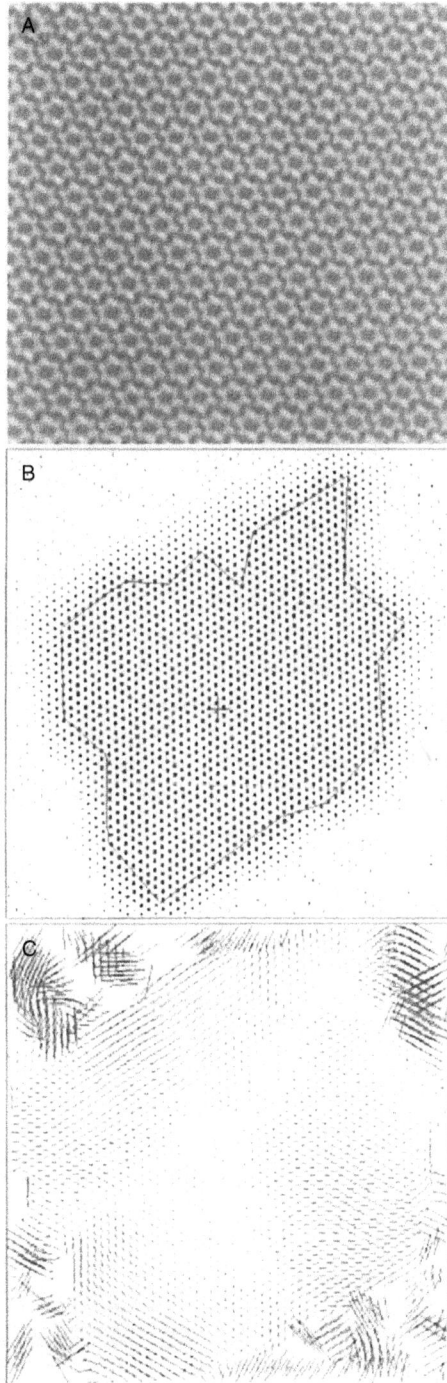

Fig. 18 (continued)

8. After boxing the area defined in step 6j, FFTRANS is then used to generate a transform of the corrected and boxed image, and MMBOX provides a digital readout of the amplitudes and phases that can be used to assess the significance of the final data list and the success of unbending (Fig. 19).

9. Effects on the phases due to the contrast transfer function of the microscope are now corrected using CTFAPPLY (Fig. 20). The parameters needed in the correction are the electron wavelength (which will be dependent on the kilovolts of the microscope), the spherical aberration coefficient of the objective lens, the defocus level of the image, the dimensions of the Fourier transform, the crystal lattice parameters, the magnification, and the densitometer scanning step size. Astigmatism will distort the Thon rings into ellipses. Examination of the Fourier transform using XIMDISP can be used to determine the approximate pixel location of the nodes in the Fourier transform along the major and minor axes of the ellipse with respect to the x-axis of the transform.

10. In determining a 2D projection or a 3D map, all of the corrected Fourier transforms have to be refined to a common phase origin. Furthermore, for an image of a tilted crystal, the tilt geometry needs to be known. An estimate for the location of the tilt axis can be determined by optical diffraction (Fig. 14). The negative is examined using a large aperture to identify the axis along which the Thon rings do not vary in diameter. For images with a high degree of tilt, the tilt axis and tilt angle can be determined using EMTILT, which computes the tilt geometry based on the distortion in the lattice. However, the accuracy of this approach decreases as the tilt angle decreases below $\sim$30 °C, and for small tilt angles also depends on the symmetry of the specimen. For images with small tilt angles, the tilt angle can be estimated by noting the shift in the nodes of the CTF on opposite sides of the tilt axis.

11. The final merging of the amplitude and phase data from multiple images of tilted crystals is performed using ORIGTILTD.

12. If the amplitudes are collected from electron diffraction patterns, then no corrections need to be made for CTF effects on the amplitude (Fig. 21), and only a background due to inelastic scattering needs to be subtracted.

Fig. 18 Correction of crystal lattice distortions in a recombinant gap junction 2D crystal by the method of Henderson *et al.* (1986). Strong reflections in the Fourier transform that are in good agreement with the lattice parameters are masked and used to calculate a filtered image at $\sim$15-Å resolution (A). A region in the filtered image that appears to have the least distortion is selected as a reference area (6 × 6 channels in this case), and the cross-correlation function is calculated. The darker symbols in the cross-correlation map (B) indicate regions of high correlation between the reference and the region being compared in the crystal. The error map in (C) shows the translational offsets (20×) with respect to the reference area that are applied to "unbend" the image. Based on the cross-correlation map, the original image is reboxed to include only those areas that have a high cross-correlation coefficient with the reference area of $\sim$75% or greater for generating the final corrected Fourier transform.

Resolution range	Before unbending									After unbending									IQ	Before	After
80–15Å	10	12	13	8	8	7	7	7	8	10	10	16	8	8	6	7	7	8	1	11	14
	6	10	20	17	13	8	7	7	6	8	8	13	13	9	8	6	6	8	2	4	6
	11	12	22	22	29	16	8	21	9	9	11	22	27	10	6	8	13	7	3	4	3
	10	9	19	271	359	73	30	17	9	7	10	11	168	312	45	32	16	12	4	5	0
	10	19	37	319	692	284	37	14	10	9	18	32	319	978	242	28	13	8	5	2	2
	9	15	18	42	96	115	34	12	11	14	11	33	39	260	144	12	11	8	6	3	1
	6	7	8	6	24	25	20	15	10	7	7	7	5	12	34	17	14	6	7	2	2
	7	8	8	9	9	9	13	6	11	6	7	9	8	10	12	13	7	8	8	2	6
	5	6	7	8	8	11	9	9	8	6	6	7	7	8	12	11	8	8	9	3	2
15–10Å	8	8	9	11	12	9	8	8	9	6	7	11	8	7	8	6	7	10	1	0	6
	8	10	11	12	17	13	12	10	10	10	6	11	14	15	11	9	10	10	2	3	8
	10	8	9	11	15	15	13	9	9	10	7	10	13	11	14	17	12	9	3	9	7
	8	11	15	20	41	19	21	12	6	10	9	13	19	27	17	10	10	7	4	3	7
	7	7	17	15	29	26	13	10	8	9	9	12	35	68	33	12	8	6	5	3	4
	7	7	11	14	15	17	14	12	7	8	8	11	16	20	18	13	8	12	6	3	0
	8	7	8	11	13	15	16	14	6	7	7	11	15	11	16	13	10	11	7	2	1
	8	8	7	8	9	12	12	11	7	6	6	8	9	11	17	13	10	10	8	12	5
	8	4	8	6	7	7	8	9	6	8	6	6	6	8	10	11	7	7	9	10	7
9.9–7.0Å	7	9	9	9	9	8	9	8	7	7	7	9	8	7	8	8	9	7	1	0	3
	8	10	10	9	11	9	10	7	7	7	8	9	10	9	10	9	8	7	2	3	12
	8	8	8	11	22	18	14	10	8	9	8	9	7	9	8	8	8	8	3	2	9
	8	10	11	11	24	32	16	14	11	8	10	10	20	41	17	8	9	8	4	8	5
	7	8	9	13	18	25	17	13	9	7	9	10	25	68	36	10	8	8	5	11	6.
	8	9	7	9	11	14	14	11	10	8	8	10	21	15	9	8	8	2	6	12	2
	6	8	8	9	10	9	10	10	9	7	8	10	7	9	10	9	8	8	7	5	7
	8	8	8	10	7	6	6	8	9	7	7	8	8	9	9	9	7	8	8	23	23
	7	6	7	8	7	8	9	10	8	8	7	7	9	8	7	8	8	10	9	11	8
6.9–5.8Å	8	8	7	7	8	6	6	7	6	8	8	8	8	6	6	7	7	6	1	0	0
	8	8	7	7	8	6	7	6	7	7	7	8	8	9	9	8	8	6	2	0	1
	8	8	8	6	7	7	8	7	7	7	7	8	8	6	8	8	8	6	3	2	5
	7	7	7	7	7	8	7	8	7	7	6	8	10	8	7	6	5	6	4	4	3
	7	5	7	8	9	8	7	6	5	9	6	7	10	10	7	8	8	6	5	8	14
	7	6	7	6	8	9	6	9	6	5	7	8	7	9	8	8	8	7	6	5	2
	6	8	9	6	6	8	7	9	7	7	7	7	8	8	8	7	7	6	7	5	3
	6	7	8	5	7	7	5	8	8	6	7	6	7	7	7	7	7	6	8	21	21
	7	6	7	7	7	8	7	6	7	6	6	6	7	7	6	7	7	10	9	24	20

Fig. 19 Comparison of the average reflection intensities before and after correction for crystal lattice distortions in specific resolution bands (80–15, 15–10, 9.9–7.0, and 6.9–5.8 Å). Note that the unbending procedure increases the signal intensity and improves the sharpness of the reflections. After unbending, significant data are even detectable in the band from 6.9- to 5.8-Å resolution. The IQ value is a measure of signal-to-noise. An IQ value of 1 corresponds to a signal-to-noise ratio ≤ 8, and an IQ of 8 corresponds to a signal-to-noise of ~ 1. Note the increase in the number of reflections with better IQ values after correction for crystal lattice distortions.

13. If the amplitudes are computed by Fourier transformation of the images, then the amplitudes are modulated by the CTF, which is corrected using LATLIN-PRESCALE. In addition, image-derived amplitudes exhibit a resolution dependent fall-off due to drift and charging during the recording of images. This fall-off can be

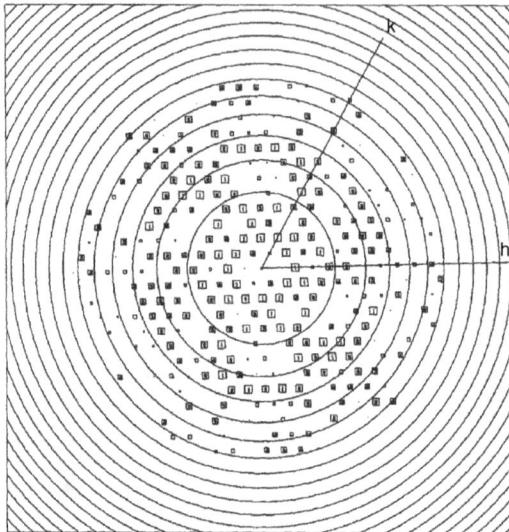

Fig. 20 Signal-to-noise ratios for all individual reflections to ~6-Å resolution derived by Fourier transformation of a corrected image of a frozen-hydrated 2D recombinant gap junction crystal. The rings represent the nodes in the CTF, determined by the level of defocus of the micrograph. The numbers are the IQ values and are inversely related to the signal-to-noise ratio (see Fig. 19).

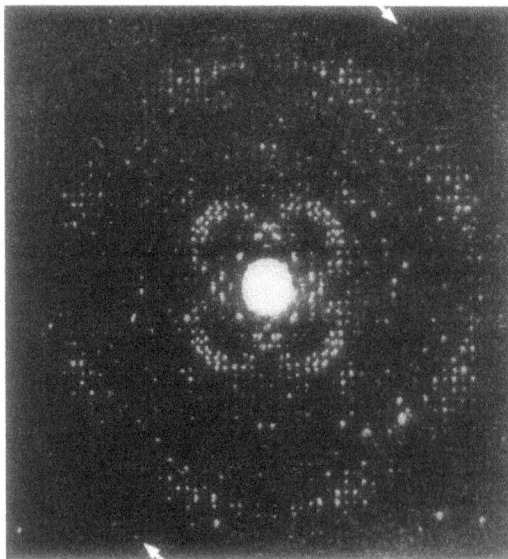

Fig. 21 The AQP1 2D crystals are sufficiently large so that electron diffraction patterns can be directly recorded. The amplitudes in the diffraction pattern are not affected by the CTF since the patterns are recorded at true focus. Hence, the derived amplitudes are much more reliable than those determined by Fourier transformation of images, as in Fig. 17. The arrows identify high-resolution reflections (3.5 Å). (From Mitra *et al.* (1995) and reproduced with permission of Nature Publishing.)

compensated by multiplying the higher resolution amplitudes with an inverse temperature factor (Havelka *et al.*, 1995; Hsu *et al.*, 1997; Schertler *et al.*, 1993).

14. Smooth curves are fitted by a least-squares method to the merged amplitude and phase data using LATLINED (Fig. 22).

15. The lattice line curves are sampled at a spacing of at least $1/T$ (where T is the estimated thickness of the specimen), and the 3D map is computed by inverse Fourier transform using symmetry constraints defined by the lattice.

16. The map can be visualized using a variety of software packages (CCP4 (CCP4, 1994), O (Jones and Kjeldgaard), and AVS (Sheehan *et al.*, 1996)).

III. Interpretation of Maps

The first milestone in the structure analysis is to derive a projection map from images of untilted, unstained crystals. The next step is to derive a 3D map by recording and processing images of tilted crystals. This is particularly challenging because drift and charging will degrade the resolution of the images when the crystals are tilted. The ultimate goal is to derive a map at near-atomic resolution. However, as noted in the introduction, crystal quality may preclude determination of a structure at atomic resolution.

A. 15–30 Å Resolution

Even at such a low resolution, important molecular details can be revealed: crystal symmetry, quaternary structure, general molecular boundary, protein/protein interactions, and the location of the aqueous pore in channels. For instance, in our analysis of 2D crystals of rat heart gap junctions (Yeager, 1994; Yeager and Gilula, 1992), the projection maps showed that the channels are formed by a hexameric cluster of subunits with a central channel. A projection map of AQP1 in this resolution range demonstrated that the protein subunits assemble as tetramers (Mitra *et al.*, 1994). Functionally important sites can be identified by labeling with ligands such as undecagold clusters and antibodies. Furthermore, difference map analysis can also be used to locate specific proteins in a complex (Yeager *et al.*, 1994).

B. 10–15 Å Resolution

In this range, there is not much information to be gleaned. The molecular boundary will be somewhat better defined, but secondary structure is typically not visualized.

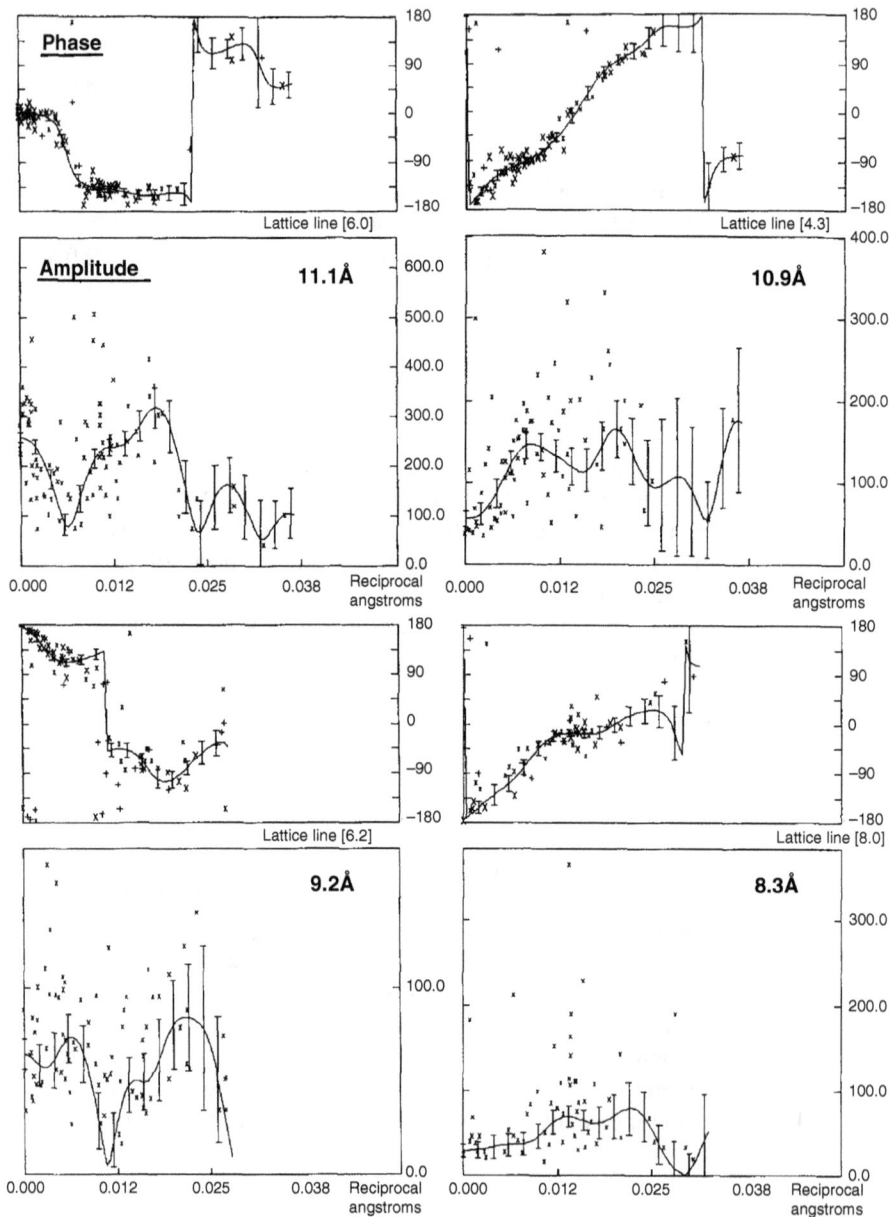

Fig. 22 Examples for phase (top) and amplitude (bottom) modulation along lattice lines of tilted 2D crystals. A least-squares method is used to generate smooth curves through the data points (Agard, 1983). Inverse Fourier transformation is used to generate a 3D map by sampling the smooth curves at an increment of at least $1/D$, where D is the thickness of the specimen.

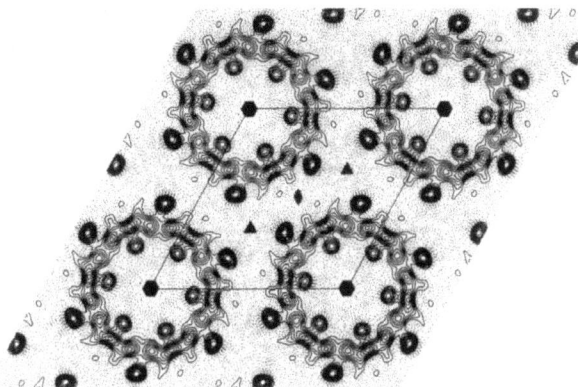

Fig. 23 Contour map of the projected density of recombinant gap junction channels at 7-Å resolution computed with p6 symmetry. The channel is formed by a hexameric cluster of subunits, each of which has three major features: (i) a ring of circular densities centered at 17-Å radius interpreted as α helices that line the channel, (ii) a ring of densities centered at 33-Å radius interpreted as α helices that are most exposed to the lipid, and (iii) a continuous band of density at 25-Å radius separating the two groups of helices. The hexagonal lattice had parameters $a = b = 79.4 \pm 0.3$ Å and $\gamma = 120 \pm 0.3°$. The map was scaled to a maximum density of 250 and contoured in steps of 25.0 (rms density = 87.4). The solid lines indicate density above the mean. (From Unger *et al.* (1997c) and reproduced with permission of Nature Publishing.)

C. 5–10 Å Resolution

A resolution of ~10 Å is just at the boundary for visualizing α helices, which tend to pack with a spacing of about 10 Å. Since packed α helices roughly perpendicular to the bilayer are a common motif for the transmembrane domains of membrane proteins, even a projection map at 5–7-Å resolution may reveal important structural details. For instance, a projection map of a recombinant gap junction channel at 7-Å resolution showed that each subunit in the hexameric connexon contains two α helices that appear roughly perpendicular to the membrane (Fig. 23). In addition, the projection map demonstrated that the docking of the two connexons in forming the complete intercellular channel involves a 30 °C rotational stagger between the apposed connexons (Unger *et al.*, 1997c). A 3D map of tetrameric AQP1 channels revealed that each monomer is formed by a barrel of tilted α helices that enclose an aqueous vestibule leading to the selective water channel (Fig. 24, color plate). However, at 5–10-Å resolution, it will be difficult to define β sheets, and α helices may be difficult to delineate if they are highly tilted.

D. Resolution of 3.5 Å or Better

Such a resolution will permit tracing of the Cα backbone and definition of the amino acid side chains (Henderson *et al.*, 1990; Kimura *et al.*, 1997; Kühlbrandt *et al.*, 1994). For bacteriorhodopsin, the level of structural detail was comparable to high-resolution maps derived by X-ray crystallography (Pebay-Peyroula *et al.*, 1997).

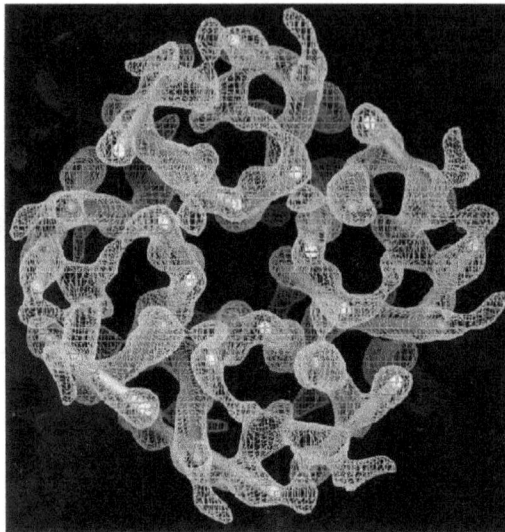

Fig. 24 3D density map of human erythrocyte AQP1 (previously named CHIP28) at 7-Å in-plane resolution, viewed approximately perpendicular to the plane of the bilayer. The molecules assemble as a tetramer, and each monomer contains a water-selective channel. The rods positioned in the density map trace the approximate paths of six tilted α helices, which form a right-handed barrel that surrounds the aqueous pathway. (From Cheng *et al.* (1997) and reproduced with permission of Nature Publishing.)

In this volume focused on ion channels, let us suppose, for example, that we have successfully derived a density map of a voltage-regulated K^+ channel. At 15–20 Å of resolution, a 3D map would confirm whether the channels are tetrameric in the crystal and reveal the external shape of the oligomer. The vestibule to the ion conduction pathway will be detected if it is more than ~15 Å in diameter. At 5–7-Å in-plane resolution, the secondary structure of the transmembrane domains could be assessed. If the vertical resolution is comparable, the folding of the cytoplasmic and extracellular loops might be visible if they are ordered to this resolution. In addition, analysis of complexes could reveal the binding sites for toxins and antibodies. To visualize the binding sites for low molecular weight drugs, the reagents would have to be labeled with a heavy atom cluster in order to increase the density of the drug. In addition, time-resolved cryomicroscopy might allow trapping of the open, closed, and inactivated conformations. Insight into the following specific aspects of K^+ channel structure would potentially be revealed in a 6-Å resolution 3D map: (1) What is the molecular boundary of the tetrameric channel? (2) Does each K^+ channel monomer contain six transmembrane α helices and a β hairpin between S5 and S6? (3) What is the molecular topography of the vestibule of the pore? (4) Where is the selectivity filter located? (5) Where is the S4 voltage sensor located? (6) What is the packing arrangement between helices of adjacent subunits that confers stability in the tetrameric assembly? (7) What is the degree of tilt of the putative α-helices and their interaction with the S5–S6 loop? At 3.5-Å resolution these points would be answered to the level of individual side chains. Indeed, a recent X-ray structure at 3.2 Å resolution of a bacterial homolog of an inward rectifying K$^+$ channel has provided answers to some of these questions (Doyle *et al.*, 1998).

IV. Conclusions

The use of electron microscopy for the structural characterization of 2D crystals of membrane proteins began in 1975 with the pioneering work of Henderson and Unwin in the analysis of glucose-embedded 2D crystals of bacteriorhodospin (Henderson and Unwin, 1975). The 6-Å resolution map revealed seven-transmembrane α helices and serves as a paradigm for membrane protein structure in the same way that the X-ray structure of myoglobin served as a paradigm for the structure of soluble proteins. Other notable events in the development of electron cryocrystallography were a projection map of bacteriorhodopsin at 3.5-Å resolution (Henderson *et al.*, 1986), a 3D map at 3.5-Å resolution (Henderson *et al.*, 1990), and most recently a map at 3.0-Å resolution (Kimura *et al.*, 1997). To date, LHCII (Kühlbrandt *et al.*, 1994) and tubulin (Nogales *et al.*, 1998) are the only other proteins determined at atomic resolution by electron cryocrystallography. For several membrane proteins, density maps with an in-plane resolution ranging from 6 to 9 Å have resolved protein secondary structure: bacterial porin (Jap *et al.*, 1991), LHCI (Karrasch *et al.*, 1995), halorhodopsin (Havelka *et al.*, 1995), aquaporin I (Cheng *et al.*, 1997; Li *et al.*, 1997; Walz *et al.*, 1997), gap junctions (Unger *et al.*, 1997c), glutathione transferase (Hebert *et al.*, 1997), vertebrate rhodopsin (Unger *et al.*, 1997b), and a bacterial H$^+$-ATPase (Auer *et al.*, 1998). In addition, tubular crystals of membrane proteins that manifest helical symmetry offer the advantage of providing all views of the molecule, thereby obviating the need to tilt the specimen. Hence, there is no missing cone of data as is inherent in datasets obtained from tilted 2D crystals. Tubular assemblies of the acetylcholine receptor (Unwin, 1995, 1993) and the Ca^{2+}-ATPase of sarcoplasmic reticulum (Zhang *et al.*, 1998) have yielded complete 3D maps at 9 and 8 Å resolution, respectively. Given all of these achievements, electron cryomicroscopy and image analysis have together emerged as a powerful strategy for the structural characterization of membrane proteins. The electron cryomicroscopes and computational methods that are currently available have the routine potential to achieve a resolution that is near atomic. Specific for membrane protein structure analysis, the critical barriers are obvious—obtaining sufficient protein and growing high-resolution 2D crystals. Given the substantial effort that is being devoted to the technology of membrane protein overexpression and the continuing list of proteins for which high-resolution 2D crystals are available, the future for this field is indeed bright.

Acknowledgments

We thank Michael Whittaker and Alan McPhee for photography and Anchi Cheng and Brian Adair for help with figure preparation. The writing of this review has been supported by grants from the National Institutes of Health (M. Y. and A. K. M.), a Grant-in-Aid from the American Heart Association (A. K. M.), the Donald E. and Delia B. Baxter Foundation (M. Y.), and a postdoctoral fellowship from the American Heart Association (V. M. U.). M. Y. is an established investigator of the American Heart Association and Bristol-Myers Squibb.

References

Adair, B., Wilson, E., Weiner, S., Kunicki, T., and Yeager, M. (unpublished observations).

Agard, D. A. (1983). *J. Mol. Biol.* **167,** 849.

Akey, C. W., Szalay, M., and Edelstein, S. J. (1984). *Ultramicroscopy* **13,** 103.

Amos, L. A., and Baker, T. S. (1979). *Nature* **279,** 607.

Amos, L. A., Henderson, R., and Unwin, P. N. T. (1982). *Prog. Biophys. Mol. Biol.* **39,** 183.

Asturias, F. J., and Kornberg, R. D. (1995). *J. Struct. Biol.* **114,** 60.

Auer, M., Scarborough, G. A., and Kühlbrandt, W. (1998). *Nature* **392,** 840.

Baldwin, J., and Henderson, R. (1984). *Ultramicroscopy* **14,** 319.

Barklis, E., McDermott, J., Wilkens, S., Schabtach, E., Schmid, M. F., Fuller, S., Karanjia, S., Love, Z., Jones, R., Rui, Y., Zhao, X., and Thompson, D. (1997). *EMBO J.* **17,** 1199.

Beeston, B. E. P., Horne, R. W., and Markham, R. (1972). *In* "Electron Diffraction and Optical Diffraction Techniques—Practical Methods in Electron Microscopy," (A. M. Glauert, ed.), p. 318. North-Holland Publishing, Amsterdam.

Bellare, J. R., Davis, H. T., Scriven, L. E., and Talmon, Y. (1988). *J. Electron Microsc. Tech.* **10,** 87.

Berriman, J., and Unwin, N. (1994). *Ultramicroscopy* **56,** 241.

Boekema, E. J. (1990). *Electron Microsc. Rev.* **3,** 87.

Brink, J., and Chiu, W. (1994). *J. Struct. Biol.* **113,** 23.

Brisson, A., and Unwin, P. N. T. (1985). *Nature* **315,** 474.

Brisson, A., Olofsson, A., Ringler, P., Schmutz, M., and Stoylova, S. (1994). *Biol. Cell* **80,** 221.

Bullough, P. A. (1990). *Electron Microsc. Rev.* **3,** 249.

CCP4. (1994). (Collaborative Computational Project, No. 4) .

Cheng, A., van Hoek, A. N., Yeager, M., Verkman, A. S., and Mitra, A. K. (1997). *Nature* **387,** 627.

Chiu, W. (1986). *Annu. Rev. Biophys. Biophys. Chem.* **15,** 237.

Chiu, W., and Schmid, M. F. (1993). *Curr. Opin. Biotech.* **4,** 397.

Chiu, W., Schmid, M. F., and Prasad, B. V. (1993). *Biophys. J.* **64,** 1610.

Chiu, W., Avila-Sakar, A. J., and Schmid, M. F. (1997). *Adv. Biophys.* **34,** 161.

Crowther, R. A., and Sleytr, U. B. (1977). *J. Ultrastruct. Res.* **58,** 41.

Crowther, R. A., Henderson, R., and Smith, J. M. (1996). *J. Struct. Biol.* **116,** 9.

Czajkowsky, D. M., and Shao, Z. (1997). *Biophys. J.* **72,** A139.

Darst, S. A., Ahlers, M., Meller, P. H., Kubalek, E. W., Blankenburg, R., Ribi, H. O., Ringsdorf, H., and Kornberg, R. D. (1991). *Biophys. J.* **59,** 387.

De Vos, A. M., Ultsch, M., and Kossiakoff, A. A. (1992). *Science* **255,** 306.

Dorset, D. L. (1996). *Acta Cryst.* **B52,** 753.

Downing, K. H. (1991). *Science* **251,** 53.

Downing, K. H. (1992). *Ultramicroscopy* **46,** 199.

Doyle, D. A., Cabral, J. M., Pfuetzner, R. A., Kuo, A., Gulbis, J. M., Cohen, S. L., Chait, B. T., and MacKinnon, R. (1998). *Science* **280,** 69.

Dubochet, J., Groom, M., and Mueller-Neuteboom, S. (1982). *Adv. Opt. Electron Microsc.* **8,** 107.

Dubochet, J., Adrian, M., Chang, J. J., Homo, J. C., Lepault, J., McDowall, A. W., and Schultz, P. (1988). *Q. Rev. Biophys.* **21,** 129.

Engel, A., Hoenger, A., Hefti, A., Henn, C., Ford, R. C., Kistler, J., and Zulauf, M. (1992). *J. Struct. Biol.* **109,** 219.

Fuller, S. D., Berriman, J. A., Butcher, S. J., and Gowen, B. E. (1995). *Cell* **81,** 715.

Garavito, R. M., and White, S. H. (1997). *Curr. Opin. Struct. Biol.* **7,** 533.

Garavito, R. M., Picot, D., and Loll, P. J. (1996). *J. Bioenerg. Biomembr.* **28,** 13.

Glaeser, R. M., Jubb, J. S., and Henderson, R. (1985). *Biophys. J.* **48,** 775.

Gogol, E., and Unwin, N. (1988). *Biophys. J.* **54,** 105.

Grisshammer, R., and Tate, C. G. (1995). *Q. Rev. Biophys.* **28,** 315.

Hatefi, Y., and Hanstein, W. G. (1974). *Methods Enzymol.* **31,** 770.

Havelka, W. A., Henderson, R., Heymann, J. A. W., and Oesterhelt, D. (1993). *J. Mol. Biol.* **234,** 837.

Havelka, W. A., Henderson, R., and Oesterhelt, D. (1995). *J. Mol. Biol.* **247,** 726.

Hebert, H., Schmidt-Krey, I., Morgenstern, R., Murata, K., Hirai, T., Mitsuoka, K., and Fujiyoshi, Y. (1997). *J. Mol. Biol.* **271,** 751.

Hemming, S. A., Bochkarev, A., Darst, S. A., Kornberg, R. D., Ala, P., Yang, D. S., and Edwards, A. M. (1995). *J. Mol. Biol.* **246,** 308.

Henderson, R., and Unwin, P. N. T. (1975). *Nature* **257,** 28.

Henderson, R., Baldwin, J. M., Downing, K. H., Lepault, J., and Zemlin, F. (1986). *Ultramicroscopy* **19,** 147.

Henderson, R., Baldwin, J. M., Ceska, T. A., Zemlin, F., Beckmann, E., and Downing, K. H. (1990). *J. Mol. Biol.* **213,** 899.

Hsu, G. G., Bellamy, A. R., and Yeager, M. (1997). *J. Mol. Biol.* **272,** 362.

Jap, B. K., and Li, H. (1995). *J. Mol. Biol.* **251,** 413.

Jap, B. K., Walian, P. J., and Gehring, K. (1991). *Nature* **350,** 167.

Jap, B. K., Zulauf, M., Scheybani, T., Hefti, A., Baumeister, W., Aebi, U., and Engel, A. (1992). *Ultramicroscopy* **46,** 45.

Jones, T. A., and Kjeldgaard, M. "O Version 5.9, The Manual." Uppsala University, Sweden.

Karrasch, S., Bullough, P. A., and Ghosh, R. (1995). *EMBO J.* **14,** 631.

Kimura, Y., Vassylyev, D. G., Miyazawa, A., Kidera, A., Matsushima, M., Mitsuoka, K., Murata, K., Hirai, T., and Fujiyoshi, Y. (1997). *Nature* **389,** 206.

Kubalek, E. W., Le Grice, S. F., and Brown, P. O. (1994). *J. Struct. Biol.* **113,** 117.

Kühlbrandt, W. (1988). *Q. Rev. Biophys.* **21,** 429.

Kühlbrandt, W. (1992). *Q. Rev. Biophys.* **25,** 1.

Kühlbrandt, W., Wang, D. N., and Fujiyoshi, Y. (1994). *Nature* **367,** 614.

Kumar, N. M., Friend, D. S., and Gilula, N. B. (1995). *J. Cell Sci.* **108,** 3725.

Lacapere, J. J., Stokes, D. L., Mosser, G., Ranck, J. L., Leblanc, G., and Rigaud, J. L. (1997). *Ann. N. Y. Acad. Sci.* **834,** 9.

Li, H., Lee, S., and Jap, B. K. (1997). *Nat. Struct. Biol.* **4,** 263.

Lynn, P. A., and Fuerst, W. (1994). *In* "Introductory Digital Signal Processing with Computer Applications," p. 8. John Wiley & Sons, Chichester, UK.

Mannella, C. A. (1984). *Science* **224,** 165.

Mannella, C. A. (1989). *Biochim. Biophys. Acta* **981,** 15.

McPherson, A. (1982). Preparation and Analysis of Protein Crystals. John Wiley and Sons, New York.

Mellema, J. E. (1980). *In* "Computer Processing of Electron Microscope Images," (P. W. Hawkes, ed.), p. 89. Springer, Berlin.

Michel, H. (ed.) (1991). "Crystallization of Membrane Proteins," CRC Press, Boca Raton, FL.

Milligan, R. A., Brisson, A., and Unwin, P. N. T. (1984). *Ultramicroscopy* **13,** 1.

Misell, D. L. (1978). *In* "Image Analysis, Enhancement and Interpretation—Practical Methods in Electron Microscopy," (A. M. Glauert, ed.), p. 40. North-Holland Publishing, Amsterdam.

Mitra, A. K., Yeager, M., van Hoek, A. N., Wiener, M. C., and Verkman, A. S. (1994). *Biochemistry* **33,** 12735.

Mitra, A. K., van Hoek, A. N., Wiener, M. C., Verkman, A. S., and Yeager, M. (1995). *Nat. Struct. Biol.* **2,** 726.

Moody, M. (1990). *In* "Biophysical Electron Microscopy, Basic Concepts and Modern Techniques," (P. W. Hawkes, and U. Valdrè, eds.), p. 145. Academic Press, London.

Nogales, E., Wolf, S. G., and Downing, K. H. (1997). *J. Struct. Biol.* **118,** 119.

Nogales, E., Wolf, S. G., and Downing, K. H. (1998). *Nature* **391,** 199.

Olofsson, A., Kavéus, U., Thelestam, M., and Hebert, H. (1988). *J. Ultrastruct. Mol. Struct. Res.* **100,** 194.

Opella, S. J. (1977). *Nat. Struct. Biol.* **4**(Suppl.), 845.

Ostermeier, C., and Michel, H. (1997). *Curr. Opin. Struct. Biol.* **7,** 697.

Pebay-Peyroula, E., Rummel, G., Rosenbusch, J. P., and Landau, E. M. (1997). *Science* **277,** 1676.

Reiss-Husson, F. (1992). *In* "Crystallization of Nucleic Acids and Proteins: A Practical Approach," (A. Ducruix, and R. Giegé, eds.), p. 175. IRL Press, Oxford.

Ribi, H. O., Ludwig, D. S., Mercer, K. L., Schoolnik, G. K., and Kornberg, R. D. (1988). *Science* **239,** 1272.

Ringler, P., Muller, W., Ringsdorf, H., and Brisson, A. (1997). *Chem. Eur. J.* **3,** 620.

Russ, J. C. (1992). *In* "The Image Processing Handbook," p. 165. CRC Press, Boca Raton, FL.

Salmon, E. D., and DeRosier, D. (1981). *J. Microsc.* **123,** 239.

Sayre, D. (1952). *Acta Cryst.* **5,** 843.

Schertler, G. F., Villa, C., and Henderson, R. (1993). *Nature* **362,** 770.

Shapiro, L., Fannon, A. M., Kwong, P. D., Thompson, A., Lehmann, M. S., Grübel, G., Legrand, J. F., Als-Nielsen, J., Colman, D. R., and Hendrickson, W. A. (1995). *Nature* **374,** 327.

Sheehan, B., Pique, M. E., and Yeager, M. (1996). *J. Struct. Biol.* **116,** 99.

Sherman, M. B., Brink, J., and Chiu, W. (1996). *Micron* **27,** 129.

Sikerwar, S. S., and Unwin, N. (1988). *Biophys. J.* **54,** 113.

Song, L., Hobaugh, M. R., Shustak, C., Chelay, S., Bayley, H., and Gouaux, J. E. (1996). *Science* **274,** 1859.

Stewart, M. (1988). *J. Electron Microsc. Tech.* **9,** 301.

Subramaniam, S., Gerstein, M., Oesterhelt, D., and Henderson, R. (1993). *EMBO J.* **12,** 1.

Taylor, K. A., and Glaeser, R. M. (1976). *J. Ultrastruct. Res.* **55,** 448.

Thon, F. (1966). *Z. Naturforschg.* **21a,** 476.

Unger, V. M., Entrikin, D. W., Guan, X., Cravatt, B., Kumar, N. M., Lerner, R. A., Gilula, N. B., and Yeager, M. (1997a). *Biophys. J.* **72,** A291.

Unger, V. M., Hargrave, P. A., Baldwin, J. M., and Schertler, G. F. (1997b). *Nature* **389,** 203.

Unger, V. M., Kumar, N. M., Gilula, N. B., and Yeager, M. (1997c). *Nat. Struct. Biol.* **4,** 39.

Unger, V. M., Cheng, A., and Yeager, M. (in preparation).

Unwin, P. N. (1975). *J. Mol. Biol.* **98,** 235.

Unwin, N. (1986). *Ann. N.Y. Acad. Sci.* **483,** 1.

Unwin, N. (1993). *J. Mol. Biol.* **229,** 1101.

Unwin, N. (1995). *Nature* **373,** 37.

Unwin, P. N. T., and Ennis, P. D. (1984). *Nature* **307,** 609.

Valpuesta, J. M., Carrascosa, J. L., and Henderson, R. (1994). *J. Mol. Biol.* **240,** 281.

Walker, M., Trinick, J., and White, H. (1995). *Biophys. J.* **68,** 87S.

Walz, T., Smith, B. L., Agre, P., and Engel, A. (1994). *EMBO J.* **13,** 2985.

Walz, T., Typke, D., Smith, B. L., Agre, P., and Engel, A. (1995). *Nat. Struct. Biol.* **2,** 730.

Walz, T., Hirai, T., Murata, K., Heymann, J. B., Mitsuoka, K., Fujiyoshi, Y., Smith, B. L., Agre, P., and Engel, A. (1997). *Nature* **387,** 624.

Wang, J., Yan, Y., Garrett, T. P. J., Liu, J., Rodgers, D. W., Garlick, R. L., Tarr, G. E., Husain, Y., Reinherz, E. L., and Harrison, S. C. (1990). *Nature* **348,** 411.

Watt, A., and Watt, M. (1992). *In* "Advanced Animation and Rendering Techniques, Theory and Practice," p. 112. Addison-Wesley, Wokingham, UK.

Williams, R. C., and Fisher, H. W. (1970). *J. Mol. Biol.* **52,** 121.

Wilson-Kubalek, E. M., Brown, R. E., Celia, H., and Milligan, R. A. (1998). *Proc. Natl. Acad. Sci. USA* **95,** 8040.

Yeager, M. (1994). *Acta Cryst.* **D50,** 632.

Yeager, M. (1995). *Microsc. Res. Tech.* **31,** 452.

Yeager, M., and Gilula, N. B. (1992). *J. Mol. Biol.* **223,** 929.

Yeager, M., Berriman, J. A., Baker, T. S., and Bellamy, A. R. (1994). *EMBO J.* **13,** 1011.

Zampighi, G., and Unwin, P. N. (1979). *J. Mol. Biol.* **135,** 451.

Zhang, P., Toyoshima, C., Yonekura, K., Green, N. M., and Stokes, D. L. (1998). *Nature* **392,** 835.

CHAPTER 14

Voltage–Clamp Biosensors for Capillary Electrophoresis

Owe Orwar, Kent Jardemark, Cecilia Farre, Ingemar Jacobson, Alexander Moscho, Jason B. Shear, Harvey A. Fishman, Sheri J. Lillard, and Richard N. Zare

Department of Chemistry
Stanford University, Stanford
California, USA

I. Introduction

During the past two decades, there have been significant developments in micro-analytical separation and sampling techniques for low molecular weight bioactive compounds. However, improvements in highly selective and sensitive detection

DOI: 10.1016/B978-0-12-382204-8.00014-X

strategies for these compounds are still needed. The heart of the problem is that many biologically active species do not contain molecular features to allow sensitive detection. Exceptions to this behavior do exist, and such compounds as catecholamines and indoleamines, which are easily oxidized, have been detected using ultra-microelectrodes in single synaptic vesicles following exocytosis (Wightman *et al.*, 1991). The exocytotic release of serotonin and tryptophan-containing proteins has also been detected using native UV fluorescence following separation (Lillard *et al.*, 1996). On the other hand, amino acid neurotransmitters such as glutamate, aspartate, glycine, and γ-aminobutyric acid (GABA) are not easily oxidized, do not have significant absorption cross sections above the water cutoff wavelength, and are virtually nonfluorescent. For their sensitive quantitation, incorporation of chemical reagents such as fluorogenic or fluorescent dyes, electroactive or highly absorbing moieties, or radioactive isotopes are required. Generally, drawbacks with such labeling schemes include lack of selectivity (similar functional groups will be altered in the same way) and uncertainties in reaction efficiencies. Furthermore, with certain labeling schemes, the biological system under study can be perturbed, possibly to the extent that the data collected are biased. Most importantly, however, any detection technique that measures singularly a physical or chemical property of the analyte, separated from its biological context, will not reveal its biological function.

Low molecular weight neuroactive compounds bind to membrane receptors. Ligand binding can lead to many different events, such as opening of ion channels or activation of G-protein-coupled intracellular cascades. Many times these events can be detected using optical or electrical recording techniques, and the receptor and its coupled reactions form the basis of a biological detector or biosensor (McConnell *et al.*, 1992; Wijesuriya and Rechnitz, 1993). The advantages of using biosensors to detect biologically active species include (1) high specificity owing to molecular recognition, (2) high sensitivity because ligand binding triggers biological amplification cascades, (3) no chemical derivatization step, and (4) biological meaning of the detector response.

Fast-acting neurotransmitters activate ligand-gated ion channels. These receptors confer a high degree of selectivity because their binding sites have been genetically engineered to suit a single or a few structurally related ligands. They also offer high sensitivity because, on ligand binding, approximately 10^4 ions will either enter or leave the cell per millisecond. This flow of ions can be measured with voltage- and patch-clamp techniques, sometimes down to the level of a single receptor. In contrast to traditional detection techniques for microseparations, this biosensor detector measures an electrical current resulting from the interaction between a receptor and a ligand binding to it, leading to meaningful biological information.

Capillary electrophoresis–patch-clamp detection (CE–PC) is a novel technique that can be used for the fractionation and detection of receptor agonists and receptor antagonists in complex and extremely small (10^{-9}–10^{-18} l) sample solutions. CE–PC is a combination of two well-established techniques. A particularly interesting aspect of CE is that it can be useful in solution nanochemistry applications because the inlet of the electrophoresis capillary can be pulled to narrow (submicron) diameters, and can be used to inject extremely small volumes (10^{-15}–10^{-18} l) or even single biopolymers, such as DNA oligomers (Chiu *et al.*, 1997). Because of this high sampling resolution it is our hope that CE–PC technology can be used to probe neurotransmitter release and dynamics in biological microdomains. Here we present information about how to optimize the union between CE and PC, especially to study ligands that act on receptors belonging to the glutamate–receptor family. Also, the use of *Xenopus laevis* oocytes that have been injected with mRNA and total rat brain RNA, as two-electrode voltage-clamp (VC) detectors in CE, is discussed.

Recently, patch-clamped neuronal cell membranes have been used as *in situ* sensors, so-called "sniffer-patch detectors," to demonstrate release of acetylcholine and glutamate-like compounds from neuronal growth cones and turtle photoreceptors (Allen, 1997). The sniffer-patch technique offers extremely good spatial resolution, down to ~1 μm^3, which allows detection of quantal release of neurotransmitters. A major limitation with this technique, however, is its inability to discriminate between multiple ligands that activate the same type of receptor. To achieve this goal without sacrificing spatial resolution, we have coupled both patch- and two-electrode VC biosensors online to capillary electrophoresis (Jardemark *et al.*, 1997; Orwar *et al.*, 1996; Shear *et al.*, 1995). In these systems, ligands are separated electrophoretically, delivered onto the surface of a cell, and detected at a characteristic migration time through the capillary. Such techniques can be optimized for a specific analyte or class of analytes. This optimization is accomplished by using cell types and cell lines that express native or recombinant receptors specific for the analyte of interest. Additionally, it is feasible to identify bioactive components from their electrophoretic migration times in such a manner that the migration time and response of the detector provide a unique signature of the bioactive component.

II. Capillary Electrophoresis

Free-solution capillary electrophoresis in its present configuration was introduced by Jorgenson and Lukacs (1981). They showed that highly efficient electrophoretic separations could be performed in narrow-bore (5–75-μm i.d., 100–400-μm o.d.) fused silica capillaries (usually 20–100 cm in length). Because thin-walled capillaries efficiently dissipate Joule heat, minimal distortion is imposed on the separated analyte bands, and under ideal circumstances, theoretical plate numbers in excess of 10^6 can be accomplished.

In CE, the capillary is filled with buffered electrolyte solution (i.e., running buffer) and both ends are immersed in reservoirs that typically contain the same solution. Most commonly, the anode and cathode are placed in the inlet and outlet reservoirs, respectively, and are connected to a high-voltage source (Jorgenson and

Lukacs, 1981; Li, 1992). Above about pH 2, the bare fused-silica capillary wall is negatively charged. Cations in the electrolyte are attracted to the charged fused silica surface and an electrical double layer is created. When an electric field is applied across the capillary, a sheath of positive ions drags the bulk solution in the cathodic direction, which results in electroosmotic flow. This electroosmotic flow acts as a noise-free pump.

The separation mechanism for free-solution CE (termed capillary zone electrophoresis, CZE) is based on the differential electroosmotic and electrophoretic migration rates of charged species. The electrophoretic mobility μ_{ep} (m^2/V s), can be expressed as

$$\mu_{ep} = q_i(6\pi\eta r_i)^{-1} \tag{1}$$

where η(N s/m^2) is the kinematic viscosity of the separation medium, and r_i (m) and q_i (A s) are the effective radius and charge, respectively, of the migrating species. The electrophoretic migration velocity in m/s, is given by

$$v_{ep} = \mu_{ep}E \tag{2}$$

where E is the electric field strength (V/m). The electroosmotic mobility, μ_{eo}, can be expressed as

$$\mu_{eo} = \varepsilon\zeta(4\pi\eta)^{-1} \tag{3}$$

where ε (C/V m) is the dielectric constant of the separation medium and ζ (V) is the zeta potential, which is the potential difference between the Stern plane and the buffer solution. The electroosmotic migration velocity is given by

$$v_{eo} = \mu_{eo}E \tag{4}$$

and the net migration velocity for a species subjected to both electroosmosis and electrophoresis is

$$v_{net} = (\mu_{eo} + \mu_{ep})E = v_{eo} + v_{ep} \tag{5}$$

Under normal operating conditions $|v_{eo}| > |v_{ep}|$, so that all species—whether possessing positive, neutral, or negative charges—migrate in the same direction past a single detector.

CE is an excellent match to cell-based biosensors because physiologic buffers can be employed as electrolytes and the dimensions of the separation capillary correspond well with those of the sensor. CE can also be used as an injection technique for administering organic and inorganic ionic species into single cells (Nussberger et al., 1996).

In CE–PC and CE–VC, detection is based on the activation of ligand-gated ion channels. On ligand binding to such receptors, ions flow either into or out of the cell, resulting in a measurable current. The transmembrane flux of ions is created by concentration and electrical driving forces, and can be controlled by changing the transmembrane potential and ionic composition of the intracellular and extra-cellular solutions. When the CE effluent is directed onto the surface of a cell, the electrophoresis electrolyte effectively becomes the extracellular solution. There-fore, proper ionic composition of the CE running buffer is critical because of the requirements set forth by the Nernst equation and the permeability of the ion channel. According to the Nernst equation, the equilibrium potential of an ionic species, V_{eq}, can be written as

$$V_{eq} = \frac{RT}{zF} \ln\left(\frac{[X]_o}{[X]_i}\right) \qquad (6)$$

where R is the gas constant (8.314 J/mol K), T is the temperature (K), z is the valence of the ion, F is Faraday's constant (9.6487×10^4 A s/mol), and $[X]_o$ and $[X]_i$ are the extracellular and intracellular concentrations, respectively, of the ionic species. At 20 °C, RT/zF is about 25 mV for a monovalent cation. For simplicity, external and internal ion activities are assumed to be equal and thus cancel. By knowing the membrane (holding) potential, V_m (which can be set by a command voltage in the patch-clamp amplifier), and the reversal potential (i.e., V_{eq}) given by the Nernst equation, the ionic current, I, through an open channel can be calculated from

$$I = \gamma(V_m - V_{eq}) \qquad (7)$$

where γ (S) is the conductance of the ion channel. Thus, if the reversal potential is determined for a specific ligand-gated ion-channel system (by selection of internal and external ionic compositions) and a proper holding potential is chosen, then it is possible to optimize the direction and amplitude of the transmembrane current.

In our experiments, HEPES–saline buffers are used as electrolytes in the CE capillaries, vials, and as cell bath media. For the CE–PC experiments, the buffer consists of 140 mM NaCl, 5 mM KCl, 1 mM $CaCl_2$, 1 mM $MgCl_2$, 10 mM HEPES, and 10 mM D-glucose (pH 7.4, NaOH). In the CE–VC experiments, the buffer consists of 95 mM NaCl, 2 mM KCl, 2 mM $CaCl_2$, and 10 mM HEPES (pH 7.5, NaOH). Other buffers, depending on application or preference, can be used. There are some receptor ligands that require certain molecules or ions as cofactors to activate the ion channel. In these cases, the electrolyte composition must be modified to include such species. For example, in the detection of N-methyl-D-aspartate (NMDA)-evoked receptor responses, a Mg^{2+}-free glycine-supplemen-ted HEPES–saline buffer is used.

III. Fabrication of Fractured Electrophoresis Capillaries

The voltage applied to a CE capillary is typically 10–30 kV, which creates electric field strengths of several hundred volts per centimeter. As mentioned earlier, typical CE setups have the outlet reservoir as the cathodic end, which is usually ground. When cell-based biosensors are used, the CE outlet reservoir is a chamber, which contains the cells and the bath solution, positioned on a microscope. To avoid high-offset potentials operating on the cell membrane during electrophoresis, the CE capillary is fractured and connected to ground 3–5 cm above the outlet. Effectively, this connection to ground functions as a voltage divider and the residual potential is compensated by an offset potential to the patch-clamp amplifier system (Jardemark *et al.*, 1997). Fractured CE capillaries were originally developed for ultrasensitive electrochemical detection (Wallingford and Ewing, 1987). Alternatives to cracking the capillaries might include laser-drilling holes in the capillary walls, as has been used for online chemical derivatization in CE (Pentoney *et al.*, 1988) or the use of a floating (virtual) ground that keeps the cell bath close to zero potential (Vaughan and Trotter, 1982).

We have developed a simple technique to fracture CE capillaries that efficiently decreases the CE outlet potential. These assemblies are robust and can be used for several months. Fused silica capillaries 10–50 cm long with 20–50-μm i.d. and 370-μm o.d. (Polymicro Technologies Inc., Phoenix, AZ) are used. The outlet end of the capillary is first fixed to a microscope coverslip with a small piece of adhesive tape. A 0.5-in. section of glass tubing (e.g., disposable microliter pipette glass capillaries; World Precision Instruments, Saratoga, FL) with an inner diameter that matches the outer diameter of the CE capillary is threaded coaxially onto the CE capillary from the outlet. This tubing is immobilized on both sides with a small droplet of cyanoacrylate glue (which is also called Super Glue or Crazy Glue) so that the outlet end of the glass tubing is located exactly at the position where the fracture is to be (Fig. 1A). The glue is allowed to cure, and the capillary is then filled with distilled water. A 1-ml syringe, in which the needle has been inserted into Teflon or medical tubing, serves this purpose. The inlet of the capillary is inserted into the tubing (whose inner diameter matches the outer diameter of the capillary) and the capillary is easily filled. The coverslip mount is then placed on the stage of a stereomicroscope.

A ceramic capillary cutter (Polymicro) is used to make a vertical cut through both the outer polyimide coating and the fused silica wall. The size of the incision is difficult to control exactly, but should allow just a small droplet of water to emerge from the capillary interior. By pushing water through the capillary with the syringe, it is possible to check the split ratio between the incision and the outlet. We have found, empirically, that if approximately 50% of the injected volume elutes from the CE outlet, then cracks with good grounding characteristics are produced. Following this, another 0.5-in. section of glass tubing is threaded onto the CE capillary. This piece is placed within 1–2 mm from the other glass tubing, centered around the incision, and immobilized at both ends using cyanoacrylate glue (Fig. 1B). When the glue has cured, two sections of 0.5-in. glass tubing are aligned in parallel, centered around the incision, and fixed with cyanoacrylate glue to the coaxially mounted pieces of glass tubing (Fig. 1C).

When the glue is cured, tapping around the incision using tweezers or a small metallic rod completes the fracture (Fig. 1D). The assembly is then lifted from the coverslip and embedded with a two-component epoxy glue (Fig. 1E). Application

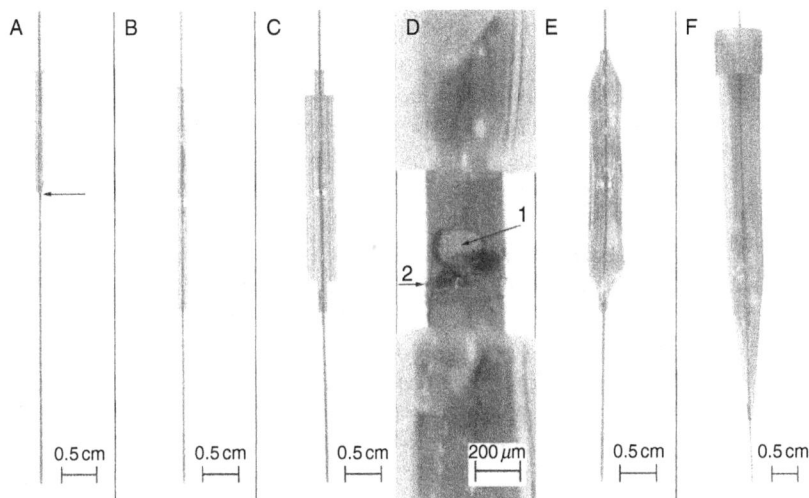

Fig. 1 Construction of a fractured CE capillary. The outlet of the capillary is at the bottom of each frame. (A) Glass tubing is threaded onto the CE capillary and immobilized on both sides. A small incision is made at the position of the arrowhead using a capillary cutter. (B) Another piece of 0.5-in-long glass tubing is threaded onto the CE capillary and immobilized 1–2 mm from the first piece. (C) Two sections of glass tubing are aligned, centered around the incision, and fixed to the pieces of coaxially mounted glass tubing. The area near the fracture is tapped and the fracture is completed. (D) Photomicrograph of the incision produced by the capillary cutter (1) and the crack (2). (E) The assembly is embedded with epoxy. (F) The mount is inserted and fixed with epoxy into a 1-ml plastic pipette tip. (Photographs courtesy of Susanne Orwar.)

of glue into the fractured area should be avoided. When the epoxy has cured, the mount is inserted and fixed with epoxy resin into a 1-ml plastic pipette tip (Fig. 1F). The apical end of the pipette tip is glued with epoxy onto the capillary. The pipette reservoir is then filled with the same electrolyte used as CE running buffer and connected to ground with a platinum wire. Offset potentials with our system (+12-kV, 40-cm-long, 50-μm-i.d. capillaries) are about 10–20 mV, which can easily be compensated with the VC amplifier (Jardemark *et al.*, 1997).

IV. Cell Preparation

A. Patch-Clamp Detection

Because virtually any cell type can be used in CE–PC, we have used acutely isolated neurons, primary cultures, or cultured cell lines as detectors. The cells are cultured and maintained according to standard tissue culture protocols. Suspended cells are preferred for whole-cell recordings and immobilized cells are preferred for outside-out patch recordings. Glass substrates coated with poly(L-lysine) or poly (D-ornithine), for example, can be used to immobilize acutely isolated or nonadherent cells.

Although the use of transfected cells expressing a single recombinant receptor type would allow for novel analyte selection, acutely isolated interneurons from the rat olfactory bulb have been used typically in our studies. These cells can be used to detect inhibitory and excitatory amino acid neurotransmitters and transmitter mimetics. Among other receptors, these cells express inhibitory $GABA_A$ and Gly receptors and excitatory receptors belonging to the glutamate receptor family, such as the (R,S)-α-amino-3-hydroxy-5-methyl-4-isoxazole propionate (AMPA), kainate, and NMDA types. The procedure for isolating olfactory bulb interneurons is as follows: newborn or adult rats (10–200 g) are anesthetized in halothane (ISC Chemicals Ltd., Avonmouth, UK) and decapitated. The olfactory bulbs are then dissected from the brains, sliced into four pieces, and placed in an incubation chamber. The incubation chamber contains proteases from *Aspergillus oryzae* (2.5 mg/ml), which are dissolved in prewarmed (32 °C) HEPES–saline buffer containing 140 mM NaCl, 5 mM KCl, 10 mM HEPES, 10 mM glucose, 2 mM $MgCl_2$ (pH 7.4, NaOH). After 25–30 min, the slices are washed with protease-free buffer solution for 20 min. During both the enzymatic treatment and washing, the solutions are continuously perfused with a gas mixture containing 95% (v/v) O_2 and 5% (v/v) CO_2.

Following enzymatic treatment and washing, the slices are kept in a HEPES–saline buffer held at 20 °C, containing 1 mM $CaCl_2$, that is continuously perfused with 95% O_2 and 5% CO_2. The slices are then disintegrated by shear forces with gentle suction through the tip of a fire-polished Pasteur pipette. The cell suspension is then placed in a Petri dish and diluted by Ca^{2+}-containing (1 mM) HEPES–saline buffer. To detect NMDA receptor-mediated responses, a Mg^{2+}-free and Ca^{2+}-containing HEPES–saline buffer supplemented with 10 μM glycine is used. The Petri dish is transferred to the microscope stage. Viable interneurons can be harvested up to 6 h after the interruption of the enzymatic treatment. Chemicals and enzymes are obtained from Sigma Chemical Co. (St. Louis, MO).

B. Two-Electrode Voltage-Clamp Detection

Oocytes from the South African clawed frog (*X. laevis*) are widely used as expression systems for a broad spectrum of proteins (Gurdon *et al.*, 1971). Oocyte detectors with tailormade selectivities and response characteristics can be produced (Shear *et al.*, 1995) by injection of *in vitro* transcribed mRNA that encode for ligand-gated ion channels (Melton *et al.*, 1984; Swanson and Folander, 1992). The sparse expression of self-made ion channels in these oocytes makes them well suited for functional studies of ligand-gated and voltage-dependent ion channels. Because the literature on the expression of ion channels in *Xenopus* oocytes is comprehensive, this chapter is limited to the procedures used for CE–VC detection in our laboratories.

X. laevis frogs are available from several companies including Xenopus Ltd. (Nutfield, UK), Nasco Biologicals, Inc. (Fort Atkinson, WI), and Dipl. Biol.-Dipl.Ing., Horst Kähler Institut für Entwicklungsbiologie (Hamburg,

Germany). The procedure to prepare *Xenopus* oocytes for two-electrode voltage-clamp experiments includes isolation of oocytes from the frogs, (preparation and) injection of RNA, and defolliculation. The oocytes have to rest for at least 24 h between each procedure. In our protocol, *Xenopus* frogs are anesthetized on ice, and the oocytes are surgically removed from the ovarium with forceps. The incision is closed with a suture.

Many different methods exist for isolation of RNA from brain tissue. The method described below, which uses a strong denaturant, was originally devised to isolate RNA from cells that cannot be separated easily into cytoplasm and nuclei (e.g., frozen fragments of tissue), or from cells that are particularly rich in RNases (e.g., pancreatic cells). The method is modified from Glisin *et al.* (1974) and Ullrich *et al.* (1974). Pieces (0.3 g) of rat (Harlan UK Ltd., Blackthorn, UK) brains are placed in prechilled Falcon tubes containing 4 M guanidine isothiocyanate, 25 mM sodium acetate (pH 6.0), and 0.14 M 2-mercaptoethanol. The cell lysates are homogenized with a Polytron homogenizer (Brinkman Instruments Inc., Westbury, NY) and centrifuged (400 rpm, 2 min) at room temperature. Homogenization shears the nuclear DNA and prevents the formation of an impenetrable mat, which can hinder sedimentation of the RNA. The supernatant is then layered onto a cushion of 5.7 M CsCl and 10 mM ethylenediaminetetraacetic acid (EDTA) (pH 7.5), prepared in RNase-free water, in a clear RNase-free ultracentrifuge tube (Beckman Instruments Inc., Fullerton, CA). The position of the top of the cushion is marked on the outside of the tube, and the contents are centrifuged at 21 °C for 21 h at a speed of 40,000 rpm (SW 55-rotor, Beckman Instruments Inc.). A swinging bucket rotor is preferred to a fixed-angle rotor so that the RNA is deposited at the bottom of the tube, rather than along the walls (where it can come into contact with the cell lysate). The centrifuge tube should be handled carefully to avoid any mixing of its contents.

A line is drawn on the outside of each tube 0.5 cm from the bottom. The fluid above the level of the cushion (upper mark) is then removed and discarded. With a fresh, RNase-free pipette, the fluid above the lower mark is removed and discarded, and the bottom of the tube is cut off, just above the level of the remaining fluid, with a heated razor blade or a scalpel. The tube is inverted to allow the fluid to drain onto a pad of Kimwipes, then returned to an upright position (and the RNA pellet verified). The pellet is washed with 70% (v/v) ethanol to remove remaining CsCl, dissolved in 300 μl of 0.3 M sodium acetate (pH 6.0), and the RNA-containing solution is transferred to an Eppendorf tube. To precipitate the RNA, 750 μl of ice-cold ethanol (99.7%) is added to the tube and stored overnight at -20 °C. To collect the RNA precipitate, the solution is centrifuged (13,000 rpm, 10 min, 4 °C). The pellet is then washed with 70% ethanol, centrifuged (13,000 rpm, 5 min, 4 °C), and the ethanol is removed. Because all ethanol has to be removed to avoid poisoning of the oocytes, the RNA pellet is allowed to dry in a vacuum. The RNA is then dissolved in a small volume of RNase-free water and stored at -80 °C until used.

Stage V or VI oocytes are microinjected with 50 nl total rat brain RNA (Goldin, 1992) or *in vitro* transcribed mRNA ($\sim$0.1 μg/μl) expressing cloned rat serotonin

(5-hydroxytryptamine, 5-HT) 5HT1c receptors (Shear *et al.*, 1995). Stage II or III oocytes can also be used if a smaller membrane capacitance is desired, which speeds up the clamping process of the oocyte's membrane potential (Kraffe and Lester, 1992). A microdispenser (Drummond Scientific Company, Brookmall, PA), which is mounted on a micromanipulator, is used for injection of the RNA. The oocyte is penetrated by the microdispenser tip and injected at the vegetal (yellowish) hemisphere. Asymmetrical ("tongue-formed") tips are preferred and the microdispenser needle should be heated in an oven (200 °C, 16 h) to be free from RNases. Following injection of RNA, small clusters of oocytes are defolli-culated by gentle agitation in 2 mg/ml collagenase (type IA; Sigma) for 2 h. Alternatively, oocytes can be gently treated with collagenase (1 mg/ml) for 1 h, then the next day placed in a solution of high osmolarity. Following this step (after 30 s), the follicular cell layer becomes visible through a microscope and can be removed manually with forceps. For proper translation of the injected RNA, the oocytes are then incubated 3–6 days in modified Barth's solution: 88 mM NaCl, 1 mM KCl, 2.7 mM $NaHCO_3$, 10 mM HEPES, 0.82 mM $MgSO_4$, 0.33 mM Ca $(NO_3)_2$, 0.41 mM $CaCl_2$, 100 μg/ml gentamicin (pH 7.5, NaOH) at 19 °C.

V. Capillary Electrophoresis–Patch-Clamp Recording

A. Patch Clamping

In a patch-clamp experiment a cell is attached to the tip of a glass microelectrode by controlled suction to yield a high-resistance seal (Neher and Sakmann, 1976; Sakmann and Neher, 1995). It is then possible to choose one of several modes such as outside-out, inside-out, or whole-cell recording. For detailed explanations on the theory and practice of patch clamping, the reader is referred to Neher and Sakmann (1976) and Sakmann and Neher (1995).

B. Pipette Solutions

Patch-clamp pipette internal solutions serve to mimic the cytosol. By including or excluding certain ionic species in the pipette electrode solution, and by choosing a proper holding potential, the direction and magnitude of the transmembrane ionic flow can be controlled for a given ion channel. In our experiments we use two kinds of pipette solutions. For detection of ligands (e.g., glutamate, kainate, and NMDA) that activate receptor or ion-channel systems permeable to Na^+, K^+, and Ca^{2+}, a solution consisting of 100 mM KF, 2 mM $MgCl_2$, 1 mM $CaCl_2$, 11 mM ethylene glycol bis(β-aminoethyl ether)-N,N,N',N'-tetraacetic acid (EGTA), and 10 mM HEPES (pH 7.2, KOH) is used. When preparing this solution, KF must be added gradually to avoid precipitation. For ligands that activate Cl^--mediating receptor–ion-channel complexes (e.g., $GABA_A$ and strychnine-sensitive Gly recep-tors) a buffer containing 140 mM CsCl, 1 mM $MgCl_2$, 1 mM $CaCl_2$, 11 mM

EGTA, and 10 mM HEPES (pH 7.4, KOH) is used. To maintain cellular energy and redox status, 2 mM Mg-ATP and 5 mM reduced glutathione can be added to the patch-clamp pipette internal solution (Marty and Neher, 1995). Chemicals can be obtained from Sigma.

C. Regeneration of Capillary and Buffer Vial Maintenance

To obtain efficient electroosmotic flow and high precision in migration times, it is necessary to keep the silanol groups ionized. The capillary can be regenerated by flushing 1 M NaOH followed by 0.1 M NaOH, distilled water, and running buffer. It is sometimes advantageous to include an organic solvent such as methanol or ethanol in the washing step, especially if hydrophobic substances have been injected into the CE capillary.

It is important to change the solution of the reservoir covering the crack in the CE capillary periodically—preferably after each run. Otherwise, receptor ligands might diffuse into the CE capillary and depending on the concentration, either increase the background noise or cause distinctive response features. Likewise, the solution of the inlet buffer vial should be changed periodically, especially if the sample contains high concentrations of ligands. The risk for analyte crossover and ubiquitous injection (Fishman *et al.*, 1994) can cause a continuous supply of receptor ligands into the capillary, again causing activation of receptors in an uncontrollable way.

D. Instrumentation

In our experiments, signals are recorded with a patch-clamp amplifier (Model List L/M EPC, List-Electronics, Darmstadt, Germany or Axopatch 200B, Axon Instruments, Foster City, CA), digitized (20 kHz, PCM 2 A/D VCR adapter, Medical Systems Corp., New York, NY) and then stored on videotape until data analysis. For the production of complete electropherograms, the signals from the videotape are digitized at 2 Hz.

The experimental setup is shown in Fig. 2. An inverted microscope is used for viewing the cells and controlling the experiment. The microscope is placed inside a Faraday cage, and all power supplies or voltage sources are placed outside the cage. The Faraday cage is constructed from 2-mm-thick aluminum plates, and may include sections of aluminum screen to allow easier viewing of the microscope region. The Faraday cage serves to shield the experiment from electronic and magnetic fields that may interfere with the patch-clamp recordings. Shielding becomes especially critical if these experiments are performed in atypical electrophysiology laboratories. Specifically, if the patch-clamp setup is in the vicinity of mechanical pumps and high-powered pulsed lasers, then additional precautions are essential. These include shielding (and grounding) the headstage and pipette holder with aluminum foil, and having an isolated circuit breaker for the patch-clamp amplifier. Also, the microscope lamp should be powered by a voltage source

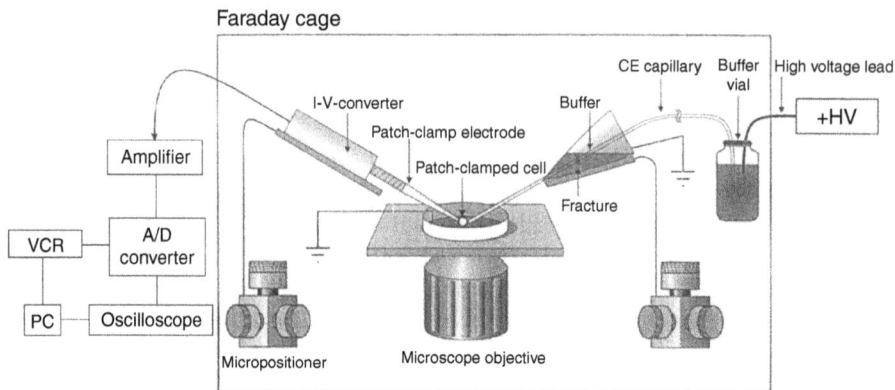

Fig. 2 Patch-clamp detection system for capillary electrophoresis (CE). The inlet of the separation capillary is inserted into a buffer vial, which is connected to a high-voltage power supply. The CE capillary is fractured and connected to ground 3 cm above the outlet. The tip of the patch-clamp electrode is positioned ~5–25 μm from the capillary outlet with a micropositioner.

(preferably DC to avoid interference from the line frequency) that is a separate unit from the microscope (thus placed outside the Faraday cage). When these issues are addressed, the rms noise decreases by nearly 10-fold (with lasers on). Further details of shielding patch-clamp experiments from electronic and magnetic noise also can be found elsewhere (Levis and Rae, 1992; Sakmann and Neher, 1995).

Electrophoresis is performed by applying a positive potential (up to 30 kV) to the inlet of the CE capillary with a high-voltage supply (LKB, Bromma, Sweden). The CE capillary and the high-voltage lead (i.e., inlet side) are inserted into a buffer vial contained in a polycarbonate holder with a 1-in. wall thickness. The polycarbonate vial holder is equipped with an interlock for safety precautions against electrical shock. The outlet of the CE capillary is mounted on a micromanipulator, and positioned into the cell-containing buffer at an angle of ~30 °C with respect to the microscope stage. Injection of sample solution into the CE capillary is performed hydrodynamically by raising the inlet of the capillary 10 cm above the level of the capillary outlet for 5–10 s. The volume, v_s, of a sample hydrodynamically injected into a CE capillary can be calculated using the Hagen–Poiseuille equation:

$$v_s = \Delta P r^4 \pi t (8\eta L)^{-1} \tag{8}$$

where ΔP is the pressure drop over the capillary, r the internal radius of the capillary, t the injection time, η the viscosity of the injected sample, and L is the total length of the capillary. Typically, our injection volumes are in the low-nanoliter range (Fishman *et al.*, 1994).

The patch-clamp pipette with the cell attached to the tip is placed 5–25 μm from the center of the outlet of the CE capillary. The pipette is fixed by a headstage that is connected to the patch-clamp amplifier (Levis and Rae, 1992; Sakmann and Neher, 1995). A chlorided silver wire, functioning as a reference electrode, is placed in the cell bath proximal to the CE capillary outlet. It is important to keep this wire well chlorided in order to effectively keep the cell bath at a low potential. This, in turn, will reduce the effect of the capillary-induced potential on the cells. The headstage is firmly mounted on a micromanipulator. It is essential that at least the micromanipulator holding the headstage and patch-clamp pipette be of a high-graduation type to allow exact positioning of the patch-clamped cell at the CE outlet. Other components included in our patch-clamp setup are an oscilloscope, a videotape recorder, an A/D converter, Butterworth and Bessel filters, and a computer system for sampling of data. In addition, a pipette puller for production of patch pipettes and a microforge for fire-polishing the electrode tips are required.

E. Example of CE–PC Data

Figure 3 shows the CE separations of GABA, Glu, and NMDA detected with outside-out patch-clamp detection. In addition to migration times, the characteristics of the current responses elicited by the separated agonists further confirm their identities. For example, analysis (see below) of single-channel conductances, distribution of conductance states, and ion-channel open and shut times yields information that can identify the ligand–receptor interaction (Jardemark *et al.*, 1997). In essence, it is feasible to detect indirectly a single analyte molecule with this technique (Orwar *et al.*, 1996). When whole-cell patch-clamp detection is employed, current traces of separated components give mean single-channel conductance levels, γ (in pS) and corner frequencies, f_{ci} (in Hz).

F. Data Analysis

A strength of the CE–PC detection technique is that it generates a high amount of information that can be used to identify a certain type of receptor interaction and the ligand that caused it. Electrophoretic migration rates together with several parameters pertaining to the electrical properties of the receptor in its activated and quiescent state are used for this purpose. In what follows, a brief account of patch-clamp data analysis that is pertinent to CE–PC is given. It is important to note that rather than characterize receptor physiology, we exploit receptors as detectors in CE and use a few key parameters for coupled receptor–ligand identification. More in-depth aspects on analysis of patch-clamp recorded current traces can be found in Sakmann and Neher (1995).

Electrophoretic migration rates are obtained by analyzing the time it takes from sample injection to detector response. Because peak responses obtained for desensitizing receptors do not track the physical presence (Gaussian concentration profile) of the analyte, the onset of the response is used as an index of a ligand's migration time. For nondesensitizing ligand–receptor interactions, the migration time is noted as the center of the normal-distributed detector response.

Various analyses can be performed on patch-clamp recorded currents. Spectral analysis of whole-cell currents yields information about ion-channel kinetics and

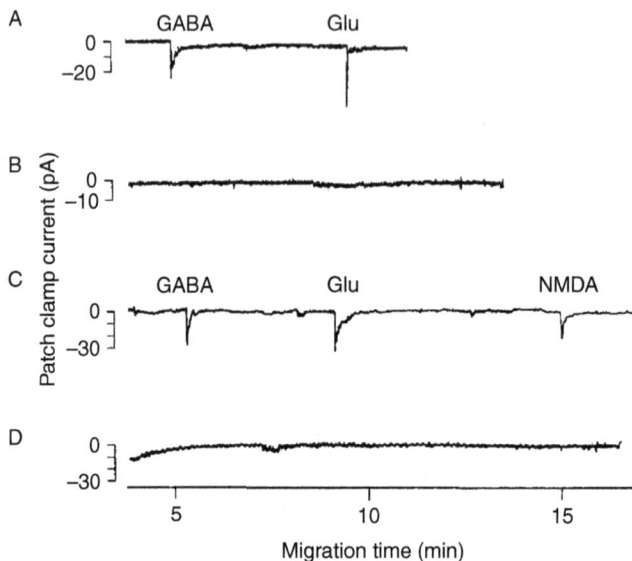

Fig. 3 Inward currents recorded from an outside-out patch following separation of GABA, Glu, and NMDA (250 µM of each) by capillary electrophoresis: (A) separation of GABA and Glu; (B) separation of HEPES–saline onto same patch used in (A); (C) separation of GABA, Glu, and NMDA; (D) separation of HEPES-saline onto same patch used in (C). Current traces were sampled at 2 Hz. The patch-clamp electrode contained CsCl solution. The holding potential was −70 mV and the applied high voltage was +12 kV. Different capillaries were used for the separations and their respective controls. (Reprinted with permission from Orwar *et al.* (1996). Copyright 1996 American Association for the Advancement of Science.)

the mean single-channel conductance. These parameters correlate to a specific ligand–receptor interaction. For spectral analysis of whole-cell currents, the signal from the videoadapter is filtered with an eight-pole Butterworth filter (bandwidth 1 kHz, −3 dB) and is digitized at 2 kHz. Records are divided into 0.5-s blocks prior to calculation of the spectral density. The mean power spectrum is calculated by averaging all power spectra obtained (at least 20) from these blocks. Agonist-induced power spectra are subtracted from power spectra obtained during membrane resting conditions. The resulting power spectrum is fitted by a single or double Lorentzian function using a least-squares Levenberg–Marquardt algorithm with proportional weighting (Colquhoun and Hawkes, 1997). Apparent single-channel conductances can be estimated according to the following equation:

$$\gamma = \frac{\sigma^2}{(E - E_r)I_m} \tag{9}$$

where σ^2 is the current variance, E the holding potential, E_r the reversal potential, and I_m is the mean current.

Unitary ion-channel events can be resolved using outside-out patch-clamp recording (Neher and Sakmann, 1976; Sakmann and Neher, 1995). Because of their small sizes, these patches have gigaohm ($G\Omega$)-resistance glass-to-membrane seals and extremely low membrane capacitances. In contrast to mean single-channel conductances that are calculated from a population of ion channels, intrinsic properties of the receptor complex such as distributions of conductance levels, transition between different conductance states, and channel open and closed times can be analyzed. The distribution of conductance states in a single receptor type can be displayed in amplitude histograms representing a "finger-print" of the activated channel (Sakmann and Neher, 1995).

Current-to-voltage (I–V) relationships can be obtained from current responses evoked by continuous superfusion of agonists, and on-the-fly from responses caused by separated analytes. Together with knowledge about the ionic composition of the intracellular (i.e., pipette) and extracellular solutions, reversal potentials and recti-fication of the I–V curve yield information about ion-channel permeability and identity of the activated receptor (Hille, 1992). For elimination of responses evoked by voltage-dependent ion channels, the I–V curve obtained between the responses is subtracted from the ramp obtained during the agonist-activated responses.

VI. Two-Electrode Voltage-Clamp Recording

Figure 4 shows a two-electrode voltage-clamped *Xenopus* oocyte during a CE experiment. The configuration includes voltage-recording (Fig. 4A) and current-injecting (Fig. 4B) electrodes (both of which penetrate the oocyte), a CE capillary

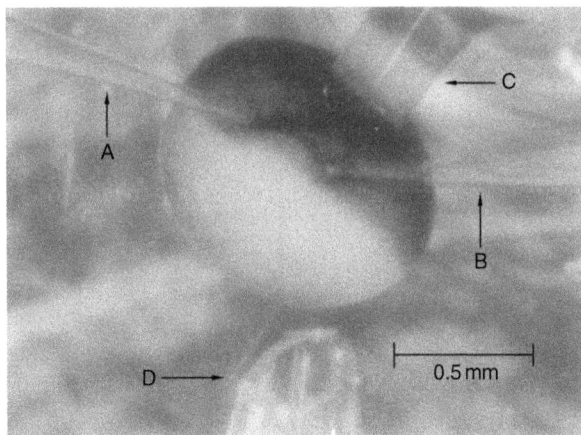

Fig. 4 A photomicrograph showing a *Xenopus* during a capillary electrophoresis experiment. The voltage-recording electrode (A) and a current-injecting electrode (B) penetrate the oocyte membrane. The capillary outlet (C) is placed ~20 μm from the surface of the oocyte's animal hemisphere. (D) Reference electrode. (Photograph courtesy of Susanne Orwar.)

positioned at a short distance from the oocyte membrane (Fig. 4C), and a reference electrode (Fig. 4D) (Stühmer, 1992). The current-passing and voltage-measuring electrodes are made from borosilicate capillaries and are back-filled with 0.3–3 M KCl. The electrodes are mounted on headstages, which, in turn, are connected to a VC amplifier. The headstages and electrodes are shielded by a Faraday cage. The oocyte for recording is placed in a small polycarbonate chamber filled with a HEPES buffer solution. The capillary outlet is placed between the electrodes, proximal to the animal hemisphere (dark area) of the oocyte. Procedures for capillary and buffer vial maintenance are as described in the previous section.

A. Example of CE–VC Data

Electropherograms of 5-HT and a buffer blank are shown in Fig. 5. Detection was performed using two-electrode voltage-clamped *Xenopus* oocytes expressing the cloned rat 5HT1c receptor (Shear *et al.*, 1995).

The *Xenopus* oocyte detection system is more mechanically stable than the CE–PC system, the latter of which is limited by the relatively short lifetime of the patch pipettes (no longer than 30 min). In addition, an oocyte in the two-electrode VC

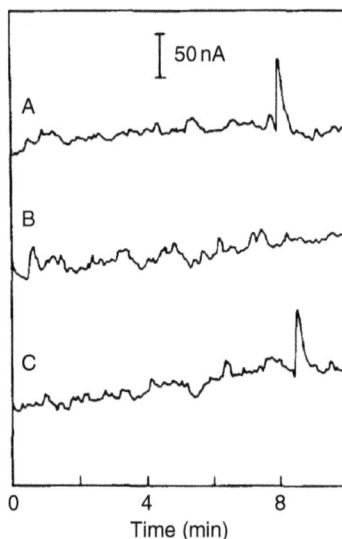

Fig. 5 Electropherograms demonstrating detection of serotonin (5-HT) using CE–VC. The membrane potential of a *Xenopus* oocyte, expressing the cloned rat 5HT1c receptor, was clamped at −70 mV, and the signal current was measured. The traces have been inverted to show typical positive-going CE peaks. (A) 5-HT (100 μM) was injected and a peak is detected at ~8 min. (B) Blank buffer was injected, and gives no response. (C) 5-HT was injected again and the electropherogram demonstrated a reproducible detected response at approximately the same migration time as in (A). The solutions were separated at ~250 V/cm in a 40-μm-i.d. capillary. (Reprinted with permission from Shear *et al.* (1995). Copyright 1995 American Association for the Advancement of Science.)

system can be used for several hours and lends itself for quantitative measurements, since multiple doses can be injected onto the same oocyte and reliable dose–response curves can be created. Also, when using preferably stage II and III oocytes with an appropriate sampling rate, power spectra can be obtained and analyzed to yield information about the mean single-channel conductance for a specific ion channel.

VII. Summary

By coupling CE to VC biosensors, a highly selective and sensitive means for analyzing biologically active components in complex mixtures with minimum sample handling is obtained. In contrast to traditional techniques for chemical analyses, this system can be engineered for sensitivity to specific biological species, thereby improving the response selectivity and greatly simplifying the challenge of chemical analysis. CE–PC and CE–VC are complementary techniques. CE–PC detection is a highly sensitive and information-rich technique that enables the study of electrophoretically separated ligands activating a single ion channel. The technique suffers from short lifetimes of the patch pipettes and is therefore mainly qualitative. CE–VC detection, on the other hand, might become a quantitative technique because the detector cells are viable for several hours. Even if mean single-channel conductances and I–V relationships can be obtained with CE–VC, the trade-off with this format is that much less information can be extracted from the current traces compared to patch-clamp recorded data.

Acknowledgments

This work is supported by the Swedish Natural Science Research Council (NFR), the Swedish Foundation for Strategic Research (SSF), and the U.S. National Institute of Drug Abuse (NIDA). S. J. L. thanks the U.S. National Institutes of Health (NIH) for a postdoctoral fellowship.

References

Allen, T. G. J. (1997). *Trends Neurosci.* **20**, 192.

Chiu, D. T., Hsiao, A. H., Gaggar, A., Garza-López, R. A., Orwar, O., and Zare, R. N. (1997). *Anal. Chem.* **69**, 1801.

Colquhoun, D., and Hawkes, A. (1997). *Proc. R. Soc. Lond., B, Biol. Sci.* **199**, 231.

Fishman, H. A., Amudi, N. M., Lee, T. T., Scheller, R. H., and Zare, R. N. (1994). *Anal. Chem.* **66**, 2318.

Glisin, V., Crkvenjakov, R., and Byus, C. (1974). *Biochemistry* **13**, 2366.

Goldin, A. L. (1992). *Methods Enzymol.* **207**, 266.

Gurdon, J. B., Lane, C. D., Woodland, H. R., and Marbaix, G. (1971). *Nature* **233**, 177.

Hille, B. (1992). "Ionic Channels of Excitable Membranes." Sinauer, Sunderland, MA.

Jardemark, K., Orwar, O., Jacobson, I., Moscho, A., and Zare, R. N. (1997). *Anal. Chem.* **69**, 3427.

Jorgenson, J. W., and Lukacs, K. D. (1981). *Anal. Chem.* **53**, 1298.

Kraffe, D. S., and Lester, H. A. (1992). *Methods Enzymol.* **207**, 340.

Levis, R. A., and Rae, J. L. (1992). *Methods Enzymol.* **207,** 14.

Li, S. F. Y. (1992). "Capillary Electrophoresis: Principles, Practice and Applications." Elsevier, Amsterdam.

Lillard, S. J., Yeung, E. S., and McCloskey, M. A. (1996). *Anal. Chem.* **68,** 2897.

Marty, A., and Neher, E. (1995). "Single-Channel Recording," 2nd edn. p. 45. Plenum Press, New York.

McConnell, H. M., Owicki, J. C., Parce, J. W., Miller, D. L., Baxter, G. T., Wada, H. G., and Pitchford, S. (1992). *Science* **257,** 1906.

Melton, D. A., Kreig, P. A., Rebagliati, M. R., Maniatis, T., Zinn, K., and Green, M. R. (1984). *Nucleic Acids Res.* **12,** 7035.

Neher, E., and Sakmann, B. (1976). *Nature* **260,** 799.

Nussberger, S., Foret, F., Hebert, S. C., Karger, B. L., and Hediger, M. A. (1996). *Biophys. J.* **70,** 998.

Orwar, O., Jardemark, K., Jacobson, I., Moscho, A., Fishman, H. A., Scheller, R. H., and Zare, R. N. (1996). *Science* **272,** 1779.

Pentoney, S. L., Huang, X., Burgi, D. S., and Zare, R. N. (1988). *Anal. Chem.* **60,** 2625.

Sakmann, B., and Neher, E. (1995). "Single-Channel Recording," 2nd edn. Plenum Press, New York.

Shear, J. B., Fishman, H. A., Allbritton, N. L., Garigan, D., Zare, R. N., and Scheller, R. H. (1995). *Science* **267,** 74.

Stühmer, W. (1992). *Methods Enzymol.* **207,** 319.

Swanson, R., and Folander, K. (1992). *Methods Enzymol.* **207,** 310.

Ullrich, A., Shine, J., Chirgwin, J., Pictet, R., Tischer, E., Rutter, W. J., and Goodman, H. M. (1974). *Science* **196,** 1313.

Vaughan, P., and Trotter, M. (1982). *Can. J. Physiol. Pharmacol.* **60,** 604.

Wallingford, R. A., and Ewing, A. G. (1987). *Anal. Chem.* **59,** 1762.

Wightman, R. M., Jankowski, J. A., Kennedy, R. T., Kawagoe, K. T., Schroeder, T. J., Leszczyszyn, D. J., Near, J. A., Diliberto, E. J., and Viveros, O. H. (1991). *Proc. Natl. Acad. Sci. USA* **88,** 10754.

Wijesuriya, D. C., and Rechnitz, G. A. (1993). *Biosens. Bioelectron.* **8,** 155.

CHAPTER 15

Ion Channels as Tools to Monitor Lipid Bilayer–Membrane Protein Interactions: Gramicidin Channels as Molecular Force Transducers

O. S. Andersen,[*] C. Nielsen,[*] A. M. Maer,[*] J. A. Lundbæk,[†] M. Goulian,[‡] and R. E. Koeppe II[§]

[*]Department of Physiology and Biophysics
Weill Cornell Medical College
New York, USA

[†]Department of Neuropharmacology
Novo-Nordisk A/S
Måløv, Denmark

[‡]Center for Studies in Physics and Biology
The Rockefeller University
New York, USA

[§]Department of Chemistry and Biochemistry
University of Arkansas, Fayetteville
Arkansas, USA

ESSENTIAL ION CHANNEL METHODS
DOI: 10.1016/B978-0-12-382204-8.00015-1

I. Recent Developments

A. Testing the Elastic Bilayer Model

The elastic bilayer model has been tested in two independent studies.

Goulian *et al.* (1998) used giant unilamellar vesicles to examine how changes in bilayer tension alter gramicidin channel appearance rates and lifetimes. Changes in bilayer tension alter lipid bilayer thickness:

$$\frac{\Delta d}{d_0} = -\frac{\sigma}{K_a},$$ (1)

where Δd denotes the change in bilayer thickness, d_0 the equilibrium thickness of the unperturbed bilayer, σ the bilayer tension, and K_a the bilayer area compression modulus. The changes in bilayer thickness alter the gramicidin monomer $\leftrightarrow$ dimer equilibrium, and thus the channel appearance frequency and lifetime. The elastic bilayer model provides good agreement with the experimental results using reasonable values for the bilayer material parameters.

Lundbæk and Andersen (1999) examined how changes in bilayer thickness alter the gramicidin channel lifetime. The elastic bilayer model, using independently determined material parameters, provides good agreement with the experimental results.

Though one must question the use of continuum models to describe molecular events, such as the energetic coupling between membrane proteins and their host bilayer, these two studies suggest that the elastic bilayer model provides a useful framework for examining the lipid bilayer regulation of membrane protein function.

B. Theoretical Developments

The role of intrinsic lipid, or monolayer, curvature, c_0, was incorporated into the elastic bilayer model by Nielsen and Andersen (2000), who also showed that the bilayer deformation energy can be expressed as a biquadratic function of the protein–bilayer hydrophobic mismatch, $l - d_0$, where l denotes the channel length, and c_0, cf. their Eqs. (17) and (28):

$$\Delta G_{\text{def}}^{\circ} = H_B(l - d_0)^2 + H_X(l - d_0)c_0 + H_C c_0^2,$$ (2)

where the three coefficients H_B, H_X, and H_C can be expressed as functions of the bilayer material properties (d_0, c_0, and the associated elastic compression and bending moduli) and the protein radius.

The elastic bilayer model was extended by Partenskii and Jordan (2002), who showed how it was possible to relax the assumption of spatially uniform bilayer material properties (elastic moduli). The model was further extended by Andersen and Koeppe (2007), who showed how Eq. (2) can be derived as a series expansion of $\Delta G^\circ_{\text{def}}$ in $l - d_0$ and c_0:

$$
\begin{aligned}
\Delta G^\circ_{\text{def}}(l - d_0, c_0) = {} & \Delta G^\circ_{\text{def}}(0,0) + \frac{\partial(\Delta G^\circ_{\text{def}})}{\partial(l - d_0)}(l - d_0) + \frac{\partial(\Delta G^\circ_{\text{def}})}{\partial c_0} c_0 \\
& + \frac{1}{2}\frac{\partial^2(\Delta G^\circ_{\text{def}})}{\partial(l - d_0)^2}(l - d_0)^2 + \frac{\partial^2(\Delta G^\circ_{\text{def}})}{\partial(l - d_0)\partial c_0}(l - d_0)c_0 \qquad (3) \\
& + \frac{1}{2}\frac{\partial^2(\Delta G^\circ_{\text{def}})}{\partial c_0^2}c_0^2 + \cdots
\end{aligned}
$$

where the first-order derivatives for reasons of symmetry are zero. The biquadratic expression for $\Delta G^\circ_{\text{def}}$ (Eq. (2)) thus is likely to apply rather generally.

In Eqs. (2) and (3) the hydrophobic mismatch and curvature contributions are coupled in the $(l - d_0)c_0$ term, which complicates attempts to evaluate the relative importance of these contributions to $\Delta G^\circ_{\text{def}}$. It is possible to separate the effects of $l - d_0$ and c_0 by noting the bilayer responds to the deformation that is associated with a protein–bilayer hydrophobic mismatch with a restoring force that, in the case of gramicidin channels, will tend to promote dimer dissociation. This disjoining force, F_{dis}, can be expressed as

$$
F_{\text{dis}} = -\frac{\partial \Delta G^\circ_{\text{def}}}{\partial(l - d_0)} = 2H_{\text{B}}(d_0 - l) - H_{\text{X}}c_0. \qquad (4)
$$

That is, the changes in gramicidin channel lifetimes can, in principle, be separated into contributions from the hydrophobic mismatch and the curvature, in each case multiplied by the appropriate elastic coefficient.

C. Experimental Developments

The two contributions to F_{dis} can be separated in experiments with gramicidin channels of different lengths (e.g., Lundbæk et al., 2005; Bruno et al., 2007). This became important in experiments that explore the bilayer-modifying effects of amphiphiles that alter the intrinsic monolayer curvature—capsaicin (Lundbæk et al., 2005) and polyunsaturated fatty acids (Bruno et al., 2007). Both capsaicin and polyunsaturated fatty acids cause negative changes in intrinsic curvature, and would thus be expected to have effects opposite to those of micelle-forming amphiphiles, such as Triton X-100. That was not the case; like Triton X-100, both capsaicin and polyunsaturated fatty acids stabilize gramicidin channels and increase the channel appearance frequencies and lifetimes.

It was possible to determine the mechanistic basis for these results by in experiments with gramicidin channels of different lengths, which showed that both capsaicin and polyunsaturated fatty acids cause larger changes in the lifetimes of the shorter gramicidin channels. These reversibly adsorbing amphiphiles thus alter the hydrophobic mismatch-dependent term in Eq. (4)—meaning that they alter bilayer elasticity, as predicted by Evans *et al.* (1995), and that the energetic consequences of the changes in bilayer elasticity dominate the changes in curvature.

D. Recent Reviews

Andersen and Koeppe (2007) summarize studies on the bilayer regulation of integral membrane protein function. Lundbæk *et al.* (2010) summarize recent developments in the regulation of membrane protein function and the use of gramicidin channels as probes for changes in bilayer properties.

II. Introduction

Numerous studies show that membrane protein function depends on the bilayer lipid composition (Bienvenüe and Marie, 1994; Devaux and Seigneuret, 1985). Specifically, the function of integral membrane proteins varies with lipid bilayer thickness (Baldwin and Hubbell, 1985; Brown, 1994; Caffrey and Feigenson, 1981; Criado *et al.*, 1984; Johannsson *et al.*, 1981) and intrinsic monolayer curvature (Hui and Sen, 1989; Jensen and Schutzbach, 1984; McCallum and Epand, 1995; Navarro *et al.*, 1984). In most cases there is only modest chemical specificity in these membrane lipid–protein interactions (Devaux and Seigneuret, 1985). This lack of chemical specificity, together with the large number of lipid types that are found in the membranes of any given cell, has caused difficulties for attempts to understand how the function of integral membrane proteins is affected by the bilayer lipid composition. These difficulties arose in part because the results were interpreted within the framework of the Singer–Nicolson *fluid-mosaic membrane* model (Singer and Nicolson, 1972).

The fluid-mosaic membrane model evolved from thermodynamic considerations relating to the organization of the main membrane components—phospholipids, cholesterol, and proteins (Singer and Nicolson, 1972). The guiding principles of the model were: first, that the lipids are organized in a liquid-crystalline bilayer in which integral membrane proteins are embedded; and second, the need to maximize hydrophobic and hydrophilic interactions. These principles have influenced all subsequent work—and led to the notion of hydrophobic coupling between the hydrophobic exterior surface of the membrane-spanning part of integral membrane proteins and the bilayer hydrophobic core. If the hydrophobic coupling were sufficiently strong, the thickness of the bilayer core adjacent to the protein would equal the length of the protein's exterior, hydrophobic surface. A weakness of the fluid-mosaic membrane model was that the lipid bilayer component was assumed to be a passive entity only—a permeability barrier that separated the extra- and intracellular aqueous phases. This point of view was strengthened by numerous studies on the permeability properties of lipid bilayers to small polar solutes (e.g., Walter and Gutknecht, 1986), which showed that the lipid bilayer could be approximated as being a ~5 nm thin sheet of liquid hydrocarbon.

The view of the lipid bilayer as a sheet of liquid hydrocarbon led to the notion of bilayer fluidity as an important determinant of protein function. The limitations of this notion were exposed by Lee (1991) who pointed out that the changes in protein function often were associated with changes in the protein's ligand affinity, which suggests that the membrane lipids somehow affect the conformational preference of bilayer-embedded proteins. Such lipid-dependent effects on the conformation of integral membrane proteins are difficult/impossible to explain within the scope of models that invoke changes in membrane fluidity as being a primary determinant of protein function. If not fluidity, however, what then?

An important, but neglected consequence of the liquid-crystalline organization of lipid bilayers is that one needs to incorporate the bilayer material properties (thickness and compression modulus, curvature, and bending modulus) into descriptions of membrane protein organization and function (Evans and Hochmuth, 1978; Helfrich, 1973; Mouritsen and Bloom, 1984). Similarly, one needs to consider specifically the importance of geometric packing criteria (Israelachvili, 1977) for lipid–protein interactions—and protein function.

III. Protein Conformational Changes and Bilayer Perturbations

The bilayer material properties can affect membrane protein function because the sequence of protein (conformational) state changes that are associated with normal protein function may involve changes in protein structure that affect the protein/lipid interface (Unwin and Ennis, 1984; Unwin et al., 1988). The thermodynamic need to maximize hydrophobic interactions between the membrane-spanning domain of integral membrane proteins and the bilayer core will couple these protein conformational changes to changes in the structure of the immediately surrounding bilayer (Fig. 1).

The liquid-crystalline organization of the lipid bilayer means that this perturbation of the bilayer structure has an energetic cost—the bilayer deformation energy $(\Delta G_{\text{def}}^{\circ})$ (Helfrich and Jakobsson, 1990; Huang, 1986). This becomes important for protein function because the free energy difference $(\Delta G_{\text{tot}}^{\text{I} \rightarrow \text{II}})$ between two protein conformations (I and II, e.g., the closed and open channel conformation in Fig. 1) is the sum of contributions from the protein per se $(\Delta G_{\text{prot}}^{\text{I} \rightarrow \text{II}})$ and terms that arise from the protein's interactions with the environment, which include changes in bilayer deformation energy contribution to $\Delta G_{\text{tot}}^{\text{I} \rightarrow \text{II}}$, $\Delta G_{\text{bilayer}}^{\text{I} \rightarrow \text{II}}$, which can be expressed as

$$\Delta G_{\text{bilayer}}^{\text{I} \rightarrow \text{II}} = \Delta G_{\text{def}}^{\text{II}} - \Delta G_{\text{def}}^{\text{I}}, \tag{5}$$

the difference in bilayer deformation energy between conformations I and II. The material properties of the bilayer, and thus $\Delta G_{\text{bilayer}}^{\text{I} \rightarrow \text{II}}$, will vary as a function of the membrane lipid composition, which provides a mechanism for the control of the protein conformational preference and function by the bilayer.

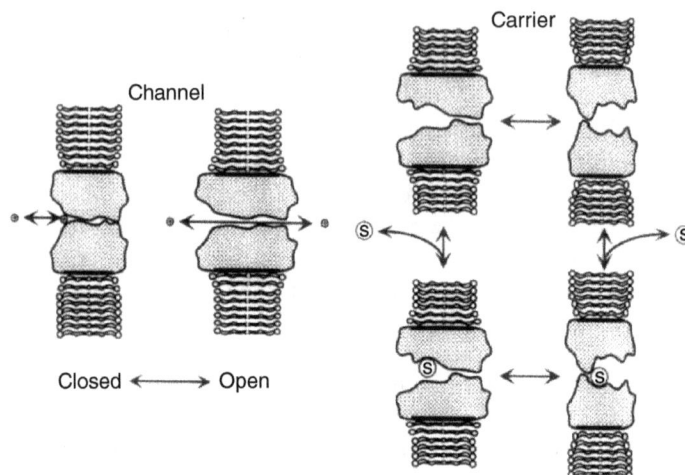

Fig. 1 Solute transfer by membrane-spanning channels and conformational carrier. Solute (ion) transfer through membrane-spanning channels does not involve major changes in channel structure. Solute transfer by conformational carriers are intimately coupled to protein conformational changes, which in turn may perturb the structure of the surrounding bilayer.

A protein conformational change that involves a change in the hydrophobic length of the membrane-spanning segment of the protein will cause both compression and bending of the two monolayers. The relative contribution of these two independent modes of membrane deformation to $\Delta G^\circ_{\text{def}}$ depends on the bilayer material properties (thickness and compression modulus; intrinsic monolayer curvature, and bending modulus). The intrinsic curvature of a lipid monolayer is determined by the variation of the intermolecular lateral interactions along the molecular axis, which usually is expressed in terms of the effective "shape" of the lipids in the monolayer (Cullis and de Kruijff, 1979) (Fig. 2).

Lipids that have a cylindrical "shape" form planar monolayers with no intrinsic curvature. If the effective cross-sectional area of the polar head group region is larger than that of the acyl chains, the monolayer will have a positive intrinsic curvature. If the effective cross-sectional area is less than that of the acyl chains, the monolayer will have a negative curvature.

In bilayers, the two monolayers must have complementary curvatures. Thus, the formation of a (planar) bilayer by lipids that by themselves would tend to form curved monolayers will cause a change in the effective shape of the lipid molecules—because of the requirement for a uniform cross-sectional area/molecule across a planar bilayer. The energy that is required to change the lipid "shape" causes a stress in the bilayer. The attractive interactions between the two monolayers in a bilayer means that one must distinguish between the intrinsic curvature of the monolayer, which is determined by the average lipid shape, and the curvature of the bilayer, which is determined by the coupled monolayers. Whenever the

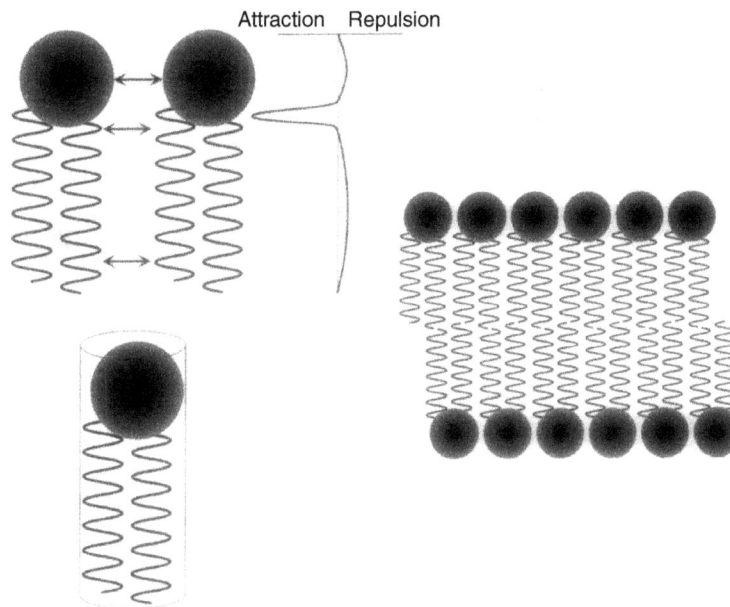

Fig. 2 Shape of lipid molecules and bilayer curvature stress. (Left) Formation of "relaxed" bilayers from molecules that in isolation have a "cylindrical shape"—as described by the profile of intermolecular interactions along the molecule length: top and bottom left panels. The isolated monolayer will have zero curvature, and two monolayers form a relaxed bilayer. (Right) Formation of bilayers that are under stress from molecules that in isolation have a "cone shape"—as described by the profile of intermolecular interactions along the molecule length: top and bottom left panels. The isolated monolayer will have positive curvature, and two monolayers will form a frustrated bilayer—in which the individual molecules are forced into an approximately cylindrical shape.

intrinsic monolayer curvature differs from the bilayer curvature, the bilayer will under a curvature-induced stress.

IV. Membrane Perturbations and Channel Function

Pharmacological modification of lipid bilayers that affect their material properties, as monitored by changes in the energetics of gramicidin channel–lipid bilayer interactions, also affect the function of integral membrane proteins (Lundbæk *et al.*, 1996). This is shown in Fig. 3, which shows the reversible effect of Triton X-100 on voltage-dependent sodium channels (μ1 subtype) expressed in HEK293 cells.

Triton X-100 promotes a positive monolayer curvature, which on *a priori* grounds would be expected to stabilize gramicidin channels—as is the case (Lundbæk *et al.*, 1996; Sawyer *et al.*, 1989). Triton also stabilizes, reversibly, one or more inactivated states in voltage-dependent calcium (Lundbæk *et al.*, 1996) and sodium channels—and the

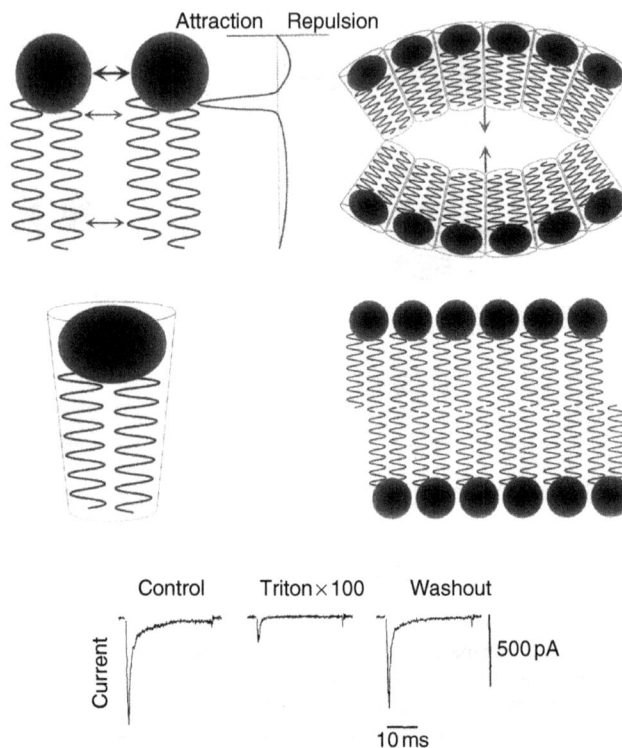

Fig. 3 Effect of Triton X-100 on sodium channel currents. Rat muscle μ1 sodium channel α-subunits were expressed in HEK293 cells. Membrane currents were measured using the whole-cell configuration following a depolarization from -80 to -10 mV. The current signal was corrected for leak conductance and filtered at 3 kHz. The three panels show sodium channel currents before, during, and 8 min after superperfusion with 100 μM Triton X-100. The plasmid carrying the μ1 message was a generous gift from Dr. Gail Mandel (SUNY at Stony Brook).

desensitized state of nicotinic acetylcholine receptors (Kasai *et al.*, 1970). This commonality suggests that Triton X-100 alters channel function by altering some general membrane property—such as the membrane deformation energy associated with the transition from closed or open channel states to inactivated or desensitized states.

V. The Membrane Deformation Energy

In the simplest description, for a given protein conformation, the magnitude of ΔG_{def}° depends on the bilayer compression–expansion and splay–distortion moduli, the bilayer thickness, and the intrinsic monolayer curvature. The different energy contributions can be unified using the theory of liquid-crystal elastic deformations (Helfrich and Jakobsson, 1990; Huang, 1986; Nielsen *et al.*, submitted), which allows the interdependent (compression, bending, etc.) contributions to

$\Delta G^{\circ}_{\mathrm{def}}$ to be evaluated. The interdependence among the components arises because the bilayer responds to an imposed distortion by minimizing the overall deformation free energy by varying both the compression–expansion and splay–distortion, etc. contributions to $\Delta G^{\circ}_{\mathrm{def}}$. Each of these contributions vary as functions of the bilayer deformation profile—and a change in the profile that minimizes, say the compression–expansion contribution, will also affect the magnitude of the splay–distortion contribution.

Subject to the choice of boundary conditions, the $\Delta G^{\circ}_{\mathrm{def}}$ associated with a local change in bilayer thickness can be approximated as a quadratic function of the extent of the membrane deformation—the difference between the hydrophobic thickness of the unperturbed bilayer (d_0) and the protein's hydrophobic exterior length (l) (Huang, 1986; Lundbæk et al., 1996; Nielsen et al., 1998):

$$\Delta G^{\circ}_{\mathrm{def}} = H_{\mathrm{B}}(l - d_0)^2, \qquad (6)$$

where H_{B} is a phenomenological spring coefficient associated with the bilayer deformation; it is a function of the bilayer material properties and the protein geometry. [Eq. (6) is exact only when the intrinsic curvature is zero, and the value of H_{B} depends on the boundary conditions at the protein/bilayer boundary (Nielsen et al., 1998).]

$\Delta G^{\circ}_{\mathrm{def}}$ is comprised of several contributions. Figure 4 shows the variation in $\Delta G^{\circ}_{\mathrm{def}}$ and its two major components, ΔG_{CE} (the compression–expansion component)

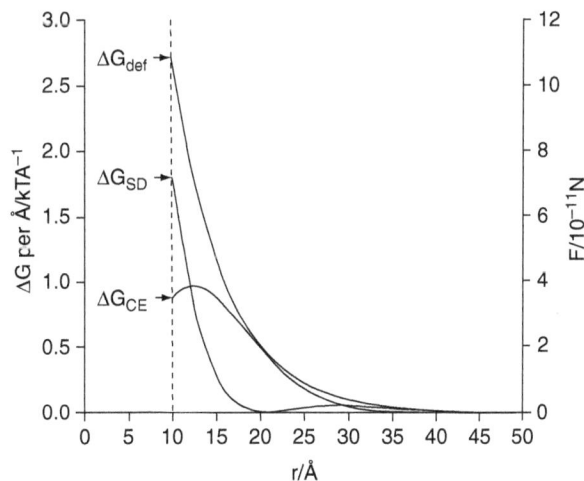

Fig. 4 Membrane deformation energy components per unit length as a function of radial distance from the bilayer/channel boundary. The membrane deformation $u = d - l = 4$ Å. ΔG_{def} denotes the total deformation energy, ΔG_{SD} the splay–distortion component, and ΔG_{CE} is the compression–expansion component. The splay–distortion modulus K_{c}, the compression–expansion modulus K_{a} and the membrane thickness d are connected by scaling laws (e.g., Nielsen et al., 1998). The example was calculated with $K_{\mathrm{c}} = 2.85 \times 10^{-10}$ N/Å, $K_{\mathrm{a}} = 2.85 \times 10^{-11}$ N/Å, and $d = 28.5$ Å.

and ΔG_{SD} (the splay–distortion component), as a function of distance from a membrane-spanning gramicidin dimer.

The splay–distortion contribution dominates close to the channel–bilayer interface, whereas the compression–expansion contribution is dominant further away from the channel. The relative contributions of the ΔG_{CE} and ΔG_{SD} components to ΔG_{def}° are comparable. Consequently, a pharmacologically induced change in the magnitude of ΔG_{def}° cannot be attributed solely to a change in ΔG_{CE} or ΔG_{SD}, which has implications for understanding how the lipid bilayer composition (and material properties) affect protein function. The membrane perturbation extends $\sim$30 Å from the dimer; but most of the deformation energy (about 75%) arises in the region between $r = 10$ and 20 Å, which corresponds to the annulus of lipid molecules that are in immediate contact with the dimer. This implies that one may be able to effect large changes in the membrane deformation energy by rather modest changes in the composition of this boundary layer—which could have implications for understanding how lipid-soluble, or amphipathic, substances affect the conformational preference of (and function) of integral membrane proteins.

The question thus becomes: Is the bilayer deformation energy of sufficient magnitude to affect the protein conformational preference and function? The total deformation free energy for the situation described in Fig. 4 is 11.9 $K_B T$, where K_B is Boltzmann's constant, and T is temperature in Kelvin. The quaternary conformational changes in gap junction channels (with radius $r_0 = 30$ Å) involve changes in the channels' hydrophobic length of approximately 0.3 Å (Unwin and Ennis, 1984). If the channels (with $l = 30$ Å) are embedded in a bilayer with $d_0 = 30$ Å (perfect hydrophobic match), the $\Delta G_{bilayer}^{C \rightarrow O}$ contribution to $\Delta G_{tot}^{C \rightarrow O}$ for the close $\leftrightarrow$ open (C $\leftrightarrow$ O) transition will be $\sim$0.8 kT. For the same transition in a membrane with $d_0 = 32$ Å, the ΔG_{def}° contribution will be $\sim$6 kT. Even larger deformation energies can result if the intrinsic monolayer curvature is different from zero. These estimates depend, however, on the choice of boundary conditions—and the assumption that the material constants that are determined in "macroscopic" experiments on isolated bilayers are valid to describe events close to the protein/bilayer boundary. It is therefore necessary to have quantitative measurements that probe how the lipid bilayer affects structurally well-defined conformational transitions in membrane proteins (Gruner, 1991).

VI. Carrier Versus Channel—Choice of Reporter Protein

The energetic coupling between proteins and bilayers will, in principle, affect the function of all bilayer embedded proteins. Membrane-spanning channels and conformational carriers, however, catalyze the transmembrane transfer of selected solutes by fundamentally different mechanisms (Fig. 1). In channels, the control of function arises from conformational changes between nonconducting (closed) and conducting (open) states. The individual catalytic events (the transfer of a solute/ion across the membrane) are uncoupled from the protein conformational changes associated with channel gating. In carriers, or ATP-driven pumps, the catalytic event(s) are inextricably coupled to protein conformation changes. Moreover, the continued operation of such membrane proteins reflects the continued cycling through the different kinetic states.

This difference in catalytic mechanism has implications for how the function of channels and carriers is affected by the lipid bilayer. In either case, a change in bilayer material properties (in $\Delta G^{\circ}_{\mathrm{def}}$) will alter the equilibrium distribution between different protein conformers. In the case of membrane-spanning channels, the distribution between conducting and nonconducting states (as well as the kinetics of the transitions between these states) will be altered. One can monitor directly the distribution between nonconducting (closed) and conducting (open) channel states (conformations) by measuring the changes in the channel-mediated ionic current, which provides for a direct readout of the number of conducting channels. In the case of conformational carriers, or pumps, a change in the equilibrium constant between the major conformers (the binding site exposed toward the left or the right in Fig. 1) will affect the rate constants for both the left → right and right → left transitions (usually in opposite directions). Systematic studies on how the protein function (turnover rate) is affected by the membrane lipid composition thus may show that the turnover rate is a nonmonotonic function of lipid composition (of $\Delta G^{\circ}_{\mathrm{def}}$) because of the opposing effects on the left → right and the right → left rate constants. This complicates attempts to understand how a change in bilayer material properties will affect the carrier function. Ion conducting channels therefore offer advantages not enjoyed by the carriers for attempts to elucidate the basis for bilayer control of protein function. Similar advantages apply to ligand-gated receptors, where one can monitor the distribution between different receptor states (conformations).

VII. Molecular Force Transducers

Among ion conducting channels, the gramicidin monomer ↔ dimer equilibrium that is associated with the formation of membrane-spanning gramicidin channels constitutes a reasonably well-defined structural transition in a membrane inclusion (Fig. 5).

Standard gramicidin channels are mini-proteins formed by the transmembrane assembly (O'Connell et al., 1990) of two $\beta^{6.3}$-helical monomeric subunits (He et al., 1994), which join at their formyl-NH-termini to form the conducting channels (see Andersen and Koeppe, 1992; Killian, 1992; Koeppe and Andersen, 1996 for reviews). Most, if not all, membrane-spanning gramicidin dimers are conducting

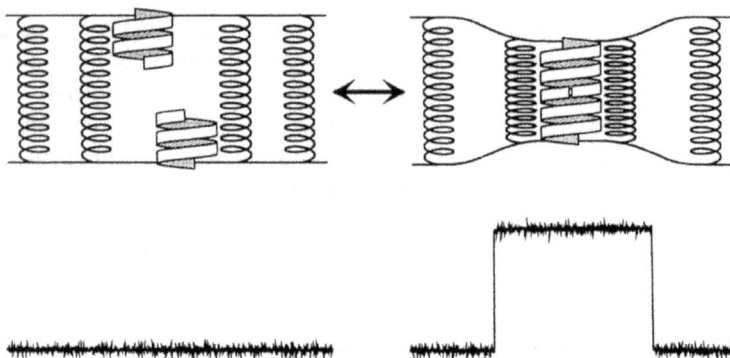

Fig. 5 Schematic representation of gramicidin channel formation by two membrane-inserted $\beta^{6.3}$-helical subunits. The average membrane thickness is larger than the length of the membrane-spanning dimer, and channel formation is associated with a membrane "dimpling." Channel formation can be monitored electrophysiologically—by the appearance of the single-channel current events (channel formation is upward, channel dissociation is downward).

channels (Veatch *et al.*, 1975), and there is no evidence for chemical specificity in the interactions between gramicidin channels and their host bilayer (Girshman *et al.*, 1997; Providence *et al.*, 1995). These properties make the gramicidins suitable to be used as molecular force transducers for investigating the mechanical properties of lipid bilayers.

Gramicidin channels can be used as force transducers because channel formation in lipid bilayers with a hydrophobic thickness, d_0, that is different than the hydrophobic length, l, of the gramicidin dimer forces the bilayer to "dimple" or "pucker" as it adapts its hydrophobic thickness to the channel length. Alterations in the ability of the bilayer to adjust to the channel will alter the equilibrium constant for channel formation. A membrane perturbant, such as the odorant limonene, 1-methyl-4-(1-ethylethenyl)-cyclohexene, does not interact specifically with gramicidin channels. Nevertheless, limonene has significant effects upon gramicidin channels, which can be observed as changes in the average channel lifetime (Fig. 6).

This result can be rationalized by considering the effective "shape" of this strongly hydrophobic molecule (cf. Fig. 2). Limonene has no polar moiety so, when incorporated in bilayers, these molecules will induce a lateral pressure in the bilayer interior—and the limonene molecules can be considered to be cone shaped with the base of the cone at the hydrophobic core of the bilayer. A planar membrane formed from phospholipids, doped with limonene, will be in a state of (curvature) stress, as the presence of the limonene will tend to drive the overall membrane shape toward concave surfaces. This curvature stress causes the twofold change in average lifetime that is observed.

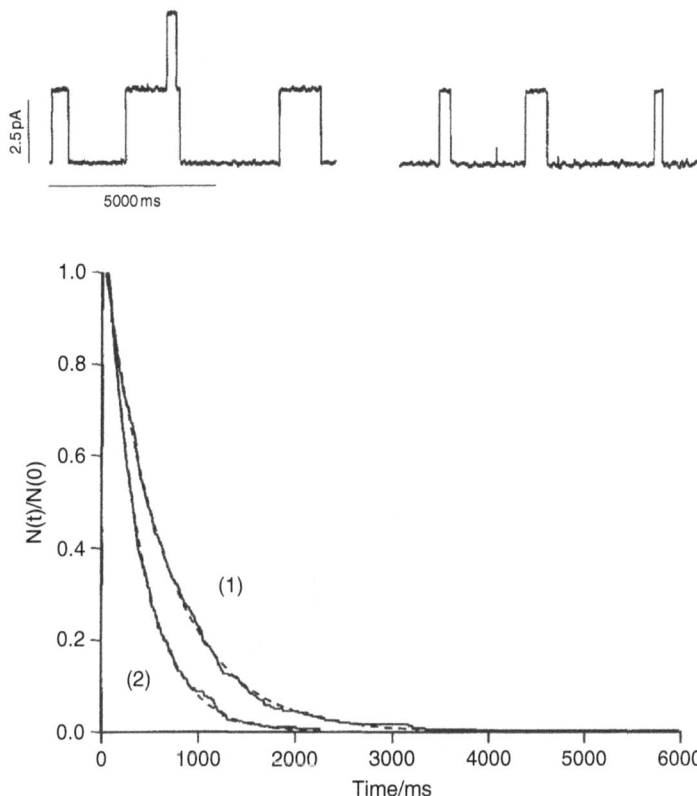

Fig. 6 The effect of limonene on the behavior of gramicidin A channels in dioleoylphosphatidylcholine/n-decane bilayers. *Top*: Single-channel current traces before (left) and a few minutes after (right) the addition of 1 mM limonene to the aqueous solution on both sides of the bilayer. *Bottom*: Normalized survivor histograms. Curve (1) denotes results in the absence of limonene; curve (2) denotes results obtained in the presence of 1 mM limonene. The interrupted curves denote the best fit of a single exponential distribution to the results: $N(t)/N(0) = \exp(-t/\tau)$, where $N(0)$ and $N(t)$ denote the number of channels with lifetimes longer than time zero and time t. In the absence of limonene, $\tau = 630$ ms, $N(0) = 365$; in the presence of limonene, $\tau = 370$ ms, $N(0) = 328$. The reduction in $N(0)$ reflects a reduction in the channel appearance rate. Experimental conditions as in Andersen (1983); applied potential 200 mV, current signal filtered at 200 Hz, 1 M NaCl, and 22 °C.

VIII. Measuring $\Delta G_{\text{bilayer}}^{\text{I}\rightarrow\text{II}}$ and the Phenomenological Spring Constant

The principle underlying the use of gramicidin channels as molecular force transducers is simple: to monitor how the gramicidin monomer $\leftrightarrow$ dimer equilibrium is affected by maneuvers that alter the bilayer properties. The practical implementation can be done in several different ways: by measuring the equilibrium constant for channel formation—as a change in membrane conductance, by measuring the disjoining force the bilayer imposes on the channels—as a change in channel lifetime, and by measuring the equilibrium distribution between channels of different length.

A. Equilibrium Constant Approach

The gramicidin dimerization constant K_D is given by

$$K_D = \frac{[D]}{[M]^2}, \tag{7}$$

where [D] and [M] denote the surface densities of gramicidin dimers and monomers, respectively. Assuming that $\Delta G_{\text{bilayer}}^{M \to D}$ is the only extrinsic contribution to $\Delta G_{\text{tot}}^{M \to D}$:

$$
\begin{aligned}
K_D = \frac{[D]}{[M]^2} &= \exp\left(\frac{-\Delta G_{\text{tot}}^{M \to D}}{k_B T}\right) = \exp\left(-\frac{\Delta G_{\text{prot}}^{M \to D} + \Delta G_{\text{bilayer}}^{M \to D}}{k_B T}\right) \\
&= K_D^{\text{prot}} \exp\left(\frac{-\Delta G_{\text{bilayer}}^{M \to D}}{k_B T}\right),
\end{aligned}
\tag{8}
$$

where k_B is Boltzmann's constant, T the temperature in Kelvin, and

$$K_D^{\text{prot}} = \exp\left(\frac{-\Delta G_{\text{prot}}^{M \to D}}{k_B T}\right). \tag{9}$$

For practical use, Eq. (8) is rewritten as

$$\Delta G_{\text{bilayer}}^{M \to D} = -k_B T \ln\left(\frac{[D]}{K_D^{\text{prot}} [M]^2}\right). \tag{10}$$

The gA channel-associated membrane conductance G is proportional to the number of gramicidin channels in the membrane:

$$G = [D]g, \tag{11}$$

where g is the single-channel conductance. Combining Eqs. (10) and (11),

$$\Delta G_{\text{bilayer}}^{M \to D} = -k_B T \ln\left(\frac{G/g}{K_D^{\text{prot}} [M]^2}\right), \tag{12}$$

which provides the desired link between bilayer energetics and electrophysiological measurements.

In practice, it is most convenient to use Eq. (12) to measure *changes* in $\Delta G_{\text{bilayer}}^{\text{M}\to\text{D}}$ ($\Delta\Delta G_{\text{bilayer}}^{\text{M}\to\text{D}}$) in the limit when $[\text{D}] \ll [\text{M}]$ (Lundbæk and Andersen, 1994; Lundbæk et al., 1997):

$$\Delta\Delta G_{\text{bilayer}}^{\text{M}\to\text{D}} = -k_{\text{B}}T \ln\left(\frac{G_{\text{test}}/g_{\text{test}}}{G_{\text{cntl}}/g_{\text{cntl}}}\right), \qquad (13)$$

where the subscripts (test and cntl) denote the two different experimental situations that are being compared (test vs. control). $\Delta G_{\text{bilayer}}^{\text{M}\to\text{D}}$ can be modified by pharmacological means (Lundbæk and Andersen, 1994; Lundbæk et al., 1997), for example, by the addition of compounds that alter the intrinsic monolayer curvature (cf. Lundbæk and Andersen, 1994; Lundbæk et al., 1996; Fig. 6). Relatively modest modifications of the bilayer properties can change $\Delta G_{\text{bilayer}}^{\text{M}\to\text{D}}$ by 10–15 kJ/mol (Lundbæk and Andersen, 1994; Lundbæk et al., 1997)—indicating that the bilayer deformation energy may be of sufficient magnitude to affect protein function.

The phenomenological spring constant H_{B} (cf. Eq. (6)) scales with the protein radius (Nielsen et al., submitted). Thus, once H_{B} is determined for any protein one can estimate its value for other proteins—assuming that the boundary conditions at the protein/lipid contact are similar for the two proteins. H_{B} can be estimated in experiments where the extent of the membrane deformation, $l - d_0$, is varied systematically. One can vary the channel length, l, at a constant membrane thickness, d_0, by changing the length of gramicidin's amino acid sequence—or vary the membrane thickness at a constant channel length by changing the length of the phospholipid acyl chains. The latter maneuver is preferable because it maintains the subunit interface invariant in the different experiments in which case $\Delta G_{\text{prot}}^{\text{M}\to\text{D}}$ should be constant. Using Eq. (6):

$$\Delta G_{\text{tot}}^{\text{M}\to\text{D}} = \Delta G_{\text{prot}}^{\text{M}\to\text{D}} + \Delta G_{\text{bilayer}}^{\text{M}\to\text{D}} = \Delta G_{\text{prot}}^{\text{M}\to\text{D}} + H_{\text{B}}(l - d_0)^2 - \Delta G_{\text{def}}^{\text{M}}, \qquad (14)$$

where $\Delta G_{\text{def}}^{\text{M}}$ denotes the bilayer deformation energy for the monomer state. When the bilayer thickness is changed, from d_0 (the reference bilayer thickness) to $d_0 + \delta$ (the test thickness), the change in $\Delta G_{\text{bilayer}}^{\text{M}\to\text{D}}$ becomes

$$\Delta\Delta G_{\text{bilayer}}^{\text{M}\to\text{D}} = H_{\text{B}}[(l - (d_0 + \delta))^2 - (l - d_0)^2] = H_{\text{B}}\delta[\delta - 2(l - d_0)]. \qquad (15)$$

Assuming that the single-channel conductance does not vary as a function of $l - (d_0 + \delta)$, H_{B} can be determined as:

$$\Delta\Delta G_{\text{bilayer}}^{\text{M}\to\text{D}} = -k_{\text{B}}T \ln\left(\frac{G_{l-(d_0+\delta)}}{G_{l-d_0}}\right) = H_{\text{B}}\delta[\delta - 2(l - d_0)]. \qquad (16)$$

It is important to note that, as the magnitude of $l - (d_0 + \delta)$ increases, the notion of strong hydrophobic coupling eventually will fail (meaning that the bilayer deformation at the channel/bilayer boundary will differ from $l - d_0$) because $\Delta G_{\text{def}}^{\circ} = H_{\text{B}}[l - (d_0 + \delta)]^2$ becomes so large that it becomes advantageous to allow hydrophobic residues to be in direct contact with water. When the bilayer deformation at the channel/bilayer boundary differs from $l - d_0$, when there is hydrophobic slippage, Eq. (16) may give rise to an erroneous determination of H_{B}.

This problem applies to integral membrane proteins as well as gramicidin channels. For example, the effective spring constant for membrane deformations adjacent to an integral membrane protein of radius 30 Å is ∼4 kJ/(mol Å^2) (Nielsen *et al.*, 1998). For the same protein, the hydrophobic penalty associated with a hydrophobic mismatch is ∼20 kJ/(mol Å). The membrane deformation energy increases as a quadratic function of $l - d_0$, cf. Eq. 6 (and Eq. 2); the hydrophobic energy increases only as a linear function of the hydrophobic mismatch. This means the incremental deformation energy eventually will exceed the incremental hydrophobic energy, which in this example will occur when $|l - d_0| > 5$ Å. Strong hydrophobic coupling thus will fail for larger membrane deformations, and there could be slippage even for rather modest deformations.

Another limitation is the assumption that a change in $l - d_0$ will affect only $\Delta G_{\text{bilayer}}^{\text{M}\to\text{D}}$. Even though the nonconducting $\beta^{6.3}$i-helical gramicidin subunits are inserted into the bilayer (He *et al.*, 1994), there may be an energetic penalty associated with subunit insertion. That may affect the equilibrium distribution between folded (membrane-inserted) and unfolded (membrane-adsorbed) subunit conformations (Mobashery *et al.*, 1997). To the extent this energy penalty varies as a function of the thickness–length mismatch $l - d_0$, it will affect the determination of $\Delta\Delta G_{\text{bilayer}}^{\text{M}\to\text{D}}$ and thus H_B.

B. Channel Lifetime Approach

Rather than the membrane conductance, one can measure the disjoining force the bilayer imposes on the membrane-spanning gramicidin dimers, which affect both the association (k_1) and dissociation (k_{-1}) rate constants. k_{-1} is of primary interest because $k_{-1} = 1/\tau$, where τ is the average dimer (channel) lifetime, which is directly measurable.

$$k_{-1} = \frac{1}{\tau_0}\exp\left(\frac{-\Delta G_{\text{tot}}^{\ddagger,\text{D}\to\text{M}}}{k_\text{B}T}\right),\qquad(17)$$

where $\Delta G_{\text{tot}}^{\ddagger,\text{D}\to\text{M}}$ is the activation energy for dimer dissociation and $1/\tau_0$ is a frequency factor (in Eyring's transition state theory, $1/\tau_0 = k_\text{B}T/h$, where h is Planck's constant). The transition state for dimer dissociation occurs when the subunits move a distance Δ apart, and $\Delta G_{\text{tot}}^{\ddagger,\text{D}\to\text{M}}$ is the sum of the intrinsic activation energy $\Delta G_{\text{prot}}^{\ddagger,\text{D}\to\text{M}}$ and the difference in bilayer deformation energy for deformations of $[(l+\Delta) - d_0]$ and $(l - d_0)$, respectively, $\Delta G_{\text{bilayer}}^{\ddagger,\text{D}\to\text{M}}$. Using Eq. (6),

$$\begin{aligned}\Delta G_{\text{tot}}^{\ddagger,\text{D}\to\text{M}} &= \Delta G_{\text{prot}}^{\ddagger,\text{D}\to\text{M}} + \Delta G_{\text{bilayer}}^{\ddagger,\text{D}\to\text{M}} = \Delta G_{\text{prot}}^{\ddagger,\text{D}\to\text{M}} + H_\text{B}[((l+\Delta) - d_0)^2 - (l - d_0)^2]\\ &= \Delta G_{\text{prot}}^{\ddagger,\text{D}\to\text{M}} + H_\text{B}\Delta[\Delta + 2(l - d_0)],\end{aligned}$$

$$(18)$$

cf. Lundbæk *et al.* (1996), and

$$\tau = \tau_{\text{prot}}\exp\left(\frac{\Delta G_{\text{def}}^{\ddagger,\text{D}\to\text{M}}}{k_{\text{B}}T}\right) = \tau_{\text{prot}}\exp\left(\frac{H_{\text{B}}\Delta[\Delta + 2(l - d_0)]}{k_{\text{B}}T}\right), \quad (19)$$

where $\tau_{\text{prot}} = \tau_0 \exp(\Delta G_{\text{prot}}^{\ddagger}/k_{\text{B}}T)$. When strong hydrophobic coupling pertains, H_{B} thus can be determined from the variation of τ as a function of d_0 (Lundbæk and Andersen, in preparation) because:

$$H_{\text{B}} = -\frac{k_{\text{B}}T}{2\Delta}\frac{\text{d}(\ln(\tau))}{\text{d}(d_0)}. \quad (20)$$

H_{B} is large, 25–30 $k_{\text{B}}T/\text{nm}^2$, again indicating that the bilayer deformation energy associated with a hydrophobic mismatch may be of sufficient magnitude to affect protein function.

The advantage of the lifetime approach is that it is more convenient to measure channel lifetimes than the large membrane conductance. In addition, one does not need to consider changes in the energetic cost of subunit insertion into the monolayers. The disadvantage is that the precise value of Δ is not known. The transition state most likely is when the dimer is stabilized by only four intermolecular hydrogen bonds (Lundbæk et al., 1996), in which case $\Delta = 1.6$ Å. But that estimate could be off by a factor of 2.

C. Heterodimer Formation

A third approach to measure $\Delta\Delta G_{\text{bilayer}}^{\text{M}\to\text{D}}$, or H_{B}, is to determine the relative stabilization of heterodimers formed between two gramicidin analogues that differ in length by, say two amino acid residues. (The gramicidin sequences must differ in length by an even number of residues, in order to ensure that the heterodimers are stabilized by six hydrogen bonds (Durkin et al., 1993).) The formation of the heterodimeric channels can be described by the reaction (cf. Greathouse et al., 1998):

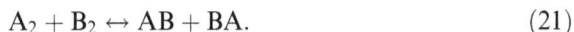

$$A_2 + B_2 \leftrightarrow AB + BA. \quad (21)$$

Let the channels formed by two gramicidin analogs, for example, a 14- and 16-residue gramicidin (A and B, respectively), differ in length by 2λ. The membrane deformations associated with the formation of A_2 and B_2 will be $(l - d_0)$ and $[(l - 2\lambda) - d_0]$, where l is the length of the 16-residue gramicidin channel, and the deformation associated with heterodimer formation will be $(l - \lambda) - d_0$. Assuming there are no subunit-specific interactions, that is, that $\Delta G_{\text{prot}}^{\text{M}\to\text{D}}$ is the same for all three channel types.

$$\Delta\Delta G_{\text{bilayer}}^{\text{M}\to\text{D}} = H_{\text{B}}\frac{2[(l - \lambda) - d_0]^2 - [(l - 2\lambda) - d_0]^2 + (l - d_0)^2}{2} = -H_{\text{B}}\lambda^2. \quad (22)$$

The advantage of this approach is that it is a one-parameter equation, one does not need to know $l - d_0$ and λ is well-determined (1.6 Å per L-D-dipeptide).

The disadvantage is that the assumption that ΔG°_{prot} is invariant among the different channel types is difficult to verify.

IX. Conclusions

The bilayer and its embedded proteins exert reciprocal effects upon each other:

$$protein\ conformational\ change \rightleftharpoons bilayer\ deformation\ energy.$$

This reciprocity emphasizes the dynamic implications of the hydrophobic coupling between bilayer and proteins, and the importance of the bilayer contribution to the free energy of membrane protein conformational changes. That is, in addition to serving as an organizing principle for the folding and insertion of membrane proteins, the need to minimize the exposure of hydrophobic groups to water (Singer and Nicolson, 1972) also provides a means for the regulation of protein function by the bilayer. The control of protein function by the membrane lipids is, *to a first approximation*, a "simple" energetic question, which can be addressed using the continuum theory of liquid-crystal deformations with minimal chemical specificity—and measured by monitoring changes in channel open probability. Chemical specificity will be important in some cases; but that is likely to be the exception rather than the rule. This simplifies the interpretational problems considerably, because one can disregard specific chemical identity of the numerous lipid types that are present in biological membranes. The situation becomes similar to that for electrified interfaces, where the Gouy–Chapman theory of the diffuse double layer serves as a major organizing principle (e.g., McLaughlin, 1989)—and where chemical specific interactions are introduced only when absolutely needed.

Acknowledgments

This work was supported by NIH grants GM21342 (O. S. A.) and GM34968 (R. E. K.), the Danish Research Council (C. N.), NSF (A. M. M.), the William Keck Foundation (M. G.), and the Norman and Rosita Winston Foundation (C. N.).

References

Andersen, O. S. (1983). *Biophys. J.* **41**, 119.
Andersen, O. S., and Koeppe, R. E., II (1992). *Physiol. Rev.* **72**, S89–S158.
Andersen, O. S., and Koeppe, R. E., II (2007). *Annu. Rev. Biophys. Biomol. Struct.* **36**, 107–130.
Baldwin, P. A., and Hubbell, W. L. (1985). *Biochemistry* **24**, 2633.
Bienvenüe, A., and Marie, J. S. (1994). *Curr. Top. Membr.* **40**, 319.
Brown, M. F. (1994). *Chem. Phys. Lipids* **73**, 159.
Bruno, M. J., Koeppe, R. E., II, and Andersen, O. S. (2007). *Proc. Natl. Acad. Sci. USA* **104**, 9638–9643.
Caffrey, M., and Feigenson, G. W. (1981). *Biochemistry* **20**, 1949.

Criado, M., Eibl, H., and Barrantes, F. J. (1984). *J. Biol. Chem.* **259,** 9188.

Cullis, P. R., and de Kruijff, B. (1979). *Biochim. Biophys. Acta* **559,** 399.

Devaux, P. F., and Seigneuret, M. (1985). *Biochim. Biophys. Acta* **822,** 63.

Durkin, J. T., Providence, L. L., Koeppe, R. E., II, and Andersen, O. S. (1993). *J. Mol. Biol.* **231,** 1102.

Evans, E. A., and Hochmuth, R. M. (1978). *Curr. Top. Membr. Transp.* **10,** 1.

Evans, E., Rawicz, W., and Hofmann, A. F. (1995). *In* "Bile Acids in Gastroenterology: Basic and Clinical Advances," (A. F. Hofmann, G. Paumgartner, and A. Stiehl, eds.), pp. 59–68. Kluwer Academic Publishers, Dordrecht.

Girshman, J., Greathouse, J. V., Koeppe, R. E., II, and Andersen, O. S. (1997). *Biophys. J.* **73,** 1310–1319.

Goulian, M., Mesquita, O. N., Fygenson, D. K., Nielsen, C., Andersen, O. S., and Libchaber, A. (1998). *Biophys. J.* **74,** 328–337.

Gruner, S. M. (1991). *In* "Biologically Inspired Physics," (L. Peliti, ed.), pp. 127–135. Plenum Press, New York.

He, K., Ludtke, S. J., Wu, Y., Huang, H. W., Andersen, O. S., Greathouse, D., and Koeppe, R. E. I. (1994). *Biophys. Chem.* **49,** 83.

Helfrich, P., and Jakobsson, E. (1990). *Biophys. J.* **57,** 1075.

Helfrich, W. (1973). *Z. Naturforsch* **28C,** 693.

Huang, H. W. (1986). *Biophys. J.* **50,** 1061.

Hui, S. W., and Sen, A. (1989). *Proc. Natl. Acad. Sci. USA* **86,** 5825.

Israelachvili, J. N. (1977). *Biochim. Biophys. Acta* **469,** 221.

Jensen, J. W., and Schutzbach, J. S. (1984). *Biochemistry* **23,** 1115.

Johannsson, A., Smith, G. A., and Metcalfe, J. C. (1981). *Biochim. Biophys. Acta* **641,** 416.

Kasai, M., Podleski, T. R., and Changeux, J. P. (1970). *FEBS Lett.* **7,** 13.

Killian, J. A. (1992). *Biochim. Biophys. Acta* **1113,** 391.

Koeppe, R. E. I., and Andersen, O. S. (1996). *Annu. Rev. Biophys. Biomol. Struct.* **25,** 231.

Lee, A. G. (1991). *Prog. Lipid Res.* **30,** 323.

Lundbæk, J. A., and Andersen, O. S. (1994). *J. Gen. Physiol.* **104,** 645.

Lundbæk, J. A., and Andersen, O. S. (1999). *Biophys. J.* **76,** 889–895.

Lundbæk, J. A., Birn, P., Girshman, J., Hansen, A. J., and Andersen, O. S. (1996). *Biochemistry* **35,** 3825.

Lundbæk, J. A., Maer, A. M., and Andersen, O. S. (1997). *Biochemistry* **36,** 5695.

Lundbæk, J. A., Birn, P., Tape, S. E., Toombes, G. E., Søgaard, R., Koeppe, R. E., II, Gruner, S. M., Hansen, A. J., and Andersen, O. S. (2005). *Mol. Pharmacol.* **68,** 680–689.

Lundbæk, J. A., Collingwood, S. A., Ingólfsson, H. I., Kapoor, R., and Andersen, O. S. (2010). *J. R. Soc. Interface* **7,** 373–395.

McCallum, C. D., and Epand, R. M. (1995). *Biochemistry* **34,** 1815.

McLaughlin, S. (1989). *Annu. Rev. Biophys. Biophys. Chem.* **18,** 113.

Mobashery, N., Nielsen, C., and Andersen, O. S. (1997). *FEBS Lett.* **412,** 15.

Mouritsen, O. G., and Bloom, M. (1984). *Biophys. J.* **46,** 141.

Navarro, J., Toivio-Kinnucan, M., and Racker, E. (1984). *Biochemistry* **23,** 130.

Nielsen, C., and Andersen, O. S. (2000). *Biophys. J.* **79,** 2583–2604.

Nielsen, C., Goulian, M., and Andersen, O. S. (1998). *Biophys. J.* **74,** 1966–1983.

O'Connell, A. M., Koeppe, R. E., II, and Andersen, O. S. (1990). *Science* **250,** 1256.

Partenskii, M. B., and Jordan, P. C. (2002). *J. Chem. Phys.* **117,** 10768–10776.

Providence, L. L., Andersen, O. S., Greathouse, D. V., Koeppe, R. E., II, and Bittman, R. (1995). *Biochemistry* **34,** 16404.

Sawyer, D. B., Koeppe, R. E. I., and Andersen, O. S. (1989). *Biochemistry* **28,** 6571.

Singer, S. J., and Nicolson, G. L. (1972). *Science* **175,** 720.

Unwin, P. N. T., and Ennis, P. D. (1984). *Nature* **307,** 609.

Unwin, N., Toyoshima, C., and Kubalek, E. (1988). *J. Cell. Biol.* **107,** 1123.

Veatch, W. R., Mathies, R., Eisenberg, M., and Stryer, L. (1975). *J. Mol. Biol.* **99,** 75.

Walter, A., and Gutknecht, J. (1986). *J. Membr. Biol.* **77,** 255.

PART VII

Purification and Reconstitution

CHAPTER 16

Purification and Reconstitution of Epithelial Chloride Channel Cystic Fibrosis Transmembrane Conductance Regulator

Mohabir Ramjeesingh, Elizabeth Garami, Kevin Galley, Canhui Li, Yanchun Wang, Paul D.W. Eckford, and Christine E. Bear

Programme in Molecular Structure and Function
Research Institute
Hospital for Sick Children
Toronto, Canada

337

DOI: 10.1016/B978-0-12-382204-8.00016-3

I. Update

The field of cystic fibrosis transmembrane conductance regulator (CFTR) research has advanced considerably in the 10 years since this chapter was written (see Riordan, 2008 for a recent review on CFTR), yet the methods described herein remain relevant today. With the continual development of a variety of small molecules that may interact directly with CFTR and rescue aberrant mutant protein trafficking and/or function at the membrane surface, the need for effective methods to purify, reconstitute, and assess the activity of the protein have increased. The purification and reconstitution methods described in this chapter remain important to the study of the function of CFTR in our laboratory. The PFO purification and reconstitution procedure described here is now employed almost exclusively in our work to rapidly and effectively isolate active wild-type and mutant CFTR protein expressed in Sf9 cells. While the SDS (sodium dodecyl sulfate) purification method is rarely used in our studies, the procedure has been retained in this chapter for completeness. Though a more lengthy and complex purification protocol, the SDS method remains suitable for the purification of CFTR expressed in Sf9 cells for applications where the introduction of a His_{10} tag to the protein or use of metal affinity chromatography may not be desirable.

This chapter describes several techniques to assess the function of purified and reconstituted CFTR protein, including measurement of its chloride channel activity in proteoliposomes via a $^{36}Cl^{-}$ uptake assay. Measurements of single-channel conductance and gating are performed by the planar bilayer techniques described herein. These methods are still relevant today and are effective in demonstrating the single molecule and ensemble activity of the protein as well as in probing the potential interaction of CFTR with many prospective small molecule therapeutics that are currently being developed.

The low ATPase activity of CFTR is monitored by radioactive ATP hydrolysis. The ATPase assay described in this chapter has been updated to correspond to our current methodology using $[\gamma\text{-}^{32}P]ATP$. ATPase assays with this method were used to show recently that CFTR in fact mediates ATPase rather than adenylate kinase activity (Ramjeesingh *et al.*, 2008), and to probe the interaction of the small molecule VRT-532 with mutant G551D (Pasyk *et al.*, 2009) and ΔF508 (Wellhauser *et al.*, 2009) CFTR. This technique will continue to be an invaluable tool to study the ATP-binding and hydrolysis functions of both wild-type and mutant CFTR under a variety of conditions and in the absence and presence of potential small molecule therapeutics.

II. Introduction

When the cystic fibrosis (CF) gene was first discovered, its protein product, the cystic fibrosis transmembrane conductance regulator (CFTR) was thought to act either as a chloride channel or as a chloride channel regulator (Riordan *et al.*, 1989).

Eventually, the chloride channel activity of CFTR was confirmed using a variety of experimental approaches. First, expression of recombinant CFTR in heterologous cell systems confers the appearance of cAMP-regulated chloride channels (Anderson *et al.*, 1991; Kartner *et al.*, 1991; Tabcharani *et al.*, 1991). Second, mutagenesis of amino acid residues thought to reside in putative membrane-spanning domains causes alterations in single-channel conductance and/or anion selectivity of the conductance conferred with CFTR expression (Anderson and Welsh, 1992; Tabcharani *et al.*, 1992). Finally, reconstitution of purified CFTR in planar phospholipid bilayers causes the appearance of cAMP-activated chloride channels, exhibiting biophysical properties identical to those observed in patch-clamp studies of epithelial cell membranes (Bear *et al.*, 1992). The chloride channel function of CFTR is currently thought to be critical for the elaboration of salt and water secretion across the epithelial cell lining of the airways, pancreatic ductules, gastrointestinal tract, and reproductive tract (Welsh *et al.*, 1995).

In this chapter, we describe the method we employ to purify and functionally reconstitute CFTR in model membranes (Bear *et al.*, 1992; Li *et al.*, 1996a,b; Ramjeesingh *et al.*, 1997). As previously mentioned, this experimental system has allowed us to define some of the functional properties of the protein which are intrinsic to CFTR by direct biophysical assays. Our most recent studies of the coupling of the catalytic and channel functions of CFTR best illustrate the utility of our reconstitution system. Using purified CFTR, we find that this molecule is not only a chloride channel, but it is also an ATPase (Li *et al.*, 1996b). Furthermore, we find that CFTR utilizes the energy released by ATP hydrolysis to fuel the opening and closing of the channel gate. To date, studies of the enzymatic activity of CFTR in cellular membranes have not been possible because of the difficulty in eliminating the background activity of other membrane ATPases. Consequently, future studies of the structural basis for CFTR ATPase activity and the coupling of this catalytic activity with channel gating must be performed using purified protein.

The body of this chapter addresses the methods we use to express, purify, and reconstitute CFTR. Furthermore, we describe the procedures we use to study the function of the reconstituted molecule. We compare two different strategies for CFTR purification from Sf9 cells; our original method, which employs conventional chromatographic techniques (Bear *et al.*, 1992), and a novel procedure, which applies metal affinity chromatography to purify a CFTR molecule engineered to possess a polyhistidine tag at its carboxy terminus (CFTR-His) (Ramjeesingh *et al.*, 1997; Fig. 1). In the original method, recombinant CFTR is extracted from Sf9 cells using sodium dodecyl sulfate (SDS). This strong anionic detergent is used because of the difficulty in solubilizing membrane incorporated CFTR using milder detergents. The use of SDS obligates the application of multistep purification and reconstitution protocols (Bear *et al.*, 1992). Our novel method capitalizes on the development of a novel family of fluorinated surfactants which can be used in conjunction with metal affinity chromatography. This one-step purification procedure is rapid and leads to the effective purification (>95%) of CFTR-His. Further, following reconstitution, the

SDS purification

Pellet of Sf9 cells
containing CFTR

Day 1

Solubilize cell pellet in
2% Triton X–100

(S) Centrifuge at
100,000*g*

Solubilize (P) in
2% SDS

(P) Centrifuge at
60,000*g*

Day 2

Apply (S) to
hydroxyapatite column

Hydroxyapatite
chromatography

Wash and elute the bound
protein from column
using phosphate gradient

Day 3

Identification and analysis of
fractions containing CFTR by:
• Dot blot
• SDS PAGE
• Silver staining
• Immunoblotting

Day 4

Concentration of fractions
containing CFTR

Day 5

Gel-filtration
chromatography

Identification and analysis of
fractions containing CFTR by:
• Dot blot
• SDS PAGE
• Silver staining
• Immunoblotting

Day 6

NaPFO purification

Pellet of Sf9 cells containing
CFTR-His

Solubilize cell pellet in buffer
containing 8% NaPFO

Centrifuge at
100,000*g* (P)

Apply (S) to nickel column

Wash and elute the bound
protein from nickel column
using pH gradient

Nickel
chromatography

Identification and analysis of
fractions containing CFTR-His by:
• Dot blot
• SDS PAGE
• Silver staining
• Immunoblotting

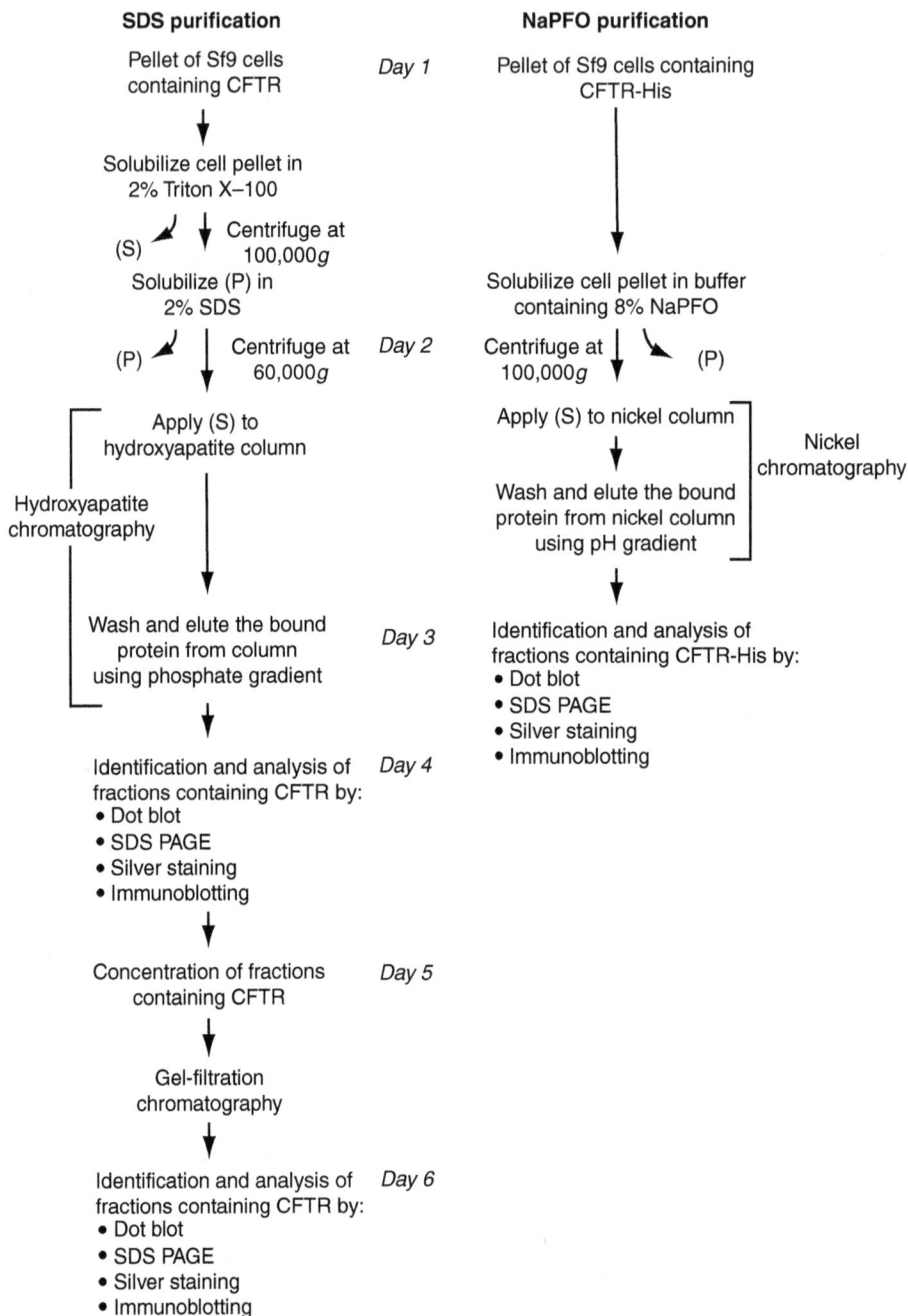

Fig. 1 Flow diagram of two different purification procedures for CFTR. *Left*: Purification scheme for SDS-solubilized CFTR protein. *Right*: Purification scheme for NaPFO-solubilized CFTR-His protein.

functional properties of CFTR-His are identical to those described for CFTR protein purified using the original protocol. We predict that this new method for CFTR purification may be applicable to other ion channels and will expedite studies of the structure–function relationships of these membrane proteins.

III. Expression of CFTR in Sf9–Baculovirus System

A. Rationale

We use the *Spodoptera frugiperda* fall armyworm ovary (Sf9)–baculovirus expression system for production of CFTR and CFTR variants for several reasons. First, functional expression of CFTR can be rapidly confirmed in patch-clamp studies of transfected cells (Kartner *et al.*, 1991). Second, the yield of recombinant protein is very high: close to 1% of total cellular protein of infected cells is CFTR (Bear *et al.*, 1992). Further, since Sf9 insect cells appear to possess relatively permissive quality controls with respect to protein processing, CFTR variants that are misprocessed and fail to reach the cell surface of mammalian cells, such as CFTRΔF508, can reach high levels of surface expression in Sf9 cells. Differences in glycosylation between Sf9–CFTR and CFTR produced in mammalian cells may be a harbinger of the quality control mechanisms that regulate membrane protein expression. Like other proteins produced in Sf9 cells (Kartner *et al.*, 1991), Sf9–CFTR possesses only core glycosylation, not the complex glycosylation that has been observed for the mature CFTR protein produced in mammalian cells. Because there is no evidence that differences in the degree of glycosylation affect the structure or function of the protein, we use the Sf9–baculovirus system to optimize our chances to obtain large quantities of membrane incorporated CFTR and CFTR mutant proteins.

B. Procedures

1. Construction of Transfer Vectors

The CFTR open reading frame (ORF) has been subcloned into the baculoviral transfer vector pBlueBac4 (Invitrogen, San Diego, CA) for the purpose of expression in the Sf9 insect cell system. This construct, pBlueBac4-CFTR ORF, and derivatives thereof, are used for all of our recent expressions of CFTR protein. To generate CFTR protein with a polyhistidine tag, a PCR (polymerase chain reaction) product corresponding to the C terminus of CFTR plus the polyhistidine tag (H_{10}) was ligated into the pBlueBac4-CFTR ORF construct so as to replace the wild-type 3' end of the ORF with the polyhistidine tagged 3' end. The reverse oligonucleotide 5'–CTGACGGTACCACTAGTGATGATGATGATGATGAT-GATGATGATGAAGCCTTGTATCTTG CACC-3' was used in conjunction with the forward oligonucleotide 5'-ATG GTGTGTCTTGGGATTCA-3' to create this PCR product, which was then subcloned back into pBlueBac4-CFTR

ORF. The entire amplified region was then confirmed by sequencing. Whereas the C terminus of CFTR is QDTRL, the C terminus of CFTR-His is QDTRLH$_{10}$.

2. Gene Transfer into Sf9 Cells and Production of Protein

Recombinant baculovirus is produced in Sf9 cells as previously described (Kartner *et al.*, 1991), incorporating recent modifications. Sf9 cells are cotransfected with the baculoviral transfer construct (pBlueBac4 encoding CFTR or CFTR-His) and linear baculoviral DNA (Bac-N-Blue DNA, Invitrogen) to produce recombinant CFTR or CFTR-His containing viruses. The supernatant generated from this transfection is used to infect cells and recombinant events are detected as blue plaques. Recombination is confirmed and clones selected for further study on the basis of purity, that is, lack of contamination by wild-type virus as assessed by PCR analysis of viral supernatant and by detection of CFTR protein as assessed by Western blot analysis of the infected cells. Working stocks of recombinant virus are produced and titered as previously described (Kartner *et al.*, 1991).

Sf9 cells are grown from frozen stocks purchased from Invitrogen. Initially, cells are grown as a monolayer in TMN-FH complete media [Grace's insect media with L-glutamine hydrolyzate (GIBCO-BRL, Gaithersburg, MD) and yeastolate (GIBCO)], at 27 °C in a CO_2-free incubator. When 90% confluency is reached, cells from 3×75-cm^2 flasks are transferred to 100 ml of the suspension culture media consisting of 50% (v/v) TMN-FH, 50% Excell-401 (JRH Biosciences, Lenexa, KS) in 250-ml shaker flasks (Bellco, NJ). The cells are then incubated at 27 °C with constant shaking at 150 rpm. To maintain cultures continuously, the cells are split when their density reaches 2–3×10^6 cells/ml, and reseeded into fresh media (1:1 TMN-FH: Excell 401) at approximately 1×10^6 cells/ml. Typically, CFTR and CFTR-His protein are produced in 1 l of suspension culture in 2-l shaker flasks. For optimal Sf9 cell infection, cell density is approximately 2×10^6 cell/ml with a viability of 98%. These cells are infected by high-titer viral stocks at a multiplicity of infection of 5 for 48 h. After this period, approximately 5% of the Sf9 cells are lysed due to infection. Infected cells are harvested from 1 l of media by centrifugation at $2000 \times g$ at 4 °C for 20 min and the cell pellet washed once with PBS. The cell pellets are then stored at -80 °C. Protein has been purified from cell pellets that have been stored in this manner for more than 1 year with no significant reduction in activity.

IV. Solubilization and Purification of CFTR

The detergent solubilization of hydrophobic integral membrane proteins is an absolute requirement for their purification. This initially involves the incorporation of the protein into water-soluble micelles by displacing the lipids and other associated molecules bound to the protein with detergent molecules. The protein–detergent micelles can subsequently be manipulated by many of the conventional techniques of protein purification. The choice of detergent is dictated by the degree of hydrophobicity of the protein. Recombinant CFTR expressed in Sf9 cells is poorly soluble in all detergents tested except for SDS, and the monovalent salts of pentadecafluorooctanoic acid (Ramjeesingh *et al.*, 1997). The purification of CFTR solubilized in these two detergents is described in the following paragraphs.

A. Solubilization and Purification of CFTR in Sodium Dodecyl Sulfate

1. Rationale

SDS is a strong anionic detergent which is very effective at extracting membrane proteins. Unfortunately, SDS is also a very powerful dissociating membrane detergent that usually results in loss of biological activity of solubilized proteins. It has been shown, however, that integral membrane proteins are resistant to complete denaturation in SDS. For example, bacteriorhodopsin maintains 50% α-helical content in SDS (Huang *et al.*, 1981) and the detergent seems to promote α-helical structure in less complex hydrophobic molecules (Dawson *et al.*, 1978; Steele and Reynolds, 1979; Wu and Yang, 1980). Because many integral membrane proteins contain a large amount of α-helical content and no disulfide bonds, the use of SDS for purification may offer distinct advantages because of its ability to disaggregate hydrophobic proteins during chromatography (Hjerten *et al.*, 1988). Hydroxyapatite, cation-exchange, and gel-filtration chromatographic methods of purification are amenable to the presence of SDS. Successful reactivation of SDS-solubilized proteins has been reported for a number of membrane proteins including bacteriorhodopsin (Huang *et al.*, 1981; London and Khorana, 1982), P-glycoprotein (Dong *et al.*, 1996), nicotinic acetylcholine receptor (Hanke *et al.*, 1990), 5′–nucleotidase, and neuraminidase (Hjerten *et al.*, 1988).

2. Procedures

a. Sample Preparation in Sodium Dodecyl Sulfate

1. Frozen Sf9 cell pellets expressing recombinant CFTR are thawed and then resuspended in 150 ml of PBS containing 2% Triton X-100 and a cocktail of protease inhibitors [leupeptin 10 μg/ml, aprotinin 10 μg/ml, E-64 (*trans*-epoxysuccinyl-L-leucylamido(4-guanidino)butane) 10 μM, benzamidine 1 mM, dithiothreitol (DTT) 2 mM, and MgCl$_2$ 5 mM] and DNase I 20 U/ml.

2. The suspension is nutated at room temperature for 1 h, and then centrifuged at 100,000 × g for 2 h at 4 °C. The supernatant is discarded and the pellets are transferred into 200 ml of 10 mM sodium phosphate, 2% SDS (w/v), and 3% mercaptoethanol (v/v), pH 7.4, and stirred overnight with a magnetic stirrer at room temperature.

3. Undissolved material is pelleted at 60,000 × g for 1 h and the supernatant is filtered using a 500-ml filtering unit (0.2 or 0.45 μm) with prefilters to prevent clogging. The filtered sample is then applied to the ceramic hydroxyapatite column.

b. Hydroxyapatite Chromatography

1. SDS-solubilized CFTR sample is applied to the column at a flow rate of 1 ml/min at room temperature.

2. The column is washed with 100 ml of buffer A, followed by 100 ml of buffer B. The column is then washed with 200 ml of a buffer containing 40% of buffer B and 60% of buffer C (400 mM final phosphate concentration). This is followed by 100 ml of a 400–600 mM phosphate gradient (100% buffer C) to elute CFTR. Washing of the column is continued with an additional 100 ml of buffer C.

3. Three-milliliter fractions from the gradient and subsequent washes are collected and analyzed by dot blot. Immunopositive fractions are further analyzed by Western blot (10 μl of each fraction) and by silver-stained protein gel (40 μl of each fraction). CFTR usually elutes as a broad band (about 10 fractions) between 540 and 600 mM phosphate. The silver-stained gel validates the quality of separation. A good separation yields a single band corresponding to CFTR and some lower molecular weight bands (see Fig. 2). A less than optimal separation may necessitate the sacrifice of a few of the initial CFTR-containing fractions to ensure the purity of the preparation. The CFTR-containing fractions are pooled and concentrated in a Centriprep 50 concentrator (Amicon, Danvers, MA) to a final volume of 500 μl.

c. Gel-Filtration Chromatography

1. The concentrated CFTR sample (500 μl) is applied to a Superose 12 column that has previously been equilibrated with 100 ml of buffer D.

2. The column is eluted with the buffer at a flow rate of 0.5 ml/min and 1-ml fractions are collected (Fig. 3). A Western blot and a silver-stained protein gel of fractions eluting at V_e/V_o of 1.4–1.6 are done to verify the protein and its purity. Pure CFTR-containing fractions are pooled for reconstitution and concentrated on a Centricon 100 concentrator (Amicon) at 15 °C to a final volume of 500 μl. The yield of CFTR is quantitated by amino acid analysis. The yield can vary from 15 to 500 μg of protein depending on the expression level of the protein in the infected Sf9 cells.

3. Reagents and Equipment

Hydroxyapatite column. The column is composed of four Econo-Pac CHT-II cartridges (5 ml each) attached in series (Bio-Rad, Richmond, CA). Initially, the column is washed with 100 ml of buffer C and equilibrated with 100 ml of buffer A before use. Between runs, an additional washing step using 100 ml of 1 N NaOH is employed.

Gel-filtration column. Preparative Superose 12 column (Pharmacia, Uppsala, Sweden).

Equipment. Fast protein liquid chromatography (FPLC; Pharmacia, Piscataway, NJ) or any liquid chromatography system with a pump capable of generating a linear gradient.

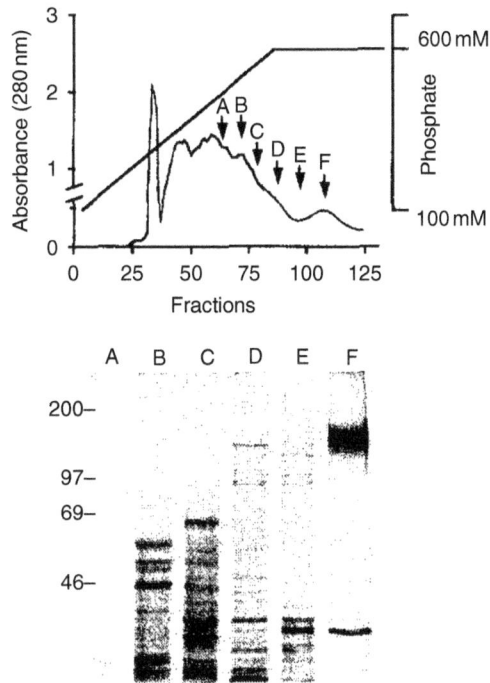

Fig. 2 Hydroxyapatite column chromatography of SDS-solubilized Sf9-CFTR. *Top*: Elution profile with a phosphate gradient generated from two buffers of 100 and 600 mM phosphate, each containing 0.15% SDS and 5 mM DTT, pH 6.8. *Bottom*: SDS–polyacrylamide gel (6%) of the peaks as indicated in the legend. Protein bands are visualized by silver staining. CFTR protein is eluted in peak F at 600 mM phosphate.

4. Buffers

Buffer A: 10 mM sodium phosphate, 0.15% LiDS (lithium dodecyl sulfate), 5 mM DTT, 0.025% NaN$_3$ (sodium azide), pH 6.8. Note that this buffer is degassed under vacuum prior to addition of LiDS. The buffer is filtered through a 0.22-μm Steritop-GP filter (Millipore, Bedford, MA).

Buffer B: 100 mM sodium phosphate, 0.15% SDS, 5 mM DTT, 0.025% NaN$_3$, pH 6.8, prepared as above.

Buffer C: 600 mM sodium phosphate, 0.15% SDS, 5 mM DTT, 0.025% NaN$_3$, pH 6.8, prepared as above. Some heating may be required to solubilize all of the reagents.

Buffer D: 10 mM Tris, 5 mM DTT, 100 mM NaCl, 0.5 mM EGTA and 0.25% LiDS, pH 7.8.

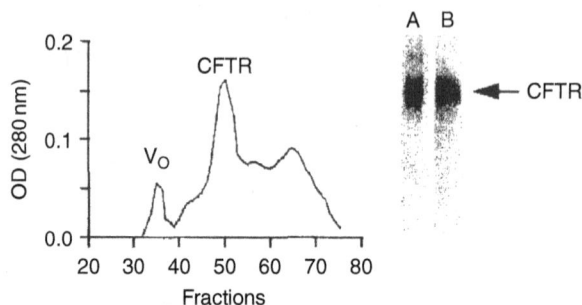

Fig. 3 Superose 12 gel-filtration chromatography of CFTR-containing fractions from a hydroxyapatite column. CFTR elutes at V_e/V_o (eluted volume relative to void volume) of 1.4. Purified CFTR runs as a single band on a silver-stained gel (A) and gives a single immunopositive band on a Western blot (B) with CFTR monoclonal antibody (MAb), M3A7. (Generously provided by Dr. N. Kartner, Pharmacology, University of Toronto.)

B. Solubilization and Purification of CFTR in Sodium Pentadecafluorooctanoic Acid

1. Rationale

The monovalent salts of pentadecafluorooctanoic acid (PFO) are derived from a novel family of fluorinated surfactants that are much more active than ordinary surfactants due to the hydrophobicity of the fluorocarbon chain. Furthermore, the compatibility of these detergents with biological assays and their ease of removal make them good candidates for purification and functional reconstitution of membrane proteins (Shepherd and Holzenburg, 1995). More importantly, sodium pentadecafluorooctanoate, unlike SDS does not abolish protein affinity interactions, thereby allowing the purification of a recombinant polyhistidine-tagged CFTR by single-step nickel affinity chromatography (Ramjeesingh *et al.*, 1997).

2. Procedures

a. Sample Preparation in NaPFO

1. Frozen Sf9 cell pellets expressing recombinant CFTR-His are thawed and then resuspended in 150 ml of phosphate-buffered saline (PBS) containing 2% Triton X-100 and a cocktail of protease inhibitors (leupeptin 10 μg/ml, aprotinin 10 μg/ml, E64 10 μM, benzamidine 1 mM, DTT 2 mM, and $MgCl_2$ 5 mM) and DNase I 20 U/ml.

2. The suspension is nutated for 1 h at room temperature, and then centrifuged at $100,000 \times g$ for 2 h at 4 °C. The supernatant is discarded and the pellets are transferred into 100 ml of 8% NaPFO in 20 mM phosphate, pH 8.0, and stirred overnight with a magnetic stirrer.

3. Undissolved material is pelleted at $60,000 \times g$ for 1 h and the supernatant filtered using a 500-ml filtering unit (0.2 or 0.45 μm) with prefilters to prevent clogging. The filtered sample is then applied to the nickel column.

b. Nickel Affinity Chromatography

1. The freshly regenerated 25-ml nickel column is attached to the FPLC and the CFTR-His containing sample is applied at 1 ml/min.

2. The column is then washed with 60 ml of buffer A. A pH gradient is applied to the column titrating buffer A with buffer B, from 0% buffer B to 100% buffer B in 100 ml.

3. Three-milliliter fractions are collected and analyzed by dot blot. Immuno-positive fractions are analyzed by Western blot (10 μl of each fraction) and by silver-stained protein gel (40 μl of each fraction) (Fig. 4). CFTR protein that elutes below pH 6.8 is pooled and concentrated in a Centriprep 50 concentrator (Amicon) to a final volume of 500 μl. The yield of CFTR is quantitated by amino acid analysis.

3. Reagents and Buffers

Pentadecafluorooctanoic acid is from Fluorochem (Old Glossop, UK).

Column. Twenty-five milliliters of packed nickel chelating resins (Qiagen, Valencia, CA) are poured and washed with 100 ml of buffer A before use.

Equipment. FPLC (Pharmacia) or any liquid chromatography system capable of generating a linear gradient.

Buffers. Buffers containing sodium pentadecafluorooctanoate (NaPFO) are prepared by adding the free acid, PFO, to the buffer and titrating with NaOH to the required pH.

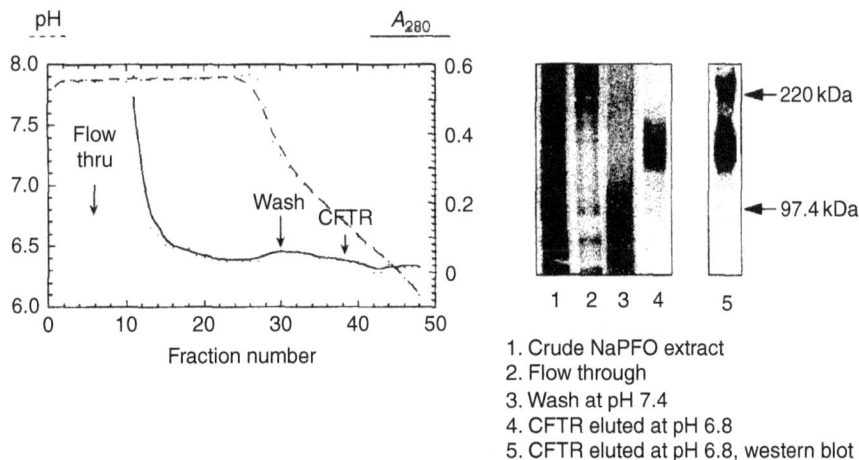

1. Crude NaPFO extract
2. Flow through
3. Wash at pH 7.4
4. CFTR eluted at pH 6.8
5. CFTR eluted at pH 6.8, western blot

Fig. 4 Immobilized metal ion affinity chromatography of Sf9–CFTR in NaPFO. *Left*: Elution profile with a pH gradient indicated, generated from two buffers of pH 7.4 and 4.0, each containing 25 mM phosphate, 100 mM NaCl, and 4% NaPFO. *Right*: Silver-stained protein bands after SDS–PAGE (6%) of crude NaPFO membrane extract and fractions as described in the legend. Lane 5 shows immunoreactivity of protein from fraction labeled as lane 4 with CFTR MAb, M3A7.

○ *Buffer A*: 20 mM phosphate, 4% NaPFO, pH 7.8. Note that the buffer is degassed under vacuum prior to NaPFO addition. The buffer is filtered through a Steritop-GP 0.22 μm filter (Millipore).

○ *Buffer B*: 20 mM phosphate, 4% NaPFO, pH 4.0, prepared as in buffer A.

V. Reconstitution of CFTR

A. Rationale

Reconstitution of hydrophobic membrane proteins involves the reinsertion of the protein into a phospholipid bilayer or vesicles of known composition. Phospholipid added to the detergent-solubilized protein initially forms mixed lipid-detergent micelles. The critical step in reconstruction is detergent removal from the mixed micelles. As detergent concentration falls below a critical minimum, phospholipid micelles spontaneously change conformation to form bilayer vesicles. Simultaneously, phospholipid displaces detergent at the exposed hydrophobic surfaces of the protein, which subsequently becomes incorporated across the vesicle bilayer. Several methods are available for detergent removal depending on the type of detergent used, such as hydrophobic adsorption chromatography (Moriyama *et al.*, 1984), gel-exclusion chromatography (Furth, 1980), dilution (Racker *et al.*, 1975), or dialysis. Detergent dialysis is the simplest of the methods and relies on the detergent monomer concentration being substantially high compared to the very low concentration of lipid monomers, allowing the detergent to be removed considerably faster than the lipid (Razin, 1972; Rhoden and Goldin, 1979). Detergents with high critical micellar concentration (CMC) and low aggregation number are the easiest to remove by this method (Furth, 1980). This dialysis method has been adopted for reconstituting SDS-solubilized proteins despite the low CMC and moderately high aggregation number by utilizing a mild detergent exchange step before vesicle reconstitution. Exchange can occur as a separate step prior to the addition of the lipid provided that the protein is soluble in its new detergent (Dong *et al.*, 1996). Alternatively, the detergent exchange can occur in the presence of lipids during dialysis (Hjerten *et al.*, 1988; Huang *et al.*, 1981). We have adopted the latter technique because of the insolubility of CFTR in other detergents. In addition, to further facilitate detergent removal, the SDS concentration is diluted below its CMC prior to dialysis, and a dialysis membrane cutoff of 50 kDa, large enough for SDS micelles to permeate, is used to ensure complete exchange of SDS for cholate.

The choice of lipids for reconstitution of integral membrane proteins is loosely based on (1) ability of the lipids to form liposomes, (2) biochemical requirements of the protein for specific lipids, and (3) capacity to form fusogenic vesicles, suitable for bilayer assays for single-channel activity. The effect of lipids on function of membrane proteins is well documented. P-Glycoprotein, a member of the ABC superfamily of proteins to which CFTR belongs, shows significantly higher

ATPase activity in the presence of phosphatidylethanolamine (PE) and phosphatidylserine (PS) than in the presence of phosphatidylcholine (PC) (Doige *et al.*, 1993). PE vesicles have been reported to be more fusogenic, possibly because the PE head groups are less hydrated than PC head groups (Lis *et al.*, 1982; Sundler *et al.*, 1981; White and King, 1985). However, pure PE does not form vesicles at neutral pH and physiologic ionic strength (Papahadjopoulos and Watkins, 1967). On the other hand, vesicle formation can be induced, when PE lipids are doped with other naturally occurring, bilayer-forming lipids such as PS and PC (Cullis and de Kruijff, 1987; Ho and Huang, 1985; Taraschi *et al.*, 1982) and when the reconstitution is performed at low ionic strength (Papahadjopoulos and Watkins, 1967). PS, a negatively charged phospholipid, helps to maintain liposome integrity by decreasing the tendency of liposomes to aggregate and has been shown to promote fusion with cultured cells (Damen *et al.*, 1982; Fraley *et al.*, 1981).

B. Procedures

1. Reconstitution of CFTR from LiDS

1. Fifty micrograms of CFTR protein (after Superose 12 column elution) in LiDS is concentrated with a Centricon 100 concentrator (Amicon) to a final volume of 100 μl. A 10-fold dilution with buffer B and reconcentration to 100 μl yields CFTR protein with a final LiDS concentration of 0.92 mM.

2. The protein is then added to 200 μl (200 μg) of the liposome preparation (PE:PS:PC:ergosterol, 5:2:1:1 by weight) and allowed to incubate at room temperature for 1 h.

3. The lipid–protein mixture is transferred to a dialysis bag (Spectra/Por membrane, molecular weight cutoff 50,000) and dialyzed against 2 l of buffer A for 17 h, followed by a 24-h dialysis against 4 l of buffer B and another 24 h of dialysis against buffer C.

4. The reconstituted proteoliposomes are aliquoted and stored under argon at $-80\,^\circ$C.

2. Reconstitution of CFTR from NaPFO

1. CFTR sample (100 μg) in 25 mM phosphate, 4% NaPFO, pH 6.2, is diluted 1:10 with buffer B and concentrated in a Centriprep 50 concentrator (Amicon) to a final volume of 500 μl (molecular weight cutoff 50,000) to 600 μl. The final NaPFO concentration is 0.4%.

2. Four hundred microliters of liposome preparation containing 4 mg of lipid (PE:PS:PC:ergosterol, 5:2:1:1 by weight) is added to 100 μg (500 μl) of CFTR. The mixture is dialyzed (Spectra/Por membrane, molecular weight cutoff 50,000) for 18 h against 4 l of buffer B containing 0.025% NaN_3 followed by a final dialysis against 4 l of buffer C. The reconstituted liposomes are aliquoted and stored under argon at $-80\,^\circ$C.

3. Liposome Preparation

     1. Three milligrams (300 μl) of the lipid mixture (PE:PS:PC:ergosterol, 5:2:1:1 by weight) in chloroform is dried in a 20-ml Pyrex test tube by argon gas. The tube is rotated during the drying process to allow the bottom of the test tube to be evenly coated with the lipid. Residual moisture is removed by a gentle stream of argon for 1 h.

     2. Buffer B (300 μl) is added to the lipid-coated tube, which is then incubated on ice for 10 min to rehydrate the lipid mixture. The latter is then sonicated in a bath sonicator until the solution is translucent. This usually requires 2–5 min of intermittent sonication. The test tube is kept on ice between sonications to prevent overheating of the lipid.

C. Reagents

1. Lipids

     PE (from egg, Avanti Polar Lipids, Birmingham, AL), PS (from brain, Avanti), and PC (from egg yolk, Avanti), ergosterol (Sigma, St. Louis, MO). Ergosterol is recrystallized from ethanol, dried under vacuum, and stored at −20 °C. A stock solution of a lipid mixture containing PE:PS:PC:ergosterol (5:2:1:1 by weight) is prepared at a concentration of 10 mg/ml of lipid in chloroform, aliquoted in 1-ml vials with Teflon stoppers (Reacti-Vial, Pierce, Rockford, IL), and stored under argon at −80 °C. Dialysis membranes (Spectra/Por) are purchased from Spectrum (Houston, TX).

2. Buffers

     Buffer A: 8 mM HEPES, 0.5 mM EGTA, 0.025% NaN_3, and 1.5% sodium cholate, pH 7.2.
     Buffer B: 8 mM HEPES, 0.5 mM EGTA, 0.025% NaN_3, pH 7.2.
     Buffer C: 8 mM HEPES, 0.5 mM EGTA, pH 7.2.

3. Equipment

     A bath sonicator (model G112SP1G, Laboratory Supplies Co., Hicksville, NY).

VI. Assessment of Functional Properties of Reconstituted CFTR Channels

     As mentioned previously, we have shown that purified, reconstituted CFTR functions as a chloride channel with biophysical and regulatory features identical to those reported for a chloride channel located on the apical membrane of epithelial cells (Bear *et al.*, 1992). Like the chloride channel in epithelial cells, purified, reconstituted CFTR requires phosphorylation by protein kinase A (PKA) and ATP hydrolysis for activity (Anderson *et al.*, 1991; Baukrowitz *et al.*,

1994; Gunderson and Kopito, 1995). Once activated, the CFTR chloride channel exhibits a low unitary conductance and slow gating kinetics (Kartner et al., 1991; Tabcharani et al., 1991). More recently, we have been using purified, reconstituted CFTR protein in our studies of the structural basis for its regulation by nucleotides (Li et al., 1996a). CFTR is a unique ion channel in that it hydrolyzes ATP and utilizes this energy to open and close the gate through which the flux of chloride ion occurs. Currently, we are assessing the effect of mutations within the Walker consensus sequences for nucleotide binding in order to determine the structural basis for CFTR ATPase activity and nucleotide-dependent gating. Each purified, reconstituted CFTR variant is currently being studied with respect to its ATPase activity and channel function.

Measurements of the ATPase activity of a suspension of proteoliposomes provide a rough estimate of the number of CFTR molecules capable of ATPase activity and of the catalytic activity of each CFTR molecule (Fig. 5). The ion-channel activity of each CFTR variant is examined using two different assays: radioisotopic flux assay (Li et al., 1996a) and single-channel studies following fusion with planar lipid bilayers (Bear et al., 1992; Li et al., 1996a,b). Similar to the ATPase assay, the electrogenic $^{36}Cl^-$ flux assay permits a macroscopic view of the number of channel competent CFTR molecules and their regulation by phosphorylation (Fig. 6). On the other hand, planar lipid bilayer studies permit a detailed examination of

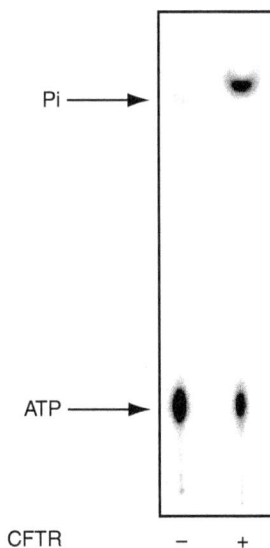

Fig. 5 ATPase activity of purified, reconstituted CFTR measured as the production of Pi from ATP. TLC separation of [γ-^{32}P]ATP and [γ-^{32}]Pi, where the upper arrow indicates [γ-^{32}]Pi, and the lower arrow [γ-^{32}P]ATP. The left lane indicates lack of [γ-^{32}]P$_i$ production (−) by liposomes alone, incubated with [γ-^{32}P]ATP for 2 h at 32 °C, and the right lane indicates [γ-^{32}]Pi production by liposomes containing CFTR-His (100 ng) (+).

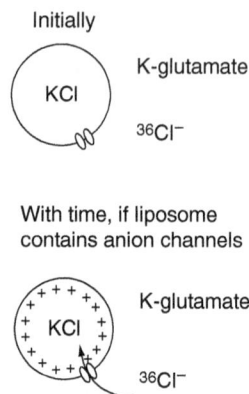

Fig. 6 Schematic representation of proteoliposome flux assay of electrogenic chloride uptake by purified CFTR. In this diagram, a single purified CFTR molecule provides the influx path for accumulation of $^{36}Cl^-$. The driving force for $^{36}Cl^-$ uptake is the positive intraliposome potential difference created by the initial creation of a chemical gradient for chloride ion.

the conductance properties and ATP-dependent gating of individual CFTR molecules on phosphorylation. A potential limitation in the study of mutant forms of CFTR using this assay relates to the difficulty in discerning between failure of proteoliposomes to fuse to the bilayer and protein dysfunction. A technique described by Woodbury (1998) helps to alleviate this problem.

Addition of the channel forming peptide, nystatin, renders all proteoliposomes fusogenic with the planar lipid bilayer and also permits the detection and confirmation of these fusion events (Li *et al.*, 1996b). Conveniently, nystatin only forms an active channel in the presence of the lipids cholesterol or ergosterol. On fusion of the nystatin- and ergosterol-containing proteoliposomes with planar lipid bilayers, a transient conductance spike is observed, providing a marker for each fusion event. On the other hand, because the bilayer phospholipid mixture contains only PS and PE, ergosterol will diffuse away from nystatin following fusion of the proteoliposomes; hence, the nystatin-channel forming complex will dissipate and will not further contaminate the activity generated by CFTR, the ion channel of interest. This technique has proven to be particularly useful in our study of CFTR variants which we suspect to have low channel open probability. In these cases, the detection of repeated proteoliposome fusion spikes in the absence of channel opening events provides convincing evidence that the variant CFTR protein is not functioning normally. We have provided an example of the utility of the nystatin-mediated fusion technique in our studies of the disease-causing CFTR mutation, CFTRG551D. In Fig. 7, it is possible to see that unlike wild-type CFTR, frequent spikes are observed, indicating the fusion of CFTRG551D proteoliposomes with no evidence of accompanying single-channel openings and closings. Experiments of this type convinced us that the CFTRG551D channel protein had difficulty in opening.

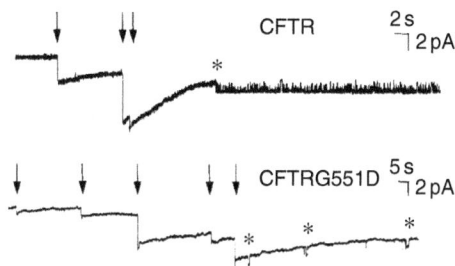

Fig. 7 Examples of the utility of nystatin-mediated liposome fusion for comparison of the channel activity of wild-type CFTR with variant CFTR proteins, that is, CFTRG551D. Arrows indicate conductance spikes associated with nystatin-mediated fusion events; asterisks indicate appearance of CFTR or CFTRG551D channel openings. For wild-type CFTR protein, three fusion spikes precede the appearance of a channel opening. For CFTRG551D, fusion of several liposomes is verified by the appearance of fusion spikes. The fifth fusion is associated with channel activity which is clearly altered from that observed for the wild-type CFTR protein.

A. Procedures

1. Phosphorylation of CFTR

Both CFTR channel and catalytic activities are dependent on phosphorylation by PKA. Hence, in our studies of CFTR, the catalytic subunit of PKA is added to the proteoliposomes to phosphorylate the protein. The enzyme must then be removed so that it will not contaminate our subsequent assays of ATPase and channel activity.

1. CFTR (10–30 μg) reconstituted in phospholipid liposomes is phosphorylated by incubation for 1 h at room temperature in a reaction mixture comprised of 200 nM catalytic subunit of PKA (Promega, Madison, WI), 0.5 mM ATP in 50 mM Tris, 50 mM NaCl, and 5 mM MgCl$_2$ at pH 7.5.

2. To remove PKA after the phosphorylation reaction, CFTR proteoliposomes are airfuged twice at 100,000 × g for 30 min. Between spins, the proteoliposomes are washed with buffer containing 50 mM Tris and 50 mM NaCl at pH 7.2. Alternatively, catalytic subunit of PKA was separated from the proteoliposomes by spin-column chromatography using Sephadex G-50 (Pharmacia).

B. Equipment

Airfuge ultracentrifuge (Beckman, Palo Alto, CA).

C. Assay of CFTR Activity as an ATPase

ATPase activity is measured as the production of [γ-^{32}]Pi from [γ-^{32}P]ATP by purified, reconstituted CFTR (Fig. 5). This ratio is corrected for spontaneous hydrolysis by subtracting the [γ-^{32}]Pi/[γ-^{32}P]ATP ratio of control liposomes (no CFTR) from the experimental ratio. Radiolabeled Pi and ATP are separated by polyethyleneimine (PEI) chromatography.

1. Proteoliposomes containing CFTR are dispersed by sonicating twice briefly for approximately 10 s.
2. The ATPase assay is carried out in a 15-μl reaction mixture containing 100 ng CFTR in phospholipid liposomes, 50 mM Tris, 50 mM NaCl, 2.5 mM MgCl$_2$, 1 mM DDT, 10 μCi of [γ-^{32}P]ATP (3 Ci/μmol), and 1 mM nonradioactive ATP at pH 7.5. Reaction mixture is sonicated for approximately 6 s and then incubated at 32 °C for 2 h.
3. The reaction is stopped by addition of 5 μl of 10% SDS or 2 μl of 1 M formic acid.
4. One-microliter samples from ATPase reaction vials are spotted on a PEI-cellulose plate and developed in 1 M formic acid–0.5 M LiCl.
5. The position and quantity of the radioactive Pi and ATP are ascertained using a Molecular Dynamics PhosphorImager. The data are analyzed using the ImageQuant software package (Molecular Dynamics, Sunnyvale, CA).

1. Reagents

Radiolabeled ATP: [γ-^{32}P]ATP 10 mCi/ml (Perkin Elmer, Woodbridge, Canada).

PEI cellulose plates: 20 × 20-cm PEI cellulose TLC plates with UV indicator (Merck KGaA, Darmstadt, Germany).

D. Assays of CFTR Activity as Chloride Channel

1. Electrogenic Tracer Uptake Assay for Study of CFTR Function

A concentrative tracer uptake assay is used to characterize the chloride conductance properties of reconstituted CFTR, as developed by Garty *et al.* (1983) and modified by Goldberg and Miller (1991) (Fig. 6).

a. Procedures
1. Proteoliposomes (100 μl) are preloaded with 150 mM KCl and centrifuged through Sephadex G-50 columns equilibrated with glutamate-containing salts; potassium glutamate (125 mM), sodium glutamate (25 mM), glutamic acid (10 mM), Tris–glutamate (20 mM) at pH 7.6, to replace external chloride. The eluted liposomes are diluted to 600 μl with the above glutamate buffer.
2. Uptake is initiated and quantified by addition of 1.0 μCi/mol of ^{36}Cl$^-$.
3. Intravesicular ^{36}Cl$^-$ is assayed at various time points following separation of 100-μl liposomes from the external media using a Dowex 1 anion-exchange minicolumn (Sigma).

b. Reagents

Sephadex G-50 (Sigma), $^{36}Cl^-$ (ICN, CA, USA), Dowex 1 ion exchanger, glutamate form (Sigma).

c. Buffers

Buffer A: Potassium glutamate (125 mM), sodium glutamate (25 mM), glutamic acid (10 mM), Tris–glutamate (20 mM), pH 7.6.

Buffer B: KCl (50 mM), MOPS [3-(N-morpholino)propanesulfonic acid] (10 mM), pH 7.2.

2. Planar Bilayer Studies of Liposomes Containing Purified CFTR

a. Procedures

Nystatin incorporation into proteoliposomes. As in our previous studies, proteo-liposome fusion with planar lipid bilayers is facilitated and detected by the introduction of nystatin (120 μg/ml), a technique originally described by Woodbury and Miller (1990).

1. Ten microliters of nystatin stock solution (1 mg/ml in methanol) is added to 1 mg (100 μl) of the lipid mixture in chloroform.

2. The mixture is dried in a 5-ml Pyrex test tube by argon gas. The tube is rotated during the drying process to allow the bottom of the test tube to be evenly coated with the lipid. Residual moisture is removed by spraying with a gentle stream of argon for 1 h.

3. Buffer B (100 μl) is added to the lipid-coated tube with incubation on ice for 10 min to rehydrate the lipid mixture.

4. This lipid mixture is then sonicated intermittently in the bath sonicator until the solution is translucent to light. The tube is kept on ice between sonications.

5. An aliquot of the nystatin-containing liposomes is added to an equal volume of the CFTR containing proteoliposomes (from either preparation) in a small Pyrex test tube. The mixture is frozen for 5 min in a dry ice–ethanol slurry, thawed at room temperature and sonicated for 15–20 s. This freeze–thaw cycle is repeated twice before the nystatin-containing proteoliposomes are studied in the planar bilayer chamber.

Bilayer formation and proteoliposome fusion. Planar lipid bilayers are formed by painting a 10 mg/ml solution of phospholipid (PE:PS at a ratio of 1:1) in *n*-decane over a 200-μm aperture in a bilayer chamber. Bilayer formation is monitored electrically by observation of the increase in membrane capacitance. In all experiments, bilayer capacitance is greater than 200 pF. Fusion of liposomes is potentiated with the establishment of an osmotic gradient across the lipid bilayer; the *cis* compartment of the bilayer chamber, defined as that compartment to which liposomes are added, contains 300 mM KCl, and the *trans* compartment, connected to ground, contains 50 mM KCl. Fusion events of nystatin-containing liposomes are indicated by the appearance of transient "nystatin spikes" in bilayer conductance (Woodbury and Miller, 1990) (Fig. 7).

E. CFTR Channel Detection and Analysis

CFTR channel activity is detected using a bilayer amplifier. Data are recorded and analyzed using pCLAMP 6.0.2 software (Axon Instruments, Burlingame, CA). Prior to analysis of open probability and dwell times, single-channel data are digitally filtered at 100 Hz. Ideal records are created by use of a half-height transition protocol.

1. Reagents

Nystatin (Sigma), *n*-decane (Sigma).

2. Equipment

Bilayer amplifier (custom made by M. Shen, Physics Laboratory, University of Alabama) and bilayer chamber (Warner Instruments, Hamden, CT).

References

Anderson, M., and Welsh, M. (1992). *Science* **257,** 1701.

Anderson, M., Gregory, R., Thompson, S., Souza, D., Paul, S., Mulligan, R., Smith, A., and Welsh, M. (1991). *Science* **253,** 202.

Baukrowitz, T., Hwang, T. C., Nairn, A., and Gadsby, D. (1994). *Neuron* **12,** 473.

Bear, C., Li, C., Kartner, N., Bridges, R., Jensen, T., Ramjeesingh, M., and Riordan, J. (1992). *Cell* **68,** 809.

Cullis, P. R., and de Kruijff, B. (1987). *Biochim. Biophys. Acta* **559,** 399.

Damen, J., Regts, J., and Scherphof, G. (1982). *Biochim. Biophys. Acta* **712,** 44.

Dawson, C. R., Drake, A. F., Helliwell, J., and Hider, R. C. (1978). *Biochim. Biophys. Acta* **510,** 75.

Doige, C. A., Yu, X., and Sharom, F. J. (1993). *Biochim. Biophys. Acta* **1146,** 65.

Dong, M., Penin, F., and Baggetto, L. G. (1996). *J. Biol. Chem.* **271,** 28875.

Fraley, R., Straubinger, R. M., Rule, G., Springer, E., and Papahadjopoulos, D. (1981). *Biochemistry* **20,** 6978.

Furth, A. J. (1980). *Anal. Biochem.* **109,** 207.

Garty, H., Rudy, B., and Karlish, S. J. (1983). *J. Biol. Chem.* **258,** 13094.

Goldberg, A. F., and Miller, C. (1991). *J. Membr. Biol.* **124,** 199.

Gunderson, K. L., and Kopito, R. R. (1995). *Cell* **82,** 231.

Hanke, W., Andree, J., Strotmann, J., and Kahle, C. (1990). *Eur. Biophys. J.* **18,** 129.

Hjerten, S., Sparrman, M., and Liao, J. (1988). *Biochim. Biophys. Acta* **939,** 476.

Ho, R. J., and Huang, L. (1985). *J. Immunol.* **134,** 4035.

Huang, K. S., Bayley, H., Liao, M. J., London, E., and Khorana, H. G. (1981). *J. Biol. Chem.* **256,** 3802.

Kartner, N., Hanrahan, J., Jensen, T., Naismith, L., Sun, S., Ackerley, C., Reyes, E., Tsui, L. C., Rommens, J. M., Bear, C. E., and Riordan, J. R. (1991). *Cell* **64,** 681.

Li, C., Ramjeesingh, M., and Bear, C. E. (1996a). *J. Biol. Chem.* **271,** 11623.

Li, C., Ramjeesingh, M., Wang, W., Garami, E., Hewryk, M., Lee, D., Rommens, J. M., Galley, K., and Bear, C. E. (1996b). *J. Biol. Chem.* **271**, 28463.

Lis, L. J., McAlister, M., Fuller, N., Rand, R. P., and Parsegian, V. A. (1982). *Biophys. J.* **37**, 667.

London, E., and Khorana, H. G. (1982). *J. Biol. Chem.* **257**, 7003.

Moriyama, R., Nakashima, H., Makino, S., and Koga, S. (1984). *Anal. Biochem.* **139**, 292.

Papahadjopoulos, D., and Watkins, J. C. (1967). *Biochim. Biophys. Acta* **135**, 639.

Pasyk, S., Li, C., Ramjeesingh, M., and Bear, C. E. (2009). *Biochem. J.* **418**, 185.

Racker, E., Chien, T. F., and Kandrach, A. (1975). *FEBS Lett.* **57**, 14.

Ramjeesingh, M., Li, C., Garami, E., Huan, L. J., Hewryk, M., Wang, Y., Galley, K., and Bear, C. (1997). *Biochem. J.* **327**, 17.

Ramjeesingh, M., Ugwu, F., Stratford, F. L., Huan, L. J., Li, C., and Bear, C. E. (2008). *Biochem. J.* **412**, 315.

Razin, S. (1972). *Biochim. Biophys. Acta* **265**, 241.

Rhoden, V., and Goldin, S. M. (1979). *Biochemistry* **18**, 4173.

Riordan, J. R. (2008). *Annu. Rev. Biochem.* **77**, 701.

Riordan, J., Rommens, J., Kerem, B. S., Alon, N., Rozmahel, R., Grzelczak, Z., Zielenski, J., Lok, S., Plavsic, N., Jia-ling, C., Drumm, M., Iannuzzi, M., *et al.* (1989). *Science* **245**, 1066.

Shepherd, F. H., and Holzenburg, A. (1995). *Anal. Biochem.* **224**, 21.

Steele, J. C., Jr., and Reynolds, J. A. (1979). *J. Biol. Chem* **254**, 1633.

Sundler, R., Duzgunes, N., and Papahadjopoulos, D. (1981). *Biochim. Biophys. Acta* **649**, 751.

Tabcharani, J. A., Chang, X. B., Riordan, J. R., and Hanrahan, J. W. (1991). *Nature* **352**, 628.

Tabcharani, J. A., Chang, X. B., Riordan, J. R., and Hanrahan, J. W. (1992). *Biophys. J.* **62**.

Taraschi, T. F., van der Steen, A. T., de Kruijff, B., Tellier, C., and Verkleij, A. J. (1982). *Biochemistry* **21**, 5756.

Wellhauser, L., Kim, C. P., Pasyk, S., Li, C., Ramjeesingh, M., and Bear, C. E. (2009). *Mol. Pharmacol.* **75**, 1430.

Welsh, M., Tsui, L. C., Boat, T. F., and Beaudet, A. L. (1995). *In* "The Metabolic and Molecular Basis of Inherited Disease," (C. R. Scriver, A. L. Beaudet, W. S. Sly, and D. Valle, eds.), pp. 3799–3876. McGraw-Hill, New York.

White, S. H., and King, G. I. (1985). *Proc. Natl. Acad. Sci. USA* **82**, 6532.

Woodbury, D. J. (1998). *Methods Enzymol.* 294, Chapter 17.

Woodbury, D. J., and Miller, C. (1990). *Biophys. J.* **58**, 833.

Wu, C. S., and Yang, J. T. (1980). *Biochemistry* **19**, 2117.

CHAPTER 17

Reconstitution of Native and Cloned Channels into Planar Bilayers

Isabelle Favre, Ye-Ming Sun, and Edward Moczydlowski
Department of Biology
Clarkson University, Potsdam
New York, USA

I. Update

Since our previous summary of popular methods for reconstitution of ion channels into planar bilayer membranes first appeared (Chapter 15 in *Methods in Enzymology*, Vol. 294), this area of research has continued to advance toward increasing sophistication of applications and functional analysis of cloned channel proteins. With the advent and availability of three-dimensional structures of channel proteins solved by X-ray crystallography and NMR over the last decade, there has been increasing emphasis on the use of planar bilayer recording to characterize the single-channel behavior of bacterial K^+ channels such as KcsA and MthK which are practically inaccessible to electrophysiological recording in their native cell membranes. An important technique which has emerged for single-channel recording of recombinant protein channels is the horizontal bilayer

DOI: 10.1016/B978-0-12-382204-8.00017-5

technique as first described by Chen and Miller (1996) for the *Torpedo* ray CLC$_0$ chloride channel and Heginbotham *et al.* (1999) for the *Streptomyces lividans* KcsA K$^+$ channel. This method is an offshoot of the earlier bilayer recording method of Wonderlin *et al.* (1990) that utilizes small holes ($\sim$50 μm in diameter) formed in a plastic partition. After forming a painted bilayer membrane from a solution of phospholipids in decane on a horizontally supported hole in a plastic film, it is possible to reproducibly incorporate purified channel proteins by pipetting reconstituted proteoliposomes directly over the bilayer or by brushing the bilayer with a small glass rod dipped in the liposome suspension. Major advantages of this technique are the ability to obtain long-lived single-channel recordings at high voltages up to 200 mV and accessibility of both sides of the membrane to solution exchange by perfusion (Chen and Miller, 1996; Heginbotham *et al.*, 1999; LeMasurier *et al.*, 2001; Zadek and Nimigean, 2006).

A particularly interesting and potentially revolutionary technique for recording of cloned channels is the recent description of bilayer insertion of heterotetramers of differentially tagged subunits of KcsA from channel protein produced by cell-free *in vitro* transcription and translation that is eluted as purified protein from bands of an SDS–PAGE gel (Rotem *et al.*, 2009). Using the latter methodology, Rotem *et al.* (2009) characterized the conductance and gating behavior of wild-type and mutant KcsA channels with different tetramer combinations of mutant E71A and native subunits. It is also noteworthy that gel-eluted KscA channels can be directly incorporated from aqueous protein solution into solvent-free bilayers of diphytanoyl-phosphatidylcholine formed by the Montal–Mueller bilayer folding technique (Rotem *et al.*, 2009). If this approach to *in vitro* expression, reconstitution, and bilayer recording of cloned ion channel proteins is broadly applicable, it would greatly simplify the process of functional channel characterization by eliminating the need for cellular expression and traditional steps of protein purification.

As a parting observation on the state of bilayer reconstitution, current literature exhibits a strong trend toward the development new microfabrication techniques and automated bilayer recording methods. Aligned with the recent development of automated patch-clamp and planar patch-clamp methods for cellular electrophysiology, we note with interest the description of ion channel reconstitution in bilayers formed on a microfabricated silicon chip (Pantoja *et al.*, 2001) and in droplet-interface bilayers (Bayley *et al.*, 2008). The emergence of these new technologies suggest that the future of ion channel research will find single-channel recording performed by robotic instrumentation to be a commonplace laboratory activity in combination with single-molecule fluorescence (Blunck *et al.*, 2008) and other ingenious assaults on protein molecules of cellular excitability that are yet to be imagined.

II. Introduction and General Overview

A planar bilayer is an artificial membrane formed across a small hole, $\sim$50 μm or larger in diameter. The hole on which the membrane is formed is usually placed in a thin plastic partition separating two aqueous compartments, but artificial bilayers may also be formed on a glass micropipette tip. Insertion or incorporation of a channel-forming molecule into such a membrane provides a simple experimental system for electrical recording of channel-mediated currents. Planar bilayer recording of ion channels is practiced for a number of reasons. Frankly, it is a technique that yields incredibly rich mechanistic information on a relatively low budget while offering kaleidoscopic displays of single-channel fluctuations that some workers find delightful, even soothing.

More formally, the following applications and research stratagems have emerged as the principal rationales and justifications for forming planar membranes: (1) assessment of the functional channel activity of peptides, purified membrane proteins, or other classes of membrane-active molecules; (2) as a biochemically defined system to investigate single-channel mechanisms and pharmacology at a basic biophysical level; (3) investigation of the influence of lipids and other membrane constituents on channel behavior; (4) as a model system for elementary studies of membrane fusion, capacitance changes, and other membrane-associated phenomena; (5) as a system for routine study of various channels in cellular membranes that are technically difficult to access (e.g., endoplasmic reticulum, bacterial and mitochondrial membranes; (6) use as a membrane dilution approach for obtaining long (>1 h) single-channel recordings of some channels that are often clustered in high density on cell membranes; (7) for experiments where simultaneous open access to solutions of both sides of the membrane is required; (8) as an assay system for screening complex biological mixtures such as scorpion venoms for the presence of channel-specific toxins, inhibitors, and activators that can be subsequently isolated by purification; and (9) use as a companion system to cellular expression for analyzing the unitary behavior of cloned channels and channel mutants produced by recombinant DNA methodology.

There are a number of excellent books (Hanke and Schlue, 1993; Miller, 1986) and chapters (Ehrlich, 1992; Labarca and Latorre, 1992; Williams, 1995) that review the historical development of the field of artificial membranes and provide detailed protocols for diverse bilayer techniques that form the basis of contemporary methodology. Our intent in this chapter is to highlight technical aspects of selected examples where the bilayer approach has been widely applied to functional analysis of major classes of ion channel proteins. Identification of techniques and preparations that have been found to be most generally applicable may facilitate extension of the bilayer approach to other classes of channel proteins that have not yet been thoroughly domesticated.

Following this preview, we first give a detailed protocol for the preparation of plasma membranes from rat skeletal muscle (Guo *et al.*, 1987; Moczydlowski and Latorre, 1983). For unknown reasons, this preparation is noted for particularly robust incorporation of maxi-Ca^{2+}-activated K^+ channels (K_{Ca}^+ channels) and voltage-sensitive Na^+ channels (Na_v^+ channels). The latter channels may be recorded in planar bilayers in a mode lacking inactivation in the presence of batrachotoxin, a method pioneered by Krueger *et al.* (1983) and French *et al.* (1984). For workers who may be just beginning to delve into bilayer techniques, this preparation is recommended as a demonstration experiment or positive control to "make sure that things are working properly" before venturing into unknown territory. Next, we provide a representative compilation of methods for bilayer incorporation of channel proteins from native tissues. Finally, we summarize new approaches to the study of cloned channels in planar bilayers. We view this last topic as a promising avenue for future research, because it offers the possibility of extending the functional analysis of specific mutations of many types of channel proteins to the bilayer assay system. This would be analogous to single-channel studies of gramicidin derivatives that have provided important molecular insights to the gating and conductance behavior of this simple channel molecule (Andersen *et al.*, 1998; Greathouse *et al.*, 1998).

Before proceeding further, it is worthwhile to briefly survey the landscape of bilayer work as applied to channels. An experimenter faces three major hurdles that must be surmounted in order to routinely collect usable data from a bilayer apparatus. These impediments are (1) electrical noise and high membrane capacitance, (2) membrane formation, and (3) channel incorporation.

Seal resistance, a critical noise factor in patch-clamp recording (Sigworth, 1983), is generally not a problem in bilayer work since the electrical resistance of artificial membranes properly formed on plastic partitions usually exceeds 100 GΩ. The major limitation of planar bilayers is "voltage noise," which arises from the high capacitance that is proportional to the membrane area. Low-pass filtering of high-frequency noise is absolutely necessary to resolve single-channel currents from a planar bilayer. Such filtering also attenuates the shortest fluctuations of channels under study and generally means that fast processes cannot be resolved as well in planar bilayer recordings versus patch-clamp recordings. The only available solution to this noise problem is to reduce the bilayer area by using as small a hole as practical. The trade-off in using small bilayers is that they can be more difficult to form and much less amenable to channel incorporation. Relatively low-noise bilayer recordings have been achieved by using specially designed holes (25–80 μm in diameter) on plastic partitions (Wonderlin *et al.*, 1990) and also by forming bilayers directly on small-diameter glass micropipettes using painting (Sitsapesan *et al.*, 1995), "tip-dip" (Coronado and Latorre, 1983; Hanke *et al.*, 1984; Schuerholz and Schindler, 1983; Suarez-Isla *et al.*, 1983), or "bilayer punch" (Andersen, 1983) methods. For routine work in our laboratory we use a bilayer diameter of $\sim$200 μm. This permits recording of well-resolved unitary currents of maxi-K^+_{Ca} channels with conductance of $\sim$250 pS at $\sim$1-kHz filtering or batrachotoxin-activated Na^+_V channels with a conductance of $\sim$20 pS at $\sim$200-Hz filtering.

Planar bilayers have enjoyed their widest application in steady-state recording of channel currents at constant voltage or by using linear voltage ramps (Eisenman *et al.*, 1986) to monitor the current–voltage behavior of open channels. The study of fast, transient, voltage-activated currents using pulsed, voltage-step protocols presents special problems for bilayer work due to large capacitative transients that are difficult to compensate. This problem can be largely overcome by the use of a specially designed headstage and amplifiers such as those marketed by Axon CNS

(Molecular Devices, Sunnyvale, CA), Dagan Corp. (Minneapolis, MN), and Warner Instrument Corp. (Hamden, CT). At least two groups of researchers have successfully recorded transient currents of voltage-activated Ca^{2+} channels and K^+ channels in planar bilayers by using such equipment (Rosenberg *et al.*, 1986; Sherman-Gold, 2008; Wonderlin *et al.*, 1990).

Contrary to its undeserved reputation as a sorcerer's ritual, the art of forming planar bilayers is not difficult to master. However, it must be said that some students experience frustration when working with bilayers despite noble efforts. For those with the right combination of patience and dexterity, there are two types of basic techniques for forming bilayers (White, 1986) known as "painting" or "folding." In the painting method (Mueller *et al.*, 1962), the partition containing the hole initially separates two chambers filled with an appropriately buffered electrolyte solution (e.g., 150 mM KCl, 10 mM, HEPES–KOH, pH 7.2). A solution of phospholipids dissolved in an alkane solvent, commonly decane, is spread over the hole with a brush, glass capillary rod, or other implement, and the resulting film is allowed to thin. Depending on the size of the hole and particular lipids, such films spontaneously thin over a time course as short as 1 min to produce a lipid bilayer that is sealed to the hole by an annulus of lipid and solvent. The term *black lipid membrane* or BLM is often used to describe such membranes, because by monitoring the thinning process optically, one first observes an iridescent film that transforms into a black hole as the membrane thins to bilayer thickness and no longer reflects light. Bilayer formation is also commonly monitored by measuring the membrane capacitance, which increases to a stable value as the membrane thins. In the folding technique (Montal and Mueller, 1972), a bilayer is assembled from a lipid monolayer formed on the surface of the solution by depositing a drop of lipid in a volatile solvent such as pentane. The bilayer is formed by raising the solution above the hole in the partition several times (Montal and Mueller, 1972) or by dipping the tip of a micropipette through the air–monolayer–water interface several times (Coronado and Latorre, 1983; Hanke *et al.*, 1984; Schuerholz and Schindler, 1983; Suarez-Isla *et al.*, 1983). Folded membranes generally contain much less solvent than painted membranes; however, there is much evidence to suggest that decane in the bilayer phase of painted membranes does not interfere with the function of many types of ion channels. As judged by the number of laboratories routinely using painted versus folded membranes, it appears that painted membrane technology is relatively easier to apply to incorporation of most types of channel proteins.

The final hurdle, incorporation of channel proteins into planar bilayers, is undoubtedly the most problematic and technically the most difficult to overcome. Although bilayer insertion of small lipophilic molecules and peptides, such as gramicidin, alamethicin, and some detergent-solubilized membrane proteins, can occur directly from the aqueous phase, most large channel proteins must be transferred from a native cellular membrane or a liposome membrane to the planar bilayer.

One approach is to first deposit the protein of interest at the air–water interface by spreading reconstituted proteoliposomes and then to use the folded membrane technique to form the bilayer and thus insert channels. Examples of the successful application of this method have included the nicotinic acetylcholine receptor channel as purified from the *Torpedo* ray electric organ and insect muscle (Hanke and Breer, 1987; Montal *et al.*, 1986).

The second technique makes use of the well-known but incompletely understood process of membrane fusion (not to be confused with cold fusion). This method, which seems to be more generally applicable, involves the fusion of native membrane vesicles or reconstituted liposomes containing the channel of interest to the planar membrane (Miller and Racker, 1976). While bilayer fusion clearly works (Chernomordik *et al.*, 1995), a person using this method somehow always has the Goldilockian perception that there is too little fusion or too much fusion and it is never "just right." Several parameters that can be exploited to control the fusion process include Ca^{2+} concentration, particular lipid mixtures (PE and acidic phospholipids favor fusion, PC inhibits), and osmotic gradients (hypertonic on the cis, vesicle-containing side favors fusion). However, incorporation conditions that work best for each type of preparation must be determined empirically. Woodbury and Miller (Woodbury, 1998; Woodbury and Miller, 1990) have also introduced a general fusion technique involving the use of liposomes containing nystatin and ergosterol. In principle, this method can be applied to any membrane preparation or channel protein and potentially offers a systematic approach for controlling and monitoring the fusion process.

III. Rat Muscle T-Tubule Membranes: A Reliable Source of K_{Ca}^+ Channels and Na_v^+ Channels

Following Miller and Racker's breakthrough (Miller and Racker, 1976) in 1976 of achieving reproducible incorporation of K^+-selective channels from sarcoplasmic reticulum vesicles into painted lipid bilayers, similar fusogenic membrane preparations were eagerly sought for biophysical studies of channels that mediate the signaling currents of electrically excitable cells. This goal was largely met in 1982 when Latorre *et al.* (1982) used a preparation of membrane vesicles derived from the transverse tubule system of rabbit skeletal muscle to incorporate and record unitary behavior of large conductance K_{Ca}^+ channels. Similarly, in 1983, Krueger *et al.* (1983) described a bilayer preparation for studying single Na_v^+ channels in the presence of batrachotoxin, a natural toxin that prevents Na_v^+ channel inactivation. These accomplishments led to the recognition that, given the right kind of preparation, the bilayer system could be applied to the *in vitro* study of native ion channels resident in all kinds of cellular membranes. Although numerous applications of planar bilayer reconstitution have been described, such preparations are known to vary considerably in their reproducibility and ease of handling. The key factors for the most successful applications seem to be the purity and homogeneity of the membrane preparation, the inherent "fusability" of the preparation as measured by channel incorporation frequency, and the experimenter's ability to identify and select for the channel of interest by controlling the ionic conditions and exploiting pharmacology.

As mentioned earlier, plasma membrane vesicles from rat skeletal membrane (Guo *et al.*, 1987; Moczydlowski and Latorre, 1983) work very well for the routine study of K_{Ca}^+ channels and batrachotoxin-activated Na_v^+ channels. The time-tested procedure for purifying such vesicles essentially involves differential centrifugation of homogenized skeletal muscle followed by separation of plasma membranes (probably a mixture of T tubules and surface sarcolemma) and SR membranes on a sucrose density gradient. The relatively rapid method given below is derived from the more sophisticated procedures and studies (Lau *et al.*, 1977; Rosemblatt *et al.*, 1981) of membrane biochemists, who developed methodology for the separation of muscle membranes by means of painstaking biochemical and electron microscopic analyses. The protocol for this preparation currently used in our laboratory is given next. The excruciating details are given for the convenience of new students, course instructors, or researchers who may just be entering the world of the planar bilayer reconstitution.

A. T-Tubule Membrane Preparation

Perform the following steps at 4 °C and keep all solutions ice-cold, except the phenylmethylsulfonyl fluoride (PMSF) solution. Prechill the solutions overnight, especially the large volume of initial homogenization buffer. For a large membrane preparation, use 8–10 adult male Sprague–Dawley rats, each weighing ~300 g, which will provide ~500 g of starting skeletal muscle tissue.

1. Euthanize each rat by exposure to CO_2 in a plastic desiccator chamber containing dry ice before decapitation and exsanguination. Lay the animal on its abdomen and start the dissection with a dorsal incision of the skin, from tail to neck along the spine. Continue with a pair of scissors and scalpel to remove the skin from the back and hind limbs. Dissect skeletal muscle from the rear legs, thighs, and back while removing as much fat and connective tissue as possible. Store the freshly dissected muscle in a beaker containing ice-cold homogenization buffer. Rinse and blot the tissue on paper towels, weigh, and divide it into batches of 100 g. Mince each batch of muscle by cutting it repeatedly with scissors. If an electric chopper is available, process the 100 g batches to ground/chopped meat.

2. In a standard, high-speed blender, add 100 g ground muscle to 300 ml of homogenization buffer. Just before turning on the blender, add 0.2 mM PMSF. Homogenize the tissue at the highest speed for 30 s. Stop the blender and repeat for another 30 s at high speed.

3. Centrifuge the homogenate in 250-ml bottles for 10 min at 5000 rpm in a Beckman JA14 rotor ($2500 \times g$). Pour the supernatant through several layers of cheesecloth and save it on ice.

4. Combine each pellet in the blender with an original volume of homogenization buffer and 0.2 mM PMSF. Rehomogenize at the highest speed twice for 30 s and centrifuge the homogenate as in step 3. Pour the supernatant through cheesecloth and combine it with the supernatant from step 3 in a large beaker. Measure the total volume of supernatant and add solid KCl to a final concentration of 0.6 M. Dissolve the KCl completely and solubilize contractile proteins by continuous stirring for 30 min at 4 °C. With a starting material of ~500 g muscle the supernatant volume may be close to 1.5 l at this stage.

5. Pellet the membranes by ultracentrifuging the preparation for 30 min at 40,000 rpm in a Beckman 45Ti or 50.2Ti rotor (~100,000 × g). After each centrifuge run, discard the clear supernatant and save the pellets. Considerable time can be saved by using one or two sets of tubes and adding homogenate on top of membrane pellets from the previous run.

6. After all the supernatant has been processed by ultracentrifugation, pool and resuspend the pellets in homogenization buffer, but do not exceed a final volume of 50 ml. Homogenize the membrane preparation thoroughly in a glass Dounce homogenizer and centrifuge it for 10 min at 8000 rpm in a Beckman JA20 rotor (5000 × g). Save the white supernatant, but exclude the brown mitochondrial pellet. This centrifugation step may be repeated to further reduce mitochondrial contamination.

7. Prepare six tubes containing 30 ml of 32% (w/v) sucrose buffer for centrifugation in a Beckman SW28 swinging bucket rotor. Carefully layer ~8 ml of the whitish supernatant from step 6 on top of the 32% sucrose layer, filling nearly to the top of the tube. Centrifuge the step gradients overnight (12–18 h) at 25,000 rpm (85,000 × g).

8. Collect and pool the whitish band at the 32% sucrose interface. Dilute this pool with one volume of dilution buffer and pellet the plasma membranes for 60 min in an ultracentrifuge at 100,000 × g. Discard the supernatant and resuspend the final pellet in a small volume (~2 ml) of resuspension buffer. Transfer the preparation into a small glass Dounce homogenizer and homogenize thoroughly. The final preparation should be homogeneously thick and white. Aliquot the membrane preparation into 50-μl portions in small microcentrifuge tubes and quick-freeze them by immersion in liquid nitrogen. When stored at −70 °C, this preparation remains active for incorporation of K_{Ca}^+ channels and Na_v^+ channels for as long as 1 year.

B. Solutions

Homogenization buffer: 300 mM sucrose, 20 mM MOPS–KOH, pH 7.4, 0.02% NaN$_3$.

32% sucrose gradient: 32% (w/v) sucrose in 20 mM MOPS–KOH, pH 7.4.
Dilution buffer: 20 mM MOPS–KOH, pH 7.4.
Resuspension buffer: 300 mM sucrose, 20 mM MOPS–KOH, pH 7.4.

PMSF solution: 200 mM PMSF in acetone treated with molecular sieves to remove water. It is best to prepare this stock solution just before use. It must be kept at room temperature to prevent precipitation.

Note: In addition to PMSF, other standard protease inhibitors may be added to the homogenization buffer. However, we have found that addition of the calcium chelator, ethylenediaminetetraacetic acid (EDTA), to the solutions dramatically reduces the incorporation activity of K_{Ca}^+ channels. The use of calcium chelators in the preparation should be avoided for studies of this channel.

IV. Preparations for Reconstituting Diverse Types of Channels from Native Tissues

Many different types of ion channels have been studied in planar bilayers with the degree of rigorous characterization that is necessary to ensure that the bilayer data reflect native function. Table I is a partial list of channel preparations that have been studied extensively. Typically, such projects start with a well-characterized membrane preparation from a favorable tissue source and progress toward the routine incorporation of a particular type of channel that can be readily identified. There are also many examples of traditional purification and reconstitution of channel proteins into liposomes followed by fusion with planar bilayers. The examples of Table I are a good source of methods that may be applied to development of new bilayer preparations.

Membrane reconstitution is indispensable for investigating and demonstrating the transport function of membrane proteins. However, there is always the hazard that native functional properties have been altered in the course of the reconstitution process. For single-channel assays, there is also the danger that current fluctuations might represent the activity of contaminants in the preparation rather than the protein of interest. Thus, in bilayer research it is imperative that the activity of reconstituted channels be thoroughly characterized with respect to established functional characteristics of the native preparation. For channels, this means that the properties of the channel in a planar bilayer must be carefully evaluated in comparison to those measured by cellular electrophysiology or biochemical assays of function. This responsibility may be difficult to carry out for channels such as Ca^{2+}-release channels that reside in intracellular membranes and cannot easily be recorded by patch clamping of the native membrane. Nevertheless, in well-documented systems, investigators have found ways of determining whether bilayer results genuinely reflect physiological behavior. One of the most powerful approaches involves the use of specific pharmacologic agents. For example, the unique blocking effect of charybdotoxin (Miller *et al.*, 1985) on maxi-K_{Ca}^+ channels assayed in planar bilayers or in cellular preparations (Wallner *et al.*, 1995) provides a good tool for cross-validation. In the case of Ca^{2+}-release channels, which are under intensive study in planar bilayers (Bezprozvanny and

Table I

Membrane Preparations for Incorporating Various Classes of Ion Channels into Planar Bilayers from Native Tissues

Channel type	Membrane preparations	References
Sarcoplasmic reticulum (SR) K^+ channel	SR vesicles from mammalian skeletal muscle	Miller and Racker (1976), Labarca *et al.* (1980)
ClC_0 Cl^- channel	Plasma membrane vesicles from noninnervated face of *Torpedo* electroplax	White and Miller (1979), Middleton *et al.* (1994)
K^+_{Ca} channels (BK or maxi-K^+_{Ca})	Plasma membrane vesicles from mammalian skeletal muscle, brain; smooth muscle from intestine, uterus, and aorta	Latorre et al. (1982), Reinhart *et al.* (1989), Cecchi et al. (1986), Toro *et al.* (1990, 1991), this work
Na^+_v channels (activated by batrachotoxin)	Plasma membrane vesicles from mammalian brain, skeletal muscle, and heart	Guo *et al.* (1987), Krueger *et al.* (1983), this work
Ca^{2+}_v channels (dihydropyridine-activated)	Plasma membrane vesicles from mammalian skeletal and cardiac muscle	Affolter and Coronado (1985), Rosenberg *et al.* (1986)
Ryanodine-receptor/ Ca^{2+}-release channel	Junctional SR from skeletal muscle, cardiac SR	Copello *et al.* (1997), Lai *et al.* (1988), Lindsay and Williams (1991), Ma *et al.* (1988)
$InsP_3$-gated Ca^{2+}-release channel	Endoplasmic reticulum vesicles from mammalian cerebellum	Bezprozvanny and Ehrlich (1994)
ATP-sensitive K^+ channels	Plasma membranes from mammalian skeletal muscle and vascular smooth muscle	Parent and Coronado (1989), Kovacs and Nelson (1991), Ottolia and Toro (1996)
Amiloride-sensitive epithelial Na^+ channel	A6 kidney cell line from *Xenopus laevis*, mammalian kidney membranes	Sariban-Sohraby *et al.* (1984), Ismailov *et al.* (1995)

Ehrlich, 1994; Copello *et al.*, 1997; Coronado *et al.*, 1992; Lai *et al.*, 1988; Lindsay and Williams, 1991), sensitivity to the plant alkaloid, ryanodine, or specific activation by inositol 1,4,5-trisphosphate ($InsP_3$) provides an unmistakable identification. Pharmacology is just one of the tools for the molecular fingerprinting of reconstituted channels. Other signature properties that must be cross-verified between cells and bilayers are unitary conductance, ionic selectivity, gating kinetics, and biochemical modulation.

V. Methods for Reconstituting Cloned and Heterologously Expressed Channels into Planar Bilayers

In view of the unique advantages of the planar bilayer recording system, one may consider its applications in connection with the heterologous expression of cloned channel proteins. This particular combination of techniques could be very useful for the functional analysis of channels that are naturally expressed at very low levels or in cases where native tissues do not provide an adequate source for protein purification. Also, the incorporation of purified mutant channels into planar bilayers offers the possibility of conducting structure–function investigations in a pristine *in vitro* environment devoid of endogenous cytoplasmic and membrane-associated constituents inherent to cellular expression systems such as *Xenopus* oocytes or eukaryotic cell lines. The particular combination of a controlled membrane environment and site-specific channel mutations may be especially applicable to questions related to channel regulation.

The purpose of this section is to show that such considerations have not escaped the attention of astute channelologists, since a number of examples can be cited where substantial progress has been made in this direction. A rather unique demonstration of the principle is given by the reconstitution of the mitochondrial voltage-dependent anion channel (VDAC) into planar lipid bilayers (Colombini *et al.*, 1992). The cloned yeast VDAC gene can be subjected to site-directed mutagenesis and such mutants can be expressed in a VDAC-deleted strain of yeast. The recombinant mutant VDAC protein can then be purified from mitochondria of transformed yeast and reconstituted into planar bilayers. This approach has been used to identify amino acid residues that affect the ionic selectivity of the VDAC channel (Colombini *et al.*, 1992). To our knowledge, the first application to involve a nonorganellar channel protein was the work of Rosenberg and East (1992), reporting planar bilayer incorporation of a recombinant *Shaker* K_v^+ channel that was originally cloned from *Drosophila*. Their unique experimental approach involved the use of a cell-free expression system. *Shaker* K_v^+ channel mRNA was translated *in vitro* using a rabbit reticulocyte lysate in the presence of avian microsomal membranes. In this method a source of rough endoplasmic reticulum vesicles is required for proper cotranslational synthesis of membrane proteins. Fusion of the resulting microsomal vesicles containing *in vitro* synthesized *Shaker* protein with planar bilayers resulted in the observation of K_v^+ channel activity with functional behavior expected for *Drosophila Shaker* channels (Rosenberg and East, 1992).

Clearly, the reconstitution of recombinant channel proteins requires the identification of a good expression system where functional channels are made in high yield and a good method for isolating the channels and efficiently inserting them into bilayers. Table II summarizes various examples of progress toward this goal as gleaned from the recent literature. Aside from the *in vitro* translation method (Rosenberg and East, 1992), these examples make use of several different systems for heterologous cellular expression of cloned channels and use two different approaches for membrane reconstitution that mirror those discussed earlier for native channels. One method is to directly isolate microsomal vesicles containing recombinant channels on a sucrose gradient after homogenization of transfected cells. The second method involves solubilization and purification of the recombinant channel protein followed by reconstitution into liposomes in a functionally active form. In the remainder of this chapter, we discuss relevant details of these

Table II
Recombinant Ion Channels Reconstituted into Planar Lipid Bilayers

Recombinant channel	Heterologous expression system	Sucrose gradient fractionation of membrane vesicles	Protein purification and incorporation into liposomes	References
Shaker K_v^+ channel	*In vitro* translation	−	−	Rosenberg and East (1992)
	COS cells	+	−	Sun et al. (1994)
CFTR[a] Cl^- channel	Sf9 cells	−	+	Bear et al. (1992)
	HEK293 cells	+	−	Xie et al. (1995, 1996)
maxi-K_{Ca}^+ channel	*Xenopus* oocytes	+	−	Pérez et al. (1994)
	COS cells	+	−	Müller et al. (1996)
	HEK293 cells	+	−	Moss et al. (1996)
K_{IR}^+ channel[b] (IRK1)	*Xenopus* oocytes	+	−	Aleksandrov et al. (1996)
ClC_0 *Torpedo* Cl^- channel	HEK293 cells	−	+	Middleton et al. (1996)
Ryanodine receptor/Ca^{2+} -release channel	Sf9 cells	−	+	Ondrias et al. (1996)

[a]Cystic fibrosis transmembrane regulator.
[b]Inwardly rectifying K^+ channel.

procedures that may ultimately provide a basis for the systematic study of many different types of recombinant channels in planar bilayers.

The widespread popularity and success of *Xenopus laevis* oocytes as an expression system for channels, transporters, and receptors (Shih *et al.*, 1998) has led to the consideration of the oocyte membrane as a source of membrane vesicles for planar bilayer incorporation. Thus far two types of cloned channels have been successfully reconstituted after expression in *Xenopus* oocytes: the *Drosophila* K_{Ca}^+ channel, Dslo (Pérez *et al.*, 1994) and the murine inwardly rectifying K^+ channel, IRK1 (Aleksandrov *et al.*, 1996). In the latter example, Aleksandrov *et al.* (1996) employed preexisting methods (Bretzel *et al.*, 1986; Kinsey *et al.*, 1980) for the oocyte membrane preparation, whereas Pérez *et al.* (1994) used the following simple procedure. In brief, oocytes expressing recombinant channels were manually disrupted in a high-K^+ potassium buffer (400 mM KCl, 5 mM PIPES, pH 6.8) in the presence of a mixture of proteinase inhibitors. The homogenate was then layered onto a discontinuous sucrose gradient (0.75 ml 20% sucrose and 0.75 ml 50% sucrose in the same high-K^+ buffer) and centrifuged at $30,000 \times g$ for 30 min to allow the separation of the membrane fraction from the yolk. A membrane band at the 20–50% sucrose interface was collected, diluted threefold with the high-K^+ buffer and membrane vesicles were recovered in a final centrifugation step. Clean separation of a membrane fraction from fatty acid contamination appeared to be critical for stability of the planar lipid bilayer after fusion with oocyte microsomal vesicles (Pérez *et al.*, 1994).

A major drawback with the use of *Xenopus* oocyte membranes is the abundant presence of endogenous Ca^{2+}-activated Cl^- channels (Barish, 1983; Takahashi *et al.*, 1987). This latter background conductance is well known to electrophysiologists who use frog oocytes for two-electrode voltage clamping. It has been overcome in two ways, either by intracellular injection of Ca^{2+} chelators or by the elimination of Cl^- from the bath solution whenever possible (Ellinor *et al.*, 1995; Schlief *et al.*, 1996; Stühmer and Parekh, 1995). Substitution of Cl^- salts by salts of the impermeant anion, methansulfonate, was used to mask insertion of the contaminating Cl^- conductance when Dslo K_{Ca}^+ channels (Pérez *et al.*, 1994) or IRK1 K^+ channels (Aleksandrov *et al.*, 1996) were studied. However, this endogenous conductance does handicap the oocyte expression system and may render it inappropriate for certain applications, for example, cloned Cl^- channels. Other factors, such as the time-consuming injection and handling of oocytes and the small yield of microsomal vesicles, may also mitigate against the routine use of oocytes for planar bilayer work.

As an alternative approach, a number of researchers have expressed various cloned ion channels in mammalian cell lines such as COS cells and HEK293 cells as a starting source for planar bilayer reconstitution. Patches are easily excised from the plasma membrane of mammalian cells and, in contrast to oocytes, these latter cell lines do not have a high background of endogenous currents that complicate electrophysiological measurements. The first example of this approach was functional reconstitution of an inactivation-deficient mutant of the *Shaker* K_v^+ channel transiently expressed in COS cells (Sun *et al.*, 1994). This study was followed by reconstitution of several other kinds of cloned channels: human cystic fibrosis transmembrane conductance regulator (CFTR) (Xie *et al.*, 1995, 1996), maxi-K_{Ca}^+ channels from mouse (Müller *et al.*, 1996), and *Drosophila* (Moss *et al.*, 1996), and more recently, in our laboratory, the $\mu 1$ Na_v^+ channel from rat skeletal muscle (see Fig. 1). The basic experimental procedure used by Sun *et al.* (1994) to recover membranes from COS cells transiently transfected with the *Shaker* K_v^+ channel first involved harvesting the cells in an alkaline buffer containing 150 mM KCl, 2 mM $MgCl_2$, 5 mM ethyleneglycol-bis(β-aminoethylether)-N,N,N',N'-tetra-acetic acid (EGTA), adjusted to pH 10.6 with NH_4OH. The harvested cells are then lysed by forcing the cell suspension several times through a small-gauge syringe needle. This homogenate is next layered on a prechilled sucrose step gradient (10-ml sample, 14 ml 20% (w/v) sucrose, 14 ml 38% (w/v) sucrose in 20 mM MOPS–KOH, pH 7.1) and centrifuged in a Beckman SW28.1 rotor for 45 min at 25,000 rpm. The turbid band at the 20/38% sucrose interface was collected, diluted fourfold with cold water, and pelleted. The final membrane pellet fraction was resuspended in a buffer containing 250 mM sucrose, 10 mM HEPES–KOH, pH 7.3, aliquoted, quickly frozen, and stored at $-70\,°C$. Similar methods have also been successfully used to recover microsomal vesicles from HEK293 cells expressing K_{Ca}^+ channels (Moss *et al.*, 1996; Müller *et al.*, 1996) and rat skeletal muscle Na_v^+ channels (Y. Sun and E. Moczydlowski, unpublished results, 1998; Fig. 1).

In another application involving the reconstitution of CFTR mutants, a more detailed approach to the membrane fractionation of transiently transfected HEK293 cells was undertaken by Xie *et al.* (1995, 1996). The question of membrane origin is especially relevant in the case of CFTR, since the biosynthetic processing of some CFTR mutants is defective. Such mutations result in a failure of newly synthesized CFTR to be properly transported and expressed in the surface plasma membrane leading to an accumulation of the protein in intracellular membranes. By using discontinuous sucrose gradient fractionation, Western blot analysis, and immunoprecipitation, Xie *et al.* (1995, 1996) showed that wild-type CFTR is found in plasma membrane vesicles that mainly band at interfaces of 28/33% sucrose and 33/36% sucrose, whereas mutant CFTR protein is found in intracellular membrane vesicles that band at 36/38.7% and 38.7/43.7% sucrose interfaces. This study demonstrated the feasibility of separating plasma membranes from intracellular membranes and should be useful to researchers attempting to express and reconstitute channel proteins that are located in either of these membrane fractions. In summary, the use of transient or stable transfection in COS or HEK293 cells combined with the isolation of semipurified membrane fractions (Moss *et al.*, 1996; Müller *et al.*, 1996; Sun *et al.*, 1994; Xie *et al.*, 1995, 1996) offers an attractive approach for expression, membrane vesicle isolation, and incorporation of cloned and mutant channels into planar lipid bilayers.

While the preceding methods may satisfy ordinary demands of most applications, a true bilayer reductionist insists on the ultimate degree of compositional control. This purist philosophy requires that the cloned channel be heterologously

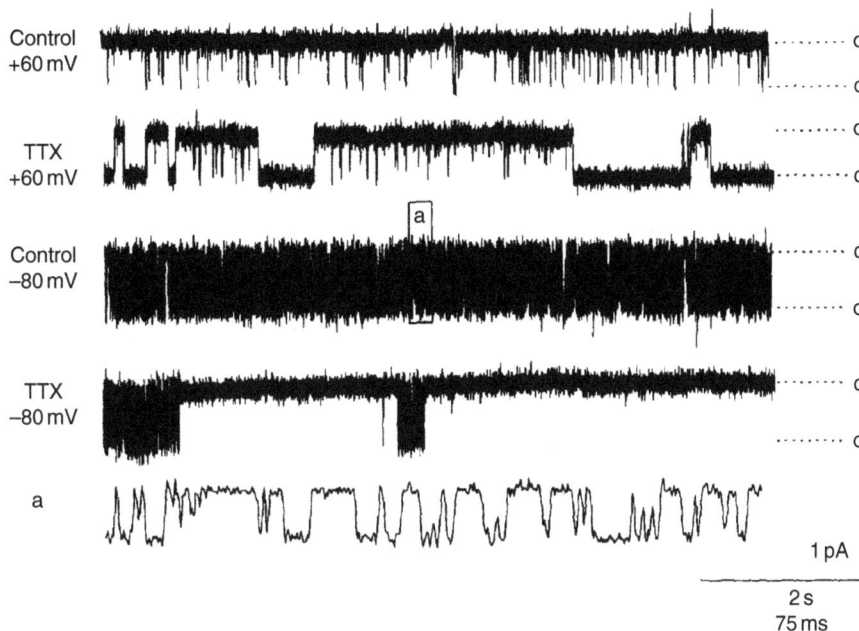

Fig. 1   Planar bilayer recording of a recombinant $\mu1$ Na$_v^+$ channel. cDNA coding for the rat skeletal muscle $\mu1$ Na$_v^+$ channel was subcloned into the mammalian cell expression vector pcDNA3 (Invitrogen, Carlsbad, CA). Human fibroblast HEK293 cells were transfected with the $\mu1$/pcDNA3 vector using the calcium phosphate precipitation method (Chen and Okayama, 1987). A cell line of stably transfected cells expressing the $\mu1$ Na$_v^+$ channel ($\sim$5-nA typical peak whole-cell Na$^+$ current) was selected by growth in the presence of the antibiotic G418 (700 μg/ml). Ten flasks of cells (50% confluent, 150 cm^2) were used to prepare membrane vesicles by modifying the procedure of Sun *et al.* (1994) previously described for COS cells. The major modifications of this latter method were that harvested cells were fragmented using a glass Dounce homogenizer followed by mild sonication and that KCl in the solutions was substituted by NaCl. Single Na$_v^+$ channels from this preparation were incorporated into planar bilayers in the presence of symmetrical 200 mM NaCl, 10 mM MOPS–NaOH, pH 7.4, and 0.5 μM batracho-toxin using methods similar to those described previously (Guo *et al.*, 1987). The figure shows selected traces from a long ($\sim$1-h) recording of a single recombinant Na$_v^+$ channel (filter frequency 100 Hz). The first and third traces from the top, labeled "control," were respectively taken at holding voltages of +60 and −80 mV (physiological convention, extracellular ground) before the addition of tetrodotoxin (TTX). The rapid flickering at −80 mV corresponds to voltage-dependent closing of the channel associated with voltage-activation (c, closed level; o, open level). The second and fourth traces from the top were taken after addition of 20 nM TTX to the external side. Long-lived closed states in these traces represent individual TTX-blocking events (French *et al.*, 1984; Guo *et al.*, 1987; Krueger *et al.*, 1983) characteristic of batrachotoxin-modified Na$_v^+$ channels. The bottom trace corresponds to a boxed segment of the third trace labeled "a" and is shown at an expanded time scale.

expressed at high levels, purified to biochemical homogeneity, incorporated into liposomes, and then fused into the bilayer. Examples of this connoisseur approach may be cited for CFTR (Bear *et al.*, 1992), the ryanodine receptor Ca^{2+}-release channel (Ondrias *et al.*, 1996), and the ClCl$_0$ Cl$^-$ channel (Middleton *et al.*, 1996).

The last example (Middleton *et al.*, 1996) is especially instructive because its methods seem to be broadly applicable. In this case, the recombinant $ClCl_0$ Cl^- channel was first expressed by large-scale transient transfection in HEK293 cells (Middleton *et al.*, 1996). Membrane protein from these cells was solubilized with CHAPS detergent and the $ClCl_0$ Cl^- channel protein was purified by immunoaffinity column chromatography. The detergent was removed by a gel filtration procedure resulting in the formation of reconstituted liposomes, which could then be fused to planar bilayers using the nystatin-mediated fusion technique (Woodbury, 1998; Woodbury and Miller, 1990). The relatively straightforward procedures used in this example were developed as an extension of work on the purification and reconstitution of the native voltage-gated Cl^- channel from *Torpedo* electric organ (Goldberg and Miller, 1991; Middleton *et al.*, 1994) and later applied to cloned wild-type and mutant $ClCl_0$ channel expressed in HEK293 cells (Middleton *et al.*, 1996). A cautiously optimistic outlook envisions that similar strategies may lead the way to the functional reconstitution of other purified channel proteins and also contribute toward two- and three-dimensional crystallization and protein structural determination.

Acknowledgments

This work was supported by grants to E. M. from the National Institutes of Health (GM51172) and the American Heart Association (95008820). Isabelle Favre was supported by a James Hudson Brown–Alexander Brown Coxe postdoctoral fellowship and an award from the CIBA-GEIGY-Jubiläums-Stiftung.

References

Affolter, H., and Coronado, R. (1985). *Biophys. J.* **48**, 341.
Aleksandrov, A., Velimirovic, B., and Clapham, D. E. (1996). *Biophys. J.* **70**, 2680.
Andersen, O. (1983). *Biophys. J* **41**, 119.
Andersen, O. S., Nielsen, C., Maer, A. M., Lundbaek, J. A., Goulian, M., and Koeppe, R. E. (1998). *Methods Enzymol* 294 Chapter 10 (this volume).
Barish, M. E. (1983). *J. Physiol. (Lond.)* **342**, 309.
Bayley, H., Cronin, B., Heron, A., Holden, M. A., Hwang, W., Syeda, R., Thompson, J., and Wallace, M. (2008). Droplet interface bilayers. *Mol. Biosyst.* **4**, 1191–1208.
Bear, C. E., Li, C. H., Kartner, N., Bridges, R. J., Jensen, T. J., Ramjeesingh, M., and Riordan, J. R. (1992). *Cell* **68**, 809.
Bezprozvanny, I., and Ehrlich, B. E. (1994). *J. Gen. Physiol.* **104**, 821.
Blunck, R., McGuire, H., Hyde, H. C., and Bezanilla, F. (2008). Fluorescence detection of the movement of single KcsA subunits reveals cooperativity. *Proc. Natl. Acad. Sci. USA* **105**, 20263–20268.
Bretzel, G. J., Janeczek, J., Born, M., Tiedemann, J. H., and Tiedemann, H. (1986). *Roux's Arch. Dev. Biol.* **195**, 117.
Cecchi, X., Alvarez, O., and Wolff, D. (1986). *J. Membr. Biol.* **91**, 11.
Chen, T. Y., and Miller, C. (1996). Nonequilibrium gating and voltage dependence of the ClC-0 Cl^- channel. *J. Gen. Physiol.* **108**, 237–250.
Chen, C., and Okayama, H. (1987). *Mol. Cell. Biol.* **7**, 2745.

Chernomordik, L., Chanturiya, A., Green, F., and Zimmerberg, J. (1995). *Biophys. J.* **69**, 922.

Colombini, M., Peng, S., Blachly-Dyson, E., and Forte, M. (1992). *Methods Enzymol.* **207**, 432.

Copello, J. A., Berg, S., Onoue, H., and Fleischer, S. (1997). *Biophys. J.* **73**, 141.

Coronado, R., and Latorre, R. (1983). *Biophys. J.* **43**, 231.

Coronado, R., Kawano, S., Lee, C. J., Valdivia, C., and Valdivia, H. H. (1992). *Methods Enzymol.* **207**, 699.

Ehrlich, B. E. (1992). *Methods Enzymol.* **207**, 463.

Eisenman, G., Latorre, R., and Miller, C. (1986). *Biophys. J.* **50**, 1025.

Ellinor, P. T., Yang, J., Sather, W. A., Zhang, J. P., and Tsien, R. W. (1995). *Neuron* **15**, 1121.

French, R. J., Worley, J. F., III, and Krueger, B. K. (1984). *Biophys. J.* **45**, 301.

Goldberg, A. F., and Miller, C. (1991). *J. Membr. Biol.* **124**, 199.

Greathouse, D. V., Koeppe, R. E., Providence, L. L., Shobana, S., and Andersen, O. S. (1998). *Methods Enzymol.* 294 Chapter 28 (this volume).

Guo, X., Uehara, A., Ravindran, A., Bryant, S. H., Hall, S. H., and Moczydlowski, E. (1987). *Biochemistry* **26**, 7546.

Hanke, W., and Breer, H. (1987). *J. Gen. Physiol.* **90**, 855.

Hanke, W., and Schlue, W. R. (1993). Planar Lipid Bilayers: Methods and Applications Academic Press, New York.

Hanke, W., Methfessel, C., Wilmsen, U., and Boheim, G. (1984). *Biochem. Bioeng. J.* **12**, 329.

Heginbotham, L., LeMasurier, M., Kolmakova-Partensky, L., and Miller, C. (1999). Single *Streptomyces lividans* K$^+$ channels: Functional asymmetries and sidedness of proton activation. *J. Gen. Physiol.* **114**, 551–559.

Ismailov, I. I., Berdiev, B. K., and Benos, D. J. (1995). *J. Gen. Physiol.* **106**, 445.

Kinsey, W. H., Decker, G. L., and Lennarz, W. J. (1980). *J. Cell Biol.* **87**, 248.

Kovacs, R. J., and Nelson, M. T. (1991). *Am. J. Physiol.* **261**, H604.

Krueger, B. K., Worley, J. F., III, and French, R. J. (1983). *Nature* **303**, 172.

Labarca, P., and Latorre, R. (1992). *Methods Enzymol.* **207**, 447.

Labarca, P., Coronado, R., and Miller, C. (1980). *J. Gen. Physiol.* **76**, 397.

Lai, F. A., Erikson, H. P., Rousseau, E., Liu, Q. Y., and Meissner, G. (1988). *Nature* **331**, 315.

Latorre, R., Vergara, C., and Hidalgo, C. (1982). *Proc. Natl. Acad. Sci. USA* **79**, 805.

Lau, Y. H., Caswell, A. H., and Brunschwig, J. P. (1977). *J. Biol. Chem.* **252**, 5565.

LeMasurier, M., Heginbotham, L., and Miller, C. (2001). KcsA: It's a potassium channel. *J. Gen. Physiol.* **118**, 303–313.

Lindsay, A. R. G., and Williams, A. J. (1991). *Biochim. Biophys. Acta* **1064**, 89.

Ma, J., Fill, M., Knudson, M., Campbell, K. P., and Coronado, R. (1988). *Science* **242**, 99.

Middleton, R. E., Pheasant, D. J., and Miller, C. (1994). *Biochemistry* **33**, 13189.

Middleton, R. E., Pheasant, D. J., and Miller, C. (1996). *Nature* **383**, 337.

Miller, C. (ed.), (1986). "Ion Channel Reconstitution," p. 577. Plenum Press, New York.

Miller, C., and Racker, E. (1976). *J. Membr. Biol.* **30**, 283.

Miller, C., Moczydlowski, E., Latorre, R., and Phillips, M. (1985). *Nature* **313**, 316.

Moczydlowski, E. G., and Latorre, R. (1983). *Biochim. Biophys. Acta* **732**, 412.

Montal, M., and Mueller, P. (1972). *Proc. Natl. Acad. Sci. USA* **69**, 3561.

Montal, M., Anholt, R., and Labarca, P. (1986). *In* "Ion Channel Reconstitution," (C. Miller, ed.), p. 157. Plenum Press, New York.

Moss, G. W., Marshall, J., Morabito, M., Howe, J. R., and Moczydlowski, E. (1996). *Biochemistry* **35**, 16024.

Mueller, P., Rudin, D., Tien, H. T., and Wescott, W. C. (1962). *Circulation* **26**, 1167.

Müller, M., Madan, D., and Levitan, I. B. (1996). *Neuropharmacology* **35**, 877.

Ondrias, K., Brillantes, A. M., Scott, A., Ehrlich, B. E., and Marks, A. R. (1996). *Soc. Gen. Physiol. Ser.* **51**, 29.

Ottolia, M., and Toro, L. (1996). *J. Membr. Biol.* **153**, 203.

Pantoja, R., Sigg, D., Blunck, R., Bezanilla, F., and Heath, J. R. (2001). Bilayer reconstitution of voltage-dependent ion channels using a microfabricated silicon chip. *Biophys. J.* **81**, 2389–2394.

Parent, L., and Coronado, R. J. (1989). *J. Gen. Physiol.* **94**, 445.

Pérez, G., Lagrutta, A., Adelman, J. P., and Toro, L. (1994). *Biophys. J.* **66,** 1022.

Reinhart, P. H., Chung, S., and Levitan, I. B. (1989). *Neuron* **2,** 1031.

Rosemblatt, M., Hidalgo, C., Vergara, C., and Ikemoto, N. (1981). *J. Biol. Chem.* **256,** 8140.

Rosenberg, R. L., and East, J. E. (1992). *Nature* **360,** 166.

Rosenberg, R. L., Hess, P., Reeves, J. P., Smilowitz, H., and Tsien, R. W. (1986). *Science* **231,** 1564.

Rotem, D., Mason, A., and Bayley, H. (2009). Inactivation of the KcsA potassium channel explored with heterotetramers. *J. Gen. Physiol.* **135,** 29–42.

Sariban-Sohraby, S., Latorre, R., Burg, M., Olans, L., and Benos, D. (1984). *Nature* **308,** 80.

Schlief, T., Schonerr, R., Imoto, K., and Heinemann, S. H. (1996). *Eur. Biophys. J.* **25,** 75.

Schuerholz, T., and Schindler, H. (1983). *FEBS Lett.* **152,** 187.

Sherman-Gold, R. (ed.) (2008). "The Axon Guide: A Guide for Electrophysiology and Biophysics Laboratory Techniques," 3rd edn. MDS Analytical Technologies, Sunnyvale, CA.

Shih, T., Smith, R., Toro, L., and Goldin, A. (1998). *Methods Enzymol.* **293,** 529.

Sigworth, F. J. (1983). *In* "Single-Channel Recording," (B. Sakmann, and E. Neher, eds.), p. 3. Plenum Press, New York.

Sitsapesan, R., Montgomery, R. A. P., and Williams, A. J. (1995). *Pflügers Arch. Eur. J. Physiol.* **430,** 584.

Stühmer, W., and Parekh, A. B. (1995). *In* "Single-Channel Recording," 2nd edn. (B. Sakmann, and E. Neher, eds.), p. 341. Plenum Press, New York.

Suarez-Isla, B. A., Wan, K., Lindstrom, J., and Montal, M. (1983). *Biochemistry* **22,** 2319.

Sun, T., Naini, A. A., and Miller, C. (1994). *Biochemistry* **33,** 9992.

Takahashi, T., Neher, E., and Sakmann, B. (1987). *Proc. Natl. Acad. Sci. USA* **84,** 5063.

Toro, L., Ramos-Franco, J., and Stefani, E. (1990). *J. Gen. Physiol.* **96,** 373.

Toro, L., Vaca, L., and Stefani, E. (1991). *Am. J. Physiol.* **260,** H1779.

Wallner, M., Meera, P., Ottolia, M., Kaczorowski, G. J., Latorre, R., Garcia, M. L., Stefani, E., and Toro, L. (1995). *Recept. Channels* **3,** 185.

White, S. H. (1986). *In* "Ion Channel Reconstitution," (C. Miller, ed.), p. 3. Plenum Press, New York.

White, M. M., and Miller, C. (1979). *J. Biol. Chem.* **254,** 10161.

Williams, A. J. (1995). *In* "Ion Channels: A Practical Approach," (R. H. Ashley, ed.), p. 43. IRL Press, Oxford.

Wonderlin, W. F., Finkel, A., and French, R. J. (1990). Optimizing planar lipid bilayer single-channel recordings for high resolution with rapid voltage steps. *Biophys. J.* **58,** 289.

Woodbury, D. J. (1998). *Methods Enzymol.* **294,** Chapter 17, 319.

Woodbury, D. J., and Miller, C. (1990). *Biophys. J.* **58,** 833.

Xie, J., Drumm, M. L., Ma, J., and Davis, P. B. (1995). *J. Biol. Chem.* **270,** 28084.

Xie, J., Drumm, M. L., Zhao, J., Ma, J., and Davis, P. B. (1996). *Biophys. J.* **71,** 3148.

Zadek, B., and Nimigean, C. M. (2006). Calcium-dependent gating of MthK, a prokaryotic potassium channel. *J. Gen. Physiol.* **127,** 673–685.

PART VIII

Second Messengers and Biochemical Approaches

CHAPTER 18

Protein Phosphorylation of Ligand–Gated Ion Channels

Andrew L. Mammen, Sunjeev Kamboj, and Richard L. Huganir

HHMI, Department of Neuroscience
John Hopkins University School of Medicine
Baltimore, Maryland, USA

DOI: 10.1016/B978-0-12-382204-8.00018-7

I. Introduction

Neurotransmitter receptors mediate signal transduction at the postsynaptic membrane of chemical synapses in the nervous system. The major excitatory and inhibitory neurotransmitter receptors in the brain are ligand-gated ion channels. These receptors directly bind neurotransmitters, resulting in the opening of an intrinsic ion channel. The predominant ligand-gated ion channels in the nervous system are the nicotinic acetylcholine receptor, the glutamate receptors, the γ-aminobutyric acid ($GABA_A$) receptors and the glycine receptors. Biochemical and electrophysiologic studies of these receptors have demonstrated that they are multiply phosphorylated by a variety of protein kinases (Roche *et al.*, 1994). Phosphorylation of these receptors regulates many functional properties, including desensitization, open channel probability, open time, and subcellular targeting (Roche *et al.*, 1994). Because of the central role of ligand-gated ion channels in synaptic transmission, protein phosphorylation of these receptors is a major mechanism in the regulation of synaptic transmission and may underlie many forms of synaptic plasticity (Nicoll and Malenka, 1995; Raymond *et al.*, 1993b; Roche *et al.*, 1994). In this chapter, we review a variety of techniques to examine the role of protein phosphorylation in the regulation of ligand-gated ion channel function. We review general strategies and methods for characterizing the phosphorylation state of ligand-gated ion channels, identifying phosphorylation sites on these channels, and analyzing the physiologic consequences of channel phosphorylation. To facilitate this discussion, we use the glutamate receptor subunit GluR1 as an example throughout this review.

II. Biochemical Characterization of Phosphorylation of Ligand-Gated Ion Channels

Ideally, the phosphorylation state of a ligand-gated ion channel should be investigated in the tissue(s) where it is endogenously expressed. For example, the phosphorylation state of glutamate receptors is often analyzed in brain slice preparations or in primary cultures of central neurons (Blackstone *et al.*, 1994). To identify and characterize specific phosphorylation sites at the biochemical and functional level, however, it is often necessary to study phosphorylation of wild-type and mutant recombinant channels. Because most primary cultures are difficult to transfect and, more importantly, have a background of wild-type channels, wild-type and mutant channels are typically expressed separately in heterologous systems such as the HEK293 or COS cell lines or in *Xenopus* oocytes. These heterologous expression systems simplify the study of channel phosphorylation by allowing biochemical and electrophysiologic comparisons between wild-type and mutant channels. However, to confirm the physiologic relevance of channel phosphorylation and its regulation in heterologous systems, it is always important to compare these findings with results obtained in cells where the channels are natively expressed.

The first step in characterizing the phosphorylation state of a given channel is to determine whether the protein of interest is indeed a substrate for protein kinases. To accomplish this, primary cultures or transfected cells expressing the channel are incubated with ortho[^{32}P]phosphate. The cells will incorporate the labeled phosphate into adenosine triphosphate (ATP) at the γ-phosphate position where it can be transferred to proteins by protein kinases. Care must be taken to allow enough time for the ortho[^{32}P]phosphate to reach equilibrium with the intracellular pools of ATP and the protein kinase substrates. For most cell types, this takes around 4 h and can be monitored by examining the ^{32}P incorporation into the substrate of interest. Following incubation of the cells with ortho[^{32}P]phosphate, the cells can be treated with activators of specific protein kinases for short periods of time or left untreated to examine the basal phosphorylation state of the substrate protein. After the cells are isolated, the ion channels are solubilized from the membrane with detergents and immunoprecipitated from the labeled cell extracts. The isolated channel is then analyzed by sodium dodecyl sulfate–polyacrylamide gel electrophoresis (SDS–PAGE), and the dried gels exposed to film to detect ^{32}P incorporation into the channel.

Below is a protocol for the immunoprecipitation of ^{32}P-labeled glutamate receptors from transiently transfected HEK293 cells. The same protocol is used for labeling and immunoprecipitating glutamate receptors from primary cultures by eliminating the transfection step. This protocol may also be easily modified for the labeling and immunoprecipitation of other ligand-gated ion channels from a variety of cell types. Because of the relatively large amounts of radioactivity used in these experiments, special care must be taken to avoid ^{32}P contamination and exposure. We recommend practicing the immunoprecipitations with unlabeled cultures prior to working with the labeled material.

A. Labeling and Cell Extract Preparation

1. Reagents

Immunoprecipitation buffer (IPB): 10 mM sodium phosphate (pH 7.0), 100 mM NaCl, 10 mM sodium pyrophosphate, 50 mM NaF, 1 mM sodium orthovanadate, 5 mM ethylenediaminetetraacetic acid (EDTA), 5 mM EGTA, 1 μM okadaic acid, 1 μM microcystin-LR, and appropriate protease inhibitors

Day 1

1. Split HEK293 cells onto 100-mm tissue culture plates.

Day 2

1. Transfect HEK293 cells with the appropriate cDNA. We typically use the calcium phosphate precipitation method to transfect HEK293 cells with 20 μg cDNA/plate (Blackstone et al., 1992).

Day 4

1. Remove the media from the plates and wash the cells twice with 5 ml phosphate-free MEM (Sigma St Louis, MO).

2. Add 2.5 ml phosphate-free MEM to each plate.

3. Prelabel each plate with 2 mCi/ml of ortho[^{32}P]phosphate. We order ortho [^{32}P]phosphate as a 50-mCi/ml solution from NEN Dupont and add 100 μl to each plate. Return the plates to the incubator for 4 h to allow the label to reach steady-state levels.

4. Activate endogenous protein kinases by treating cells with phorbol esters, forskolin, calcium ionophores, etc. Leave one plate untreated to examine basal phosphorylation.

5. Remove "hot" media and rinse each plate twice with 3–4 ml room temperature phosphate-buffered saline (PBS). Carefully dispose of the liquid radioactive waste from these washes.

6. Add 150 μl of room temperature IPB with 1% SDS to each plate and scrape with a cell scraper. Leave the extracts in the plates at this stage.

7. Dilute extracts by adding 750 μl of ice-cold IPB with 1% Triton X-100 to each plate and mix with a cell scraper.

8. Transfer the labeled extract to a 15-ml conical tube on ice.

9. Carefully sonicate each sample with a probe sonicator at setting 6 for 20 s, making sure not to contaminate area with ^{32}P, and return to ice. Sonication breaks up the DNA, making the solution less viscous and easier to work with. Samples can be frozen at this stage, if desired.

B. Immunoprecipitation of Protein from Labeled Extracts

1. Transfer labeled extract to a 1.5-ml screwtop Eppendorf tube containing 200 μl of a 1:1 slurry of protein A-Sepharose beads (Sigma) suspended in IPB with 1% bovine serum albumin (BSA), 50 μl preimmune serum, 15 units DNase, and 150 μg RNase. The addition of DNase and RNase minimizes the contamination of the sample with ^{32}P-labeled DNA or RNA. Rotate tubes for 1 h at 4 °C. This step preabsorbs any protein that may stick nonspecifically to the protein A-Sepharose beads.

2. Centrifuge at 2000 rpm for 1 min and add supernatant to a 1.5-ml screwtop Eppendorf tube containing 200 μl of the 1:1 slurry of protein A-Sepharose beads and the precipitating antibody. Rotate tubes for 2 h at 4 °C.

3. Wash the beads with 1 ml each of IPB with 1% Triton X-100 ($\times$2), IPB with 1% Triton X-100 and 500 mM additional NaCl ($\times$3), and IPB ($\times$2). Carefully dispose of these radioactive washes.

4. Elute immunoprecipitated material from beads with 150 μl SDS–PAGE sample buffer. Analyze samples by SDS–PAGE and stain the gel with Coomassie blue. Then, dry the gel between two sheets of cellophane (Bio-Rad, Richmond, CA) and expose to film.

In the experiment shown in Fig. 1, we transfected HEK293 cells with wild-type GluR1, a mutant form of GluR1 in which the serine at residue 845 has been converted to an alanine (GluR1 S845A), and used mock transfected cells as controls. These cells were labeled with ortho[^{32}P]phosphate, treated with phorbol ester (a PKC activator), forskolin (which indirectly stimulates PKA activity), and IBMX, and immunoprecipitate with anti-GluR1 antibodies as described above. Figure 1A shows that both the wild-type and mutant receptors are phosphorylated under these conditions. Although we have previously shown that serine-845 is a substrate for phosphorylation PKA (Roche *et al.*, 1996) (and see below), in this experiment, the mutant receptor appears to be phosphorylated even more robustly than the wild type. This is because GluR1 contains phosphorylation sites for other protein kinases (including PKC; Roche *et al.*, 1996) and the mutant receptor was expressed at higher levels than the wild-type receptor in this experiment. This result exemplifies that careful controls to examine the expression level of your protein of interest need to be performed in order to quantitate phosphorylation. No labeled receptor was immunoprecipitated from mock transfected cells.

C. Phosphopeptide Mapping of Ligand–Gated Ion Channels

The next step in analyzing the phosphorylation of a ligand-gated ion channel involves subjecting the ^{32}P-labeled phosphoprotein to two-dimensional phospho-peptide map and phosphoamino acid analysis. In the first procedure, the labeled material is excised from the gel and digested with protease. The resulting peptides

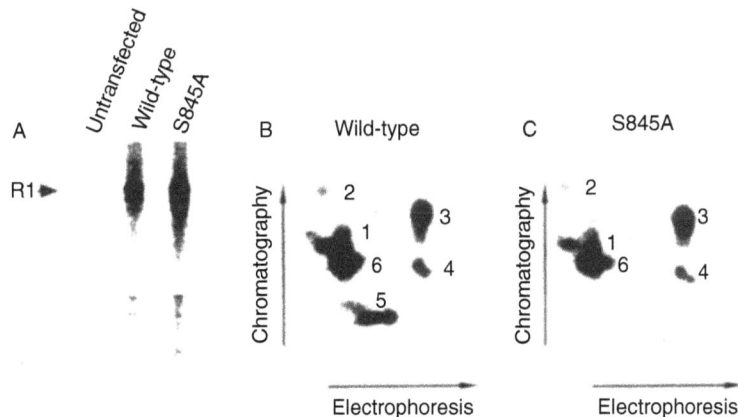

Fig. 1 *Immunoprecipitation and mapping of [^{32}P]-labeled channels from transfected cells.* QT6 cells were transfected with wild-type GluR1, mutant GluR1 (S845A), or mock transfected. These plates were incubated with ortho[^{32}P]phosphate for 4 h, treated with 100 nM phorbol 12-myristate 13-acetate (PMA), 10 μM forskolin (FSK), and 100 μM IBMX (IBMX) for 15 min. GluR1 was immunoprecipi-tated from detergent extracts of each plate (A). Two-dimensional phosphopeptide maps were generated from immunoprecipitated GluR1 (B) and GluR1 S845A (C).

are spotted onto a cellulose thin-layer chromatography (TLC) plate and subjected to electrophoresis. This separates the proteolytic fragments according to charge. Next, the plate is subjected to ascending chromatography where the peptides are separated according to their solubility in the chromatography buffer. The TLC plates are then exposed to film in order to visualize the phosphopeptides. Each ligand-gated ion channel subunit will yield a unique map depending on the location of its proteolytic cleavage sites and phosphorylated residues.

Below is a protocol for the two-dimensional phosphopeptide map analysis of glutamate receptors. To achieve optimal separation of phosphopeptides generated from other ligand-gated ion channels, it may be necessary to employ different proteases, electrophoresis times, and ascending chromatography buffers. These issues—and phosphopeptide mapping in general—are discussed thoroughly elsewhere (Boyle *et al.*, 1991).

D. Proteolytic Digestion

1. Reagents

> Destain solution: 25% methanol/10% acetic acid in H_2O
> 1× Trypsin solution: 0.3 mg/ml trypsin TPCK (Sigma) dissolved in 50 mM NH_4HCO_3
> 10× Trypsin solution: 3 mg/ml trypsin TPCK dissolved in 50 mM NH_4HCO_3

2. Procedure

1. Cut the ^{32}P-labeled ligand-gated ion channel out of the dried acrylamide gel and place in a glass scintillation vial.
2. Wash each protein-containing gel fragment with 20 ml destain solution for 30 min (×3).
3. Wash with 20 ml 50% methanol for 30 min (×2).
4. Transfer the gel fragment to an Eppendorf tube and dry down in a Speed-Vac, approximately 2 h.
5. To each tube, add 1 ml of 1× trypsin solution and incubate overnight at 37 °C. The next morning add 100 μl of a 10× trypsin solution and incubate an additional 3–4 h at 37 °C.
6. Remove the supernatant from the tube and save. Add 1 ml H_2O to the gel fragment-containing tube and incubate 1 h at 37 °C.
7. Combine the supernatants and dry down in a Speed-Vac. Resuspend the dried material in 0.5 ml H_2O and dry down again; repeat three times to completely sublimate the NH_4HCO_3.
8. Resuspend the dried material in 10 μl H_2O and microcentrifuge at 14,000 rpm for 10 min.
9. Remove the supernatant to a fresh Eppendorf tube and count in a scintillation counter (Cerenkov method).

E. Two-Dimensional Thin-Layer Chromatography

1. Reagents

> Electrophoresis buffer (pH 3.4): acetic acid:pyridine:H_2O at 19:1:89
> Chromatography buffer: pyridine:butanol:acetic acid:H_2O at 15:10:3:12
> Basic fuchsin solution: 1 mg/ml solution
> Phenol red solution: 1 mg/ml solution

2. Procedure

1. On a 20×20-cm Kodak (Rochester, NY) 13,255 cellulose TLC sheet, make a single pencil mark at the sample origin, 10 cm from each side and 4 cm from the bottom of the sheet. Make two additional pencil marks 5 cm from each side and 4 cm from the bottom of the sheet.

2. Spot at least 250 cpm of the sample at the sample origin. However, do not spot more than about 2000 cpm of the sample per plate. Use only a fraction of the 10 μl sample if possible because overloading of the sample can cause streaking of the phosphopeptides. The sample should be spotted 1 μl at a time with drying in between (a hair dryer works much faster than air drying). Avoid gouging the TLC plate when spotting the sample.

3. Spot 1 μl each of basic fuchsin and phenol red solutions at the origin.

4. Prepare two sheets of 25×25-cm Whatman paper. Place a pencil mark 6.5 cm from the bottom and 12.5 cm from each side of one piece of Whatman paper; using the mark as its center, cut a 3-cm-diameter hole in this piece of Whatman paper. Place the spotted TLC plate on top of the intact piece of Whatman paper. Prewet the Whatman paper with the hole with pH 3.4 electrophoresis buffer. Lay the prewetted Whatman paper on the TLC plate so that the spotted sample appears in the middle of the hole. With a Pasteur pipette, dribble pH 3.4 electrophoresis buffer on the top piece of Whatman paper until the entire TLC plate underneath is wet. To avoid movement of the sample away from the origin, one should dribble electrophoresis buffer around the perimeter of the hole such that the diffusing buffer reaches the spotted sample from all sides simultaneously.

5. While still wet, place the TLC plate in an electrophoresis tank containing the electrophoresis buffer with the spotted sample in the middle between the two electrodes. Run at 500 V until the dyes reach the marks 5 cm from the edges of the plate (about 1.5 h).

6. Remove the plate and allow to dry completely.

7. Place the plate in an ascending chromatography chamber with the bottom of the plate submerged in the chromatography buffer. When the buffer is 1 cm from the top of the plate (4–6 h) remove it and allow to dry completely.

8. Wrap the TLC plate in Saran wrap and expose to film. For faster results place the wrapped plate in a PhosphoImager cassette.

The two-dimensional phosphopeptide map generated for each ligand-gated ion channel represents a highly reproducible "phosphopeptide fingerprint" of that channel. In the case of wild-type GluR1, phosphopeptide maps of the immuno-precipitated receptor from phorbol ester and forskolin treated HEK293 cells include seven distinct phosphopeptides (Fig. 1B). However, this does not necessarily imply that there are seven distinct phosphorylation sites on GluR1; incomplete protease digestion is common and often generates multiple phosphopeptides, which include the same phosphorylation site. In fact, we have shown that GluR1 phosphopeptides 3, 4, and 6 all include the same PKC phosphorylation site (Roche *et al.*, 1996).

F. Phosphoamino Acid Analysis of Ligand–Gated Ion Channels

Serine, threonine, and tyrosine are each substrates for protein phosphorylation, and determining which of these amino acids are phosphorylated is an important step in characterizing the phosphorylation of a ligand-gated ion channel. To perform phosphoamino acid analysis, follow the protocol for proteolytic digestion of the sample described above and then use the isolated sample for acid hydrolysis to individual amino acids.

G. Phosphoamino Acid Analysis Protocol

1. Reagents

Electrophoresis buffer (pH 1.9): formic acid:acetic acid:H_2O at 1:10:89
Electrophoresis buffer (pH 3.4): acetic acid:pyridine:H_2O at 19:1:89

2. Procedure

1. Add at least 500 cpm of the sample to a 16×75-mm black-topped Kimax tube with a Teflon screw cap containing 0.5 ml 6 N HCl. Blow N_2 gently over the liquid before screwing the lid on tightly.

2. Place in a 105 °C oven for 1–2 h and then transfer to a microcentrifuge tube and Speed-Vac until dry.

3. Resuspend the sample in 0.5 ml H_2O and redry.

4. Resuspend the sample in 10 μl H_2O and vortex 30 s. Spin the tube at 14,000 rpm for 10 min and transfer the supernatant to a second microcentrifuge tube.

5. Prepare a Kodak TLC 13,255 cellulose sheet as follows: Mark the two side edges of the sheet with a pencil 4 cm from the bottom. Lay a ruler across the sheet between the two marks and, starting 2 cm from the side, make pencil marks every 4 cm. Up to five samples may be spotted at these five origins. Make additional marks 5 and 14 cm above where the samples will be spotted.

6. Spot the samples 1 μl at a time with drying in between.

7. Prepare fresh phosphoserine, phosphothreonine, and phosphotyrosine standards at 10 mg/ml in H_2O. Spot 1 μl of each on top of each spotted sample. Also spot 1 μl of phenol red on each sample.

8. Cut three pieces of Whatman paper: 25×25 cm, 25×17 cm, and 25×5 cm. Lay the TLC plate on the large piece of Whatman paper and prewet the others with the pH 1.9 electrophoresis buffer. Lay the 25×17-cm Whatman paper over the TLC plate such that its lower edge is 1.5 cm above the spotted samples. Lay the 25×5-cm Whatman paper such that its upper edge is 1.5 cm below the spotted samples. Drip pH 1.9 solution onto the prewetted Whatman papers until the TLC plate underneath is wet. The buffer should gradually diffuse from the upper and lower pieces of Whatman paper to wet the samples without moving them.

9. While wet, place the TLC plate in an electrophoresis tank containing pH 1.9 electrophoresis buffer. The samples should be nearest the cathode. Electrophorese the samples at 500 V until the phenol red dye reaches the first pencil mark, 5 cm above where the samples were spotted.

10. Without allowing it to dry, transfer the plate to an electrophoresis tank containing pH 3.5 electrophoresis buffer and electrophorese at 500 V until the phenol red dye reaches the next pencil mark, 14 cm above where the samples were spotted.

11. Dry the TLC plate, then dip in ninhydrin (1% in acetone) and allow to dry. The phosphoamino acid standards will turn purple in about 15 min. Phosphoserine migrates at the front, followed by phosphothreonine and, finally, phosphotyrosine.

12. Cover the plate with Saran wrap and expose to film.

H. Fusion Protein Phosphorylation Studies

Ligand-gated ion channels are composed of relatively large polypeptides which may contain many serines, threonines, and tyrosines. Thus, determining which residues are phosphorylated may seem like a daunting task. However, if the protein kinase that phosphorylates the channel has been identified, one can begin by searching the protein for the appropriate protein kinase consensus sites (Pearson and Kemp, 1991). Furthermore, if the topology of the channel in the membrane has been determined, candidate sites can be narrowed to those that are present on intracellular regions of the channel. Once regions likely to include phosphorylation sites have been identified, it is often useful to make fusion proteins corresponding to these regions and phosphorylate them *in vitro* with purified protein kinases. If the channel of interest has large continuous domains which might contain phosphorylation sites, it is helpful to generate several smaller fusion proteins (50–100 residues long) which span the length of the domain. These fusion proteins can then be analyzed by two-dimensional phosphopeptide mapping and the resulting maps compared to those generated from native channels or full-length recombinant channels expressed *in vivo*.

I. *In Vitro* Fusion Protein Phosphorylation

1. Reagents

5× PKA reaction buffer: 50 mM HEPES, pH 7.0, 100 mM $MgCl_2$
5× PKC and CaMKII reaction buffer: 50 mM HEPES, pH 7.0, 50 mM $MgCl_2$,
5 mM $CaCl_2$
ATP mixture: 250 μM ATP (2000 cpm/pMol)

2. Procedure

1. Add 10 μl (1 μg) of a 0.1-mg/ml solution of fusion protein substrate (dissolved in PBS) to a 1-ml microcentrifuge tube on ice.

2. Add 20 μl of the appropriate 5× reaction buffer to each tube.

3. For PKC reactions add 10 μl of a 500-μg/ml phosphatidylserine, 50 μg/ml diolein solution in H_2O. For CaMKII reactions add 10 μl of a 0.3-mg/ml calmodulin solution in H_2O.

4. Add H_2O to bring volume to 90 μl.

5. Add 9 μl ATP mixture.

6. Begin reaction by adding 1 μl of a 100-μg/ml solution of the desired protein kinase and incubate at 30 °C for 30 min. Terminate the reactions by adding 50 μl of 3× sample buffer and analyze by SDS–PAGE. Stain the gel with Coomassie blue, dry the gel between two sheets of cellophane, and expose to film. Cut out the appropriate band from the gel to perform phosphopeptide map and phosphoamino acid analysis as described above.

When comparing fusion protein maps with maps from full-length channels, one should look for phosphopeptides that are of similar shape and comigrate on TLC plates. Figure 2B shows the map of a fusion protein corresponding to the C terminus of GluR1 phosphorylated *in vitro* with purified PKA. Note the presence of a phosphopeptide cluster that closely resembles and seems to comigrate with phosphopeptide 5 from full-length receptor (Fig. 1B). This finding suggests that phosphopeptide 5 may be contained within this fusion protein and that it may be a substrate for PKA phosphorylation. (To confirm that phosphopeptides from the two preparations actually comigrate, it is often useful to spot both on the same TLC plate and look to see whether they overlap when processed together.) Also note the presence of an additional phosphopeptide not observed in maps of the full-length receptor (indicated by an arrow). Such spurious phosphopeptides are often seen when proteins are phosphorylated *in vitro* and may represent phosphorylation at sites that are normally extracellular or otherwise inaccessible to protein kinases.

J. Site-Specific Mutagenesis of Phosphorylation Sites

Once phosphorylation sites have been narrowed to small domains within a channel and the identity of the phosphorylated amino acid has been determined by phosphoamino acid analysis, candidate serines, threonines, and tyrosines can be

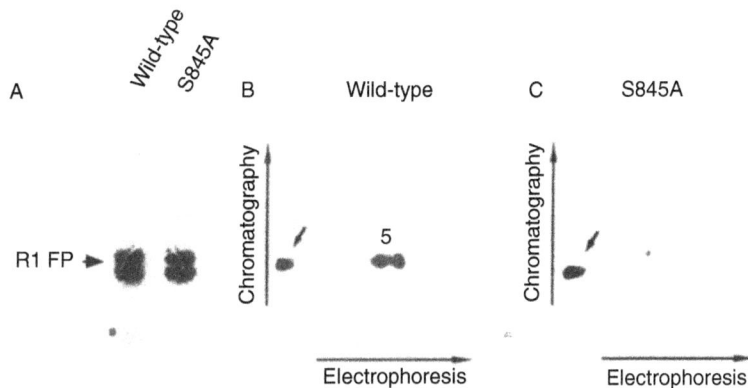

Fig. 2 *In vitro phosphorylation and mapping of fusion proteins.* Fusion proteins corresponding to the C termini of GluR1 and GluR1 S845A were phosphorylated *in vitro* with purified PKA and analyzed by SDS–PAGE (A). Two-dimensional phosphopeptide maps were generated from the PKA phosphorylated GluR1 (B) and GluR1 S845A (C) C-terminal fusion proteins.

targeted for point mutation. Typically, alanines are substituted for serines and threonines whereas phenylalanines are substituted for tyrosines. The resulting mutant recombinant receptors are then expressed in transfected cells and phosphopeptide maps generated as previously described. In the case of GluR1, for example, mutating Ser-845 to Ala abolished phosphopeptide 5 (compare Fig. 1B and C). Mutant fusion proteins may also be created, phosphorylated *in vitro*, and screened by two-dimensional map analysis. Figure 2C shows that a phosphopeptide map of the S845A C-terminal GluR1 fusion protein phosphorylated *in vitro* with purified PKA does not include phosphopeptide 5. Taken together with the finding that phosphopeptide maps of GluR1 from cells which are not treated with forskolin do not include phosphopeptide 5 (not shown), these results suggest that PKA phosphorylates Ser-845 on GluR1. In general, the absence of wild-type phosphopeptides from maps of mutant channels is taken as strong evidence that the mutated residues are sites of protein phosphorylation.

K. Phosphorylation Site–Specific Antibodies

After a specific channel residue has been identified as a phosphorylation site, its phosphorylation state in a number of different preparations and under a wide variety of conditions is often of interest. To facilitate such studies, phosphorylation site-specific antibodies which recognize the channel only when the residue of interest is phosphorylated can be generated. These antibodies reduce the difficulty and expense associated with phosphopeptide mapping and allow the rapid screening of many samples by Western blot analysis. To generate such antibodies, phosphopeptides with chemically phosphorylated serine, threonine, or tyrosine residues must be synthesized and injected into rabbits. We have found that

12-mers with the phosphorylated residue at the sixth position work well. Including a lysine at the end of the phosphopeptide to facilitate coupling to the carrier thyroglobulin is also helpful (see Tingley *et al.*, 1997 for a detailed discussion of phosphopeptide synthesis methods).

Phosphorylation site-specific antibodies can be separated from antibodies that recognize the nonphosphorylated peptide by loading the serum on an affinity column containing the nonphosphorylated equivalent of the phosphopeptide antigen. The phosphorylation site-specific antibodies should be contained in the flow-through along with other nonspecific antibodies which fail to bind the nonphosphorylated peptide. Often, the serum contains very few antibodies that recognize the nonphosphorylated peptide, and this first purification step may not be necessary. To purify the antibody of interest away from nonspecific antibodies, the flow-through from the first column (or crude serum) may be loaded onto an affinity column containing the phosphopeptide antigen. After washing the column, the phosphorylation site-specific antibodies can be eluted with standard methods and used for Western blot analysis.

Figure 3 demonstrates the utility of a phosphorylation site-dependent antibody that recognizes GluR1 only when Ser-845 is phosphorylated (GluR1 845-P). QT6 cells were transfected with wild-type or mutant GluR1 and treated with phorbol

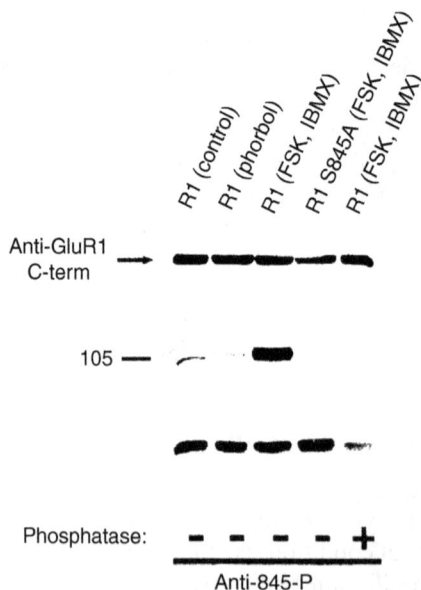

Fig. 3 *Specificity of a phosphorylation site-specific antibody.* QT6 cells expressing GluR1 or GluR1 S845A were treated with control solution (Control), 100 nM phorbol 12-myristate 13-acetate (PMA), or 10 μM forskolin and 100 μM IBMX (FSK, IBMX) for 15 min as indicated. Membranes were prepared and run on SDS–PAGE, transferred to PVDF membranes, and immunoblotted with the phosphorylation site-specific antibody GluR1 845-P or an antibody to the C terminus of GluR1.

ester, forskolin and IBMX, or vehicle. Membrane extracts were prepared from each sample, separated by SDS–PAGE, blotted onto PVDF membrane, and analyzed by Western blot. Forskolin treatment caused an increase in GluR1 845-P labeling of wild-type channel, confirming that PKA phosphorylates GluR1 on Ser-845. However, when the blot was treated with lambda phosphatase prior to application of the antibody, GluR1 845-P did not label GluR1 from forskolin-treated cells. Furthermore, when QT6 cells were transfected with the GluR1 S845A channel and treated with forskolin, the GluR1 845-P antibody did not recognize the mutant channel. These two results demonstrate the phosphorylation state dependence of the GluR1 845-P antibody. A different antibody raised against the carboxy terminus of GluR1 shows that equal amounts of the channel were loaded in each lane.

Phosphorylation site-specific antibodies are especially useful for examining the phosphorylation state of ligand-gated ion channels in preparations that are not amenable to labeling with ortho[^{32}P]phosphate. For example, the GluR1 845-P antibody may be used to compare the degree of GluR1 Ser-845 phosphorylation in the brains of rats that have undergone different treatments. These antibodies are also potentially useful for examining the subcellular distribution of phosphorylated channels by immunocytochemistry.

III. Functional Effects of Ligand–Gated Ion Channel Phosphorylation

A variety of electrophysiologic studies have provided evidence that ligand-gated ion channels are modulated by protein phosphorylation. However, it is not clear from these studies whether this modulation occurs via the direct phosphorylation of the channel itself. To address this question, patch-clamp studies using wild-type and mutant receptors expressed in heterologous systems are necessary. The physiologic properties of recombinant wild-type receptors are studied before and after phosphorylation and compared to various phosphorylation site mutant receptors.

A. Stimulation of Phosphorylation in Whole-Cell Recordings

The functional effects of phosphorylation have been investigated by applying pharmacologic agents that activate kinases (e.g., phorbol esters or forskolin) to primary cultures or transfected cells. This approach has been particularly useful in establishing roles for PKA- and PKC-mediated phosphorylation in modulating ligand-gated ion channel function (e.g., see Courtney and Nicholls, 1992; Greengard et al., 1991; Markram and Segal, 1992). An alternative approach is to apply the kinase itself to the intracellular side of the channel (Chen and Huang, 1992; Moss et al., 1992; Raymond et al., 1993a; Roche et al., 1996; Wang et al., 1993; Yakel et al., 1995). This requires the isolation of a constitutively active form of the kinase or coapplication of the native kinase with appropriate activators.

We have found that simply supplementing the intracellular solution with the kinase can be problematic since it may diffuse via the tip into the cell before a control response can be established. Furthermore, it is often difficult to obtain a gigaseal when a protein is present in the intracellular solution. This problem can be circumvented by tip-filling the electrode with a control intracellular solution (lacking the kinase) and back-filling the pipette with the intracellular solution containing kinase. Consistent volumes of tip-filling and back-filling solutions should be used to preventing intercell variability in the measured effect of the kinase and the latency of its onset.

Intracellular patch perfusion is a preferred method for introducing kinases to the inside of cells. It involves injection of intracellular solution containing the kinase (and a suitable carrier protein) directly to the inside of cells via tubing placed close to the tip of the patch electrode. We have found this method to be particularly successful: It avoids problems in gigaohm seal formation, since intracellular injection always occurs after "breaking through" into the whole-cell configuration. Furthermore, it allows an unambiguous measurement of a baseline response in the absence of kinase. Placing the intracellular perfusion tubing near the tip of the electrode (<2 mm) should allow relatively rapid entry of solution into the cell, due to diffusion and the fluid turbulence created by pressure injection. For example, dye (trypan blue) can be seen near the tip of the electrode within seconds of injection into a patch electrode. Measurable changes in GluR1 whole-cell currents after intracellular application of PKA (Roche *et al.*, 1996) or a constitutively active form of CaMKII (unpublished data) are seen in as little as 2 min.

The internal patch-perfusion tubing is easily fabricated from polypropylene tubing, which can be melted and gently drawn out to the desired diameter and length. For these experiments we use special electrode holders with an additional inlet for the patch-perfusion tubing purchased from Adams and List Associates, Ltd. Several other manufacturers produce suitable holders. These electrode holders are similar in most respects to standard holders, however, due to the noise created by the proximity of the perfusion tubing and the electrode wire, it may be difficult to obtain low-noise single-channel recordings using this method.

The solution containing the kinase can be expelled from the intracellular perfusion tubing by applying pressure via a syringe. The final 50–100 μl of this tubing (leading into the electrode holder) should be detachable. This part of the tubing is filled with the kinase containing solution. The rest of the tubing that connects this 50–100 μl section to the syringe should be filled with mineral oil. The latter is noncompressible and prevents volume changes within the intracellular perfusion tubing from occurring when the patch electrode experiences changes in pressure. Thus, intracellular perfusion occurs smoothly and without formation of bubbles.

B. "Intracellular" Solution Composition

Because various intracellular components are required for phosphorylation/dephosphorylation reactions, it is important that they be preserved or at least "mimicked" during extended recordings. For this reason, some experimenters use an "ATP-regenerating solution," (Forscher and Oxford, 1985) which, as the name implies, mimics the cellular ATP-generating machinery. In addition to ATP (2–4 mM), this solution is supplemented with creatine phosphokinase (50 U/ml) and phosphocreatine (20 mM). Use of this solution has allowed NMDA and AMPA-receptor mediated responses to be recorded over extended durations without appreciable "rundown," which appears to be mediated by dephosphorylation (Wang *et al.*, 1993).

The intracellular solution should be kept on ice to prevent degradation of labile contents. As far as is practical, the kinase containing intracellular solution should also be kept on ice. It is difficult to keep the kinase in the patch-perfusion tubing cool, therefore only small volumes should be used to fill this tubing. The tube should frequently be flushed and replenished with newly thawed and diluted kinase. How often the kinase containing solution is replaced depends on its stability at room temperature. For our experiments using constitutively active CaMKII or PKA, we limit the use of a batch of kinase to 2 h.

An additional consideration for these experiments is that proteins bind to glass by way of multiple nonspecific interactions with the charged glass surface of the patch electrode. Thus, in addition to adding the kinase of interest to the intracellular solution, it is necessary to saturate most of the nonspecific glass–protein interactions by adding a "carrier protein" such as BSA or creatine phosphokinase.

C. Basal Phosphorylation of Ligand–Gated Ion Channels

Sites of basal phosphorylation in glutamate receptors have been detected in neurons and HEK293 cells (Blackstone *et al.*, 1994; Roche *et al.*, 1996). The level of basal phosphorylation may depend on the cell type in which the receptor is expressed and factors such as the composition of the tissue culture serum (e.g., serum often contains thrombin, which can activate PKC in many cell types). Basal phosphorylation may make the detection of any functional effect of an exogenous kinase difficult or impossible. If, on the other hand, sites are already stoichiometrically phosphorylated by basal kinase activity, functional changes may be detected in the presence of phosphatases.

In the case of GluR1 for example, the C-terminal S831 site was found to be basally phosphorylated in HEK293 cells, making it difficult to investigate the functional effect of phosphorylation at this site. However, S831 was not phosphorylated in the QT6 quail fibroblast cell line. Detailed biochemical experiments showed that this site was phosphorylated by CaMKII and PKC. Furthermore, electrophysiologic studies revealed that receptors expressed in these cells can be strongly potentiated when exposed to a constituitively active form of CaMKII (unpublished observations).

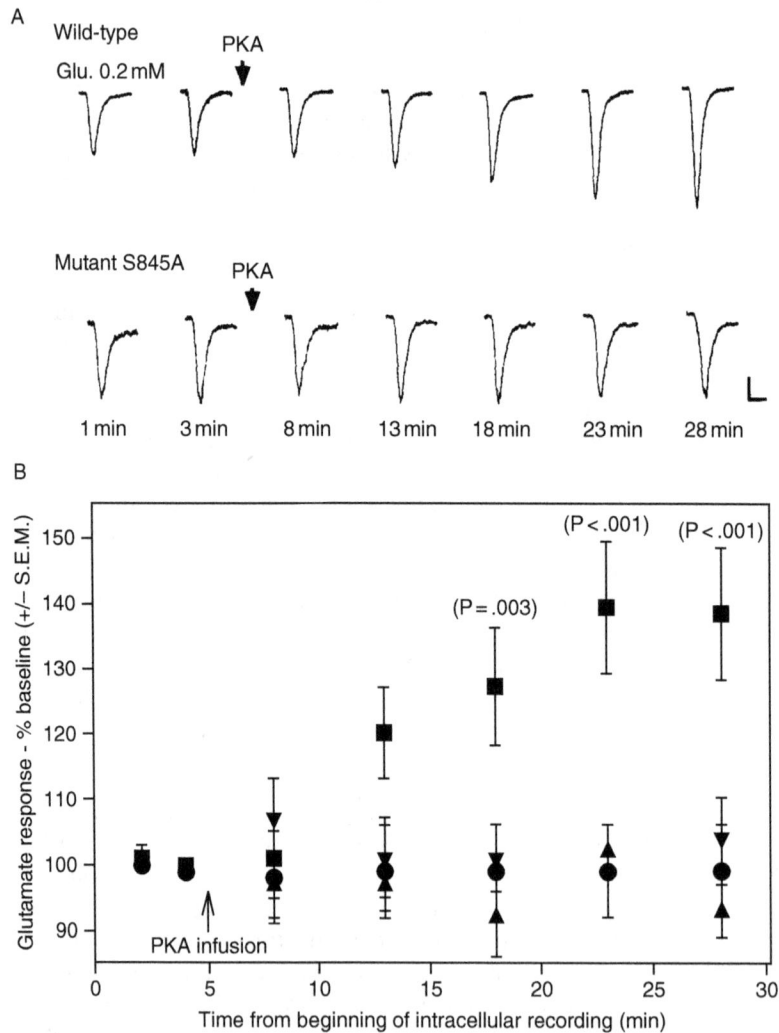

Fig. 4 *Functional effects of GluR1 phosphorylation by PKA.* Wild-type and mutant GluR1 responses activated by glutamate in HEK293 cells (duration of glutamate application is indicated by solid line). The calibrations bars indicate 100 pA and 10 ms. PKA (20 μg/ml) was infused after holding the cells for ~5 min. In wild-type GluR1-expressing cells, the responses become significantly larger after ~15 min (A and B, ■). This effect appeared to be specifically mediated by PKA: the PKA buffer caused no potentiation (B, ●) and PKA inhibitor peptide (PKI$_{5-24}$ amide; 200 μg/ml) blocked potentiation (B, ▲). In addition, the GluR1 S845A mutant was not potentiated by PKA (A and B, ▼). Symbols and vertical lines indicate data gathered from 4–12 cells ± SEM.

D. Determining Functional Effect of Phosphorylation at Specific Sites

Biochemical analysis using wild-type and mutant receptors has allowed phosphorylation sites to be specifically identified. To establish a physiologic role for phosphorylation at particular sites, responses of wild-type and mutant receptors are measured. For example, the effect of PKA phosphorylation of GluR1 at S845 has been determined using internal kinase perfusion. This site is robustly phosphorylated when the receptor is exposed to forskolin and IBMX. Furthermore, infusion of PKA into the cell caused a significant potentiation of the peak response of the channel, while the S845A mutant was neither phosphorylated by PKA, nor underwent potentiation in the presence of PKA (see Fig. 4). That the observed increased in the GluR1-mediated response was due to the specific action of PKA was confirmed using a peptide inhibitor of PKA which blocked the potentiation (Fig. 4B). These results strongly suggest that PKA phosphorylation in Ser-845 directly potentiates GluR1 function.

IV. Conclusion

In this article we have attempted to describe a biochemical and electrophysiologic approach to investigating ligand-gated ion channel phosphorylation. As a model we have used phosphorylation of the GluR1 receptor subunit, but the principles and concepts should be applicable to other classes of ligand-gated ion channels. These studies provide us with a better understanding of the regulation of ligand-gated ion channels by protein phosphorylation, which may be a major mechanism for the modulation of synaptic function.

References

Blackstone, C. D., Moss, S. J., Martin, L. J., Levey, A. I., Price, D. L., and Huganir, R. L. (1992). *J. Neurochem.* **58,** 1118.

Blackstone, C., Murphy, T. H., Moss, S. J., Baraban, J. M., and Huganir, R. L. (1994). *J. Neurosci.* **14,** 7585.

Boyle, W. J., Geer, P. V. D., and Hunter, T. (1991). *Methods Enzymol.* **201,** 110.

Chen, L., and Huang, L. Y. (1992). *Nature* **356,** 521.

Courtney, M. J., and Nicholls, D. G. (1992). *J. Neurochem.* **59,** 983.

Forscher, P., and Oxford, G. S. (1985). *J. Gen. Physiol.* **85,** 743.

Greengard, P., Jen, J., Nairn, A. C., and Stevens, C. F. (1991). *Science* **253,** 1135.

Markram, H., and Segal, M. (1992). *J. Physiol. (Lond.)* **457,** 491.

Moss, S. J., Smart, T. G., Blackstone, C. D., and Huganir, R. L. (1992). *Science* **257,** 661.

Nicoll, R. A., and Malenka, R. C. (1995). *Nature* **377,** 115.

Pearson, R. B., and Kemp, B. E. (1991). *Methods Enzymol.* **200,** 62.

Raymond, L. A., Blackstone, C. D., and Huganir, R. L. (1993a). *Nature* **261,** 637.

Raymond, L. A., Blackstone, C. D., and Huganir, R. L. (1993b). *Trends Neurosci.* **16,** 147.

Roche, K. W., Tingley, W. G., and Huganir, R. L. (1994). *Curr. Opin. Neurosci.* **4,** 383.

Roche, K. W., O'Brien, R. J., Mammen, A. L., Bernhardt, J., and Huganir, R. L. (1996). *Neuron* **16,** 1179.

Tingley, W. G., Ehlers, M. D., Kameyama, K., Doherty, C., Ptak, J. B., Riley, C. T., and Huganir, R. L. (1997). *J. Biol. Chem.* **272,** 5157.

Wang, L. Y., Taverna, F. A., Huang, X. D., MacDonald, J. F., and Hampson, D. R. (1993). *Science* **259,** 1173.

Yakel, J. L., Vissavajjhala, P., Derkach, V. A., Brickey, D. A., and Soderling, T. R. (1995). *Proc. Natl. Acad. Sci. USA* **92,** 1376.

CHAPTER 19

Analysis of Ion Channel Associated Proteins

Michael Wyszynski and Morgan Sheng

Howard Hughes Medical Institute
Mass General Hospital
Boston, Massachusetts, USA

I. Introduction

Ion channels are typically heterooligomeric protein complexes composed of several distinct subunits. Ion channels in turn interact with a variety of intracellular proteins including clustering proteins, cytoskeletal proteins, protein kinases, and G proteins. The distinction between a true channel "subunit" and an ion channel "associated protein" can be somewhat arbitrary, especially if the primary function of the protein is not known. For the purposes of this article, however, we define ion

DOI: 10.1016/B978-0-12-382204-8.00019-9

channel associated proteins as those that are not directly involved in channel function (e.g., they do not contribute to the formation of the channel pore or alter significantly the biophysical properties of the channel). Rather ion channel associated proteins include those that are involved in localization, cytoskeletal anchoring, or regulation of their associated ion channels.

Ion channel associated proteins can be initially identified by many means, including biochemical copurification (e.g., acetylcholine receptor (AChR) and rapsyn, glycine receptors and gephyrin, *Shaker*-type K^+ channels and their β-subunits) (Sheng and Kim, 1996), genetic interactions (e.g., degenerin channels and their interacting proteins in *Caenorhabditis elegans*; reviewed in Garcia-Anoveros and Corey, 1997), and the yeast two-hybrid system (Niethammer and Sheng, 1998). In this chapter, we consider the approaches for confirming and analyzing an interaction between an ion channel and its putative interacting protein, using as a primary example the association between PSD-95 and *Shaker* K^+ channels and *N*-methyl-D-aspartate (NMDA) receptors (see Fig. 1).

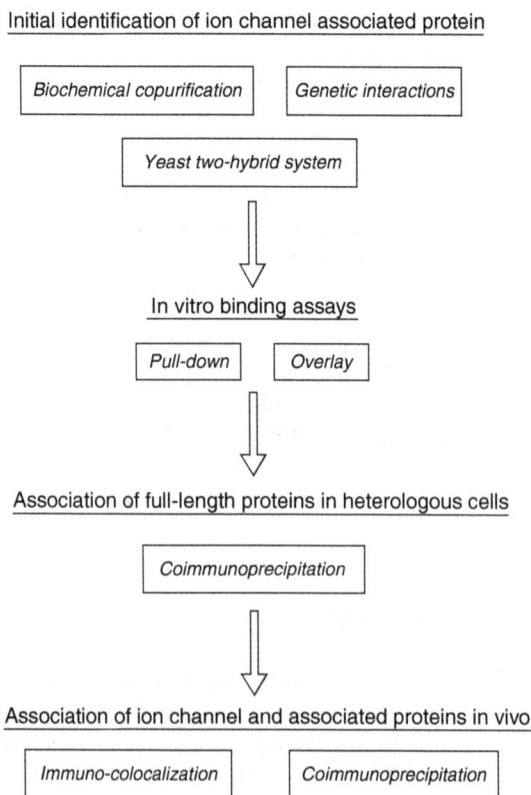

Initial identification of ion channel associated protein

| Biochemical copurification | Genetic interactions |

| Yeast two-hybrid system |

⇓

In vitro binding assays

| Pull-down | Overlay |

⇓

Association of full-length proteins in heterologous cells

| Coimmunoprecipitation |

⇓

Association of ion channel and associated proteins in vivo

| Immuno-colocalization | Coimmunoprecipitation |

Fig. 1 A scheme that summarizes a general approach for analyzing putative ion channel associated proteins.

This interaction was first identified by the yeast two-hybrid system (Kim *et al.*, 1995; Kornau *et al.*, 1995; Niethammer *et al.*, 1996).

II. General Considerations

After a protein is identified as a putative channel associated protein via yeast two-hybrid screening or other kinds of affinity-based screening, the most important task is to confirm the authenticity of the interaction *in vivo*. The most significant tests of *in vivo* association of two proteins in common use are coimmunoprecipitation (a convenient and stringent way to show biochemical copurification) and double-label immunofluorescent colocalization (a direct way to show that they are in the same subcellular location, a prerequisite if they are to interact directly). Ultimately, genetic experiments provide critical evidence for an *in vivo* interaction and offer important clues to its functional significance. For instance, genetic deletion of the channel-associated protein could result in mislocalization of the ion channel or in abnormal regulation of channel properties. Of course, genetic approaches are essential to determine the function of the interaction at the organismal level. Before undertaking the more laborious *in vivo* studies, however, it is reasonable to conduct *in vitro* binding experiments to confirm an interaction between an ion channel and its suspected partner.

III. *In Vitro* Binding of Recombinant Proteins

In vitro binding is useful to confirm an interaction initially identified by the yeast two-hybrid system. Although interactions detected by the yeast two-hybrid system almost invariably involve direct binding between the interacting proteins, *in vitro* binding assays using purified proteins can confirm that the interaction requires no other ancillary protein. More importantly, an *in vitro* binding assay offers the opportunity to measure directly the affinity of binding.

Typically, *in vitro* binding assays are performed with the relevant parts of the proteins produced in bacteria as recombinant fusion proteins. We describe here two commonly used *in vitro* binding assays: a filter overlay assay (sometimes known as a Far-Western) and a "pull-down" or coprecipitation assay. In filter overlay assays, one protein (protein X) is subjected to SDS–PAGE and transferred to nitrocellulose filters. The other protein (protein Y) is applied in solution to the filter and its binding to protein X is visualized (through radioactive or indirect chemiluminescent labeling and autoradiography) as a "band on a gel." Thus, in filter overlays, protein Y is used as a probe in analogous fashion to antibodies in a Western blot. In the "pull-down" assay, protein X is attached to a solid support (e.g., agarose bead) and incubated with protein Y in solution.

If proteins X and Y associate, pelleting of the beads by centrifugation will result in the precipitation of both X and Y. The presence of Y can then be assayed by immunoblotting the precipitated proteins.

A. Filter Overlay Assay

The example of a filter overlay assay provided here is that of PSD-95 binding to the C-terminus of NMDA receptor NR2 and *Shaker* K^+ channel subunit proteins (Niethammer *et al.*, 1996). The first two PDZ domains of PSD-95 were produced as a hexahistidine-fusion protein (using pRSETB, Invitrogen) in bacteria and purified by nickel-nitrilotriacetic acid (Ni-NTA) chromatography (using Probond resin, Invitrogen, Carlsbad, CA). A hexahistidine (H_6)-labeled fusion protein of PSD-95 (H_6-PSD-95) is here used as the probe. The NR2 C-terminal tails were fused to glutathione *S*-transferase (vector pGEX, Pharmacia, Piscataway, NJ) and over-expressed in bacteria. It is not necessary to purify the GST fusions since they will be separated by SDS–PAGE and transferred to a nitrocellulose filter to act as targets for binding by H_6-PSD95. Specificity of binding is then determined by the position on the filter where the hexahistidine-fusion protein binds.

1. Electrophorese total bacterial lysates containing GST-fusion proteins of the NR2 C-terminus ($\sim$40 kDa) and GST-fusion proteins of the Kv1.4 C-terminus ($\sim$37 kDa) in a 10% SDS–PAGE gel. Usually we electrophorese 1–5 μg of lysate protein and aim to have approximately 100 ng of target protein available for binding. To confirm the specificity of the interaction, GST alone or GST fused to unrelated protein sequences should also be used as targets.

2. Transfer the resolved proteins from the acrylamide gel to a nitrocellulose filter at 35 V/cm for 2 h in transfer buffer [50 mM Tris–base, 40 mM glycine, 1 mM ethylenediaminetetraacetic acid (EDTA), 20% (v/v) methanol, pH 8.3]. To facilitate the transfer of proteins larger than 100 kDa, 0.1% (w/v) SDS may also be included in the transfer buffer.

3. To check that the transfer was successful, visualize the proteins transferred to the nitrocellulose filter by incubating the filter in Ponceau S protein stain. The molecular weight markers should also be identified and marked at this stage.

4. The denatured filter-bound proteins may require partial or complete renaturation of their conformations for successful binding to the recombinant fusion proteins serving as probe. To renature the filter-bound proteins, incubate the nitrocellulose filter in buffer A (10 mM HEPES, 60 mM KCl, 1 mM EDTA, and 1 mM 2-mercaptoethanol) containing 6 M guanidine hydrochloride at 4 °C for 10 min. Repeat the incubations sequentially at 4 °C for 10 min using decreasing concentrations of guanidine hydrochloride (3, 1.5, 0.75, 0.38, 0.19, 0.1, and 0 M). The buffer A solutions containing guanidine hydrochloride are reusable and should be stored at 4 °C.

5. Block the filter in buffer B (25 mM HEPES, 120 mM KCl, 1 mM EDTA, and 0.2% Triton X-100) containing 5% nonfat dry milk (NFDM) at 4 °C for 1 h with shaking.

6. Repeat step 5 using buffer B containing 1% NFDM.

7. For the protein binding reaction, overlay blocked filter with buffer B containing 10 mg/ml bovine serum albumin (BSA) (to suppress nonspecific binding by the protein probe) and 1 μg/ml of the purified hexahistidine-fusion protein of the first two PDZ domains from PSD-95. Typically, 3–5 ml of the overlay solution is sufficient for covering a nitrocellulose filter with a surface area of 50 cm^2. Incubate at 4 °C overnight.

8. Next day, bring the overlaid filter to room temperature by incubating at room temperature for 30 min.

9. Wash the filter with buffer B for 10 min at room temperature. Repeat wash twice.

10. Rinse the filter briefly with TBST [10 mM Tris (pH 7.4), 150 mM NaCl, 0.1% Tween 20].

11. To visualize the bound hexahistidine-fusion protein by Western blotting, incubate the filter in antibody dilution solution (TBST containing 2% BSA and 2% horse or goat serum) containing antibodies directed against an epitope tag in the hexahistidine leader sequence (1: 10,000 dilution of mouse monoclonal anti-T7-Tag antibodies, Novagen, Milwaukee, WI) at 4 °C overnight. Next morning, wash the filter with TBST containing 5% NFDM at room temperature for 10 min. Repeat the wash with TBST alone. Incubate the washed filter in antibody dilution solution containing a 1:1000 dilution of antimouse HRP-conjugated antibodies (Amersham, Arlington Heights, IL) at room temperature for 30 min, wash twice with TBST at room temperature for 10 min, and perform enhanced chemiluminescence (ECL, Amersham).

12. To show the position and relative abundance of the GST-fusion proteins in each lane, strip the filter of antibodies by incubating the filter in stripping buffer [62 mM Tris (pH 6.7), 2% (w/v) SDS, and 0.1% mercaptoethanol] at 60 °C for 30 min. Block the filter with TBST containing 5% NFDM at room temperature for 30 min, and perform Western blotting as described above using 0.5 μg/ml rabbit anti-GST antibodies (Santa Cruz Biotechnology, Santa Cruz, CA).

B. "Pull-Down" Assay

The example of a pull-down assay provided here is that of α-actinin-2 binding to NMDA receptors (Wyszynski *et al.*, 1997). The pull-down assay can be applied to extracts from native tissues (as illustrated here) or to extracts from transfected heterologous cells or even semipurified preparations of bacterial fusion proteins. Obviously, "pull-down" of protein Y by protein X from a complex mixture (such as a brain extract) does not indicate that there is a direct interaction between X and Y. On the contrary, it is typical to obtain a heterogeneous mixture of proteins by this method, and it is important to check the specificity of the precipitation and to confirm it in as stringent conditions as possible.

The coprecipitation or pull-down of ion channel associated proteins from rat brain extracts begins with the preparation of a brain membrane fraction from whole brain homogenates.

1. Dissect the whole brain from 1 adult rate (age ~6 weeks, weight ~150–200 g).

2. Mince the rat brain on ice with a razor.

3. Homogenize the minced brain tissue in 3 ml of ice-cold homogenization buffer [0.32 M sucrose, 10 mM Tris (pH 7.4), 1 mM EDTA, and various protease inhibitors] using a Dounce homogenizer. For protease inhibitors, we typically use a cocktail consisting of 1 mM phenylmethylsulfonyl fluoride (PMSF) and 10 μg/ml each of pepstatin, aprotinin, and leupeptin.

4. Transfer the homogenate to a 15-ml conical centrifuge tube, centrifuge at $1000 \times g$ for 10 min at 4 °C, and separate the supernatant from the insoluble pellet.

5. Resuspend by homogenization the insoluble pellet in 1 ml of ice-cold homogenization buffer, repeat centrifugation, and separate the supernatant from the remaining insoluble material.

6. Pool supernatants from steps 4 and 5 and perform a high-speed centrifugation at $60,000 \times g$ at 4 °C for 1 h in an ultracentrifuge.

7. The supernatant obtained from this centrifugation may be considered the crude "soluble" brain fraction.

8. Resuspend the insoluble pellet in 8 ml of ice-cold resuspension buffer [20 mM Tris (pH 7.4), 1 mM EDTA] with various protease inhibitors.

9. Incubate on ice for 10 min and pellet by centrifugation at $60,000 \times g$ at 4 °C for 30 min.

10. Resuspend the pellet in 3 ml of ice-cold resuspension buffer. This resuspension represents the crude "membrane" fraction.

Following preparation of the rat brain membrane fraction, ion channel associated proteins in the membrane fractions are made soluble by detergent extraction. Alternatively, the soluble brain fraction can be used for the pull-down if the putative ion channel associated protein is present in this fraction.

11. Dilute 200 μl of membrane fraction in 200 μl of ice-cold extraction buffer [80 mM Tris (pH 8.0), 300 mM NaCl, 1 mM EDTA, and 2% Triton X-100].

12. Allow extraction to proceed at 4 °C for 1 h with mixing by inversion.

13. Centrifuge at $60,000 \times g$ at 4 °C for 30 min.

14. Save the supernatant containing the extracted protein frozen at −80 °C.

The "pull-down" assay also requires the affinity purification of a recombinant fusion protein containing the ion channel protein of interest. This fusion protein will be used to affinity-purify (i.e., pull-down) proteins from the detergent extract prepared in step 14.

15. Prepare GST-fusion protein of the NR1 C-terminus using glutathione-Sepharose beads (Pharmacia) by standard purification protocols. Do not release the GST-fusion protein from the glutathione-Sepharose beads.

16. Incubate GST-fusion protein (~100 μg) attached to the glutathione-Sepharose beads with Triton X-100 extracts of rat brain (~200 μg protein from step 14) at 4 °C for 1 h.

17. Wash the beads three times by resuspending protein bound glutathione-Sepharose beads with phosphate-buffered saline (PBS) containing 0.1% Triton X-100.

18. After washing, elute the bound proteins from the beads with SDS sample buffer, resolve by SDS–PAGE, and visualize the α-actinin "pulled-down" by immunoblotting with α-actinin-2 specific antibodies (Wyszynski et al., 1997).

As in the overlay assay, the specificity of the precipitation needs to be determined by using glutathione-Sepharose beads coupled to GST alone or GST fused to unrelated protein sequences as control affinity matrices. Further, the relative amount of GST-fusion proteins used should be equivalent, and can be assayed by anti-GST immunoblotting as described earlier for the filter overlay assay.

IV. Association of Full-Length Proteins in Heterologous Cells

The assays described above, the filer overlay or "pull-down" assays, are essentially tests for *in vitro* association between proteins, one or both of which is typically produced as a recombinant bacterial fusion protein containing only one fragment of the protein. These are useful assays when beginning to confirm a putative interaction between two proteins because they are relatively easy and do not require specific antibodies against either protein. It is obviously important, however, to confirm that a *full-length* ion channel protein can interact with a *full-length* channel associated protein in a *cellular* environment. This can be conveniently done by coexpression in a heterologous cell by transfection, and coimmunoprecipitation of the two expressed proteins. Such an approach requires the complete coding region of the cDNAs for the two putatively interacting proteins, and also specific antibodies to both proteins. In the absence of specific antibodies, one or both proteins can be tagged with an epitope tag [such as *myc* or hemagglutinin (HA) tags] which can be recognized by a commercially available monoclonal antibody. In the following protocol, the example given is that of Kv1.4 (a *Shaker*-type K$^+$ channel) binding to PSD-95 in heterologous (COS-7) cells (Kim and Sheng, 1996).

A. Coimmunoprecipitation of Full-Length Proteins Expressed in Heterologous Cells

Coimmunoprecipitation of proteins transiently expressed in heterologous cells requires transfection of cultured cells, detergent solubilization of the expressed proteins (since ion channels are intrinsic membrane proteins), and immunoprecipitation employing antibodies that are specific for the expressed proteins.

1. Plate COS-7 cells in tissue dishes (35 mm) for a confluence of 50–70%.

2. Next day, the cells should be 70–80% confluent. Prepare a DNA transfection mix of the eukaryotic expression constructs for full-length PSD-95 and Kvl.4 by adding equivalent amounts of each plasmid (no more than 1 μg total DNA) to 200 μl OPTI-MEM (GIBCO-BRL, Gaithersburg, MD) and 5 μl of lipofectamine (GIBCO-BRL). Allow the DNA/lipofectamine complexes to form for 30 min at room temperature. Following this incubation, add 800 μl of OPTI-MEM to each transfection mix.

3. Wash the cells thoroughly with 2 ml of OPTI-MEM.

4. Remove the OPTI-MEM wash and quickly add the DNA/lipofectamine OPTI-MEM mix ($\sim$1 ml) to the washed cells.

5. Incubate the cells in the DNA/lipofectamine mix in 5% CO_2 at 37 °C for 5 h.

6. Exchange the DNA/lipofectamine mix for normal medium and incubate the cells in 10% CO_2 at 37 °C. Depending on the efficiency of the eukaryotic expression vector employed and on the protein expressed, efficient expression (up to 1% of total cellular protein) of the protein takes 24–48 h.

Prior to immunoprecipitation, the transfected cells are harvested from the cell culture dish, and the expressed proteins extracted with nondenaturing detergents.

7. Wash the cells twice with 2 ml of ice-cold PBS.

8. Following the last wash, quickly add to the dish 350 μl of ice-cold extraction buffer [50 mM Tris (pH 7.4), 150 mM NaCl, 1 mM EDTA, plus the extraction detergent of choice]. For the solubilization of PSD-95 and Kvl.4 from heterologous cells we use RIPA [1% (v/v) Nonidet P-40 (NP-40), 0.5% (w/v) deoxychlorate, and 0.1% (w/v) SDS].

9. On ice, flush cells off the bottom of the dish by triturating the extract buffer over the surface of the dish.

10. Transfer the cell suspension to a 1.5-ml microcentrifuge tube. Allow extraction to proceed with shaking for 1 h at 4 °C.

11. Pellet insoluble material by centrifugation at 60,000 × g at 4 °C for 30 min. The supernatant contains the detergent-extracted proteins (soluble fraction). The pellet contains the detergent-insoluble proteins (insoluble fraction). Prior to immunoprecipitation, perform Western blotting analysis to determine the efficiency of solubilization by comparing the amount of expressed protein in the insoluble fraction to the amount of protein in the soluble fraction. Efficient immunoprecipitation requires that the protein be efficiently solubilized by detergent (at least 10–30%).

The immunoprecipitation of protein complexes from cell extract requires either the expression of epitope-tagged proteins or antibodies which specifically recognize the proteins of interest.

12. To 100-μl aliquots of the detergent extract (input), add the immunoprecipitating antibodies to a final concentration of 5 μg/ml.

In this example, guinea pig polyclonal antibodies specific for either PSD-95 (Kim *et al.*, 1995) or rabbit polyclonal antibodies to Kv1.4 (Sheng *et al.*, 1992) are used. Reciprocal immunoprecipitation using antibodies specific for either of the binding partners is useful in confirming a positive result; for example, immunoprecipitation of PSD-95 leads to the coimmunoprecipitation of Kv1.4 and immunoprecipitation of Kv1.4 leads to the coimmunoprecipitation of PSD-95.

Perform a negative control for the immunoprecipitation with the final 100-μl aliquot of detergent extract by immunoprecipitating with a nonspecific antibody such as rabbit IgG or the same quantity of an unrelated antibody. The nonspecific antibody should not immunoprecipitate either PSD-95 or Kv1.4 from the extracted fraction.

13. Incubate primary antibody and detergent extract at 4 °C for 1–2 h with mixing by inversion.

14. Add 20 μl (20% of immunoprecipitation reaction volume) of protein A-Sepharose conjugated beads (Pharmacia) preequilibrated in extraction buffer (50% slurry) and incubate for an additional 2 h with mixing by inversion.

15. Pellet protein A-Sepharose beads by spinning in a microcentrifuge at maximum speed for 10 s.

16. Wash protein A-Sepharose conjugated beads with 1 ml of ice-cold extraction buffer three times.

17. Elute bound proteins from the beads with SDS sample buffer, resolve eluted proteins by SDS–PAGE, and visualize components of immunoprecipitated protein complexes by immunoblotting using antibodies specific for PSD-95 (Kim et al., 1995) or Kv1.4 (Sheng et al., 1992). To strip and reprobe the filter, incubate the filter in stripping buffer [62 mM Tris (pH 6.7), 2% SDS, and 0.1% mercaptoethanol] at 60 °C for 30 min, block the filter with TBST containing 5% NFDM at room temperature for 30 min. Perform Western blotting as described earlier for the filter overlay assay.

V. Colocalization of Ion Channel and Associated Protein *In Vivo*

A prerequisite for two proteins interacting *in vivo* is that they be expressed in the same cells and in the same subcellular location (at least in part). If the ion channel of interest has a specific expression pattern and a targeted subcellular localization, a similar distribution of its putative associated protein greatly supports the validity of the interaction *in vivo*. By far the best way to show colocalization is with double-labeling immunocytochemistry using specific antibodies in the two interacting proteins X and Y. Typically this is achieved using indirect immunofluorescence, with the secondary antibodies tagged by distinct fluorophores such as fluorescein (green emission) or Cy-3 (red). This requires that antibodies derived from different species (rabbit or mouse, for instance) be available for proteins X and Y; otherwise, the secondary antibodies cannot distinguish between the respective primary antibodies. At the electron microscopy (EM) level, the species-specific secondary antibodies can be labeled with gold particles of distinct size. In the available antibodies to the interacting proteins are from the same species, then the primary antibodies to X and Y can be directly tagged with different fluorophores or gold particles, but this is not an easy procedure.

Immunocolocalization can be achieved at various levels of spatial resolution. For instance, immunofluorescence confocal microscopy can provide submicron resolution, while immunogold EM can resolve to ~10–20 nm. Obviously, the higher the resolution, the more meaningful the colocalization. For instance, two proteins can "colocalize" by light microscopy but one can be in the presynaptic membrane and the other in the postsynaptic membrane. Generally, it is impossible to distinguish pre- and postsynaptic localization at a given synapse by light microscopy alone—EM is required. Another point worth emphasizing is that colocalization does not necessarily indicate that the two proteins interact directly. Direct interaction can only be concluded from biochemical experiments *in vitro*. Nevertheless, colocalization of proteins X and Y by immunocytochemistry offers powerful support for the idea that X and Y interact *in vivo*, partly because immunostaining does not have some of the inherent biases of the biochemical approaches to showing association *in vivo* (see below). Here we provide a protocol for immunocytochemical colocalization of NMDA receptors and α-actinin-2 in rat brain.

A. Immunocolocalization of Ion Channel and Associated Protein in Rat Brain

1. Using a Vibratome tissue sectioning system (Technical Products International, Inc., St Louis, MO), prepare floating brain sections (50 μm thick) from anesthetized adult rats (~6 weeks age) perfused transcardiacally with 4% paraformaldehyde.

2. Permeabilize sections with PBS containing either 0.1–0.3% Triton X-100 or 50% (v/v) ethanol at room temperature for 15 min.

3. Wash sections in PBS at room temperature for 10 min three times.

4. Block potential nonspecific antibody binding sites by incubating the sections in antibody dilution solution (PBS containing 3% horse or goat serum and 0.1% crystallized BSA) at room temperature for 1 h. If detergent permeabilization was employed, include 0.1–0.3% Triton X-100 in the antibody dilution solution.

5. Add primary antibodies that specifically recognize NR1 receptor subunits (2 μg/ml of mouse monoclonal antibodies 54.1; PharMingen, San Diego, CA) and α-actinin-2 (1 μg/ml of rabbit polyclonal antibodies 4B2[6]) together to the tissue sections floating in the antibody dilution solution. Incubate sections at room temperature overnight.

6. Next day, wash the sections thoroughly three times in PBS at room temperature for 10 min.

7. Submerge the primary-stained sections in antibody dilution solution containing species-specific fluorophore-conjugated secondary antibodies (1:1000 dilution of antimouse Cy3- and 1:200 dilution of antirabbit FITC-conjugated antibodies;

Jackson ImmunoResearch, West Grove, PA). Incubate at room temperature for 2 h.

8. Wash sections three times in PBS at room temperature for 10 min.

9. To mount sections, transfer stained sections to microscope slides, add one or two drops of fluorescence mounting medium, and secure section with a glass coverslip. The fluorescently stained sections can be visualized by epifluorescence light microscopy (e.g., using a Zeiss Axioskop microscope, Thornwood, NY) or by confocal microscopy (e.g., using a confocal Bio-rad Richmond, CA, MRC 1000 microscope).

VI. Coimmunoprecipitation of Ion Channel and Associated Protein from Native Tissue

The most direct way to confirm that two proteins are associated *in vivo* is to purify them as an intact complex from a native tissue. Coimmunoprecipitation using antibodies specific for either protein is the most convenient way to achieving this. For ion channels and their associated proteins, this approach presents some difficulties. First, ion channels are typically not abundant proteins, thus their purification by any method is tricky. A partial purification to enrich for the ion channel may be needed prior to the immunoprecipitation to increase the yield. Second, ion channels are integral membrane proteins and they need to be solubilized by detergents before they can be purified by immunoprecipitation. Some ion channels are highly insoluble—NMDA receptors, for instance, are tightly bound to the postsynaptic density and are difficult to extract with anything less than SDS. These solubilization protocols may disrupt the interaction between the ion channel and their association proteins, leading to loss of the complex.

Typically, coimmunoprecipitation is performed using antibodies to protein X to immunoprecipitate the complex, and antibodies to protein Y to detect coimmunoprecipitated protein Y by Western blotting. Ideally, this coimmunoprecipitation should be performed in both directions, that is, antibodies to either protein should precipitate the other binding partner. It is also important to assay for the amount of cognate antigen that is precipitated (i.e., the amount of protein X brought down by anti-X antibodies) and to compare it with the amount of protein Y that is coimmunoprecipitated (i.e., the amount of protein Y brought down by anti-X antibodies). This will give you an idea of the efficiency of coimmunoprecipitation, that is, what fraction of Y is stably associated with X, etc. For this, it will be essential to include on the immunoblots control lanes that contain known or starting amounts of proteins X and Y in the extract to be immunoprecipitated. It is rare that the quantities of protein that can be isolated by immunoprecipitation are adequate to visualize by nonspecific means such as Coomassie blue staining. In any case, the immunoglobulins (Igs) used for the immunoprecipitation constitute a major portion of the precipitated proteins and their heterogeneity or impurity will complicate the interpretation of nonspecific protein staining (especially when using sensitive methods such as silver staining). In general then, the detection of the coimmunoprecipitated protein relies on its prior identification as a putative associated polypeptide and on its detection by specific antibodies in immunoblotting.

Because of this inherent "bias," and because of the high sensitivity of immunoblotting, it is critical to apply a large number of negative controls to ensure that the coimmunoprecipitation is specific. For instance, a small amount of protein X and Y may nonspecifically stick to the Igs or to the protein A-Sepharose beads used to precipitate the Igs, and be detected by immunoblotting and interpreted as coimmunoprecipitation of these proteins. Good controls include leaving out the immunoprecipitating antibody and using unrelated immunoprecipitating antibodies. In addition, control antibodies should be used to probe the immunoblots of the precipitated proteins, since the presence of unexpected extraneous proteins in these precipitates will raise concerns about the specificity of the coimmunoprecipitation. This latter problem is especially pertinent with ion channels, which are integral membrane proteins and sometimes difficult to solubilize. Detergent treatment of membranes may release large macromolecular aggregates or even small membrane domains, both of which may include proteins not even peripherally associated with the ion channel of interest. These proteins may be falsely coimmunoprecipitated with the ion channel and interpreted to be interacting with the ion channel. To reduce this problem, it is essential to clear the detergent extracts with high-speed ultracentrifugation ($100,000 \times g$ for 30 min at 4 °C to pellet partly solubilized membranes and large protein aggregates), and to use negative control antibodies to probe the immunoblots of the immunoprecipitated proteins.

In conclusion, the most meaningful results are those in which the coimmunoprecipitations (1) succeed in both directions (antibodies to protein X bring down protein Y, and antibodies to protein Y bring down protein X); (2) are highly specific (immunoprecipitates contain as little as possible other than proteins X and Y); and (3) are highly efficient (the ratio of amounts of X and Y in the immunoprecipitates is close to unity). Due to the inherent difficulties of working with nonabundant membrane proteins, it is extremely challenging to obtain convincing coimmunoprecipitation results approaching the ideal ones listed above. This is especially true when trying to do immunoprecipitation studies from the highly complex mixture of proteins in the brain.

A. Methods of Solubilization

Since solubilization is critical for any form of purification (including immunoprecipitation), it is worthwhile to optimize this step prior to immunoprecipitation. For membrane proteins such as ion channels, we generally try three types of detergent: Triton X-100 (1–2%); RIPA (1% NP-40, 0.5% deoxycholate, 0.1% SDS); and SDS (1–2%), followed by excess Triton X-100 to sequester the SDS prior to addition of antibodies. In addition, other types of detergent such as CHAPS or digitonin can also be tried. These detergents are typically used in buffers of physiologic pH (pH 7.4–8.0) containing 150 mM salt. Finding the right detergent is largely empirical. The SDS approach is only used when other methods fail to solubilize the ion channel, since SDS is a relatively harsh denaturing detergent.

Next, we present a protocol for showing coimmunoprecipitation of NMDA receptors and PSD-95. NMDA receptors and PSD-95 are highly insoluble except in SDS detergent.

B. Coimmunoprecipitation of Ion Channel and Associated Protein from Native Tissue

Due to the relative insolubility of NMDA receptors and PSD-95, the coimmunoprecipitation of NMDA receptors and PSD-95 complexes from rat brain membranes requires SDS solubilization of these proteins. Surprisingly, the complex remains associated in these conditions. The following protocol is adapted from Lau *et al.* (1996)

1. Prepare a rat brain membrane fraction as described earlier for the pull-down assay.
2. For each immunoprecipitation reaction add 25 μl of the membrane fraction (~200 μg protein) to an equal volume of 2× ice-cold extraction buffer [100 mM Tris (pH 8.0), 300 mM NaCl, 5 mM KCl, 10 mM EDTA, 10 mM EGTA, and various protease inhibitors] containing 2% SDS.
3. Incubate at 4 °C for 2 h with shaking.
4. Pellet insoluble materials at 60,000 × g at 4 °C for 1 h.
5. Remove the supernatant (extracted fraction) and "neutralize" the SDS by diluting in 5 volumes (250 μl) of 1× ice-cold extraction buffer [50 mM Tris (pH 8.0), 150 mM NaCl, 2.5 mM KCl, 5 mM EDTA, 5 mM EGTA, and various protease inhibitors] containing 2% Triton X-100.
6. Immunoprecipitate using 5 μg/ml of immunoprecipitating antibodies specific for either PSD-95 (Kim *et al.*, 1995) or SAP97 (Kim *et al.*, 1996). The rest of the protocol is as described earlier for coimmunoprecipitation of full-length proteins expressed in heterologous cells.

VII. Conclusion

The analysis of a putative ion channel associated protein involves numerous steps ranging from *in vitro* to *in vivo* studies. Proving the validity of such an interaction *in vivo* is extremely difficult, and requires genetic as well as biochemical and histological approaches. The most important point is that multiple lines of evidence are necessary to build up a case for a biologically significant direct interaction between an ion channel and its putative associated proteins.

References

Garcia-Anoveros, J., and Corey, D. (1997). *Annu. Rev. Neurosci.* **20,** 567.

Kim, E., and Sheng, M. (1996). *Neuropharmacology* **35,** 993.

Kim, E., Niethammer, M., Rothschild, A., Jan, Y. N., and Sheng, M. (1995). *Nature* **378,** 85.

Kim, E., Cho, K. O., Rothschild, A., and Sheng, M. (1996). *Neuron* **17,** 103.

Kornau, H. C., Schenker, L. T., Kennedy, M. B., and Seeburg, P. H. (1995). *Science* **269,** 1737.

Lau, L. F., Mammen, A., Ehlers, M. E., Kindler, S., Chung, W. J., Garner, C. C., and Huganir, R. L. (1996). *J. Biol. Chem.* **271,** 21622.

Niethammer, M., and Sheng, M. (1998). *Methods Enzymol.* **293**(7), .

Niethammer, M., Kim, E., and Sheng, M. (1996). *J. Neurosci.* **16,** 2157.

Sheng, M., and Kim, E. (1996). *Curr. Opin. Neurobiol.* **6,** 602.

Sheng, M., Tsaur, M. L., Jan, Y. N., and Jan, L. Y. (1992). *Neuron* **9,** 271.

Wyszynski, M., Lin, J., Rao, A., Nigh, E., Beggs, A. H., Craig, A. M., and Sheng, M. (1997). *Nature* **385,** 439.

CHAPTER 20

Secondary Messenger Regulation of Ion Channels/Plant Patch Clamping

Sarah M. Assmann and Lisa Romano

Department of Biology
Pennsylvania State University
University Park, Pennsylvania

I. Update

The original chapter on this topic, published in 1999, describes two major sets of techniques: methods for patch clamping of plant cells and methods for discerning secondary messenger regulation of ion channels, with particular emphasis on regulation by cytosolic calcium concentrations. The information provided in the original chapter concerning methods of plant patch clamping is still quite relevant, as the major advances in this area have not been technological but rather consist of application of standard methods to greater numbers of plant species and plant cell

411

DOI: 10.1016/B978-0-12-382204-8.00020-5

types. By contrast, as briefly described below, there have been major technical advances in the past 11 years in methods for measuring cytosolic secondary messengers, particularly ions.

A. Plant Patch Clamping

In the original chapter, there was an emphasis on the patch clamping of stomatal guard cells. This is still one of the plant cell types that is most heavily studied using patch clamp techniques, in part because osmotic regulation of cell volume, mediated primarily by ion fluxes, is central to guard cell function. In order for the patch pipette to access the cell membrane of any plant cell, the cell wall must be removed, resulting in a plant protoplast. Several alternative methods of guard cell protoplast isolation have been developed since 1999, with an emphasis on proto-plast isolation from *Arabidopsis thaliana* (Leonhardt *et al.*, 2004; Pandey *et al.*, 2002). *Arabidopsis* is the first plant species to have its genome sequenced, and remains the plant species with the largest genetic toolkit. Methods for isolation of *Arabidopsis* guard cell protoplasts on a large scale have allowed identification of guard cell transcriptomes and proteomes, and this omics information has been mined for applications to the study of guard cell ion channels and their regulation (Jammes *et al.*, 2009; Leonhardt *et al.*, 2004; Zhao *et al.*, 2008). In *Arabidopsis*, genetic knockout of specific ion channel genes has allowed elegant assessment of those genes' contributions to cellular function (Hirsch *et al.*, 1998; Hosy *et al.*, 2003; Kwak *et al.*, 2001; Lebaudy *et al.*, 2008; Mouline *et al.*, 2002).

Other *Arabidopsis* cell types which have joined the list of "patchable" cells include sieve tube elements (Hafke *et al.*, 2007) and companion cells (Ivashikina *et al.*, 2003) of the phloem (the plant vascular system responsible for distributing carbohydrates throughout the plant body), various types of root cells (Demidchik *et al.*, 2004; Kiegle *et al.*, 2000), hypocotyl cells (Cho and Spalding, 1996; Colcombet *et al.*, 2005), pollen (Becker *et al.*, 2004; Mouline *et al.*, 2002; Wang *et al.*, 2004), and crown gall tumor cells induced by *Agrobacterium* (Deeken *et al.*, 2003).

In addition to *Arabidopsis*, the chapter published in 1999 reported on patch clamp protocols for broad bean, pea, maize, barley, and rice. Since then, several additional species have been assessed by this technique. In particular, numerous patch clamp studies on various types of plant root cells have provided insights on how plants use ion channels to take up mineral nutrients from the soil. A non-comprehensive list of species in which various types of root cells have been patch clamped includes many studies on the cereals maize, barley, rice, and wheat, as well as poplar (Arend *et al.*, 2005), carrot (Bregante *et al.*, 2008; Costa *et al.*, 2004), common bean (Etherton *et al.*, 2004), and the symbiosome membrane of legume root nodules (Roberts and Tyerman, 2002), the structures in which fixation of atmospheric N_2 occurs.

Patch clamp studies on unusual plant species such as *Thellungiella halophila* which is a salt-tolerant *Arabidopsis* relative, *Thlaspi caerulescens* (Piñeros and Kochian, 2003), which is adapted to grow on natural soils with high heavy metal content, *Alyssum bertolonii*, which is adapted to serpentine soils of high nickel, cobalt, and cadmium content (Corem *et al.*, 2009), and white lupin (Zhang *et al.*, 2004), which forms unusual cluster roots in response to low soil phosphate, are revealing ion transport-based mechanisms utilized by plants to cope with adverse soil conditions. In addition, patch clamp analyses of ion channels in marine higher plants (Carpaneto *et al.*, 1999), algae (Binder *et al.*, 2003), and phytoplankton (Taylor and Brownlee, 2003) are uncovering unique ion channel physiologies that underlie these plants' existence in the saline marine environment.

As indicated in the original chapter, essentially every new plant cell type and species will require some tailoring of standard protocols in order to obtain high quality protoplasts for patch clamping. The best approach remains to simply survey and then test available protocols, such as those described in the articles referenced above, and then modify the most promising protocol according to observations made concerning protoplast health, yield, and susceptibility to gigaseal formation.

When the original chapter was written in 1999, green fluorescent protein (GFP) and its variants had not yet revolutionized cell biology (Dixit *et al.*, 2006). It is now well-established in many organisms, including plants, that expression of GFP driven by a cell-specific promoter can be used to mark a specific cell type. For plant patch clamping, the advantage of this method is that when protoplasts are isolated from a plant tissue containing many different types of cells, it is possible to pick out and patch solely those protoplasts of the GFP-marked cell type (Ivashikina *et al.*, 2003; Maathuis *et al.*, 1998). GFP tagging of ion channel proteins has also been employed to study the spatial distribution and redistribution of ion channels within the plant cell (Bregante *et al.*, 2008; Sutter *et al.*, 2007).

B. Secondary Messenger Regulation of Ion Channels

As in animal cells, secondary messenger regulation of plant ion channels includes regulation by channel protein modification events such as de/phosphorylation, by metabolites, by cellular redox status, and by pCa^{2+} and pH. Transgenic over- or underexpression of relevant enzymes and signaling proteins (e.g., kinases/phosphatases, Jammes *et al.*, 2009; Kwak *et al.*, 2002; Li *et al.*, 2000; Mori *et al.*, 2006; Xu *et al.*, 2006; NADPH oxidases, Kwak *et al.*, 2003; and heterotrimeric G proteins, Fan *et al.*, 2008; Wang *et al.*, 2001) has expanded greatly in the past decade as a powerful approach to elucidate signal transduction chains that target plant ion channels.

As discussed in the original chapter and as remains true today, "pipette biochemisty" also provides a powerful approach to the study of secondary messenger regulation of ion channels, particularly with regard to regulation by small molecules and metabolites. Secondary messengers such as ATP or cyclic nucleotides (Ma *et al.*, 2009), introduced via the patch pipette solution to the cytosol (in whole cell patch clamping), or to the cytosolic face of the membrane (in patching of outside-out membrane patches), can be used to dissect ligand-based modulation of ion channel activity.

Among the potent regulators of ion channel behavior in plant as well as animal cells is cytosolic Ca^{2+}, thus, measurement of cytosolic Ca^{2+} concentrations is of great interest. The original chapter focused on the available technology of Ca^{2+} sensor dyes such as Calcium Green-1, Indo-1, and Fura-2. In order to utilize such dyes to image cytosolic Ca^{2+} concentrations in intact plant or animal cells, the dyes must be either microinjected into individual cells or modified such that they are readily membrane-permeant. Patch clamping offers some work-arounds to this problem, as provision of the Ca^{2+} dye in the patch pipette results in its introduction to the cytosol via the whole-cell configuration, allowing simultaneous Ca^{2+} measurement and electrophysiological recording (Pei et al., 2000; Romano et al., 2000; Schroeder and Hagiwara, 1990). One of the most recent developments in Ca^{2+} measurement technology is FLEP (Fluorescence Combined with Excised Patch), in which the ratiometric Ca^{2+} dye, Fura-2, is included in the patch pipette solution, and a photomultiplier is focused at the tip of the patch pipette to record the Ca^{2+} signal (Gradogna et al., 2009). The method provides sufficient sensitivity to measure efflux of Ca^{2+} into the pipette via the small number of ion channels located in the circumscribed patch of membrane (Gradogna et al., 2009).

A major advance in Ca^{2+} imaging technology has been modification of GFP for use as a Ca^{2+} sensor (Miyawaki et al., 1999). Such GFP-based "cameleon" Ca^{2+} sensors have the advantage that, since they are proteins, the organism of interest can be transformed with a construct encoding the sensor. Cameleons can be targeted to specific cell types according to which promoter is chosen to drive expression, and cameleons can also be produced as fusion proteins with peptide sequences that target the cameleon to different organelles. Like Indo-1 and Fura-2, cameleons have the advantage of being ratiometric Ca^{2+} sensors, thereby allowing corrections for photobleaching and inhomogeneity of expression levels. The cameleon methodology was first applied to animal systems, but the technology readily transferred to plant systems (Allen et al., 1999), and has provided detailed information on temporal changes in cellular Ca^{2+} concentrations (Allen et al., 2001; Monshausen et al., 2009).

At the cell membrane, plants utilize primarily H^+ ATPases, rather than $2Na^+/K^+$ ATPases, to maintain a resting membrane potential that is typically -100 to -200 mV, considerably more negative than the membrane potential of the typical animal cell. As such, the regulation of cytosolic and extracellular pH and in turn the regulation of ion channels by protons is of particular interest in plant cell biology. GFP has been modified to also serve as a pH sensor, and this tool has been utilized in extracellular and intracellular pH imaging in plants (Fasano et al., 2001; Monshausen et al., 2009), although it has not yet been combined with simultaneous electrophysiological measurements of ion channel activity.

In conclusion, due primarily to the now routine ability in *Arabidopsis* and some other plant species to manipulate expression levels of specific ion channel and regulatory genes and to the explosion of GFP-based methodologies, there has been a great expansion in the methods available to the researcher interested in the study of plant ion channels and their regulation by secondary messengers.

II. Introduction

Early in the history of patch clamping it was recognized that the whole-cell and inside-out configurations of the technique offered a way to manipulate signals acting from within the cell, so-called "secondary messengers," and record consequent effects on electrophysiologic responses (Cahalan and Neher, 1992). In recent years, with the advent of sophisticated tools such as caged compounds and fluorescent ion indicators, the types of information that can be garnered by experimenters who combine electrophysiology and biochemical manipulations has grown tremendously. The purpose of this chapter is to provide an overview of the techniques available to manipulate and measure secondary messengers in living cells while simultaneously performing patch-clamp recording. We assume that the reader already has a basic familiarity with the patch-clamp technique. In this overview, we describe general information that will be of use regardless of the biological system under study. In addition, however, we also provide pointers on aspects of the techniques and special problems that are unique to plant systems. Many of our examples are based on the guard cell system. Guard cells have proved to be a premier system for the study of secondary messenger regulation of ion channels (Assmann, 1993). Guard cells are specialized cells located in the outermost tissue layer of aerial plant organs. Pairs of guard cells border microscopic pores called stomata. Through osmotic swelling and shrinking, based largely on transmembrane ion fluxes, guard cells control stomatal apertures, and thereby regulate both water vapor loss and photosynthetic CO_2 uptake, which occur through the stomatal pores.

III. Membrane Access

To study secondary messenger regulation of ion channels via the patch-clamp technique, the first requirement is that the experimenter be able to achieve the high-resistance "gigohm" seal between the patch pipette and the cell membrane that is essential to record low-noise patch-clamp data. Below, we describe methods of membrane access that have been successfully applied to plant cells.

A. Enzymatic Access

Each plant cell is completely encased in a cell wall. Therefore, one of the major differences between the patch clamping of animal and plant cells is the necessity in the latter case of removing the wall in order to gain membrane access. Several methods are available to achieve this. The first and still the most widespread approach is the use of cell-wall degrading enzymes (Raschke and Hedrich, 1989) to digest the wall, resulting in production of what is called a plant "protoplast." This use of enzymes parallels enzymatic treatments in some animal systems, where collagenase and/or trypsin are used to break down connective tissue and expose a clean membrane surface (Trube, 1983), but of course the particular enzymes involved are different. Primary plant cell walls contain carbohydrate polymers such as cellulose and pectin as major constituents, thus the enzyme mixtures used typically contain cellulases and pectinases. Secondary plant cell walls, and primary walls of some cell types in later stages of development contain, in addition to cellulose and pectin, significant amounts of lignin. This phenolic polymer is very difficult to degrade. Saprophytic organisms such as fungi do produce ligninases that might be of use in protoplasting lignified plant cells. However, to our knowledge, this approach has yet to be investigated by plant electrophysiologists, with the corollary that most patch-clamp investigations to date have been performed on protoplasts from cells possessing only primary walls.

Cellulases and pectinases are commercially available (see Appendix for list of suppliers with addresses) from several sources, including Calbiochem, Worthington, Sigma, and the Japanese companies Yakult and Seishin. In our own laboratory, the enzymes from Japan are preferred. Yakult produces two cellulases, Onozuka R-10 and Onozuka RS. Anecdotally, we have found that it is easier to obtain gigohm seals on guard cell protoplasts that have been produced using Onozuka RS than on those obtained with R10. Seishin produces a very effective pectinase, Pectolyase Y-23, which we use in conjunction with cellulase RS.

The specific concentrations and ratios of cellulase and pectinase that are optimal for obtaining protoplasts differ with both species and cell type. For example, guard cell protoplasts from *Arabidopsis thaliana* are prepared using Onozuka R10 and Macerozyme R10 in a 16-h digestion (Pei *et al.*, 1997) while guard cells of *Vicia faba* require Onozuka RS and Pectolyase Y-23 in an approximately 1-h digestion (Miedema and Assmann, 1996). In contrast, when preparing mesophyll cell protoplasts from *V. faba* we use R10 cellulase and Macerozyme R10 for 1 h (Li and Assmann, 1993). Because of such variations in wall digestibility, a significant amount of time may be spent trying to find an enzyme cocktail that can break down the cell wall of cells of interest, and yet leave the resulting protoplasts in a condition suitable for patch clamping. Other factors that influence the effectiveness of the enzyme solutions include the presence or absence of ascorbic acid, digestion temperature, pH, presence and strength of agitation during tissue digestion, and extent of trituration to release protoplasts from the tissue. In addition, the removal of the cell wall leaves the protoplast vulnerable to osmotic stress. The optimum osmolality of the isolation buffers must be assessed so that the protoplasts are round, but not swollen or easily ruptured. For example, while we utilize the same enzyme mixture for isolation of both *Arabidopsis* and *V. faba* mesophyll cell protoplasts, we have found that it is best to utilize a 0.45-M sorbitol solution for *Arabidopsis* and a 0.6-M sorbitol solution for *V. faba* (see Table I). Unfortunately, there is little substitute for trial and error in determining the proper enzyme recipe and protocol. As a starting point, sample recipes for protoplasting solutions for several different types of plant cells are given in Table I.

Table I
General Protoplast Isolation Protocols[a]

Species/tissue	Solutions	Enzyme 1[b]	Enzyme 2[b]	Digestion conditions	References
Vicia faba (broad bean) guard cells (blended epidermal peels of 3, youngest, fully expanded leaves of plants 3–4 weeks old)	Blending medium: 10 mM MES, 5 mM CaCl₂, 0.5 mM ascorbic acid, 0.1% PVP40, pH 6 (KOH) Basic medium: 0.45 M sorbitol, 0.5 mM CaCl₂, 0.5 mM MgCl₂, 0.5 mM ascorbic acid, 100 μM KH₂PO₄, 5 mM MES, pH 5.5 (KOH)	0.7% Cellulysin cellulase, 0.1% PVP40, 0.25% BSA, 0.5 mM ascorbic acid, in 55% basic medium, 45% distilled water (v/v), pH 5.5 (KOH), 10 ml total volume	1.5% RS cellulase, 0.02% Pectolyase Y-23, 0.25% BSA, 0.5 mM ascorbic acid 100% basic medium, pH 3.5 for 5 min (HCl), final pH 5.5 (KOH), filter through 0.45-μm filter, 10 ml total volume	Enzyme 1: 30 min/28 °C; shaking: 162 excursions/min Enzyme 2: 50–80 min at 17 °C; shaking: 108 excursions/min	Miedema and Assmann (1996)
Arabidopsis guard cells (blended epidermal strips of 2, 2-cm rosette leaves)	Blending medium: distilled water Incubation medium: 0.5 M mannitol, 0.1 mM KCl, 0.1 mM CaCl₂, 10 mM ascorbic acid, 0.1% kanamycin sulfate, pH 5.5 (Tris)	1.3% Cellulase R10, 0.7% Macerozyme R10, 0.5% BSA in incubation medium	NA	16 hr at 22 °C; 70 rpm	Pei *et al.* (1997)
Arabidopsis mesophyll cells (8 rosette leaves, 2 cm in length with abaxial epidermis peeled)	Basic medium (see *V. faba* guard cell protocol above)	0.4% Macerozyme R10, 1% Cellulase R10, 0.1% PVP40, 0.2% BSA, 10 ml basic medium, pH 5.5	Same as enzyme 1, 10 ml	Enzyme 1: 2 min vacuum infiltration, followed by 5 min shaking (160 excursions/min) Enzyme 2: 45 min at 24 °C shaking: 108 excursions/min	Romano and Assmann (1997)
Vicia faba mesophyll cells (1 bifolate leaf from a plant 3–4 weeks old with abaxial epidermis peeled)	0.6 M Sorbitol, 1 mM CaCl₂	Enzymes same as *Arabidopsis* mesophyll, in 0.6 M sorbitol, 1 mM CaCl₂ solution	Same as enzyme 1	Enzyme 1: same as *Arabidopsis* mesophyll Enzyme 2: same as *Arabidopsis* mesophyll but for 1–1.5 hr	Li and Assmann (1993)
Hordeum vulgare (barley) xylem parenchyma (minced stelar tissue from roots stripped of cortex)	0.5 M Mannitol, 1 mM CaCl₂	2% Cellulase R10, 0.02% Pectolyase Y-23, 2% BSA, 10 mM sodium ascorbate, pH 5.5 (H₂SO₄)	NA	2–2.5 hr at 20 °C; shaking: 100 excursions/min	Wegner and Raschke (1994)

(continues)

Table I *(continued)*

Species/tissue	Solutions	Enzyme 1[b]	Enzyme 2[b]	Digestion conditions	References
Pisum sativum (pea) stem (epidermal strips from 7-day-old etiolated plants) and leaf (lower epidermis of young leaves in light grown plants) epidermal cells	Wash solution: 2.325% Gamborg's B5 (GIBCO), 2 mM CaCl₂, 10 mM MES mannitol to 0.61 M, pH 5.5 (KOH) Bath solution: 5 mM CaCl₂, 2 mM MgCl₂, 10 mM potassium-citrate, mannitol to 0.21 M, pH 5.5	1.7% RS cellulase, 1.7% Cellulysin cellulase, 0.026% Pectolyase Y-23, 0.2% BSA, in wash solution	NA	5–10 min in enzyme solution, 2 × 5 min in wash solution (all steps at 30° on rotary shaker at 50 rpm) Place in bath solution (in which patch-clamp experiments are to be done)	Elzenga *et al.* (1991)
Zea mays (maize) root cortex cells (chopped cortex tissue)	0.5 M sorbitol, 1 mM CaCl₂, 5 mM MES, pH 6 (KOH)	0.5% PVP, 10,000 MW, 0.5% BSA, 0.8% RS cellulase, 0.08% pectolyase in 0.5 M sorbitol, 1 mM CaCl₂, 5 mM MES, pH 6.0 solution	NA	3 h at 28 °C with agitation	Roberts and Tester (1995)
Oryza sativa (rice) root-hair protoplasts (0.5 cm of root tip of 3-day-old rice seedlings)	Basic medium (see above) with 0.40 M sorbital	1.5% RS cellulase, 0.1% Pectolyase Y-23, 0.2% BSA in basic medium	NA	Place roots in a 100-μm pore nylon filter. Fix filter into 50-ml centrifuge tube which contains enzyme solution. After 10 min at room temperature, add 40 ml basic medium and centrifuge. Repeat wash step with basic medium.	Yu and Wu (1997)

[a]All percent values are in w/v unless indicated. NA, Not applicable; BSA, bovine serum albumin; PVP40, polyvinylpyrrolidone, MW 40,000; MES, 2-(N-Morpholino)ethanesulfonic acid.

[b]Addresses of sources of enzymes are given in the Appendix: Cellulysin cellulase (Calbiochem), RS cellulase, Cellulase R10, Macerozyme R10 (Yakult Pharmaceutical Ind. Co. Ltd), Pectolyase Y-23 (Seishin Corporation), pectolyase (Sigma Chemical Co.).

Although the specific recipes for protoplast isolation may vary from tissue to tissue, a general procedure can be followed for most protoplast isolations. The first step is to prepare the tissue from which the protoplasts will be isolated. For example, to isolate guard cells or epidermal cells, the epidermis of leaves can be peeled either manually or with the aid of a blender, and these peels subjected to further treatment. This procedure may be performed using a buffer or distilled water. The second step is to place the prepared tissue in the appropriate enzyme solution. In some cases, the tissue may be subjected to two different enzyme cocktails in order to obtain a population that is enriched for one type of protoplast. For example, the guard cell isolation procedure commonly employed by our laboratory (Miedema and Assmann, 1996) uses two enzyme solutions, the first of which releases epidermal cells. The osmolality of the first enzyme solution is low (<300 mOsm kg^{-1}) so that the released epidermal protoplasts burst. The peels, which still contain the intact guard cells, are then rinsed and placed in a second enzyme solution which preferentially releases guard cells. During incubation in the enzyme solution, the protoplasts should be gently shaken and held at an appropriate temperature. In general cooler temperatures are preferred for optimal protoplast health, but this also slows the digestion process. After digestion is complete, the protoplasts are filtered through nylon mesh which allows released protoplasts to pass through freely, thus separating them from debris. The protoplasts can then be rinsed free of the enzyme solution by centrifugation in the presence of a "washing" medium. The centrifugation speed and time should be the minimum necessary to allow collection of the protoplasts. After the final centrifugation, the protoplasts should be allowed to recover for at least 1 h on ice in the dark.

One drawback to the use of enzymes is that the commercially available enzymes are impure. Contaminating proteases may influence ion channel intactness and activity, yet such effects are difficult to assess, simply because it is impossible to patch clamp a completely walled plant cell. In the case of guard cells, double-barreled voltage-clamp recordings on intact cells have confirmed, at least qualitatively, current responses seen in protoplasts (Armstrong and Blatt, 1995), but voltage clamping is not practical for all cell types. For example, it cannot be readily used for interior cells within plant organs (such cells can be accessed as protoplasts since the protoplasting procedure results in dissociation and digestion of the entire tissue). Another drawback to the enzymatic digestion of the cell wall is that protoplasting may remove signaling compounds such as oligosaccharins that originate from the wall itself. Finally, removal of the wall may significantly alter protoplast physiology and thus influence signal transduction chains that regulate channel activity. Most obviously, plant cells differentiate into a variety of shapes as a result of physical constraints imposed by the cell wall. When the wall is removed the resulting protoplast is inevitably spherical. Cytoskeletal rearrangements that ensue as, for example, wall–membrane–cytoskeletal linkers are removed, may have significant effects on ion channel activity (Kim et al., 1995; Thion et al., 1996).

B. Alternatives to Standard Enzymatic Protoplasting: Osmotic Shock and Laser–Assisted Patch Clamping

For the above reasons, it would be helpful to have a method of membrane access that minimized or avoided enzymatic treatments. Two such approaches have recently been developed. The first involves the short-term exposure of tissue to relatively high concentrations of enzymes, followed by hypoosmotic shock (Elzenga *et al.*, 1991). The cells of the tissue are first plasmolyzed in a high-osmolality solution, and then briefly (5–10 min) exposed to a cocktail of hydrolytic enzymes. This enzyme exposure weakens the cell wall. The tissue is then placed in a low-osmolality solution, which causes the protoplasts to swell and pop out of the weakened cell walls. This technique produces viable protoplasts with a high seal success rate from several tissue types, including epidermis, mesophyll, and coleoptile cortex (see Table I, *Pisum sativum* protocol).

Depending on the system under study, the experimenter can also take advantage of inherent weaknesses in the cell wall to minimize enzyme exposure. More than a decade ago, procedures were described to obtain "subprotoplasts" from root hairs by brief enzymatic digestion (Cocking, 1985). Root hairs are outgrowths from root epidermal cells and their cells walls are weakest at the growing tip. Thus, enzymatic exposure results in the bulging of exposed membrane from the tip of the root hair. With gentle centrifugation, these "subprotoplasts" pinch off and can be collected. Alternatively, the exposed membrane can simply be "patched" *in situ*, thus allowing the recording of patch-clamp data from the intact root. Figure 1A shows protoplasts emerging from rice root hairs (see Table I), and Fig. 1B gives an example of the whole-cell K^+ currents recorded from such subprotoplasts.

Recently, our laboratory has refined a protocol that allows membrane access for patch-clamp experiments in the complete absence of enzymatic digestion of the cell wall (Henriksen *et al.*, 1996). The technique was originally applied to large algal cells (De Boer *et al.*, 1994; Taylor and Brownlee, 1992), but if adequate precision is employed, it can also be applied to the relatively small cells of higher plants. A microbeam (diameter of 0.3–1 μm) from a 337-nm UV laser is focused through the microscope onto a portion of the cell wall (Fig. 2A). The high-energy laser light ablates the wall, thus exposing the membrane. Focusing is achieved by passing the laser light through a Galilean type beam expander that has been empirically adjusted such that the laser beam is parfocal with the microscope focal plane. Once this is achieved, the beam will remain in focus as the microscope focus is changed during the course of an experiment, provided that the microscope is designed such that the carrier containing the dichroic mirror (Fig. 2A) does not change its vertical position when the microscope focus is changed. Detailed information on the equipment and optics required for the setup is given in a recent methodology (Henriksen and Assmann, 1997).

Wall ablation is enhanced by prior exposure of the tissue to Calcofluor White (Sigma), a UV-absorbing dye that binds to cell walls (De Boer *et al.*, 1994). Damage to the protoplast from the laser beam is avoided by the precise focusing

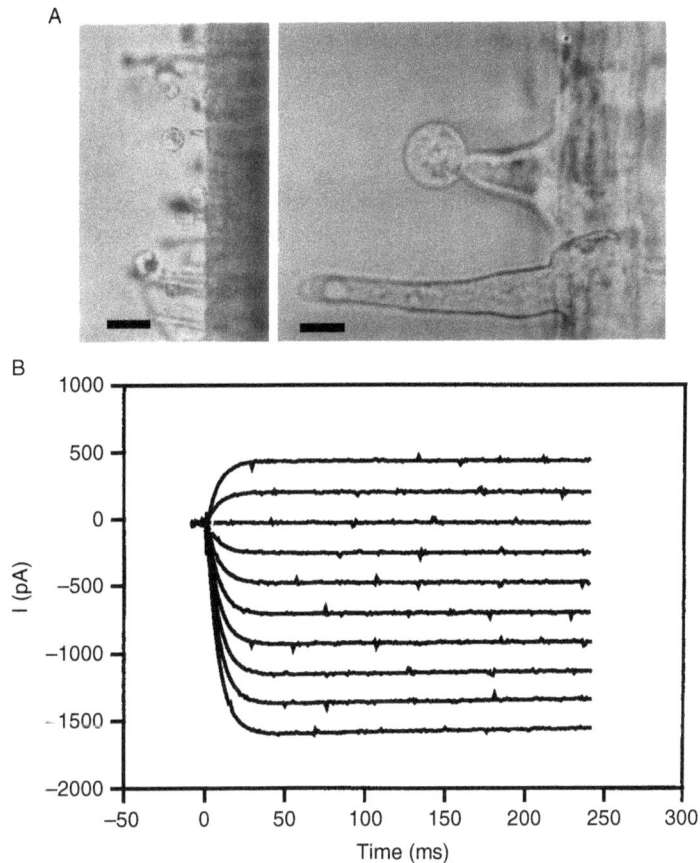

Fig. 1 (A) Emerging protoplasts from rice (*Oryza sativa*) root hairs. Bars: 70 μm (left) and 10 μm (right). (B) Whole-cell recording from rice root hair protoplast. Voltage protocol: The membrane potential was held at −50 mV and pulsed to potentials ranging from −180 to 0 mV. Pipette solution: 98 mM potassium glutamate, 2 mM KCl, 2 mM EGTA, 0.3 M sorbitol, 10 mM HEPES, pH 7.2 (KOH). Bath solution: 10 mM potassium glutamate, 2 mM $MgCl_2$, 5 mM $CaCl_2$, 0.4 M sorbitol, 10 mM HEPES, pH 6.0 (KOH)(Yu and Wu, 1997).

of the laser beam in the *z* dimension, and by temporarily plasmolyzing the protoplast with high osmoticum so as to remove the protoplast from the laser light path (Fig. 2B). On deplasmolysis, the plasma membrane "blebs" through the hole in the wall and is accessible to the patch pipette (Fig. 2C). We have found that a slow rate of deplasmolysis, as provided by use of a linear gradient maker (De Boer *et al.*, 1994), is essential to expose the membrane without rupturing the protoplast. In our experience with guard cells of *V. faba*, it is easy to obtain high-resistance seals on the exposed membrane with little or no application of suction, although this is apparently not to be true for all systems (De Boer *et al.*, 1994). While technically

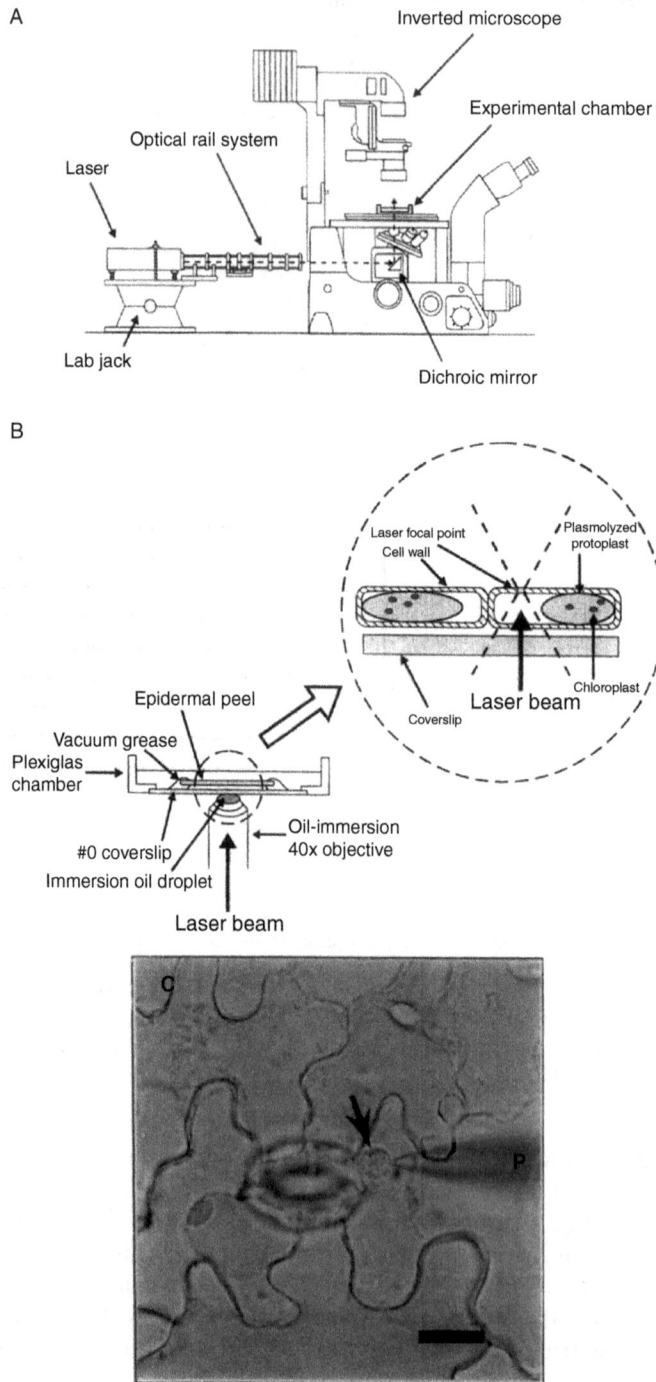

Fig. 2 (A) Schematic of the setup used for laser-assisted patch clamping. (B) Schematic illustrating laser-targeting of the upper wall of a plasmolyzed guard cell. (C) Laser-accessed membrane "bleb" from a guard cell of *Vicia faba*. Bar: 20 μm. (A, B, reproduced with permission from Henriksen and Assmann, 1997; C, reproduced with permission from Henriksen *et al.*, 1996.)

more demanding than membrane exposure via protoplasting, laser-assisted patch clamping provides an alternative approach when enzymatic methods of protoplast isolation fail. Additionally, laser-assisted patch clamping allows recording of patch-clamp data *in situ*. The technique also has the potential to provide information on the spatial distribution of ion channels that cannot be obtained from spherical protoplasts.

IV. Patch Clamping of Plant Versus Animal Cells

The technique of patch clamping is similar for all types of cells, and the reader is referred to general reviews on this topic (Hamill *et al.*, 1981; Rudy and Iverson, 1992; Sakmann and Neher, 1995). However, there are subtle differences between the patch clamping of animal cells and plant protoplasts. This section focuses on such differences.

Isolated animal cells from primary cultures can usually be maintained in nutrient media at the appropriate temperature for up to 1 week and used in patch-clamp experiments (Trube, 1983). Plant protoplasts must generally be used within 12 hr of isolation. This is because protoplasts, no matter how healthy, are still missing a vital organelle, the cell wall, and are therefore fragile and more susceptible to physical, osmotic, and temperature stress. The life of protoplasts can be greatly prolonged if they are handled gently and kept on ice in the dark. In addition to the frailty of the protoplasts, protoplasts of many plant cell types are able to synthesize new cell walls after isolation, and this may eventually render the membrane too "dirty" to form tight seals.

Another difference between patch clamping of plant and animal cells is the solutions which are commonly used, especially bath solutions. In the majority of animal studies, the bath solution contains sodium, in order to mimic the extracellular environment that prevails *in vivo*. The Na^+ electrochemical gradient is an important source of energy for the uptake of sugars and amino acids and for pH regulation, while the resting membrane potential of animal cells is largely determined by the K^+ electrochemical gradient. Placing many animal cell types in a high K^+ bath solution could depolarize the membrane to unphysiologic levels. This could trigger the activation of voltage-gated channels, especially Ca^{2+} permeable channels, and set in motion a number of Ca^{2+}-dependent processes, such as secretion. In contrast, even moderate Na^+ levels are toxic to plants, and Na^+ plays a more minor role in plant membrane physiology. Therefore, many plant patch-clamp experiments are performed in solutions where K^+ salts are the predominant ions. Depending on the types of experiments, KCl, potassium glutamate, or potassium gluconate are commonly used. In our laboratory, we focus primarily on K^+ currents, and a typical bath solution consists of 10 or 100 mM KCl, 1 mM $CaCl_2$, 1 mM $MgCl_2$, 10 mM HEPES (or MES), pH 5.6 or 7.2, with the osmolality adjusted to 460 mOsm kg^{-1} with sorbitol or mannitol. If anion channels are being studied, CsCl can be substituted for KCl. It is also important to note that in many animal patch-clamp experiments, pharmacologic blockers are used to isolate specific ion currents. This is also possible in plant systems, with the caveat that the "specificity" of the blocker must be assessed. For example, verapamil, which is used in animal systems to block Ca^{2+} currents, blocks outward rectifying K^+ currents in many types of plant cells (L. A. Romano personal observation; Thomine et al., 1994).

The pipette solution commonly used in our laboratory to study K^+ currents consists of 80 mM potassium glutamate, 20 mM KCl, 2 mM $MgCl_2$, 2 mM Mg-ATP (in Tris), 10 mM HEPES, pH 7.2 or 7.8. The final osmolality is adjusted to 500 mOsm kg^{-1} with sorbitol or mannitol. Ethyleneglycol-*bis*(β-aminoethyl)-*N*, *N*,*N'*,*N'*-tetraacetic acid (EGTA) or 1,2-*bis*(o-Aminophenoxy)ethane-*N*,*N*,*N'*,*N'*-tetraacetic acid (BAPTA) is added in various concentrations depending on the amount of Ca^{2+} buffering desired for a particular experiment.

The pipettes used for patch-clamping plant protoplasts, in general, are smaller than those used for animal cells. Typically for patch-clamping animal cells in the whole-cell configuration, pipette resistances of 1–5 $M\Omega$ are used. With plant protoplasts, pipette resistances of as high as 10–30 $M\Omega$ (in the solutions described above) are used for whole-cell recordings. The reason that smaller sized pipettes are required to form tight seals is unknown, but may arise from the fact that the cytoskeletal structure of the plant cell is disrupted when the cell wall is removed. This may allow a large amount of cytoplasm to be sucked into the pipette when seal formation is attempted, which tends to lower seal success rates. This problem may also be addressed by lowering the osmolality of the bath solution, which causes the protoplasts to swell (Thomine et al., 1994).

When patch clamping was a new technique, it was widely accepted that it was extremely difficult to patch clamp plant cells. Now it is possible, at least with some types of plant cells, for example, *V. faba* guard cell protoplasts, to obtain similar seal success rates as those enjoyed by animal patch clampers. However, after obtaining a high-resistance seal, there are a few problems a plant patch clamper may specifically encounter. The main difficulty we have found is increases in series resistance. There are two main causes of such increases. The first has to do with the membrane properties of the protoplasts. Some types of protoplasts, for example, guard cells but not mesophyll cells of *V. faba*, have membranes that can best be described as "gooey." This can be especially noticed when attempting to excise a patch for single-channel recordings. The protoplast seems to ooze along with the pipette as it is moved instead of a patch of membrane being plucked off. This membrane consistency may also promote a gradual increase in series resistance if the whole-cell configuration is maintained for prolonged periods. It is extremely important to monitor and correct for changes in series resistance during the course of an experiment, especially when a cell serves as its own control, where an increase in series resistance may be mistaken for a secondary messenger effect. The second cause of increases in series resistance is the partial or complete blocking of the pipette by a chloroplast. This problem can be extremely frustrating in cells such as mesophyll cells, which are packed with chloroplasts just the right size to plug the end of a patch pipette. The chloroplasts can be moved out of the pipette tip by suction or blowing, but this may result in loss of the seal.

Protoplasts also seem to be sensitive to ambient temperature. Most patch-clamp experiments using plant cells are conducted at room temperature. This is usually not an issue, except when room temperature becomes too high (e.g., 28 °C). At these high temperatures protoplast health becomes compromised.

After overcoming the technical difficulties of getting and keeping a seal, it would seem that patch clamping of plant protoplasts would be almost identical to the patch clamping of animal cells. For the most part, it is. However, it is important to note that by nature, plant cells respond to virtually every environmental stimulus. For example, while in mammals vision is restricted to particular cell types, almost every cell type in a plant contains some type of photoreceptor. Since in patch clamping the experimenter seeks to gain insight into the physiologic function of ion channels, the hope is that isolated protoplasts maintain their responsiveness to these stimuli. However, the very act of patch clamping also exposes the protoplast to a barrage of "stimuli" that the researcher must try to minimize or at best keep constant from cell to cell. For example, simply viewing cells under the microscope introduces two stimuli: light and heat. Several studies have shown effects of different wavelengths of light on various transport processes, including ion channel (Cho and Spalding, 1996) and H^+-ATPase (Assmann et al., 1985) activity. Such effects can be overcome by installing a green filter in the microscope light path (since most plant cells do not respond to this region of the spectrum) or by minimizing light exposure by turning the microscope light off after the seal is obtained. Microscope lights also generate a fair amount of heat, and at least one study reports an effect of temperature on plant ion channels (Ilan et al., 1995). Again, this problem can be overcome simply by installing heat filters and/or minimizing the amount of exposure to light. While performing simultaneous patch clamping and Ca^{2+} imaging (see below) we also noticed that on days when the ambient temperature was high due to warm outdoor temperatures and large amounts of heat generated by equipment, the Ca^{2+} level of patch-clamped guard cell protoplasts was abnormally high. Whether this high level of Ca^{2+} was in response to the signal of high temperature or a sign of poor protoplast health is unclear. As mentioned previously, high temperature may also compromise the health of the protoplasts, even before patch clamping is initiated.

Finally, the biggest difference between patch-clamping plant and animal cells is the amount of patience required of the experimenter. In many animal systems, seal formation is almost instantaneous. The plant cell patch clamper must be willing to spend sometimes up to 10 min or even longer nursing a patch along until a gigohm seal is formed. The time required varies from cell type to cell type, and also within a preparation. The key is to not abandon a cell that does not seal immediately, but to experiment to find different tricks to get the cells to seal. Sometimes doing the exact opposite of what intuition suggests works. For example, after several attempts to apply inordinate amounts of suction to get a particularly unruly batch of mesophyll protoplasts to seal, we changed our strategy to that of applying almost no suction at all, and this was successful. In our experience, if it is possible to achieve seal resistances of about 200 MΩ, then the protoplasts are "patchable" and successful seal formation should result following suitable experimentation by the investigator to find the proper pipette shape and size, suction protocol, etc. However, if seal resistances as high as 200 MΩ cannot be achieved, some modification of the enzymatic isolation protocol will probably be required.

V. Secondary Messenger Regulation of Ion Channels

Once a high resistance seal and the appropriate patch-clamp configuration have been obtained, there are two basic approaches toward connecting secondary messengers with ion channel behavior. The researcher can quantify endogenous levels of a signaling compound, and correlate changes in the level of this signal with alterations in channel activity. Or, the experimenter can deliberately manipulate levels of a putative secondary messenger and observe the electrophysiologic result. The following sections provide technical background on each of these topics. Such approaches have been used extensively by researchers studying animal systems, and are currently being incorporated into plant research as well. In the guard cell system, for example, electrophysiologic studies of the signal cascades underlying ion channel responses to the plant hormones abscisic acid and auxin have suggested involvement of secondary messengers also common to animal systems, including Ca^{2+}, pH, G proteins, inositol 1,4,5-trisphosphate (IP_3), and a calcineurin-like phosphatase. In addition, plant-specific signals such as a calcium-dependent protein kinase with a calmodulin-like domain have also been implicated in ion channel regulation (Pei et al., 1996). Several recent reviews (Assmann, 1993; Blatt and Thiel, 1993; Ward et al., 1995) on the guard cell system provide a good source of references for the reader seeking examples of how the techniques described below are actually capitalized on to address biological questions.

A. Measuring Endogenous Secondary Messengers

The availability of fluorescent indicator dyes for Ca^{2+}, H^+, and other ions, coupled with patch clamping, provides the means to measure in vivo changes in endogenous secondary messengers concomitantly with changes in cell electrophysiology. In our laboratory we perform simultaneous Ca^{2+} measurements and whole-cell recording. Therefore, our discussion focuses on the use of Ca^{2+} indicator dyes. The general principles discussed would hold equally for imaging of other secondary messengers, for example, pH.

The ability to monitor two vitally important aspects of cellular physiology, the activity of Ca^{2+}-regulated ion channels via patch-clamp recording (Hille, 1992) and changes in intracellular Ca^{2+} concentration via ratiometric Ca^{2+} imaging (Grynkiewicz et al., 1985), has led to enormous insight into the impact of Ca^{2+}

metabolism on ion channels in animals systems (Lledo *et al.*, 1992; van den Pol *et al.*, 1996). The combination of these two powerful tools has, to our knowledge, been applied previously only once to plant protoplasts, in a study employing photometric Ca^{2+} measurements with whole-cell patch clamp (Schroeder and Hagiwara, 1990). Grabov and Blatt (1997) have also recently applied the techniques of intracellular Ca^{2+} measurement and voltage clamp to intact guard cells still situated within an epidermal peel. The reason for the relative dearth of studies may simply be that the long time "inaccessibility" of plant protoplasts to patch-clamp studies has caused an overall lag in the types of experimental techniques that have been applied to them.

1. Overview of Ca^{2+} Measurement in Living Cells

The first decision the experimenter must make when setting up for simultaneous electrophysiology and Ca^{2+} quantification is what type of Ca^{2+} measurement system to employ. Three options are available. From most to least expensive, these are confocal Ca^{2+} imaging, standard Ca^{2+} imaging, and photometry. Confocal microscopy offers extremely high resolution imaging in the *x, y,* and *z* planes. It is usually the method of choice when spatial information is important (Hernández-Crus *et al.*, 1990), unless quite fast measurements of cellular Ca^{2+} changes are required, in which case photometry may be the only option. However, because of their high cost, confocal microscopes are not uniformly available, and if available they are often part of multiuser facilities, with limitations on the amount of time that the equipment is accessible to any one user.

Standard Ca^{2+} imaging provides spatial information in the *x* and *y* dimensions. While the spatial resolution is poorer than with confocal microscopy, standard Ca^{2+} imaging is also significantly less expensive. Finally, in photometry, a single fluorescent signal integrated over the entire cell or region of interest is recorded, using one or two photomultipliers (PMTs) as detectors of the emitted light. Other than the option to select a region of interest, photometry provides no spatial information, but it requires the least amount of time per measurement because essentially one "giant" pixel (i.e., the entire protoplast or selected region of interest) is instantaneously measured (as opposed to the finite scan time required for a confocal image), and because the use of highly sensitive PMTs as detectors reduces or eliminates the need for temporal signal averaging, as may be required for standard Ca^{2+} imaging. Because each signal comprises many photons, photometry may also be the only workable method when signal strength is weak.

The second decision facing the researcher is which Ca^{2+} indicator dye to use. The catalog from Molecular Probes provides a wealth of information on this topic (Haugland, 1996). One issue apparently unique to plants is that some indicator dyes partition into the large central vacuole of plant cells, and are therefore useless for measuring cytoplasmic Ca^{2+} concentrations (Bush and Jones, 1987; Elliott and Petkoff, 1990). Dyes that are conjugated to dextran ("dextranated") avoid this problem; other dyes must be evaluated for their usefulness on a case-by-case basis.

When possible, it is always preferable to use a Ca^{2+} indicator that is ratiometric, such as Indo-1 or Fura-2. A ratiometric dye is one that has a shift in either the excitation peak (Fura-2) or the emission peak (Indo-1) on binding of Ca^{2+}, instead of the increase in fluorescence intensity that is characteristic of nonratiometric dyes such as Calcium Green-1. The use of ratiometric dyes significantly reduces the potential artifacts of measuring intracellular Ca^{2+} concentrations. Because the fluorescence intensities of two wavelengths are ratioed, the effects of photobleaching, inhomogeneous dye distribution, and nonuniform excitation intensity are canceled. In some cases it may be impossible to use Indo-1 or Fura-2, either because there is not an appropriate UV light source available, or because the experimental protocol precludes the use of UV light, for example, if caged compounds are to be used (see below). Under these circumstances it may be possible to use pseudoratiometric dyes such as Calcium Green–Texas Red dextran (Molecular Probes). Calcium Green is a single-wavelength dye whose fluorescence intensity increases when Ca^{2+} is bound. Texas Red fluorescence is totally Ca^{2+} independent. These two molecules are linked via dextran molecules so that they are *de facto* equally distributed. The Ca^{2+}-independent fluorescence intensity of Texas Red can be used as an indicator of photobleaching and dye concentration, and the Ca^{2+}-dependent Calcium Green fluorescence can then be pseudoratioed against the Ca^{2+}-independent background of Texas Red to reduce the artifacts associated with single-wavelength Ca^{2+} indicators.

2. Simultaneous Patch-Clamp and Confocal Ca^{2+} Imaging in Plant Protoplasts

Recently, our laboratory has combined patch clamping and Ca^{2+} quantification in the study of guard cell physiology using confocal laser scanning microscopy (Schild, 1996) to perform the Ca^{2+} imaging. In general, the setup to perform the techniques is similar to that used in animal systems. Therefore, a relatively brief overview of the system and methods will be provided, with emphasis given to the modifications in both electrophysiology and imaging necessary for application to plant protoplasts.

The preparation of protoplasts for combined patch clamping and confocal Ca^{2+} imaging is identical to the protocol for "regular" patch-clamp experiments (see above). The solutions used are also the same except that EGTA is omitted from the pipette solution, and 60 μM Indo-1 pentapotassium salt (Molecular Probes) is included. A 1-mg ml^{-1} stock solution of Indo-1 is almost exactly 1 mM, so we simply add 1 ml of pipette solution to 1 mg, store it at $-80\,^{\circ}$C in 60-μl aliquots, and add one aliquot to 1 ml total of pipette solution. This concentration of Indo-1 allows for a good fluorescence signal, without adding excessive exogenous Ca^{2+} buffering, which may alter the Ca^{2+} signal that one is trying to measure. The pipette solution is filtered just before use with a 0.2-μm syringe filter and stored on ice covered with foil during the experiment.

The resistance of the pipettes used for simultaneous patch-clamp and Ca^{2+} imaging is critical. Typically with guard cell protoplasts, it is easier to obtain a high-resistance seal with a pipette resistance of 15–20 MΩ (in 100 mM KCl bath solution/20 mM KCl–80 mM potassium glutamate pipette solution). However, this size of pipette is not suitable for Ca^{2+} imaging because the resulting series resistance is about 30–40 MΩ. The high series resistance leads to a low diffusion rate of dye into the protoplasts and, as a result, a poor fluorescence signal. Typically for Ca^{2+} imaging, we use pipettes with resistances of 6–10 MΩ. With this size of pipette it is still possible to have a high success rate of obtaining the initial seal (about 80%), and the dye loading is complete and stable within 5–10 min. There is one drawback to using large pipettes, however. The faster diffusion rate of dye into the cell also means that there is a faster rate of diffusion of soluble intracellular factors out of the cell (Horn and Korn, 1992). Therefore, if the currents being studied are susceptible to rundown, the rate of rundown may also increase. With guard cells we have routinely been able to maintain cells with robust currents and Ca^{2+} signals for 30 min to 1 h.

To perform simultaneous patch-clamp measurements and confocal Ca^{2+} imaging, we simply attach a patch-clamp setup to an existing Zeiss confocal laser scanning microscope. A schematic of the confocal setup is shown in Fig. 3A. For Ca^{2+} imaging, Indo-1 is excited with an 8-s scan of a 364-nm UV laser. An 80/20 beamsplitter in the laser path allows 20% of the UV laser light to reach the specimen and 80% of the fluorescence signal from the specimen to be transmitted through the light path. The fluorescence signal is then split by a 460-nm dichroic mirror, and detected simultaneously by two PMT detectors that receive light of 450–505 and 400–435 nm through the use of appropriate emission filters. An important issue in imaging plant cells is the fluorescence of chlorophyll, which is in the red (>600-nm) range. The fluorescence of Indo-1 peaks at 400 and 480 nm. If a longpass filter (i.e., one that passes wavelengths longer than 460 nm) is placed in the position of emission filter 1, as is commonly done for Indo-1 imaging, the fluorescence from chloroplasts will contaminate the signal received by the detector. Therefore, a 450- to 505-nm bandpass filter is installed for emission filter 1, which transmits the Indo-1 signal but not the chlorophyll fluorescence. With this modification, contamination from autofluorescence on either PMT is negligible in guard cell protoplasts.

During the course of an experiment, it is also desirable to view a transmitted light image of the cell under study, to adjust the pipette position or simply to monitor the cell's health. In our system, this is accomplished by using either a red (633-nm) or a green (543-nm) laser and a transmission detector. As mentioned earlier, isolated protoplasts may respond to light of various wavelengths, so it is important to conduct appropriate controls to make sure that whatever sources of light one uses do not produce an effect. This can be accomplished by comparing data obtained in "regular" patch-clamp experiments with data from patch-clamp/Ca^{2+} imaging experiments. Also, simply subjecting patch-clamped cells to the same laser regime as used during an imaging experiment but in the absence of Indo-1 and treatment will confirm that the laser exposure is without effect.

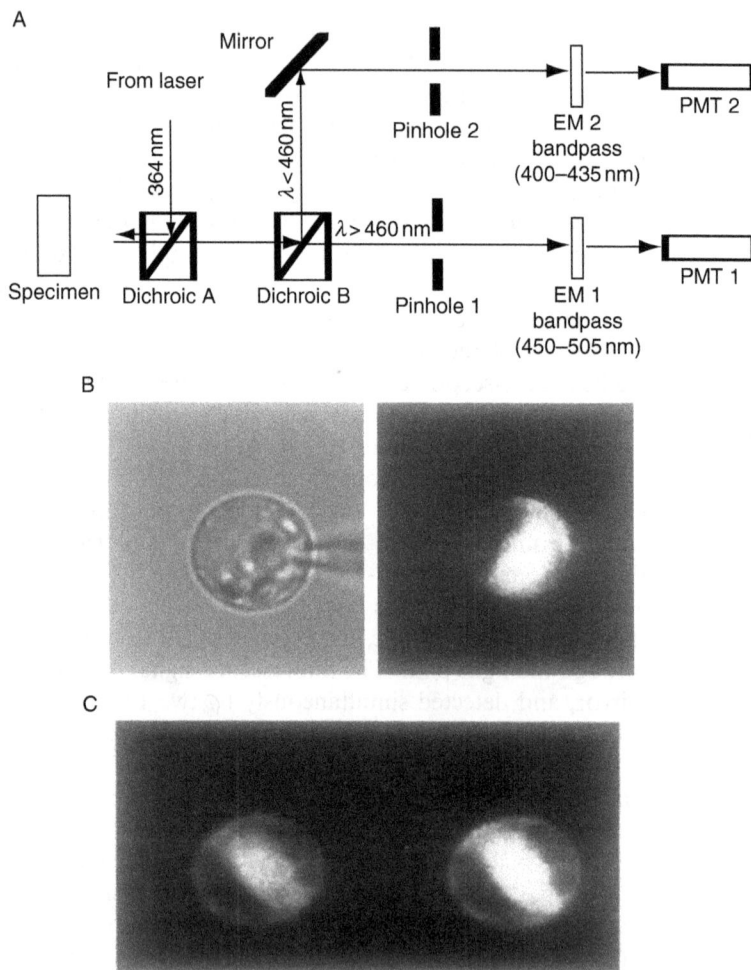

Fig. 3 (A) Diagram of confocal laser scanning microscope used for simultaneous patch clamping and Ca^{2+} imaging. The same setup, with modifications, is used to release caged Ca^{2+} (nitr-5). See text for details. (B) Bright-field (left) and fluorescence (right) images of a guard cell protoplast from *V. faba* loaded with Indo-1 pentapotassium salt via the patch pipette. Note that the fluorescence signal is isolated to the cytoplasmic region, indicated by the presence of chloroplasts. Pipette solution: 80 mM potassium glutamate, 20 mM KCl, 2 mM $MgCl_2$, 2 mM Mg-ATP (in Tris), 60 μM Indo-1 pentapotassium salt, sorbitol to 500 mOsm kg^{-1}, 10 mM HEPES, pH 7.8 (KOH). Bath solution: 100 mM KCl, 1 mM $CaCl_2$, 1 mM $MgCl_2$, sorbitol to 460 mOsm kg^{-1}, 5 mM HEPES, 5 mM MES, pH 5.6 (Tris) (L. A. Romano, S. Gilroy, S. M. Assmann, unpublished.). (C) Increase in Calcium Green fluorescence showing a qualitative increase in Ca^{2+} after UV photolysis of nitr-5. Pipette solution as in (B), except Indo-1 is omitted and 50 μM Calcium Green dextran (10,000 molecular weight) and 2 mM nitr-5 charged with 1 mM Ca^{2+} (as $CaCl_2$) are added. Bath solution same as in (B). (L. A. Romano, S. Gilroy, S. M. Assmann, unpublished).

As mentioned previously, a potential problem with measuring intracellular Ca^{2+} with indicator dyes in plant protoplasts is compartmentalization of the dye into the vacuole (Bush and Jones, 1987; Elliott and Petkoff, 1990). Examination of transmitted light and fluorescence images of the same guard cell protoplast under study clearly shows that Indo-1 pentapotassium salt loaded via the patch pipette does not concentrate in the vacuole of this cell type (Fig. 3B). The cytoplasm is clearly distinguished in bright field from the vacuole by the presence of chloroplasts, whereas the vacuole is clear. The corresponding fluorescence image of the same cell loaded with Indo-1 shows a similar pattern.

When studying secondary messenger systems in patch clamping, the investigator tries to maintain the cell in a true physiologic state. The use of the whole-cell patch-clamp technique gives the experimenter the advantage of being able to control the contents of the cell cytosol via equilibration with the patch pipette solution. However, this also results in the loss of intracellular factors that may be crucial to signal transduction pathways. The loss of these components can be reduced or eliminated by using the perforated-patch technique, although this technique has not yet been widely applied to plant cells. To combine perforated patch recording with Ca^{2+} imaging, the indicator dye must be preloaded into the cells. Investigators studying animal systems have utilized the acetoxymethyl (AM) esters of Ca^{2+} indicator dyes (Lledo et al., 1992; van den Pol et al., 1996). However, the use of AM esters in plant systems has been difficult at best. The main problems associated with AM loading of plant protoplasts are external hydrolysis of the ester, lack of internal ester hydrolysis, and compartmentalization into the vacuole (Bush and Jones, 1987; Elliott and Petkoff, 1990). Some plant cell types can be loaded with Indo-1 in a pH-dependent manner. Bush and Jones (1987) developed a method to load the pentapotassium salt of Indo-1 into barley aleurone protoplasts at pH 4.5. This method has also been successful in loading Arabidopsis roots (Legué et al., 1997). In this method, cells are incubated for approximately 1 h with 25 µM Indo-1 pentapotassium salt in a buffer held at pH 4.5 with 25 mM dimethylglutaric acid (DMGA). The cells are then washed twice to remove Indo-1 from the bathing media. Using this method, there is no detectable accumulation of Indo-1 into the vacuole. We have yet to combine this approach with confocal Ca^{2+} imaging/ electrophysiology because, in our hands, attempts to "acid load" guard cells of V. faba have been unsuccessful. It is possible that because of the strong pH regulation of this cell type, either the environment surrounding the cells may be of different pH than the bulk media, or the cells may be sensitive to the high concentration of DMGA needed for acid loading.

Calibration of the fluorescence signals is routinely performed in vitro (Gilroy, 1996; Legué et al., 1997). This is done on a daily basis using Ca^{2+} calibration standards from Molecular Probes and 6-µm polystyrene beads (Bangs Laboratories). In this procedure, 10 µM Indo-1 (1 µl of a 1-mM stock) and a small volume of bead suspension (less than 0.5 µl) is added to 100 µl of calibration solution. A small amount of this solution is placed between two coverslips and mounted on the microscope. The midplane of the beads is then focused on and imaged as described above. Focusing on the beads ensures that there are neither artifacts from imaging close to the coverslip nor edge effects from a droplet of calibration solution. *In vivo* calibrations are performed to confirm the validity of the *in vitro* approach (Lledo *et al.*, 1992). This is accomplished by patch clamping the cell as described above, except that pipette Ca^{2+} concentration is altered by adding either $CaCl_2$ (5 or 10 mM) or EGTA (5–20 mM). These conditions correspond to totally Ca^{2+}-saturated dye and totally Ca^{2+}-free dye, respectively.

The methods of acquisition of patch-clamp and Ca^{2+} data are largely determined by the types of experiments to be performed. The two acquisition systems on our setup are not linked, that is, they are independently triggered. The data obtained from simultaneous patch-clamp/confocal Ca^{2+} imaging experiments are analyzed as follows. The electrophysiology data are acquired and analyzed using standard software such as pClamp version 6.0.3 (Axon Instruments). The Indo-1 imaging data are analyzed using IPLab Spectrum (Signal Analytics Corp.), and analysis consists of the following steps. The fluorescence intensity data from the two emission wavelengths are separated, and the background, quantified as a confocal scan on a region of the dish lacking cells, substracted. If significant autofluorescence is present, one can obtain an average value for autofluorescence from a number of cells, and subtract this value from the data. This, however, is imprecise because each cell will have its own specific level of autofluorescence, and autofluorescence intensity will also vary from compartment to compartment (e.g., cytoplasm vs. vacuole) within the cell. Therefore, the best strategy is to load sufficient dye into the cells such that the contribution to the signal from autofluorescence is negligible. After any corrections, the fluorescence intensity values are then ratioed, and the results are presented as a map of Ca^{2+} concentration across the cell. Pseudocolor is usually applied as a visual aid in distinguishing spatial distributions in Ca^{2+} concentration. The Ca^{2+} concentration can be derived in two ways. The first approach is to convert the ratio value using the expression:

$$[Ca^{2+}] = K_d Q[(R - R_{min})/(R_{max} - R)]$$

where K_d is the dissociation constant for Indo-1, R is the ratio value (λ_2/λ_1 where λ_2 is 400–435 nm and λ_1 is 450–505 nm), R_{min} is the ratio value of a calibration solution with zero free Ca^{2+}, R_{max} is the ratio value of a calibration solution with saturating free Ca^{2+}, and Q is the ratio of F_{max}/F_{min} for λ_2. The second approach is to construct an empirical *in vitro* Ca^{2+} calibration curve using standards (Molecular Probes) that provide several free Ca^{2+} concentrations, and then to plot the ratio values obtained experimentally onto this calibration curve.

When Ca^{2+} imaging is performed, it is possible to obtain information about spatial changes in intracellular Ca^{2+}. However, it may also be beneficial to obtain overall cell averages in Ca^{2+} concentration. In plant cells, a large portion of the cell volume is taken up by the vacuole, and this may present a problem. To obtain an average value of cytoplasmic Ca^{2+} concentration, the largest available contiguous area of cytoplasm is measured, and black areas that correspond to the vacuole are excluded. It is important to avoid the area directly adjacent to the pipette, since it may produce artifacts due to reflection from the glass, the fact that the seal, although gigohm, is not infinite, and the fact that cellular buffering of supplied Ca^{2+} diffusing from the patch pipette is not instantaneous.

3. Standard Ca^{2+} Imaging and Photometry

Several companies, including Axon Instruments, Photon Technology International, and Universal Imaging, offer various configurations of imaging systems. Axon Instruments, with a history as a vendor of electrophysiology equipment, appears to be progressing most rapidly toward offering an integrated patch-clamp/ standard imaging system.

The two indicators that plant biologists most commonly use at present for standard Ca^{2+} imaging are Fura-2-dextran and Indo-1. Fura-2 is a dual excitation/single emission ratiometric dye, meaning that the emission signal at a single wavelength is measured following excitation of the dye by two different wavelengths. Since both excitation wavelengths for Fura-2 are in the UV region (340 and 360–380 nm) a common method to achieve the two different excitation wavelengths is to install a filter wheel with appropriate filters downstream of the standard fluorescence light source for the microscope. Excitation wavelengths can also be produced using a monochromator, if sufficient funds are available for purchase of this expensive option. Excitation light is directed to the specimen by the use of an appropriate dichroic in the microscope filter holder, and emission light (510 nm for Fura-2) is similarly directed through an appropriate filter to a detector. If spatial information is not required, the detector can be a PMT. Alternatively, if somewhat lower sensitivity and temporal resolution are permissible, and spatial information is important, the detection apparatus is a camera. A bewildering variety of options is available, including cooled charge-coupled device (CCD) cameras, intensified CCD cameras, integrating video cameras, and silicon intensified (SIT) cameras. Individual camera models may differ considerably, but as a brief overview, Table II summarizes the various types of cameras in terms of cost, sensitivity, spatial resolution, speed, and data storage requirements. Not included in the table are frame transfer cameras, which are out of the price range of most investigators. These cameras are very fast because the images are electronically shuttled to a new region of the chip. Camera technology is evolving rapidly; therefore, no specific recommendations are made here. Before purchasing a system, the best strategy is simply to have a variety of cameras demonstrated, such that the most appropriate one for the application at hand can be purchased.

When performing standard Ca^{2+} imaging with Fura-2, the camera (or photomultiplier) can be mounted directly on the microscope port. This is advantageous because there is no possibility that the image will shift in the camera field during the

Table II
Types and Properties of Cameras Used in Imaging

Camera type	Output	Price	Sensitivity	Spatial resolution	Temporal resolution	Data storage requirement per image (approximate) (Mbyte)
Integrating video camera	Analog (video)	Inexpensive	Low	Fair	Slow because longer integrating time required to overcome low sensitivity	0.5
Cooled CCD	Digital	Relatively inexpensive	Very high	Good (varies between cameras)	Slow because of the read-off time required (this can be shortened by choosing a smaller region of interest)	1–2 or more
Intensified CCD	Analog (video) or digital	Moderate	Moderate	Good (varies between cameras)	Fast but integration may be required	0.5
	Usually analog (video)	High	High to very high	Good	Fast but integration may be required	0.5

experiment or even between collection of the two successive images that comprise a ratiometric measurement as might be the case if, for example, the detector were mounted on the vibration isolation table instead of directly on the microscope.

There are two disadvantages to the use of Fura 2-dextran. The first is that, since it is excited by UV light, this dye cannot be used to measure Ca^{2+} before and after uncaging of a second messenger: the initial Ca^{2+} measurement will itself uncage the dye, and, conversely, the high intensity light used for uncaging may photobleach the Ca^{2+} indicator. Work-arounds to this problem have been applied to animal systems. For example, a system has been described (Kirby *et al.*, 1994) that uses low intensity light from a xenon arc lamp to measure Ca^{2+} while releasing only minor amounts of caged Ca^{2+} and a laser flash to release the caged compound that is sufficiently brief so as to avoid photobleaching of the Ca^{2+} indicator. A specific problem with the use of Fura-2 in plant cells is that it emits toward the red region of the spectrum. For this reason, its use to date in plants has been largely restricted to achlorophyllous cells such as root cells and pollen tubes (Felle and Hepler, 1997; Pierson *et al.*, 1994). It may be possible to separate the Fura-2 fluorescence, with a peak at 510 nm, from the longer wavelengths of chlorophyll fluorescence by using a narrow bandpass emission filter, although we have not tested this.

The problem of chlorophyll autofluorescence can be avoided by the use instead of Indo-1, as described in the section on confocal imaging. However, Indo-1 also offers a problem for combined patch-clamp and standard imaging. Since Indo-1 is a dual emission dye, the filter wheel must be mounted directly onto the microscope on the emission side. When such a system was demonstrated in our laboratory, the amount of vibration generated by the filter wheel inevitably resulted in loss of the seal on the protoplast. We do not know whether this is a universal problem or one unique to our cell type (guard cells). To work around this problem, one could mount two cameras, and install a beamsplitter upstream, such that light of the appropriate wavelengths for Indo-1 emission is directed to each of the two cameras. Nikon and perhaps other microscopy companies as well do offer the necessary hardware, and it is indeed an option if photometry is to be done. However, for standard imaging, (1) it is likely to be prohibitively expensive to purchase two cameras, and (2) unless the cameras are precisely in register, down to the individual pixel, the Ca^{2+} measurement will be completely erroneous. This issue does not arise in confocal microscopy of ratiometric dyes, where alignment occurs simply because both of the pixel-by-pixel images are formed simultaneously, one pixel at a time.

Given the problems described above, we are in the initial stages of investigating the applicability of Calcium Green/Texas Red dextran for Ca^{2+} imaging in chlorophyllous cells. A filter wheel on the upstream side provides the two excitation wavelengths of 488 and 578 nm. A custom-made emission filter that has two windows of transmission, for the emission wavelengths of Calcium Green and Texas Red, but cuts off sharply to eliminate chlorophyll fluorescence, transmits the emitted light alternately from the two dyes, as the two dyes are alternately excited by the spinning filter wheel. We purchased this custom-made filter from Chroma, but it may also be available from other companies, such as Omega Optical.

Once decisions have been made on how to configure the imaging system and which indicator to use, simultaneous patch-clamp and standard Ca^{2+} imaging experiments follow the same basic procedures for data acquisition and analysis as described earlier in the section on confocal Ca^{2+} imaging.

B. Administering Exogenous Secondary Messengers

To unravel signal transduction chains involved in ion channel modulation, channel activity may also be compared before and after administration of a regulatory molecule. Such regulators may include plant hormones, kinases/phosphatases and other regulatory proteins, antibodies against regulatory proteins, ions such as Ca^{2+} and H^+, and synthetic regulators of endogenous signaling molecules. Examples of compounds falling in the last category are GTPγS and GDPβS, which are nonhydrolyzable analogs of GTP and GDP. GTPγS and GDPβS lock G proteins into active and inactive states, respectively, thereby allowing detection of involvement of a G protein in the signaling pathway under investigation (Fairley-Grenot and Assmann, 1991; McFadzean and Brown, 1992).

If the regulatory molecule acts from the apoplastic side of the membrane, or if it is internally acting and either membrane permeant or efficacious in the inside-out patch configuration, then it is a relatively simple task to apply the regulator in the bath solution and monitor subsequent effects. However, if the regulator is effective only from within the intact cell, or if it is membrane-permeant but available in such limited amounts or at such high expense that bath application is impractical, then it is necessary to introduce the regulator via the pipette solution. There are various means of accomplishing this, and each has advantages and disadvantages.

1. Comparison of Control and Experimental Cells

One approach is simply to make up two pipette solutions, with and without the regulatory molecule in question, and then to compare recordings made from different cells using the two different solutions. This approach has the advantages of simplicity, and of enabling the researcher to compare control and treatment data obtained at precisely the same point in time during each recording, thus avoiding successive treatments on a single cell, a situation in which rundown or other temporal alterations in cell physiology may confound data interpretation. On the other hand, interpretation of results from these types of experiments can be difficult if cells within the population have variable electrophysiologic "profiles" even within the control treatment. Under such circumstances, it is most useful to record from one and the same cell before and after application of the signaling molecule, using the techniques described below. Because the recordings made will be sequential in nature, it is crucial to determine from control cells that the "baseline" response does not change throughout the duration of the recording.

2. Repatching and Pipette Perfusion

Several methods are available for successive administration of intracellular treatments. If the cells are sufficiently robust as to withstand successive sealing attempts, one can initially obtain whole-cell access and record control currents, then withdraw that patch pipette and seal onto the cell again and achieve whole-cell access using a second pipette that contains the regulatory substance. A more elegant approach is to utilize the pipette perfusion technique, in which an additional port on the pipette holder is used to introduce solution, thus forcing a replacement of the initial pipette solution with the new solution. Pipette perfusion has the practical advantage that the equipment is inexpensive, and the methodological advantage that the regulatory molecule may be subsequently removed by further perfusion, and the cell monitored for recovery of the initial response. This technique has been a standard method in the animal patch clamper's toolkit for some time (Tang *et al.*, 1992). A recent paper (Maathuis *et al.*, 1997) describes an inexpensive perfusion apparatus that has been used in whole-cell recordings from a variety of plant cell types. In theory, pipette perfusion is simple. The main difficulty lies in fine-tuning the perfusion so as to not disrupt the seal. The only special equipment that is required is a pipette holder that has an additional port for the perfusion capillary. Such pipette holders can be purchased from Warner Instrument Corp. in a variety of sizes to fit commonly used headstages. The perfusion capillary should be made of fused silica coated with polyimide (Polymicro Technologies). The capillary is drawn out in a flame to the appropriate size (30–50 μm in diameter; Tang *et al.*, 1992). The silica capillary should be filled with the initial pipette solution to remove air bubbles. The perfusion capillary is then fit into a piece of polyethylene tubing, which is threaded through the perfusion port of the pipette holder. The polyethylene tubing does not enter the pipette shaft, but stops at the end of the perfusion port. The silica capillary is then moved close to the tip of the patch pipette, taking care not to push it through the end of the pipette. The free end of the polyethylene tubing is placed in a small reservoir filled with the starting pipette solution. A seal is formed in the usual way, and after control recordings have been made, the free end of the polyethylene tubing is repositioned in a reservoir containing the pipette solution to be perfused. Suction is then applied through the usual suction port, and the perfusion can be monitored by viewing the rise of fluid in the pipette shaft. The rate of pipette perfusion will vary depending on the diameter of the perfusion capillary and the amount of suction applied. The rate of diffusion of the perfused solution will depend on both the distance of the perfusion capillary from the pipette tip and the diameter of the tip. If fluorescence microscopy is available, the inclusion of a fluorescent dye such as Lucifer Yellow in the second pipette solution can aid in determining the rate of perfusion into the cell.

3. Caged Compounds

The use of caged compounds in combination with the patch-clamp technique is a powerful probe into the secondary messenger regulation of ion channels. Caged compounds are biologically inert until they are released from their protecting group, that is, "uncaged," by exposure to UV light, which breaks specific chemical bonds within the compound. Thus, like repatching and pipette perfusion, this technique also allows a "before" and "after" look at ion channel modulation by a chemical stimulus. A wide variety of caged compounds is available, including Ca^{2+}, amino acids, nucleotides, neurotransmitters, hormones, and enzymes (Adams and Tsien, 1993; Haugland, 1996). Kits for the synthesis of caged compounds are also available from Molecular Probes.

As with simultaneous Ca^{2+} imaging and patch clamping, the basic technique for using caged secondary messengers in combination with patch clamping is the same as for conventional patch-clamp experiments except for the inclusion of the desired caged compound in the pipette solution. There must also be a means to UV irradiate the cell under study to release the caged probe and, if possible, to monitor the uncaging. We will describe the use of caged Ca^{2+} in combination with patch clamping or guard cell protoplasts.

There are currently three types (Adams and Tsien, 1993; Ellis-Davies and Kaplan, 1994; Gurney *et al.*, 1987) of caged Ca^{2+} available from Molecular Probes or Calbiochem. We routinely use 1-[2-Amino-5-(1-hydroxy-1-[2-nitro-4,5-methylenedioxyphenyl]methyl)phenoxy]-2-(2'-amino-5'-methylphenoxy)ethane-N,N,N',N' tetraacetic acid, sodium (nitr-5), because it has been used successfully in intact guard cells (Gilroy *et al.*, 1990) and in barley aleurone protoplasts (Gilroy, 1996). Another form of caged Ca^{2+}, o-nitrophenyl EGTA (NP-EGTA), is based on EGTA, and is also Ca^{2+} selective; however, it has not been used extensively in plant systems. The other commonly used form of caged Ca^{2+}, 1-(2-Nitro-4,5-dimethoxyphenyl)-1,2-diaminoethane-N,N,N',N'-tetraacetic acid (DM-nitrophen), is based on EDTA, and therefore has a high affinity for Mg^{2+} as well, which limits its usefulness. The chemical structure of nitr-5 is based on BAPTA, and it is therefore highly selective for Ca^{2+} over Mg^{2+}. Upon exposure to UV light (360 nm), the affinity of nitr-5 for Ca^{2+} drops from a K_d of 150 nM to about 6.5 μM (Adams and Tsien, 1993). Figure 3 illustrates the elevation in cytosolic Ca^{2+} engendered following treatment of a nitr-5-loaded guard cell protoplast with UV light.

To measure the effect of a rapid increase in intracellular Ca^{2+} on ion currents in guard cell protoplasts, we use the following methods. Guard cell protoplasts are prepared as usual and the ionic compositions of the bath and pipette solutions are also the same as usual with the following exceptions. EGTA is omitted from the pipette solution and replaced with 1 or 2 mM nitr-5. We prepare a 10-mM stock solution of nitr-5 in pipette solution, which is stored at -80 °C until use. Calcium Green-1 is also added at 50 μM, from a 2-mM stock solution, also prepared in pipette solution and added to the pipette solution just prior to the experiment. Calcium Green-1 is a nonratiometric dye, and is included to give a qualitative indication of the changes in intracellular Ca^{2+} on photolysis of the caged Ca^{2+}. It is important to keep all caged probes and fluorescent indicators protected from light.

Caged Ca^{2+} is actually a Ca^{2+} chelator whose affinity for Ca^{2+} is decreased on photolysis. Therefore, Ca^{2+} must be bound to the chelator which can be released. Often, the caged Ca^{2+} is charged with additional Ca^{2+} before being added to the pipette solution. We have found that this sometimes produces abnormal whole-cell currents in guard cell protoplasts. When nitr-5 is added uncharged (i.e., with no added Ca^{2+}) whole-cell currents are normal and there is an increase in Calcium Green-1 fluorescence after exposure to UV light. This phenomenon most likely results from initial charging of the nitr-5 with abundant intracellular Ca^{2+} as the cells attempt to regulate their Ca^{2+} at normal levels (S. Gilroy, personal communication (1997) and Gurney et al., 1987). The appropriate concentrations to add of caging compound and free Ca^{2+} (if any) must be determined empirically for each experimental system.

In our laboratory, the system used for releasing caged Ca^{2+} and monitoring it with Calcium Green-1 is the same as that used for confocal Ca^{2+} imaging with Indo-1, with the following alterations. Dichroic A is a 488 beamsplitter, which allows wavelengths 488 nm and shorter to reach the specimen and reflects wavelengths longer than 488 through the light path. A 488-nm argon laser is used to excite Calcium Green-1. Dichroic B is open, and a 515- to 565-nm bandpass emission filter is placed upstream of the channel 1 PMT (Calcium Green-1 has an emission peak at 530 nm). Calcium is released by 2- to 8-s scans of the UV laser. If a UV laser is not available, a xenon flashlamp (Rapp and Guth, 1988) or even the standard UV source for fluorescence microscopy may be used. If a flashlamp is employed, care must be taken that the attendant electrical surge does not damage sensitive electrophysiology equipment.

There are several possible caveats to the use of nitr-5 or any caged compound. First, energy introduced into the biological system by the UV treatment may itself affect the cells, thus necessitating that control data be obtained on cells that contain no caged compound and are subjected to the identical UV treatment as the experimental cells. As for previous procedures, it is important to be certain that the indicators and lasers/light sources used in the experimental protocol do not have an ancillary, unknown effect on the cellular physiology. For example, the 488-nm light used to excite Calcium Green-1 is in the blue region of the spectrum, and plants possess several blue light receptors, which have been implicated in a variety of signaling pathways (Ahmad and Cashmore, 1993; Assmann et al., 1985; Cho and Spalding, 1996). In our experience, the laser wavelengths and intensities used for confocal Ca^{2+} imaging with Indo-1 or for caged Ca^{2+} release have no effect on whole-cell currents of guard cells. However, Calcium Green-1 is potentially phototoxic (Eilers et al., 1995), thus the dye concentrations and excitation intensities must be carefully tested to ensure that the cell is not damaged. Second, by-products from the uncaging event may themselves have a physiologic effect. For example, Blatt et al. hypothesized that a decrease in outward K^+ current that they observed on photorelease of caged IP_3 in guard cells actually resulted from a decrease in cytosolic pH stemming from H^+ release during uncaging (Assmann,

1993; Blatt *et al.*, 1990). Because of such possibilities, it is valuable, when feasible, to compare results obtained from the same compound chemically caged in more than one way. In the case of caged Ca^{2+}, the control experiment can alternatively consist of performing the uncaging in the presence of BAPTA, so that the Ca^{2+} that is released is quickly sequestered. In practice, this requires high amounts of BAPTA (10–20 mM; Bates and Gurney, 1993) to abolish the rise in Ca^{2+}, and this may make seal formation difficult. A third issue is that, unlike the case with pipette perfusion, the signaling compound, once uncaged, cannot be subsequently removed. Finally, the caged compound approach is, of course, only available if the requisite molecule can be purchased or synthesized. Despite these potential roadblocks, the elegance and simplicity of the caged compound approach dictates that use of this technique will continue to increase.

VI. Concluding Remarks

After publication of the seminal paper on the patch-clamp method in 1981, it was 3 years until plant membrane biologists capitalized on this technique (Moran *et al.*, 1984; Schroeder *et al.*, 1984). Since then, however, the gap in progress between studies of animal versus plant ion transport systems has been steadily narrowing. It is our hope that this chapter will, in the realm of secondary messenger regulation of ion channels, contribute toward this trend.

Appendix

Addresses of major vendors cited in this article are listed here.

Axon Instruments, Inc., 1101 Chess Drive, Foster City, CA 94404; phone: 650-571-9400, fax: 650-571-9500

Bangs Laboratories Inc., 979 Keystone Way, Carmel, IN 46032; phone: 317-844-7176

Calbiochem, P.O. Box 12087. La Jolla, CA 92039; phone: 619-450-9600, fax: 619-453-3552

Chroma Technology Corp., 72 Cotton Mill Hill, Unit A-9, Brattleboro, VT 05301; phone: 802-257-1800, fax: 802-257-9400

GIBCO BRL, Grand Island, NY; phone: 800-828-6686, fax: 800-331-2286

Molecular Probes Inc., 4849 Pitchford Avenue, Eugene, OR 97402; phone: 541-465-8300, fax: 541-344-6504

Omega Optical, P.O. Box 573, 3 Grove Street, Brattleboro, VT 05302; phone: 802-254-2690, fax: 802-254-3937

Photon Technology International, 1 Deerpark Drive, Suite F, South Brunswick, NJ 08852; phone: 908-329-0910, fax: 908-329-9069

Polymicro Technologies Inc., 18019 North 25 Avenue, Phoenix, AZ 85023; phone: 602-375-4100, fax: 602-375-4110

Seishin Corporation, 4-13 Koamicho, Nihonbashi, Chuo-ku, Tokyo, Japan; phone: 03-3669-2876, fax: 03-3669-1684

Sigma, St. Louis, MO; phone: 800-325-3010, fax: 800-325-5052

Signal Analytics Corp., 440 Maple Avenue East, Suite 201, Vienna, VA 22180; phone: 703-281-3277, fax: 703-281-2509

Universal Imaging Corp., 502 Brandywine Parkway, West Chester, PA 19380; phone: 610-344-9410, fax: 610-344-9515

Warner Instruments Corp., 1125 Dixwell Avenue, Hamden, CT 06514; phone: 203-776-0664, fax: 203-776-1278

Yakult Pharmaceutical Ind. Co., Ltd., 1-1-19 Higashi-Shinbashi, Minato-ku, Tokyo, 105 Japan; phone: 03-3574-6766 fax: 03-3574-7254

Acknowledgments

We thank Dr. Simon Gilroy (University of Wisconsin) for many valuable discussions on Ca^{2+} measurement and imaging techniques, and Dr. Henk Miedema (Pennsylvania State University) for helpful comments on the text. We also thank Dr. Wei-hua Wu (Beijing Agricultural University), for access to the data depicted in Fig. 1. Research in S.M. Assmann's laboratory on plant cell signal transduction and ion channel regulation is supported by grants from NSF and USDA.

References

Adams, S. R., and Tsien, R. Y. (1993). *Annu. Rev. Physiol.* **55,** 755.

Ahmad, M., and Cashmore, A. R. (1993). *Nature* **366,** 162.

Allen, G. J., Kwak, J. M., Chu, S. P., Llopis, J., Tsien, R. Y., Harper, J. F., and Schroeder, J. I. (1999). Cameleon calcium indicator reports cytoplasmic calcium dynamics in *Arabidopsis* guard cells. *Plant J.* **19,** 735.

Allen, G. J., Chu, S. P., Harrington, C. L., Schumacher, K., Hoffmann, T., Tang, Y. Y., Grill, E., and Schroeder, J. I. (2001). A defined range of guard cell calcium oscillation parameters encodes stomatal movements. *Nature* **411,** 1053.

Arend, M., Stinzing, A., Wind, C., Langer, K., Latz, A., Ache, P., Fromm, J., and Hedrich, R. (2005). Polar-localised poplar K^+ channel capable of controlling electrical properties of wood-forming cells. *Planta* **223,** 140.

Armstrong, F., and Blatt, M. R. (1995). *Plant J.* **8,** 187.

Assmann, S. M. (1993). *Annu. Rev. Cell. Biol.* **9,** 345.

Assmann, S. M., Simoncini, L., and Schroeder, J. I. (1985). *Nature* **318,** 285.

Bates, S. E., and Gurney, A. M. (1993). *J. Physiol.* **466,** 345.

Becker, D., Geiger, D., Dunkel, M., Roller, A., Bertl, A., Latz, A., Carpaneto, A., Dietrich, P., Roelfsema, M. R., Voelker, C., Schmidt, D., Mueller-Roeber, B., *et al.* (2004). AtTPK4, an *Arabidopsis* tandem-pore K^+ channel, poised to control the pollen membrane voltage in a pH- and Ca^{2+}-dependent manner. *Proc. Natl. Acad. Sci. USA* **101,** 15621.

Binder, K. A., Wegner, L. H., Heidecker, M., and Zimmermann, U. (2003). Gating of Cl^- currents in protoplasts from the marine alga *Valonia utricularis* depends on the transmembrane Cl^- gradient and is affected by enzymatic cell wall degradation. *J. Membr. Biol.* **191,** 165.

Blatt, M. R., and Thiel, G. (1993). *Annu. Rev. Plant Physiol. Plant Mol. Biol.* **44,** 543.

Blatt, M. R., Thiel, G., and Trentham, D. R. (1990). *Nature* **346,** 766.

Bregante, M., Yang, Y., Formentin, E., Carpaneto, A., Schroeder, J. I., Gambale, F., Lo Schiavo, F., and Costa, A. (2008). KDC1, a carrot Shaker-like potassium channel, reveals its role as a silent regulatory subunit when expressed in plant cells. *Plant Mol. Biol.* **66,** 61.

Bush, D. S., and Jones, R. L. (1987). *Cell Calcium* **8,** 455.

Cahalan, M., and Neher, E. (1992). *Methods Enzymol.* **207,** 3.

Carpaneto, A., Cantu, A. M., and Gambale, F. (1999). Redox agents regulate ion channel activity in vacuoles from higher plant cells. *FEBS Lett.* **442,** 129.

Cho, M. H., and Spalding, E. P. (1996). An anion channel in *Arabidopsis* hypocotyls activated by blue light. *Proc. Natl. Acad. Sci. USA* **93,** 8134.

Cocking, E. C. (1985). *Biotechnology* **3,** 115.

Colcombet, J., Lelievre, F., Thomine, S., Barbier-Brygoo, H., and Frachisse, J. M. (2005). Distinct pH regulation of slow and rapid anion channels at the plasma membrane of *Arabidopsis thaliana* hypocotyl cells. *J. Exp. Bot.* **56,** 1897.

Corem, S., Carpaneto, A., Soliani, P., Cornara, L., Gambale, F., and Scholz-Starke, J. (2009). Response to cytosolic nickel of Slow Vacuolar channels in the hyperaccumulator plant *Alyssum bertolonii. Eur. Biophys. J.* **38,** 495.

Costa, A., Carpaneto, A., Varotto, S., Formentin, E., Marin, O., Barizza, E., Terzi, M., Gambale, F., and Lo Schiavo, F. (2004). Potassium and carrot embryogenesis: Are K^+ channels necessary for development? *Plant Mol. Biol.* **54,** 837.

De Boer, A. H., Van Duijn, B., Giesberg, P., Wegner, L., Obermeyer, G., Köhler, K., and Linz, K. W. (1994). *Protoplasma* **178,** 1.

Deeken, R., Ivashikina, N., Czirjak, T., Philippar, K., Becker, D., Ache, P., and Hedrich, R. (2003). Tumour development in *Arabidopsis thaliana* involves the Shaker-like K^+ channels AKT1 and AKT2/3. *Plant J.* **34,** 778.

Demidchik, V., Essah, P. A., and Tester, M. (2004). Glutamate activates cation currents in the plasma membrane of *Arabidopsis* root cells. *Planta* **219,** 167.

Dixit, R., Cyr, R., and Gilroy, S. (2006). Using intrinsically fluorescent proteins for plant cell imaging. *Plant J.* **45,** 599.

Eilers, J., Schneggenburger, R., and Konnerth, A. (1995). *In* "Single-Channel Recording," 2nd edn. (B. Sakmann, and E. Neher, eds.). Plenum, New York, NY.

Elliott, D. C., and Petkoff, H. S. (1990). *Plant Science* **67,** 125.

Ellis-Davies, G. C. R., and Kaplan, J. H. (1994). *Proc. Natl. Acad. Sci. USA* **91,** 187.

Elzenga, J. T. M., Keller, C. P., and Van Volkenburgh, E. (1991). *Plant Physiol.* **97,** 1573.

Etherton, B., Heppner, T. J., Cumming, J. R., and Nelson, M. T. (2004). Opposing effects of aluminum on inward-rectifier potassium currents in bean root-tip protoplasts. *J. Membr. Biol.* **198,** 15.

Fairley-Grenot, K., and Assmann, S. M. (1991). *Plant Cell* **3,** 1037.

Fan, L. M., Zhang, W., Chen, J. G., Taylor, J. P., Jones, A. M., and Assmann, S. M. (2008). Abscisic acid regulation of guard-cell K^+ and anion channels in Gβ- and RGS-deficient *Arabidopsis* lines. *Proc. Natl. Acad. Sci. USA* **105,** 8476.

Fasano, J. M., Swanson, S. J., Blancaflor, E. B., Dowd, P. E., Kao, T. H., and Gilroy, S. (2001). Changes in root cap pH are required for the gravity response of the *Arabidopsis* root. *Plant Cell* **13,** 907.

Felle, H. H., and Hepler, P. K. (1997). *Plant Physiol* **114,** 39.

Gilroy, S. (1996). *Plant Cell* **8,** 2193.

Gilroy, S., Read, N. D., and Trewavas, A. J. (1990). *Nature* **346,** 769.

Grabov, A., and Blatt, M. R. (1997). *Planta* **201,** 84.

Gradogna, A., Scholz-Starke, J., Gutla, P. V., and Carpaneto, A. (2009). Fluorescence combined with excised patch: Measuring calcium currents in plant cation channels. *Plant J.* **58,** 175.

Grynkiewicz, G., Poenie, M., and Tsien, R. Y. (1985). *J. Biol. Chem.* **260,** 3440.

Gurney, A. M., Tsien, R. Y., and Lester, H. A. (1987). *Proc. Natl. Acad. Sci. USA* **84,** 3496.

Hafke, J. B., Furch, A. C., Reitz, M. U., and van Bel, A. J. (2007). Functional sieve element protoplasts. *Plant Physiol.* **145,** 703.

Hamill, O. P., Marty, A., Neher, E., Sakmann, B., and Sigworth, F. J. (1981). *Pflügers Arch.* **391,** 85.

Haugland, R. P. (1996). *In* "Handbook of Fluorescent Probes and Research Chemicals," (M. T. Z. Spence, ed.). Molecular Probes, Eugene, OR.

Henriksen, G. H., and Assmann, S. M. (1997). *Pflügers Arch.* **433,** 832.

Henriksen, G. H., Taylor, A. R., Brownlee, C., and Assmann, S. M. (1996). *Plant Physiol.* **110,** 1063.

Hernández-Crus, A., Sala, F., and Adams, P. R. (1990). *Science* **247,** 858.

Hille, B. (1992). "Ionic Channels in Excitable Membranes." Sinauer Associates, Sunderland, MA.

Hirsch, R. E., Lewis, B. D., Spalding, E. P., and Sussman, M. R. (1998). A role for the AKT1 potassium channel in plant nutrition. *Science* **280,** 918.

Horn, R., and Korn, S. J. (1992). *Methods Enzymol* **207,** 149.

Hosy, E., Vavasseur, A., Mouline, K., Dreyer, I., Gaymard, F., Poree, F., Boucherez, J., Lebaudy, A., Bouchez, D., Very, A. A., Simonneau, T., Thibaud, J. B., *et al.* (2003). The *Arabidopsis* outward K$^+$ channel GORK is involved in regulation of stomatal movements and plant transpiration. *Proc. Natl. Acad. Sci. USA* **100,** 5549.

Ilan, N., Moran, N., and Schwartz, A. (1995). *Plant Physiol.* **108,** 1161.

Ivashikina, N., Deeken, R., Ache, P., Kranz, E., Pommerrenig, B., Sauer, N., and Hedrich, R. (2003). Isolation of AtSUC2 promoter-GFP-marked companion cells for patch-clamp studies and expression profiling. *Plant J.* **36,** 931.

Jammes, F., Song, C., Shin, D., Munemasa, S., Takeda, K., Gu, D., Cho, D., Lee, S., Giordo, R., Sritubtim, S., Leonhardt, N., Ellis, B. E., *et al.* (2009). MAP kinases MPK9 and MPK12 are preferentially expressed in guard cells and positively regulate ROS-mediated ABA signaling. *Proc. Natl. Acad. Sci. USA* **106,** 20520.

Kiegle, E., Moore, C. A., Haseloff, J., Tester, M. A., and Knight, M. R. (2000). Cell-type-specific calcium responses to drought, salt and cold in the *Arabidopsis* root. *Plant J.* **23,** 267.

Kim, M., Hepler, P. K., Eun, S. O., Ha, K. S., and Lee, Y. (1995). *Plant Physiol.* **109,** 1077.

Kirby, M. S., Hadley, R. W., and Lederer, W. J. (1994). *Pflügers Arch* **427,** 169.

Kwak, J. M., Murata, Y., Baizabal-Aguirre, V. M., Merrill, J., Wang, M., Kemper, A., Hawke, S. D., Tallman, G., and Schroeder, J. I. (2001). Dominant negative guard cell K$^+$ channel mutants reduce inward-rectifying K$^+$ currents and light-induced stomatal opening in *Arabidopsis. Plant Physiol.* **127,** 473.

Kwak, J. M., Moon, J. H., Murata, Y., Kuchitsu, K., Leonhardt, N., DeLong, A., and Schroeder, J. I. (2002). Disruption of a guard cell-expressed protein phosphatase 2A regulatory subunit, RCN1, confers abscisic acid insensitivity in *Arabidopsis. Plant Cell* **14,** 2849.

Kwak, J. M., Mori, I. C., Pei, Z. M., Leonhardt, N., Torres, M. A., Dangl, J. L., Bloom, R. E., Bodde, S., Jones, J. D., and Schroeder, J. I. (2003). NADPH oxidase AtrbohD and AtrbohF genes function in ROS-dependent ABA signaling in *Arabidopsis. EMBO J.* **22,** 2623.

Lebaudy, A., Vavasseur, A., Hosy, E., Dreyer, I., Leonhardt, N., Thibaud, J. B., Very, A. A., Simonneau, T., and Sentenac, H. (2008). Plant adaptation to fluctuating environment and biomass production are strongly dependent on guard cell potassium channels. *Proc. Natl. Acad. Sci. USA* **105,** 5271.

Legué, V., Blancaflor, E., Wymer, C., Perbal, G., Fantin, D., and Gilroy, S. (1997). *Plant Physiol.* **114,** 789.

Leonhardt, N., Kwak, J. M., Robert, N., Waner, D., Leonhardt, G., and Schroeder, J. I. (2004). Microarray expression analyses of *Arabidopsis* guard cells and isolation of a recessive abscisic acid hypersensitive protein phosphatase 2C mutant. *Plant Cell* **16,** 596.

Li, W., and Assmann, S. M. (1993). *Proc. Natl. Acad. Sci. USA* **90,** 262.

Li, J., Wang, X. Q., Watson, M. B., and Assmann, S. M. (2000). Regulation of abscisic acid-induced stomatal closure and anion channels by guard cell AAPK kinase. *Science* **287,** 300.

Lledo, P. M., Somasundaram, B., Morton, A. J., Emson, P. C., and Mason, W. T. (1992). *Neuron* **9,** 943.

Maathuis, F. J. M., Taylor, A. R., Assmann, S. M., and Sanders, D. (1997). *Plant J.* **11,** 891.

Maathuis, F. J., May, S. T., Graham, N. S., Bowen, H. C., Jelitto, T. C., Trimmer, P., Bennett, M. J., Sanders, D., and White, P. J. (1998). Cell marking in *Arabidopsis thaliana* and its application to patch-clamp studies. *Plant J.* **15,** 843.

Ma, W., Qi, Z., Smigel, A., Walker, R. K., Verma, R., and Berkowitz, G. A. (2009). Ca^{2+}, cAMP, and transduction of non-self perception during plant immune responses. *Proc. Natl. Acad. Sci. USA* **106,** 20995.

McFadzean, I., and Brown, D. A. (1992). *In* "Signal Transduction: A Practical Approach," (G. Milligan, ed.). Oxford University Press, Oxford, UK.

Miedema, H., and Assmann, S. M. (1996). *J. Membr. Biol.* **154,** 227.

Miyawaki, A., Griesbeck, O., Heim, R., and Tsien, R. Y. (1999). Dynamic and quantitative Ca^{2+} measurements using improved cameleons. *Proc. Natl. Acad. Sci. USA* **96,** 2135.

Monshausen, G. B., Bibikova, T. N., Weisenseel, M. H., and Gilroy, S. (2009). Ca^{2+} regulates reactive oxygen species production and pH during mechanosensing in *Arabidopsis* roots. *Plant Cell* **21,** 2341.

Moran, N., Ehrenstein, G., Iwasa, K., Bare, C., and Mischke, C. (1984). *Science* **226,** 835.

Mori, I. C., Murata, Y., Yang, Y., Munemasa, S., Wang, Y. F., Andreoli, S., Tiriac, H., Alonso, J. M., Harper, J. F., Ecker, J. R., Kwak, J. M., and Schroeder, J. I. (2006). CDPKs CPK6 and CPK3 function in ABA regulation of guard cell S-type anion- and Ca^{2+}-permeable channels and stomatal closure. *PLoS Biol* **4,** e327.

Mouline, K., Very, A. A., Gaymard, F., Boucherez, J., Pilot, G., Devic, M., Bouchez, D., Thibaud, J. B., and Sentenac, H. (2002). Pollen tube development and competitive ability are impaired by disruption of a Shaker K^+ channel in *Arabidopsis*. *Genes Dev.* **16,** 339.

Pandey, S., Wang, X. Q., Coursol, S. A., and Assmann, S. M. (2002). Preparation and applications of *Arabidopsis thaliana* guard cell protoplasts. *New Phytol.* **153,** 517.

Pei, Z. M., Ward, J. M., Harper, J. F., and Schroeder, J. I. (1996). *EMBO J.* **15,** 6564.

Pei, Z. M., Kuchitsu, K., Ward, J. M., Schwarz, M., and Schroeder, J. I. (1997). *Plant Cell* **9,** 409.

Pei, Z. M., Murata, Y., Benning, G., Thomine, S., Klusener, B., Allen, G. J., Grill, E., and Schroeder, J. I. (2000). Calcium channels activated by hydrogen peroxide mediate abscisic acid signalling in guard cells. *Nature* **406,** 731.

Pierson, E. S., Miller, D. D., Callaham, D. A., Shipley, A. M., Rivers, B. A., Cresti, M., and Hepler, P. K. (1994). *Plant Cell* **6,** 1815.

Piñeros, M. A., and Kochian, L. V. (2003). Differences in whole-cell and single-channel ion currents across the plasma membrane of mesophyll cells from two closely related *Thlaspi* species. *Plant Physiol.* **131,** 583.

Rapp, G., and Guth, K. (1988). *Pflügers Arch.* **411,** 200.

Raschke, K., and Hedrich, R. (1989). *Methods Enzymol.* **174,** 312.

Roberts, S. K., and Tester, M. (1995). *Plant J.* **8,** 811.

Roberts, D. M., and Tyerman, S. D. (2002). Voltage-dependent cation channels permeable to NH_4^+, K^+, and Ca^{2+} in the symbiosome membrane of the model legume *Lotus japonicus*. *Plant Physiol.* **128,** 370.

Romano, L. A., and Assmann, S. M. (1997). Unpublished observations.

Romano, L. A., Jacob, T., Gilroy, S., and Assmann, S. M. (2000). Increases in cytosolic Ca^{2+} are not required for abscisic acid-inhibition of inward K^+ currents in guard cells of *Vicia faba* L. *Planta* **211,** 209.

Rudy, B., and Iverson, L. E. (eds.) (1992). "Methods Enzymology," Vol. 207. Academic Press, San Diego, CA.

Sakmann, B., and Neher, E. (eds.) (1995). "Single-Channel Recording," 2nd edn. Plenum, New York, NY.

Schild, D. (1996). *Cell Calcium* **19,** 281.

Schroeder, J. I., and Hagiwara, S. (1990). Repetitive increases in cytosolic Ca^{2+} of guard cells by abscisic acid activation of nonselective Ca^{2+} permeable channels. *Proc. Natl. Acad. Sci. USA* **87,** 9305.

Schroeder, J. I., Hedrich, R., and Fernandez, J. M. (1984). *Nature* **312,** 361.

Sutter, J. U., Sieben, C., Hartel, A., Eisenach, C., Thiel, G., and Blatt, M. R. (2007). Abscisic acid triggers the endocytosis of the *Arabidopsis* KAT1 K$^+$ channel and its recycling to the plasma membrane. *Curr. Biol.* **17**, 1396.

Tang, J. M., Wang, J., and Eisenberg, R. S. (1992). *Methods Enzymol* **207**, 176.

Taylor, A. R., and Brownlee, C. (1992). *Plant Physiol.* **99**, 1686.

Taylor, A. R., and Brownlee, C. (2003). A novel Cl$^-$ inward-rectifying current in the plasma membrane of the calcifying marine phytoplankton *Coccolithus pelagicus*. *Plant Physiol.* **131**, 1391.

Thion, L., Mazars, C., Thuleau, P., Graziana, A., Rossignol, M., Moreau, M., and Ranjeva, R. (1996). *FEBS Lett.* **393**, 13.

Thomine, S., Zimmermann, S., Van Duijn, B., Barbier-Brygoo, H., and Guern, J. (1994). *FEBS Lett.* **340**, 45.

Trube, G. (1983). *In* "Single Channel Recording," (B. Sakmann, and E. Neher, eds.). Plenum, New York, NY.

van den Pol, A. N., Obrietan, K., and Chen, G. (1996). *J. Neurosci.* **16**, 4283.

Wang, X. Q., Ullah, H., Jones, A. M., and Assmann, S. M. (2001). G protein regulation of ion channels and abscisic acid signaling in *Arabidopsis* guard cells. *Science* **292**, 2070.

Wang, Y. F., Fan, L. M., Zhang, W. Z., Zhang, W., and Wu, W. H. (2004). Ca^{2+}-permeable channels in the plasma membrane of *Arabidopsis* pollen are regulated by actin microfilaments. *Plant Physiol.* **136**, 3892.

Ward, J. M., Pei, Z. M., and Schroeder, J. I. (1995). *Plant Cell* **7**, 833.

Wegner, L. H., and Raschke, K. (1994). *Plant Physiol.* **105**, 799.

Xu, J., Li, H. D., Chen, L. Q., Wang, Y., Liu, L. L., He, L., and Wu, W. H. (2006). A protein kinase, interacting with two calcineurin B-like proteins, regulates K$^+$ transporter AKT1 in *Arabidopsis*. *Cell* **125**, 1347.

Yu, C. J., and Wu, W. H. (1997). Unpublished observations .

Zhang, W. H., Ryan, P. R., and Tyerman, S. D. (2004). Citrate-permeable channels in the plasma membrane of cluster roots from white lupin. *Plant Physiol.* **136**, 3771.

Zhao, Z., Zhang, W., Stanley, B. A., and Assmann, S. M. (2008). Functional proteomics of *Arabidopsis thaliana* guard cells uncovers new stomatal signaling pathways. *Plant Cell* **20**, 3210.

PART IX

Special Channels

CHAPTER 21

ATP–Sensitive Potassium Channels

**M. Schwanstecher, C. Schwanstecher, F. Chudziak, U. Panten,
J. P. Clement IV, G. Gonzalez, L. Aguilar-Bryan, and J. Bryan**
Department of Cell Biology
Baylor College of Medicine
One Baylor Plaza
Houston, Texas, USA

ESSENTIAL ION CHANNEL METHODS
Copyright © 1999 by Elsevier Inc. All rights reserved.

DOI: 10.1016/B978-0-12-382204-8.00021-7

I. Introduction

ATP-sensitive potassium channels, or K_{ATP} channels, couple changes in cellular metabolism with membrane electrical activity. In pancreatic β cells these channels set the resting membrane potential. Increased glucose metabolism changes the ATP/ADP ratio reducing the opening of K_{ATP} channels causing membrane depolarization and activation of voltage-gated Ca^{2+} channels. The resulting increase in $[Ca^{2+}]_i$ stimulates insulin exocytosis. Pharmacologically distinct types of K_{ATP} channels have been identified in muscle cells, where their opening would be expected to reduce electrical activity. K_{ATP} channels are assembled from sulfonylurea receptors, SURs, members of the ATP-binding cassette superfamily, and members of the inwardly rectifying potassium channel family, $K_{IR}6.x$. Genetic, biochemical, and electrophysiologic data establish that the β-cell channel is assembled from SUR1 (OMIM 600509) and $K_{IR}6.2$ (OMIM 600937) (Aguilar-Bryan and Bryan, 1996; Aguilar-Bryan *et al.*, 1998b; Babenko *et al.*, 1998; Bryan and Aguilar-Bryan, 1997). $K_{IR}6.2$ forms the pore of the channel, while SUR1 regulates channel activity and confers responsiveness to channel openers like diazoxide, pinacidil, and cromakalim, and to channel blockers, sulfonylureas, like tolbutamide and glibenclamide. SUR1 and $K_{IR}6.2$ are present in the channel complex in a 1:1 ratio in a tetrameric stoichiometry $(SUR1/K_{IR}6.2)_4$. Loss of function mutations in SUR1 and $K_{IR}6.2$ have been shown to result in persistent hyperinsulinemic hypoglycemia of infancy (OMIM256450), a neonatal disorder characterized by unregulated insulin release despite severe hypoglycemia. The β-cell K_{ATP} channel is inhibited by nanomolar concentrations of glibenclamide, and can be activated by diazoxide. The SUR1/$K_{IR}6.2$ channels have also been identified in neuronal tissue (Jonas *et al.*, 1991; Ohno-Shosaku and Yamamoto, 1992; Schwanstecher and Bassen, 1997).

SUR2A and $K_{IR}6.2$ assemble a potassium selective, inwardly rectifying channel with the electrical and pharmacologic properties of the cardiac K_{ATP} channel (Inagaki *et al.*, 1996). These channels are less sensitive to glibenclamide, IC_{50} $\sim$0.3 μM, and are not responsive to diazoxide, but can be activated by pinacidil at micromolar concentrations, IC_{50} $\sim$20–100 μM. A splice variant of SUR2A, designated SUR2B, differing only in the splicing of the C-terminal 45 amino acid exons (Aguilar-Bryan *et al.*, 1998b), has been reported to pair with both $K_{IR}6.2$ and $K_{IR}6.1$ to form K_{ATP} channels similar to those found in vascular smooth muscle (Isomoto *et al.*, 1996). Like the cardiac channels, the SUR2B channels are less sensitive to glibenclamide than the β-cell channel, but are activated by diazoxide and can be activated by lower concentrations of pinacidil than the cardiac channel. The SUR2B/K_{IR}6.1 channels are reported to be activated by nucleoside diphosphates, but not inhibited by ATP (Beech *et al.*, 1993; Isomoto *et al.*, 1996).

This article describes methods used in the characterization of K_{ATP} channels, including synthesis and radioiodination of two derivatives of glibenclamide that photolabel SUR1.

II. Drug Synthesis

Two derivatives of glibenclamide, iodoglibenclamide (Aguilar-Bryan *et al.*, 1990) and azidoiodoglibenclamide (Chudziak *et al.*, 1994; Schwanstecher *et al.*, 1994b), have been critically important in the isolation and characterization of SUR1, the high-affinity sulfonylurea receptor. Although the radioiodinated species are available commercially at great cost, their synthesis, including radioiodination, can be accomplished easily by the average laboratory. We have detailed the synthesis of both derivatives here.

A. Synthesis of Iodoglibenclamide

1. Chemicals

Chemicals (pure grade) are purchased from Aldrich Chemical Company Inc. (Wilwaukee, WI) (unless stated) and are used as received. Solvents are dried and distilled before use.

The individual steps for making and radiolabeling iodoglibenclamide are described in detail below. Figure 1 provides an overview of the synthesis and the structures.

Step 1: Synthesis of 4-[β-(2-Hydroxybenzenecarboxamido)ethyl]benzene-sulfon-amide. Salicyclic acid 4.1 g (0.031 mol) is dissolved in 50 ml of dry acetone, 4.3 ml (0.031 mol) of triethylamine is added, and the solution is cooled to $-20\,°C$. To this solution 3 ml (0.031 mol) of ethyl chloroformate is added dropwise while stirring. The resulting mixed anhydride is held at $-20\,°C$ for 15 min with occasional stirring.

A suspension is prepared from 6 g (0.031 mol) of *p*-(β-aminoethyl)benzenesulfonamide, 4.3 ml of triethylamine, and 50 ml of dry acetone and cooled to $-20\,°C$. This suspension is added rapidly, while stirring, to the mixed anhydride prepared above. The mixture is stirred for 2 h at $4\,°C$ and then at room temperature ($23\,°C$) for 4 h. The acetone is removed by vacuum distillation and the residue acidified with 0.2 N HCl. The resulting precipitate is collected by filtration, washed with H_2O, air dried, and recrystallized from ethanol.

Step 2: Synthesis of N-{4-[β-(2-Hydroxybenzenecarboxamido)ethyl]benzenesul-fonyl}-N'-cyclohexylurea. 4[β-(2-Hydroxybenzenecarboxamido) ethyl]benzenesulfonamide (12 g) is dissolved in 90 ml of acetone at $4\,°C$. NaOH (1.8 g), dissolved in

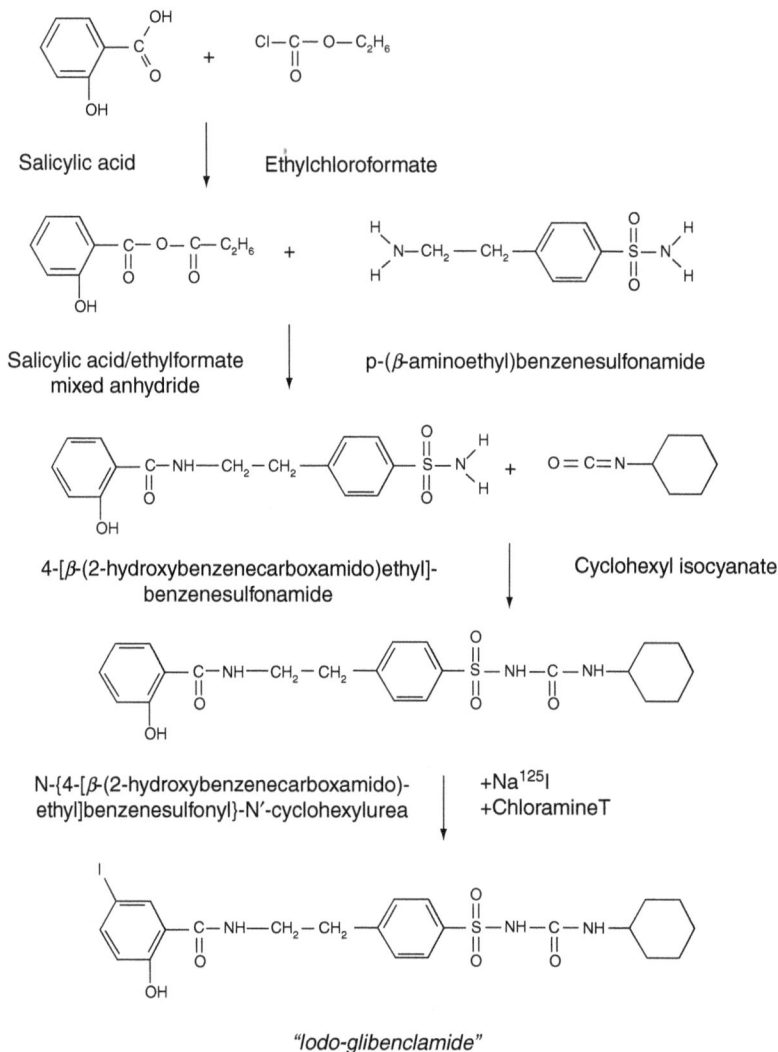

Fig. 1 Synthesis of iodoglibenclamide.

10 ml of H_2O, is added, with 5.75 ml of cyclohexyl isocyanate in four equal aliquots over a period of 80 min. After stirring overnight, 150 ml of H_2O is added, stirred for 10 min, then filtered to remove a small amount of precipitate. Acidification by the addition of 75 ml of 2 N HCl produces a precipitate that is collected by filtration after 60 min, air dried, and recrystallized from ethanol.

Step 3: Synthesis of N-{4-[β-(2-Hydroxy-5-iodobenzenecarboxamido)ethyl]benzenesulfonyl}-N'-cyclohexylurea. N-{4-[β-(2-Hydroxybenzenecarboxamido)ethyl] benzenesulfonyl}-N'-cyclohexylurea, 0.5 g (1.13 mmol), and 0.26 g of NaI are dissolved in 5 ml of dimethylformamide. Chloramine-T (0.39 g), dissolved in 2 ml of dimethylformamide, is added to start the reaction. After stirring for 1 h at 23 °C, the reaction is terminated by addition of 25 ml of 0.1 N HCl. The iodinated compound (iodoglibenclamide) is extracted into ethyl acetate and washed sequentially with equal volumes of 50 mM sodium bisulfite, H_2O, 0.5 M NaCl (2×), and H_2O (2×). The volume of the ethyl acetate phase is reduced to approximately 5–10 ml and the iodinated product is collected and recrystallized from ethyl acetate with an approximate yield of 70%.

Step 4: Radioiodination of N-{4[β-(2-Hydroxybenzenecarboxamido)ethyl]benzenesulfonyl}-N'-cyclohexylurea. One microliter of a 10-mM solution of N-{4-[β-(2-hydroxybenzenecarboxamido)ethyl]benzenesulfonyl}-N'-cyclohexylurea dissolved in dimethylformamide is added to 2–4 μl of $Na^{125}I$ (1–4 mCi) in a solution of dilute NaOH. Iodination is initiated by the addition of 1 μl of 10 mM chloramine-T dissolved in dimethylformamide. The reaction is terminated after 10 min at 23 °C by addition of 1 μl of 14 M 2-mercaptoethanol and 20 μl of 50% (v/v) methanol in H_2O. This mixture is absorbed to a μBondapak C_{18} column (Waters/Millipore, Bedford, MA) equilibrated with 50% methanol, and then separated by high-performance liquid chromatography (HPLC) using a 50–90% methanol gradient with a flow rate of 1 ml/min. A slightly concave gradient was used for increased resolution. The uniodinated and iodinated species are well resolved with retention times of 9.9 and 13.3 min, respectively. The fractions (0.5 ml) containing the ^{125}I-labeled iodoglibenclamide are pooled and stored at −20 °C. The concentration of the drug, based on the specific activity of the ^{125}I, is determined immediately after purification. The material gives a single spot on reversed-phase thin-layer chromatography when analyzed on Whatman $KC_{18}F$ plates using 80% methanol/20% 0.5 M NaCl as a solvent. The iodinated material cochromatographs with the unlabeled iodinated compound synthesized as described in step 3.

B. Synthesis of Azidoiodoglibenclamide

The individual steps for the synthesis and radiolabeling of azidoiodoglibenclamide are described in detail below. Figure 2 provides an overview of the synthesis and the structures.

Step 1: Synthesis of N-Acetyl-2-phenylethylamine. 2-Phenylethylamine (91.7 g) and 91.7 g of 2,4,6-trimethylpyridine are dissolved in 600 ml of dry dichloromethane. Acetyl chloride (59.4 g) is added dropwise with stirring at room temperature. The resulting precipitate is collected by filtration, washed with 0.1 M HCl, 0.1 M NaOH, and H_2O, and finally dried over anhydrous $MgSO_4$. Dichlormethane is removed by vacuum distillation.

Step 2: Synthesis of 4-(2-Acetamidoethyl)benzenesulfonamide. To 50 g of N-acetyl-2-phenylethylamine 17.5 g of chlorosulfonic acid is added drop-wise with stirring at 50 °C. The reaction mixture is allowed to cool to room temperature, stirred for another 30 min, and then poured into ice-cold H_2O. The precipitated

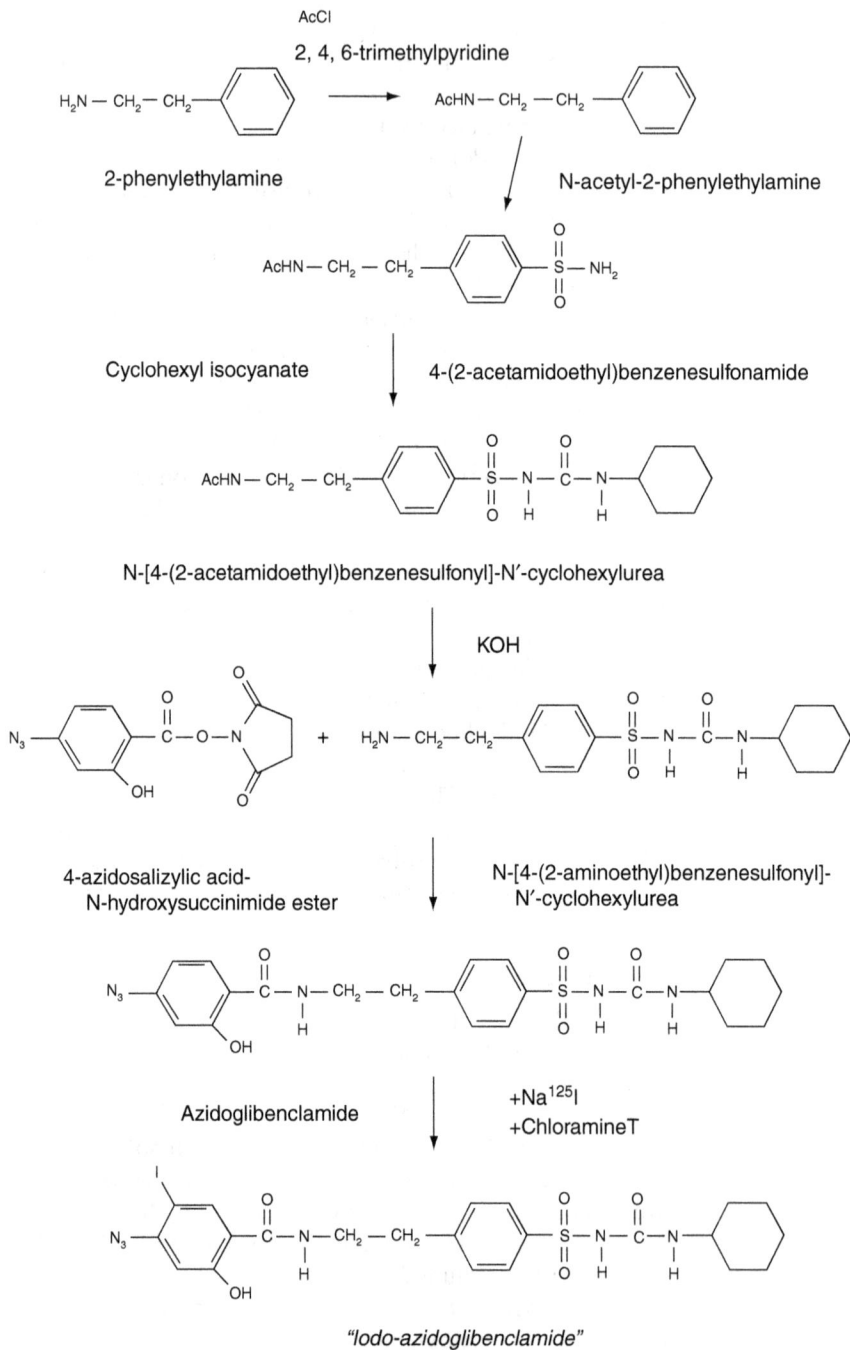

Fig. 2 Synthesis of azido[125I]iodoglibenclamide.

sulfochloride is collected by filtration and refluxed 15 min with a mixture of 90 ml concentrated NH_3 plus 135 ml H_2O. After heating with activated carbon and stirring overnight at 5 °C the resulting precipitate is collected by filtration, washed with ice-cold H_2O, and dried by vacuum distillation.

Step 3: Synthesis of N-[4-(2-Acetamidoethyl)benzenesulfonyl]-N'-cyclohexylurea. To 7.6 g of 4-(2-acetamidoethyl)benzenesulfonamide in 40 ml acetone 12.7 ml of 2 M NaOH is added and the solution cooled to 4 °C. At this temperature 5.1 g of cyclohexyl isocyanate is added dropwise while stirring. The mixture is then stirred at room temperature (23 °C) for 3 h. The acetone is removed by vacuum distillation and the solid residue obtained dissolved in 0.1 M NaOH. The solution is adjusted to pH 9 and filtrated to remove dicyclohexylurea. The filtrate is acidified with 2 M HCl and the resulting precipitate collected by filtration and recrystallized from ethanol/H_2O.

Step 4: Synthesis of N-[4-(2-Aminoethyl)benzenesulfonyl]-N'-cyclohexylurea. Hydrolysis of the *N*-acetyl group is carried out by boiling 5.0 g of *N*-[4-(2-acetamidoethyl)benzenesulfonyl]-*N'*-cyclohexylurea with 7.6 g KOH for 8 h using 34 ml of 90% (v/v) ethanol as solvent. Ethanol is removed by vacuum distillation and the residue dissolved in H_2O. The product is precipitated by neutralization with 2 M HCl, collected by filtration, and washed with ethyl acetate.

Step 5: Synthesis of N-[4-(2-{4-Azido-2-hydroxybenzamido}ethyl)benzenesulfonyl]-N'-cyclohexylurea (N₃-GA). To 6 ml dimethylformamide containing 32.5 mg *N*-[4-(2-aminoethyl)benzenesulfonyl]-*N'*-cyclohexylurea and 28 μl triethylamine 27 mg of 4-azidosalicyclic acid *N*-hydroxysuccinimide ester (Sigma, St. Louis, MO) is added at 0 °C. After stirring in the dark for 2 h at 0 °C and then overnight at room temperature, the reaction mixture is poured into ice-cold H_2O, followed by acidification with 2 M HCl. The resulting precipitate is collected by filtration and washed with ice-cold H_2O.

Step 6: Radioiodination of N₃-GA. To a NENSURE-vial containing 5 mCi of $Na^{125}I$ (2.35 nmol) in 50 μl of 10 μM NaOH (NEZ-033A, NEN, Dreieich, Germany) is added N₃-GA (9.38 nmol, 37.5 μl of a 0.25-mM solution). The latter solution is prepared by dissolving 1.22 mg of N₃-GA in 8 ml of 10 mM NaOH, adding 1.95 ml of 0.5 M sodium phosphate buffer (pH 7.4), and adjusting to pH 7.4 with 2 M HCl. Iodination is started by addition of 37.5 μl of 0.25 mM chloramine-T dissolved in 0.1 M sodium phosphate buffer (pH 7.4). After 15 min at room temperature in the dark, the reaction is quenched by addition of 2 μl of 0.14 M 2-mercaptoethanol and mixing. Separation of the reaction mixture is carried out immediately by HPLC (ODS-Hypersil 5-μm column). The mobile phase (60% acetonitrile, 40% water, 0.07% trifluoroacetic acid) is used isocratically with a flow rate of 2 ml/min. The retention times of the radioactive peaks representing *N*-[4-(2-{4-azido-2-hydroxy-5-[^{125}I]iodobenzamido}ethyl)benzenesulfonyl]-*N'*-cyclohexylurea ($^{125}IN_3$-GA) and the 3,5-diiodinated derivative are 4.6 and 9.8 min, respectively. The radioactivity of the diiodinated derivative is 4% of the radioactivity of $^{125}IN_3$-GA. A minor radioactive peak is eluted just ahead of $^{125}IN_3$-GA and is excluded from the fractions containing $^{125}IN_3$-GA. The retention time of N_3-GA is 3.0 min, and nonradioactive UV peaks are not detected between 3.0 and 15.0 min. The fractions (333 μl) containing $^{125}IN_3$-GA are collected, counted (2.8 mCi), and stored at 4 °C (used for up to 2 months). Because monitoring of the eluate did not reveal chemical or radiochemical impurities in the fractions containing $^{125}IN_3$-GA, a specific activity of approximately 2100 Ci/ml can be assumed. The radiochemical yield based on ^{125}I was 56%.

III. Tissue Culture

A. Maintenance of Tissue Culture Cells

Tissue culture cells producing SUR1 include the hamster insulin secreting cell line, HIT-T15 (passage 65–75; CRL1777 ATCC, Rockville, MD), the rat insulinoma cell line, RINm5f, and the glucagon secreting cell line, αTC-6. Chinese hamster ovary (CHO) and COS (1, m6, and 7) cell lines do not produce SUR1 and are used for receptor and channel expression. Cells are maintained in T-175 culture flasks (Falcon) as monolayers in Dulbecco's modified Eagle's medium with high glucose (DMEM)-HG medium supplemented with 10% fetal bovine serum (FBS), 100 U/ml penicillin, and 0.1 mg/ml streptomycin. Cells are grown in 5% CO_2 at 37 °C, maintained in subconfluent cultures, fed three times a week, and subcultured as needed. To subculture, confluent cells are detached with 0.05% trypsin/EDTA, resuspended in supplemented DMEM-HG medium and replated at one-tenth the original density. For expression experiments, COS cells are plated at 50–60% confluence prior to transient transfection.

IV. Transfection Protocols

Cells are transfected using either a DEAE-dextran or a lipofectamine protocol.

A. DEAE-Dextran Protocol

For $^{86}Rb^+$ efflux studies, 3-day-old cultures of COS cells are trypsinized and replated at a density of 2.0×10^5 cells per 35-mm well (six-well dish) and allowed to attach overnight. Typically, 5 μg of a SUR1 plasmid is mixed with 5 μg of a K_{IR} plasmid and brought up to 7.5 μl final volume in TBS (8 g/l NaCl; 0.38 g/l KCl; 0.2 g/l Na_2HPO_4; 3.0 g/l Tris base; 0.15 g/l $CaCl_2$; 0.1 g/l $MgCl_2$, pH 7.5) before addition of DEAE-dextran (30 μl of a 5-mg/ml solution in TBS). The samples are vortexed, collected by briefly spinning in a microfuge, then incubated for 15 min at room temperature before addition of 500 μl 10% NuSerum (Collaborative Biomedical Products, Twin Oak Park, Bedford, MA) in TBS. Cells are washed twice with Hanks' balanced salt solution (HBSS), the DNA mix is added, and the cells are maintained in a 37 °C CO_2 incubator. After 4 h the DNA mix is decanted and the cells shocked for 2 min in 1 ml HBSS + 10% dimethyl sulfoxide (DMSO), then placed in 1.5 ml of DMEM-HG + 2% FBS + 10 μM chloroquine and kept in a 37 °C CO$_2$ incubator. After 4 h, the cells are washed twice with HBSS and incubated in normal growth media until assayed (usually 36–48 h posttransfection).

B. Lipofectamine Protocol

COS cells are plated in six-well dishes as described above and are 70–80% confluent when transfected. Typically 1 μg of a SUR1 plasmid and 1 μg of a K$_{IR}$6.2 plasmid are mixed with 375 μl of Opti-Mem reduced serum medium (Life Technologies, Inc.) then added to 375 μl of Opti-Mem containing 9 μl of lipofectamine. After a 45-min incubation, the mixture is brought to 1 ml with Opti-Mem and added to the cells that had been washed once with 3 ml of Opti-Mem. Transfections are allowed to proceed for 5 h, at which time the Opti-Mem is replaced with DMEM-HG medium plus 10% FBS.

Transfections are scaled based on the area of the plates used. For example, 150-mm plates used for membrane isolations are transfected with 100 μg of each plasmid using the DEAE-dextran protocol.

V. Rubidium Efflux Assays

Twenty-four hours posttransfection, cells are placed in fresh media containing approximately 1 μCi/ml ^{86}RbCl, incubated for an additional 12–24 h, and assayed as follows. Cells are incubated for 30 min at 25 °C in Krebs–Ringer solution under one of three conditions: no additions (basal), with oligomycin (2.5 μg/ml) and 2-deoxy-D-glucose (1 mM) (metabolically inhibited) or with oligomycin and deoxyglucose plus 1 μM glibenclamide (K$_{ATP}$ channels inhibited). Cells are washed once in ^{86}Rb$^+$-free Krebs–Ringer solution, with or without the added inhibitors, then time points are taken by removing all the medium from the cells and replacing it with fresh medium at the indicated times. Equal portions of the medium from each time point are counted, and the values summed to determine flux. Total ^{86}Rb$^+$ is defined as the sum of counts from each time point plus the counts released by addition of 1% sodium dodecyl sulfate (SDS) to the cells at the end of the experiment. Results can be presented either as the percentage of total cellular ^{86}Rb$^+$ released or as percent glibenclamide inhibitable efflux. The latter measure represents specific K$_{ATP}$ channel activity and is defined as the percent ^{86}Rb$^+$ efflux from metabolically inhibited cells minus the percent ^{86}Rb$^+$ efflux from metabolically inhibited plus glibenclamide inhibited cells (efflux through K$_{ATP}$ channels defined as % glibenclamide inhibitable efflux = % metabolically inhibited efflux − % metabolically and glibenclamide inhibited efflux).

VI. Membrane Isolation

Membranes are prepared 60–72 h posttransfection from 10–20 150-mm dishes. Transfected cells are washed twice in phosphate-buffered saline (PBS), pH 7.4, then scraped in PBS and collected in 10-ml plastic tubes. The cells are pelleted, resuspended in 10 ml of hypotonic buffer (5 mM Tris–HCl, pH 7.4, 2 mM EDTA), and allowed to swell for 45 min on ice. Cells are then homogenized, transferred to a 15-ml glass tube, and spun at 1000 g for 20 min at 4 °C to remove nuclei and unbroken cells. The supernatant is then transferred to a polycarbonate centrifuge tube and membranes are collected by ultracentrifugation at 40,000 rpm in an 80Ti fixed-angle rotor for 2 h. The pelleted membranes are resuspended in 3–500 μl of membrane buffer (50 mM Tris–HCl, pH 7.4, 5 mM EDTA) and stored at -80 °C. Typical protein concentrations are 2–5 mg/ml.

VII. Photolabeling Protocols

A. Photolabeling of Membranes

[^{125}I]Iodoglibenclamide (Aguilar-Bryan *et al.*, 1990) or azido[^{125}I]iodoglibenclamide (Schwanstecher *et al.*, 1994b) is added to samples (typically to a concentration of 5–10 nM) and the samples incubated at 23 °C for 30 min. Sample concentration was at 5–10 mg protein/ml. Samples are then transferred onto Parafilm and irradiated at 312 nm in a UV cross-linker (Model FB-UVXL-1000, Spectronics Corp., Westbury, NY) or at 356 nm by use of a hand lamp (Camag, Berlin, Germany). The photolabeling reaction is optimized in terms of energy required. In typical experiments, 1.0–1.5 J/cm^2 is used for iodoglibenclamide and 0.2 J/cm^2 for azidoiodoglibenclamide. Nonspecific labeling is evaluated by addition of unlabeled glibenclamide at a concentration of 0.1 μM.

B. Photolabeling of Live Cells

Photolabeling is carried out as described for isolated membranes (Aguilar-Bryan *et al.*, 1992; Nelson *et al.*, 1992). Living cells, grown in in six-well dishes, are washed three times with PBS, then incubated in the dark with 10 nM azido[^{125}I] iodoglibenclamide or 10 nM [^{125}I]iodoglibenclamide in Krebs–Ringer solution supplemented with 10 mM glucose. After 30 min at room temperature, the cells are irradiated in a UV cross-linker (Model FB-UVXL-1000 Spectronics Corp., Westbury, NY) at a setting of 0.9 J/cm^2. Excess unbound drug is removed by three 5-ml washes with PBS (1 min each). The cells are then solubilized in 250–500 μl of 2× SDS sample buffer (Laemmli, 1970). Aliquots are separated on 8% polyacrylamide gels, stained with Coomassie blue, dried, and placed on X-ray film (Aguilar-Bryan *et al.*, 1998a; Nelson *et al.*, 1992).

These protocols label the core and complex glycosylated SUR1 receptors when [^{125}I]iodoglibenclamide is used as the probe. When azido[^{125}I]iodoglibenclamide is

I. Introduction

ATP-sensitive potassium channels, or K_{ATP} channels, couple changes in cellular metabolism with membrane electrical activity. In pancreatic β cells these channels set the resting membrane potential. Increased glucose metabolism changes the ATP/ADP ratio reducing the opening of K_{ATP} channels causing membrane depolarization and activation of voltage-gated Ca^{2+} channels. The resulting increase in $[Ca^{2+}]_i$ stimulates insulin exocytosis. Pharmacologically distinct types of K_{ATP} channels have been identified in muscle cells, where their opening would be expected to reduce electrical activity. K_{ATP} channels are assembled from sulfonylurea receptors, SURs, members of the ATP-binding cassette superfamily, and members of the inwardly rectifying potassium channel family, $K_{IR}6.x$. Genetic, biochemical, and electrophysiologic data establish that the β-cell channel is assembled from SUR1 (OMIM 600509) and $K_{IR}6.2$ (OMIM 600937) (Aguilar-Bryan and Bryan, 1996; Aguilar-Bryan *et al.*, 1998b; Babenko *et al.*, 1998; Bryan and Aguilar-Bryan, 1997). $K_{IR}6.2$ forms the pore of the channel, while SUR1 regulates channel activity and confers responsiveness to channel openers like diazoxide, pinacidil, and cromakalim, and to channel blockers, sulfonylureas, like tolbutamide and glibenclamide. SUR1 and $K_{IR}6.2$ are present in the channel complex in a 1:1 ratio in a tetrameric stoichiometry $(SUR1/K_{IR}6.2)_4$. Loss of function mutations in SUR1 and $K_{IR}6.2$ have been shown to result in persistent hyperinsulinemic hypoglycemia of infancy (OMIM256450), a neonatal disorder characterized by unregulated insulin release despite severe hypoglycemia. The β-cell K_{ATP} channel is inhibited by nanomolar concentrations of glibenclamide, and can be activated by diazoxide. The SUR1/$K_{IR}6.2$ channels have also been identified in neuronal tissue (Jonas *et al.*, 1991; Ohno-Shosaku and Yamamoto, 1992; Schwanstecher and Bassen, 1997).

SUR2A and $K_{IR}6.2$ assemble a potassium selective, inwardly rectifying channel with the electrical and pharmacologic properties of the cardiac K_{ATP} channel (Inagaki *et al.*, 1996). These channels are less sensitive to glibenclamide, IC_{50} ~0.3 μM, and are not responsive to diazoxide, but can be activated by pinacidil at micromolar concentrations, IC_{50} ~20–100 μM. A splice variant of SUR2A, designated SUR2B, differing only in the splicing of the C-terminal 45 amino acid exons (Aguilar-Bryan *et al.*, 1998b), has been reported to pair with both $K_{IR}6.2$ and $K_{IR}6.1$ to form K_{ATP} channels similar to those found in vascular smooth muscle (Isomoto *et al.*, 1996). Like the cardiac channels, the SUR2B channels are less sensitive to glibenclamide than the β-cell channel, but are activated by

CHAPTER 21

ATP-Sensitive Potassium Channels

M. Schwanstecher, C. Schwanstecher, F. Chudziak, U. Panten,
J. P. Clement IV, G. Gonzalez, L. Aguilar-Bryan, and J. Bryan
Department of Cell Biology
Baylor College of Medicine
One Baylor Plaza
Houston, Texas, USA

ESSENTIAL ION CHANNEL METHODS
Copyright © 1999 by Elsevier Inc. All rights reserved.

DOI: 10.1016/B978-0-12-382204-8.00021-7

used, $K_{IR}6.2$ is cophotolabeled in addition to the core and complex glycosylated receptors (Clement *et al.*, 1997; Schwanstecher *et al.*, 1994a).

VIII. Receptor Solubilization

A. Preparation of Digitonin

Digitonin (Sigma) is purchased as a powder and is not further purified. For receptor solubilization, 20% (w/v) digitonin is prepared fresh each day by addition of the detergent to deionized water, vortexing for a few seconds, and boiling in a closed test tube for 2 min. Boiling results in a relatively clear detergent solution diluted as outlined below.

B. Digitonin Solubilization

For receptor purification membranes are thawed, rehomogenized using a Teflon–glass homogenizer, and mixed with ice-cold digitonin to a final protein concentration of 5 mg/ml and 1% digitonin. Subsequent steps are performed at room temperature in the presence of a cocktail of protease inhibitors [0.1 mM phenylmethylsulfonyl fluoride (PMSF), 0.1 mM phenanthroline, and 0.1 mM iodoacetamide]. Membranes are homogenized, then solubilized for 15 min; the solubilized receptor is separated from insoluble material by centrifugation at 100,000 g for 1 h at 4 °C.

C. Triton X-100 or CHAPS Solubilization

Membranes are thawed, rehomogenized using a Teflon–glass homogenizer, and mixed with ice-cold solubilization buffer to give the final concentrations: 10 mg/ml membrane protein; 0.2% (w/v) phosphatidylcholine (egg); 10% (w/v) glycerol; 115 mM KCl; 1 mM ethylenediaminetetraacetic acid (EDTA); 100 μM PMSF; 20 mM N-2-hydroxyethylpiperazine-N'-2-ethanesulfonic acid (HEPES), pH 7.4. Triton X-100 or CHAPS (3-[(3-cholamidopropyl)dimethylammonio]-1-propane sulfonate) is added to 2% (w/v). The solution is gently stirred at 4 °C (1 h) and centrifuged at 120,000 g and 4 °C for 1 h. The supernatant is then collected and stored at −80 °C until binding studies are performed.

IX. Partial Purification of SUR1

A. Lectin Affinity Chromatography

1. Concanavalin A

For purification of the 140-kDa core glycosylated species of the sulfonylurea receptor 4-ml aliquots of digitonin-solubilized receptor are cycled four times over a 1-ml concanavalin A (con A)-Sepharose (Sigma) column equilibrated with 25 mM

Tris (pH 7.5), 0.1 M NaCl, 2 mM EDTA, 1% (w/v) digitonin. The column is washed with 8 ml of the equilibrating buffer and eluted with 4 ml of the equilibrating buffer containing 0.5 M methyl α-D-mannopyranoside. The eluted protein can be stored at this stage at $-80\,°C$. See also Aguilar-Bryan *et al.* (1998a).

2. Wheat Germ Agglutinin

Wheat germ agglutinin (WGA; Sigma)-Sepharose was used instead of con A-Sepharose for the purification of the 150-kDa complex glycosylated receptor. The receptor was eluted with 0.3 M N-acetylglucosamine. All other manipulations are as described for the purification of the 140 kDa protein. See also Aguilar-Bryan *et al.* (1998a).

B. Ni^{2+}-Agarose

1. Histidine–Tagged Receptor

$Sur1_{N\text{-}6X\text{-}HIS}$ and $SUR1_{N\text{-}6X\text{-}HIS}/K_{IR}6.x$ complexes can be partially purified by chromatography on a column of Ni^{2+}-agarose (Qiagen Corp., Santa Clarita, CA) equilibrated in solubilization buffer (1.0% (w/v) digitonin, 150 mM NaCl, 25 mM Tris, pH 7.4) plus 4 mM imidazole. For example, to demonstrate that SUR1 and $K_{IR}6.2$ form a complex, approximately 150 μg of membranes prepared from COS cells transfected with $K_{IR}6.2$ and SUR1 plasmids is photolabeled with 10 nM azido [^{125}I]iodoglibenclamide, pH 6.5. The labeled membranes are solubilized in 200 μl of solubilization buffer for 30 min on ice, then spun for 30 min at 100,000 g at 4 $°C$ (Model TI-100, Beckman Instruments, Fullerton, CA) to remove insoluble material before passing over the Ni^{2+} column four times. The column is washed with 20 ml (40 times column volume) of digitonin wash buffer (0.2% (w/v) digitonin, 150 mM NaCl, 25 mM Tris, pH 7.4, plus 4 mM imidazole), then eluted with wash buffer containing 100 mM imidazole, pH 7.4. Aliquots of each fraction are analyzed by polyacrylamide gel electrophoresis as described above.

X. Additional Purification Steps

A. Reactive Green-19 Affinity Chromatography

The eluate from con A-Sepharose is cycled twice over a 1-ml column of Reactive Green 19-agarose (Sigma) equilibrated with 50 mM HEPES, pH 8.5, 2 mM EDTA, 0.2% (w/v) digitonin. After washing with 8 ml of the equilibrating buffer, and 8 ml of the equilibrating buffer plus 0.4 M NaCl, the protein is eluted with 4 ml of 1.5 M NaCl in the equilibrating buffer.

B. Phenylboronate-10 Affinity Chromatography

The eluate from the Reactive Green-19 purification step is diluted 1:1 with 50 mM HEPES, pH 8.5, 2 mM EDTA, 0.2% digitonin to reduce the ionic strength, then cycled twice over a 1-ml phenylboronate-10 Sepharose (Amicon, Danvers, MA) column. The phenylboronate column is washed with 8 ml of the HEPES buffer, followed by 2 ml of 0.1 M Tris (pH 7.5), 2 mM EDTA, and 0.1% (w/v) digitonin. Protein is eluted with 4 ml of 0.1 M Tris (pH 7.5), 2 mM EDTA, 0.1% (w/v) SDS.

C. Concentration of Receptor

Pooled samples from the various column steps are concentrated by centrifugation (3000 g for 30 min at 4 °C) to 0.5 ml using Amicon 100,000 molecular weight cutoff filters. Filters are pretreated overnight at 4 °C with 5% (v/v) Tween 20 to prevent loss of protein.

Additional steps used in purification of SUR1 have been described (Aguilar-Bryan et al., 1998; Aguilar-Bryan et al., 1995).

XI. Sedimentation

Sucrose gradient centrifugation is used to estimate the molecular weight of receptor and receptor/K_{IR} complexes. Membranes from transfected cells are isolated, photolabeled with azido[^{125}I]iodoglibenclamide, and solubilized as described above. Twelve milliliter 5–20% linear sucrose gradients (in 0.1% digitonin, 100 mM NaCl, 50 mM Tris, pH 7.4) are poured in SW41 tubes using a BioComp Gradient Master 106 (BioComp Instruments, Inc., Fredericton, NB, Canada). The solubilized proteins are loaded on top of the gradient and sedimented at 36,000 rpm in a SW-41 Ti rotor using a Beckman L8–80 M ultracentrifuge for 9 h at 4 °C. Fractions (0.5 ml) are collected using a Bio-Rad (Richmond, CA) model 2110 fraction collector. Seventy microliters of each fraction are combined with 15 μl of 2-mercaptoethanol and 25 μl of 5× SDS sample buffer. This mixture (100 μl) is separated on a 1.5-mm-thick 7.5% polyacrylamide gel. Gels are stained, dried, and visualized by autoradiography. The markers used are immunoglobulin M (IgM) (950 kDa), thyroglobulin (660 kDa), urease (hexamer, 545 kDa), urease (trimer, 272 kDa), catalase (240 kDa), and adolase (160 kDa).

XII. Binding Assays

A. Glibenclamide Binding to Particulate Receptors

Stored membranes are thawed and rehomogenized at 4 °C in 50 mM Tris buffer, pH 7.4, either by using a glass homogenizer with Teflon pestle (10 strokes at 500 rpm) or by transferring the suspension to a 0.7-ml polypropylene tube and pressing the stoppered tube in ice-cold H_2O for 40 s against the 12-mm tip of a sonifier (Branson type B15P, pulsed mode with 40% duty cycle). Measurement of [³H]glibenclamide binding is performed in incubation mixtures consisting of 0.7 ml Tris buffer (50 mM, pH 7.4), 100 μl [³H]glibenclamide (0.1–250 nM), and 200 μl membrane suspension (final protein concentration 2–500 μg/ml). To reach equilibrium, incubations are carried out for 1 h at room temperature and terminated by rapid filtration of aliquots through Whatman (Clifton, NJ) GF/B filters (25-mm diameter, soaked in ice-cold Tris buffer before use) under reduced pressure. The filters are washed three times with ice-cold Tris buffer. Filtration and washing take less than 15 s. The ³H content of the filters is determined in a liquid scintillation counter after addition of 4 ml of scintillation fluid (Quickszint 402, Zinsser, Frankfurt, Germany) and equilibration for 24 h at room temperature. Nonspecific binding is determined by incubations in the additional presence of 0.1–10 μM unlabeled glibenclamide. Specific binding is determined by subtracting nonspecific from total binding.

B. Glibenclamide Binding to Solubilized Receptors

The incubation mixture consists of 0.7 ml Tris buffer (50 mM, pH 7.4), 100 μl [³H]glibenclamide (0.1–250 nM), and 200 μl detergent extract (final protein concentration 10–500 μg/ml). To reach equilibrium, incubations are carried out for 2 h at room temperature and terminated by addition of 1 ml of an ice-cold mixture containing 0.4% (w/v) bovine γ-globulin, 20% (w/v) polyethylene glycol, 0.4% (w/v) Triton X-100, and 50 mM Tris (pH 7.4). The incubation tubes are vortexed immediately after the addition of this mixture and were kept on ice for 10–20 min. Aliquots are filtered (Whatman GF/C filters, 25-mm diameter, soaked in ice-cold wash buffer before use) under vacuum and washed rapidly three times with 4 ml of a Tris buffer [50 mM Tris, 8% (w/v) polyethylene glycol, pH 7.4]. The [³H]content of the filters and specific binding is determined as described above.

These protocols can be used to assay binding of azido[¹²⁵I]iodoglibenclamide and [¹²⁵I]iodoglibenclamide.

C. [³H]P1075 Binding to Particulate SUR2B

Incubations are performed in 1-ml aliquots containing 3 nM [³H]P1075, 100 μg/ml of resuspended COS membranes, 1 mM free Mg^{2+}, 100 μM ATP, and 50 mM Tris buffer (pH 7.4). Incubations are carried out for 1 h at room temperature and terminated by rapid filtration through Whatman GF/B filters as described above (see Glibenclamide binding to particulate receptors). Nonspecific binding is defined by 100 μM pinacidil and amounted to $19 \pm 3\%$ ($n = 20$) of total binding.

Acknowledgments

This work was supported by grants from the DFG (M.S.), the NIH NIDDK (J.B.), the JDFI (J.B.), the ADA (L.A-B.), and the Houston Endowment (L.A.-B.).

References

Aguilar-Bryan, L., and Bryan, J. (1996). *Diabetes Rev.* **4,** 336.

Aguilar-Bryan, L., Nelson, D. A., Vu, Q. A., Humphrey, M. B., and Boyd, A. E., III (1990). *J. Biol. Chem.* **265,** 8218.

Aguilar-Bryan, L., Nichols, C. G., Rajan, A. S., Parker, C., and Bryan, J. (1992). *J. Biol. Chem.* **267,** 14934.

Aguilar-Bryan, L., Nichols, C. G., Wechsler, S. W., Clement, J. P., IV, Boyd, A. E., III, Gonzalez, G., Herrera-Sosa, H., Nguy, K., Bryan, J., and Nelson, D. A. (1995). *Science* **268,** 423.

Aguilar-Bryan, L., Clement, J. P., IV, and Nelson, D. A. (1998a). *Methods Enzymol.* (in press).

Aguilar-Bryan, L., Clement, J. P., IV, Gonzalez, G., Kunjilwar, K., Babenko, A., and Bryan, J. (1998b). *Physiol. Rev.* (in press, January issue).

Babenko, A. P., Aguilar-Bryan, L., and Bryan, J. (1998a). *Ann. Rev. Physiol.* **78,** 227.

Beech, D. J., Zhang, H., Nakao, K., and Bolton, T. B. (1993). *Br. J. Pharmacol.* **110,** 573.

Bryan, J., and Aguilar-Bryan, L. (1997). *Curr. Opin. Cell Biol.* **9,** 553.

Chudziak, F., Schwanstecher, M., Laatsch, H., and Panten, U. (1994). *J. Labelled Comp. Radiopharm.* **34,** 675.

Clement, J. P., IV, Kunjilwar, K., Gonzalez, G., Schwanstecher, M., Panten, U., Aguilar-Bryan, L., and Bryan, J. (1997). *Neuron* **18,** 827.

Inagaki, N., Gonoi, T., Clement, J. P., IV, Wang, C. Z., Aguilar-Bryan, L., Bryan, J., and Seino, S. (1996). *Neuron* **16,** 1011.

Isomoto, S., Kondo, C., Yamada, M., Matsumoto, S., Higashiguchi, O., Horio, Y., Matsuzawa, Y., and Kurachi, Y. (1996). *J. Biol. Chem.* **271,** 24321.

Jonas, P., Koh, D. S., Kampe, K., Hermsteiner, M., and Vogel, W. (1991). *Pflügers Arch.* **418,** 68.

Laemmli, U. K. (1970). *Nature* **227,** 680.

Nelson, D. A., Aguilar-Bryan, L., and Bryan, J. (1992). *J. Biol. Chem.* **267,** 14928.

Ohno-Shosaku, T., and Yamamoto, C. (1992). *Pflügers Arch* **422,** 260.

Schwanstecher, C., and Bassen, D. (1997). *Br. J. Pharmacol.* **121,** 193.

Schwanstecher, M., Loser, S., Chudziak, F., and Panten, U. (1994a). *J. Biol. Chem.* **269,** 17768.

Schwanstecher, M., Loser, S., Chudziak, F., Bachmann, C., and Panten, U. (1994b). *J. Neurochem.* **63,** 698.

CHAPTER 22

Simplified Fast Pressure-Clamp Technique for Studying Mechanically Gated Channels

Don W. McBride, Jr. and Owen P. Hamill

University of Texas Medical Branch
301 University Blvd
Galveston, Texas
USA

I. Introduction

The pressure-clamp technique allows the application of repetitive and reproducible steps of mechanical stimuli to small patches of membrane tightly sealed in a patch recording pipette. This is the only method currently available for characterizing the dynamic properties of individual mechanosensitive (MS) channels. Knowledge of these properties including the latency, turn-on and turn-off times and the adaptation, and/or desensitization kinetics provides not only the performance limits of the channel but also insight into the type of mechanism (direct or indirect) that may confer mechanosensitivity on a channel.

Since the first descriptions of the pressure-clamp method and its applications (Hamill and McBride, 1997; McBride and Hamill, 1992, 1995) there have been several new improvements to the piezo valve based prototype (Besch *et al.*, 2002). This chapter describes the simplified fast and reliable pressure clamp based on an inexpensive piezoelectric valve that can generate suction and pressure steps with ~1 ms rise times. We also describe the recent modifications into this design that have been introduced into a commercially available pressure clamp (http://www.alascience.com/).

A fundamental feature of the tight-seal patch-clamp technique is the existence of a suction port in the patch pipette holder. This is critical for applying suction to the membrane patch to form a tight seal. This feature also enables, after seal formation, the application of pressure/suction to stimulate the membrane patch mechanically. Initially, mouth-applied suction was used to obtain the seal and provide the stimulation. Although this is convenient, it lacks precision in terms of the magnitude and duration of the mechanical stimulus. To overcome this deficiency, we have developed a system that can apply controlled pressure waveforms to the suction port of the pipette holder. We refer to these prototypes as "pressure clamps." Although there have been previous descriptions of earlier prototypes (McBride and Hamill, 1992, 1995), we now describe a simpler but fast system and provide more details for its construction.

The basic strategy of the pressure clamp is that the desired pressure applied to the patch pipette is achieved by a balancing of pressure and suction. Central to this balancing is the use of a proportional piezoelectric valve whose opening is proportional to the applied voltage. Through feedback control of this valve, the amount of pressurized N_2 allowed to enter a mixing chamber can be regulated to balance the constant outflow due a continual vacuum efflux and thus achieve the desired pressure.

II. Mechanical Arrangement of Simplified Pressure Clamp

Figure 1 is a schematic illustrating the mechanical arrangement of the pressure clamp. Suppliers (and addresses) are listed in the Appendix. The basic principle of the valve action has been illustrated. The system volume whose pressure is being controlled is represented by the lightly shaded region, which includes the mixing chamber and the patch pipette and holder as well as the tubing to the holder. Simple analysis of the system using the ideal gas law ($PV = nRT$, where V is the system volume, and other terms have their usual meanings; see also McBride and Hamill, 1992) reveals that the speed with which the pressure can be changed is directly proportional to the change in the input or output flux and inversely proportional to the volume of the system [$dP/dt = (RT/V)(dn/dt)$]. Ideally, it would be preferable to control the input and output flux reciprocally, that is, as the input flux increases the output flux decreases and *vice versa*. In the high-speed modified version of the pressure clamp we included two values (McBride and Hamill, 1995). This gave rise/fall step transition times of less than 0.5 ms. Unfortunately, during the course of several experiments, we realized that it is nearly impossible to prevent solution from entering the valve on the vacuum side and shorting it out, thus ruining it. (The piezo element in each valve has a constant 160 V_{DC} across it.) Since on the pressure side dry N_2 is always blowing through the value into the mixing chamber, pipette solution never enters that valve. We lost

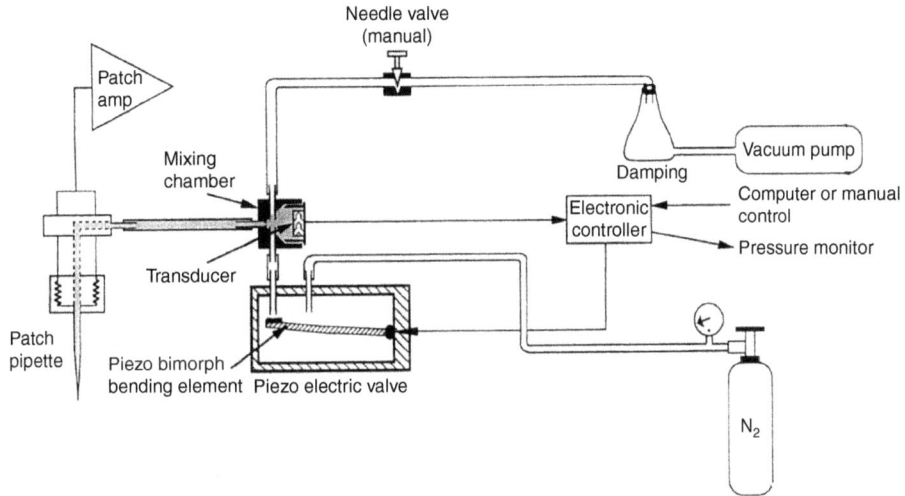

Fig. 1 Schematic diagram of the mechanical arrangement of the pressure clamp. The patch pipette holder and amplifier are shown on the left. The suction port of the pipette holder is connected by a tube to the mixing chamber. The mixing chamber contains the transducer as well as the ports leading to either a piezoelectric valve for pressure injection or to the constant vacuum. The basic mechanism of the piezoelectric valve is illustrated in which the position of the bimorph is dependent on the applied voltage. The upper port is connected to the vacuum pump through a needle valve (to adjust the suction or out flux) and damping flask (necessary to filter out cycle-to-cycle pump fluctuations). Note that the vacuum source is not critical. We have found that either house vacuum or an inexpensive commercially available vacuum pump (e.g., KNF-Neuberger) is adequate. The piezo valve is connected via tubing to the pressure regulator on the N_2 tank. Here again, house compressed air may be adequate if it is sufficiently clean and dry and if the pressure can be regulated appropriately (10 20 psi).

several valves before we realized what was occurring. To avoid this "solution-shorting" problem, we use only a single Lee valve controlling the pressure injection into a constant vacuum output flux. The amount of output flux is adjusted by a manual needle valve or a similar type valve. We find that the speed of the pressure clamp using one piezo valve still produces a step change of ∼1 ms.

Because the speed of the clamp is also inversely proportional to the volume, we have sought to minimize the system volume. In addition to the small size of the Lee valve aiding in this, we have also, as described below, incorporated the transducer into the mixing chamber (i.e., the sensor element of the transducer forms one wall of the mixing chamber). These two modifications permit closer proximity of the pressure-clamp mixing chamber to the pipette holder and thus minimize the connecting tubing. Tubing length from the vacuum and pressure is less critical, although it can affect input/output flux.

To consolidate the transducer/mixing chamber we started with a Sen-Sym transducer, which is enclosed in a metal, transistor type TO-39 package. The actual metal can has a diameter of 0.325 in. and a height of 0.265 in. Normally the top of the can has a small-diameter hole that gives access to the sensor. By cutting off the top portion of the can (Thorlabs sells a convenient diode "can-opener"), the sensor element itself is exposed and the remaining bottom portion of the can containing the leads may be conveniently press fitted into the end of a cylindrically shaped mixing chamber. The mixing chamber is made from a piece of 0.5-in. diameter plexiglass rod ~0.5 in. long. Into one end a hole is drilled to accept the sensor (e.g., letter O drill bit and ~0.25 in. deep). The chamber has three additional ports in it. We use #16 or #18 size hypodermic needles cut to a suitable length. Two of the ports, diametrically opposed, are perpendicular to the axis of the chamber and are drilled so as to intersect the large central transducer hole near the tip of the cone formed during its drilling. One port connects to the piezo valve and the other to the constant vacuum. The third port is along the axis of the cylinder and opposite the sensor and connects to the pipette holder. With regard to the third port care must be taken to prevent the cutoff hypodermic needle from damaging the sensor element. The total volume of the mixing chamber is about 250 μl. However, as described below, to increase the sensitivity of the pressure clamp large (either longer or bigger diameter) mixing chambers could be used.

III. Electronic Control of Pressure Clamp

The basic design of the electronic controller of the pressure clamp has been retained (McBride and Hamill, 1992). Figure 2 shows the schematic of the electronic controller and is divided into four sections. The first section is the voltage command. There are two internal inputs and one external input (A5). The internal control includes both the manual (A6) (i.e., potentiometer) and an adjustable test pulse, which can be switched in or out. The test pulse is used during the adjustment procedure for optimizing the step response. This pulse could be supplied externally. However, because it is frequently used it has been incorporated into the controller. An LM555 timer used as an astable multivibrator (National Semiconductor, Santa Clara, CA) was used to generate a repetitive ~50-ms pulse at ~2 Hz. This pulse was then scaled by multiplying it by the output of a potentiometer using an AD633 multiplier chip (Analog Devices, available from Newark, also data sheet). The result was a test square pulse that could be varied between +5 and −5 V. The second section of the controller (Fig. 2) is responsible for pressure measurement. The transducer used is the SCC05GSO (SenSym, Inc., Milpitas, CA). The modifications in incorporating this into the mixing chamber have been discussed above. An LM334 current regulator (National Semiconductor, available from Newark, also see data sheet) was used to excite the transducer bridge and was set at a current of ~1 mA. The AD620 (Analog Devices) instrumentation amplifier with a gain of ~500 (see data sheet for AD620) was used to measure the output of the bridge. The AD620 is an 8-pin chip which is convenient to use. The output of the AD620 is further amplified (A4) to give the desired gain of 100 mV/mmHg. As another simplification, the Bessel filter that followed A4 has been removed. The subsequent integration was found to be sufficient. The third section is the feedback control

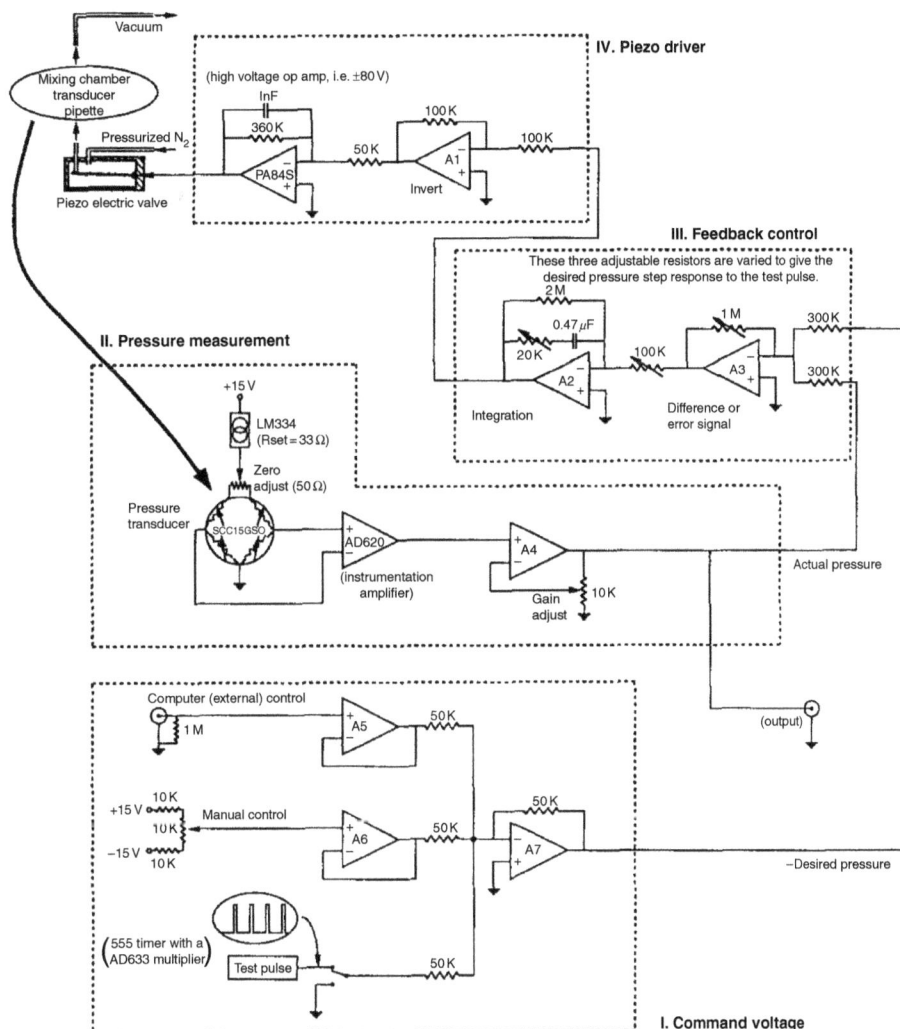

Fig. 2 Electronic control of the pressure clamp. The circuit can be divided into four blocks. (I) Command voltage. This represents the desired pressure and includes internal and external sources. The internal sources include a manual control for setting the steady-state values (A6) and a variable test pulse, which can be turned on and off by a toggle switch. The test pulse is used in the step response adjustment procedure. It could have been supplied externally but because it is frequently used, for convenience it has been incorporated into the circuitry. (II) Pressure measurement. This measures the pressure in the mixing chamber. A constant current source excites the transducer bridge whose output is measured by the instrumentation amplifier. A voltage follower (A4) is included to enable gain adjustment. (III) Feedback control. This section is composed of an initial summing amplifier (A3), which generates an error signal followed by an integrating amplifier (A2). In this section there are three adjustable resistors. While to some extent there may be some redundancy they do allow flexibility in terms of shaping the desired waveform (i.e., step response). (IV) Piezo driver. This section includes a high-voltage operational amplifier operated with a gain of about 7 and a low-pass filter. The integrated signal from section III is inverted (A1) before the high-voltage driving amp. This is necessary to provide negative feedback for the feedback loop. (All op-amps used were AD712, available from Newark.)

section and is composed of a summation amplifier (A3) used to generate the error signal (between the desired and actual pressures) followed by integration or averaging (with compensation) of this error signal (A2). Three adjustable resistors have been included to adjust the response. While this may be somewhat redundant, it does allow flexibility in adjusting the desired waveform. The final stage is the piezo driver section. This includes a high-voltage op-amp (PA84S from Apex, Tucson, AZ), which is powered at ±80 V. The control signal for the pressure injection from the N_2 tank has been inverted (A1) so as to give overall negative feedback.

IV. Some Practical Tips for Construction

In the construction, we suggest that Section II (Fig. 2), pressure measurement, be assembled first. The data sheet for the transducer gives some good tips, especially with regard to adjusting the offset. The most critical part of this section is the calibration of the overall gain of 100 mV/mmHg. To do this, a calibrated pressure transducer must be used with which to compare the pressure clamp transducer. The mixing chamber can be constructed with the transducer in place. With the sensor open to the atmosphere, adjust the zero trim potentiometer appropriately. To adjust the gain, block one of the three ports of the mixing chamber. Connect one port to a syringe and the other to the calibrated sensor. The pressure in the system can then be changed by moving the syringe plunger and monitored via the calibrated sensor. Simply adjust the gain on amplifier A4 until the two sensors agree. Two calibrated transducers (SenSym) that can be used are the 143SC series and a digital manometer. The 143SC is a simple three-pin device (power, ground, and output) which gives a ±analog signal output centered at 3.5 V. The more accurate digital manometer only has a digital readout.

A ±80 V_{DC} power supply is needed to drive the piezo valve. One can be easily and inexpensively built using a dual isolation (1:1) transformer such as the Magnetek FP230-50 available from Newark Electronics. The dual transformer is necessary to obtain ±80 V. Appropriate full-wave bridges and capacitors can be used in conjunction with the TL783C, a three-pin, 125-V voltage regulator (Texas Instruments, available from Newark Electronics) to build two +80 V supplies (see the TL783C data sheets). The voltage output of the TL783C is determined by the ratio of two resistors. We use values for R_1 of 130 Ω and R_2 of 2200 Ω. Since both supplies are isolated, they can be connected in series (i.e., the positive of one connected to the ground of the other) to give one ±80 V power supply.

At the present gain of 100 mV/mmHg the practical pressure limit of the clamp is ±100 mmHg (i.e., ±10 V input). In studies using standard sized patch pipettes (i.e., 1–2 μm) this upper limit value is sufficient for determining stimulus–response relations of mechanically gated (MG) channels since most MG channels studied saturate at <50 mmHg. Furthermore, this limit exceeds typical patch rupture pressures. However, there are some reports of MG channels that have saturation pressure exceeding ~100 mmHg. To achieve higher pressures, the gain of the pressure measurement section could be reduced to, for instance, 50 V/mmHg [i.e., the gain adjust on A4 (Fig. 2)]. This would double the practical range to ±200 mmHg. However, note that at high suctions the membrane can be decoupled from the underlying cytoskeleton and alter (i.e., either increase or decrease) the mechanosensitivity of a channel (Hamill and McBride, 1997).

When considering noise limitations and sensitivity of the pressure clamp itself (i.e., the minimal distinguishable step size), there is a reciprocal relationship between these and the speed of the clamp with regard to the volume of the system. On the one hand, larger volumes decrease the speed (i.e., for a given flux it takes longer to change the pressure in a larger volume). On the other hand, a larger volume stabilizes the pressure of the system making it less sensitive to fluctuations in input or output flux. This decreases the pressure noise of the system and also, because it requires a larger change in flux for a given change in pressure, the control can be more sensitive albeit slower. Our preference has been to optimize the time response of the clamp since oocyte and muscle MG channels are activated by moderate pressures (~10–20 mmHg) by minimizing the mixing chamber volume. In the present configuration, the pressure rms (root mean square) noise is ~0.1 mmHg and this presumably represents the limitation for the minimal distinguishable step size that can be applied by the clamp for the given volume of the system. This sensitivity is adequate for many MG channels, which show half saturation pressure of 10–15 mmHg. However, some MG channels have been reported to have half-saturation pressure as low as 1–2 mmHg. For these more sensitive MG channels the minimal pressure increments could be reduced by increasing the mixing chamber volume (i.e., making it longer or larger diameter).

Several factors influence the step response of the pressure clamp. There are three potentiometers in the feedback control section. Varying these potentiometers will affect the speed and shape of a pressure step. In addition, the setting of the needle valve on the vacuum side and the regulator pressure from the N_2 tank (or other pressure source) in front of the piezo valve (5–20 psi) both greatly influence the step response. Indeed, all five of these possible adjustment points determine the step response. With some experimentation, one can determine how each adjustment point affects the speed and shape of the step response so that the optimum response can be obtained.

Since the development of the above described pressure clamp Besch *et al.* (2002) have provided an elegant solution for overcoming the "saline short circuiting" problem, by introducing a photosensitive water detector—consisting of a photodiode/photodetector pair (OP299/OP535A, Optek, Carrollton, TX, USA)—that detects saline entering the outlet tube and switches the valve to positive pressure. In addition, they have utilized a single piezoelectric bimorph (Piezo Systems, Cambridge, MA, USA) to control both pressure and suction. By also incorporating the pressure sensor within the valve and further reducing the dead volume they were able to significantly reduce the latency between the command voltage and initial pressure change to 120 μs, while maintaining the pressure step rise times of

1–3 ms for 20–200 mmHg steps (Besch *et al.*, 2002), which are only slightly slower than submillisecond rise times reported for the first fast pressure-clamp (McBride and Hamill, 1995). The Besch *et al.* (2002) prototype is currently marketed by ALA Scientific Instruments (http://www.alascience.com/).

Appendix

We indicate here the suppliers we have used in constructing the pressure clamp.

High voltage op-amp, as well as the socket for the op-amp [Apex Microtechnology Corporation, 5980 North Shannon Road, Tucscon, AZ 85741-5230, phone: (800) 546-2739].

Piezo valve [The Lee Company, 2 Pettipaug Road, P.O. Box 424, Westbrook, CT 06498-0424, phone: (800) 533-7584].

Pressure transducer [SynSym Inc., 1804 McCarthy Boulevard, Milpitas, CA 95035, phone: (800) 392-9934].

Can opener for pressure transducer [Thorlabs, Inc., P.O. Box 366, Newton, NJ 07860-0366, phone: (201) 579-7227].

Power supply parts and electronic parts [Newark Electronics, 4801 North Ravenswood Avenue, Chicago, IL 60640-4496, phone: (773) 784-5100].

Vacuum pumps [KNF-Neuberger, 2 Black Forest Road, Trenton, NJ 08691, phone: (609) 890-8889].

Acknowledgments

We acknowledge the NIH (RO1-AR42782), the NSF (Instrument Development for Biological Research) and the MDA for their support.

References

Besch, S. R., Suchyna, T., and Sachs, F. (2002). *Pflügers Arch.* **445,** 161.
Hamill, O. P., and McBride, D. W., Jr. (1997). *Annu. Rev. Physiol.* **59,** 621.
McBride, D. W., Jr., and Hamill, O. P. (1992). *Pflügers Arch.* **421,** 606.
McBride, D. W., Jr., and Hamill, O. P. (1995). *In* "Single Channel Recording," 2nd edn. (B. Sakmann, and E. Neher, eds.), p. 329. Plenum, New York.

CHAPTER 23

Purification and Heterologous Expression of Inhibitory Glycine Receptors

Bodo Laube and Heinrich Betz

Max-Plank Institute
Hirnforschung
Deutschordenstrasse
Frankfurt, Germany

473
DOI: 10.1016/B978-0-12-382204-8.00023-0

I. Introduction

Glycine is a major inhibitory neurotransmitter in the vertebrate central nervous system, most prominently in brain stem and spinal cord. Its postsynaptic actions are mediated via a pentameric chloride channel protein, the inhibitory glycine receptor (GlyR). Strychnine, a convulsive poison in man and animals, antagonizes glycine activation of the GlyR and has proved to be a highly useful tool to characterize this membrane protein (Kuhse et al., 1995). In adult spinal cord, the GlyR is composed of three copies of ligand binding α_1 and two copies of structural β subunits. During development and in peripheral tissues, additional α-subunit genes (α_2–α_4) are expressed that display distinct pharmacologic and functional properties (Kuhse et al., 1995). Heterologous expression has allowed their detailed characterization. Here, we describe methods for the purification and expression in heterologous cell systems of the GlyR.

II. Purification of Glycine Receptor from Mammalian Spinal Cord

Affinity purification of the GlyR has been successfully achieved from rat (Pfeiffer et al., 1982), mouse (Becker et al., 1986), and porcine (Graham et al., 1985) spinal cord tissue as well as from cells expressing recombinant receptors (Hoch et al., 1989). All procedures involve adsorption of the detergent-solubilized receptor on aminostrychnine-agarose followed by biospecific elution with the agonist glycine. Routinely this procedure results in highly enriched (>80%) preparations containing the postsynaptic GlyR complex, that is, its α and β subunits together with the anchoring protein gephyrin (Prior et al., 1992; Schmitt et al., 1987). All steps have to be performed at 4 °C in the presence of a protease inhibitor mix (Pfeiffer et al., 1982). After solubilization, phospholipids (phosphatidylcholine) have to be included in all detergent buffers to stabilize the native conformation of the GlyR (Pfeiffer and Betz, 1981; Pfeiffer et al., 1982).

A. Preparation of Membranes

Spinal cord, medulla oblongata, and pons are removed from adult (100–200 g) Wistar rats (or mice or pigs), frozen in liquid nitrogen, and stored at −70 °C. Crude synaptic membranes are prepared from the frozen tissue after homogenization in >20 volumes of buffer A consisting of 25 mM potassium phosphate buffer, pH 7.4, containing 5 mM ethylenediaminetetraacetic acid (EDTA), 1 mM dithiothreitol (DTT), and protease inhibitors [100 mM benzethonium chloride, 1 mM benzami-dinehydrochloride, 100 mM phenylmethylsulfonyl fluoride (PMSF), and 16 mU/ml of aprotinin] using a motor-driven Teflon or Polytron homogenizer. After optional removal of debris and nuclei by centrifugation at $1000 \times g$ for 30 min, membranes are collected at $30,000 \times g$ for 30 min and washed once with buffer A following the protocol described earlier. The resulting membrane pellet is rehomogenized in buffer A containing 0.6 M sucrose and centrifuged at $30,000 \times g$ for 60 min. After resuspending the membrane fraction in 10 volumes of buffer A and centrifugation at $48,000 \times g$ for 20 min, the membranes are suspended in 3 volumes of buffer A, frozen in liquid nitrogen, and stored at $-70\,°C$. Preparation of membranes from transfected tissue culture cells (Hoch *et al.*, 1989; Sontheimer *et al.*, 1989) basically follows the same protocol. However, the sucrose buffer step is not required.

B. Solubilization of Glycine Receptor

The GlyR has been solubilized and purified from spinal cord membranes using either the nonionic detergent Triton X-100 (Becker *et al.*, 1986; Pfeiffer *et al.*, 1982) or the ionic detergent cholate (Graham *et al.*, 1985). Solubilization efficiencies with cholate are usually better; however, the GlyR associated protein gephyrin is largely lost (Graham *et al.*, 1985).

For routine purification (Pfeiffer *et al.*, 1982), membranes are incubated in 25 mM KP_i, pH 7.4, containing final concentrations of 3% (w/v) Triton X-100, 1 M KCl, 5 mM neutralized EDTA, 5 mM neutralized ethylene glycol bis(β-aminoethyl ether)-N,N,N',N'-tetraacetic acid (EGTA), 5 mM DTT, and protease inhibitors. Protein concentrations may range from 1 to 10 mg/ml. The mixture is held on ice and repeatedly homogenized during 1 h. After centrifugation at $100,000 \times g$ for 1 h at 4 °C, the supernatant is carefully decanted, filtered through a porous glass sieve to retain myelin lipids, and used as source of the solubilized GlyR.

C. Preparation of Aminostrychnine–Agarose

2-Aminostrychnine had been commercially available previously; however, no supplier is presently known. The compound therefore has to be synthesized from strychnine by nitration followed by reduction with Sn/HCl as described (Mackerer *et al.*, 1977). For coupling to Affi-Gel 10 (Bio-Rad, Richmond, CA), 20 ml of the gel is washed three times with 1 volume of dimethylform-amide. To the washed resin in 20 ml of dimethylformamide, 500 mg of solid 2-aminostrychnine are added, and the mixture is gently shaken for 2 days at 4 °C in a brown bottle. After the addition of 200 ml of ethanolamine and incubation for another 2 h, the gel is washed with 20 ml of dimethylformamide followed by 20 ml of 50 mM potassium phosphate buffer, pH 7.4, containing 0.02% sodium azide. After transfer into a column the gel is washed with an additional 400 ml of this buffer and stored in a brown bottle at 4 °C for up to 2 years (Pfeiffer *et al.*, 1982). Aliquots are washed extensively with 25 mM potassium phosphate buffer (KP_i), pH 7.4, containing 1% (w/v) Triton X-100 and 1 M KCl (buffer B), before being used for affinity chromatography.

D. Affinity Chromatography

An aminostrychnine-agarose column (1–5 ml) is washed with 10–20 column volumes of buffer B, and then solubilized GlyR (up to ca. 300 ml) is applied to the column with a maximal flow rate of about 2.5 ml/h of affinity resin. The column is washed extensively with about 20 column volumes of buffer B containing 5 mM EDTA, 5 mM EGTA, 5 mM DTT, 2.5 mM phosphatidylcholine, and protease inhibitors (buffer C). For elution, 1 column volume of buffer C containing 200 mM glycine (buffer D) is continuously pumped through the column for several hours (Pfeiffer *et al.*, 1982). After elution, a second addition of 1 column volume of buffer D results in a secondary eluate that also contains native GlyR. A final elution with 1 column volume buffer D containing 5 M urea yields denatured GlyR polypeptides and regenerates the column for future use in additional purifications. We have been able to repeatedly use the affinity matrix for purification over periods greater than 1 year.

III. Heterologous Expression of Glycine Receptor

Expression in heterologous cell systems has been widely used to investigate functional properties of the GlyR by both electrophysiologic recording and biochemical methods (Kuhse *et al.*, 1995). Injection into *Xenopus* oocytes of poly(A)$^+$ RNA prepared from spinal cord generates glycine-gated chloride channels in the oocyte membrane whose properties closely correspond to those detected in neurons. Functional GlyRs have also been generated by injection of *in vitro* transcribed rat and human α- and β-subunit mRNAs. Expression of the α_1, α_2, or α_3 subunits alone and in combination with the β subunit has shown that both homo- and heterooligomeric GlyRs exist (Kuhse *et al.*, 1995). Transient expression of the recombinant GlyRs has also been achieved by transfection of cultured mammalian cells or using the baculovirus system.

A. Oocyte Expression System

Since Miledi and Sumikawa (1982) demonstrated that various types of ion channels and receptors can be expressed in *Xenopus laevis* oocytes injected with poly(A)$^+$ RNA isolated from appropriate tissue, the oocyte expression system has proved highly useful to investigate various aspects of the pharmacology and function of GlyR proteins. These include (1) identification of structure–function relationships by mutational analysis (Schmieden *et al.*, 1993), (2) posttranslational processing and assembly of this multisubunit receptor (Kuhse *et al.*, 1993), (3) comparison of the pharmacologic properties of native and recombinant GlyRs (Schmieden *et al.*, 1989), (4) receptor modulation by effectors (Laube *et al.*, 1995), and (5) analysis of mutations causing hereditary animal and human motor disorders (Langosch *et al.*, 1994). The *Xenopus* oocyte expression system is ideally suited for electrophysiology, since the large cells are easy to handle and current amplitudes are high (routinely up to 5 μA). Biochemical studies in contrast are tedious since large numbers of cells have to be injected individually.

Heterologous expression of the GlyR in *Xenopus* oocytes has been achieved by both cytoplasmic injection of poly(A)$^+$ RNA or *in vitro* synthesized cRNA encoding different subunits and nuclear injection of cDNAs.

B. Isolation of Poly(A)$^+$ RNA from Neural Tissue

Poly(A)$^+$ RNA suitable for injection can be prepared from spinal cord and brain according to standard methods (Goldin and Sumikawa, 1992). However, pitfalls have been encountered with both isolated and *in vitro* transcribed RNA samples due to contamination with RNases. The following general comments on the handling of RNA should help to eliminate this problem: (1) Wear gloves throughout all manipulations (RNases are everywhere). (2) Prepare all buffers with filtrated, autoclaved H_2O. All tubes and pipette tips should also be autoclaved. (3) Clean all gel tanks and glass plates thoroughly. A common source of RNase on gel electrophoresis equipment stems from DNA preparations that have been treated with RNase A. (4) Templates used for transcription must be RNase free. DNA minipreparations are suitable if care is taken to remove contaminating RNases. Generally the template is linearized with a restriction endonuclease that cleaves downstream of the RNA polymerase promoter and the DNA inserted into the multiple cloning site. The digested DNA should be purified by proteinase K followed by two phenol–chloroform (1:1, v/v) steps and one chloroform extraction followed by ethanol precipitation. (5) It is recommended that RNase inhibitors be included in the transcription reaction.

C. *In Vitro* Transcription

RNA suitable for expression in *Xenopus* oocytes can be readily prepared by *in vitro* transcription of GlyR subunit cDNAs with bacteriophage T3 or T7 RNA polymerase in the presence of a 5′cap [P-1,5′-(methyl)guanosine P-3,5-guanosine triphosphate] analog (Kuhse *et al.*, 1993). Addition of the 5′cap structure appears to be essential for RNA stability and efficient expression of transcribed RNA in oocytes (Ben-Azis and Soreq, 1990). Receptor cDNAs should be inserted into suitable cloning vectors, that is, pBluescript II M13$^{+/-}$ KS/SK (Stratagene, La Jolla, CA), pRc/CMV, or pcDNAI/AMP (Invitrogen) plasmids, which provide T3/T7 promoters. Before transcription, the plasmid has to be linearized at a convenient restriction site as follows.

D. Specific Procedures

Linearization of the Vector

1. Mix 4 pmol of vector DNA, required enzyme, 10 μl buffer (10×), and H_2O to 100 μl

2. Incubate for 1 h at 37 °C (or at the temperature required by the enzyme). Check digest by running 2 μl on a 1% agarose gel together with undigested and linearized plasmid

3. Add 2 μl of proteinase K (50 μg/ml) and incubate for another 30 min at 37 °C

4. Extract twice with 1 volume phenol/chloroform (1:1), once with 1 volume chloroform/isoamyl alcohol (24:1), and precipitate with 0.1 volume 3 M sodium-acetate, pH 5.2, and 2 volumes of ethanol (100%) for 0.5 h at −20 °C (or at −70 °C for 5 min)

5. Wash once with 1 ml of 70% (v/v) ethanol and resuspend in 10 μl TE buffer (10 mM Tris, pH 7.5, 0.1 mM EDTA)

6. Dot 1 μl on an ethidium bromide agarose plate or measure the OD at 260 nm for quantitation

Transcription

1. Mix

	At final concentration
1 μg linearized plasmid (1 μg/μl)	50 ng/μl
2 μl rATP-, rCTP-, rUTP-, and rGTP-mix (5 mM each)	500 μM
2 μl CAP (5 mM)	500 μM
4 μl transcription buffer 5× (200 mM Tris–HCl, pH 7.5, 250 mM NaCl, 40 mM $MgCl_2$, 10 mM spermidine)	
1 μl RNase inhibitor II	1 U/μl
2 μl SP6, T3, or T7 polymerase (Boehringer, Mannheim, Germany)	2 U/μl
H_2O	To 20 μl

2. Incubate for 1 h at 37 °C

3. Add 0.5 μl DNase 1 (10 U/μl) and incubate for 15 min at 37 °C

4. Add 1 μl of 0.5 M EDTA, pH 8.0, and 80 μl of TE buffer (10 mM Tris, pH 7.5, 0.1 mM EDTA) or diethyl pyrocarbonate (DEPC)-treated water

5. Extract once with 1 volume phenol/chloroform (1:1) and 1 volume chloroform/isoamyl alcohol (24:1)

6. Add 1 volume 4 M ammonium acetate and 2 volumes of ethanol. Incubate and precipitate the RNA for 15 min at −70 °C in a dry ice–ethanol mixture

7. Spin for 30 min at 13,000 rpm and 4 °C in an Eppendorf centrifuge.

8. Drain the pellet and rinse with 80% ethanol cooled to −20 °C

9. Spin for 10 min at 13,000 rpm, dry in a Rotovap (Fischer, Frankfurt, Germany), and resuspend in 10 μl H_2O

10. Determine the concentration of RNA in the sample and store at $-70\,^\circ$C in 1-μl aliquots

E. Preparation of Oocytes

X. laevis oocytes are egg precursors that on hormonal stimulation become fertilizable eggs. They are stored as paired gonades in the abdominal cavity and can be surgically removed and maintained in Barth's medium [88 mM NaCl, 1 mM KCl, 0.4 mM $CaCl_2$, 0.3 mM $Ca(NO_3)_2$, 8.4 mM $MgSO_4$, 2.4 mM $NaHCO_3$, 10 mM Tris, pH 7.2]. Only the large stage V and VI oocytes (Dumont, 1972) with a diameter of approximately 1–1.2 mm should be used for electrophysiological experiments. Several ovarian lobes can be removed from female frogs anesthetized by immersion in 1% (w/v) urethane (Sigma, St Louis, MO) by making a 1- to 2-cm incision in the abdomen. This incision can be rapidly sutured under urethane narcosis with sterile surgical thread and needles; the animals recover rapidly and may be used again as oocyte donors in a later experiment. The ovarial lobes are then mechanically separated in pieces of about 10–50 oocytes by using forceps. For the isolation of single oocytes, connective tissue and blood vessels are removed by using forceps and a pair of microscissors (Dösch, Heidelberg). Subsequently, the layer of follicle cells surrounding the oocyte must be removed carefully (see below) because it constitutes a potential source of problems due to electrical coupling to the oocyte via gap junctions. The surrounding vitelline membrane, a glycoprotein matrix that gives the oocyte structural rigidity, does not impair binding of compounds to the oocyte surface up to a molecular mass of about 8000 Da and can therefore be left on the oocyte for routine electrophysiology. All dissection procedures should be done in Barth's medium.

There are two ways to remove the follicle cell layers:

1. Extensive collagenase treatment that completely strips off the layers is achieved by immersing the oocyte for 1–3 h in calcium-free saline containing 2 mg/ml of collagenase Type IIA (Sigma)

2. A less extensive collagenase treatment followed by manual removal of the follicular layer with fine forceps (Inox 5) under binocular control (M3Z; Zeiss, Oberkochen, Germany) may be the better method, since oocytes undergoing the extensive collagenase treatment often show reduced survival times

A striking feature of *Xenopus* oocytes is their highly characteristic polarity: the animal pole region is of dark green-brownish color whereas the vegetal pole has a bright yellow-white appearance. Polarity is maintained inside the oocyte: the nucleus consistently is found at the animal pole, whereas receptors expressed from exogenous mRNAs accumulate preferentially in the vegetal pole region of the plasma membrane. This nonhomogeneity has no consequences for whole-cell two-electrode voltage clamping, but great importance when recording GlyRs in different patch-clamp configurations (Hamill *et al.*, 1981). For achieving high-resistance seals, a clean oocyte surface is essential. Because devitellinized oocytes are extremely fragile and tend to disintegrate on contact with air–water interfaces, the vitelline membrane should only be removed for single-channel recordings requiring a gigaseal. This can be achieved by placing the oocyte in a hypertonic stripping solution (200 mM potassium aspartate, 20 mM KCl, 1 mM MgCl$_2$, 10 mM EGTA, 10 mM HEPES, pH 7.4) and allowing it to shrink. During shrinking, the vitelline membrane detaches from the plasma membrane and forms a transparent sphere around the oocyte that can be removed with fine forceps. Devitellinized oocytes should be stored in small culture dishes containing a bottom layer of 2% agarose to avoid sticking to the plastic.

F. Injection of Oocytes

Oocytes of healthy appearance with clearly visible vegetal and animal poles are transferred into a culture dish and allowed to recover for >2 h. Thereafter damaged oocytes can be easily identified by visual inspection and discarded. Oocyte injections are performed under a fixed-stage binocular (Zeiss, Leitz) with objective lenses of 6.5–40× and eye pieces of 10× magnification. The binocular is mounted to an injection table with a movable injection chamber and a micromanipulator (Narishige, East Meadow, NY, or Bachofer, Reutlingen, Germany) for positioning the injection capillary, which is connected to a pressure generator (microinjector).

1. Pulling and Calibration of Injection Capillaries

Pull glass capillaries (borosilicate, 2 mm outer diameter; Hilgenberg) with a total volume of 500 nl for multiple injections by using a List electronic two-step puller L/M-3P-A or a similar device. When using the List puller, the capillary is fixed in the upper and lower holders for maximal distance. The real distance of the first pull is given by a disk with a height of 7 mm and made with a current of 20 A. Changing the disk to the 9-mm one allows pulling of the fine tip in the second step with a current of 16.5 A resulting. Pressure injection is useful for injecting volumes >1 nl in a reproducible manner. For RNA injection, volumes of 10–50 nl per oocyte have to be applied. Two different positive pressure modes should be available: a high one for ejection and a low pressure to prevent backfilling of the pipette by capillary action or diffusion (microinjector PV820; Bachofer). One of the most critical factors for injecting reproducible volumes is experience in manufacturing suitable micropipettes. To prevent backfilling of the pipette by capillary action and to avoid difficulties during injection, the pipette tip must be inspected visually and corrected to a size of about 4–8 μm. With tip sizes of less than 1 μm, pressure ejection becomes increasingly difficult; this can be partially overcome by cleaning the micropipette with chromic acid solutions or silanization to decrease surface tension. The volume of fluid ejected is strongly dependent on tip size because a reduction or increase in tip size results in remarkable changes in the flow rate. To calculate the volume ejected per pressure application, deposit a drop of fluid on the tip of the micropipette by a single pressure increase. The volume of this drop may be calculated by measuring the radius and assuming the drop to be spherical (100 μm diameter correspond to $\approx$500 pl). For injection, we use a manipulator (MD4, Bachofer) where 20 oocytes fixed on an injection table can be injected consecutively. A desirable injection volume is about 50 nl corresponding to 10 ng of RNA per oocyte when using a solution containing 200 ng RNA/μl. The injection capillary holds a total volume of 500 nl; therefore up to 10 oocytes can be injected with one filled capillary. For transferring the RNA from the tube into the injection needle, 10-μl siliconized (dichloromethylsilane; Merck, Darmstadt, Germany) and sterilized 10-μl micropipettes (Brand, Wortheim, Germany) are required.

G. Cytoplasmic Injection

1. Fix the injection capillary in the holder of the micromanipulator and divide its tip into 10 equal sections using a small overhead marker
2. Fill a 10-μl micropipette (Brand) with 1 μl of RNA solution by using a micrometer syringe
3. Connect a 5-ml injection syringe to the freshly pulled injection capillary and position it close to the RNA containing micropipette
4. Form a drop of RNA solution at the tip of the micropipette by increasing pressure using the microsyringe
5. When the pipette is inserted into the fluid, the meniscus can be seen to raise in the tip of the capillary. Create negative pressure to fill the injection needle
6. Position the oocytes on the injection table
7. Create a slight positive pressure in the injection capillary and penetrate the oocyte membrane by using the micromanipulator
8. Increase the pressure until the RNA solution starts to flow
9. Remove the needle by using the micromanipulator after 50 nl of solution have been injected and proceed to the next oocyte
10. After injection, incubate all oocytes at 19 °C (Heraeus, Hanau, Germany) in 35-mm petri dishes (Greiner, Frickenhausen, Germany) containing Barth's medium and change the medium daily

GlyR subunits express efficiently; thus electrophysiological recording is possible already after 24 h of injection. Maximal expression levels are reached after 2 days and maintained for up to 3 days. After incubation at 19 °C, oocytes may be stored at 4 °C for up to 2 weeks without significant changes in channel behavior and membrane properties. Before recording, any remaining follicular cells should be removed mechanically under the microscope (M3Z, Zeiss) by using forceps.

H. Nuclear Injection

Expression of GlyRs via nuclear injection of cDNAs has the advantage that no preparation and handling of RNA is required. In addition, higher expression rates may be achieved (Taleb and Betz, 1994). However, the survival of oocytes after nuclear injection is generally much lower than that obtained on cytoplasmic injection of RNA.

Nuclear injection can be easily performed provided the nucleus of the oocyte is visible. To this end, the nucleus is moved close to the cell surface by gentle centrifugation. This can be achieved by putting the oocytes with the dark side up into the wells of a 96-well microtiter plate, adding Barth's medium, and centrifuging the plates at $1000 \times g$ for 15 min at room temperature. Nuclei then will appear as white spots in the dark animal pole region. The plasmid pCis (Gorman $et\ al.$, 1990), pRc/CMV, or pcDNAI/Amp (Invitrogen) containing the appropriate GlyR cDNA insert is diluted in water to concentrations of 25–50 ng/μl; when using an injection volume of $\sim$10 nl, this corresponds to a total amount of 250 pg of plasmid. Injection capillaries are made from Drummond microcaps in two steps using a vertical puller (LM-3P-A; List Electronics) as described before. Then the capillary tip is broken using optical control (microforge; Bachofer) until the desired tip diameter is obtained ($\approx$5 μm). The capillary is inserted into a micromanipulator (MD4; Bachofer) and connected to a pressure generator (microinjector PV 820; Bachofer) that allows the generation of different injection parameters by varying the time and pressure of injection. The protocols for the calibration of the needle and the injection procedure are similar to those used for cytoplasmic injection. Only the inside volume of the capillary for the nuclear injection is lowered to 100 nl, thus allowing 10 injections of 10 nl. Sometimes the oocytes have to be turned under binocular control (40$\times$) to orient the visible nucleus perpendicularly to the tip of the capillary. Then the tip is introduced into the nucleus, and pressure is activated until the above volume is injected. Subsequently the cells are incubated in Barth's medium at 19 $^{\circ}$C as above.

IV. Transient Expression in HEK293 Cells

Coprecipitates composed of "calcium phosphate" (hydroxyapatite) and purified DNA have been used for >20 years for the transfer to and expression of genetic information in mammalian cells in culture. This technique has become one of the preferred methods to express recombinant GlyRs in mammalian cells for analyzing their functional properties (Langosch $et\ al.$, 1994). We use a popular immortalized cell line, human embryonic kidney 293 cells (HEK 293; CRL 1573, ATCC, Rockville, MD), and a transfection procedure adapted from Chen and Okayama (Chen and Okayama, 1987). Routinely, transfection efficacies of ca. 50% can be achieved. The expression vectors contain a promoter that interacts with factors in the cell line, that is, a cytomegalovirus (CMV) promoter or other viral promoters (SV40) to provide high transcript levels. Poly(A) addition sites, which are contained in many eukaryotic expression vectors, enhance in general the production of receptors. The quality of the plasmid DNA is also important; phenol/chloroform extraction and ethanol precipitation should be used for plasmid preparation. DNA purity can be checked by measuring optical densities at 260 nm versus 280 nm; the ratio should be ≈ 2.

Expression by transfecting cells provides some advantages over the *Xenopus* expression system: (1) Biochemical studies are possible, since large numbers of cells can be rapidly obtained. (2) No *in vitro* translation of RNA is required. (3) Control over the intracellular solution is possible during electrophysiologic recording. (4) Fast kinetics can be recorded easily in the whole-cell mode because cell sizes are much smaller than for oocytes.

A. Cultivation of HEK293 Cells

HEK293 cells can be kept for 20–30 passages when replated before confluence. For electrophysiologic recording, the cells can be seeded onto fibronectin-coated (20 μg/ml fibronectin) plastic or glass coverslips (Thermanox, 13 mm; Nunc, Rochester, NY) in standard 24-well culture dishes with minimum essential medium (MEM). Cultures for biochemical experiments are grown in 30- or 100-mm uncoated tissue culture dishes. The cells are transfected by the calcium phosphate method 48 h after plating and can be analyzed 48 h after the transfection. For efficient transfection, the cells must be preconfluent because they have to divide in order to take up and express DNA. One inherent problem is the presence of nontransfected cells; even in good experiments, only about 50% of the cells are transfected. Thus, identification of transfected cells becomes important for electrophysiology; cotransfection with green fluorescence protein (Marshall *et al.*, 1993) (GFP) cDNA can be used for visualization of transfected cells. In addition, under differential interference contrast optics transfected cells often show a granular or vacuolated appearance. To eliminate electrical coupling of the HEK293 cells, an inhibitor of mitosis (14.3 mM uridine) can be added to the medium 24 h after transfection; in addition, isolated cells should be selected for whole-cell recording. For electrophysiology the coverslips with the transfected cells are transferred to the recording chamber where the culture medium can be exchanged by the appropriate extracellular solution for recording.

B. Preparing HEK293 Cells for Transfection

Routinely, HEK293 cells are cultivated in minimal essential medium (MEM) (GIBCO-BRL, Gaithersburg, MD), containing 10% (v/v) heat-inactivated fetal calf serum (FCS), 10 mM L-glutamine, and 100 U/ml penicillin/100 μg/ml streptomycin in an incubator (Heraeus) maintaining a 5% CO_2 atmosphere at 37 °C. Cells are replated every 3 days as follows.

1. Harvest 293 cells from one or several 10-cm plates (Becton Dickinson) containing 8×10^6 cells by removing the medium and adding 15 drops of trypsine (GIBCO-BRL) per plate for 2 min
2. Collect the cells with medium and distribute them either into 24-well plates (Becton Dickinson) to obtain 5×10^4 cells/well, or into 10-cm dishes to obtain 2×10^6 cells/dish
3. Let the cells grow for 48 h
4. Check one plate to see that cells cover $\approx 50\%$ of the plate surface

C. Transfection

1. For the transfection of 4 wells in a 24-well plate (or one 10-cm dish), put into a clear polycarbonate-type tube 1 μg (10 μg) of DNA from a 1-μg/μl stock and add 150 μl (375 μl) water
2. Then add first 50 μl (125 μl) of 1 M $CaCl_2$ and then 200 μl (500 μl) of $2\times$ BBS (see below)
3. Wait 1–3 min and add 1.6 ml (4 ml) MEM^{+++} (precipitation will form if you wait long enough and will be visible with a very faint haze in the solution)
4. Gently add 0.5 ml (5 ml) to each well (dish); avoid dislodging of the cells
5. Put the plates in 3% CO_2 incubator for 24 h
6. Remove and discard media, wash with Puck's saline A [PSA/Ca^{2+}; EGTA 1 mM, $CaCl_2 \cdot 2H_2O$ 0.1 mM, glucose 5 mM, KCl 5 mM, NaCl 150 mM, NaHCO$_3$ 3 mM, sterilize using 0.2 μM bottle top filter (Becton Dickinson)] and add 0.5 ml (5 ml) fresh MEM^{+++} per well (dish)
7. Place plates in a 5% (v/v) CO_2 incubator at 37 °C
8. Forty-eight hours after transfection cells should be optimal for recording or harvesting

To increase transfection efficiency, a glycerol shock may be applied 5 h after the transfection:

1. Remove transfection medium
2. Incubate the cells for 3 min with a sterile 10% glycerol solution in water and wash $3\times$ with phosphate-buffered saline (PBS, 80 mM; $Na_2HPO_4 \cdot 2H_2O$; 20 mM $NaH_2PO_4 \cdot H_2O$; 1.4 M NaCl; $10\times$)
3. Add new medium and continue as described before

V. $2\times$ BBS

1. Weight into beaker using fine balance; 1.07 g BES (N,N-bis[2-hydroxyethyl-2-aminoethansulfonic acid) (Sigma), 50 mM final; 1.63 g NaCl, 280 mM final; 0.0267 g $Na_2HPO_4 \cdot 2H_2O$, 1.5 mM final
2. Add H_2O to 100 ml, stir for 15 min

3. Add NaOH (from 2 N stock) to pH 6.95 at room temperature and wait for 15 min

4. Check if pH is at 6.95 (add HCl if necessary), sterilize solution using 0.2-μm filter, and store at $-20\ °C$

VI. Baculovirus System

Baculoviruses have proved to be powerful and versatile eukaryotic expression systems, which produce recombinant proteins at high levels of the total insect cell protein. The α_1 subunit of the GlyR has been successfully expressed in the baculovirus/Sf9 (*Spodoptera frugiperda* fall armyworm ovary) insect cell system (PharMingen, San Diego, CA) and shown to be functional by both biochemical and electrophysiologic criteria (Cascio *et al.*, 1993; Morr *et al.*, 1995). The protocols used are very similar to those described in the supplier's manual and are therefore not dealt with in this article. However, much of the protein produced seems to be localized in intracellular compartments (Morr *et al.*, 1995), suggesting that posttranslational processing and/or sorting may constitute a limiting factor of GlyR production in this versatile expression system.

References

Becker, C. M., Hermanns-Borgmeyer, I., Schmitt, B., and Betz, H. (1986). *J. Neurosci.* **6,** 1358.
Ben-Azis, R., and Soreq, H. (1990). *Nucleic Acids Res.* **18,** 3418.
Cascio, M., Schoppa, N. E., Grodzicki, R. L., Sigworth, F. J., and Fox, R. O. (1993). *J. Biol. Chem.* **268,** 22135.
Chen, C., and Okayama, H. (1987). *Mol. Cell. Biol.* **8,** 2745.
Dumont, J. N. (1972). *J. Morphol.* **136,** 153.
Goldin, A. L., and Sumikawa, K. (1992). *Methods Enzymol.* **207,** 279.
Gorman, C. M., Gies, R. D., and McCray, G. (1990). *DNA Prot. Eng. Tech.* **2,** 3.
Graham, D., Pfeiffer, F., Simler, R., and Betz, H. (1985). *Biochemistry* **24,** 990.
Hamill, O. P., Marty, A., Neher, E., Sakmann, B., and Sigworth, F. J. (1981). *Pflügers Arch.* **391,** 85.
Hoch, W., Betz, H., and Becker, C. M. (1989). *Neuron* **3,** 339.
Kuhse, J., Laube, B., Magalei, D., and Betz, H. (1993). *Neuron* **11,** 1049.
Kuhse, J., Betz, H., and Kirsch, J. (1995). *Curr. Opin. Neurobiol.* **5,** 318.
Langosch, D., Laube, B., Rundstroem, N., Schmieden, V., Bormann, J., and Betz, H. (1994). *EMBO J.* **13,** 4223.
Laube, B., Kuhse, J., Runstroem, N., Kirsch, A., Schmieden, V., and Betz, H. (1995). *J. Physiol. (Lond.)* **483,** 613.
Mackerer, C. R., Kochman, R. L., Shen, T. F., and Hershenson, F. M. (1977). *J. Pharmacol. Exp. Ther.* **201,** 326.
Marshall, J., Molloy, R., Moss, G., Howe, J. R., and Hughes, T. E. (1993). *Neuron* **14,** 211.
Miledi, R. E., and Sumikawa, K. (1982). *Biomed. Res.* **3,** 390.
Morr, J., Rundstroem, N., Betz, H., Langosch, D., and Schmitt, B. (1995). *FEBS Lett.* **368,** 495.
Pfeiffer, F., and Betz, H. (1981). *Brain Res.* **226,** 273.
Pfeiffer, F., Graham, D., and Betz, H. (1982). *J. Biol. Chem.* **257,** 9389.

Prior, P., Schmitt, B., Grenningloh, G., Pribilla, I., Multhaup, G., Beyreuther, K., Maulet, Y., Werner, P., Langosch, D., Kirsch, A., and Betz, H. (1992). *Neuron* **8,** 1161.

Schmieden, V., Grenningloh, G., Schofield, P., and Betz, H. (1989). *EMBO J.* **8,** 695.

Schmieden, V., Kuhse, J., and Betz, H. (1993). *Science* **262,** 256.

Schmitt, B., Knaus, P., Becker, C. M., and Betz, H. (1987). *Biochemistry* **26,** 805.

Sontheimer, H., Becker, C. M., Prittchett, P. R., Schofield, P. R., Grenningloh, G., Kettenmann, H., Betz, H., and Seeburg, P. H. (1989). *Neuron* **2,** 1491.

Taleb, O., and Betz, H. (1994). *EMBO J.* **13,** 1318.

CHAPTER 24

Functional Analyses of Aquaporin Water Channel Proteins

Peter Agre, John C. Mathai, Barbara L. Smith, and Gregory M. Preston

Department of Biological Chemistry
John Hopkins University School of Medicine
Baltimore, MD, USA

DOI: 10.1016/B978-0-12-382204-8.00024-2

I. Introduction

Water is the most abundant component of all living organisms, so the entry and release of water from cells must be considered a fundamental process of life. Red blood cells and multiple "leaky" epithelia such as renal tubules, salivary glands, and choroid plexus have long been thought to contain water-selective channels or pores (Finkelstein, 1987), which have been studied by several biophysical techniques (Solomon, 1989). Although the last 20 years have featured major advances in the recognition and molecular cloning of numerous salt, ion, and sugar transport proteins, the molecular identity of the membrane water transporters has eluded investigators until recently. This in part has resulted from the relatively high diffusional permeability of water through simple lipid bilayers, causing a significant background water permeability that foiled attempts to clone water channels by expression. Efforts to isolate candidate water transport molecules from leaky epithelia were also without success, and the ubiquity of water and its simple molecular structure has precluded chemical modifications for affinity labeling.

Identification of the first recognized water transport molecule resulted from the serendipitous observation made possible by the identification, purification, and cDNA cloning of a previously unknown 28-kDa membrane protein from red cells and renal tubules (Denker *et al.*, 1988; Preston and Agre, 1991; Smith and Agre, 1991). The distribution, abundance, and physical nature of this protein suggested that it might be the sought-after water channel, and this was demonstrated by expression of the cRNA in *Xenopus laevis* oocytes, which then became highly permeable to water (Preston *et al.*, 1992). Now designated AQP1, this protein is the archetypal member of the aquaporins (Agre *et al.*, 1993a,b), a large family of water transport molecules found in vertebrates, invertebrates, plants, and microorganisms (Chrispeels and Agre, 1994). A large body of aquaporin literature is emerging, but most experimental methods are derived from those established for AQP1, which are outlined here.

II. Purification of Red Cell AQP1 Protein

AQP1 is unique among membrane channels, since the native protein may be purified in milligram amounts from human red cells. Purification is made simple by the limited solubility of this protein in *N*-lauroylsarcosine (Denker *et al.*, 1988; Smith and Agre, 1991). Thus, while AQP1 makes up ~2.4% of the total red cell membrane protein, it is >50% pure after extraction of cytoskeleton-stripped membrane vesicles in *N*-lauroylsarcosine. Our current preparation technique is modified from published methods (Denker *et al.*, 1988; Nielsen *et al.*, 1993; Smith and Agre, 1991; Zeidel *et al.*, 1992, 1994).

A. Materials

Two units of anticoagulated human red blood cells (American Red Cross, local affiliate)

BPF4 high-efficiency leukocyte removal filters (Pall Biomedical, East Hills, NY)

Wash buffer: 150 mM NaCl, 7.5 mM sodium phosphate (pH 7.4), 1 mM NaEDTA

Lysis buffer: 7.5 mM sodium phosphate (pH 7.4), 1 mM NaEDTA, 10 μM leupeptin, 0.5 mM diisopropyl fluorophosphate (DFP)

1 M potassium iodide

Sarcosine buffer: 1% (w/v) N-lauroylsarcosine, 1 mM NH_4HCO_3, 1 mM NaN_3, 1 mM dithiothreitol (DTT) (pH 7.8)

Triton X-100 (except where noted, reagents were from Sigma, St Louis, MO)

Octylglucoside (n-octyl-β-D-glucopyranoside, Calbiochem, San Diego, CA)

Chromatography buffer: 20 mM Tris–HCl (pH 7.8), 1 mM NaN_3, 1 mM DTT

0.22-μm Millex GV membranes (Millipore, Bedford, MA)

10-mm × 100-mm POROS Q/F column and 10-mm × 100-mm POROS HQ/F column (PerSeptive Biosystems, Framingham, MA).

B. Procedures

Preparation of Red Cell Membrane Vesicles. Two units of anticoagulated human red cells (~200 ml packed red cells/unit) obtained within 5 days after blood donation are passed through leukocyte removal filters, and washed with wash buffer. Membranes are prepared by lysing the washed red cells in hypotonic lysis buffer, and membrane skeleton proteins are extracted with 1 M potassium iodide (Bennett, 1983). The extracted membrane vesicles from the 400 ml of red cells are further extracted by shaking for 1 h at 22 °C in 800 ml of sarcosine buffer and pelleted at 30,000 × g for 4 h at 4 °C in a Beckman JA-14 rotor. The pellet is washed once in 7.5 mM sodium phosphate (pH 7.4) and solubilized by shaking for 1 h at 22 °C in 1.2 l of chromatography buffer containing 4% Triton X-100 (v/v). After a 4-h centrifugation at 4 °C as above, the supernatant is filtered through 0.22-μm Millex GV membranes. AQP1 is the major protein present at this stage when analyzed by SDS–PAGE with silver staining (Fig. 1).

1. Chromatographic Procedures

A 600-ml aliquot of the filtrate is loaded onto a 10-mm × 100-mm POROS Q/F column equilibrated with chromatography buffer containing 0.1% Triton X-100 driven at 3 ml/min by an FPLC apparatus (Pharmacia, Piscataway, NJ). The UV monitor is used to measure A_{280}, but the baseline must be adjusted to compensate for absorption by 0.1% Triton X-100. The column is washed with 40 ml of chromatography buffer and eluted with a 120 ml 0.2–0.6 M NaCl gradient at 4 ml/min while A_{280} is recorded (elution no. 1). Peaks eluted at ~0.3 M NaCl from

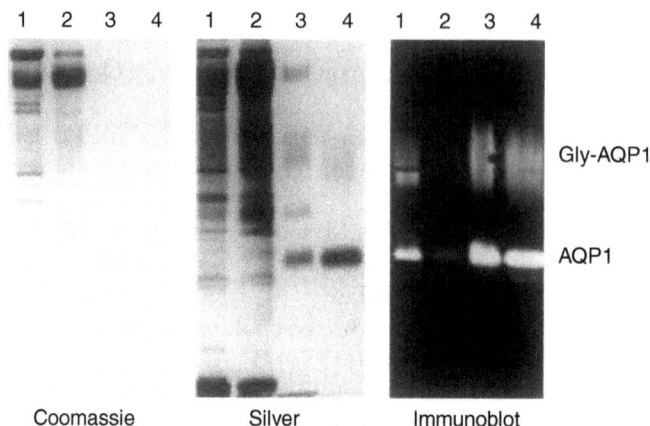

Fig. 1 *Purification of AQP1 protein from human red blood cells.* Panels correspond to membranes analyzed by SDS–PAGE with Coomassie staining (left), silver staining (middle), or after transfer to an immunoblot incubated with anti-AQP1 IgG and visualized by ECL (Amersham). Individual lanes: 1, whole red cell membranes (6 μg); 2, N-lauroylsarcosine-soluble proteins (6 μg); 3, N-lauroylsarcosine-insoluble proteins (0.4 μg); 4, pure AQP1 (0.4 μg). Note that AQP1 is a homotetramer comprised of three nonglycosylated subunits (AQP1) and one subunit with a polylactosaminoglycan attached (Gly-AQP1). (Denker *et al.*, 1988; Smith and Agre, 1991).

two runs are combined, diluted 7- to 10-fold with chromatography buffer containing 1.2% octylglucoside and loaded onto a 10-mm × 100-mm POROS HQ/F column equilibrated with the same buffer while running at 2 ml/min. The column is washed until the A280 baseline is recovered, and then eluted at 1 ml/min with a 40-ml gradient of 0–0.6 M NaCl in the same buffer (elution no. 2). The peak fractions eluting at ~0.3 M NaCl are combined and diluted 15- to 20-fold in chromatography buffer containing 1.2% octylglucoside and reloaded onto the same POROS HQ/F column at 2 ml/min, washed with 40 ml, and concentrated by elution at 1 ml/min with a 0–0.6 M NaCl gradient of 15 ml (elution no. 3). Peak fractions collected in 250-μl volumes elute at ~0.3 M NaCl and range in concentration from 1 to 3 mg/ml (BCA protein method, Pierce, Rockford, IL). The fractions are snap frozen in dry ice and stored at −80 °C until reconstitution. SDS–PAGE analysis with silver staining reveals the final product to be ~99% pure (Fig. 1), and a two-unit purification yields approximately 5 mg of AQP1.

C. Comments

If the highest purity is not essential or final elution into octylglucoside is not required, substantially larger yields of less concentrated AQP1 may be obtained by eliminating anion-exchange elution no. 2 or no. 3. Because the critical step in the purification comes at the N-lauroylsarcosine extraction, the subsequent anion-exchange steps may be varied. We have found that the POROS columns have excellent flow properties but may deteriorate after repeated exposure to detergents and often cannot be regenerated. Also, somewhat different elution profiles have been noted for columns with different lot numbers. Thus, chromatography fractions are routinely analyzed by SDS–PAGE with visualization by silver staining, since the protein is poorly visible after Coomassie blue staining (Fig. 1). Other prepacked, high-performance anion-exchange columns may also be used (e.g., Mono Q HR, Pharmacia). Alternatively, less expensive anion-exchange beads may be purchased in bulk, packed into columns, and periodically replaced with fresh material (e.g., Source 15Q or 30Q, Q-Sepharose HP or Fast Flow, Pharmacia).

III. Expression of AQP1 in Yeast

Temperature-sensitive *sec6–4* mutant strains of yeast are a powerful system for producing membrane proteins for functional and structural studies. Although AQP1 can be purified from red cells (Section II), other aquaporins are less abundant and are expressed in complex tissues, making their purifications impractical. The yeast expression system also offers some advantages to the oocyte system (Section VII), since failure of AQP1 mutants to increase the water permeability when expressed in oocytes often reflects defective targeting to the plasma membrane (Jung *et al.*, 1994). Quantities of up to 1 mg of native AQP1 and mutant forms of AQP1 can be generated with these methods, although for most studies smaller scale preparations are convenient and provide approximately 100 μg yields. The same approach is used for preparation of AQP2 protein (Coury *et al.*, 1998).

A. Materials

Yeast strains: temperature-sensitive SY1 strain of *Saccharomyces cerevisiae* (*MATα, ura3–52, leu2–3, 112, his4–619, sec6–4, GAL* (Nakamoto *et al.*, 1991) (Source: Dr. C. W. Slayman, Department of Genetics, Yale University School of Medicine, New Haven, CT)

Defined minimal media: yeast nitrogen base, 6.7 g/l (Bio 101, Vista, CA) 20 mg/l histidine, 30 mg/l leucine, 2% raffinose

YEP–galactose medium: 0.5% yeast extract; 1% Bacto-peptone (Difco Labs, Detroit MI); 2% galactose (w/v)

Spheroplasting medium: 1.4 M sorbitol, 50 mM K_2HPO_4 (pH 7.5), 10 mM NaN_3, 40 mM 2-mercaptoethanol

Lysis buffer: 0.8 M sorbitol, 10 mM triethanolamine (pH 7.2), 1 mM NaEDTA

Lyticase and concanavalin A (Con A) (Sigma Chemical Co., St Louis MO)

100 mM $MnCl_2$ and 100 mM $CaCl_2$

Dounce homogenizer with pestle A (Wheaton, Millville, NJ).

B. Procedures

1. pYES2-AQP1 Plasmid Construction and Yeast Cultures

Aquaporin cDNAs are cloned into the pYES2 plasmid (Invitrogen, Carlsbad, CA) using standard molecular biological methods, with orientation confirmed by restriction digestions (978-bp DNA fragment containing human *AQP1* is released with *BclI* and *XbaI*) and dideoxynucleotide sequencing. The SY1 strain is transformed with pYES2-AQP1 by the lithium acetate method (Ito *et al.*, 1983) and is maintained in uracil-deficient defined media. One liter of culture (200 ml media in five 1000-ml culture flasks) is grown at room temperature to midlog phase ($OD_{600} \sim 1$) and cells are harvested by centrifugation at $2000 \times g$ for 10 min at room temperature. Production and accumulation of secretory vesicles is induced by transfer to 1 l of rich YEP-galactose medium during overnight incubation at 37 °C. Sodium azide is added to a final concentration of 10 mM 10 min prior to centrifugation at $2000 \times g$ for 10 min at 4 °C.

2. Spheroplast Preparation and Secretory Vesicle Isolation

The yeast cells are suspended to a density of 50 OD_{600} units/ml in spheroplasting medium containing lyticase (0.2 mg/ml) and incubated at 37 °C for 45 min. Spheroplasts are pelleted by low-speed centrifugation ($1000 \times g$, 15 min, 4 °C) and resuspended in the same volume of spheroplasting media containing 1 mM $MnCl_2$ and 1 mM $CaCl_2$ and incubated with Con A (25 mg/1600 OD_{600} units) for 15 min at 0 °C. Spheroplasts are again pelleted by centrifugation, resuspended at 80 OD_{600} units/ml in cold lysis buffer, and lysed at 4 °C with 25 strokes of the pestle A in a Dounce homogenizer. The lysate is spun at $10,000 \times g$ for 10 min at 4 °C to remove unbroken cells, nuclei, mitochondria, and cell debris. Secretory vesicles are collected by centrifugation of supernatant at $100,000 \times g$ for 1 h in a Beckman Type 50 Ti rotor at 4 °C (Nakamoto *et al.*, 1991). Protein expression is confirmed by immunoblotting with anti-AQP1 (Smith and Agre, 1991). For many studies, the isolated membranes may be used directly; however, AQP1 may be partially purified from isolated secretory membrane vesicles by extraction with 0.6% *N*-lauroylsarcosine (w/v) yielding aquaporin proteins that are $\sim 50\%$ pure at up to 1 mg/l of culture. The protein may be solubilized in 4% Triton X-100 (v/v) or 1.2% octylglucoside (w/v) using conditions outlined in Section II.

IV. Reconstitution of AQP1 into Proteoliposomes

Investigators studying bacterial transport proteins have routinely characterized the functions of purified membrane proteins after reconstitution into lipid bilayers. Although it is expensive, the detergent octylglucoside is particularly useful, since it has a relatively high micellar concentration. By rapidly lowering the octylglucoside concentration, solubilized phospholipids will form liposomes with integral membrane proteins imbedded in the lipid bilayer (Ambudkar and Maloney, 1986a,b). We have adapted this method to reconstitute purified human red cell AQP1 into proteoliposomes (Zeidel *et al.*, 1992, 1994).

A. Reconstitution of Purified Red Cell AQP1

1. Materials

Octylglucoside (octyl-β-D-glucopyranoside, Calbiochem, San Diego, CA), 15% (w/v) in water, stored at $-80\,°C$

Escherichia coli lipids (bulk phospholipid, Avanti Polar Lipids, Birmingham AL) depleted of neutral lipid by acetone/ether wash (Ambudkar and Maloney, 1986a), 50 mg/ml in 2 mM 2-mercaptoethanol; stored at $-70\,°C$ under nitrogen

100 μg AQP1 protein in 1.2% octylglucoside buffer (Section II)

Carboxyfluorescein (Molecular Probes, Eugene, OR)

Millipore filter: Pore size 1.2 μm, filter type RA (Millipore Co., Bedford, MA)

Individual stocks solutions: 500 mM Tris–HCl (pH 7.5), 10 mM NaN$_3$, 100 mM DTT, and 500 mM MOPS [3-(N-morpholino)propanesulfonic acid, pH 7.5]

Reconstitution buffer: 50 mM MOPS (pH 7.5), 150 mM N-methyl-D-glucamine chloride (neutralized with HCl to pH 7.0), 10–15 mM carboxyfluorescein, 1 mM DTT, and 0.5 mM phenylmethylsulfonyl fluoride (PMSF).

2. Procedure

The reconstitutions are performed in a final volume of 1.0 ml of 50 mM Tris–HCl (pH 7.5), 1 mM NaN$_3$, 1 mM DTT, 1.25% (w/v) octylglucoside, 100 μg pure AQP1 protein (Section II), and 9 mg of bath-sonicated, purified *E. coli* phospholipid. The AQP1 and phospholipid mixture is vortex mixed and incubated for 20 min at $0\,°C$. Proteoliposomes form at room temperature by rapidly injecting the mixture through a 23-gauge needle into 25 ml of reconstitution buffer. Liposomes with simple lipid bilayer membranes are prepared identically, except AQP1 protein is not included. Proteoliposomes or liposomes are then collected by centrifugation for 1 h at 123,000 $\times g$ in a Beckman Type 42.1 rotor at $4\,°C$. The pellets are resuspended in 8 ml of reconstitution buffer and then centrifuged at 10,000 $\times g$ for 15 min in a Beckman Type 50 Ti rotor to remove any lipid aggregates. The supernatant is then centrifuged for 1 h at 152,000 $\times g$ in the same rotor. Proteoliposomes or liposomes are resuspended in 0.3 ml of reconstitution buffer for up to 24 h at $4\,°C$.

Proteoliposome and liposome sizes are determined by negative staining electron microscopy or by dynamic light scatter in a DynaPro-801 (Protein Solutions, Charlottesville, VA). Most preparations yield vesicles 0.1–0.15 μm in diameter. Proteoliposomes contain multiple intramembranous particles (each representing an AQP1 tetramer) which may be examined by freeze-fracture electron microscopy (Fig. 2). Protein was measured by a modification of the Schaffner and Weissman method (Schaffner and Weissman, 1973). The final concentration of sodium

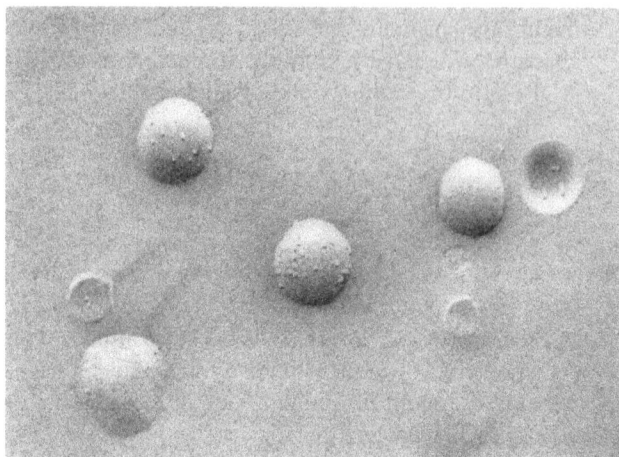

Fig. 2 Freeze-fracture electron micrographs of proteoliposomes reconstituted with AQP1 at a protein:lipid ratio of 1:50. Intramembranous particles represent AQP1 tetramers. Magnification: ×90,000. [Modified and reprinted from Zeidel *et al.* (1994) with permission. Micrograph provided by Søren Nielsen and Arvid Maunsbach, Department of Cell Biology, Institute of Anatomy, University of Aarhus, Denmark.]

dodecyl sulfate (SDS) in the assay is increased by 1%, and the protein is precipitated with 12% trichloroacetic acid and vacuum filtered onto a 48-mm-diameter Millipore filter disk (Ambudkar and Maloney, 1986b). Phospholipid is estimated by assuming 70% recovery in the proteoliposomes but may be accurately determined by the method of Hallen (Hallen, 1980).

3. Comments

In a typical reconstitution, about one-half of the purified red cell AQP1 protein and two-thirds of the pure phospholipid are incorporated into proteoliposomes. The protein:lipid ratios may be varied from 1:125 to 1:25 (or even higher) and yield correspondingly increased water permeabilities (Zeidel *et al.*, 1994). The purified phospholipid is comprised of phosphatidylethanolamine (70%), phosphatidylglycerol (15%), and cardiolipin (15%) (Chen and Wilson, 1984); however, experiments with other phospholipids yield proteoliposomes that are functionally similar (Zeidel *et al.*, 1994). Internal volumes of 1 μl/mg phospholipid are obtained for proteoliposomes (Ambudkar and Maloney, 1986b).

B. Reconstitution of AQP1 Expressed in Yeast

1. Materials

Secretory vesicles or membranes (8–12 mg/ml) containing AQP1 (Section II)

Solubilization buffer: 1.5% (w/v) octylglucoside, 50 mM MOPS (pH 7.5), 20% (v/v) glycerol, 1 mM DTT, 0.4% *E. coli* phospholipids, 0.5 mM PMSF
Reconstitution buffer: (see Section IV).

2. Procedures

a. Solubilization of Secretory Vesicles

Solubilization buffer should be mixed well before addition of octylglucoside. One milliliter of secretory vesicles (8–12 mg/ml) is mixed with 3 ml of solubilization buffer, incubated for 20 min on ice, and centrifuged at $100,000 \times g$ for 1 h at 4 °C. The supernatant is used for reconstitution.

b. Reconstitution of Liposomes

Reconstitution is performed in a final volume of 1.14 ml containing 1.4% octylglucoside, 6 mg of bath-sonicated *E. coli* phospholipid, and 1.0 ml of the above supernatant. This mixture is incubated on ice for 20 min and proteoliposomes are formed by rapid injection of the mixture into 25 ml of reconstitution buffer at room temperature. Proteoliposomes are collected and washed by centrifugation as described in Section IV.

3. Comments

Although the methods for reconstitution of heterologously expressed AQP1 and other aquaporins are still being established, the technique promises an approach for physical analysis of these proteins. Note that as a cost-effective alternative, carboxyfluorescein may also be loaded into proteoliposomes by overnight incubation of proteoliposomes at 0 °C in 1 ml of reconstitution buffer containing 15 mM of carboxyfluorescein. The carboxyfluorescein-loaded vesicles are then washed just prior to use by centrifugation as described in Section IV.

V. Water Permeability of AQP1 Proteoliposomes

The water permeability of AQP1 has been determined by comparing wild-type red cells and AQP1-deficient red cells (Mathai *et al.*, 1996). The possibility of measuring water permeability of purified, reconstituted AQP1 should be of greater general usefulness and is described here. By directly measuring the coefficients of osmotic water permeability of proteoliposomes containing known concentrations of AQP1, the unit permeability may be calculated for the AQP1 tetramer, which is comprised of four independently functional subunits. Moreover, by substituting other solutes for sucrose, the permeability of a variety of osmolytes may be measured.

A. Materials

Antifluorescein antibody (Molecular Probes): Isotonic solution: reconstitution buffer without carboxyfluorescein (Section IV); measured osmolality 356 mOsm by using an osmometer such as the Vescor 5100C vapor pressure osmometer (Vescor Inc., Logan, UT).

Hypertonic solution: add pure sucrose to isotonic solution to increase total osmolality of the solution threefold. For example, add 19.0 g of sucrose to 100 ml of reconstitution buffer (736 mOsm + 356 mOsm = 1092 mOsm) and mix 1:1 with isotonic solution by stopped flow. The final osmolality will be approximately twice the original osmolality (724 mOsm). Note that osmolalities of sucrose solutions are not directly related to calculated molarities but are listed in *The Handbook of Chemistry and Physics*, CRC Press, Boca Raton, FL.

AQP1 inhibitors: 0.5 mM $HgCl_2$ and 1 mM *p*-chloromercuribenenesulfonate.

B. Procedure

Measurements of the coefficient osmotic water permeability (P_f) are performed on AQP1 proteoliposomes and control liposomes (Section IV). Extravesicular carboxyfluorescein is removed by centrifugation (Section IV) just before the water permeability measurements, if carboxyfluorescein is loaded into proteoliposomes by overnight incubation. P_f is measured as described (Zeidel *et al.*, 1992) by abruptly exposing proteoliposomes to an increase in extravesicular osmolality using a stopped-flow fluorimeter (SF.17 MV, Applied Photophysics, Leatherhead, UK) with a measured dead time of 0.7 ms. Excitation wavelength is 490 ± 1.5 nm, using a monochrometer, and emission wavelength is >515 nm using a cut-on filter (Oriel Corp., Stratford, CT). Extravesicular carboxyfluorescein fluorescence is completely quenched by using antifluorescein antibody of 1:1000 dilution. Vesicles act as perfect osmometers, and relative volume (absolute volume divided by initial volume) is linearly related to relative fluorescence (absolute fluorescence divided by initial fluorescence). Averaged data from multiple determinations are fitted to single exponential curves using software provided by Applied Photophysics (Zeidel *et al.*, 1992). Fitting parameters are then used to determine P_f exactly as described (Priver *et al.*, 1993; Zeidel *et al.*, 1992) using the equation:

$$dV(t)/dt = (P_f)(\text{SAV})(\text{MVW})\{[C_{in}/V(t)] - C_{out}\}$$

where $V(t)$ is the relative volume as a function of time, SAV is the surface area to volume ratio, MVW is the molar volume of water (18 cm^3/mol), and C_{in} and C_{out} are the initial concentrations of total solute inside and outside the vesicle.

C. Comments

At 37 °C, typical measurements yield estimates of $P_f \sim 500 \times 10^{-4}$ cm/s (for AQP1 proteoliposomes) versus $\sim 100 \times 10^{-4}$ cm/s (for control liposomes). Note that the background permeability rapidly declines at lower temperatures, yielding a relatively larger value of P_f to background. By measuring the P_f at a range of temperatures (8–22 °C), the Arrhenius activation energies may be computed from the slope of a plot of P_f and absolute temperature (In P_f vs. $1/T$, K). The slope multiplied by gas constant R (1.986 cal/K/mol) will give the activation energy. AQP1 proteoliposomes are routinely found to have $E_a < 5$ kcal/mol, which is equivalent to the diffusion of water in bulk solution. In contrast, control liposomes yield values of >10 kcal/mol, which represents a significant barrier to water flow. When incubated for 30 min at 37 °C in submillimolar concentrations of the known inhibitors $HgCl_2$ or p-chloro-mercuribenzene sulfonate, the P_f of AQP1 proteoliposomes is reduced to $\sim 20\%$ of the original level, indicating that this method may be very useful for screening for new potential inhibitors.

The number of AQP1 tetramers per milliliter of suspension is calculated from the amount of protein/ml in each preparation and the molecular weight of the AQP1 tetramer (4×28.5 kDa). The measured total entrapped volume/mg phospholipid, the calculated volume of each proteoliposome, and the total number of proteoliposomes/ml of suspension are calculated from the total phospholipid content. The number of AQP1 tetramers/ml divided by the number of proteoliposomes/ml yields the numbers of AQP1 tetramers/proteoliposome. Thus the subunit permeability may be calculated by multiplying the P_f (cm/s) by the surface area of each proteoliposome ($6.2 \times 10_{-10}$ cm$_2$) and dividing by four times the number of AQP1 tetramers reconstituted per proteoliposome. This typically yields values of $\sim 5 \times 10^{-10}$ cm/s/AQP1 subunit (equivalent to 2×10^9 water molecules/subunit/s).

Heterologous expression of AQP1 in yeast will permit the assessment of osmotic water permeability of wild-type or site-directed AQP1 mutants. This may be performed directly on the secretory vesicles isolated from yeast or reconstituted vesicles containing expressed AQP1. Osmotic water permeability is measured and calculated as describe above by abrupt exposure of vesicles to increased external osmolality using the stopped-flow fluorimeter (Fig. 3).

VI. Homology Cloning of Aquaporins by Degenerate Oligonucleotide PCR

All members of the aquaporin family of water channels are encoded by DNA sequences that represent the duplication of an ancestral gene. Each tandem repeat contains the highly conserved NPA motif (asparagine–proline–alanine) and other conserved amino acids (Preston and Agre, 1991). Since the cloning and expression of AQP1, multiple aquaporin homologs have been cloned using degenerate oligonucleotide primers corresponding to the most highly conserved regions of the

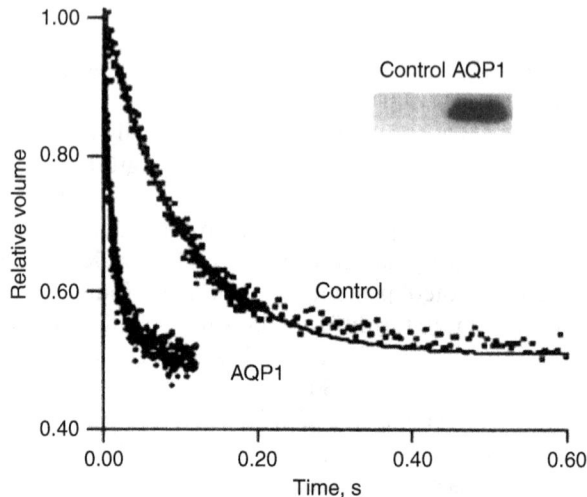

Fig. 3 *Measurement of osmotic water permeability (P_f) of reconstituted proteoliposomes.* Tracings are from reconstituted membranes analyzed by stopped-flow transfer to hyperosmolar solution (100% increase in osmolality). AQP1 proteoliposomes reconstituted with vesicle protein from AQP1 transformed *sec⁻* yeast (P_f = 70 μm/s) are compared to control proteoliposomes reconstituted with vesicle protein from untransformed *sec⁻* yeast (P_t = 6.2 μm/s). *Inset*: Immunoblot (2 μg protein per lane) of control proteoliposomes and AQP1 proteoliposomes.

AQP1 molecule for amplification by polymerase chain reaction (PCR) of the intervening sequence (Fig. 4). The amplified cDNAs are then used to probe the appropriate libraries for full-length cDNAs. Examples where this approach was used include the cloning of *AQP4* from brain (Jung *et al.*, 1994), *AQP5* from salivary gland (Raina *et al.*, 1995), and *aqpZ* from *E. coli* (Calamita *et al.*, 1995).

A. Materials

10× PCR reaction buffer: 100 mM Tris–HCl (pH 8.3 at 25 °C), 500 mM KCl, 15 mM MgCl2, 0.1% w/v gelatin. Incubate at 50 °C to melt the gelatin, filter sterilize, and store at −20 °C in aliquots.

dNTP stock solution (1.25 mM dATP, dGTP, dCTP, dTTP) made by diluting commercially available deoxynucleotides with sterile water.

Thermostable DNA polymerase (such as AmpliTaq DNA polymerase, Perkin-Elmer-Cetus Norwalk, CT) supplied at 5 units/μl.

Mineral oil.

A programmable thermal cycler machine (available from Perkin-Elmer-Cetus, MJ Research Inc., Stratagene, and other manufacturers).

The following degenerate oligonucleotide primers should be purified by reversed-phase HPLC or by elution from acrylamide gels. The primers should be resuspended at 20 pmol/μl in sterile water, and stored at −20 °C in aliquots.

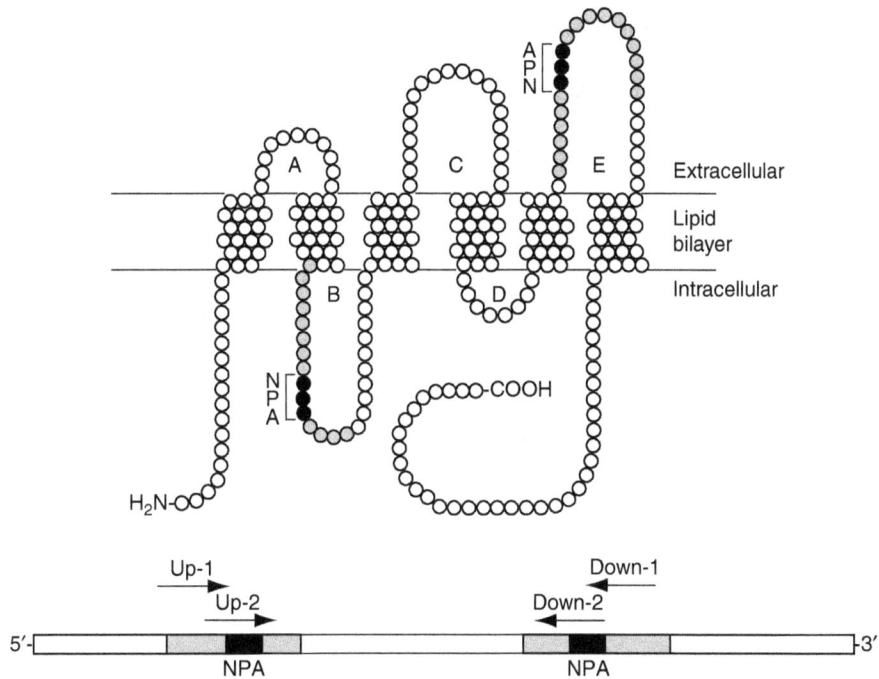

Fig. 4 Highly conserved domains with aquaporins are used for homology cloning by PCR. *Top*: Membrane topology model of AQP1 showing areas of high conservation among different members of the aquaporin family (loop B and loop E). Residues shared among most aquaporins (shaded circles) and NPA motifs absolutely conserved among all aquaporins (black circles). *Bottom:* Model of aquaporin cDNA showing sites for hybridization of forward (Up-1 and Up-2) and reverse (Down-1 and Down-2) degenerate oligonucleotide primers (see text).

Forward primers:
Up-1 5′-STB GGN CAY RTB AGY GGN GCN CA-3′
Up-2 5′-G GGA TCC GCH CAY NTN AAY CCH GYN GTN GTN AC-3′
Reverse primers:
Down-1 5′-GC DGR NSC VAR DGA NCG NGC NGG-3′
Down-2 5′-CGG AAT TCG DGC DGG RTT NAT NSH NSM NCC-3′
Degenerate codes: B = G, C, or T (not A); D = A, G, or T (not C); H = A, C, or T (not G); K = G or T (keto); M = A or C (amino); N = any (A, G, C, or T); R = A or G (purine); S = G or C (strong); V = A, C, or G (not T); W = A or T (weak); Y = C or T (pyrimidine). The underlined sequences correspond to *Bam*HI and *Eco*RI restriction sites. A related series of degenerate oligonucleotide primers with broader reactivity for nonmammalian aquaporins has also been published (Preston, 1997).

The DNA template can be almost any DNA sample, including a single-stranded cDNA from a reverse transcription reaction, DNA from a phage library, or genomic DNA. The DNA is heat denatured at 99 °C for 10 min and stored at −20 °C.

Chloroform.

Tris–saturated phenol: prepared using ultrapure redistilled crystalline phenol as recommended by the supplier (GIBCO-BRL, Gaithersburg, MD). Use polypropylene or glass tubes for preparation and storage.

PC9: mix equal volumes of buffer-saturated phenol (pH > 7.2) and chloroform, extract twice with an equal volume of 100 mM Tris (pH 9.0), separate phases by centrifugation at room temperature for 5 min at 2000 × *g*, and store at 4 to −20 °C for up to 1 month.

7.5 M Ammonium acetate is preferred to sodium acetate for precipitation of DNA because nucleotides and primers generally do not precipitate. Dissolve ammonium acetate in water, filter through 0.2-μm membrane, and store at room temperature.

For the elution of specific PCR-amplified DNA products from agarose gels, we have observed consistent results with the QIAEX gel extraction kit (Qiagen Inc., Santa Clarita, CA) for products from 50 bp to 5 kbp, however, several other kits are available.

B. Procedures

1. PCR Amplification Reaction

The DNA template should be PCR amplified with all four possible pair-combinations of forward and reverse primers (Up-1 and Down-1, Up-1 and Down-2, Up-2 and Down-1, Up-2 and Down-2), as well as the individual degenerate primers to determine if any of the bands amplified are derived from one of the degenerate primer pools. A DNA-free control is required to assess if there is any contaminating DNA in any of the other reagents. Pipette into 0.5-ml microcentrifuge tubes in the following order: 58.5 μl sterile water (0.2 μm filtered and autoclaved); 10 μl 10× PCR reaction buffer; 16 μl 1.25 mM dNTP stock solution; 5.0 μl primer Up-1 or Up-2; 5.0 μl primer Down-1 or Down-2; 5.0 μl heat denatured DNA (1–100 ng). If several reactions are being set up concurrently, a master reaction mix can be made up of the reagents used in all of the reactions. Samples are briefly vortex mixed and spun for 10 s in a microfuge. Each sample is overlaid with two to three drops of mineral oil. Amplify by hot-start PCR by pausing the thermocycler at step 4—cycle 1 and add 0.5 μl AmpliTaq DNA polymerase to each tube. Cycle in the following protocol: step 1, 95 °C for 3 min; step 2, 95 °C for 10 s (denaturation); step 3, 50 °C for 30 s (annealing); step 4, 72 °C for 30 s (extension); step 5, cycle 29 or 34 times to step 2; step 6, 72 °C for 4 min; step 7, 10 °C hold.

2. DNA Isolation and Gel Electrophoresis Analysis

The reaction tubes are removed from the thermal cycler, 200 μl chloroform is added, and the tubes are spun for 10 s in a microfuge to separate oil–chloroform layer from the aqueous layer. The aqueous layers are transferred to clean microfuge tubes, and the AmpliTaq DNA polymerase is removed by extracting the aqueous phase twice with 100 μl PC9. The tubes are spun for 2 min to separate the lower organic layer from the upper aqueous layer, and the aqueous layers are transferred to clean microfuge tubes. Ammonium acetate–ethanol precipitation of cDNA in 100-μl samples is performed by adding 50 μl 7.5 M ammonium acetate plus 350 μl 100% ethanol. The samples are vortex mixed for 15 s, placed on ice for 15 min, and the DNA pelleted at 12,000 $\times$ g for 15 min at 4 °C in a microfuge. The aqueous waste is decanted, replaced with 250 μl 70% ethanol, and briefly vortex-mixed and spun 5 min at 4 °C. The ethanol is decanted and the pellets are allowed to dry inverted at room temperature before resuspending in 20 μl sterile water. Aliquots of the PCR fragments (2–10 μl) may be resolved by gel electrophoresis through 2–3% NuSieve agarose gels (FMC BioProducts, Rockland, ME). The DNA is stained by soaking the gel for 5–30 min in about 10 volumes of water containing 1 μg/ml ethidium bromide and photographed under UV light. Since the NPA motifs are separated by 110–140 amino acids within all known members of the aquaporin gene family, PCR amplification of the known aquaporin cDNAs using these internal degenerate primers will yield products of approximately 400 bp (Fig. 4).

3. Secondary PCR Amplifications and DNA Purification

Based on the results from gel electrophoresis of the PCR amplified DNA products, multiple subsequent approaches may be made: (1) the initial DNA sample may be PCR amplified under different conditions (altered MgCl$_2$ concentration, annealing temperature, or primers); (2) a different DNA template may be used for PCR amplification; (3) one or more bands from the agarose gel may be eluted for subcloning or reamplification by PCR; (4) the product may be reamplified by PCR with an internal pair of degenerate primers. Options 1 and 2 are self-explanatory. Depending on whether no band is visible on the agarose gel or if multiple product bands are detected, the amount of template DNA, the MgCl$_2$ concentration, or the primer annealing temperatures may be altered. The goal is to obtain cDNA products that are detectable as bands on the agarose gels which are in the 340- to 420-bp range. If some products are detected in this size range, they may be purified with the QIAEX gel extraction kit (Qiagen Inc.) and either reamplified with the same primer set, an internal primer set (to be designed), or the cDNAs may be directly subcloned. When attempting to identify an aquaporin cDNA homolog from a tissue that is known to express recognized aquaporins, the final PCR sample may be enriched for novel homologs. Because the degenerate oligonucleotide primers are designed from the sequences of known members of the aquaporin family, these primers will likely be biased for those homologs. For example, AQP1 is abundant in the capillaries in soft tissues throughout the body, but AQP1 is absent within the salivary gland. To identify a salivary homology of the aquaporin gene family, a rat salivary gland cDNA library was used which also contained AQP1 cDNAs presumably from the surrounding capillaries

(Raina *et al.*, 1995). The rat cDNA library was first amplified with the external set of degenerate primers, Up-1 and Down-1. The PCR-amplified products were then digested with the restriction enzyme PstI, which cuts between the NPA motifs of rat AQP1. This was followed by reamplification with the internal pair of primers (Up-2 and Down-2). The PCR fragments of approximately 400 bp were cloned and sequenced. Obviously this strategy would not work if the resulting cDNA (*AQP5*) also contained a *Pst*I site. By trying different restriction enzymes that cut DNA infrequently (6- to 8-bp recognition sites), multiple novel homologs may be identified.

After identifying a novel aquaporin cDNA by cloning and sequencing the PCR products, the resulting PCR clone should be used to screen a cDNA library for a recombinant with a full-length insert. Alternatively the resulting DNA sequence can be used to design primers for cloning the 5′ and 3′ ends of the cDNA by anchor PCR (Preston, 1997). A disadvantage of anchor PCR is that it yields fragmented cDNA sequences, which need to be spliced together for expression studies.

C. Comments

All PCR reactions should be set up in either a sterile laminar flow hood or a dead air box using pipette tips containing filters (aerosol-resistant tips) or positive displacement pipettors to prevent the contamination of samples, primers, nucleotides, and reaction buffers by DNA. If the PCR reaction is going to be reamplified by PCR, all possible intervening steps should also be performed in a sterile hood with the same precautions to prevent DNA contamination. These precautions should also be extended to all extractions and reactions on the nucleic acid (RNA or DNA) through the last PCR reaction. Likewise, all primers, nucleotides, and reaction buffers for PCR should be made up and aliquoted using similar precautions. All buffers for PCR should be made with great care using sterile disposable plastic or baked glass, and restricted for use with aerosol-resistant pipette tips. The surfaces and pipettors should be cleaned with 10% bleach (made fresh daily) and rinsed with distilled deionized water between amplifications to destroy contaminating DNA, which could be a source of amplification artifact. Standard PCR reaction buffers contain 15 mM MgCl2 (1.5 mM final concentration). In many cases, changing the MgCl2 concentration will have significant consequences on the amplification of specific bands, and various concentrations between 0.5 and 5.0 mM MgCl2 have been successful. Note that organic solvents (phenol and chloroform) and ethidium bromide are hazardous materials and should always be handled with caution while wearing gloves and eye protection. The institutional hazardous waste department should be contacted regarding the proper disposal procedures.

VII. Water Permeability of AQP1 Expressed in Oocytes

Female *X. laevis* lay their eggs in freshwater ponds, so the water permeability of the oocytes is extremely low. This observation prompted Professor Erich Windhager (Cornell University Medical School, New York, NY) to propose this system for functional analysis of cDNAs encoding putative water channels. The suggestion was followed during the first functional expression of AQP1 (Preston *et al.*, 1992) and all other aquaporins.

A. Materials

An oocyte expression construct, such as pSP64 T (Krieg and Melton, 1984)

Various restriction endonucleases

T4 DNA polymerase and ligase

Calf intestinal alkaline phosphatase

DH5 α-competent *E. coli*

DEPC–water (ribonucleases are inactivated in diethylpyrocarbonate-treated water)

T7 RNA polymerase (50 U/μl) and 5 × T3/T7 buffer (GIBCO-BRL) ACUG mix: 10 mM ATP, 10 mM CTP, 10 mM UTP, and 2 mM GTP 50 mM DTT, made up in DEPC–water, stored in aliquots at $-20\,°C$

10 mM G(5′)ppp(5′)G cap (Pharmacia) in DEPC–water, stored in aliquots at $-20\,°C$

RNasin (40 U/μl) and RQ1 DNase (1 U/μl) (Promega, Madison, WI)

Female *X. laevis* and storage facilities

Modified Barth's solution: 88 mM NaCl, 1.0 mM KCl, 2.4 mM NaHCO$_3$; 15 mM Tris–HCl (pH 7.6); 0.3 mM Ca(NO$_3$)$_2$, 0.4 mM CaCl$_2$, 0.8 mM MgSO$_4$, sodium penicillin 0.1 mg/ml, streptomycin sulfate, 0.1 mg/ml, 0.5 mM theophylline; make up to 1000 ml with water

B. Procedures

1. Preparation of cDNA Expression Constructs

The human *AQP1* cDNA is available in a *Xenopus* oocyte expression construct from the American Type Culture Collection (Rockville, MD) and plasmids containing rat *AQP4*, rat *AQP5*, and *E. coli aqpZ* are also available (Table I). The entire coding regions for other aquaporin cDNAs must be spliced into the *Bgl*II site of pSP64T or another suitable *Xenopus* oocyte expression construct (Krieg and Melton, 1984). For blunt-end ligation reactions, the *Bgl*II restriction site overhangs are filled in with T4 DNA polymerase, followed by calf intestinal alkaline phosphatase to remove the 5′-phosphates from the DNA. For cohesive ligation reactions, the plasmid is treated with the phosphatase following *Bgl*II digestion. Control ligations are set up with the vector alone to assess efficiency of phosphatase

Table I
Aquaporin cDNAs[a]

Aquaporin	Species	Catalog No.
AQP1	Human	95674, 95675, 99538[b]
AQP4	Rat	87184
AQP5	Rat	87185
AqpZ	*E. coli*	87386

[a]Available through American Type Tissue Collection (Tel: 800-638-6597; fax: 301-816-4361; e-mail: sales@atc-c.org; Internet: http://www.atcc.org).
[b]In *X. laevis* expression vector.

treatment. The ligation reactions are transformed into DH5 α-competent bacteria. Plasmid DNA minpreparations are prepared from several of the resulting colonies, and a clone is identified with the correct cDNA insert orientation relative to the 5′ and 3′ UTRs of the expression construct by digesting the plasmid DNA with different restriction enzymes followed by agarose gel electrophoresis and DNA sequencing. A clone with the insert in the reverse orientation may be used as a negative control in the expression experiments.

2. Preparation of cRNAs for Oocyte Expression

Approximately 50 μg of each expression cDNA is digested with a restriction enzyme that cuts on the 3′ end of the construct, producing either a blunt end or a 3′ overhang. This is extracted twice with PC9, once with chloroform, and is precipitated by ammonium acetate/ethanol. The precipitate is resuspended at 1 mg/ml in DEPC–water. The following cRNA synthesis reaction is set up at room temperature in a 0.5-ml sterile microfuge tube with the following reagents added in order: 44.5 μl DEPC–water, 20 μl 5 × T3/T7 buffer, 10 μl ACUG mix, 10 μl 50 mM DTT, 6 μl 10 mM GppG cap, 5 μl linearized DNA, 2.5 μl RNasin, 2 μl T7 RNA polymerase. These are mixed gently and incubated at 37° for 90 min. A 2.5-μl aliquot of RNasin and 4 μl of RQ1 DNase I are added, mixed, and incubated at 37 °C for 30 min. The DNA digestion reactions are stopped with 1 μl of 500 mM NaEDTA and are extracted twice with 100 μl PC9 and once with chloroform. The aqueous supernatants are transferred to a clean sterile tube, and precipitated on ice with ammonium acetate/ethanol. The pellet is washed with 500 μl 70% ethanol, spun 5 min at 4 °C, decanted, and residual fluid is removed with a dry paper tissue. The cRNA is resuspended in 20 μl DEPC–water with 2 μl RNasin. (*Note:* Vacuum drying RNA is faster, but overly dried RNA pellets are often difficult to resuspend. All cRNAs must be kept on ice.) The absorbancy is read at 260 and 280 nm on a spectrophotometer with 0.5–2 μl RNA. Concentrations and purity are calculated

(1 U OD$_{260}$ = 40 μg/ml ssRNA; OD$_{260}$/OD$_{280}$ ~ 2.0 for pure ssRNA). The material is diluted with DEPC–water to 1 mg/ml and stored in 2-μl aliquots at −80 °C. A small aliquot of the cRNA may be visualized by an RNA gel with known amounts of other in vitro synthesized RNA molecules to verify concentration, purity, and lack of degradation.

3. Preparation and Microinjection of Xenopus Oocytes

The amphibia are anesthetized on ice, and stage V and VI oocytes are isolated (Lu et al., 1990) and stored overnight at 18 °C in modified Barth's buffer (MBS). Oocytes are injected with 50 nl of water (control injected oocytes) or 50 nl of water containing 0.1–50 ng cRNA. Injected oocytes are incubated at 18 °C in MBS for 1–5 days with daily changes of the buffer to promote good viability.

4. Measurement of Osmotic Water Permeability

A computer-interfaced videomicroscope is needed for determination of oocyte swelling (available from Nikon and other manufacturers (see Fig. 5). The hypotonic buffer (70 mOsm) is prepared by diluting one volume of MBS with two parts distilled water. Single oocytes are transferred from isotonic MBS (200 mOsm) to hypotonic MBS (at 22 °C), and digitized images are collected by computer (Universal Imaging Corporation, West Chester, PA) for up to 2 min or whenever the oocyte ruptures. The surface area of sequential images is calculated (Image-1 software, version 4.01B, Universal Imaging) assuming the oocytes are spheres without microvilli:

$$V = (4/3) \times (\text{area}) \times (\text{area}/\pi)^{1/2}$$

The relative changes in oocyte volume with time for up to 2 min, $d(V/V_o)/dt$, are fit by computer to a quadratic polynomial, and the initial rate of swelling is calculated. The osmotic water permeability (P_f, cm/s $\times 10^{-4}$) is determined between 10 and 20 s, using the average initial oocyte volume ($V_o = 9 \times 10^{-4}$ cm^3), the average initial oocyte surface area (S = 0.045 cm2), the molar ratio of water ($V_w = 18$ cm^3/mol), and the following formula (Zhang et al., 1990):

$$P_f = V_o[d(V/V_o)/dt]/[SV_w(\text{osm}_{in} - \text{osm}_{out})]$$

In addition to water permeability measurements, other small uncharged molecules such as glycerol have also been shown to permeate AQP3 (Ishibashi et al., 1994). This may be measured by incubating the oocytes in buffer containing isotopically labeled glycerol in the absence of an osmotic gradient followed by solubilization of the washed oocytes in SDS and scintillation counting. Aquaporins are not believed to conduct ions or protons, (Agre et al., 1997) so electrophysiologic techniques are not described here.

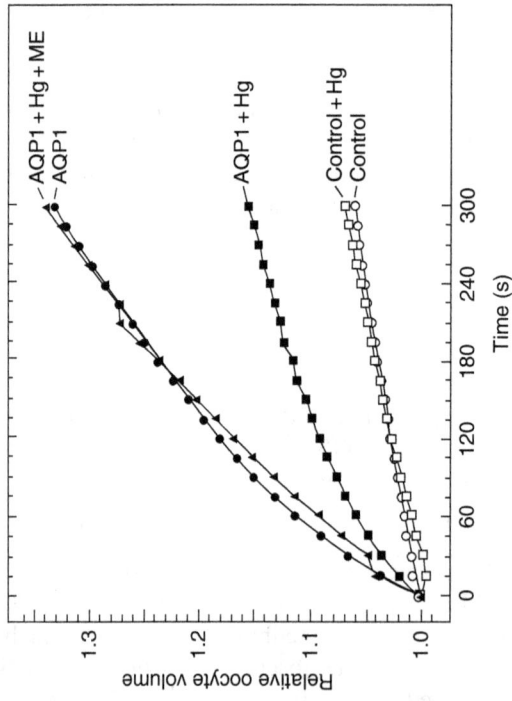

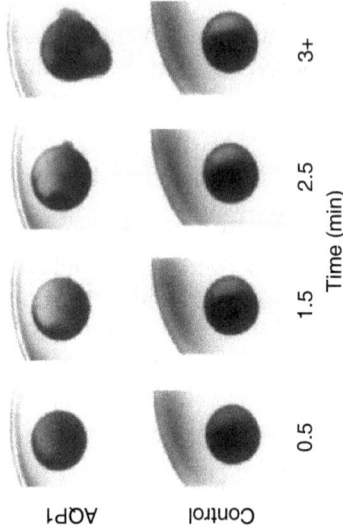

Fig. 5 *Time course and Hg^{2+} inhibition of osmotic swelling of oocytes expressing AQP1. Left:* Oocytes were injected with 50 nl of water (control) or 50 nl of water containing 1 ng of cRNA (AQP1). After 72 h, the oocytes were transferred from 200 to 70 mOsm modified Barth's buffer, and changes in size were measured by video microscopy. This permitted determination of P_f (AQP1, $\sim 200 \times 10^4$ cm/s; control, $\sim 20 \times 10^{-4}$ cm/s). Where indicated (+Hg), the oocytes were incubated for 5 min in buffer containing 0.3 mM $HgCl_2$ followed by osmotic swelling. After incubation for 5 min in 0.3 mM $HgCl_2$, other oocytes (AQP1 + Hg + ME) were then incubated for 15 min in buffer containing 5 mM 2-mercaptoethanol followed by osmotic swelling. *Right:* Series of still photographs of oocytes during osmotic swelling from video micrographs taken at indicated intervals. [Modified and reprinted from Preston *et al.* (1992) and Preston *et al.* (1993) with permission.]

5. Oocyte Membrane Immunoblot Analysis

It is often useful to confirm that the oocytes expressed the aquaporin of interest, particularly when testing site-directed mutant forms (Preston *et al.*, 1993). Groups of 5–10 oocytes are transferred with modified Barth's buffer into 1.5-ml microcentrifuge tubes on ice. After chilling for 5 min, the buffer is removed and the oocytes are lysed in 0.5–1 ml ice-cold hypotonic lysis buffer (7.5 mM Na_2HPO_4, pH 7.4, 1 mM NaEDTA) in the presence of protease inhibitors (20 μg/ml PMSF, 1 μg/ml pepstatin A, 1 μg/ml leupeptin, 1:2000 DFP) by repeatedly vortex agitating and pipetting. The yolk and cellular debris are pelleted at 750 $\times$ g for 5 min at 4 °C, and membranes are then pelleted from the supernant at 16,000 $\times$ g for 30 min at 4 °C. The floating yolk is removed from the top of the tubes with a cotton applicator, and the supernatant is removed. The membrane pellets are gently washed once with an equal volume of ice-cold hypotonic lysis buffer and resuspended in 10 μl of 1.25% (w/v) SDS per oocyte and electrophoresed into a 12% SDS–polyacrylamide gel, transferred to nitrocellulose, incubated with a 1:1000 dilution of affinity-purified anti-AQP1 (Smith and Agre, 1991) or another anti-aquaporin immunoglobulin G (IgG), and visualized by ECL (enhanced chemiluminescence) Western blotting detection system (Amersham Corp., Arlington Heights, IL).

Acknowledgments

This work was supported in part by research grants from the National Institutes of Health to P.A. and a postdoctoral fellowship from the American Heart Association, Maryland Affiliate, to J.C.M. We thank our colleagues Suresh V. Ambudkar, Mark L. Zeidel, Larry A. Coury, and William B. Guggino for their efforts in developing these methods.

References

Agre, P., Sasaki, S., and Chrispeels, M. J. (1993a). *Am. J. Physiol. Renal.* **265**, F461.

Agre, P., Preston, G. M., Smith, B. L., Jung, J. S., Raina, S., Moon, C., Guggino, W. B., and Nielsen, S. (1993b). *Am. J. Physiol. Renal.* **265**, F463.

Agre, P., Lee, M. D., Devidas, S., and Guggino, W. B. (1997). *Science* **275**, 1490.

Ambudkar, S. V., and Maloney, P. C. (1986a). *Methods Enzymol.* **125**, 558.

Ambudkar, S. V., and Maloney, P. C. (1986b). *J. Biol. Chem.* **261**, 10079.

Bennett, V. (1983). *Methods Enzymol.* **96**, 313.

Calamita, G., Bishai, W. R., Preston, G. M., Guggino, W. B., and Agre, P. (1995). *J. Biol. Chem.* **270**, 29063.

Chen, C. C., and Wilson, T. H. (1984). *J. Biol. Chem.* **259**, 10150.

Chrispeels, M. J., and Agre, P. (1994). *Trends Biochem. Sci.* **19**, 421.

Coury, L. A., Mathai, J. C., Prasad, B. V. R., Brodsky, J. L., Agre, P., and Zeidel, M. L. (1998). *Am. J. Physiol. Renal.* **274**, F34.

Denker, B. M., Smith, B. L., Kuhajda, F. P., and Agre, P. (1988). *J. Biol. Chem.* **263**, 15634.

Finkelstein, A. (1987). "Water Movement Through Lipid Bilayers, Pores, and Plasma Membranes: Theory and Reality." Wiley and Sons, New York.

Hallen, R. M. (1980). *J. Biochem. Biophys. Methods* **2**, 251.

Ishibashi, K., Sasaki, S., Fushimi, K., Uchida, S., Kuwahara, M., Saito, H., Furukawa, T., Nakajima, K., Yamaguchi, Y., Gojobori, T., and Marumo, F. (1994). *Proc. Natl. Acad. Sci. USA* **91,** 6269.

Ito, H., Fukuda, Y., Murata, K., and Kimura, A. (1983). *J. Bacteriol.* **153,** 163.

Jung, J. S., Bhat, R. V., Preston, G. M., Guggino, W. B., Baraban, J. M., and Agre, P. (1994). *Proc. Natl. Acad. Sci. USA* **91,** 13052.

Krieg, P. A., and Melton, D. A. (1984). *Nucleic Acids Res.* **12,** 7057.

Lu, L., Montrose-Rafizadeh, C., Hwang, T. C., and Guggino, W. B. (1990). *Biophys. J.* **57,** 117.

Mathai, J. C., Mori, S., Smith, B. L., Preston, G. M., Mohandas, N., Collins, M., van Zijl, P. C. M., Zeidel, M. L., and Agre, P. (1996). *J. Biol. Chem.* **271,** 1309.

Nakamoto, R. K., Rao, R., and Slayman, C. W. (1991). *J. Biol. Chem.* **266,** 7940.

Nielsen, S., Smith, B. L., Christensen, E. I., Knepper, M., and Agre, P. (1993). *J. Cell Biol.* **120,** 371.

Preston, G. M. (1997). *In* "Methods in Molecular Biology" (B. A. White, ed.), Vol. **67,** pp. 433–449. Human Press Inc, Totowa, NJ.

Preston, G. M., and Agre, P. (1991). *Proc. Natl. Acad. Sci. USA* **88,** 11110.

Preston, G. M., Carroll, T. P., Guggino, W. B., and Agre, P. (1992). *Science* **256,** 385.

Preston, G. M., Jung, J. S., Guggino, W. B., and Agre, P. (1993). *J. Biol. Chem.* **268,** 17.

Priver, N., Rabon, E. C., and Zeidel, M. L. (1993). *Biochemistry* **21,** 2459.

Raina, S., Preston, G. M., Guggino, W. B., and Agre, P. (1995). *J. Biol. Chem.* **270,** 1508.

Schaffner, W., and Weissman, C. (1973). *Anal. Biochem.* **56,** 502.

Smith, B. L., and Agre, P. (1991). *J. Biol. Chem.* **266,** 6407.

Solomon, A. K. (1989). *Methods Enzymol.* **173,** 192.

Zeidel, M. L., Ambudkar, S. V., Smith, B. L., and Agre, P. (1992). *Biochemistry* **31,** 7436.

Zeidel, M. L., Nielsen, S., Smith, B. L., Ambudkar, S. V., Maunsbach, A. B., and Agre, P. (1994). *Biochemistry* **33,** 1606.

Zhang, R., Logee, K. A., and Verkman, A. S. (1990). *J. Biol. Chem.* **265,** 15375.

PART X

Toxins and Other Membrane Active Compounds

CHAPTER 25

Conus Peptides as Probes for Ion Channels

**J. Michael McIntosh, Baldomero M. Olivera, and
Lourdes J. Cruz**
Department of Biology
University of Utah
Salt Lake City, Utah, USA

I. Introduction

Conus peptides are increasingly used as tools for investigating ion channels. The 500 species of predatory cone snails each produces a complex venom that has a large number of biologically active peptides. The majority of *Conus* peptides characterized to date appear to be targeted to different types of ion channels. It is estimated that the venom of each *Conus* species has between 50 and 200 peptides. Because of the remarkable divergence that occurs when cone snails speciate, the complement of venom peptides in any one *Conus* species is distinct from that of any other. Thus,

DOI: 10.1016/B978-0-12-382204-8.00025-4

many thousands of peptides that affect ion channel function are present in *Conus* venoms but only a miniscule fraction of these have been characterized biochemically. An even smaller number have been used as tools in neurobiology. However, there is little doubt that as more of these peptides become available to the neurobiological community, an increasing number will be used as ligands for characterizing ion channel structure and function. Because of their relatively small size, most of these peptides can be chemically synthesized, and thus be made widely available.

II. Biochemical Overview of *Conus* Peptides

The *Conus* venom peptides can be divided into two general groups: (1) multiply disulfide-bonded peptides from 12 to 50 amino acids in length (most under 30 residues). Generically, these are called conotoxins, and (2) other peptidic venom components that are not disulfide-rich; these either completely lack disulfide bonds or have a single disulfide linkage. The latter are a heterogeneous group of peptides with several distinct families.

In the following sections, we focus first on *Conus* peptides that are targeted to ligand-gated ion channels, followed by peptides that are targeted to voltage-gated ion channels. The last section discusses practical considerations for using *Conus* peptides. It should be noted parenthetically that in much of the literature of the late 1980s and early 1990s, the term *conotoxin* was routinely used to refer to one specific molecule out of the many tens of thousands of *Conus* peptides—this was ω-conotoxin GVIA, the first natural toxin known to inhibit voltage-gated calcium channels. Given the very large number of *Conus* peptides, it is no longer appropriate to use the term *conotoxin* for this one peptide. In this review, *conotoxin* will be used generically for all multiply disulfide-bonded *Conus* peptides.

For neurobiologists, the major interest in *Conus* peptides is that they are highly subtype-specific ligands. For several ion channel targets, *Conus* peptides are the most specific ligands known. For example, among ligands that target voltage-gated sodium channels, μ-conotoxin GIIIA has unprecedented specificity for the skeletal muscle subtype. This isoform is among the set of sodium channels that are tetrodotoxin and saxitoxin sensitive. However, μ-conotoxin GIIIA is much more specific than either of the guanidinium toxins; it has a preference for the skeletal muscle isoform by at least three orders of magnitude over other tetrodotoxin-insensitive subtypes (Cruz *et al.*, 1985; Gonoi *et al.*, 1987). This high subtype selectivity is proving to be a general feature of *Conus* peptides. As a consequence, with more isoforms of ion channel families being cloned and characterized, and the need for subtype-specific ligands increasing, *Conus* peptides will undoubtedly be increasingly used to discriminate functionally between closely related molecular forms of ion channels. In many ways, having a very highly subtype-specific *Conus* peptide ligand provides a complementary approach to having a gene knockout of one particular ion channel isoform.

Table I
Classes of *Conus* Peptides and Their Macromolecular Targets

Peptide class	Characteristic structural features (number of amino acids)	Mode of action
α-Conotoxins	CC–C–C (12–19)	Competitive inhibitor of nicotinic ACh receptor
αA-conotoxins	CC–C–C–C–C (25–30)	Competitive inhibitor of nicotinic ACh receptor
ψ-Conotoxins	CC–C–C–CC (24)	Noncompetitive inhibitor of nicotinic ACh receptor
Conantokins	γ-Carboxylate residues, Cys residues not necessary (17–27)	Noncompetitive inhibitor of NMDA receptor
μ-Conotoxins	CC–C–C–CC (22)	Sodium channel blocker; competes with saxitoxin and tetrodotoxin for site I
μO-conotoxins	C–C–CC–C–C (31)	Sodium channel blocker; does not compete with saxitoxin for site I binding
μ-Conotoxins	CC–C–C–C–C (17)	Blocks molluscan sodium channels
δ-Conotoxins	C–C–CC–C–C (27–31)	Delays sodium channel inactivation; binds to site VI of the channel
κ-Conotoxins	C–C–CC–C–C (27)	Potassium channel blocker
ω-Conotoxins	C–C–CC–C–C (24–29)	Calcium channel blocker

An overview of the *Conus* peptides known to affect ion channel function is given in Table I.

A. *Conus* Peptides Targeting Ligand–Gated Ion Channels

Four families of *Conus* peptides are known to target ligand-gated ion channels; three of these target nicotinic acetylcholine receptors (nAChRs). These include the α-conotoxins, the αA-conotoxins, and the ψ-conotoxins. The first two families are believed to be competitive antagonists of the nicotinic receptor, while the ψ-conotoxins have recently been shown to be noncompetitive antagonists. To date, peptides in all three families have been found that target the skeletal muscle subtype of nicotinic receptors. However, all *Conus* peptides characterized so far that preferentially inhibit *neuronal* nicotinic receptors belong to the α-conotoxin family.

The other group of peptides that target ligand-gated ion channels is the conantokins; these are unusual *Conus* peptides that have been shown to antagonize the NMDA (*N*-methyl-D-aspartate) subclass of glutamate receptors.

Preliminary evidence for *Conus* peptides that target other ligand-gated ion channels such as the 5HT3 receptor has been obtained, but a complete biochemical characterization of these peptides is not yet published (England *et al.*, 1997).

B. *Conus* Peptides Targeting Skeletal Muscle Subtype of Nicotinic Acetylcholine Receptors

1. α-Conotoxins

One group of α-conotoxins is known to target the skeletal muscle subtype of nicotinic receptors (the "α3/5 subfamily"). Characteristically, these have three amino acids between the second and third cysteine residues, and five amino acids between the third and fourth cysteine residues of the peptide. The sequences of all α-conotoxins of this subfamily are shown in Table II. Among these is the very first *Conus* peptide that was biochemically characterized, α-conotoxin GI. Certain members of a second subfamily of α-conotoxins, the "α4/7 subfamily," also target the muscle receptor. One example is α-conotoxin EI (Martinez *et al.*, 1995).

The α3/5 subfamily of α-conotoxins is the best characterized with respect to high targeting specificity for the muscle receptor. α-Conotoxin MI has been shown to discriminate between the α/δ and the α/γ interface of the mammalian nAChR

Table II
Structure and Specificity of α-Conotoxins

Disulfide bond arrangement : CC————C————C

Conotoxin	Source	Primary structure	Site preference	Reference
Targeted to skeletal muscle nAChR				
α3/5 subfamily				
GI	*Conus geographus*	ECCNPACGRHYSC[a]	Mouse: α/δ subunit interface	Gray *et al.* (1981), Groebe *et al.* (1995), Hann *et al.* (1994), Kreienkamp *et al.* (1994), Utkin *et al.* (1994)
GIA	*Conus geographus*	ECCNPACGRHYSCGK[a]		Gray *et al.* (1981)
GII[c]	*Conus geographus*	ECCHPACGKHFSC[a]		Gray *et al.* (1981)
MI	*Conus magus*	GRCCHPACGKNYSC[a]	Mouse: α/δ subunit interface	Groebe *et al.* (1995), Hann *et al.* (1994), Kreienkamp *et al.* (1994), McIntosh *et al.* (1982), Utkin *et al.* (1994)
SI	*Conus striatus*	ICCNPACGPKYSC[a]		Zafaralla *et al.* (1988)
SIA[c]	*Conus striatus*	YCCHPACGKNFDC[a]	*Torpedo*: α/γ subunit interface	Myers *et al.* (1991), Hann *et al.* (1997)
SII[c]	*Conus striatus*	GCCCNPACGPNY GCGTSCS[b]		Ramilo *et al.* (1992)
α4/7 subfamily				
EI	*Conus ermineus*	RDOCCYHPTCNM SNPQIC[a]	*Torpedo*: α/δ subunit interface	Martinez *et al.* (1995)
Targeted to neuronal nAChRs				
α4/7 subfamily				
MII	*Conus magus*	GCCSNPVCHLEHSNLC[a]	Rat: $\alpha_3\beta_2$ subunit interface	Cartier *et al.* (1996)
PnIA	*Conus pennaceus*	GCCSLPPCAANNPDYC[a]	*Aplysia*: neuronal nAChR	Fainzilber *et al.* (1994a,b)
PnIB	*Conus pennaceus*	GCCSLPPCALSNPDYC[a]	*Aplysia*: neuronal nAChR	Fainzilber *et al.* (1994a,b)
AuIA/B/C	*Conus aulicus other*	Unpublished	Rat: $\alpha_3\beta_4$ subunit interface	Luo *et al.* (1997)
ImI	*Conus imperialis*	GCCSDPRCAWRC[a]	Rat: α_7 nAChR; *Aplysia*: neuronal nAChR	Johnson *et al.* (1995), Kehoe *et al.* (1996), McIntosh *et al.* (1994), Pereira *et al.* (1996)

[a]C-terminal α-carboxyl group is amidated.
[b]C-terminal α-carboxyl group is the free acid.
[c]Disulfide bond arrangement has not been determined for GII, SIA, or SII, but very likely is conserved.

receptor by approximately 10^4. When the nicotinic receptor from *Torpedo* is used, α-conotoxin SIA has been shown to discriminate totally between the two ligand-binding sites (in this case targeting to the α/γ interface of the *Torpedo* receptor). α-Conotoxins MI and GI have been shown to be inactive at neuronal nAChRs including $\alpha_2\beta_2$, $\alpha_2\beta_4$, $\alpha_3\beta_2$, $\alpha_3\beta_4$, $\alpha_4\beta_2$, and $\alpha_4\beta_4$ subtypes. Additionally, they do not block α_7 and α_9 homomers in contrast to the long α-neurotoxins from elapiid snakes, such as α-bungarotoxin. Thus, compared to α-bungarotoxin, peptides such as α-contoxin MI appear to be much more highly specific.

It is noteworthy that a small subset of the α3/5 family shows a much greater differential affinity for teleost nicotinic receptors versus mammalian nicotinic receptors. The majority of the peptides in this subfamily (α-conotoxins GI, MI, and SIA) have high affinity for all skeletal muscle nicotinic receptors; in contrast, peptides such as α-conotoxin SI have a dramatically lower affinity for the mammalian skeletal muscle nicotinic receptors (Groebe *et al.*, 1995).

In contrast to the α3/5 conotoxins which have high affinity for the mammalian α/δ but not the α/γ interface in mammalian muscles, but not the α/γ interface in *Torpedo* (Hann *et al.*, 1994; Utkin *et al.*, 1994), the α4/7 conotoxin EI shows high affinity for the α/δ interface in both systems and can be used as a selective probe for the α/δ site in *Torpedo* (Martinez *et al.*, 1995). The structures of several α-conotoxins have been solved both by nuclear magnetic resonance (NMR) techniques and, more recently, by X-ray crystallography.

2. αA-Conotoxins

Like α-conotoxins of the α3/5 subfamily, αA-conotoxins are believed to be competitive antagonists of skeletal muscle nicotinic receptors (Table III). It has been demonstrated that in contrast to α-conotoxin MI, αA-conotoxin EIVA from the fish-hunting species *Conus ermineus* has almost equal affinity for the two ligand-binding sites of the nicotinic receptor. Indeed, αA-conotoxin EIVA exhibited a higher affinity than any other *Conus* peptide for the α/γ ligand-binding site of the mouse skeletal muscle nicotinic receptor. Thus, α-conotoxins and αA-conotoxins that target the skeletal muscle nicotinic receptor subtype have different specificity for the two ligand-binding sites of mammalian receptors. Clearly, the different structures reflect different "microsite" interactions (Olivera *et al.*, 1990) even though both groups of peptides are competitive antagonists. The structures of two αA-conotoxins have been solved by NMR.

3. ψ-Conotoxins

A novel noncompetitive nicotinic receptor antagonist has been described, ψ-conotoxin PIIIE, from *Conus purpurascens*. At least two other peptides belonging to this family have been discovered (R. Jacobsen and B. Olivera, unpublished results). ψ-Conotoxin PIIIE has been shown to inhibit the skeletal muscle subtype of nicotinic receptors expressed in oocytes, although it has a significantly higher

TABLE III

Structure of αA-Conotoxins, ψ-Conotoxins, and Conantokins

Conotoxin	Source	Primary structure[a]	Reference
Competitive muscle nAChR antagonists			
Disulfide bond arrangement:		CC————C——C—C————C	
αA-EIVA	*Conus ermineus*	GCCGPYONAACHOCGCKVGROOYCDROSGG[b]	Jacobsen *et al.* (1997)
αA-EIVB	*Conus ermineus*	GCCGKYONAACHOCGCTVGROOYCDROSGG[b]	Jacobsen *et al.* (1997)
αA-PIVA	*Conus purpurascens*	GCCGSYONAACHOCSCKDROSYCGQ[b]	Hopkins *et al.* (1995)
Noncompetitive muscle nAChR antagonists			
Disulfide bond arrangement:		CC————C————C————CC	
ψ-PIIIE	*Conus purpurascens*	HOOCCLYGKCRRYOGCSSASCCQR[b]	Shon *et al.* (1997a,b)
Noncompetitive NMDA receptor antagonists			
Conantokin-G	*Conus geographus*	GEγγLQγNQγLIRγKSN[b]	Mena *et al.* (1990), Cruz *et al.* (1985)
Conantokin-T	*Conus tulipa*	GEγγYQKMLγNLRγAEVKKNA[b]	Haack *et al.* (1990)

[a] γ, γ-Carboxyglutamate; O, *trans*-4-hydroxyproline.
[b] C-terminal α-carboxyl group is amidated.

affinity for the *Torpedo* receptor compared to the homologous mouse receptor. The structure of ψ-conotoxin PIIIE has been determined by multidimensional NMR.

C. *Conus* Peptides Targeted to Neuronal Subtypes of Nicotinic Receptors

All *Conus* ligands for neuronal subtypes of nicotinic receptors in mammalian systems belong to the α-conotoxin family. The most specific of such peptides described to date is α-conotoxin MII. This peptide has a very high affinity and target specificity for the $\alpha_3\beta_2$ subtype of neuronal nicotinic receptors. This peptide was used to demonstrate that at least two presynaptic subtypes of neuronal nicotinic receptor are involved in striatal dopamine release, one of which contains an $\alpha_3\beta_2$ interface (Kulak *et al.*, 1997). Additionally, MII has been used to pharmacologically dissect nicotinically mediated synaptic transmission in chick parasympathetic ciliary ganglion. At this ganglion, MII selectively inhibits the slowly decaying versus rapidly decaying current (Ullian *et al.*, 1997). A combination of MII and IMI has been used to distinguish subpopulations of nAChRs in frog sympathetic ganglion (Tavazoie *et al.*, 1996). The NMR structure of α-conotoxin MII has recently been solved. A variety of data suggest that α-conotoxin MII is a Janus ligand, with two interacting interfaces. One interface is proposed to specifically cause rapid association with the β_2 subunit, and the other to cause functional block and very slow dissociation from the α_3 subunit.

A variety of *Conus* peptides have also been shown to target the α_7 subtype of nicotinic receptors. The first one of these characterized was α-conotoxin IMI from *Conus imperialis* venom. In addition to its specificity for α_7 in mammalian systems, this peptide has been used to discriminate between different types of nicotinic receptors in molluscan systems. Other α-conotoxins have recently been discovered that target the α_7 subtype with significantly higher affinity than α-conotoxin ImI (J. M. McIntosh, unpublished results).

A number of peptides from *Conus aulicus* venom (α-conotoxins AUIA, AuIB, and AuIC), which prefer the $\alpha_3\beta_4$ subtype of neuronal nicotinic receptor, have been characterized. However, the sequences of these peptides have not yet been published.

Some of the α-conotoxins have been shown to act potently at molluscan nAChRs. The first reported peptides were α-conotoxins PnIA and PnIB from *Conus pennaceus* (Fainzilber *et al.*, 1994a,b). The peptides block the nAChR of cultured *Aplysia* neurons. More recently, α-conotoxin ImI was shown to be a selective antagonist of subpopulations of *Aplysia* nAChRs (Kehoe *et al.*, 1996).

D. *Conus* Peptides Targeting NMDA Receptors

The conantokins, which are perhaps the most novel family of *Conus* peptides have been shown to be NMDA receptor antagonists (Mena *et al.*, 1990). In contrast to the conotoxins, conantokins are not multiply disulfide-bonded but instead have a very unusual posttranslational modification, the γ-carboxylation of glutamate residues to γ-carboxyglutamate (Gla). The discovery of the first member of this family, conantokin-G, established that this unusual posttranslational modification could occur outside mammalian systems (McIntosh *et al.*, 1984).

Three conantokins have been characterized so far, conantokin-G from *Conus geographus* (Olivera *et al.*, 1984), conantokin-T from *Conus tulipa* (Haack *et al.*, 1990), and conantokin-R from *Conus radiatus* (White *et al.*, 1997). These peptides were purified from venom by following an unusual *in vivo* activity in mammals: the ability to induce a sleep-like state in young mice (under 2 weeks of age). Thus, in the earlier papers describing these peptides (before they were found to be NMDA receptor antagonists), they are referred to as "sleeper peptides."

The conantokins are the only natural peptides known to inhibit NMDA receptors. So far, all natural conantokins tested cause inhibition of a variety of NMDA receptor isoforms, albeit with very different affinities. No other subclass of glutamate receptors that have been examined are inhibited by the conantokin peptides. A report has demonstrated that conantokins have potential as anticonvulsant compounds, exhibiting great potency in an audiogenic seizure mouse model, with a very high protective index compared to commercial anticonvulsant compounds (White *et al.*, 1997).

Several structural investigations have been carried out on the conantokins using circular dichroism and NMR techniques (Myers *et al.*, 1990; Rigby *et al.*, 1997; Skjaebaek *et al.*, 1997). These studies are in general agreement that conantokins are highly structured peptides, with α-helical structure as well as a distorted 3_{10} helix. For conantokin-G at least, the peptide becomes more structured in the presence of divalent cations. Like the Gla-containing peptides of the blood clotting cascade, conantokin-G binds acidic membranes in the presence of Ca^{2+} ions (Myers *et al.*, 1990).

It has recently been shown that the conantokins are initially translated as a large prepropeptide precursor; the mature peptide is found in the C-terminal end in a single copy. In the excised region, which is N terminal to the mature conantokin-encoding C-terminal region, a recognition signal sequence is present that facilitates vitamin K-dependent carboxylation of selected glutamate residues in the mature peptide region (Bandyopadhyay *et al.*, 1997). Thus, in contrast to the conotoxins where structure is largely stabilized by the multiple disulfide cross-links, in the conantokin family of peptides the structure is stabilized by the presence of multiple γ-carboxyglutamate (Gla) residues, appropriately spaced for a helical configuration to be assumed. Sequences in the prepropeptide precursor that do not appear in the mature peptide play an important role in the posttranslational conversion of Glu to Gla.

E. *Conus* Peptides That Target Voltage–Gated Ion Channels

1. Overview

The most widely used *Conus* peptides in neurobiology are those that target voltage-gated calcium channels; these all belong to the ω-conotoxin family (see Table IV). Several different *Conus* peptide families target voltage-gated sodium channels; the first of these discovered were the μ-conotoxins, which are Na^+ channel blockers (Cruz *et al.*, 1985). The δ-conotoxins are a family of *Conus* peptides that inhibit sodium channel inactivation (Shon *et al.*, 1994). Finally, the μO-conotoxins are also sodium channel antagonists (Fainzilber *et al.*, 1994a,b), but do not appear to act on the same site as the μ-conotoxins and have a different structural motif (see Table V). The first *Conus* peptide that targets a voltage-gated potassium channel, κ-conotoxin, has been characterized (Shon *et al.*, 1997a,b).

F. *Conus* Peptides That Target Voltage–Gated Calcium Channels

The literature on the ω-conotoxins that target voltage-gated calcium channels is very extensive, but in this chapter, only a very brief overview is presented. For a more comprehensive review, the reader is referred to Olivera *et al.* (1994) and Dunlap *et al.* (1995).

The first ω-conotoxin that was biochemically characterized was ω-conotoxin GVIA from *C. geographus* venom, followed by ω-conotoxin MVIIA from *Conus magnus* venom. These were the first natural peptide toxins that inhibited voltage-gated calcium channels. In mammalian systems, these two peptides are very highly subtype-specific, targeting voltage-gated calcium channel complexes that contain an α_{1B} subunit (which correspond to what is known as the "N-type" Ca current).

Note that these peptides may have broader selectivity in lower vertebrates (see a discussion in Olivera *et al.*, 1994). In the literature, there has been a tendency to

Table IV

Structure and Specificity of the Calcium Channel Blockers. ω-Conotoxins

Disulfide linkages: CC————C————C————CC

ω-Conotoxin	Source	Primary structure	Specificity	Reference[b]
GVIA	*Conus geographus*	CKSOGSSCSOTSYNCCRSCNOYTKRCY[a]	N-type calcium channels (α_{1B} subunit)	Nishiuchi and Sakakibara (1982), Nishiuchi et al. (1986)
GVIIA	*Conus geographus*	CKSOGTOCSRGMRDCCTSCLLYSNKCRRY[a]		Olivera et al. (1984)
MVIIA	*Conus magus*	CKGKGAKCSRLMYDCCTGSCRSGKC[a]	N-type calcium channels (α_{1B} subunit)	Olivera et al. (1985)
MVIIB	*Conus magus*	CKGKGASCHRTSYDCCTGSCNRGKC[a]		Olivera et al. (1987)
MVIIC	*Conus magus*	CKGKGAPCRKTMYDCCSGSCGRRGKC[a]	P/Q- and N-type calcium channels (α_{1B} and $\alpha 1A$)	Olivera et al. (1987) Hillyard et al. (1992)
MVIID	*Conus magus*	CQGRGASCRKTMYNCCSGSCNRGRC[a]	P/Q- and N-type calcium channels (α_{1B} and $\alpha 1A$)	Monje et al. (1993)
SVIA	*Conus striatus*	CRSSGSPCGVTSICCGRCYRGKCT[a]		Ramilo et al. (1992)
SVIB	*Conus striatus*	CKLKGQSCRKTSYDCCSGSCGRSGKC[a]	N- and P/Q-type calcium channels (α_{1A} and $\alpha 1B$)	Ramilo et al. (1992)
TxVIIA	*Conus textile*	CKQADEPCDVFSLDCCTGICLGVCMV[c]	Dihydropyridine-sensitive currents in *Aplysia*	Fainzilber et al. (1996)

[a] C-terminal α-carboxyl group is amidated.
[b] See also reviews for primary references (Bean, 1989; Dunlap et al., 1995; Hess, 1990; Miller, 1987; Olivera et al., 1994; Tsien et al., 1988, 1991; Snutch, 1992).
[c] C-terminal amide is the free acid.

Table V
Structure and Specificity of Sodium and Potassium Channel Ligands from *Conus*

Conotoxin	Source	Primary structure	Specificity	Reference
		Disulfide linkages: μ-conotoxins (III-family)		Hidaka et al. (1990), Shon et al. (1994, 1997a,b), Terlau et al. (1996a,b)

μ-conotoxins (III-family)

μO-, δ- and κ-conotoxins

Blockers of voltage-sensitive sodium channels

Conotoxin	Source	Primary structure	Specificity	Reference
μ-GIIIA	*Conus geographus*	RDCCTOOKKCKDRQCKOQRCCA[a]	Skeletal muscle Na channel (μl subtype); binds to site I	Cruz et al. (1985), Moczydlowski et al. (1986), Sato et al. (1983), Stephan et al. (1994)
μ-GIIIB	*Conus geographus*	RDCCTPPRKCKDRRCKPMKCCAGR[a]	Same	Cruz et al. (1985), Gonoi et al. (1987), Sato et al. (1983), Stephan et al. (1994)
μ-GIIIC	*Conus geographus*	RDCCTPPRKCKDRRCKPMKCCAGR[a]	Same	Cruz et al. (1985)
μO-MrVIA	*Conus marmoreus*	ACRKKWEYCIVPIIGFIYCCPGLICG PFVCV[b]	Molluscan neurons (~100 nM); type II Na+ channels and Na+ channels in cultured rat hippocampal cells; block of rapidly inactivating Ca2+ current at higher concentrations (>1 μM)	Fainzilber et al. (1995a,b,c), McIntosh et al. (1995a,b), Terlau et al. (1996a,b)

Name	Species	Sequence	Action	Reference
μO-MrVIB	*Conus marmoreus*	ACSKKWEYCIVPILGFVYCCPGLIC GPFVCV[a]	Same	Same
μ-PnIVA	*Conus pennaceus*	CCKYGWTCLLGCSPCGC[b]	Tetrodotoxin-insensitive molluscan $Na+$ channels	Fainzilber et al. (1995a,b,c)
μ-PnIVB	*Conus pennaceus*	CCKYGWTCWLGCSPCGC[b]	Same	Fainzilber et al. (1995a,b,c)

Ligands that delay inactivation of voltage-sensitive sodium channels

Name	Species	Sequence	Action	Reference
δ-GmVIA	*Conus gloriamaris*	VKPCRKEGQLCDPIFQNCCRGWN CVLFCV[b]	Molluscan neurons; shifts voltage-dependent activation curve to more negative potentials and inactivation curve to more positive potentials	Shon et al. (1994)
δ-TxVIA	*Conus textile*	WCKQSGEMCNLLDQNCCDGYC IVLVCT[b]	Molluscan neurons; binding to mammalian Na^+ channels with no apparent physiologic effects and acts to protect against toxic effects of other toxins binding to the same site	Fainzilber et al. (1991, 1994a,b), Woodward et al. (1990)
δ-PVIA	*Conus purpurascens*	EACYAOGTFCGIKOGLCCSEFCL PGVCFG[b]	Rat brain type II $Na+$ channel; rat hippocampal neurons; vertebrate neuromuscular junction	Shon et al. (1995), Terlau et al. (1996a,b)
NgVIA	*Conus nigropunctatus*	SKCFSOGTFCGIKOGLCCSVRCF SLFCISFE[b]	Molluscan and vertebrate $Na+$ channels; δ-TxVIA is a partial antagonist of NgVIA	Fainzilber et al. (1995a,b,c)

Potassium channel blocker

Name	Species	Sequence	Action	Reference
κ-PVIIA	*Conus purpurascens*	CRIONQKCFQHLDDCCSRKCN RFNKCV[b]	*Shaker* K^+ channel	Shon et al. (1997a,b), Terlau et al. (1996a,b)

[a]C-terminal α-carboxyl group is amidated.
[b]C-terminal α-carboxyl group is the free acid.

assume that any voltage-gated calcium channel that is sensitive to ω-conotoxin GVIA or MVIIA must be α_{1B} containing (i.e., an N-type calcium channel), while any voltage-gated calcium channel resistant to these peptides must be of a different subtype. Although there are no known exceptions so far to this generalization in mammalian systems, there is reason to suspect that the correlation will not hold in lower vertebrates, and almost certainly does *not* apply to invertebrates.

The structures of both ω-conotoxins GVIA and MVIIA have been reported by several laboratories, using multidimensional NMR techniques. Some structure–function studies have been carried out. Both peptides have been radiolabeled, and used productively in binding experiments, and in autoradiographic studies (e.g., see Filloux *et al.*, 1994).

In electrophysiologic experiments, ω-conotoxin GVIA is used to inhibit α_{1B}-containing complexes irreversibly, while ω-conotoxin MVIIA is the ligand of choice when a high-affinity but reversible block is desired. Several other homologs of these peptides have been described in the literature (see Table IV).

A second group of ω-conotoxins inhibits both α_{1B}- and α_{1A}-containing calcium channel complexes. These have broader specificity than the α_{1B}-specific ω-conotoxins described above. The most widely used of these peptides is ω-conotoxin MVIIC, which has been used to discriminate between different subclasses of voltage-gated calcium channels. Both ω-conotoxins MVIIC and MVIID clearly inhibit the so-called "P/Q subclasses" of voltage-gated calcium channels, which are widely believed to contain an α_{1A} subunit, although the precise correspondence of P- and Q-type calcium currents as described by electrophysiologic investigations to α_{1A}-containing calcium channel complexes is still uncertain.

The structure of ω-conotoxin MVIIC has been reported (Hillyard *et al.*, 1992). This peptide has been radiolabeled and used for binding studies. Tsien *et al.* have proposed that ω-conotoxin MVIIC can serve as a key reagent in discriminating between P- and Q-type calcium currents, but this view has not been universally accepted (Dunlap *et al.*, 1995).

Additional ω-conotoxins which inhibit voltage-gated Ca^{2+} channels in invertebrate systems, particularly in mollusks, have been reported (Fainzilber *et al.*, 1996). However, although these peptides have been biochemically characterized, their specificity for particular calcium channel subtypes has not yet been established. In certain cases, peptides that were originally isolated as being voltage-gated calcium channel antagonists have proved to be more potent as sodium channel inhibitors.

G. *Conus* Peptides That Target Voltage-Gated Sodium Channels

1. μ-Conotoxins

The μ-conotoxins were the first polypeptide toxins to compete for the same site on Na^+ channels as the well-established guanidinium toxins which target sodium channels, tetrodotoxin, and saxitoxin. In the nomenclature of Catterall (1992), all of these toxins bind to site I, which is believed to be the outer vestibule of the ion channel pore. The μ-conotoxins were originally characterized from *C. geographus* venom, but more recently another μ-conotoxin was isolated and characterized from *C. purpurascens*. As noted earlier, the μ-conotoxins have narrower subtype specificity than the guanidinium toxins. Like the critical guanidinium moiety in saxitoxin and tetrodotoxin, there is believed to be a key arginine in all μ-conotoxins that have been characterized. However, it has been suggested that the guanidinium group of arginine does not in fact interact with the same residues on the voltage-gated ion channel as does the guanidinium group on tetrodotoxin. The structure of several μ-conotoxins, including some analogs, has been described by several groups using NMR techniques.

2. δ-Conotoxins

The first δ-conotoxin was originally called a "King-Kong peptide" from *Conus textile* venom, because it elicited a peculiar symptomatology when injected into lobsters. It was subsequently shown using electrophysiological methods that the peptide delayed inactivation of voltage-gated sodium channels in *Aplysia* ganglion cells (Fainzilber *et al.*, 1994a,b; Woodward *et al.*, 1990). Another δ-conotoxin from a snail-hunting *Conus*, δ-conotoxin GmVIA, has also been characterized (Shon *et al.*, 1994).

A δ-conotoxin from a fish-hunting cone snail, δ-conotoxin PVIA from *C. purpurascens* venom, has been isolated and chemically synthesized. This peptide has been shown to be important for the very rapid stunning effect of *C. purpurascens* venom on prey (Shon *et al.*, 1995; Terlau *et al.*, 1996a,b). This peptide is believed to play a key role in the prey capture strategy of this fish-hunting cone snail (Terlau *et al.*, 1996a,b). Like the δ-conotoxins from snail-hunting *Conus* venoms, δ-conotoxin PVIA also causes a delay in inactivation.

A conotoxin, NgVIA, that delays inactivation of molluscan and vertebrate sodium channels has been isolated (Fainzilber *et al.*, 1995a,b,c) and appears to act on a receptor site distinct from that of δ-TXVIA.

It is notable that although the δ-conotoxins have the same disulfide bonding pattern as the ω-conotoxins, they differ strikingly in the type of amino acids found in the loop regions between disulfide linkages. While ω-conotoxins largely have hydrophilic and positively charged amino acids, in all δ-conotoxins there is a preponderance of hydrophobic residues. It was proposed that the δ-conotoxins bind to a unique site on voltage-gated sodium channels, which has been called site VI. Given the very hydrophobic nature of these peptides, this site may be at least partially in the lipid bilayer (Fainzilber *et al.*, 1994a,b).

Because fast inactivation of voltage-gated sodium channels is generally believed to be mediated by a cytoplasmic "ball" region of the ion channel complex, the δ-conotoxins present an intriguing mechanistic puzzle in that they cause an inhibition of fast inactivation from the extracellular side of the membrane.

3. μO-Conotoxins

Two peptides from the snail-hunting species *Conus marmoreus*, μO-conotoxins MrVIA and MrVIB, were shown to block voltage-gated sodium channels (Fainzilber *et al.*, 1995a,b,c; McIntosh *et al.*, 1995a,b). They differ from the μ-conotoxins in being more closely related to the δ-conotoxins than to the μ-conotoxins, and also in being the first polypeptide inhibitors that inhibit conductance through Na^+ channels that do not compete for binding with tetrodotoxin/saxitoxin, and clearly target a different site (Terlau *et al.*, 1996a,b). Furthermore, in contrast to the μ-conotoxins, these peptides act more broadly on different voltage-gated sodium channel subtypes, and a wide variety of different voltage-gated sodium channels are inhibited.

Two conotoxins from *C. pennaceus*, μ-PnIVA and μ-PnIVB, were found by Fainzilber *et al.* (1995a,b,c) to block the tetrodotoxin-insensitive molluscan sodium channels. These peptides are structurally distinct from the originally described μ-conotoxins (e.g., μ-conotoxin GIIIA) and are named with a Roman numeral IV to indicate this difference.

H. *Conus* Peptides That Target Voltage-Gated Potassium Channels

So far, only one *Conus* peptide has been shown to inhibit a voltage-gated potassium channel, κ-conotoxin PVIIA from *C. purpurascens* venom. This peptide has a disulfide bonding pattern generally similar to the ω-conotoxins, but instead of inhibiting voltage-gated calcium channels it targets potassium channels. Although the peptide is active both in lower vertebrate systems (where together with δ-conotoxin PVIA, it appears to be responsible for the very fast stunning effect of venom injection on the prey), and shows activity in mammalian systems as well, no vertebrate potassium channel subtype has yet been identified as being targeted by κ-conotoxin PVIIA. However, the well-characterized *Drosophila Shaker* channel is a κ-conotoxin PVIIA target (Shon *et al.*, 1997a,b).

There is preliminary evidence for a number of peptides unrelated in structure to κ-conotoxin PVIIA which also inhibit voltage-gated potassium channels. However, the biochemical characterization of these peptides is still in progress, and has not been published. It will be interesting to compare the subtype specificity of these peptides with κ-conotoxin PVIIA. Given the vast diversity of potassium channels, it seems likely that the *Conus* venom system will provide many novel peptides that target potassium channels in the future.

III. Some Practical Considerations in Handling *Conus* Peptides

A. Solubility

Conus peptides are soluble in aqueous solutions. In general, a stock concentration of 500 μM may be prepared without difficulty. Some peptides are soluble at higher concentration. Care should be taken, however, to ensure that peptide is actually in solution. Adding buffer to lyophilized peptide often gives the appearance of dissolving the peptide, when, in fact, a suspension has been created. This usually can be detected by holding the mixture up to a light and inspecting for particulates or cloudiness. Examining the solution under a dissecting microscope is often helpful. Certain peptides such as the µO- and δ-conotoxins are much less soluble and require the addition of organic solvents such as dimethyl sulfoxide (DMSO) or acetonitrile to achieve higher micromolar stock concentrations.

B. Storage

Conus peptides are most stable in lyophilized form. For transport over a few days, they can be safely shipped at room temperature. For longer periods they should be stored at −20 or −80 °C. Static charge can cause the lyophilized peptide powder to "fly" out of the test tube. If static is encountered, use of an antistatic gun eliminates the problem. Particularly after transport of peptide, it is wise to centrifuge the container to ensure that peptide will not exit the tube on opening. As a side note, peptides lyophilize in a somewhat unpredictable fashion. Small quantities of peptide lyophilized side by side in a rotary evaporator often appear as either a very visible white powder, or a nearly invisible crystalline substance. The latter can easily be mistaken for "no peptide in the tube" without close inspection.

Peptides solutions can also be stored. For immediate use, solutions are generally kept at room temperature or on ice. For longer storage, solutions are frozen at −20 to −80 °C. With some peptides we have noted decreased activity after repeated freeze–thaw cycles. High-performance liquid chromatography (HPLC) of these peptide solutions suggests that loss of peptide in solution, rather than peptide breakdown, is occurring. To avoid this, we routinely make aliquots of solutions such that a given aliquot will not need to be thawed more than two or three times prior to consumption.

We often store peptides in HPLC elution buffer consisting of 0.1% trifluoroacetic acid (TFA) and acetonitrile/H_2O. We have found that with long-term storage, however, some peptides (e.g., α-conotoxin EI) undergo degradation, which is consistent with deamination as measured by mass spectrometry. We presume that this is secondary to the acidic pH, and therefore avoid long-term storage under these conditions.

C. Nonspecific Adsorption

Many *Conus* peptides are hydrophobic in nature and have a tendency to "stick" to glassware and plasticware. At nanomolar concentrations and below, this can lead to significant changes in solution concentration of peptide. To avoid this, we often add 0.1 mg/ml lysozyme or 0.1–1.0 mg/ml bovine serum albumin (BSA) to the solution.

Lyophilization of small quantities of peptide (<1 nmol) can lead to significant loss of peptide to container walls. We have found that the addition of carrier protein (e.g., 10–50 µg of lysozyme) to the solution prior to lyophilization largely circumvents this problem. Conodipine-M (McIntosh *et al.*, 1995a,b) (a phospholipase A_2 from *Conus magus*) is a particularly striking example. The apparent IC_{50} shifts by two orders of magnitude to the right without the utilization of carrier protein.

The use of carrier protein is not always sufficient to prevent nonspecific adsorption, particularly at low peptide concentrations. We have found, for example, that static bath application of α-conotoxins to *Xenopus* oocyte recording chambers leads to an apparent 10-fold decrease in potency compared to preparations where the solution is applied as a continuous flow (Harvey *et al.*, 1997).

Radioiodinated *Conus* peptides may be particularly sticky. We routinely siliconize (Sigmacote, Sigma, St. Louis, MO) pipette tips and test tubes (including the caps) when using iodinated peptides and assess radioactivity after solution transfer (e.g., pipette tips) using a gamma counter. We also gamma count final reaction tubes as a measure of true radioactivity concentration. Iodinated peptides may also stick to dust particles introduced into solution, for example, by pipette tips. This can lead to scatter of signal in receptor binding assays. To avoid this, stock solutions of radiolabeled peptide are centrifuged (e.g., in an Eppendorf microfuge) to pellet such particles prior to solution use.

References

Bandyopadhyay, P. K., Colledge, C. J., Walker, C. S., Zhou, L. M., Hillyard, D. R., and Olivera, B. M. (1997). *J. Biol. Chem.* submitted.

Bean, B. P. (1989). *Annu. Rev. Physiol.* **51**, 367.

Cartier, G. E., Yoshikami, D., Gray, W. R., Luo, S., Olivera, B. M., and McIntosh, J. M. (1996). *J. Biol. Chem.* **271**, 7522.

Catterall, W. A. (1992). *Physiol. Rev.* **72**, S15.

Cruz, L. J., Gray, W. R., Olivera, B. M., Zeikus, R. D., Kerr, L., Yoshikami, D., and Moczydlowski, E. (1985). *J. Biol. Chem.* **260**, 9280.

Dunlap, K., Luebke, J. I., and Turner, T. J. (1995). *Trends Neurosci.* **18**, 89.

England, L. J., Imperial, J., Jacobsen, R., Craig, A. G., Gulyas, J., Rivier, J., Julius, D., and Olivera, B. M. (1997). "Seratonin Symposium" San Francisco, CA.

Fainzilber, M., Gordon, D., Hasson, A., Spira, M. E., and Zlotkin, E. (1991). *Eur. J. Biochem.* **202**, 589.

Fainzilber, M., Hasson, A., Oren, R., Burlingame, A. L., Gordon, D., Spira, M. E., and Zlotkin, E. (1994a). *Biochemistry* **33**, 9523.

Fainzilber, M., Kofman, O., Zlotkin, E., and Gordon, D. (1994b). *J. Biol. Chem.* **269**, 2574.

Fainzilber, M., Lodder, J. C., Kits, K. S., Kofman, O., Vinnitsky, I., Van Rietschoten, J., Zlotkin, E., and Gordon, D. (1995a). *J. Biol. Chem.* **270**, 1123.

Fainzilber, M., Nakamura, T., Gaathon, A., Lodder, J. C., Kits, K. S., Burlingame, A. L., and Zlotkin, E. (1995b). *Biochemistry* **34**, 8649.

Fainzilber, M., van der Schors, R., Lodder, J. C., Li, K. W., Geraerts, W. P., and Kits, K. S. (1995c). *Biochemistry* **34**, 5364.

Fainzilber, M., Lodder, J. C., van der Schors, R. C., Li, K. W., Yu, Z., Burlingame, A. L., Geraerts, W. P., and Kits, K. S. (1996). *Biochemistry* **35**, 8748.

Filloux, F., Schapper, A., Naisbitt, S. R., Olivera, B. M., and McIntosh, J. M. (1994). *Dev. Brain Res.* **78**, 131.

Gonoi, T., Ohizumi, Y., Nakamura, H., Kobayashi, J., and Catterall, W. A. (1987). *J. Neurosci.* **7**, 1728.

Gray, W. R., Luque, A., Olivera, B. M., Barrett, J., and Cruz, L. J. (1981). *J. Biol. Chem.* **256**, 4734.

Groebe, D. R., Dumm, J. M., Levitan, E. S., and Abramson, S. N. (1995). *Mol. Pharmacol.* **48**, 105.

Haack, J. A., Rivier, J., Parks, T. N., Mena, E. E., Cruz, L. J., and Olivera, B. M. (1990). *J. Biol. Chem.* **265**, 6025.

Hann, R. M., Pagán, O. R., and Eterovic, V. A. (1994). *Biochemistry* **33**, 14058.

Hann, R. M., Pagán, O. R., Gregory, L. M., Jácome, T., and Eterovic, V. A. (1997). *Biochemistry* **36**, 9051.

Harvey, S. C., McIntosh, J. M., Cartier, G. E., Maddox, F. N., and Luetje, C. W. (1997). *Mol. Pharmacol.* **51**, 336.

Hess, P. (1990). *Annu. Rev. Neurosci.* **13**, 337.

Hidaka, Y., Sato, K., Nakamura, H., Kobayashi, J., Ohizumi, Y., and Shimonishi, Y. (1990). *FEBS Lett.* **1**, 29.

Hillyard, D. R., Monje, V. D., Mintz, I. M., Bean, B. P., Nadasdi, L., Ramachandran, J., Miljanich, G., Azimi-Zoonooz, A., McIntosh, J. M., Cruz, L. J., Imperial, J. S., and Olivera, B. M. (1992). *Neuron* **9**, 69.

Hopkins, C., Grilley, M., Miller, C., Shon, K. J., Cruz, L. J., Gray, W. R., Dykert, J., Rivier, J., Yoshikami, D., and Olivera, B. M. (1995). *J. Biol. Chem.* **270**, 22361.

Jacobsen, R., Yoshikami, D., Ellison, M., Martinez, J., Gray, W. R., Cartier, G. E., Shon, K., Groebe, D. R., Abramson, S. N., Olivera, B. M., and McIntosh, J. M. (1997). *J. Biol. Chem.* **36**, 22531.

Johnson, D. S., Martinez, J., Elgoyhen, A. B., Heinemann, S. S., and McIntosh, J. M. (1995). *Mol. Pharmacol.* **48**, 194.

Kehoe, J., Spira, M., and McIntosh, J. M. (1996). *Soc. Neurosci.* **22**, 267.

Kreienkamp, H. J., Sine, S. M., Maeda, R. K., and Taylor, P. (1994). *J. Biol. Chem.* **269**, 8108.

Kulak, J. M., Nguyen, T. A., Olivera, B. M., and McIntosh, J. M. (1997). *J. Neurosci.* **17**, 5263.

Luo, S., Yoshikami, D., Cartier, G. E., Jacobsen, R., Olivera, B. M., and McIntosh, J. M. (1997). *J. Neurosci. Abst.* **23**, 384.

Martinez, J. S., Olivera, B. M., Gray, W. R., Craig, A. G., Groebe, D. R., Abramson, S. N., and McIntosh, J. M. (1995). *Biochemistry* **34**, 14519.

McIntosh, J. M., Cruz, L. J., Hunkapiller, M. W., Gray, W. R., and Olivera, B. M. (1982). *Arch. Biochem. Biophys.* **218**, 329.

McIntosh, J. M., Olivera, B. M., Cruz, L. J., and Gray, W. R. (1984). *J. Biol. Chem.* **259**, 14343.

McIntosh, J. M., Yoshikami, D., Mahe, E., Nielsen, D. B., Rivier, J. E., Gray, W. R., and Olivera, B. M. (1994). *J. Biol. Chem.* **269**, 16733.

McIntosh, J. M., Ghomashchi, F., Gelb, M. H., Dooley, D. J., Stoehr, S. J., Giordani, A. B., Naisbitt, S. R., and Olivera, B. M. (1995a). *J. Biol. Chem.* **270**, 3518.

McIntosh, J. M., Hasson, A., Spira, M. E., Li, W., Marsh, M., Hillyard, D. R., and Olivera, B. M. (1995b). *J. Biol. Chem.* **270**, 16796.

Mena, E. E., Gullak, M. F., Pagnozzi, M. J., Richter, K. E., Rivier, J., Cruz, L. J., and Olivera, B. M. (1990). *Neurosci. Lett.* **118**, 241.

Miller, R. J. (1987). *Science* **235**, 46.

Moczydlowski, E., Olivera, B. M., Gray, W. R., and Strichartz, G. R. (1986). *Proc. Natl. Acad. Sci. USA* **83**, 5321.

Monje, V. D., Haack, J., Naisbitt, S., Miljanich, G., Ramachandran, J., Nasdasdi, L., Olivera, B. M., Hillyard, D. R., and Gray, W. R. (1993). *Neuropharmacology* **32**, 1141.

Myers, R. A., River, J., and Olivera, B. M. (1990). *J. Neurosci.* **16**, 958.

Myers, R. A., Zafaralla, G. C., Gray, W. R., Abbott, J., Cruz, L. J., and Olivera, B. M. (1991). *Biochemistry* **30**, 9370.

Nishiuchi, Y., and Sakakibara, S. (1982). *FEBS Lett.* **148**, 260.

Nishiuchi, Y., Kumagaye, K., Noda, Y., Watanabe, T. X., and Sakakibara, S. (1986). *Biopolymers* **25**, 561.

Olivera, B. M., McIntosh, J. M., Cruz, L. J., Luque, F. A., and Gray, W. R. (1984). *Biochemistry* **23**, 5087.

Olivera, B. M., Gray, W. R., Zeikus, R., McIntosh, J. M., Varga, J., Rivier, J., de Santos, V., and Cruz, L. J. (1985). *Science* **230,** 1338.

Olivera, B. M., Cruz, L. J., de Santos, V., LeCheminant, G., Griffin, D., Zeikus, R., McIntosh, J. M., Galyean, R., Varga, J., Gray, W. R., and Rivier, J. (1987). *Biochemistry* **26,** 2086.

Olivera, B. M., Rivier, J., Clark, C., Ramilo, C. A., Corpuz, G. P., Abogadie, F. C., Mena, E. E., Woodward, S. R., Hillyard, D. R., and Cruz, L. J. (1990). *Science* **249,** 257.

Olivera, B. M., Miljanich, G., Ramachandran, J., and Adams, M. E. (1994). *Ann. Rev. Biochem.* **63,** 823.

Pereira, E. F. R., Alkondon, M., McIntosh, J. M., and Albuquerque, E. X. (1996). *J. Pharmacol. Exp. Ther.* **278,** 1472.

Ramilo, C. A., Zafaralla, G. C., Nadasdi, L., Hammerland, L. G., Yoshikami, D., Gray, W. R., Kristipati, R., Ramachandran, J., Miljanich, G., Olivera, B. M., and Cruz, L. J. (1992). *Biochemistry* **31,** 9919.

Rigby, A. C., Baleja, J. D., Furie, B. C., and Furie, B. (1997). *Biochemistry* **36,** 6906.

Sato, S., Nakamura, H., Ohizumi, Y., Kobayashi, J., and Hirata, Y. (1983). *FEBS Lett.* **155,** 277.

Shon, K. J., Hasson, A., Spira, M. E., Cruz, L. J., Gray, W. R., and Olivera, B. M. (1994). *Biochemistry* **33,** 11420.

Shon, K., Grilley, M. M., Marsh, M., Yoshikami, D., Hall, A. R., Kurz, B., Gray, W. R., Imperial, J. S., Hillyard, D. R., and Olivera, B. M. (1995). *Biochemistry* **34,** 4913.

Shon, K., Grilley, M., Jacobsen, R., Cartier, G. E., Hopkins, C., Gray, W. R., Watkins, M., Hillyard, D. R., Rivier, J., Torres, J., Yoshikami, D., and Olivera, B. M. (1997a). *Biochemistry* (in press).

Shon, K., Stocker, M., Terlau, H., Stühmer, W., Jacobsen, R., Walker, C., Grilley, M., Watkins, M., Hillyard, D. R., Gray, W. R., and Olivera, B. M. (1997b). *J. Biol. Chem.* (in press).

Skjaebaek, N., Nielsen, K. J., Lewis, R. J., Alewood, P., and Craik, D. J. (1997). *J. Biol. Chem.* **272,** 2291.

Snutch, T. P. (1992). *Curr. Biol.* **2,** 247.

Stephan, M. M., Potts, J. F., and Agnew, W. S. (1994). *J. Membr. Biol.* **137,** 1.

Tavazoie, S. F., Tavazoie, M. F., McIntosh, J. M., Olivera, B. M., and Yoshikami, D. (1996). *Br. J. Pharmacol.* **120,** 995.

Terlau, H., Shon, K., Grilley, M., Stocker, M., Stühmer, W., and Olivera, B. M. (1996a). *Nature* **381,** 148.

Terlau, H., Stocker, M., Shon, K., McIntosh, J. M., and Olivera, B. M. (1996b). *J. Neurosci.*

Tsien, R. W., Lipscombe, D., Madison, D. V., Bley, K. R., and Fox, A. P. (1988). *Trends Neurosci.* **11,** 431.

Tsien, R. W., Ellinor, P. T., and Horne, W. A. (1991). *Trends Pharmacol. Sci.* **12,** 349.

Ullian, E. M., McIntosh, J. M., and Sargent, P. B. (1997). *J. Neurosci.* **17,** 7210.

Utkin, Y. N., Kobayashi, F. H., and Tsetlin, V. I. (1994). *Toxicon* **32,** 1153.

White, H. S., McCabe, R. T., Abogadie, F., Torres, J., Rivier, J. E., Paarmann, I., Hollmann, M., Olivera, B. M., and Cruz, L. J. (1997). *J. Neurosci. Abst.* **23,** 2164.

Woodward, S. R., Cruz, L. J., Olivera, B. M., and Hillyard, D. R. (1990). *EMBO J.* **1,** 1015.

Zafaralla, C. G., Ramilo, C., Gray, W. R., Karlstrom, R., Olivera, B. M., and Cruz, L. J. (1988). *Biochemistry* **27,** 7102.

CHAPTER 26

Scorpion Toxins as Tools for Studying Potassium Channels

Maria L. Garcia, Markus Hanner, Hans-Günther Knaus, Robert Slaughter, and Gregory J. Kaczorowski

Membrane Biochemistry & Biophysics
Merck Research Labs R80N-C31
Rahway, New Jersey, USA

I. Introduction

   Ion channels play a fundamental role in the control of cell excitability. Thus, their activity is largely involved in modulating the contractility of muscle cells, and in the release of hormones and neurotransmitters from endocrine and neuronal cells. Of all the families of ion channels, K^+ channels represent the largest and most diverse group of proteins. Gating of these proteins occurs through conformational changes that are controlled by voltage and/or ligand binding. Therefore, K^+ channels can be broadly divided into two groups: voltage-dependent and ligand-activated channels. A number of techniques have become available in the past few years for studying ion channel structure and function. Electrophysiology allows the determination of biophysical parameters that are inherent to each individual ion channel. With the use of molecular biology, a large amount of information regarding the structure and existence of subfamilies of K^+ channels has become available due to molecular cloning of cDNAs encoding these proteins (Wei *et al.*,

DOI: 10.1016/B978-0-12-382204-8.00026-6

1996). However, two major questions are still the subject of further investigation. These concern the molecular composition of given channels as expressed *in vivo*, as well as the physiologic role that channels play in cell function. To explore these questions, high-affinity, selective modulators for a given channel must be found. Progress in this area has occurred due to the discovery of high-affinity peptidyl inhibitors of K^+ channels in the venom of different organisms such as scorpions, snakes, bees, spiders, and sea anemones (Garcia *et al.*, 1997). These peptidyl inhibitors have been useful in the development of the pharmacology of K^+ channels, and have also been employed in binding reactions as a marker during the purification of channels from native tissues. This has allowed the determination of a channel's subunit composition, as well as identification of specific auxiliary subunits of K^+ channels. These auxiliary proteins are very important for channel function because they cause profound effects on both the biophysical and pharmacological properties of the pore-forming subunit. In addition, due to their well-understood mechanism of action, peptidyl inhibitors of K^+ channels have allowed the identification and molecular characterization of the pore-forming region of these proteins, and determination of subunit stoichiometry.

Scorpion venoms constitute a rich source of peptidyl inhibitors of K^+ channels (Garcia *et al.*, 1997). These peptides typically consist of 37–39 amino acids and display significant sequence homology. They usually contain six Cys residues that form three disulfide bridges, thus providing the peptide with a very rigid structure. Production of significant quantities of these peptides can be accomplished by either solid-phase synthesis or recombinant techniques. This has allowed determination of the three-dimensional solution structure of some of these peptides by nuclear magnetic resonance (NMR) techniques. The overall structures so far obtained are very conserved and possess an α-helical region that is linked by disulfide bonds to a three-strand antiparallel β sheet. Knowledge of the three-dimensional structure of these peptides, combined with site-directed mutagenesis, has allowed identification of those residues critical for binding to the channel pore, and has provided a picture of the interaction surface with K^+ channels. Importantly, all residues that are crucial for interaction with the channel are located on one face of the β sheet, while the α-helical region does not make contact with the channel directly. Complementary mutagenesis targeting residues in the channel pore has then allowed derivation of models describing shape and dimensions of the receptor in the vestibule of the channel.

Peptidyl inhibitors from scorpion venoms can be subdivided into different groups based on sequence conservation and specificity (Garcia *et al.*, 1997) (Fig. 1). Members of the first group inhibit Ca^{2+}-activated K^+ channels, although only iberiotoxin (IbTX) and limbatustoxin (LbTX) are specific for the high-conductance Ca^{2+}-activated K^+ channel. All other groups inhibit voltage-gated K^+ channels. However, only members of the KV1 family are sensitive to inhibition; KV2-, KV3-, and KV4-type channels are not blocked by these inhibitors. Within the KV1 family, only KV1.1, 1.2, 1.3, 1.6, and 1.7 are targets of these peptides; KV1.4 and 1.5 are refractory to inhibition.

I	ChTX	Z F T N V S C T T S K E C W S V C Q R L H N T S R G K C M N K K C R C Y S
	IbTX	Z F T D V D C S V S K E C W S V C K D L F G V D R G K C M G K K C R C Y Q
	Lq₂	Z F T Q E S C T A S N Q C W S I C K R L H N T N R G K C M N K K C R C Y S
	LbTX	V F I D V S C S V S K E C W A P C K A A V G T D R G K C M G K K C K C Y ...
II	NxTX	T I I N V K C T S P K Q C S K P C K E L Y G S S A G A K C M N G K C K C Y N N
	MgTX	T I I N V K C T S P K Q C L P P C K A Q F G Q S A G A K C M N G K C K C Y P H
	C.l.l. I	I T I N V K C T S P Q Q C L R P C K D R F G Q H A G G K C I N G K C K C Y P ...
	TyKα	V F I N A K C R G S P E C L P K C K E A I G K A A G K C M N G K C K C Y P
III	AgTX₁	G V P I N V K C T G S P Q C L K P C K D A G M R F G K C I N G K C H C T P K
	AgTX₂	G V P I N V S C T G S P Q C I K P C K D A G M R F G K C M N R K C H C T P K
	AgTX₃	G V P I N V P C T G S P Q C I K P C K D A G M R F G K C M N R K C H C T P K
	KTX	G V E I N V K C S G S P Q C L K P C K D A G M R F G K C M N R K C H C T P K
	KTX₂	V R I P V S C K H S G Q C L K P C K D A G M R F G K C M N G K C D C T P K

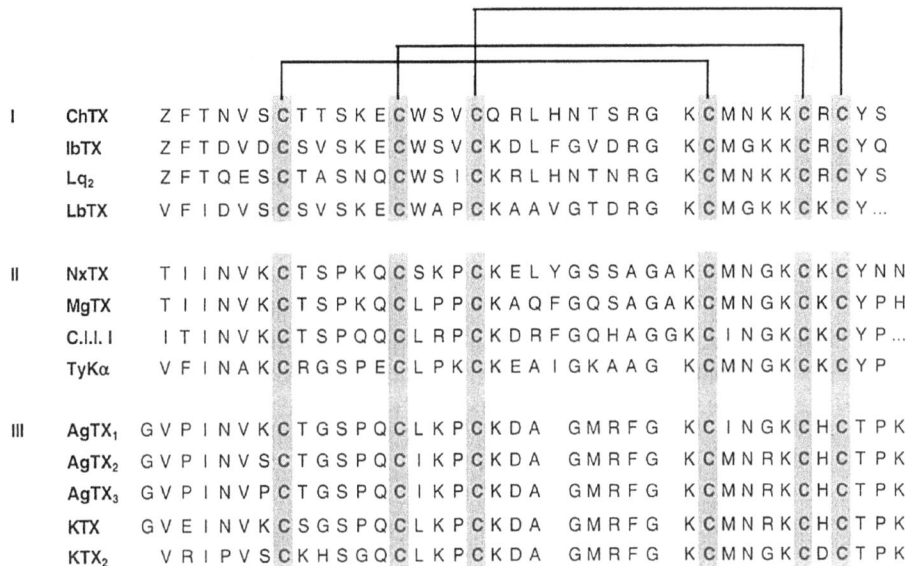

Fig. 1 Comparison of the amino acid sequences of charybdotoxin (ChTX), iberiotoxin (IbTX), *Leiurus Centruvoides quinquestriatus* toxin 2 (Lq2), limbatustoxin (LbTX), noxius-toxin (NxTX), margatoxin (MgTX), *C. limpidus limpidus* toxin I (*C.l.l.* 1), tityustoxin-Kα (TyKα), agitoxin 1 (AgTX₁), agitoxin 2 (AgTX₂), agitoxin 3 (AgTX₃), kaliotoxin (KTX), and kaliotoxin 2 (KTX₂). The sequences have been aligned with respect to the six cysteine residues which are in boldface type. The position of the disulfide bonds is indicated.

In this review, we discuss methods for (1) purifying peptides from venom sources, (2) synthesis of the peptides by recombinant techniques, (3) radio-labeling of peptides in biologically active form, and (4) use of radiolabeled peptides to characterize high-affinity receptors in native tissue.

II. Purification of Peptidyl Inhibitors of K⁺ Channels from Scorpion Venoms

Scorpion venoms contain a vast number of components with activities directed against ion channels. For example, they constitute a rich source of peptidyl Na⁺ channel modulators (Catterall, 1980). A distinguishing feature of K⁺ channel inhibitors is the presence of a large number of positively charged residues in these peptides. Thus, it is possible to take advantage of this property to achieve a simple, efficient, and highly reproducible purification procedure. For all peptides that have been purified in our laboratory [charybdotoxin (ChTX), IbTX, LbTX margatoxin (MgTX), agitoxin 1–3 (AgTX₁₋₃)], we have employed two consecutive chromatographic steps: cation-exchange chromatography on a Mono S column

(Pharmacia, Piscataway, NJ), and reversed-phase chromatography on a C_8 or C_{18} column (Galvez *et al.*, 1990; Garcia *et al.*, 1994; Garcia-Calvo *et al.*, 1993; Gimenez-Gallego *et al.*, 1988; Novick *et al.*, 1991). This has always afforded the production of pure material as judged by amino acid sequence, amino acid composition, and mass spectroscopic analyses.

The first step in peptide purification involves separation of venom components on a Mono S HR5/5 or HR 10/10 column, depending on the amount of material being processed, using a high-performance liquid chromatography (HPLC) system. Lyophilized venom is initially resuspended in 20 mM sodium borate, pH 9.0, at a final protein concentration of ca. 5 mg/ml using vortex agitation. The suspension is then subjected to centrifugation for 15 min at $27,000 \times g$ at 4 °C to remove insoluble material. Before applying the soluble material onto the Mono S column, the sample must be filtered through a Millex-GV 0.2-μm pore size filter (Millipore, Bedford, MA) to remove particulate material that could cause obstruction in the HPLC system, thereby increasing back pressure of the column. We do not recommend subjecting the pellet obtained after initial centrifugation to a second extraction. Even after processing the soluble material as described earlier, we frequently observe an increase in the pressure of the system, up to the limit tolerated by the column. If this occurs, the direction of the column must be reversed with continued pumping of buffer until the pressure decreases to normal values.

Before injecting the sample, the ion-exchange column should be equilibrated with 20 mM sodium borate, pH 9.0, at a flow rate of 0.5 ml/min (HR 5/5 column) or 2 ml/min (HR 10/10 column) depending on the Mono S column employed. The sample is then applied and absorbance monitored at 280 nm. Because of the basic pH conditions, a large amount of material absorbed at 280 nm is not retained by the column and appears in the void volume. We have never observed any biological activity in this material against the K^+ channels tested and, therefore, it can be discarded. Once the absorbance returns to baseline values, the retained material is eluted with a linear gradient of NaCl in 20 mM sodium borate, pH 9.0; 0.75 M/h for HR 5/5 or 0.5 M/h for HR 10/10 columns. Individual peaks are collected manually and used for determining their biological activity. These fractions can be stored at −70 °C until further processing is accomplished.

The second step of purification is achieved by using a C_{18} reversed-phase HPLC column (25×0.46-cm, 5-μm particle size, or 25×1-cm, 5-μm particle size, depending on the amount of material to be processed; the Separations Group). In some cases, we have also employed a C_8 reversed-phase HPLC column and have obtained nearly identical results. The column is equilibrated with 10 mM trifluoroacetic acid (TFA) at a flow rate of either 0.5 ml/min (25×0.46-cm column) or 3 ml/min (25×1-cm column), depending on the size of the column/amount of material to be loaded. Because most of the peptides are highly charged and do not contain many hydrophobic surfaces in their native conformation, it is very important to have the column well equilibrated in starting buffer. Failure to do this may lead to lack of retention of peptides by the column. Fractions of interest from the Mono S column are then directly applied, without further processing, to the reversed-phase column. Elution is achieved with a linear gradient of organic solvent. We have successfully employed a combination of 2-propanol/acetonitrile (2:1) in 4 mM TFA, although acetonitrile by itself is also an appropriate solvent for elution. The gradient can be applied from 0% to 40% over either a 30-min period (25 × 0.46-cm column) or a 60-min period (25 × 1-cm column) depending on column size. For best results, it is better to monitor absorbance of eluting material, at least at two different wavelengths (e.g., 280 and 235 nm). Some peptides have little or no aromatic amino acid content and, therefore, give very small or sometimes undetectable absorbance at 280 nm. Components are separated manually and, in most cases, well-defined peaks with good baseline separation can be obtained.

For testing biological activity, it is suggested that small amounts of material be subjected to lyophilization, and then reconstituted in any buffer containing high ionic strength (e.g., 100 mM NaCl). We also suggest including 0.1% (w/v) bovine serum albumin (BSA) in the resuspension buffer to prevent loss of peptide due to binding to glass or plastic surfaces. Remaining material can be stored at −70 °C. Once a fraction of interest is identified, and because only small amounts of material are needed for most experiments, it is desirable to lyophilize samples in small aliquots of ca. 20 μg and store them at −70 °C. Material stored in this way is usually stable for very long periods of time. When an aliquot is needed, the peptide is resuspended as indicated above and can be stored at 4 °C for several months without loss of biological activity. We do not routinely subject toxin solutions to repetitive freeze–thaw cycles as we have not investigated such procedures with respect to toxin stability. It is important to note that K⁺ channel peptides are typically highly positively charged molecules that will stick to glass surfaces unless a high ionic strength buffer is used to prevent such an interaction from taking place. This is particularly important when drying the peptides in glass tubes. If water is used as the sole resuspension agent, it is likely that most of the peptides will be lost by absorption onto the surface of the tube.

Although some material may appear to be chromatographically pure, this does not necessarily mean that it is homogeneous. For example, two different peptide entities may elute together, or a major absorbance peak may contain a component that does not have significant absorbance at the wavelength monitored. To resolve the latter situation, it is helpful to monitor the absorbance of eluting material at two different wavelengths. If more than one component has eluted together, it is often possible to ascertain that this situation exists by applying the material again to the reversed-phase column, and selecting more shallow gradient elution conditions. Despite all of these precautions, it is still necessary to characterize the material of interest in terms of its amino acid composition, amino acid sequence, and mass spectroscopic properties. When performing automated Edman degradation using most commercially available instruments, it is important to be aware that the last residue of the peptide is usually washed off the filter support, and this amino acid must, therefore, be determined by an independent means. Thus, it is important that the amino acid composition be determined after acid hydrolysis of the peptide, and that it matches the composition obtained by sequence. Furthermore, both these parameters must correlate well with results obtained by mass spectroscopy.

Finally, it is imperative that a peptide of interest be synthesized to confirm that the amino acid sequence determined corresponds to the biological activity of interest. This can be accomplished by either of two independent methods: solid-phase synthesis or biosynthesis by recombinant techniques. This latter approach is discussed in Section 3. The solid-phase synthesis of some K^+ channel inhibitory peptides has been accomplished, and fully reduced peptides have been oxidized to yield material that is indistinguishable from samples purified from crude scorpion venom (Aiyar *et al.*, 1995; Bednarek *et al.*, 1994; Drakopoulou *et al.*, 1995; Johnson and Sugg, 1992; Sugg *et al.*, 1990). This has allowed confirmation of the identities of some peptides, and has further provided a means by which to obtain different variants of these agents for structure–activity relationship studies.

III. Synthesis of K^+ Channel Inhibitory Peptides by Recombinant Techniques

The production of K^+ channel inhibitory peptides by recombinant techniques has become a very popular approach, not only for obtaining large quantities of material at relatively low cost, but also for producing peptide variants with which to carry out mechanistic studies in order to identify those residues important for channel inhibition (Aiyar *et al.*, 1995; Goldstein *et al.*, 1994; Ranganathan *et al.*, 1996; Stampe *et al.*, 1994). In general terms, a gene encoding the peptide of interest is inserted into an *Escherichia coli* expression vector. The peptide is produced as part of a fusion protein, folded, cleaved, and then purified to homogeneity by conventional methods.

The first step of this process consists of constructing an artificial gene for the peptide of interest, and inserting it into an appropriate expression vector. We have successfully employed two types of vectors. In pCSP105, the resulting construct encodes a fusion protein of the viral T7 gene 9 product with the K^+ channel inhibitory peptide, where the two proteins are separated by either a Factor X_a cleavage site, or an enteropeptidase site. In pG9, six histidine residues are inserted between the fusion protein and the Factor X_a cleavage site, at the beginning of the K^+ channel inhibitory peptide sequence (Fig. 2). This latter construct may facilitate purification of the fusion protein through application of Ni^{2+}-affinity chromatographic techniques. However, in our experience, this type of chromatographic step appears to work only with some constructs, and does not represent a very significant advantage in isolation of the fusion protein.

An appropriate strain of *E. coli*, such as BL21(DE3), is then transformed with the corresponding plasmid and a single colony is selected to inoculate the culture

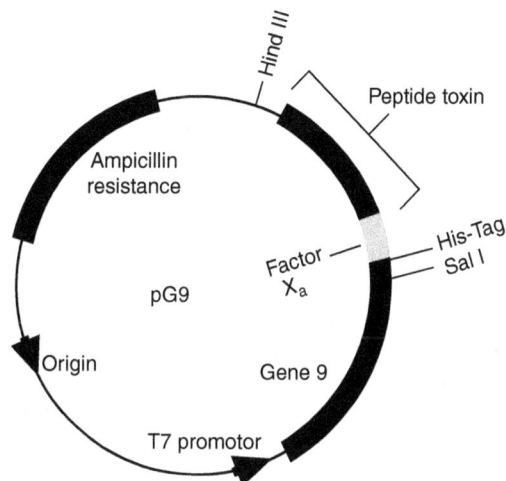

Fig. 2 Design of the synthetic peptide gene. The plasmid map of K$^+$ channel peptides shows the location of the synthetic peptide gene, the Factor X$_a$ cleavage site, the His tag, and T7 gene 9 fusion protein.

medium. The cells are grown at 30–37 °C with shaking, and when the optical density of the culture suspension at 650 nm has reached 0.6–0.7, the culture is induced with 0.5 mM isopropylthiogalactoside (IPTG), followed by further incubation for 3 h at 37 °C. Induction of the fusion protein can be assayed by employing 12% SDS–PAGE gels and subsequently staining these with Coomassie blue. After destaining, a major product at ca. 57,000 should be observed on testing the induced culture. If no induction of fusion protein is seen, one should check for proper orientation of the toxin gene and make sure that the cells do not grow above 0.7 optical density units, before addition of IPTG. If induction is successful, cells are pelleted by centifugation and washed once with 50 mM NaCl, 10 mM Tris–HCl, pH 8.0, 1 mM ethylenediaminetetraacetic acid (EDTA), and 1 mM dithiothreitol (DTT). Cells are then resuspended in 10 ml/l of culture using the buffer described earlier. Cells are frozen in liquid N$_2$ and can be stored at −70 °C until further processing. In our experience, one person can easily manage the purification of up to four different peptides at once. For this reason, we store cells at −70 °C until sufficient material is ready for purification. Cells are thawed on ice and the volume is adjusted to 50 ml/l of culture with 50 mM NaCl, 10 mM Tris–HCl, pH 8.0, 1 mM EDTA, 1 mM DTT, 100 μM PMSF (phenylmethylsulfonyl fluoride), and 0.5 mg/ml lysozyme. After incubation on ice for 1 h, the sample is subjected to sonication four times at 1-min intervals, and then subjected to centrifugation at 27,000 × g for 15 min. The supernatant is collected and the nucleic acids are removed by precipitation with streptomycin sulfate. It is recommended that at each step, samples be subjected to 12% SDS–PAGE analysis to monitor purification progress.

The next step involves purification of the fusion protein by ion-exchange chromatography. For this purpose, the supernatant is loaded onto a DEAE-Sepharose fast flow column equilibrated with 50 mM NaCl, 10 mM Tris–HCl, pH 8.0, 1 mM EDTA, 1 mM DTT, and 100 μM PMSF. After extensive washing with equilibration buffer, the fusion protein is batch eluted with 350 mM NaCl, 10 mM Tris–HCl, pH 8.0, 1 mM EDTA, 1 mM DTT, and 100 μM PMSF, followed by dialysis overnight at 4 °C against 100 mM NaCl, 20 mM Tris–HCl, pH 8.0, and 0.5 mM 2-mercaptoethanol. During the dialysis step, folding of the peptide (i.e., formation of disulfide linkages) takes place.

Purified fusion protein must undergo enzymatic cleavage in order to release the K^+ channel inhibitory peptide. To accomplish this, either of two procedures can be employed. In one of these, Factor X_a is used to release the peptide from the fusion protein. For this purpose, $CaCl_2$ is added to a final concentration of 3 mM, followed by addition of 1 μg of Factor X_a per milligram of fusion protein, and the digestion mixture is incubated overnight at room temperature. The time course of digestion can be followed by SDS–PAGE, but in most cases an overnight incubation is sufficient. We have found this procedure to give reproducible results for given constructs. However, with some peptides, Factor X_a causes a cleavage upstream from its recognition site. We interpret this occurrence as the result of either of two different phenomena: (1) inability of Factor X_a to reach its primary site of action (most likely due to steric factors), and/or (2) lack of specificity of Factor X_a. Indeed, if the released peptide containing a part of the fusion protein is purified and subsequently subjected to further incubation with Factor X_a, no further digestion is observed, suggesting that the conformation of the peptide prevents the enzyme from reaching its binding site.

A different and less expensive way of achieving release of the K^+ channel inhibitory peptide from its fusion protein is through the use of trypsin. For this purpose, $CaCl_2$ is added to a final concentration of 5 mM, and the sample is incubated with 5 μg trypsin/mg of fusion protein for 30 min at room temperature. In this case, time of incubation with enzyme is critical as longer incubations could lead to cleavage within the peptide. This procedure, however, works well with all constructs tested so far, and it is less expensive than those protocols involving Factor X_a. Even in those situations where Factor X_a produced wrong cleavage of the fusion protein, it is possible to treat the resulting sample with trypsin to release the appropriate peptide.

The K^+ channel inhibitory peptides must next be purified from the rest of the digestion mixture. This can be accomplished, as described in the previous section, by loading the sample onto a Mono S HR 10/10 column equilibrated with 20 mM sodium borate, pH 9.0. The fusion protein and various proteolytic fragments are not retained by the column under these conditions. Highly positively charged peptides, such as ChTX, MgTX, noxiustoxin (NxTX), and $AgTX_{1-3}$, can be loaded directly onto the column, whereas less positively charged peptides, such as IbTX, must first be dialyzed against 20 mM sodium borate, pH 9.0, before loading. Bound peptides are eluted in the presence of a linear gradient of NaCl in

20 mM sodium borate, pH 9.0 (0.5 M/h) at a flow rate of 2 ml/min. A major absorbance peak representing the peptide of interest is usually observed. Certain peptides, such as ChTX and IbTX, have their amino-terminal group cyclized in the form of pyroglutamine. Because cyclization is important for the biological activity of these peptides, this reaction must be accomplished before the last purification step. For this purpose, HPLC-grade acetic acid is added to the Mono S sample at a final concentration of 5%, and the mixture is incubated at either 65 °C (ChTX) or 45 °C (IbTX) until full cyclization of the N-terminal amino group is achieved. Cyclization is followed by N-terminal sequencing, as the peptide species with the blocked N-terminal is resistant to Edman degradation. For ChTX, cyclization takes place during an overnight incubation with acetic acid, whereas for IbTX, this process may take up to 2 days, perhaps due to the lower incubation temperature used. We have found that at higher temperatures, IbTX is not stable in the presence of acetic acid, as indicated by the appearance of different chromatographic species with time.

The final stage of peptide purification is achieved on reversed-phase chromatography, using a semipreparative C_{18} reversed-phase HPLC column, employing the same conditions as those described in the previous section. The composition of the purified peptide should be checked by N-terminal sequencing, amino acid hydrolysis, and/or mass spectroscopic analysis. Production of peptides through recombinant techniques is an easy way of obtaining large quantities of material at low cost. The yields of purified peptide per liter of *E. coli* culture vary depending on the construct, but it is highly reproducible for any given peptide. For instance, 10–15 mg of MgTX per liter of culture can routinely be obtained, whereas for ChTX the yields are lower: 1–5 mg of peptide per liter of culture. We believe that these differences in yield are related to the efficiency of the cleavage step as the amounts of purified fusion protein produced are similar in all cases.

IV. Radiolabeling of K^+ Channel Inhibitory Peptides

The aim of radiolabeling K^+ channel inhibitory peptides is to obtain a biologically active derivative that can be used to identify the target protein in tissues of interest, and develop the molecular pharmacology of given K^+ channels. Radiolabeled peptides at high specific activity can be produced after reaction of peptidyl Tyr or His residues with $Na^{125}I$. Alternatively, Lys residues can be covalently modified by reaction with ^{125}I-labeled Bolton–Hunter reagent. Because certain Lys residues in K^+ channel inhibitory peptides are crucial for their activity (Goldstein *et al.*, 1994; Park and Miller, 1992), this latter method has not been successfully employed to date. Labeling of Tyr residues in ChTX and MgTX has been successful and, thus, these radiolabeled peptides have played a crucial role in characterizing K^+ channels (Knaus *et al.*, 1995; Vazquez *et al.*, 1989). Other peptides such as $AgTX_{1-3}$ do not contain Tyr residues, or, as in the case of IbTX, iodination of the Tyr residue leads to complete loss of biological activity.

In these situations, it is possible to engineer amino acid residues into the sequence so that they can be subsequently modified by incorporation of a radiolabeled tag at a position that is not critical for biological activity (Aggarwal and MacKinnon, 1996; Knaus *et al.*, 1996; Koschak *et al.*, 1997; Shimony *et al.*, 1994).

Iodination of peptides such as ChTX, MgTX, and IbTX-D19Y/Y36F has been achieved using either of two methods: the Iodo-Gen (Pierce, Rockford, IL) or glucose oxidase/lactoperoxidase protocols. The latter utilizes beads to which the enzymes have been immobilized. Use of this system usually yields chromatograms displaying fewer reaction side products, due to the milder oxidizing condition of the procedure. Unfortunately, the manufacturer of these beads, Bio-Rad (Richmond, CA), has discontinued their production. In the reaction with Iodo-Gen, the water-insoluble reagent is immobilized on the surface of a glass vial. A solution of peptide and $Na^{125}I$ is added, and after a certain period of time, the reaction mixture is removed and injected onto a reversed-phase HPLC column for purification of the radiolabeled peptide.

For a typical iodination, a solution of Iodo-Gen reagent is prepared in acetone and an aliquot corresponding to 0.5–1.0 μg of reagent is placed at the bottom of a vial and dried under argon or nitrogen. Once dried, the vial can be rinsed several times with buffer to remove any aqueous soluble contaminants. Then, a solution containing 10–20 μg of peptide in 100 mM sodium phosphate, pH 7.3–8.0, is added, followed by 2–5 μCi of $Na^{125}I$ (2200 Ci/mmol). After mixing, the vial is capped and the reaction is allowed to proceed at room temperature for 10–15 min. One minute before the end of the reaction, the vial is opened, the iodination mixture is removed with a syringe, and it is injected onto a C_{18} reversed-phase HPLC column that has been equilibrated with 10 mM TFA. All these procedures should be carried out inside a hood with appropriate ventilation; special care should be taken when loading the iodination mixture onto the HPLC column because the low pH buffer employed could lead to the generation of radioactive iodine gas, which will permeate the skin and accumulate in the thyroid gland. For this reason, the loading and washing volumes from the HPLC column are always collected in 1 N NaOH to block formation of iodine gas. HPLC column elution conditions will depend on the peptide under investigation, and it is recommended that the iodination reaction be performed with unlabeled NaI first, in order to optimize the separation conditions. With ChTX, for instance, a linear gradient of 2-propanol/acetonitrile 2:1 (5–14%, 40 min) at a flow rate of 0.5 ml/min leads to a well-defined and well-separated peak corresponding to monoiodotyrosine ChTX, based on the specific activity of the radiolabeled peptide and on sequence analysis of the modified toxin. Because the amount of material used in these procedures is low, changes in absorbance should be monitored at 210 nm. To store radiolabeled peptides for extended periods of time, the fraction of interest is made 0.1% (w/v) in BSA, lyophilized, and reconstituted with 100 mM NaCl, 20 mM Tris–HCl, pH 7.4. Aliquots of this material containing ca. 5 μCi are frozen in liquid N_2 and stored at -70 °C. When needed, fractions are thawed and then stored at 4 °C. These handling conditions have been found to be optimal in that no loss of biological activity of the labeled peptide has been observed, even after several months of storage. In most cases, in addition to the monoiodotyrosine derivative, a second minor peak corresponding to the diiodotyrosine peptide is obtained. We usually do not retain this material because in many cases its biological activity is undefined.

An alternative method for introducing a radiolabel into these peptides makes use of an analog in which a Cys residue has been placed at a position in the peptide that is not critical for bioactivity. The free sulfydryl is then reacted with [^{3}H]N-ethyl-maleimide (NEM) to produce material with full biological activity, but at lower specific activity than that obtained with Na^{125}I labeling (e.g., ca. 60 Ci/mmol). This procedure has been successfully employed to radiolabel ChTX and IbTX making use of position 19 (Koschak $et\ al.$, 1997; Shimony $et\ al.$, 1994), and with AgTX$_1$ and AgTX$_2$ at position 20 (Aggarwal and MacKinnon, 1996). Note that this type of reaction is not restricted solely to radiolabeling of the peptide; other fluorescent and biotinylated derivatives of NEM can also be reacted to yield interesting peptidyl ion channel probes. However, there is one consideration to note with this approach which concerns the fact that a newly introduced Cys residue is not initially reactive on purification of the peptide, as it does not exist as a free sulfhydryl group. Instead, it exists in a disulfide bridge with either 2-mercaptoethanol or another toxin molecule. It is necessary, therefore, to selec-tively reduce this disulfide bond without disrupting the integrity of the natural disulfide bridges of the peptide that are crucial for biological activity before labeling with the sulfhydryl reagent. To accomplish this, peptides are incubated in 50 mM sodium phosphate, pH 7.0, with 1–10 mM DTT at room temperature for 1 h. The reaction mixture is then applied to a C_{18} reversed-phase HPLC column. Elution conditions must be optimized for each particular peptide preparation. The peptide containing the free sulfhydryl group should be easily identified by a change in retention time when compared with unreduced material (Fig. 3A). The change in retention time may not be very large and, therefore, appropriate elution conditions should be used to achieve good separation. In the event that the separation is suboptimal, this should not preclude continuation of the procedure as unreduced material will not react with NEM. In our experience, optimal recovery of C_{19}/C_{20} reduced peptide approximates 50%. Although either higher concentrations of DTT or increases in pH can lead to full disappearance of starting material (Fig. 3B), they also cause appearance of other forms of the peptide with much longer retention times in which additional disulfide bonds have been reduced. This material displays no biological activity and can be discarded. Reduced peptide can be concentrated to about 20 μl using a Speed-Vac, and then 100 μl of 50 mM sodium phosphate is added. It is important to use reduced peptide for reaction with [^{3}H]NEM as soon as possible after its production. We have observed that, in some cases, reduced peptide can dimerize via formation of a disulfide linkage, even when stored at $-70\,^{\circ}$C.

For the alkylating reaction, one vial containing 1 mCi of [^{3}H]NEM in pentane is opened and the content transferred to a glass tube. The solvent is evaporated under N$_2$ to about 300 μl, and then 200 μl of 50 mM sodium phosphate, pH 7.0, is added.

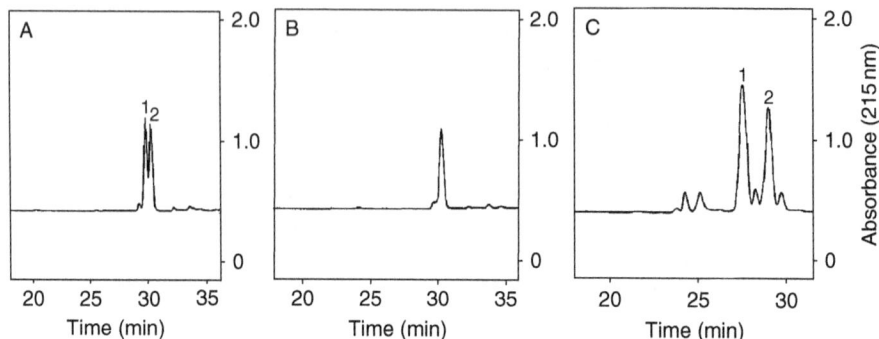

Fig. 3 Synthesis of [^{3}H]IbTX-D19C-NEM. An aliquot of IbTX-D19C in 50 mM sodium phosphate, pH 7.0, was treated with 2 mM (A) or 10 mM (B) DTT for 1 h at room temperature. Peaks 1 and 2 in (A) with retention times of 29.8 and 30.2 min, respectively, represent the unreduced and reduced forms at position 19 of IbTX-D19C. In (B) the retention time of the reduced peptide is 30.3 min. (C) Reduced IbTX-D19C prepared as in (B) was reacted with 1 mCi [^{3}H]NEM (56 Ci/mmol) for 1 h at 37 °C. Samples were loaded onto a C$_{18}$ reversed-phase column equilibrated with 10 mM TFA. Elution was achieved in the presence of a linear gradient (7–25%, 30 min) of 2-propanol/acetonitrile (2:1) in 4 mM TFA. Peaks 1 and 2 with retention times of 27.5 and 28.9 min, respectively, have identical specific activity and their biological activity is also indistinguishable.

After mixing by vortex, the remaining pentane is evaporated, the [^{3}H]NEM solution is added to the peptide solution, and the vial is rinsed with an additional 200 μl of 50 mM sodium phosphate, pH 7.0, which is then also transferred to the peptide solution. The reaction mixture is allowed to react for 1 h at 37 °C, and then injected onto a C$_{18}$ reversed-phase HPLC column equilibrated with 10 mM TFA. Elution conditions should be optimized for each individual peptide, and alkylated material can be easily identified by the change in retention time. In addition, radioactivity should be associated with this material. If good separation is achieved, the specific activity of the alkylated peptide should correspond to that of the [^{3}H]NEM employed for the reaction, ca. 60 Ci/mmol. It is worth noting that in the case of reaction of IbTX-C$_{19}$ with [^{3}H]NEM, two well-separated radiolabeled peaks are obtained in equivalent amounts, and both display identical biological activity (Fig. 3C). The exact nature of these two species is unknown, but this may correspond to the production of two isomers of IbTX-C$_{19}$-NEM.

V. Receptor Binding Studies

The aim of radiolabeling a given K$^+$ channel inhibitory peptide is to use it to characterize its receptor in native tissues. This includes characterizing the kinetics of ligand association and dissociation, the molecular pharmacology of the channel, and the tissue distribution of the channel (Knaus *et al.*, 1994, 1995; Koschak *et al.*, 1997; McManus *et al.*, 1993; Vazquez *et al.*, 1989, 1990). Additionally, radiolabeled peptides have been successfully employed in the purification of native ion channels (Garcia-Calvo *et al.*, 1994; Giangiacomo *et al.*, 1995; Parcej and Dolly, 1989; Rehm and Lazdunski, 1988). As a matter of fact, all auxiliary (*β*) subunits of ion channels have been identified and subsequently characterized through the purification of the corresponding ion channel complex using binding of channel probes as a monitor of channel purification. There are different ways in which to monitor peptide–K^+ channel interactions, such as autoradiographic techniques or binding of ligand to intact cells or tissues, but perhaps the most common studies are those in which the peptide interaction is measured in highly enriched membrane preparations.

Three major considerations should be taken into account when considering the development of a binding reaction involving K^+ channel inhibitory peptides discussed in this chapter. The first one concerns the potencies of the peptides as inhibitors of K^+ channels; under physiologic conditions, some of these peptides display nanomolar affinities, but their potency can be greatly enhanced by selecting appropriate binding reaction conditions (i.e., by lowering the ionic strength of the incubation medium, which will enhance electrostatic interaction between positively charged residues on the peptide and negatively charged residues in the channel's vestibule (Candia *et al.*, 1992; Giangiacomo *et al.*, 1992; MacKinnon and Miller, 1988). Second, problems arising from the handling of these peptides should be considered. The easiest way to separate bound from free ligands is to use filtration techniques and, therefore, the filters used must have specific characteristics, or be treated accordingly to minimize binding of the positively charged peptides. We have found that GF/C glass fiber filters (Whatman, UK) presoaked in 0.5–1.0% (w/v) polyethyleneimine (Sigma, St. Louis, MO) provide a low background binding support if high ionic strength buffer (e.g., 100 mM NaCl, 20 mM Tris–HCl, pH 7.4) is used to quench the binding reaction. This high ionic strength quench buffer is also important to induce immediate dissociation of toxin bound to low-affinity sites. We should also consider the physical properties of the test tubes in which the binding reaction is carried out. If low ionic strength media are used, then glass tubes should be avoided because peptides may adhere to these surfaces; under such conditions, use of polystyrene tubes is recommended. In addition, inclusion of 0.1% (w/v) BSA in the incubation medium would also prevent peptide from being absorbed onto the surface of tubes. Finally, the last consideration is the source of the receptor. For instance, ChTX blocks two different types of K^+ channels: high-conductance Ca^{2+}-activated K^+ channels, and voltage-dependent K^+ channels. If one considers using this peptide as a ligand, it is important to select a tissue in which either, but not both, of these channels is present. Moreover, use of highly enriched plasma membrane preparation should give a higher ratio of specific to nonspecific binding and should be employed if at all feasible.

Other parameters which are optimal for each particular ligand–receptor interaction should be determined according to each individual situation. For example, the incubation time necessary to reach equilibrium for any particular ligand can vary considerably from about 1 h for [^{125}I]MgTX or [^{125}I]ChTX (Knaus *et al.*, 1995; Vazquez *et al.*, 1990) to close to 60–72 h for [^{125}I]IbTX (Koschak *et al.*, 1997).

VI. Summary

The search for peptidyl inhibitors of K^+ channels is a very active area of investigation. In addition to scorpion venoms, other venom sources have been investigated; all these sources have yielded novel peptides with interesting properties. For instance, spider venoms have provided peptides that block other families of K^+ channels (e.g., KV2 and KV4) that act via mechanisms which modify the gating properties of these channels (Sanguinetti *et al.*, 1997; Swartz and MacKinnon, 1995, 1997a,b). Such inhibitors bind to a receptor on the channel that is different from the pore region in which the peptides discussed in this chapter bind (Swartz and MacKinnon, 1997a,b). In fact, it is possible to have a channel occupied simultaneously by both inhibitor types. It is expected that many of the methodologies concerning peptidyl inhibitors from scorpion venom, which have been developed in the past and outlined above, will be extended to the new families of K^+ channel blockers currently under development.

References

Aggarwal, S. K., and MacKinnon, R. (1996). *Neuron* **16**, 1169.

Aiyar, J., Withka, J. M., Rizzi, J. P., Singleton, D. H., Andrews, G. C., Lin, W., Boyd, J., Hanson, D. C., Simon, M., Dethlefs, B., Lee, C. L., Hall, J. E., *et al.* (1995). *Neuron* **15**, 1169.

Bednarek, M. A., Bugianesi, R. M., Leonard, R. J., and Felix, J. P. (1994). *Biochem. Biophys. Res. Commun.* **198**, 619.

Candia, S., Garcia, M. L., and Latorre, R. (1992). *Biophys. J.* **63**, 583.

Catterall, W. A. (1980). *Ann. Rev. Pharmacol. Toxicol.* **20**, 15.

Drakopoulou, E., Cotton, J., Virelizier, H., Bernardi, E., Schoofs, A. R., Partiseti, M., Choquet, D., Gurrola, G., Possani, L. D., and Vita, C. (1995). *Biochem. Biophys. Res. Commun.* **213**, 901.

Galvez, A., Gimenez-Gallego, G., Reuben, J. P., Roy-Contancin, L., Feigenbaum, P., Kaczorowski, G. J., and Garcia, M. L. (1990). *J. Biol. Chem.* **265**, 11083.

Garcia, M. L., Garcia-Calvo, M., Hidalgo, P., Lee, A., and MacKinnon, R. (1994). *Biochemistry* **33**, 6834.

Garcia, M. L., Hanner, M., Knaus, H. G., Koch, R., Schmalhofer, W., Slaughter, R. S., and Kaczorowski, G. J. (1997). *Adv. Pharmacol.* **39**, 425.

Garcia-Calvo, M., Leonard, R. J., Novick, J., Stevens, S. P., Schmalhofer, W., Kaczorowski, G. J., and Garcia, M. L. (1993). *J. Biol. Chem.* **268**, 18866.

Garcia-Calvo, M., Knaus, H. G., McManus, O. B., Giangiacomo, K. M., Kaczorowski, G. J., and Garcia, M. L. (1994). *J. Biol. Chem.* **269**, 676.

Giangiacomo, K. M., Garcia, M. L., and McManus, O. B. (1992). *Biochemistry* **31**, 6719.

Giangiacomo, K. M., Garcia-Calvo, M., Knaus, H. G., Mullmann, T. J., Garcia, M. L., and McManus, O. (1995). *Biochemistry* **34**, 15849.

Gimenez-Gallego, G., Navia, M. A., Reuben, J. P., Katz, G. M., Kaczorowski, G. J., and Garcia, M. L. (1988). *Proc. Natl. Acad. Sci. USA* **85**, 3329.

Goldstein, S. A. N., Pheasant, D. J., and Miller, C. (1994). *Neuron* **12**, 1377.

Johnson, B. A., and Sugg, E. E. (1992). *Biochemistry* **31**, 8151.

Knaus, H. G., McManus, O. B., Lee, S. H., Schmalhofer, W. A., Garcia-Calvo, M., Helms, L. M. H., Sanchez, M., Giangiacomo, K., Reuben, J. P., Smith, A. B., III, Kaczorowski, G. J., and Garcia, M. L. (1994). *Biochemistry* **33**, 5819.

Knaus, H. G., Koch, R. O. A., Eberhart, A., Kaczorowski, G. J., Garcia, M. L., and Slaughter, R. S. (1995). *Biochemistry* **34**, 13627.

Knaus, H. G., Schwarzer, C., Koch, R. O. A., Eberhart, A., Kaczorowski, G. J., Glossmann, H., Wunder, F., Pongs, O., Garcia, M. L., and Sperk, G. (1996). *J. Neurosci.* **16,** 955.

Koschak, A., Koch, R. O., Liu, J., Kaczorowski, G. J., Reinhart, P. H., Garcia, M. L., and Knaus, H. G. (1997). *Biochemistry* **36,** 1943.

MacKinnon, R., and Miller, C. (1988). *J. Gen. Physiol.* **91,** 335.

McManus, O. B., Harris, G. H., Giangiacomo, K. M., Feigenbaum, P., Reuben, J. P., Addy, M. E., Burka, J. F., Kaczorowski, G. J., and Garcia, M. L. (1993). *Biochemistry* **32,** 6128.

Novick, J., Leonard, R. J., King, V. F., Schmalhofer, W., Kaczorowski, G. J., and Garcia, M. L. (1991). *Biophys. J.* **59,** 78a.

Parcej, D. N., and Dolly, J. O. (1989). *Biochem. J.* **257,** 899.

Park, C. S., and Miller, C. (1992). *Neuron* **9,** 307.

Ranganathan, R., Lewis, J. H., and MacKinnon, R. (1996). *Neuron* **16,** 131.

Rehm, H., and Lazdunski, M. (1988). *Proc. Natl. Acad. Sci. USA* **85,** 4919.

Sanguinetti, M. C., Johnson, J. H., Hammerland, L. G., Kelbaugh, P. R., Volkmann, R. A., Saccomano, N. A., and Mueller, A. L. (1997). *Mol. Pharmacol.* **51,** 491.

Shimony, E., Sun, T., Kolmakova-Partensky, L., and Miller, C. (1994). *Prot. Eng.* **7,** 503.

Stampe, P., Kolmakova-Partensky, L., and Miller, C. (1994). *Biochemistry* **33,** 443.

Sugg, E. E., Garcia, M. L., Reuben, J. P., Patchett, A. A., and Kaczorowski, G. J. (1990). *J. Biol. Chem.* **265,** 18745.

Swartz, K. J., and MacKinnon, R. (1995). *Neuron* **15,** 941.

Swartz, K. J., and MacKinnon, R. (1997a). *Neuron* **18,** 665.

Swartz, K. J., and MacKinnon, R. (1997b). *Neuron* **18,** 675.

Vazquez, J., Feigenbaum, P., Katz, G., King, V. F., Reuben, J. P., Roy-Contancin, L., Slaughter, R. S., Kaczorowski, G. J., and Garcia, M. L. (1989). *J. Biol. Chem.* **264,** 20902.

Vazquez, J., Feigenbaum, P., King, V. F., Kaczorowski, G. J., and Garcia, M. L. (1990). *J. Biol. Chem.* **265,** 15564.

Wei, A., Jegla, T., and Salkoff, L. (1996). *Neuropharmacology* **35,** 805.

CHAPTER 27

Use of Planar Lipid Bilayer Membranes for Rapid Screening of Membrane-Active Compounds

Tajib A. Mirzabekov, Anatoly Y. Silberstein, and Bruce L. Kagan

Department of Psychiatry
UCLA, Los Angeles
California, USA

I. Update

Over the past decade, there has been an exponential increase in interest in the use of high-throughput screening techniques for exploring large-scale chemical libraries for membrane-active compounds and channel-blocking compounds. While the planar lipid bilayer chamber that we have described has received continued attention, several new platforms for high-throughput screening of channels embedded in patch-clamp membranes have been developed and used quite widely. These techniques were first pioneered by large pharmaceutical firms interested in using the power of their extensive chemical libraries to search for molecules that blocked or modified known ion channels. The standard patch-clamp electrophysiology technique was quite time and labor intensive and, thus, it was worthwhile to invest in automating the platform for high-throughput screening. This has been relatively successfully accomplished and the apparatus for performing this kind of high-throughput screening is now available commercially (Jones *et al.*, 2009).

In the planar bilayer field, parallel recording of a large number of planar bilayers has begun to be described only in the past couple of years. This includes a high-throughput lipid membrane platform (Poulos *et al.*, 2009), a 96-well planar lipid bilayer chip (Suzuki *et al.*, 2009), and a planar microelectrode cavity array for high-resolution and parallel electrical recording (Baaken *et al.*, 2008; Le Pioufle *et al.*, 2008). These approaches are quite new and have yet to be widely disseminated throughout the field. However, they show the widespread interest in adapting high-throughput screening techniques to the planar lipid bilayer electrophysiology.

A number of parameters will need to be defined with these new techniques. Most specifically, it will need to be established that they reproduce the planar lipid bilayer conditions faithfully and measurably. Furthermore, experiments on known channels will be necessary to establish the validity of these techniques. Nonetheless, they provide an exciting first step toward a future in which large chemical libraries can be screened and large libraries of samples can be screened for channel forming activities. These new techniques will likely help move the electrophysiology field into a stage long ago accomplished by biochemistry where a large number of samples could be operated on in parallel. The large amounts of data and the resulting analysis will require adaptation of existing analytical programs to handle these large amounts of data and also to reconcile the inherent variability in electrical recordings. Combining the power of being able to record single ion channels with the power being able to record from a large number of single ion channels at once will enhance the speed of progress in the field as a whole. It is still uncertain as to which of these techniques will be proved to be most reliable, most adaptable, and most useful to investigators in the field but the appearance of several new techniques at once suggests that the time is right for planar lipid bilayers to join other types of biochemical assays in the world of high-throughput screening.

Part B: How to avoid pitfalls

The planar lipid bilayer technique is plagued by a number of easily avoided traps.

• Always check and standardize electrical recording equipment. It is important to ensure that the electrical recording apparatus is recording accurately and faithfully.

• Noise reduction is critical. When measuring currents of picoampere magnitude, small amounts of noise can disrupt your signal easily. Attention to shielding, grounding and minimization of vibration are all well-established techniques for noise reduction.

• It is important to establish the conductance and capacitance of the planar membrane once it is formed and before adding external proteins or peptides. Frequent problems and artifacts can arise in the use of membranes that have not completely thinned or that possess microlenses or thickened tori. These problems can usually be easily avoided by monitoring capacitance as the membrane forms. Optical monitoring of the membrane is useful as well but the capacitance measurement is more accurate in determining the presence of a bimolecular thin membrane.

II. Introduction

Planar lipid bilayer membranes (BLMs) have been in use for more than four decades for the study of membrane-active compounds (Mueller *et al.*, 1962). BLMs provide a unique environment that allows the assay of the functional activity of carriers and channels translocating ions across membrane (Kagan and Sokolov, 1994; Miller, 1986). Although BLMs are *in vitro* systems, which have been used extensively for the measurement of biophysical properties of carriers and channels, their relevance to *in vivo* phenomena has been demonstrated repeatedly through the more recent technology of patch clamping (Sakmann and Neher, 1983) which has shown that channel properties measured in BLMs correspond qualitatively and often quantitatively to those observed in whole cells. In this review we describe a new BLM perfusion chamber that allows rapid, reversible, and quantitative assay of compounds on the bilayers.

One of the major advantages of the *in vitro* bilayer system is its remarkable sensitivity in that the ionic current flowing through a single channel can be readily observed. Furthermore, the opening and closing of this single ionic channel can be easily detected and modification of the channel properties by voltage, pH, ionic composition, blockers, mutation, and chemical reagents can also be quantified. This sensitivity also confers specificity on the assay, because identifiable channel properties, such as single-channel conductance, channel lifetime, voltage dependence, and ionic selectivity, can serve as a "fingerprint" to identify uniquely a given channel.

A second major advantage of BLMs is their freedom from many potential sources of confounding error. The system has a well-defined set of components including water, salts, buffers, lipids, and organic solvent, making it feasible to observe the effects of the introduction of new proteins and peptides into the BLM. In the absence of added membrane-active material, the system is quite predictably dull, showing little permeability to ions, no voltage dependence, and no interesting transport behavior. However, the introduction of peptide or protein channel-forming agents that spontaneously incorporate into the bilayer dramatically alters the situation. The system is readily manipulated to study the transport properties of these agents. Furthermore, a wide variety of ionic compositions, lipids, temperatures, and pH conditions can be employed to study the biophysical properties of the channels in question over a range not usually possible in physiological situations. This allows important biophysical questions to be addressed in addition to the more traditional physiologic questions.

The stability of BLMs is also an important asset in their utility. Early on, BLMs were shown to be capable of lasting as long as 120 days, the typical lifetime of a red blood cell, and it is common for BLMs to have long lifetimes even in the presence of added channels. However, it is clear that some membrane-active compounds do weaken BLM stability and can eventually lead to BLM destruction. Indeed, it is likely that for many membrane-active toxic compounds their toxic mechanism involves the disruption of membrane integrity (Kagan *et al.*, 1990; Lin *et al.*, 1997; Mirzabekov *et al.*, 1996).

Another attractive feature of BLMs is their simplicity. The number of defined parameters is small. The electrical properties are predictable and well defined, and interpretation of data is often straightforward.

BLMs do have some well-known shortcomings. The assay for channel activity in BLMs is often poorly reproducible due to a variety of factors including geometric shape of the chamber, conformation of the BLM, accessibility of the BLM, presence of a variable unstirred layer directly in front of the BLM, variability of BLM surface area, and the presence of "microlenses" of organic solvent. These factors combine to make it difficult sometimes to observe reproducibly the same quantitative channel activity for a given concentration of protein or peptide in the solution. Even with the most efficient channel formers, only a small fraction of the added protein incorporates into the membrane (approximately 1 in 10^5 molecules) (Kagan *et al.*, 1981; Schein *et al.*, 1978). This makes BLMs difficult to use as an assay for ion channel activity and has hindered their use in the purification of channels and of putative channel activities from various sources.

The poor stability of BLMs in response to hydrostatic pressure gradients is also a limiting factor. This can limit the ability to add or subtract volume from the solution, and can also hinder the ability to change the solution bathing the membrane without breaking the membrane. Furthermore, it can slow the speed at which solutions can be exchanged in the aqueous compartments. The slowness of solution changes often required can limit the ability to look at rapid transitions in the functional status of channels.

Noise is frequently a problem in BLM recording. The issues of electrical, vibrational, mechanical, and other noise have been discussed in detail elsewhere (Alvarez, 1986; Mueller *et al.*, 1962). Noise reduction is frequently an important goal of chamber and membrane design, so that smaller currents can be measured on faster time scales.

The requirement for organic solvent in the BLM may alter membrane and channel properties from the *in vivo* situation. "Solvent-free" BLMs dramatically reduce the amount of solvent present, but are significantly less stable (Montal and Mueller, 1972).

III. A New Bilayer Membrane System

In this chapter we present a new BLM system including chamber, electrodes, and membrane supports for rapid screening of new compounds for membrane activity. The system has several advantages over older designs: (1) Solutions can be exchanged rapidly and completely. (2) Membranes are stable to these solution changes. (3) The volumes employed are quite small, which allows sample size to be in the microliter range. Thus, repeated measurements can be made on material that is in very short supply. (4) The use of solution changes allows the reversibility of sample effects to be tested handily. (5) Because the system allows complete volume exchange, multiple samples can be tested for activity on a single membrane. Of course, once activity is obtained, a new membrane must be formed, but inactive samples can be simply washed out and the next sample tested on the same membrane. This system also allows positive controls with known channel formers to be used to calibrate and quantify activity. (6) Active compounds may be applied to the membrane in high concentrations (without dilution).

The chamber is also extremely easy to use. It comes apart quite rapidly, and is made from inexpensive materials. It is easily and efficiently cleaned without the use of volatile organic solvents. The system is also remarkably quiet due to the fact that the membrane itself has very little solution surface area, and thus is largely insulated from a great deal of electrical and vibrational noise that normally interferes with recording. The electrodes are made from common materials and easily customized to the requirements of the chamber. Virtually any type of standard electronic recording devices such as voltage clamps, current clamps, patch clamp amplifiers, etc., can be used with this system. Examples of studies in which this new system can be used include (1) screening fractions of protein preparations for channel activity, (2) screening for channel blockers, (3) screening peptides for channel activity, and (4) screening pharmaceuticals.

IV. Planar Lipid Membrane Setups and Chambers

A. Perfusion Setup for Painted Membranes

The formation of painted, or solvent-containing, BLMs in a rapid perfusion chamber has been described in limited detail (Mirzabekov *et al.*, 1993; Silberstein, 1989). The BLM setup used was specifically designed to allow perfusion of one chamber in a few seconds. The *cis* chamber of 30- to 50-μl total volume was a 2-mm-diameter tube bored longitudinally into a rectangular Lucite block (10 × 25 × 30 mm). The *trans* chamber was a 2-mm-diameter cylinder bored perpendicularly to the *cis* chamber (Fig. 1). The *trans* chamber was fitted with a 15-mm-long Teflon tube (with an inner diameter of 0.25, 0.5, or 1 mm; outer diameter, 2 mm) on which the membrane was formed. A flexible plastic (Tygon) tube was fitted over the Teflon tube to connect the electrode. A plastic threaded sleeve fitted to the Teflon tube and the hole allowed easy fixation of the Teflon tube to the chamber. On insertion of the *trans* electrode, the *trans* side became a closed volume, decreasing

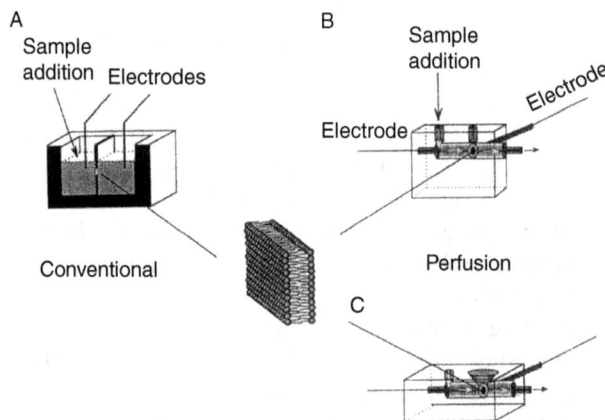

Fig. 1 Chambers for the conventional and perfusion planar membrane setups. (A) Traditional planar membrane chamber. Membrane here is made on an aperture in the Teflon film separating two 0.3- to 2-ml volumes, *cis* and *trans* sides of the chamber. Samples of compounds are added as stocks and have to be concentrated because of the increase in volume of the *cis* or *trans* side of chamber where the compound is added. Increases in the volume due to addition of a large volume of diluted compound can result in a hydrostatic pressure gradient on the membrane leading to membrane rupture. Magnetic stirrers are needed on both sides of the chamber for mixing of added compounds. Because of unstirred layers bordering the membrane, added compounds reach the membrane only after a few (3–5) minutes. Each new compound must be added to a cleaned chamber with a newly formed membrane. (B) Perfusion chamber for "painted" membranes. Membranes were made from heptane solutions of lipids by the use of a pipette inserted into the right vertical hole to the level of the hole in the Teflon tubing (black circle). Samples were premixed with a salt solution at the necessary concentration and dropped in the left vertical hole. Solution dropped out was collected in a reservoir. Membranes that had a slight concave shape were immediately accessible for the additives, and if the added solution contained membrane-active compounds they reacted with the membrane immediately (in 1–3 s). New compounds were added until a membrane-active compound was found. Then compound-free solution was added, the membrane was reformed, and screening was continued. (C) Perfusion planar membrane system for the solvent-free membrane or membranes made from native membrane vesicles. The lipid dissolved in hexane is gently spread to the surface of the right side conical hole. Membrane is made 20–30 min later after evaporation of hexane. In the case where membranes were made from vesicles or liposomes, 20–30 μl of them were added to the same hole and incubated for 30 min for formation of lipid (or lipid–protein) monolayer on the surface of the solution. The membrane was made by gentle lowering and raising of the salt solution surface level across the Teflon tubing hole.

noise and increasing stability to hydrostatic pressure because the water solution is practically noncompressible. The perfusing medium was dropped into the *cis* chamber at one end and flowed out by gravity from the other end. The *cis* chamber was also accessed vertically by three "wells," the center one directly above the *trans* chamber. These wells allow variable points for insertion of sample and escape for air bubbles in the system. The bilayer lipid membrane was formed on an air bubble of a 13–15-mg/ml solution of lipids in *n*-heptane (or *n*-decane) at the end of Teflon tubing with a 0.25-, 0.5-, or 1-mm diameter using 10-μl small pipette tips, the end of which was cut at an angle of 45°. The tip was dipped in a lipid solution, shaken, and then dipped into the solution of the *cis* chamber. The hemisphere of a bubble was pressed out of the end of the tip facing the Teflon tubing. The bubble hemisphere was gently contacted to the Teflon tubing and simultaneously the bubble was gently sucked back by the release of finger pressure on a tube connected to the pipette tip.

Use of *n*-decane as a solvent resulted in formation of membranes that have been suggested to have a smaller percentage of residual solvent in the membrane and less membrane thickness, closer to the thickness of a native membrane. Heptane offers the advantage of more rapid thinning of the membrane to a bilayer. The construction of the chamber allowed substitution of the solution in one (*cis*) compartment within several seconds (Silberstein, 1989). After formation, the brightly colored membrane becomes optically black within 20–30 s. Turning "black" is evidence that the BLM's thickness has become much smaller than the wavelength of visible light. For visual observation of the color change of the membrane we used a stereo microscope in which one eyepiece was replaced by the light source. The light reflected from the membrane was visually detected with the other eyepiece of the microscope. The solution in the *cis* side was replaced with other solutions usually of the same salt composition but containing compounds to be screened for activity. After the initial incorporation of membrane-active ingredients, the newly added solution could be washed out and substituted by the original one. Membranes were typically stable for periods >60 min and had conductances of less than 10 pS up to voltages of ±150 mV. Stability to voltage and membrane longevity depended to some extent on the specific lipids and solutions used.

B. Perfusion Chamber for Solvent-Free Membranes and Membranes Made of Native Membrane Vesicles

We also developed a perfusion planar membrane system for the formation of "solvent-free" membranes (Montal and Mueller, 1972) and membranes made from liposomes or native membrane vesicles (Schindler, 1980). The construction of this chamber is similar to the perfusion chamber for the painted (solvent-containing) planar lipid membranes. Like the solvent-containing membranes, the membrane was made on the end of Teflon tubing. The *trans* chamber became a closed volume after formation of the membrane. This kept the membrane stable and prevented it from breaking under the fast changes of hydrostatic pressure (up to 10–15 mm) on the membrane during salt solution perfusion on the *cis* side. (Note that "solvent-free" membranes formed from monolayers are not usually very sensitive to hydrostatic pressure, whereas solvent-containing BLMs are typically very sensitive.)

The chamber itself was made of Teflon and had a size of 20 × 20 × 25 mm. The 2-mm hole was made through the long axis at 15 mm from the bottom and 12 mm from the front surface (Fig. 1C). Two holes with diameters of 2 and 4 mm were made from the surface down, perpendicular to the horizontal 2-mm-diameter hole (*cis* chamber). These two holes were made at a distance of 7 mm from the right and left edges of the Teflon chamber top and served to allow addition of samples to the chamber and for the formation of membranes, respectively. The planar lipid membrane was formed from the lipid monolayer covering the 4-mm-diameter hole. For formation of the monolayer, 10 μl of 10 mg/ml lipid in hexane was applied to the surface of the salt solution in the 4-mm-hole, which was kept at approximately 2 mm above the end of the hexadecane-coated end of the Teflon tubing. Instead of hexadecane, squalene can also be used. This reduces membrane stability unless divalent cations are added at millimolar concentrations. For precoating with hexadecane, the end of the Teflon tubing was first dipped in a 20-mg/ml solution of hexadecane in pentane and air dried for 10 min to allow full evaporation of pentane. The Teflon tubing was exposed to the 4-mm hole. For membrane formation the salt solution surface near the Teflon tubing was lowered below the end of the Teflon tubing (for formation of one lipid monolayer) and again raised above it (second monolayer). After formation of the membrane, the salt solution level could be varied up to 5–10 mm above the membrane during salt solution perfusion, and the membrane remained stable. During manipulation with the salt solution levels, the hole in the Teflon tubing where the membrane was made was kept under visual observation. A low-magnification (10–12×) stereoscopic microscope with a long focus distance (about 100 mm) containing the light source in place of one eyepiece was used as before.

A rectangular oscillating voltage (1 Hz, 10 or 20 mV) was applied between the electrodes and membrane formation was monitored using capacitance measurements. Membrane capacitance was estimated using the calibration on standard 100- to 1000-pF capacitors. The amplitudes of charging–recharging currents on capacitors were measured at a standard oscillating voltage.

It is clear that when solvent-free membranes were made, the diameter of the membrane was practically equal to the inner diameter of the Teflon tubing. Two kinds of tubing were used in experiments, and they had an inner diameter of 100 and 250 μm. Tubing was purchased from Pharmacia-Upjohn, which sells Teflon tubing as kits for HPLC and FPLC chromatography.

V. Materials

A. Lipids

Lipids are purchased from Avanti Polar Lipids Inc. (Birmingham, AL). The purity of lipids used for planar lipid membrane experiments is essential for the formation of stable membranes with low baseline conductance, <10 pS. All lipids are stored at –20 °C under N_2. Most lipids are stored as 10–20 mg/ml solutions in chloroform. A few [asolectin, diphythanoyl-phosphatidylcholine (DPPC)] are stored as lyophilized powders. For membrane experiments, 1–2 mg of lipid or lipid mixture is dried under a stream of N_2 and dissolved in *n*-heptane at a final concentration of 13–15 mg/ml. In the case of membranes made from lipid mixtures, lipids dissolved in chloroform are mixed first and then dried and dissolved in heptane. It has been previously shown that the lipid composition of actual BLMs corresponds well to the composition of the lipid mixture used for formation of the bilayer (Muller and Finkelstein, 1972). Pure DPPC membranes are very stable and usually are not very sensitive to added channel formers. We use DPPC membranes for the study of channel-forming compounds with high membrane activity, such as porins or gramicidin. Membranes made from asolectin (soybean phosphatide extract, granulated, 45% (w/w) phosphocholine content), a phospholipid mixture containing a high percentage of unsaturated fatty acid hydrocarbon chains, are very sensitive to added compounds, and are used often in the screening experiments. The percentage of lipids with a net negative charge in the membrane as well as the percentage of the sterol (cholesterol) can be varied between 0% and 50%.

B. Electrodes

Ag|AgCl electrodes with agar bridges are used for current measurements on BLMs. The total volumes of both chambers with salt solution bathing the membrane are very small, around 30–50 μl for the *cis* chamber and 10–20 μl for the *trans* chamber. Therefore, it is necessary to use electrodes that have (1) small volumes which are comparable with the volumes of the chambers and (2) agar prepared in salt solutions with the same ionic strength as the solution used in the membrane experiments.

After membrane formation the *trans* side chamber is closed. Therefore, if electrodes have a volume much bigger than the *trans* side, it could lead to two unfavorable results: any small temperature increase could break the membrane due to the thermal expansion of the *trans* side chamber salt solution volume with subsequent "pressing out" of the membrane. Under these conditions it would be difficult to form stable, long-lasting membranes. On the other hand, a small temperature decrease could lead to compression of the *trans* side volume with resulting "sucking in" of the BLM from the end of Teflon tubing inside of it. As a result the BLM could be less accessible to added compounds.

The electrodes we use are simple to prepare and do not suffer from the short-coming mentioned above. Electrodes are made from the lower part of 200-μl micropipette tips filled by 2% agar prepared in the salt solutions used in experiments (usually 100 mM or 1 M KCl). Silver wire (1.5 cm in length) having a 1-mm diameter with one-half electrolytically covered by silver chloride is used to make an electrode. The chloride end is covered by 5–10 thin layers of Parafilm leaving 3–4 mm uncovered at the end of the electrode. The wider end of a 200-μl plastic pipette tip is cut to 15–20 mm of length, filled with hot agar, and the Ag|AgCl electrode pressed into it. The end of the tip on the side of the wire is covered by a couple of Parafilm layers, which prevent the drying of the agar. Electrodes are stored in the same salt solutions as those used for the experiments. One electrode, connected to the amplifier headstage, is inserted into a 10-mm soft silicon tube sleeve over the membrane supporting Teflon tube. A second electrode is inserted into the left end of the *cis* chamber. Electrode asymmetry is always less than 1 mV.

C. Chamber Cleaning

The perfusion chamber is rinsed in a stream of distilled water for 10–15 s, then the Teflon tubing is disconnected, shaken a few times in a chloroform:methanol (2:1, v/v) solution, and dried. For experiments we filter salt solutions through detergent-free antibacterial filters (e.g., Millipore, Bedford, MA; 0.22 mm) although we found this step not to be essential.

D. Recording Equipment

Virtually any standard voltage-clamp recording equipment will work with this chamber. Membrane formation is assessed by monitoring of membrane capacitance and resistance. Data are digitized and stored on VHS tape and played back for later analysis. An Axopatch 1C amplifier (Axon Instruments, Sunnyvale, CA) with head-stage CV-3B is used for measuring membrane current. For data acquisition, a digital tape recorder and video cassette recorder allow recording of large amounts of data. A storage oscilloscope is used for monitoring membrane capacitance and single-channel recordings.

VI. Applications for Perfusion Planar Lipid Membrane Technique

A. Single-Channel Reconstitution

BLMs have been intensively used in the identification and reconstitution of channel-forming proteins (Miller, 1986). The perfusion planar membrane system has several advantages over the traditional membrane system. It is approximately 20–100 times faster. For example, with the new chamber it takes *Borrelia* porin about 5–20 s to reach a steady-state conductance (hundreds of channels). At the same concentration with a standard chamber, we have to wait for 10–30 min to reach a steady-state conductance. The perfusion planar membrane setup requires sample volumes 10–100 times smaller than typical setups. A solution with a membrane-active compound immediately (within 1–3 s) reaches the planar membrane. In Fig. 2, the *cis* solution was substituted by solution containing 10 mg/ml of porin isolated from *Borrelia burgdorferi*. The resulting channel formation stops immediately after washout out of porin-containing solution. This procedure could be repeated many times. This property of the setup allows us to easily reconstitute and study the properties of a single channel in the membrane. Long-lasting single

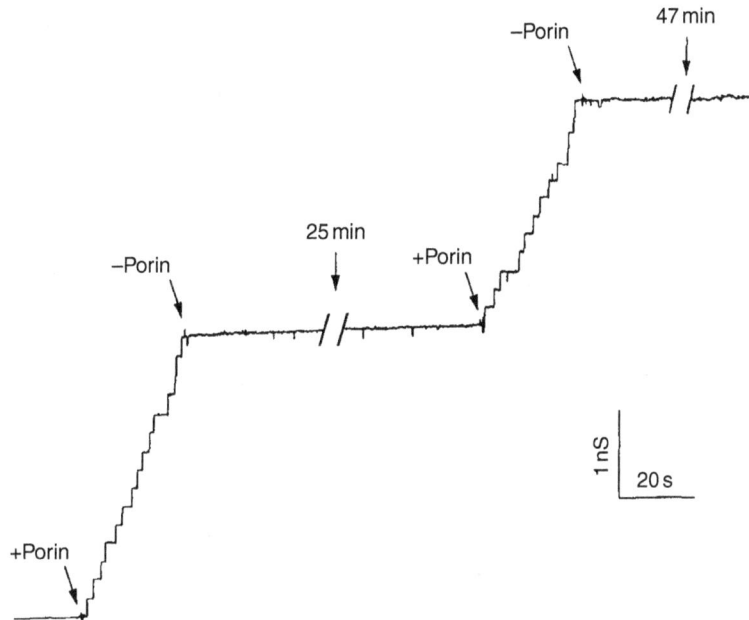

Fig. 2 Rapid initiation and termination of *Borrelia* porin channel activity with the new perfusion chamber. The record shows current as a function of time while the membrane was held at constant voltage (+10 mV). At the first arrow (+Porin), solution containing porin was introduced into the chamber. The conductance begins to increase instantaneously and insertion of single porin channels can be seen. At the second arrow (−Porin), porin-free solution was introduced and the insertion events stop at once. As the rest of the record demonstrates, the insertion of porin channels can be turned on and off rapidly and repeatedly. This immediacy and reversibility, coupled with the small sample volumes needed, make this chamber ideally suited for testing the activity of multiple samples in succession rapidly.

channels of the mitochondrial outer membrane channel, VDAC (voltage-dependent anion channel, also called mitochondrial porin), were easily reconstituted in BLMs made in the perfusion chamber and the transport properties of the channel were studied (Mirzabekov *et al.*, 1993). The new chamber also allows screening of compounds (such as blockers or other modifiers) on reconstituted ion channels as well as studying the reversibility of the interaction of these compounds. By addition of salt solutions at different temperatures or pH, these parameters can also be changed very rapidly. In our experiments we found that planar lipid membranes can be formed and remain stable at temperatures between 4 and 50 °C and in a pH range between pH 3 and 10.

B. Interactions of Amyloidogenic Peptides with Membranes

Deposition of proteins called amyloid proteins is observed in a wide variety of animal and human diseases. Amyloid proteins share certain microscopic and biochemical properties and have recently been implicated in the cell death and tissue pathology underlying these illnesses. Our recent work using the perfusion membrane system has shown that at least three of these amyloid-forming peptides can form ion channels at cytotoxic concentrations, and we have proposed that channel formation is a mechanism of toxicity.

Alzheimer's disease (AD) pathology is characterized by plaques, tangles, and neuronal cell loss. The main constituent of plaques is β-amyloid peptide (Aβ), a 39–42-residue peptide that has been linked to disruption of calcium homeostasis and neurotoxicity *in vitro*. We have demonstrated that a neurotoxic fragment of Aβ, Aβ (25–35) spontaneously inserted in planar lipid membranes to form weakly selective, voltage-dependent, ion-permeable channels. We suggest that channel formation may be involved in the pathogenesis of AD and that Aβ (25–35) may be the active channel-forming segment (Mirzabekov *et al.*, 1994).

Amylin is a 37-amino-acid cytotoxic constituent of amyloid deposits found in the islets of Langerhans of patients with type II diabetes. Extracellular accumulation of this peptide results in damage to insulin-producing β-cell membranes and cell death. We have shown that at cytotoxic concentrations, amylin forms voltage-dependent, relatively nonselective, ion-permeable channels in planar phospholipid BLMs (Mirzabekov *et al.*, 1996).

Prions cause neurodegenerative disease in animals and humans. It has been shown that a 21-residue fragment of the prion protein (106–126) could be toxic to cultured neurons. We found that this peptide forms ion-permeable channels in planar lipid BLMs. These channels are freely permeable to common physiologic ions, and their formation is significantly enhanced by "aging" and/or low pH. We suggest that channel formation is the cytotoxic mechanism of action of amyloidogenic peptides found in prion-related disease (Lin *et al.*, 1997).

C. Search for New Membrane-Active Compounds

A fast and highly sensitive planar membrane setup allows a new approach to the screening and identification of membrane-active compounds in complex organic mixtures of nature and synthetic origin.

It has been shown that the mechanism of microbicidal activity by many antibiotics is based on the formation of ion-conducting channels in the host cell membranes (Kagan *et al.*, 1990; Lin *et al.*, 1997; Schein *et al.*, 1978). Among the classes of membrane-active antibiotics are the polyenes, for example, amphotericin B (Silberstein, 1989); the host defense peptides, for example, defensins (Kagan *et al.*, 1990), magainins (Duclohier, 1994); and the ion channel forming exotoxins, yeast killer toxin (Kagan, 1983), and colicins (Schein *et al.*, 1978). All these compounds, when added to the membrane bathing aqueous solution, interacted with the BLM and induced ion currents at very low concentrations ranging from 10^{-6} to 10^{-12} g/ml. On the other hand, compounds that are not membrane related do not induce any changes in membrane currents at concentrations up to 10^{-3} g/ml. Therefore, the planar lipid membrane technique can identify membrane-active compounds present in a complex mixture even at concentrations as low as

10^{-3}–10^{-9} part of the total. Use of the traditional planar membrane system for the screening of new membrane-active compounds was practically difficult and time consuming. The speed and low sample volume of the perfusion planar membrane technique can be applied to the screening of such mixtures as bacterial metabolites, plant extracts, combinatorial peptide libraries, mutations of known membrane-active antimicrobials, and peptides synthesized on the basis of their predicted membrane-active structure.

Use of the perfusion planar lipid membrane setup in such screening could potentially result in the discovery of new bioactive compounds of bacterial, plant, animal, or synthetic origin (such as antibacterials, antivirals, toxins) in short time periods and at low cost. Automation of this system is also feasible since the same membrane can be used over and over again until an "active" sample is encountered.

D. Identification, Purification, and Characterization of Bacterial Outer Membrane Transport Proteins: Porins

The use of the perfusion planar lipid membrane technique allowed fast identification, isolation, and characterization of transport proteins (porins) in the outer membrane of bacteria. In the Lyme disease-associated spirochete *B. burgdorferi*, we identified, isolated, and characterized three different porin channels (Skare *et al.*, 1995, 1996, 1997). The method was based on solubilization of bacterial membranes, separation of solubilized membrane proteins into fractions (100–150) eluted from an HPLC column followed by immediate screening on the BLM for channel-forming activity (Fig. 2), allowing precise localization of the porin-containing protein peak. Since Triton X-100 solubilized *Borrelia* porin channels lose channel-forming activity within 20–40 h after isolation, this rapidity is essential. This methodology can be easily applied to the isolation and study of membrane proteins of other disease-related bacteria, whose porins or pathogenic toxins might be quite labile.

E. Microemulsions

Microemulsions are emulsified drug delivery systems used for improving dissolution and delivery (oral) of hydrophilic drugs. Surfactants, one of three to five basic compounds of microemulsions, are amphiphilic molecules necessary to stabilize the water-in-oil or oil-in-water conformation of the emulsion. Membrane-active amphiphilic surfactants are often cytotoxic. This toxicity strongly correlates with the ability of the surfactants to permeabilize and/or destroy the cellular membrane. Therefore, for the development of nontoxic microemulsions, it is important to know how basic compounds of microemulsions interact with membranes.

The planar lipid membrane setup is a good instrument in the search for less toxic surfactants. It can help predict quickly and quantitatively the nontoxic concentrations of surfactants. Questions that can be directed to planar membrane studies include these: How do the microemulsions interact with membranes? How do they change lipid membrane permeability and stability? What is the role of membrane lipid composition, solution pH, temperature, and ionic strength in microemulsion interactions with membranes?

F. Enveloped Viruses

The perfusion planar membrane system is a useful instrument in the study of viral fusion. Using this system we screened synthetic peptides corresponding to Moloney murine leukemia viral and human immunodeficiency viral envelope protein fusion peptides. A number of point-mutated peptides were screened for the ability to induce ion leakage and break membranes (Epand and Zhang *et al.*, 2008). Results of these studies were used for directed mutational *in vivo* analysis of viral fusion-replication.

Acknowledgments

This work was supported by grants from the Alzheimer's Association, the NIMH (MH 01174), the University of California AIDS Research Program, and the UCLA Alzheimer's Disease Center.

References

Alvarez, O. (1986). *In* "Ion Channel Reconstitution," (C. Miller, ed.), pp. 115–130. Plenum Press, New York.

Baaken, G., Sondermann, M., *et al.* (2008). Planar microelectrode-cavity array for high-resolution and parallel electrical recording of membrane ionic currents. *Lab Chip* **8**(6), 938–944.

Duclohier, H. (1994). *Toxicology* **87**, 175.

Epand, R. F., Zhang, Y. L., *et al.* (2008). Membrane activity of an amphiphilic alpha-helical membrane-proximal cytoplasmic domain of the MoMuLV envelope glycoprotein. *Exp. Mol. Pathol.* **84**(1), 9–17.

Jones, K. A., Garbati, N., *et al.* (2009). Automated patch clamping using the QPatch. *Methods Mol. Biol.* **565**, 209–223.

Kagan, B. L. (1983). *Nature (Lond.)* **302**, 709.

Kagan, B. L., and Sokolov, Yu. (1994). *Methods Enzymol.* **235**, 699.

Kagan, B. L., Colombini, M., and Finkelstein, A. (1981). *Proc. Natl. Acad. Sci. USA* **78**, 4950.

Kagan, B. L., Selsted, M. E., Ganz, T., and Lehrer, R. I. (1990). *Proc. Natl. Acad. Sci. USA* **87**, 210.

Le Pioufle, B., Suzuki, H., *et al.* (2008). Lipid bilayer microarray for parallel recording of transmembrane ion currents. *Anal. Chem.* **80**(1), 328–332.

Lin, M. C., Mirzabekov, T. A., and Kagan, B. L. (1997). *J. Biol. Chem.* **272**, 44.

Miller, C. (1986). "Ion Channel Reconstitution," Plenum Press, New York.

Mirzabekov, T. A., Ballarin, C., Zatta, P., Nicolini, M., and Sorgato, C. M. (1993). *J. Membr. Biol.* **33**, 129.

Mirzabekov, T. A., Lin, M. C., Yuan, W., Marshall, P., Carman, M., Tomaselli, K., Lieberburg, I., and Kagan, B. L. (1994). *Biochem. Biophys. Res. Commun.* **202**, 1142.

Mirzabekov, T. A., Lin, M. C., and Kagan, B. L. (1996). *J. Biol. Chem.* **271**, 1988.

Montal, M., and Mueller, P. (1972). *Proc. Natl. Acad. Sci. USA* **69**, 3561.

Mueller, P., Rudin, D. O., Tien, H. T., and Wescott, W. C. (1962). *Nature (Lond.)* **194**, 979.

Muller, R. U., and Finkelstein, A. (1972). *J. Gen. Physiol.* **60**, 285.

Poulos, J. L., Jeon, T. J., *et al.* (2009). Ion channel and toxin measurement using a high throughput lipid membrane platform. *Biosens. Bioelectron.* **24**(6), 1806–1810.

Sakmann, B., and Neher, E. (1983). "Single Channel Recording," Plenum Press, New York.

Schein, S. J., Kagan, B. L., and Finkelstein, A. (1978). *Nature (Lond.)* **276,** 159.

Schindler, H. (1980). *FEBS Lett* **122**(1), 77.

Silberstein, A. Ya. (1989). *Biol. Membr.* **6,** 1317.

Skare, J., Shang, E., Foley, D., Blanco, D. R., Champion, C. I., Mirzabekov, T. A., Sokolov, Y., Kagan, B. L., Miller, R., and Lovett, M. (1995). *J. Clin. Invest.* **96,** 2380.

Skare, J. T., Mirzabekov, T. A., Shang, E., Blanco, D. R., Erjument-Bromage, H., Tempst, P., Kagan, B. L., Miller, J. N., and Lovett, M. A. (1996). *J. Bacteriol.* **178,** 4909.

Skare, J. T., Mirzabekov, T. A., Shang, E., Blanco, D. R., Kagan, B. L., Miller, J. N., and Lovett, M. A. (1997). *Inf. Imm.* **65,** 3654.

Suzuki, H., Le Pioufle, B., *et al.* (2009). Ninety-six-wellplanar lipid bilayer chip for ion channel recording fabricated by hybrid stereolithography. *Biomed. Microdevices* **11**(1), 17–22.

CHAPTER 28

Antibodies to Ion Channels

Angela Vincent, Ian Hart, Ashwin Pinto, and F. Anne Stephenson

Neurosciences Group
Institute of Molecular Medicine
John Radcliffe Hospital
Oxford, United Kingdom

I. Introduction

Antibodies to ion channels have been of considerable use in research on the distribution of different channels and the association of ion channel subunits with other subunits or with associated proteins. Tables I–XVI describe both polyclonal and monoclonal antibodies and give an indication of the uses to which they have been put. Note that many antibodies will not necessarily have been tested for all purposes, nor will their cross-reactivity with different species be defined.

Polyclonal antibodies have been raised against purified proteins, recombinant polypeptides, or synthetic peptides, and in the latter case the antibodies can easily be affinity purified on peptide conjugated to Sepharose using conventional techniques. Most of the immunolocalization work, for instance, uses affinity-purified antibodies.

DOI: 10.1016/B978-0-12-382204-8.00028-X

Table I

Spontaneous Human Antibodies against Ion Channels (Tzartos et al., 1991)

Ion channels	Source of sera	Subunit specificity	Region	Detection of antibody by	Sources for further information	References
Muscle AChR	Patients with myasthenia gravis	Mainly α_1 but very variable; α_{67-76} represents main immunogenic region but many antibodies bind to other subunits or partly overlapping sites	Extracell	Immunoprecipitation of ^{125}I-α-BuTx-labeled AChR	Vincent (Oxford) Tzartoz (Athens)	Tzartos et al. (1991) Vincent et al. (1987)
	Fab fragments cloned from myasthenia gravis combinatorial libraries	α_1, γ_1	Extracell	Immunoprecipitation of ^{125}I-α-BuTx-labeled AChR	Vincent (Oxford)	Farrar et al. (1997)
Fetal form of muscle AChR	Mothers of babies with antibody-mediated fetal arthrogryposis	Principally γ_1	Extracell	Inhibition of fetal AChR function but not adult	Vincent (Oxford)	Riemersma et al. (1996)
GluR1	Patients with paraneoplastic disorders			IH, agonist-like activity	Rogers (Salt Lake City)	Gahring et al. (1995)
GluR3	Children with Rasussen's encephalitis		Extracell p372–395	WB, IP, IH	Rogers (Salt Lake City)	Rogers et al. (1994)
GluR4	Patients with paraneoplastic disorders			IH	Rogers (Salt Lake City)	Gahring et al. (1995)
GluR5	Patients with paraneoplastic disorders			WB	Rogers (Salt Lake City)	Gahring et al. (1995)
GluR6	Patients with paraneoplastic disorders			IH	Rogers (Salt Lake City)	Gahring et al. (1995)
VCCC	Patients with Lambert Eaton myasthenic syndrome	Mainly α_1	Extracell	Immunoprecipitation of ^{125}I-ω-conotoxin-labeled VGCC	Lang (Oxford)	Lang and Newsom-Davis (1995)
VGKC KCNA6	Patients with acquired neuromyotonia		Extracell	Immunoprecipitation of ^{125}I-dendrotoxin-labeled VGKCs or KCNA6 subunits expressed in oocytes	Hart (Liverpool) Vincent (Oxford)	Shillito et al. (1995) Hart et al. (1997a,b)
VGKC KCNA1a,2	Patients with acquired neuromyotonia		Probably extracell	Immunohistochemistry of oocytes expressing KCNA subunit	Hart (Liverpool)	Hart et al. (1997a,b)

The following abbreviations are used in Tables I–XVI: T, *Torpedo*; rt, rat; m, mouse; h, human; WB, Western blotting; IH, immunohistochemistry; IP, immunoprecipitation; IEP, immunoelectron microscopy; AChR, acetylcholine receptor; VGCC, voltage-gated calcium channel; VGKC, voltage-gated potassium channel; and GluR, glutamate receptor (usually AMPA). Extracell, extracellular epitope; Cyt, cytoplasmic epitope.

Table II

Rat Monoclonal Antibodies Raised Against *Torpedo* or Mammalian Muscle AChR

Name or number isotype (if known)	Species and source of antigen	Subunit	Region and peptide (if known)	Binding to native (N) or denatured (D) epitope	Species specificity (if known)	Uses	Commercial or academic source	References
Rat 1,2,4	*Torpedo* AChR	α_1	MIR extracell	N	T	IP, IEM	Tzartos S (Athens)	Loutrari *et al.* (1997), Tzartos and Lindstrom (1980)
6 IgG$_1$	*Torpedo* AChR	$\alpha_1 + \alpha_3$	MIR α_{67-76}	N > D	T, rt, m, h	WB, IP, IH, IEM	Tzartos S (Athens)	Tzartos and Lindstrom (1980), Tzartos *et al.* (1988)
35, 42 IgG$_1$	Eel AChR	$\alpha_1 + \alpha_3$	MIR α_{67-76}	N >>> D	T, rt, m, h	IP, IH, IEM	Tzartos S (Athens)	Papadouli *et al.* (1993), Tzartos *et al.* (1988, 1981)
198	Human AChR	$\alpha_1 + \alpha_3$	MIR α_{67-76}	N > D	T, rt, m, h	WB, IP, IH, IEM	Tzartos S (Athens)	Mamalaki *et al.* (1993), Tzartos *et al.* (1988, 1983)
202, 145, 190, 192	Human AChR	α_1	MIR	N >>> D N	rt, m, h	IP	Tzartos S (Athens)	Loutrari *et al.* (1997), Tzartos *et al.* (1981)
3, 5	*Torpedo* α	α_1	Cyt $\alpha_{351-360}$	N = D	T	WB, IP, IEM	Tzartos S (Athens)	Tzartos and Lindstrom (1980), Tzartos and Remoundos (1992)
8	*Torpedo* α	α_1	Cyt $\alpha_{370-378}$	N = D	T	WB, IP, IEM	Tzartos S (Athens)	Tzartos and Lindstrom (1980), Tzartos and Remoundos (1992)
152, 153, 155, 154	*Torpedo* α	α_1	Cyt $\alpha_{373-380}$	N = D	T, rt, m, h	WB, IP, IH, IEM	Tzartos S (Athens)	Tzartos and Remoundos (1992), Tzartos *et al.* (1986, 1988)
12	*Torpedo* AChR	α_1	Near-MIR	N	T	IP, IEM	Tzartos S (Athens)	Loutrari *et al.* (1997), Tzartos and Lindstrom (1980)
14	*Torpedo* AChR	Non-α_1	Near-MIR	N	T	IP	Tzartos S (Athens)	Loutrari *et al.* (1997), Tzartos and Lindstrom (1980)
64	Calf AChR	α_1	Extracell	N >> D	rt, m, h	IP	Tzartos S (Athens)	Loutrari *et al.* (1997), Tzartos *et al.* (1986)
73	Calf AChR	β_1	Extracell	N >>> D	rt, m, h	IP	Tzartos S (Athens)	Tzartos *et al.* (1986)

Table II *(continued)*

Name or number isotype (if known)	Species and source of antigen	Subunit	Region and peptide (if known)	Binding to native (N) or denatured (D) epitope	Species specificity (if known)	Uses	Commercial or academic source	References
111, 124, 148, 151	*Torpedo* β	β_1	Cyt $\beta_{354-359}$	N = D	T, rt, m, h	WB, IP, IH, IEM prefers nonphosphorylated	Tzartos S (Athens)	Tzartos *et al.* (1986, 1993)
125	*Torpedo* β	β_1	Cyt $\beta_{408-414}$	N = D	T	nk	Tzartos S (Athens)	Tzartos *et al.* (1986, 1993)
117	*Torpedo* β	β_1	Cyt $\beta_{343-352}$	N = D	T	nk	Tzartos S (Athens)	Tzartos *et al.* (1986, 1993)
66, 67	Calf AChR	γ_1	Extracell	N >>> D	h	IP	Tzartos S (Athens)	Tzartos *et al.* (1986)
154, 168 IgG$_1$	*Torpedo* γ	γ_1-Torp; ε_1-mammal.	Cyt $\gamma_{364-370}$	N = D	T, rt, m, h	WB, IP	Tzartos S (Athens)	Tzartos *et al.* (1986, 1995a,b)
165	*Torpedo* γ	γ_1, ε_1	Cyt $\gamma_{364-370}$	N = D	T	WB, IP	Tzartos S (Athens)	Tzartos *et al.* (1986, 1995a,b)
7	*Torpedo* γ	δ_1, ε_1	Cyt $\delta_{389-395}$ $\gamma_{382-387}$	N = D	T, m, h	WB, IP	Tzartos S (Athens)	Tzartos *et al.* (1986, 1995a,b)
137 IgG$_{2a}$	*Torpedo* γ	δ_1	Cyt $\delta_{374-391}$	N = D	T, rt, m, h	WB, IP	Tzartos S (Athens)	Tzartos *et al.* (1986, 1995a,b)
141 IgG$_{2a}$	*Torpedo* γ	δ_1	Cyt $\delta_{385-392}$	N = D	T, rt, m, h	WB, IP	Tzartos S (Athens)	Loutrari *et al.* (1997), Tzartos and Remoundos (1992)
166 IgG$_{2a}$	*Torpedo* AChR	δ_1	Cyt $\delta_{300-410}$	N + D	T, Rana, *Xenopus*	IH (amphibians)	Tzartos S (Athens)	

Table III
Mouse Monoclonal Antibodies against Muscle AChR

Name or number isotype (if known)	Species and source of antigen	Subunit specificity	Region and peptide (if known)	Binding to native (N) or denatured (D) epitope	Species specificity (if known)	Uses	Commercial or academic source	References
D6α IgG$_{2b}$	Human muscle AChR	α$_1$	MIR	N >>> D	h, m, rt, calf	WB, IP, IH	Vincent (Oxford) Serotec (**Kidlington**)	Heidenreich et al. (1998), Jacobson et al. (2010), Whiting et al. (1986)
G10α	Human muscle AChR	α$_1$	MIR	N >>> D	h, m, calf	WB, IP, IH	Vincent (Oxford) Serotec (**Kidlington**)	Whiting et al. (1986), Heidenreich et al. (1998); Jacobson et al. (2010)
C3a IgG$_1$ B3β IgG$_1$	Human muscle AChR	β$_1$	Extracell	N >>> D	h, m, calf h, calf	WB, IP, IH	Vincent (Oxford) Serotec (**Kidlington**)	Heidenreich et al. (1998), Jacobson et al. (2010), Whiting et al. (1986)
C7δ IgG$_{2a}$	Human muscle AChR	δ$_1$	Extracell	N >>> D	h, m, calf	WB, IP, IH	Vincent (Oxford) Serotec (**Kidlington**)	Heidenreich et al. (1998), Jacobson et al. (2010), Whiting et al. (1986)
G3δ IgG$_{2b}$ B8γ, C2γ, C9γ, F8γ IgG$_1$	Human muscle AChR	γ$_1$	Extracell	N >> D	h h, calf fetal-type only	WB, IP, IH	Vincent (Oxford) Serotec (**Kidlington**)	Heidenreich et al. (1998), Jacobson et al. (2010), Whiting et al. (1986)
WF6	Electric organ	α$_1$	Extracell	N	Agonist binding site of Torpedo and chick AChR	Acts as agonist and then blocks	Maelicke (Munich)	Bufler et al. (1996)

Table IV
Rabbit Polyclonal Antibodies to Human Muscle AChR

Name or number, isotype (if known)	Species and source of antigen	Subunit	Region	Native or denatured	Species specificity	Uses	Source for further information	References
Anti-α	Two overlapping α peptides	α_1	Cyt $\alpha_{309-368}$	N + D	h, m	WB, IP, IF, IH	Vincent (Oxford)	Beeson *et al.* (1996), Slater *et al.* (1997)
Anti-ε	Two overlapping ε peptides	ε_1	Cyt $\varepsilon_{341-413}$	N + D	h	WB, IP, IF, IH	Vincent (Oxford)	Beeson *et al.* (1996), Slater *et al.* (1997)
Anti-γ	Two overlapping γ peptides	γ_1	Cyt $\gamma_{337-410}$	N + D	h	WB, IP, IF, IH	Vincent (Oxford)	Beeson *et al.* (1996)

Many of the antibodies listed were originally tested by Western blotting and, therefore, were chosen more for their reactivity with denatured protein than with the intact molecule. Nevertheless, many can also be used for immunoprecipitation of solubilized proteins, or as immunohistochemical or immunoelectron microscopic reagents. Where there is sufficient information on reactivity with native or denatured protein, this has been added to the tables.

Antibodies raised against purified proteins often bind to extracellular determinants, whereas antibodies raised against synthetic peptides or recombinant polypeptides often do not bind well to the extracellular surface of native proteins. Thus, for identification of ion channels in nonpermeabilized cell lines, antisera raised against the native protein are best. A novel alternative approach is to use a specific neurotoxin to label cell surface receptors or ion channels. Recently, a monoclonal antibody against ω-agatoxin IVA has been used in conjunction with the toxin, that binds to P-type voltage-gated calcium channels (VGCCs), to immunolocalized P-type channels on cerebellar Purkinje cells (see Table XII).

Antibodies are not the only way of identifying the expression of a particular ion channel. In particular, the use of neurotoxins and specific pharmacologic ligands can be very useful in affinity purification of proteins, and in localizing channels *in situ*. *In situ* hybridization using specific probes is an obvious alternative method to immunohistochemical or autoradiographic studies.

The references given in Tables I–XVI refer to the main papers in which the antibodies were described. The name of the senior author or the commercial company is also given. Many of the companies distribute information via the Internet.

II. Spontaneous Antibodies in Human Disease

One of the intriguing aspects of ion channel research is the extent to which they are involved in human disorders. Thus not only are they targets for genetic disorders (Ptacek, 1977) but also for autoimmune diseases in which spontaneous

Table V
Monoclonal Antibodies Raised against Neuronal AChR

Name or number isotype (if known)	Species and source of antigen	Subunit	Region and peptide (if known)	Native (N) or denatured (D)	Species specificity	Uses	Source for further information	References
321–326	Chick recombinant	α_2	Cyt		Ck	nk	Lindstrom (Philadelphia)	See Lindstrom (1996) for a review and much further information
313–315 IgG$_{2a}$	Chick recombinant	α_3	Cyt $\alpha3_{315-441}$	N + D	Ck	IH, IP	Lindstrom (Philadelphia)	
286 IgM	Chick brain AChR	α_4		N + D	Ck, rt, h	WB	Lindstrom (Philadelphia)	
289 IgM	Chick brain AChR	α_4	$\alpha4_{330-511}$	N + D	Ck	WB	Lindstrom (Philadelphia)	
292 IgG$_1$	Rat brain AChR	α_4		N + D	Rt	WB	Lindstrom (Philadelphia)	
293 IgG$_{2a}$	Rat brain AChR	α_4		N + D	Ck, rt, h	WB	Lindstrom (Philadelphia)	
299 IgG$_1$	Rat brain AChR	α_4	Extracell	N + D	Ck, rt, h	WB, IH	Lindstrom (Philadelphia)	
268 IgG$_{1/2a}$	Chick brain AChR	α_5	Extracell $\alpha5_{91-100}$	D	Ch, h	WB	Lindstrom (Philadelphia)	
306, 307 IgG$_1$	a-BuTx-binding rat brain AChR	α_7	Cyt $\alpha7_{380-400}$	N (Ck) + D	Ck, rt, h	IH (but not mammalian)	Lindstrom (Philadelphia)	
320 IgG	Recombinant a7 cyt	α_7	Cyt $\alpha7_{380-400}$	N + D	Ck		Lindstrom (Philadelphia)	
308 IgG$_{2b}$	Recombinant a8 cy5	α_8	Cyt $\alpha8_{323-342}$	N + D	Ck	IH, AP	Lindstrom (Philadelphia)	
270 IgG$_{2a}$	Purified Ck brain AChR	β_2	Extracell	N.D	Ck, m, rt	IH, AP	Lindstrom (Philadelphia)	
287 IgM	Purified Ck brain AChR	β_2	Cyt	D	Ck		Lindstrom (Philadelphia)	
290 IgG$_1$	Purified rt brain AChR	β_2	Extracell	N	Ck, rt, b, h	IH, AP	Lindstrom (Philadelphia)	
295, 297, 298 IgG$_{2a}$	Purified rt brain AChR	β_2	Extracell	N	Ck, rt, b, h		Lindstrom (Philadelphia)	

Ck, chick.

Table VI
Rabbit Polyclonal Antibodies Raised Against Synthetic Peptides of Extracellular Domain of Neuronal AChRs

Subunit	Region and peptide	Native (N) or denatured (D)	Species specificity	Uses	Source for further information	References
Rat						
α_2	Extracell 68–81	D > N?	Not tested	Tested on WB Some binding to transfected cells	Patrick (Houston)	Neff *et al.* (1995)
α_3	Extracell 68–81	D > N?	Not tested	Tested on WB Some binding to transfected cells	Patrick (Houston)	Neff *et al.* (1995)
α_4	Extracell 68–81	D > N?	Not tested	Tested on WB Some binding to transfected cells	Patrick (Houston)	Neff *et al.* (1995)
α_5	Extracell 68–81	D > N?	Not tested	Tested on WB Some binding to transfected cells	Patrick (Houston)	Neff *et al.* (1995)
α_6	Extracell 68–81	D > N?	Not tested	Tested on WB Some binding to transfected cells	Patrick (Houston)	Neff *et al.* (1995)
β_2	Extracell 68–81	D > N?	Not tested	Tested on WB Some binding to transfected cells	Patrick (Houston)	Neff *et al.* (1995)
β_3	Extracell 68–81	D > N?	Not tested	Tested on WB Some binding to transfected cells	Patrick (Houston)	Neff *et al.* (1995)
β_4	Extracell 68–81	D > N?	Not tested	Tested on WB Some binding to transfected cells	Patrick (Houston)	Neff *et al.* (1995)

Table VII

Monoclonal and Polyclonal Antibodies to Glycine and GABA$_A$ Receptors

Name or number (if relevant)	Antigen: peptide (p-) or recombinant antigen (r-)	Subunit	Region	Uses	Source for further information	References
4a IgG1	Purified glycine receptor	α and β		IP, WB	Betx (Frankfurt)	Kirsch et al. (1993), Meyer et al. (1995), Pfeiffer et al. (1984)
7a IgG1	Purified glycine receptor	α		WB	Betz (Frankfurt)	Kirsch et al. (1993), Meyer et al. (1995), Pfeiffer et al. (1984)
bd17 monoclonal	Native GABA$_A$ receptor	β$_2$, β$_3$			Mohler (Zurich)	Ewert et al. (1990), Schoch et al. (1985)
bd24 monoclonal	Native GABA$_A$ receptor	Human and bovine a1 N-terminus			Mohler (Zurich)	Ewert et al. (1990)
Rabbit polyclonal sera to GABA$_A$	p324–341	α$_1$	Cyt	IP, WB	Stephenson (London)	Duggan and Stephenson (1990, 1989)
	p414–424	α$_2$	Cyt	IP, WB	Stephenson (London)	Duggan and Stephenson (1990, 1989)
	p454–467	α$_3$	Cyt	IP, WB	Stephenson (London)	Duggan and Stephenson (1990, 1989)
	p1–9	α$_1$	Extracell	WB, IH	Sieghart (Vienna)	Nusser et al. (1996a,b), Zezula and Sieghart (1991)
	p1–14 (rat)	α$_1$	Extracell	WB	Stephenson (London)	Pollard and Stephenson (unpublished)
	p328–382	α$_1$	Cyt	WB, IP	Sieghart (Vienna)	Mossier et al. (1994)
	p324–338 (rat)	α$_1$		WB, IH	Seighart (Vienna)	Kern and Sieghart (1994), Zezula and Sieghart (1991)
	p416–424	α$_2$	Cyt	WB	Sieghart (Vienna)	Kern and Sieghart (1994), Zezula and Sieghart (1991)
	p459–467	α$_3$	Cyt	WB	Sieghart (Vienna)	Kern and Sieghart (1994), Zezula and Sieghart (1991)
	p517–523	α$_4$	Extracell	WB	Sieghart (Vienna)	Kern and Sieghart (1994), Zezula and Sieghart (1991)
	p427–433	α$_5$	Cyt	WB	Sieghart (Vienna)	Kern and Sieghart (1994), Zezula and Sieghart (1991)
	p1–15 (rat)	α$_5$	Extracell	WB, IH	Stephenson (London)	Jones et al. (1997), Pollard et al. (1993)

(continues)

Table VII (*continued*)

Name or number (if relevant)	Antigen: peptide (p-) or recombinant antigen (r-)	Subunit	Region	Uses	Source for further information	References
p1–16 (bovine)		α_6-N	Extracell	IP, WB, IH	Stephenson (London)	Nusser et al. (1996a,b), Pollard et al. (1993)
p429–434 (rat)		α_6	Extracell	IP, IH	Sieghart (Vienna)	Nusser et al. (1996a,b), Togel et al. (1994)
p423–434		α_6	Extracell	IP, WB	Stephenson (London)	Pollard and Stephenson (unpublished)
Recombinant cytoplasmic loops		α_1	Cyt	WB, IP	Whiting (UK)	McKernan et al. (1991)
Recombinant cytoplasmic loops		α_2	Cyt	WB, IP	Whiting (UK)	McKernan et al. (1991)
Recombinant cytoplasmic loops		α_3	Cyt	WB, IP	Whiting (UK)	McKernan et al. (1991)
Recombinant cytoplasmic loops		α_5	Cyt	WB, IP	Whiting (UK)	McKernan et al. (1991)
p382–393		β_1		WB, IP	Stephenson (London)	Pollard and Stephenson (1997), Pollard et al. (1991)
p380–391		β_3		WB, IP	Stephenson (London)	Pollard and Stephenson (1997), Pollard et al. (1991)
p381–395		β_2		WB, IP	Stephenson (London)	Pollard and Stephenson (1997), Pollard et al. (1991)
p379–394		β_3		WB, IP	Stephenson (London)	Pollard and Stephenson (1997), Pollard et al. (1991)
p345–408 (rat)		β_3	Extracell	IP, IH, WB, IEM	Sieghart (Vienna)	Benke et al. (1994)
p1–15		γ_2	Cyt	WB, IP, IH	Mohler (Zurich)	Benke et al. (1991)
p336–350		γ_2	Cyt	WB, IP, IH	Mohler (Zurich)	Khan et al. (1994)
p1–15		γ_2	Extracell	WB, IP, IH	Stephenson (London)	Greferath et al. (1995), Stephenson et al. (1990)
p316–352		γ_2	Cyt	WB, IP	Sieghart (Vienna)	Togel et al. (1994)
p339–353 (rat)		γ_2		WB, IH	Siegel (Cleveland)	Nadler et al. (1994)
p1–9		γ_3	Extracell	WB, IP	Siegel (Cleveland)	Nadler et al. (1994)
p322–372		γ_3	Cyt	WB, IP	Siegel (Cleveland)	Nadler et al. (1994)
p1–44 (rat)		δ				Mossier et al. (1994)
p324–338 (rat)		α_1	nk	WB, IP, IH, IF	Pharmingen	Nadler et al. (1994)
p (rat)		α_3	nk	IH	Pharmingen	Nadler et al. (1994)
p (rat)		α_4	nk	IH	Pharmingen	Nadler et al. (1994)
p (rat)		α_6	nk	IH	Pharmingen	Nadler et al. (1994)

Table VIII
Monoclonal and Polyclonal Antibodies to NMDA Receptors

Name	Subunit	Region	Peptide (p-) or recombinant antigen (r-)	Uses	Source for further information	References
Monoclonal 54.4	Rat NMDAR1	Transmembrane and cytoplasmic	rNMDAR1 660–811	WB, IH	Heinemann (San Diego)	Sucher et al. (1993)
Monoclonal 54.1	Rat NMDAR1	Transmembrane and cytoplasmic	rNMDAR1 660–811	WB, IH, IEM	Heinemann (San Diego)	Brose et al. (1994a,b), Huntley et al. (1994), Siegel et al. (1994)
Monoclonal 54.2	Rat NMDAR1	Transmembrane and cytoplasmic	rNMDAR1 660–811	WB, IH, IEM	Heinemann (San Diego)	Brose et al. (1994a,b), Huntley et al. (1994), Siegel et al. (1994)
Monoclonal NR1-TM3/4	Rat NMDAR1	Transmembrane and cytoplasmic	rNMDAR1 660–811	WB, IH	Heinemann (San Diego)	Siegel et al. (1994)
Polyclonal antisera						
NR1 827 AP	Rat NMDAR1	Transmembrane and cytoplasmic	rNMDAR1 660–811	WB, IH	Heinemann (San Diego)	Brose et al. (1994a,b)
NR1-NH$_2$	Rat NMDAR1	N-terminus	p19–38	WB	Huganir (Baltimore)	Lau and Huganir (1995)
NR1-COOH	Rat NMDAR1	C-terminus	r919–938	WB	Huganir (Baltimore)	Tingley et al. (1993)
NR2A	Rat NMDAR1	C-terminus	el-1247–1464 (fusion protein)	WB	Huganir (Baltimore)	Lau and Huganir (1995)
NR2B	Rat NMDAR1	C-terminus	p1463–1482	WB, IF	Huganir (Baltimore)	Lau and Huganir (1995)
NR1A antiphosphopeptide	Rat NMDAR1	Cyt	r891–902 phosphorylated	WB, IF, WB of tryptic digests	Huganir (Baltimore)	Mei et al. (1994), Tingley et al. (1997)
	Rat NMDAR1	C-terminus	p884–895 phosphorylated	WB, IH	Wenthold (Bethesda)	Petralia et al. (1994)

Table IX

Monoclonal and Polyclonal Antibodies to AMPA Receptors[a]

Name or number isotype (if known)	Peptide (p-) or recombinant antigen (r-)	Subunit or subtype	Region	Uses	Source for further information	References
Monoclonal antibodies						
3a11 Mouse IgG$_{2a}$	17–430	GluR2 + 4	N-terminus	WB, IH, IP, IEM	Huganir	Blackstone et al. (1992a,b); Raymond et al. (1993)
4F5	N-terminal extracellular domain fusion protein	GluR5/6/7	Extra N-terminus	IH, IF, IEM	Pharmingen	Huntley et al. (1993)
Mouse IgM 1F1	C-terminal 13 aas		C-terminus	IH, IEM	Somogyi (Oxford)	Nusser et al. (1996a,b)
Rabbit polyclonal antisera						
67	877–889	GluR1	Extracell	IP, WB, IAP	Wenthold (Bethesda)	Wenthold et al. (1992)
23	369–381	GluR2	Extracell	IP, WB	Wenthold (Bethesda)	Wenthold et al. (1992)
33	834–844	GluR2	Extracell	IP, WB	Wenthold (Bethesda)	Wenthold et al. (1992)
24	372–383	GluR3	Extracell	IP	Wenthold (Bethesda)	Wenthold et al. (1992)
34	838–848	GluR3	Extracell		Wenthold (Bethesda)	Wenthold et al. (1992)
35	838–848	GluR3	Extracell	IP, WB	Wenthold (Bethesda)	Wenthold et al. (1992)
22	868–881	GluR4	Extracell	IP, WB	Wenthold (Bethesda)	Wenthold et al. (1992)
25	850–862	GluR2 + 3	Extracell	IP, WB, IAP	Wenthold (Bethesda)	Wenthold et al. (1992)
	894–907	GluR1	C-terminus	WB, IH, IF, IP (2,4,4c)	Huganir (Baltimore)	Craig et al. (1993)
		GluR2/3/4c		WB, IH, IF	Chemicon Int Iner	Martin et al. (1992)
		GluR4		WB, IH, IF	Chemicon Int Iner (London)	Martin et al. (1992)
		GluR6/7	C-terminus	WB, IH	Chemicon Int Iner (London)	Martin et al. (1992)
GluR1	p876–889			WB, IH	Huganir (Baltimore)	Puchalski et al. (1994), Rubio and Wenthold (1997)
GluR1	p872–889			WB, IH, ?IP	Wenthold (Bethesda)	Puchalski et al. (1994), Rubio and Wenthold (1997)
GluR2/3/4c	p843–862			WB, IH	Huganir (Baltimore)	Puchalski et al. (1994), Rubio and Wenthold (1997)
GluR2/3/4c	p850–862			WB (2,3)	Wenthold (Bethesda)	Puchalski et al. (1994), Rubio and Wenthold (1997)
GluR4	p862–881			WB, IH	Huganir (Baltimore)	Puchalski et al. (1994), Rubio and Wenthold (1997)
GluR4	p868–881			WB, IH	Wenthold (Bethesda)	Puchalski et al. (1994), Rubio and Wenthold (1997)

GluR6/7	p863–877		WB, IH, IP	Huganir (Baltimore)	Puchalski et al. (1994), Rubio and Wenthold (1997)
GluR6/7	p864–877		WB, IH, IP	Wenthold (Bethesda)	Puchalski et al. (1994), Rubio and Wenthold (1997)
GluR3	r245–457	p245–274, p372–395	WB, IH agonist-like activity	Rogers (Salt Lake City)	Twyman et al. (1995)

[a]Most of the antibodies raised against sequences of rat AMPA receptors. Other antibodies against synthetic peptides of glutamate receptors are available from Pharmingen.

Table X
Rabbit Polyclonal Antibodies to Voltage-Gated Calcium Channels

Name or number	Species and source of antigen	Reactivity with native (N) or denatured (D)	Subunit	Region	Species specificity	Uses	Source for further information	References
Anti-α_{1A}	Human, GST fusion protein 1041–1202	N + D	α_{1A}	Cyt	Human, rat	WB, IH	Beattie (Surrey, UK)	Beattie *et al.* (1997)
Anti-α_{1B}	Human, GST fusion protein 983–1106	N + D	α_{1B}	Cyt	Human, rat	WB, IH	Beattie (Surrey, UK)	Beattie *et al.* (1997)
Anti-α_{1E}	Human, GST fusion protein 984–1099	N + D	α_{1E}	Cyt	Human, rat	WB, IH	Beattie (Surrey, UK)	Beattie *et al.* (1997)
Anti-β_{1B}	Human, GST fusion protein 431–598	N + D	β_{1B}	Cyt	Human, rat	WB, IH	Beattie (Surrey, UK)	Beattie *et al.* (1997)
Anti-β_2	Human, GST fusion protein 554–660	N + D	β_2	Cyt	Human, rat	WB, IH	Beattie (Surrey, UK)	Beattie *et al.* (1997)
Anti-β_3	Human, GST fusion protein 418–483	N + D	β_3	Cyt	Human, rat	WB, IH	Beattie (Surrey, UK)	Beattie *et al.* (1997)
Anti-β_4	Human, GST fusion protein	N + D	β_4	aa 410–520 Cyt	Human, rat	WB, IH	Beattie (Surrey, UK)	Beattie *et al.* (1997)

Table XI
Rabbit Polyclonal Antibodies to VGCC

Name or number	Species and source of antigen	Reactivity with native (N) or denatured (D)	Subunit specificity	Region	Species specificity	Uses	Source for further information	References
Anti-CNA1	Rat, peptide 865–881	N + D	α_{1A}	II–III Cyt	Rat, mouse	IH, WB, IP	Catterall (Seattle) Alomone Labs	Sakurai et al. (1995)
Anti-CNA3	Rat, peptide 882–896	N + D	α_{1A}	II–III Cyt	Rat	IH, WB, IP	Catterall (Seattle) Alomone Labs	Sakurai et al. (1995)
Anti-CNA5	Rat, GST fusion protein 842–981	N + D	α_{1A}	II–III Cyt	Rat	IH, WB, IP	Catterall (Seattle) Alomone Labs	Sakurai et al. (1996)
Anti-CNA6	Human, GST fusion protein 569–712	N + D	α_{1A}	II–III Cyt	Rat	IH, WB	Catterall (Seattle) Alomone Labs	Sakurai et al. (1996)
Anti-NBI-1	Rabbit, peptide 845–861	N + D	α_{1A}	II–III Cyt	Rat	IH, WB	Catterall (Seattle) Alomone Labs	Sakurai et al. (1996)
Anti-NBI-2	Rabbit, peptide 904–918	N + D	α_{1A}	II–III Cyt	Rat	IH, WB	Catterall (Seattle) Alomone Labs	Sakurai et al. (1996)
Anti-CNB1	Rat, peptide 851–867	N + D	α_{1B}	II–III Cyt	Rat, mouse	IH, WB	Catterall (Seattle) Alomone Labs	Westenbroek et al. (1992a)
Anti-CNC1	Rat, peptide 815, 835	N + D	α_{1C}	II–III Cyt	Rat, mouse	IH, WB	Catterall (Seattle) Alomone Labs	Hell et al. (1993)
Anti-CND1	Rat, peptide 809–825	N + D	α_{1D}	II–III Cyt	Rat	IH, WB	Catterall (Seattle) Alomone Labs	Hell et al. (1993)
Anti-pan α_1	? species, peptide 1382–1400	N + D	α_{1S}		Rat, mouse	IH, WB	Alomone Labs	Striessnig et al. (1990)
Anti-β_{1B}	Rat, GST fusion protein 428–597	N + D	β_{1B}	Cyt	Rabbit	WB, IP	Campbell (Iowa)	Scott et al. (1996)
Anti-β_{2A}	Rat, GST fusion protein 462–578	N + D	β_2	Cyt	Rabbit	WB, IP	Campbell (Iowa)	Scott et al. (1996)
Anti-β_3	Rat, GST fusion protein 369–484	N + D	β_3	Cyt	Rabbit	WB, IP	Campbell (Iowa)	Scott et al. (1996)
Anti-β_4	Rat, GST fusion protein 419–519	N + D	β_4	Cyt	Rabbit	WB, IP	Campbell (Iowa)	Scott et al. (1996)
Anti-α_2 Anti-α_{1A}	Rabbit, peptide 839–856 aa 965–983	N + D	α_2 α_{1A}	Cyt	Rabbit	WB, IH	Campbell (Iowa) Froehner (North Carolina)	Gurnett et al. (1995) Ousley and Froehner (1994)
Anti-α_{1A}	IVS5-S6	N + D	α_{1A}	Extracell	Rabbit	WB, IH	Froehner (North Carolina)	Barry et al. (1995a,b)

Table XII

Monoclonal Antibodies to VGCC-Specific Neurotoxins for Immunolocalization Studies

Name or number	Immunogen	Epitope	Uses	Senior author	References
Anti-agatoxin IVA	Purified peptide	ω-Aga-IVA binding sites (α_{1A})	IH using ω-Aga-IVA as first layer	Beattie (Surrey)	Gillard *et al.* (1997)
Anti-conotoxin GVIA	Purified peptide	ω-CTx-GVIA binding sites (α_{1B})	IH using ω-CTx-GVIA as first layer	Beattie (Surrey)	Gillard *et al.* (1997)

antibodies lead to loss of ion channel and neurologic dysfunction (Table I). Many of these spontaneous antisera have well-defined functional effects which have generally not been achieved with experimentally produced antibodies.

III. Ligand–Gated Receptors

A. Acetylcholine Receptors

The acetylcholine receptor was the first ion channel protein to be purified, cloned, and sequenced. Polyclonal antisera and monoclonal antibodies were raised against the purified detergent-solubilized, alpha-neurotoxin-purified protein before it was fully cloned and characterized. Polyclonal antibodies to individual purified subunits were also raised and polyclonal sera to defined sequences became available as the genes were cloned. As a result of the earlier work of Jon Lindstrom and Socrates Tzartos, there are many available monoclonal antibodies against *Torpedo* and human acetylcholine (AChR), many of which cross-react with other species. Some of these bind to the main immunogenic region (MIR), a determinant on the two α subunits which is distinct from the neurotoxin-binding site (Table II). The studies performed on characterizing epitopes on AChR have provided paradigms for subsequent investigations on other ion channels.

Several antibodies raised against human AChR were also obtained by Paul Whiting, of which four are specific for the fetal form ($\alpha_2\beta\delta\epsilon$) and do not bind to the adult form ($\alpha_2\beta\gamma\delta$); these monoclonals have been shown to bind specifically to the γ subunit on Western blots of recombinant proteins (Table III). In general, though, anti-AChR antibodies bind much better to the native AChR than to subunits on Western blots, or to peptides. This phenomenon emphasizes the extent to which antibodies raised against the intact proteins do not recognize linear sequences. The exception, interestingly, is antibodies directed toward cytoplasmic epitopes. These antibodies, both monoclonal and polyclonal, often bind strongly to both denatured sequences and the intact (solubilized protein or permeabilized cells) molecule (e.g., Table IV). They can sometimes be induced by immunization against the whole recombinant peptide. Thus, it has become clear that the easiest way to induce a specific antibody to an individual AChR subunit, for instance, is to immunize against a peptide sequence of the cytoplasmic domain. These observations

Table XIII
Mouse Monoclonal Antibodies to Voltage-Gated Potassium Channels

Name	Channel subunit	Region	Epitope if known	Immunogen	Uses	Senior author or commercial company	References
K1C3 (rat)	Kv1.1	Cyt	141 aas of C-terminus	Kv1.1 (RCK1)	WB, IH	Pongs (Hamburg)	Reinhardt-Maelicke et al. (1993)
K20/78	Kv1.1		C-terminus	Peptide		Upstate Biotechnology	
Anti-Kv1.2	Kv1.2	Cyt	C-terminus	Peptide	IH, IEM	Dolly (London)	McNamara et al. (1996)
K14/16 IgG$_{2a}$	Kv1.2		C-terminus	Peptide		Trimmer (Stony Brook, NY) Upstate Biotechnology	Bekele-Arcuri et al. (1996), Shi et al. (1996)
K13/31 IgG$_1$	Kv1.4		N-terminus	Peptide		Trimmer (Stony Brook, NY) Upstate Biotechnology	Bekele-Arcuri et al. (1996), Shi et al. (1996)
K7/45 IgG$_1$	Kv1.5		C-terminus	Peptide		Trimmer (Stony Brook, NY) Upstate Biotechnology	Bekele-Arcuri et al. (1996), Shi et al. (1996)
K19/36 IgG$_3$	Kv1.6		C-terminus	Peptide		Trimmer (Stony Brook, NY) Upstate Biotechnology	Bekele-Arcuri et al. (1996), Shi et al. (1996)
D4/11 IgG$_1$	Kv2.1		C-terminus	Peptide		Trimmer (Stony Brook, NY) Upstate Biotechnology	Bekele-Arcuri et al. (1996), Shi et al. (1996)
K37/89 IgG$_{2a}$	Kv2.2		N-terminus	Peptide		Trimmer (Stony Brook, NY) Upstate Biotechnology	Bekele-Arcuri et al. (1996), Shi et al. (1996)
K9/40 IgG$_{2b}$	Kvb1		N-terminus	Peptide		Trimmer (Stony Brook, NY)	Bekele-Arcuri et al. (1996), Shi et al. (1996)
K17/70 IgG$_1$	Kvb2		N-terminus	Peptide		Trimmer (Stony Brook, NY)	Bekele-Arcuri et al. (1996), Shi et al. (1996)

Table XIV

Rabbit Polyclonal Antibodies Raised against Synthetic Peptides of Voltage-Gated Potassium Channels

Name	Channel subunit	Native (N) or denatured (D)	Region	Epitope if known	Uses	Senior author or commercial company	References
Anti-Kv1.1 (mouse)	Kv1.1	N = D	Cyt	C-terminus	WB, IP, IH	Tempel (Seattle)	Wang et al. (1993)
Anti-Kv1.1 (rat)	Kv1.1	D	Cyt	C-terminus	WB bovine	Pongs (Hamburg)	Scott et al. (1994)
Anti-Kv1.1 (rat)	Kv1.1	N = D	Cyt	C-terminus, aa 354–495	WB, IH	Pongs (Hamburg)	Veh et al. (1995)
Anti-Kv1.1	Kv1.1		Cyt	aa 458–476		Trimmer (Stony Brook, NY)	Nakahira et al. (1996)
KCN1A (human)	KCN1A	N = D	Extracell	S3–S4	IH, IP	Hart (Liverpool, UK)	Hart et al. (1997a,b)
Anti-Kv1.2	Kv1.2	N = D	Cyt	C-terminus	WB, IP, IH	Tempel (Seattle)	Wang et al. (1993)
Anti-Kv1.2	Kv1.2	D	Cyt	C-terminus	WB (bovine)	Pongs (Hamburg)	Scott et al. (1994)
Anti-Kv1.2	Kv1.2	N = D	Cyt	p468–486	WB, IH	Jan (San Francisco)	Sheng et al. (1994)
Anti-Kv1.2	Kv1.2	N = D	Cyt	p422–498	WB, IH	Pongs (Hamburg)	Veh et al. (1995)
Anti-Kv1.3	Kv1.3	D	Extracell	S1–S2 and S3–S4	WB (h, m, Drosophila, yeast)	Gutman (Irvine, CA)	Spencer et al. (1993)
Anti-Kv1.3	Kv1.3	N = D	Cyt	aa 409–525	WB, IH	Pongs (Hamburg)	Veh et al. (1995)
Anti-Kv1.3	Kv1.3	D	Cyt	aa 456–474	WB	Slaughter (Merck Research Labs)	Helms et al. (1997)
Anti-Kv1.4N	Kv1.4	N = D	Cyt	aa 13–37	WB, IH	Jan (San Francisco)	Sheng et al. (1992)
Anti-Kv1.4	Kv1.4	D	Cyt	C-terminus	WB (bovine)	Pongs (Hamburg)	Scott et al. (1994)
Anti-Kv1.4	Kv1.4	N = D	Cyt	aa 578–655	WB, IH	Pongs (Hamburg)	Veh et al. (1995)
Anti-Kv1.5	Kv1.5	D	Extracell	aa 272–312, S1–S2	WB rat heart	Levitan (Pittsburgh)	Takimoto and Levitan (1994)
Anti-Kv1.5	Kv1.5	N = D	Cyt	aa 542–602. C-terminus	WB, IH rat heart	Nerbonne (St. Louis) Trimmer (Stony Brook, NY) Upstate Biotechnology	Barry et al. (1995a,b), Trimmer (1991)
Anti-Kv1.5N	Kv1.5	N = D	Cyt	N-terminus	WB, IH human heart	Mays (Nashville)	Mays et al. (1995)
Anti-Kv1.5S1-2	Kv1.5	N = D	Extracell	S1–S2 extracellular	WB, IH human heart	Mays (Nashville)	Mays et al. (1995)
Anti-Kv1.5	Kv1.5		Cyt	aa 586–602	WB	Trimmer (Stony Brook, NY)	Trimmer (1993)
Anti-Kv1.5	Kv1.5	N	Cyt	C-terminus	WB rat brain	Roy (Connecticut)	Roy et al. (1996)
Anti-Kv1.6	Kv1.6	D	Cyt	C-terminus	WB bovine brain	Pongs (Hamburg)	Scott et al. (1994)
Anti-Kv1.6	Kv1.6	N = D	Cyt	aa 438–530, C-terminus	IH, WB rat brain	Pongs (Hamburg)	Veh et al. (1995)

Antibody	Channel	N/D	Location	Epitope	Methods	Source	Reference
Anti-Kv1.6	Kv1.6			aa 506–524		Trimmer (Stony Brook, NY)	Nakahira et al. (1996)
drk1	DRK1 (Kv2.1)	N = D		aa 506–533	WB, IH Confocal	Trimmer (Stony Brook, NY)	Sharma et al. (1993), Trimmer (1993)
KC	DRK1 (Kv2.1)	N = D	Cyt	aa 837–853	WB, IH Confocal	Trimmer (Stony Brook, NY)	Sharma et al. (1993), Trimmer (1993)
Anti-Kv2.1	DRK1 (Kv2.1)	N = D	Cyt	aa 853–857, C-terminus	WB, IH rat heart	Nerbonne (St. Louis)	Barry et al. (1995a,b)
Anti-Kv3.1b	Kv3.1b	N = D	Cyt	19 aa, C-terminus	WB, IH	Rudy (New York); Alomone Labs	Weiser et al. (1995)
Anti-Kv3.2	Kv3.2	N = D	Cyt	aa 184–204, C-terminus	IH, IP	Rudy (New York); Alomone Labs	Moreno et al. (1995)
Anti-Kv3.4	Kv3.4	N = D	Cyt	C-terminus	IH, WB	Pongs (Hamburg)	Veh et al. (1995)
Anti-Kv4.2N	Kv4.2	N = D	Cyt	aa 23–42, N-terminus	WB, IH	Jan (San Francisco)	Sheng et al. (1992)
Anti-Kv4.2C	Kv4.2	N = D	Cyt	aa 484–502, C-terminus	NB, IH	Jan (San Francisco)	Nakahira et al. (1996)
Anti-SHB N22 Drosophila	Shaker B	N = D	Cyt	N-terminus	WB, IP, IEM	Schultz (Houston)	Dubinsky et al. (1993), Li et al. (1992)
Anti-AKT1 pore plant	AKT1	D	Pore	aa 250–258	WB cross-reacts with maxi-K, DRK1, KAT1	Berkowitz (New Brunswick)	Mi and Berkowitz (1995)
Anti-minK	minK (Kvs1) slowly activating	N		Full sequence	IH guinea pig myocytes	Kass (New York)	Freeman and Kass (1993)
Anti-ERG (human)	IKr (delayed rectifier)	N = D		aa 321–335, aa 820–834	IH ferrit heart	Morales (Durham)	Brahmajothi et al. (1997)
ROMK1-GST	ROMK1 Kir1.1a (inward rectifier)	N		Not stated	IH rat kidney	White (Sheffield, UK)	Li et al. (1995)
Anti-IRK	IRK1 (Kir2.1) (inward rectifier)	N = D	Cyt	N-terminus	WB, IH	Schwartz (Stanford)	Mi et al. (1996)
C-1	IRK1 (Kir2.1) (inward rectifier)	N	Cyt	aa 376–403, C-terminus	IH	Miyashita (Tokyo)	Miyashita and Kubo (1997)
Anti-GIRK1C1	GIRK1 (kir3.1) (inward rectifier)	D > N	Cyt	aa 488–501, C-terminus	WB, IP, IH, IEM	Kurachi (Osaka, Japan)	Inanobe et al. (1995); Morishige et al. (1996)
C-2	GIRK1 (kir3.1) (inward rectifier)	N	Cyt	aa 346–375	IH	Miyashita (Tokyo)	Miyashita and Kubo (1997)
Anti-CIR	CIR (kir 3.4) (inward rectifier)	N = D	Cyt	N-terminus	WB, IH, IP	Iizuka (Nippoa Boehringer Ingelheim, Japan)	Iizuka et al. (1997)

(continues)

Table XIV (*continued*)

Name	Channel subunit	Native (N) or denatured (D)	Region	Epitope if known	Uses	Senior author or commercial company	References
Anti-KAB-2	kir4.1 KAB inward rectifier	N = D	Cyt	C-terminus, aa 366–379	WB, IH, IEM rat kidney	Ito (Yamagata)	Ito *et al.* (1996)
ROMK1	ROMK1 (inward rectifier)	N = D			WB, IH	Alomone Labs	
GIRK1	GIRK1 (inward rectifier)	N = D			WB, IH	Alomone Labs	
GIRK2	GIRK2 (inward rectifier)	N = D			WB, IH	Alomone Labs	
Anti-RACTK1	RACTK1	N = D	Extracell + Cyt	aa 268–283 cross-linked to aa 205–220	WB, IH rabbit kidney	Suzuki (Minamikawachi, Japan)	Suzuki *et al.* (1995)
Anti-Slo	Slo (voltage- and calcium-gated)	N = D		aa 913–926	WB, IP, IH	Knaus (Innsbruck)	Knaus *et al.* (1996)

Table XV
Monoclonal Antibodies to CFTR

Name	Immunized species, nature of antibody	Region	Peptide (p-) or recombinant antigen (r-)	Uses	Senior author or company	References
CF1	Mouse mAbs	Cyt	Peptides, see 1	WB, IH, IF	Banting (Bristol)	Walker et al. (1995)
CF2		Cyt			Banting (Bristol)	Walker et al. (1995)
CF3		Extracell			Banting (Bristol)	Walker et al. (1995)
CF4		Extracell			Banting (Bristol)	Walker et al. (1995)
CF5		Cyt			Banting (Bristol)	Walker et al. (1995)
CF6		Cyt			Banting (Bristol)	Walker et al. (1995)
CF7		Cyt			Banting (Bristol)	Walker et al. (1995)
CF8		Cyt			Banting (Bristol)	Walker et al. (1995)
L11E8	Mouse IgG$_1$	NBF1		IH, IP	Riordan (Toronto)	Kartner et al. (1992)
M3A7	Mouse IgG$_1$	NBF2		WB, IH, IP	Riordan (Toronto)	Kartner et al. (1992)
L12b4	Mouse IgG$_{2a}$	R domain		WB, IH, IP	Riordan (Toronto)	Kartner et al. (1992)
14H10		Mid R domain		IH, IP	Riordan (Toronto)	Kartner et al. (1992)
M13-1	IgG$_{1k}$	Not stated	p729–736	WB, IP, IF	Genzyme	
M24-1	IgG$_{2ak}$	Not stated	p1377–1480	WB, IP	Genzyme	
	Polyclonal	Pre-NBF	p415–427	WB	Genzyme	Zeitlin et al. (1992)
		R domain	p724–746	WB	Genzyme	Zeitlin et al. (1992)
MATG 1016	IgG$_{2q}$		p101–117	WB, IP	Transgene	
MATG 1016	IgG$_1$		p101–117	WB, IP	Transgene	
MATG 1061	IgG$_{2a}$	NBF	r503–5C7/509–515	IP	Transgene	
MATG 1101	IgG$_1$	R domain	p722–734	WB	Transgene	
MATG 1102	IgG$_{2a}$	R domain	p722–734	IP	Transgene	
MATG 1103	IgG$_1$	R domain	p722–734	WB	Transgene	
MATG 1104	IgG$_1$	R domain	p722–734	WB, IP	Transgene	
MATG 1105	IgG$_1$	R domain	p722–734	WB, IP	Transgene	
MATG 1106	IgG$_1$	R domain	p722–734	IP	Transgene	
MATG 1107	IgG$_1$	R domain	p722–734	WB, IP	Transgene	

Table XVI
Monoclonal and Polyclonal Antibodies to Photoreceptor cGMP-Gated Ion Channels

Name	Subunit	Region	Peptide (p-) or recombinant antigen (r-)	Immunized species, nature of antibody	Uses	Source	References
1D1	α		Purified bovine rod outer segment protein	Mouse	IEM	Molday (British Columbia)	Colville and Molday (1996), Molday et al. (1991)
2G11	α			Mouse		Molday (British Columbia)	Chen et al. (1994)
6E7	α	93–115	Bovine ROS channel	Mouse	WB, IP, IH, IF	Moday (British Columbia)	Hsu and Molday (1993)
1F6	α		Rod and cone channels	Mouse	WB	Molday (British Columbia)	Bonigk et al. (1993)
5E11	β GARP part		Purified 240-kDa protein	Mouse	WB, IH, IEM	Molday (British Columbia)	Hsu and Molday (1993), Molday et al. (1990)
4B2	β GARP part		ROS 240-kDa protein	Mouse	WB	Molday (British Columbia)	Molday et al. (1990)
3C9	β part		ROS 240-kDa protein	Mouse		Molday (British Columbia)	Korschen et al. (1995)
PPc6N	α	93–115	ROS channels		WB	Molday (British Columbia)	Molday et al. (1991)
Ab331	α	93–102	ROS channels		WB	Molday (British Columbia)	Molday et al. (1991)
PPc32K	β	1292–1334	Peptide 1292–1334	Rabbit AP		Molday (British Columbia)	Bonigk et al. (1993)
PPcCC1			Recombinant channel	Rabbit AP		Molday (British Columbia)	Bonigk et al. (1993)
63-4			Recombinant channel	Rabbit AP		Molday (British Columbia)	Bonigk et al. (1993)

on the immunogenicity of AChR have taught us useful lessons regarding the best ways to raise antibodies to other ion channels.

Animals immunized with purified AChR frequently develop signs and symptoms of experimental myasthenia gravis. Patients with myasthenia gravis have spontaneous antibodies directed toward their muscle AChRs that cross-react poorly with AChRs of other species. These antibodies are very variable in specificity between patients, however, and some MG sera show functional effects on AChR function. A particularly notable example is the complete inhibition of fetal AChR function by sera from women whose babies suffer from fetal paralysis leading to joint contractures and other abnormalities. These antibodies do not affect adult AChR function, and often the women themselves are clinically normal (see Table I).

Monoclonal and polyclonal antibodies have also been raised, the former mainly by Jon Lindstrom and colleagues, against neuronal forms of AChR (Tables V and VI).

B. Glycine and GABA$_A$ Receptors

Glycine and GABA$_A$ receptors were first purified in the 1980s. The purified receptors were first used for the production of monoclonal antibodies, and later polyclonal antibodies to peptide sequences or to fusion proteins were raised (Table VII). Since the GABA$_A$ and glycine receptor subunits are highly homologous, antibodies specific for particular isoforms required the use of sequences from the N- or C-terminals, or cytoplasmic loops. Most of the antibodies work in blotting and immunoprecipitation assays across different species, but single amino acid changes can result in the loss of reactivity. For immunocytochemical studies, particularly subcellular localization by electron microscopy, antibodies directed at extracellular determinants have been the most successful. None of the antibodies have been demonstrated to affect function.

C. NMDA and AMPA Receptors

Several groups have made monoclonal antibodies or polyclonal antisera to different peptide sequences of the various NMDA receptors (Table VIII) or AMPA receptors (Table IX). It is not clear whether there are advantages to the different preparations, and indeed a close comparison suggests that there are few differences. These antisera have been very productive in terms of immunohistochemical studies; they are not always as reliable when it comes to immunoprecipitation, and specificity for the subunit should not be inferred without direct evidence (see Mei *et al.* (1994) and Petralia *et al.* (1994) in Table VIII).

Antibodies to some of the GluR isoforms have been found in a childhood form of epilepsy and in a few patients with paraneoplastic (cancer-associated) neurologic disorders (see Table I). Their significance is not yet clear.

IV. Voltage–Gated Ion Channels

A. Voltage–Gated Calcium Channels

Attempts to raise antibodies to native VGCCs have been complicated by the difficulty in obtained sufficient quantities of purified VGCC complex for immunization. Antibodies have been successfully raised against cytoplasmic epitopes derived either from recombinant fusion proteins or peptides. These antibodies are largely generated against the highly variable II–III cytoplasmic domain and generally show marked specificity for the appropriate α_1 subunit (Tables X and XI). There appear to be few successful attempts to raise antibodies to the extracellular domains of VGCCs. An interesting technique to overcome these problems has been described by Gillard *et al.* (1997) (see Table XII). Monoclonal antibodies to the peptide neurotoxins ω-Aga-IVA, specific for the P/Q-type VGCC (α_{1A} subunit), and ω-CTx-GVIA, specific for the N-type VGCC (α_{1B} subunit), have been raised in mice. These antibodies can then be used to detect ω-Aga-IVA or ω-CTx-GVIA binding sites in paraformaldehyde-fixed tissue by incubation with the appropriate peptide neurotoxin prior to application of antibody. Specific antibodies to the different β subunits have been generated by immunization with recombinant fusion proteins derived from the hypervariable carboxy terminus of the β subunit gene.

Patients with the Lambert Eaton myasthenic syndrome have antibodies directed against ω-conotoxin-MVIIC binding VGCCs. These antibodies are highly specific for the disease state, and bind poorly to other forms of VGCC. It is not yet clear whether some of these antibodies directly inhibit function, but they lead to down-regulation of VGCC in cell lines (Table I).

B. Voltage–Gated Potassium Channels

Voltage-gated potassium channels (VGKCs) are very heterogenous, oligomeric proteins. Most of the antibodies have been raised against synthetic peptides, fusion proteins, or recombinant subunits. Monoclonal antibodies for C-terminal domains exist for many subunits (Table XIII). Polyclonal antisera can be used to demonstrate the distribution of VGKCs by immunohistochemistry and some immunoprecipitate-solubilized VGKCs. The large number of groups involved, and the interest of pharmaceutical companies, means that many antibodies are available (Table XIV).

Antibodies to VGKCs are present in some patients with acquired neuromyotonia, an autoimmune disease that results in muscle twitching and cramps as a result of neuronal hyperexcitability. It is not yet clear whether these antibodies are mainly directed against a single subtype of VGKC; early evidence suggests that they may be widely cross-reactive or that there are several different antibody specificities in some sera (Table I).

C. Voltage-Gated Sodium Channels

There are few available antibodies to voltage-gated sodium channels. Polyclonal antibodies raised in rabbit against brain type 1 Na^+ channel p 465–481 recognize the α subunit in an intracellular loop between domains I and II on WB, and may be useful for immunohistochemistry and immunoprecipitation (Gordon *et al.*, 1987; Westenbroek *et al.*, 1992b). It is marketed by Alomone Laboratories. A polyclonal antibody to the III–IV loop region is obtainable from Upstate Biotechnology.

D. cGMP-Gated Channels and CFTR

A number of monoclonal antibodies have been generated against CGMP-gated ion channels of the rod outer segment and CFTR (Tables XV and XVI).

V. Commercial Products

Several of the antibodies or similar ones are available through commercial companies as listed below. Only those which are described in the commercial literature are listed in Tables I–XVI. It is worth checking with the companies for new products, and looking for information on the Internet.

Voltage-gated ion channel antibodies:
Alomone Labs, Headquarters
Shatner Center 3, PO Box 4287
Jerusalem 91042
Israel
Tel.: 972-2-652-8002
Fax: 97202065205233
E-mail: alomone@netvision.net.il
http://www.alomone.com

CFTR antibodies:
Transgene S.A.
11 rue de Molsheim
67082 Strasbourg Cedex
France
Tel.: 33-388-279100
Fax: 33-388-279111

CFTR antibodies:
Genzyme
http://www.genzyme.com
Calcium, sodium, potassium channel and glutamate receptor antibodies:
Upstate Biotechnology
http://www.upstate biotech.com or www.biosignals.com

Glutamate and GABA receptor antibodies:
Boehringer Ingelheim Bioproducts
http://www.bi-bioproducts.de

Human acetylcholine receptor antibodies:
Serotec (Immunotec)
22 Bankside, Station Approach
Kidlington, Oxford OX5 1JE
England
Tel.: 44-1865-852700
Fax: 44-1865-373899

Potassium channels antibodies:
Merck Research Labs
Rahway, New Jersey 07065
USA

Acknowledgments

We thank S. Tzartos, J. Lindstrom, H. Betz, R. Huganir, R. S. Molday, M. Li, and J. S. Trimmer for sending information.

References

Barry, D. M., Trimmer, J. S., Merlie, J. P., and Nerbonne, J. M. (1995a). *Circ. Res.* **77,** 361.
Barry, E. L., *et al.* (1995b). *J. Neurosci.* **15,** 274.
Beattie, R. E., *et al.* (1997). *Brain Res. Prot.* **1,** 307.
Beeson, D., Omar, M., Bermudez, I., Vincent, A., and Newson-Davis, J. (1996). *Neurosci. Lett.* **207,** 57.
Bekele-Arcuri, Z., Matos, M. F., Manganas, L., Strassle, B. W., Monaghan, B. W., Rhodes, K. J., and Trimmer, J. S. (1996). *Neuropharmacology* **35,** 851.
Benke, D., Mertens, S., Trzeciak, A., Gillessen, D., and Mohler, H. (1991). *J. Biol. Chem.* **266,** 4478.
Benke, D., Fritschy, J. M., Trzeciak, A., *et al.* (1994). *J. Biol. Chem.* **269,** 27100.
Blackstone, C. D., Moss, S. J., Martin, L. J., Levey, A. I., Price, D. L., and Huganir, R. L. (1992a). *J. Neurochem.* **58,** 1118.
Blackstone, C. D., Levey, A. I., Martin, L. J., Price, D. L., and Huganir, R. L. (1992b). *Ann. Neurol.* **31,** 680.
Bonigk, W., Altenhofen, W., Muller, F., Dose, A., Illing, M., Molday, R. S., and Kaupp, U. B. (1993). *Neuron* **10,** 865.
Brahmajothi, M. V., Morales, M. J., Reimer, K. A., and Strauss, H. C. (1997). *Circ. Res.* **81,** 128.
Brose, N., Huntley, G. W., Stern-Bach, Y., Sharma, G., Morrison, J. H., and Heinemann, S. F. (1994a). *J. Biol. Chem.* **269,** 16780.
Brose, N., Huntley, G. W., Stern-Bach, Y., Sharma, G., Morrison, J. H., and Heinemann, S. F. (1994b). *J. Biol. Chem.* **269,** 16780.
Bufler, J., Kahlert, S., Tzartos, S., Toyka, K. V., Maelicke, A., and Franke, C. (1996). *J. Physiol. Lond.* **492,** 107.
Chen, T. Y., Illing, M., Molday, L. L., Hsu, Y. T., Yau, K. W., and Molday, R. S. (1994). *Proc. Natl. Acad. Sci. USA* **91,** 11757.
Colville, C., and Molday, R. S. (1996). *J. Biol. Chem.* **271,** 32968.
Craig, A. M., Blackstone, C. D., Huganir, R. L., and Banker, G. (1993). *Neuron* **10,** 1055.
Dubinsky, W. P., Mayorga-Wark, O., Garretson, L. T., and Schultz, S. G. (1993). *Am. J. Physiol.* **265,** C548.
Duggan, M. J., and Stephenson, F. A. (1989). *J. Neurochem.* **53,** 132.
Duggan, M. J., and Stephenson, F. A. (1990). *J. Biol. Chem.* **265,** 3831.
Ewert, M., Shivers, B. D., Luddens, H., *et al.* (1990). *J. Cell Biol.* **110,** 2043.

Farrar, J., Portolano, S., Willcox, N., *et al.* (1997). *Int. Immunol.* **9**, 1311.

Freeman, L. C., and Kass, R. S. (1993). *Circ. Res.* **73**, 968.

Gahring, L. C., Twyman, R. E., Greenlee, J. E., and Rogers, S. W. (1995). *Mol. Med.* **1**, 245.

Gillard, S. E., *et al.* (1997). *Neuropharmacology* **36**(3), 405.

Gordon, G., Moskowitz, H., Eitan, M., Warner, C., Catterall, W. A., and Zlotkin, E. (1987). *Proc. Natl. Acad. Sci. USA* **27**, 8682.

Greferath, U., Grunert, U., Fritschy, J. M., Stephenson, A., Mohler, H., and Wassle, H. (1995). *J. Comp. Neurol.* **353**, 553.

Gurnett, C. A., *et al.* (1995). *J. Biol. Chem.* **270**, 9035.

Hart, I. K., *et al.* (1997a). *Ann. Neurol.* **41**, 238.

Hart, I., Water, C., Vincent, A., *et al.* (1997b). *Ann. Neurol.* **41**, 238.

Heidenreich, F., Vincent, A., Roberts, A., and Newsom-Davis, J. (1998). *Autoimmunity* **1**, 285.

Hell, J. W., *et al.* (1993). *J. Cell Biol.* **123**, 949.

Helms, L. M., *et al.* (1997). *Biochemistry* **36**, 3737.

Hsu, Y. T., and Molday, R. S. (1993). *Nature* **361**, 76.

Huntley, G. W., Rogers, S. W., Moran, T., Janssen, W., Archin, N., Vickers, J. C., Cauley, K., Heinemann, S. F., and Morrison, J. H. (1993). *J. Neurosci.* **13**, 2965.

Huntley, G. W., Vickers, J. C., Janssen, W., Brose, N., Heinemann, S. F., and Morrison, J. H. (1994). *J. Neurosci.* **14**, 3603.

Iizuka, M., *et al.* (1997). *Neuroscience* **77**, 1.

Inanobe, A., *et al.* (1995). *Biochem. Biophys. Res. Commun.* **217**, 1238.

Ito, M., *et al.* (1996). *FEBS Lett.* **338**, 11.

Jacobson, L., Beeson, D., and Vincent, A. (2010). (in preparation).

Jones, A., Korpi, E. R., McKernan, R. M., *et al.* (1997). *J. Neurosci.* **17**, 1350.

Kartner, N., Augustinas, O., Jensen, T. J., Naismith, A. L., and Riordan, J. R. (1992). *Nat. Genet.* **1**, 321.

Kern, W., and Sieghart, W. (1994). *J. Neurochem.* **62**, 764.

Khan, Z. U., Guitterez, A., and De Blas, A. L. (1994). *J. Neurochem.* **63**, 371.

Kirsch, J., Wolters, I., Triller, A., and Betz, H. (1993). *Nature* **366**, 745.

Knaus, H. G., *et al.* (1996). *J. Neurosci.* **16**, 955.

Korschen, H. G., Illing, M., Seifert, R., Sesti, F., Williams, A., Gotzes, S., Colville, C., Muller, F., Dose, A., Godde, M., Molday, L., Kaupp, U. B., *et al.* (1995). *Neuron* **15**, 627.

Lang, B., and Newsom-Davis, J. (1995). *Springer Semin. Immunopathol.* **17**, 3.

Lau, L. F., and Huganir, R. L. (1995). *J. Biol. Chem.* **270**, 20036.

Li, M., Jang, Y. N., and Jan, L. Y. (1992). *Science* **257**, 1225.

Li, Q., Cope, G., Hornby, D., and White, S. (1995). *J. Physiol.* **489**, 93P.

Lindstrom, J. (1996). *Ion Channels* **4**, 377.

Loutrari, H., Kokla, A., Trakas, N., and Tzartos, S. J. (1997). *Clin. Exp. Immunol.* **109**, 538.

Mamalaki, A., Trakas, N., and Tzartos, S. J. (1993). *Eur. J. Immunol.* **23**, 1839.

Martin, L. J., Blackstone, C. D., Huganir, R. L., and Price, D. L. (1992). *Neuron* **9**, 259.

Mays, D. J., Foose, J. M., Philipson, L. H., and Tamkun, M. M. (1995). *J. Clin. Invest.* **96**, 282.

McKernan, R. M., *et al.* (1991). *Neuron* **7**, 667.

McKernan, R. M., and Whiting, P. J. (1996). *Trends Neurosci.* **19**, 139.

McNamara, N. M., *et al.* (1996). *Eur. J. Neurosci.* **8**, 688.

Mei, L., Doherty, C. A., and Huganir, R. L. (1994). *J. Biol. Chem.* **269**, 12254.

Meyer, G., Kirsch, J., Betz, H., and Langosch, D. (1995). *Neuron* **15**, 563.

Mi, F., and Berkowitz, G. A. (1995). *PNAS* **92**, 3386.

Mi, H., *et al.* (1996). *J. Neurosci.* **16**, 2421.

Miyashita, T., and Kubo, Y. (1997). *Brain Res.* **750**, 251.

Molday, L. L., Cook, N. J., Kaupp, U. B., and Molday, R. S. (1990). *J. Biol. Chem.* **265**, 18690.

Molday, R. S., Molday, L. L., Dose, A., Clark-Lewis, I., Illing, M., Cook, N. J., Eismann, E., and Kaupp, U. B. (1991). *J. Biol. Chem.* **266**, 21917.

Moreno, H., *et al.* (1995). *J. Neurosci.* **15**, 5486.

Morishige, K. I., *et al.* (1996). *Biochem. Biophys. Res. Commun.* **220**, 300.

Mossier, B., Togel, M., Fuchs, K., and Sieghart, W. (1994). *J. Biol. Chem.* **269**, 25777.

Nadler, L. S., Guirguis, E. R., and Siegel, R. E. (1994). *J. Neurobiol.* **25**, 1533.

Nakahira, K., Shi, G., Rhodes, K. K., and Trimmer, J. S. (1996). *J. Biol. Chem.* **271**, 7084.

Neff, S., Dineley-Miller, K., Char, D., Quik, M., and Patrick, J. (1995). *J. Neurochem.* **64**, 332.

Nusser, Z., Sieghart, W., Stephenson, F. A., and Somogyi, P. (1996a). *J. Neurosci.* **16**, 103.

Nusser, Z., Sieghart, W., Stephenson, F. A., and Somogyi, P. (1996b). *J. Neurosci.* **16**, 103.

Ousley, A. H., and Froehner, S. C. (1994). *Proc. Natl. Acad. Sci. USA* **91**, 12263.

Papadouli, I., Sakarellos, C., and Tzartos, S. T. (1993). *Eur. J. Biochem.* **211**, 227.

Petralia, R. S., Yokotani, N., and Wenthold, R. J. (1994). *J. Neurosci.* **14**, 667.

Pfeiffer, F., Simler, R., Grenningloh, G., and Betz, H. (1984). *Proc. Natl. Acad. Sci. USA* **81**, 7224.

Pollard, S., and Stephenson, F. A. (1997). *Biochem. Soc. Trans.* **25**, 5475.

Pollard, S., Duggan, M. J., and Stephenson, F. A. (1991). *FEBS Lett.* **295**, 81.

Pollard, S., Duggan, M. J., and Stephenson, F. A. (1993). *J. Biol. Chem.* **268**, 3753.

Ptacek, L. (1977). *Neuromusc. Dis.* **7**, 250.

Puchalski, R. B., Louis, J. C., Brose, N., Traynelis, S. F., Egebjerg, J., Kukekov, V., Wenthold, R. J., Rogers, S. W., Lin, F., Moran, T., Morrison, J. H., and Heinemann, S. F. (1994). *Neuron* **13**, 131.

Raymond, L. A., Blackstone, C. D., and Huganir, R. L. (1993). *Nature* **361**, 637.

Reinhardt-Maelicke, S., *et al.* (1993). *J. Rec. Res.* **13**, 513.

Rhodes, K. J., Kcilbaugh, S. A., Barrezueta, N. X., Lopez, K. L., and Trimmer, J. S. (1995). *J. Neurosci.* **15**, 5360.

Rhodes, K. J., Barrezueta, N. X., Monaghan, M. M., Bekele-Arcuri, Z., Nakahira, K., Nawoschlik, S., Schcchter, L. E., and Trimmer, J. S. (1996). *J. Neurosci.* **16**, 4846.

Riemersma, S., Vincent, A., Beeson, D., *et al.* (1996). *J. Clin. Invest.* **98**, 2358.

Rogers, S. W., Andrews, P. I., Gahring, L. C., *et al.* (1994). *Science* **265**, 648.

Roy, M. L., *et al.* (1996). *Glia* **18**, 177.

Rubio, M. E., and Wenthold, R. J. (1997). *Neuron* **19**, 939.

Sakurai, T., *et al.* (1995). *J. Biol. Chem.* **270**, 21234.

Sakurai, T., *et al.* (1996). *J. Cell Biol.* **134**, 511.

Schoch, P., Richards, J. G., Haring, P., *et al.* (1985). *Nature* **314**, 168.

Scott, V. E., *et al.* (1994). *Biochemistry* **33**, 1617.

Scott, V. E., *et al.* (1996). *J. Biol. Chem.* **271**, 3207.

Sharma, N., *et al.* (1993). *J. Cell Biol.* **123**, 1835.

Sheng, M., Tsaur, M. L., Jan, Y. N., and Jan, L. Y. (1992). *Neuron* **9**, 271.

Sheng, M., Tsaur, M. L., Jan, Y. N., and Jan, L. Y. (1994). *J. Neurosci.* **14**, 2408.

Shi, G., Nakahira, K., Hammond, S., Schechter, L. E., Rhodes, K. J., and Trimmer, J. S. (1996). *Neuron* **16**, 843.

Shillito, P., Molenaar, P. C., Vincent, A., *et al.* (1995). *Ann. Neurol.* **38**, 714.

Siegel, S. J., Brose, N., Janssen, W. G., Gasic, G. P., Jahn, R., and Heinemann, S. F. (1994). *Proc. Natl. Acad. Sci. USA* **91**, 564.

Slater, C. R., Young, C., Wood, S. J., *et al.* (1997). *Brain* **120**, 1513.

Spencer, R. H., Chandy, K. G., and Gutman, G. A. (1993). *Biochem. Biophys. Res. Commun.* **191**, 201.

Stephenson, F. A. (1995). *Biochem. J.* **310**, 1.

Stephenson, F. A., Duggan, M. J., and Pollard, S. (1990). *J. Biol. Chem.* **265**, 21160.

Striessnig, J., Glossmann, H., and Catterall, W. A. (1990). *Proc. Natl. Acad. Sci. USA* **87**, 9108.

Sucher, N. J., Brose, N., Deitcher, D. L., Awobuluyi, M., Gasic, G. P., Bading, H., Cepko, C. L., Greenberg, M. E., Jahn, R., Heinemann, S. F., *et al.* (1993). *J. Biol. Chem.* **268**, 22299.

Suzuki, M., *et al.* (1995). *Am. J. Physiol.* **269**, C496.

Takimoto, K., and Levitan, E. S. (1994). *Circ. Res.* **75**, 1006.

Tingley, W. G., Roche, K. W., Thompson, A. K., and Huganir, R. L. (1993). *Nature* **364**, 70.

Tingley, W. G., Ehlers, M. D., Kameyama, K., Doherty, C., Ptak, J. B., Riley, C. T., and Huganir, R. L. (1997). *J. Biol. Chem.* **272**, 5157.

Togel, M., Mossier, B., Fuchs, K., and Sieghart, W. (1994). *J. Biol. Chem.* **269,** 12993.

Trimmer, J. S. (1991). *Proc. Natl. Acad. Sci. USA* **88,** 10764.

Trimmer, J. S. (1993). *FEBS Lett.* **324,** 205.

Twyman, R. E., Gahring, L. C., Spiess, J., and Rogers, S. W. (1995). *Neuron* **14,** 755.

Tzartos, S. J., and Lindstrom, J. L. (1980). *Proc. Natl. Acad. Sci. USA* **77,** 755.

Tzartos, S. J., and Remoundos, M. S. (1992). *Eur. J. Biochem.* **207,** 915.

Tzartos, S. J., Rand, D. E., Einarson, B. E., and Lindstrom, J. M. (1981). *J. Biol. Chem.* **256,** 8635.

Tzartos, S., Langeberg, L., Hochschwender, S., and Lindstrom, J. (1983). *FEBS Lett.* **158,** 116.

Tzartos, S., Langeberg, L., Hockschwender, S., Swanson, L., and Lindstrom, J. (1986). *J. Neuroimmunol.* **10,** 235.

Tzartos, S. J., Kokla, A., Walgrave, S., and Conti-Tronconi, B. (1988). *Proc. Natl. Acad. Sci. USA* **85,** 2899.

Tzartos, S., Cung, M. T., Demange, P., *et al.* (1991). *Mol. Neurobiol.* **5,** 1.

Tzartos, S. J., Valcana, C., Kouvatsou, R., and Kokla, A. (1993). *EMBO J.* **12,** 5141.

Tzartos, S. J., Kouvatsou, R., and Tzartos, E. (1995a). *Eur. J. Biochem.* **228,** 463.

Tzartos, S. J., Tzartos, E., and Tzartos, J. S. (1995b). *FEBS Lett.* **363,** 195.

Veh, R. W., *et al.* (1995). *Eur. J. Neurosci.* **7,** 2189.

Vincent, A., Whiting, P. J., Schluep, M., *et al.* (1987). *Ann. N.Y. Acad. Sci.* **505,** 106.

Walker, J., Watson, J., Homes, C., Edelman, A., and Banting, G. (1995). *J. Cell Sci.* **108,** 2433.

Wang, H., *et al.* (1993). *Nature* **365,** 75.

Weiser, M., *et al.* (1995). *J. Neurosci.* **15,** 4298.

Wenthold, R. J., Yokotani, N., Doi, K., and Wada, K. (1992). *J. Biol. Chem.* **267,** 501.

Westenbroek, R. E., *et al.* (1992a). *Neuron* **9,** 1099.

Westenbroek, R. E., Noebels, J. L., and Catterall, W. A. (1992b). *J. Neurosci.* **12,** 2259.

Whiting, P. J., Vincent, A., Schluep, M., and Newsom-Davis, J. (1986). *J. Neuroimmunol.* **11,** 223.

Zeitlin, P. L., Crawford, I., Lu, L., *et al.* (1992). *Proc. Natl. Acad. Sci. USA* **89,** 344.

Zezula, J., and Sieghart, W. (1991). *FEBS Lett.* **284,** 15.

INDEX